Physical Geography
Great Systems and Global Environments

William M. Marsh taught physical geography and related courses at the University of Michigan for 30 years, where he founded the Department of Earth and Resource Science. He is now with the University of British Columbia where he teaches courses in landscape analysis. He is an experienced textbook author, having written three textbooks in physical geography and two in land use applications, one of which has become a standard in the field of environmental planning.

Martin M. Kaufman has taught Physical Geography and Geographic Information Systems at the high school, undergraduate, and graduate levels for over 20 years. He is also an experienced textbook author. Currently, he is a professor of Earth Science at the University of Michigan-Flint, where he teaches the introductory sequence of physical geography courses.

'Marsh and Kaufman eloquently link the science of physical geography with the impacts of human activities. As such this text is a perfect tool for encouraging students to become environmentally-informed citizens.'

Professor Dean P. Lambert, Department of Geography, San Antonio College, Texas

'At last, a textbook that successfully merges a graphic storyline with the text to describe the interconnectedness of Earth's great physical systems. The authors do a masterful job using this approach to explain the geographic character of the planet. This textbook will be understandable to both science and non-science majors.'

Dr Richard Crooker, Department of Geography, Kutztown University, Pennsylvania

'Authoritative, useful, balanced, and wise, this is more than a textbook. It is a modernized classic and comprehensive presentation of the physical geography perspective of the great natural systems operating on planet Earth. … should be successful with students and teachers alike as the scientific process and physical science fundamentals are presented with relevance to everyday life.'

Professor Dean Fairbanks, Department of Geography and Planning, California State University, Chico

'Will Marsh, an excellent scientist and a talented artist, has written the best textbooks in physical geography for three decades, and [this] new offering with Marty Kaufman… continues this tradition. The book's distinctive features include explanations that start from a comprehensible scale that the student can understand, along with gorgeous illustrations.'

Professor Jeff Dozier, Bren School of Environmental Science and Management, University of California, Santa Barbara

Physical Geography

Great Systems and Global Environments

William M. Marsh
University of British Columbia,
University of Michigan (Emeritus)

Martin M. Kaufman
University of Michigan-Flint

CAMBRIDGE
UNIVERSITY PRESS

CAMBRIDGE UNIVERSITY PRESS
Cambridge, New York, Melbourne, Madrid, Cape Town,
Singapore, São Paulo, Delhi, Mexico City

Cambridge University Press
The Edinburgh Building, Cambridge CB2 8RU, UK

Published in the United States of America by Cambridge University Press, New York

www.cambridge.org
Information on this title: www.cambridge.org/9780521764285

Printed in the United States

A catalogue record for this publication is available from the British Library

Library of Congress Cataloguing in Publication data

ISBN   978-0-521-76428-5 Hardback

Additional resources for this publication at www.cambridge.org/mk

Dedicated to

Walter A. Schroeder
of Missouri

Brief Contents

Contents
Preface
Acknowledgments

Contents

Part II Earth's Life Support Systems

Preface

If you give our planet a hard poke somewhere, it is apt to set off a chain of reactions leading to change somewhere else, maybe in many places, and often far away. We live in a broadly interconnected geographic environment, one laced with multitudes of systems, a veritable planetary network of wiring and plumbing in three-dimensional space.

These interconnections are astounding. Among them are systems linking temperatures of tropical seas with the size and number of hurricanes that strike midlatitude coastlines, dust storms over the Sahara of North Africa with the fertility of soils in the Amazon Basin of South America, soil erosion on the plains of northern China with the quality of air over Seattle, fertilizer applications on cornfields in Iowa with sick and dying ecosystems on the Mississippi Delta, and earthquakes in Indonesia with giant ocean waves capable of killing hundreds of thousands of people on the other side of the Indian Ocean more than 3000 miles away.

Yet we find it difficult to think in broad patterns and networks because we have learned to see the world in geographic compartments. To physical geography, which is interested in the distribution of natural phenomena, this is a dilemma because it implies that the nature operating in one place may have little or nothing to do with the nature operating in another place. This sort of thinking is reinforced again and again in our lives. The way we studied geography and history in school, for example, tended to signify it by drawing stiff boundaries between the pink and green patches on world maps. And international politics has also played a role by ascribing artificial significance to national borders, lines that nature can neither see nor follow. In short, we have a habit of defining Earth in terms of its subdivisions rather than its integrated whole.

This book employs a different tactic. It argues the geographic character of Earth is best understood when viewed through the window of systems. The largest of these systems, which we call great systems, operate throughout the planet. They include an energy system that begins when solar radiation enters the top of the atmosphere, systems of water and air

circulation that range over the entire planet, systems of currents coursing throughout the vastness of the oceans, systems of running water flowing over the land, and systems of organisms forming a living skin over all the Earth's lands and waters. These systems and their offspring, operating in large spaces or small ones, shape all things geographical, mountain chains, coastlines, plains, watersheds, climates, forests, lakes, and swamps. They are Earth's kinetic elements, the very foundation of physical geography.

Our planet is a geographic wonder and for centuries geographers have documented its diversity, producing maps of virtually every part of the lands, seas, and atmosphere. But we have also discovered that nothing on the planet is truly permanent. The patterns of rainfall, forest cover, river networks, coastlines, virtually everything mapped in one decade turns up different in the next, sometimes dramatically so. But knowing about change does not tell us what produces it. And this brings us to the principal objective of this book: to discover the nature of geographic change on Earth, not only how it takes place but what drives it. And this brings us back to systems, because they are the planet's big geographic drivers.

The content of physical geography is huge and the story told in textbooks can be overwhelming. The maps and diagrams alone can be daunting. The problem facing all writers in the field is how to make the medium, a book, work most effectively in support of the message, the physical geography of a changing planet. This was a real challenge for us, because with new research discoveries appearing almost daily, the message keeps getting bigger and in many ways more complicated. Among the options, we considered simplifying the message, distilling it down to its bare essence, but concluded that would sell our audience short. We also considered amplifying the medium by adding more accessories such as bigger and fancier graphics, special essays and guest commentary, items from the news media, and so on, but decided that would only detract from the central message. What we settled on was a back-to-basics approach based on two main objectives:

- The first addresses the medium, which in its simplest form is just words and graphics. We decided to write the text in a somewhat narrative style of prose, and to create a set of graphics that feature single concepts rather than a gang of concepts bunched together in a few large, complex graphics. The resultant graphics not only tend to be small and clear but easy to tie to the storyline in the text. Next, the graphic storyline and the word (text) storyline had to be woven together. This was accomplished by using an "arm-in-arm" page design, that is, one that places most graphics (figures) alongside the related passages in the text.

This facilitates learning because graphics and text are complementary, made to match, and create a simple path to help the student navigate through the large amount of information in a physical geography course.

- The second objective addresses the message, and this we reasoned had to feature a story about a planet on which geographic change is the norm rather than the exception, where systems are more significant, extensive, and interconnected than we could possibly have guessed only a few decades ago, and where geography is more central to understanding our magnificent home than at any other time in the long history of the field.

The Book as a Learning and Teaching Tool

The systems approach is capped in each chapter with an eclectic end-of-chapter summary diagram featuring key

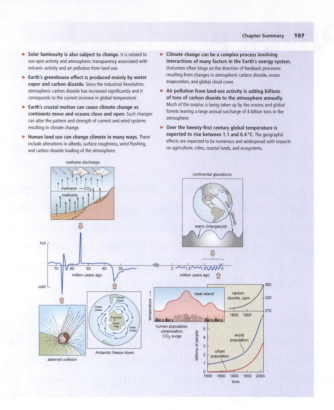

concepts in a big-picture format, a useful learning and teaching tool.

To further round out each chapter, two sets of summaries are provided: in-chapter summaries at the ends of selected sections, and chapter-end summaries made up of a concluding paragraph followed by a series of overview statements. The overview statements are topped off with a set of review questions, 10 to 15 queries to help the student gauge his/her comprehension of the chapter's main points.

Systems-based teaching can be a rewarding experience, especially if it is preceded by a little planning, and to help the teacher-scholar focus the learning experience, a companion volume is provided. The online **Instructor's Guidebook** highlights the core concepts of each chapter including the relevant graphics in the text, and suggests strategies to help teach the material from a systems perspective.

Online Resources

Online at www.cambridge.org/mk, alongside the Instructor's Guidebook, you will find Powerpoints of figures from the book, example responses for selected questions and flashcards.

WMM and MMK

Acknowledgments

Every book benefits from a host of participants and this book is no exception. Although students seldom see how they influence the character of a book, they are a major force in shaping the voice of a textbook, that is, the way the message is communicated. And so we humbly acknowledge the thousands of students who, over several decades of teaching, have helped us understand how to tell the story of physical geography.

The production of a volume such as this one requires a entire team of people and we are indebted to the team at Cambridge University Press not only for the heavy lifting with editing, graphics, and design, but for the constructive project climate they created which has fostered thoughtful exchanges and innovative thinking at many levels.

We acknowledge our colleagues in physical geography and related fields who provided manuscript reviews and suggestions on how to improve the book. The list is long and includes colleagues from colleges and universities in the United States, Canada, and Europe.

Finally, we acknowledge an inner circle of colleagues, friends, and relatives whom we used as sounding boards, who lent a hand when needed, and who willingly altered personal agendas to accommodate writing schedules. Chief among these is Alison Mewett, wife of the senior author. This circle also includes the participants in the chapter opening stories: William Steinhoff, Jeff Dozier, Alison Mewett, Earl Steinhoff, Roberta Steinhoff, Jack Goodnoe, Bruce D. Marsh, William R. Marsh, James G. Marsh, M. Leonard Bryan, John Koerner, Charles Douthitt, Ray Adair, and Vernie Anderson. It was all great fun.

WMM
MMK

Mapping Our Course of Study

We long to have nature in our lives and the field of physical geography helps fulfill that need. But it's not just natural things and where they are on maps that interests us here. We are equally interested in examining how things like glaciers and stream valleys form and change, and to do this we have to look at the systems that deliver the forces and processes of change. Systems. They are the blood and bones of the Earth, the networks of energy and matter that, like the belts of winds that deliver heat and water, lace the planet together. These global scale systems run the Earth and we call them "great systems". What are their attributes and how do they operate? And what does it all mean in terms of scientific inquiry? How do we use scientific thinking, rather than, say, religion or mythology, to explain the world around us?

Introduction

Earth is a glorious planet, a nurturing place that is home to countless billions of organisms. We humans are among the most geographically versatile of these organisms. In less than 50 thousand years – a mere wink of Earth time – our species has spread over the entire planet, rooted itself in its different geographic environments, and made them part of our heritage. We celebrate our geographic heritage in nearly every aspect of culture – in poetry, prose, song, painting, architecture, science, religion, recreation, and much more. In short, Earth's places and landscapes are central to understanding who we are as people – as nations, societies, communities, families, and individuals.

But in the modern era these geographic ties have begun to unravel as many people leave homelands because of war, famine, degradation of the environment, loss of opportunity, or the hope of finding happier and more fruitful lives in other places. Many choose the built environments of cities, and lives remote from nature. For them, farms, woodlands, prairies, and country villages are becoming secondhand places. Today, fewer than three in ten Americans and Canadians live in traditional rural settings. At the same time, many in cities are leaving for suburbia and a landscape, such as the one shown in Figure 1.1, which is neither urban nor rural, and lives connected more to automobiles and highways than to a place in the land. And the trend towards placelessness is not limited to developed countries. In poor countries

Figure 1.1 Modern suburban development, a landscape which is neither urban not rural with little local character and a poor relationship with nature where more space is provided for automobiles than for plants.

throughout the world people are abandoning the countryside and crowding into sprawling, faceless cities. Mexico City, which receives as many as 1000 immigrant families a day, will soon exceed a population of 40 million persons and Shanghai, China, has exploded to a population of more than 50 million. Roughly half of all humanity now lives in urban areas.

1.1 Our Place on the Planet

Many of us today are being threatened with the loss of our geographic roots and the opportunity, shared by hundreds of generations before us, to become intimate with a piece of Earth, to commit to memory and emotion its forms and features, to ride the rhythms of its seasons, to mark the signs of years building one upon another, to suffer the ravages of nature's destructive forces upon it, to celebrate the processes of healing, and to remember these things as a part of one's time on the planet. With life increasingly confined to massive cities, giant freeways, huge commercial malls, overgrown universities, mega corporations, and maze-like bureaucracies, a great many of us appear to be losing the privilege of knowing a parcel of land and embracing the real workings of its nature. Yet we long to have nature in our lives.

How then do we come to know nature? In our college days, some of us believed the answer lay in turning away from "the system" and returning to the land, as Thoreau did at Walden to build social arrangements close to the rudiments of nature in small, knowable, and supposedly happy places. Others of us believed the answer lay in our books, research projects, and expeditions to distant and exotic places to discover inner workings of nature. On many occasions we have weighed these two perspectives and reflected on our own good luck to have had a childhood lived close to the land, free of the rigors of formal science, followed by an adulthood lived close to science with its resources and abstract insights into nature's more hidden reaches.

Our lives today, we believe, need both a scientific and romantic acquaintance with nature. The first is necessary to becoming a responsible citizen in a very large world and the second to shaping a sense of one's place with nature in a very small world. This book is written with both objectives in mind, though it can address only the first with any confidence. In many respects, the second is far too personal to be learned in a book, and for a great many North Americans in the twenty-first century, it awaits a time in life more toward the end with retirement than at the beginning with youth. Our task in colleges and universities, then, is learning about the larger world through the windows provided by science and there is no better vehicle for approaching this challenge than physical geography.

1.2 The Great Systems of Physical Geography

Before you lies Earth, the Eden planet of the Solar System (Figure 1.2). Wonderfully diverse and constantly changing, it is a geographer's paradise. As the photograph in Figure 1.2 suggests, it is a superb place to study the patterns, processes, and products of nature. Here we have all that the other planets possess plus a lot more, most notably vast and mobile systems of life and water, which we call the biosphere and hydrosphere, wrapped in an envelope of gases, called the atmosphere, and set upon a stage of rock and soil, called the lithosphere. Yet, as impressive as the hydrosphere, biosphere, atmosphere, and lithosphere are individually, Earth's true geographic character is found by discovering their interplay, for all are intricately intertwined in great flowing systems.

Giant freeway complexes are glaring illustrations of our disconnect with the land.

Thoreau's Walden Pond where he spent time contemplating and writing about nature and life.

Connecting with the land in a modern world, part of being a responsible citizen of Earth.

Figure 1.2 A most glorious planet. A view of home from the Moon taken during the Apollo 11 mission about 40 years ago.

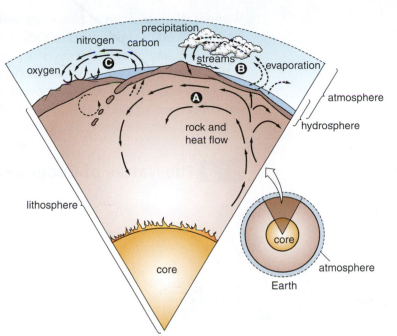

Figure 1.3 A slice of the Earth showing three important global systems, one in the lithosphere (A) that moves rock and heat; one in the hydrosphere (B) that moves water and sediment; and one in the atmosphere (C) that cycles gases between the atmosphere, biosphere, hydrosphere, and lithosphere.

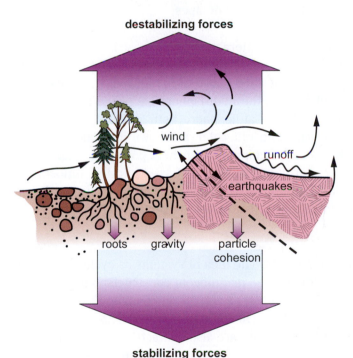

Figure 1.4 The contest between destabilizing forces, which work to change the Earth's surface, and stabilizing forces, which work to hold it together. Geographic change occurs when the threshold of stability is exceeded.

In their simplest form, Earth's systems are defined by cycles of matter and energy. Predictably, they operate at different scales and rates. Some, such as prevailing wind systems, operate almost continuously over the whole planet; others, such as weather systems, begin, expand, and contract with the seasons. Still others, such as ocean currents, flow with graceful uniformity, whereas beneath the sea molten rock in volcanoes moves with unpredictable irregularity. But they are all capable of rendering change in the Earth's surface, some slowly and gradually, some suddenly and violently.

The Drivers of Geographic Change: These vast systems – let us call them **great systems** – touch all parts of the planet and drive geographic change. Deep within the Earth, for example, a great system of heat and rock slowly mixes the mantle (see A in Figure 1.3), which in turn pushes the crust around, rearranging the continents, building mountains, bringing new rock to the surface via volcanoes, and gradually drawing older rock down along the margins of the ocean basins. On the Earth's surface, the atmosphere continuously takes in water from the oceans, and, through the mechanisms of storms of various sizes and durations, delivers a large part of it to the continents (B in Figure 1.3). The streams that result from this rainwater erode the continents and carry the resultant sediment to the ocean where it settles to the bottom, gradually building new rock on the continental edges. And above us, the atmosphere's system of gases couples with the oceans and biosphere in a massive flow system (C in Figure 1.3) that recycles carbon, nitrogen, oxygen, and other elements necessary to the chemical balance of air, water, and life itself.

These and many other geographic systems set the framework for Earth's global environment. Within them the individual forces of change or destabilization are played out while stabilizing or holding forces such as gravity, molecular cohesion among rock particles, and the binding effect of plant roots in soil struggle to hold it all together. As Figure 1.4 suggests, the relative balance between the forces of change and those of stabilization (that is, which side wins the struggle) is defined by critical limits or *thresholds* and these thresholds vary from place to place over the Earth's surface.

Without the protection of plants, for instance, sand has a low threshold to erosion by wind. In other words, it takes a wind of only modest velocity to dislodge and blow away loose sand. But with a cover of plants, the threshold of erosion is raised enormously; under a forest, winds of hurricane force can barely dislodge a particle. We also know that wind velocities vary over the Earth, so the geographic pattern of wind erosion we see on maps is actually a product of both the distribution of driving force (wind) and resisting force (vegetation).

1.3 The Nature of Geographic Systems

Any set of things linked together by some kind of interaction constitutes a system. You deal with all sorts of systems every day. Highway systems in which cities and towns interact with each other via vehicles moving over the ground; telecommunication systems in which individuals interact with others via wires and electronic signals; and water systems in which communities of people interact with a river or lake by taking in water through one set of pipes and returning it (as sewer water) through another set of pipes.

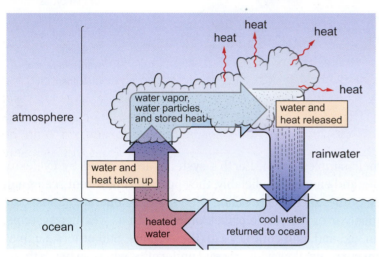

Figure 1.5 Geographic systems are characterized by interactions among different parts of the environment such as in this cycle that exchanges water between the atmosphere and the oceans.

All **systems** involve interactions and most interactions take the form of exchanges or cycles of energy and matter. In the water cycle example cited above (B in Figure 1.3), for instance, the atmosphere and the hydrosphere exchange matter when the oceans feed the atmosphere with water vapor and the atmosphere in turn feeds the oceans with rainwater, as shown in Figure 1.5. The exchange process is driven by energy, beginning with heat powering evaporation on the ocean surface. The heat energy in turn is transferred into the atmosphere with the vapor and then released into the air with the formation of clouds and rainwater. Thus we have a cycle, one of many that exists between the atmosphere and hydrosphere.

General Characteristics of Earth Systems: The idea of systems as cycles of energy and matter is central to understanding these magnificent constructs, but there is of course much more to them than this one characteristic. For example, all systems can be described in terms of their **geographic scale**, that is, how much Earth space they occupy and where they are found. Then there are also characteristics that describe their operation and organization. Five of the most important ones are:

Figure 1.6 In these diagrams the box represents a system. In an open system (a) energy and/or matter are able to move in and out. By contrast, in a closed system there are no inputs or outputs.

- *Whether they are open or closed.* **Open systems** receive inputs from outside their boundaries (see Figure 1.6a). Earth is an open system, because it receives energy (solar radiation) from the Sun and matter (meteorites) from space. The Earth also gives up energy and matter. It releases infrared radiation and molecules of gas to space from the top of the atmosphere. The balance between outgoing and incoming energy and matter currently favors incoming for both. Therefore, Earth is gaining energy and matter, but the gains are small for each. All the systems that operate on the Earth's surface, such as ocean currents, stream networks, and rainstorms, are open systems.

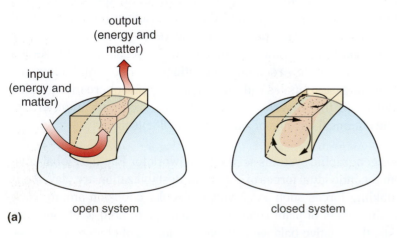

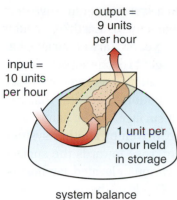

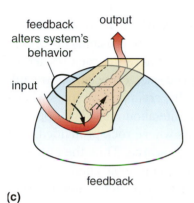

(b)

(c)

Figure 1.6 The diagram at (b) measures the inputs and outputs to determine the system's balance. The last diagram (c) represents the concept of feedback in which an outcome of the system comes back to affect the system's operation, making it slow down or speed up.

• *How much energy and/or matter they receive and release.* This characteristic, referred to as **system balance** or **budget**, addresses the relative quantities of inputs and outputs in an open system over time as illustrated in Figure 1.6b. Think of a banking system as a model. If it gains more energy or matter (that is, money) than it gives up over some period of time, it is positively balanced and the surplus goes into storage. If the reverse, it is negatively balanced and the reserve in storage is drawn down. When inputs and outputs balance, the system is in equilibrium. We know that the Earth's atmosphere currently has a positive energy balance because its temperature is rising, meaning that it is taking on or storing more energy (heat) than it is giving up. We will say a lot more about the warming of planet Earth later in the book.

• *Systems have the capacity to perform work.* When particles of sediment are carried by a river, or when moisture is moved by wind, work is being performed because matter is lifted and carried from one place to another. Energy is needed to drive systems to perform work, but when work is produced, it does not mean that energy has been consumed or lost. It means only that it has changed its form as, for example, from gasoline (chemical energy) to heat and kinetic energy as a car is propelled down the road. In a river, the kinetic energy required to do the work of moving water and its sediment is converted to heat from friction as the water and sediment rub and bounce along the channel.

• *Systems also possess the ability to change themselves.* Since they consist of exchanges of energy and matter, the outcomes from certain exchanges may affect the subsequent operation of the system itself. This is known as **feedback** (Figure 1.6c). In your body system it might work like this: You increase the intake of food which gives you more energy to undertake more physical activity and the increased activity burns more energy, which in turn causes you to eat more food. This is called positive feedback. Positive feedback occurs when the outcome of a process within a system speeds up or magnifies the system's activity or work output. Negative feedback results when an output from a process in a system slows down or dampens the overall operation of the system. If you eat bad food, for example, it curbs your appetite, causing you to eat less food which in turn dampens your activity and ultimately lowers your appetite even more.

• *Another general property of systems is the way in which they are organized.* Large systems are typically composed of an extended family of smaller systems, and these systems may consist of even smaller systems. This arrangement is called **hierarchical organization**. And here is where geography fits in well, because systems at different levels in the hierarchy operate at different geographic scales. Precipitation systems, for example, operate at several different scales as illustrated in Figure 1.7. Hurricanes are decidedly regional, covering large parts of oceans like the North Atlantic, whereas thunderstorms are typically local covering areas only the size of a city or smaller. And sometimes the systems in a hierarchical order are set or nested one inside another. Watersheds are arranged in this way. Little watersheds are nested in bigger watersheds, which in turn are nested in even bigger watersheds, and so on.

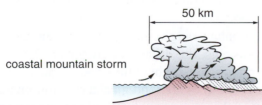

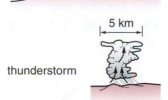

Figure 1.7 Three rainstorm systems which occur in nature at distinctly different geographic scales. All three perform basically the same function as systems in that they cycle water to and from the Earth's surface.

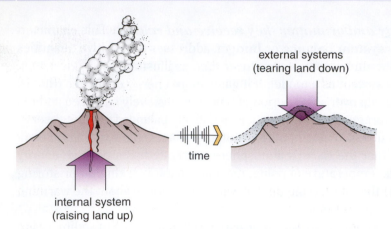

external systems
(tearing land down)

time

internal system
(raising land up)

Figure 1.8 A diagram portraying the Earth's surface as the product of the interplay between Earth's internal system that raises the land up and its external systems that tear it down.

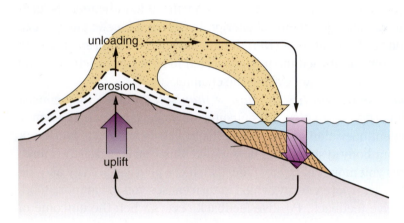

unloading

erosion

uplift

Figure 1.9 A simple feedback loop in which erosion lowers a mountain's mass which induces more uplift which in turn advances erosion.

The Systems Perspective: How can a systems perspective help us understand the geographic character of Earth? Consider this grand scenario. Over billions of years, Earth's surface has been hammered from above by the agents of one set of systems, the atmosphere and hydrosphere, which generated processes such as erosion driven by rainfall, runoff, and wind that cut, scoured, and sculpted the planet's surface. At the same time, from below, hot rock in the mantle, driven by another system, the Earth's internal energy system, rose towards the surface and lifted the crust up, creating various elevated landforms, such as plateaus and mountains. Think of these two sets of forces as they are shown in Figure 1.8.

At the Earth's surface, the processes which lifted the crust up and created mountains interacted with the erosional processes that were working to tear it down. Where the rock tended to be weak and/or the erosional processes especially powerful, the surface was worn down rapidly, creating low spots where water collected, soil formed, and life abounded. The opposite combination produced high spots where rock protruded, and soil, water, and life were relatively scarce. Thus the shape of the Earth's surface and much of its geographic character is the result of interplay between these great systems, one internal and two external.

The Role of Feedback: But there is more to the story, for at the point where the systems interact at the Earth's surface, they affect each other, forming complex feedback loops. Start with this simple principle: the greater the mass (weight) of the Earth's crust, the more energy it takes to lift it up. Therefore, as the erosional forces tear the crust down and it becomes lighter, the internal forces react by lifting it up, and when they do, they re-energize the erosional processes, as Figure 1.9 tries to show. The latter is explained by looking at a river. Where it flows down a mountain slope it has more erosive power than where it flows across a relatively flat plain.

The geographic aspect of this interplay is related to the fact that the work of neither the external nor internal systems is evenly distributed over the Earth. Where the internal forces are most concentrated, mountain ranges like the Rockies and Andes are built. As the mountain ranges grow, erosional processes grow more powerful, for the reason described above. But there is more.

The formation of the mountain ranges also influences the behavior of the hydrologic and atmospheric systems, the drivers of erosional processes. How so? First, because as mountain ranges get higher, they induce increases in precipitation. Therefore, more water is fed to rivers (and glaciers) in mountainous areas causing more erosion, and second, because mountain building also induces a reduction in atmospheric carbon dioxide, a major heat-absorbing gas, causing – if the mountains are vast enough – global cooling.

Figure 1.10 The world's largest mountain mass, the Himalayas and related mountain ranges. The formation of these mountains can induce climate cooling as carbon dioxide is drawn from the atmosphere with the exposure and breakdown of limestone in the mountains.

This change in the atmosphere's thermal balance takes place when limestone, a common rock of mountains, is broken down chemically in a process that requires carbon dioxide from the atmosphere. Therefore, imagine what happens when a huge mass of mountains like the Himalayas is raised up (Figure 1.10). The climate of south Asia changes dramatically. First, it grows wetter as the mountains get higher, and second, it grows cooler (at a scale probably affecting the whole planet) as carbon dioxide is pulled from the atmosphere. Now comes the feedback.

A Major Balancing Act: If the atmosphere is cooled with the loss of carbon dioxide, then its capacity to hold water vapor is reduced (because the capacity of air to hold water vapor is controlled mainly by air temperature). With less water vapor in the atmosphere, rainfall declines. With the decline in rainfall, stream flow and erosion decline, and so does the breakdown of limestone and with it the rate of carbon dioxide extraction from the atmosphere declines as well. With more carbon dioxide in the atmosphere, global cooling diminishes, and the planet's thermal balance returns to a more moderate state. Figure 1.11 traces this line of reasoning in seven steps. To understand these geographic changes requires, first, an understanding of individual systems, particularly the controls on their inputs and outputs, and second, how systems interact to create various forms of feedback.

Figure 1.11 Feedback from atmospheric cooling with the breakdown of limestone. The system begins at step **1** with the extraction of carbon dioxide from the atmosphere. The resultant cooling limits atmospheric water vapor content, reduces precipitation, and slows the rate of limestone breakdown, which in turn reduces carbon dioxide extraction, leading to a reduction in atmospheric cooling and perhaps a trend toward warming.

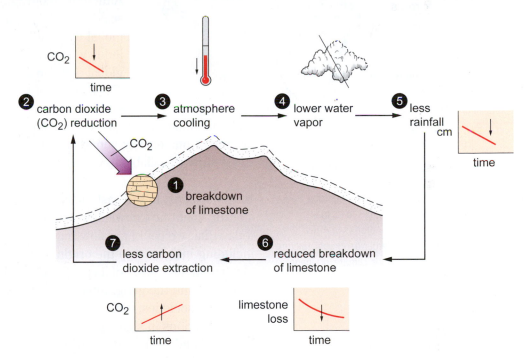

Summary on Systems: In these few pages we have argued *that* now more than ever we need to see the world as an interconnected whole in which we humans live within nature and its functions rather than apart from them; *that* in order to do this we have to take a systems perspective using the circumglobal networks, or great systems, of air, water, energy, rock, and life as the framework for our thinking; *that* in order to make sense of systems and how they affect the Earth geographically, we need to know that they:

(1) are open because they take in and release energy and matter;

(2) can be measured by their relative balance or budget based on how much energy and/or matter they take in and release;

(3) have the capacity to perform work including changing the Earth's surface;

(4) are self regulating via feedback mechanisms; and

(5) occupy space on the planet and operate at vastly different geographic scales.

1.4 The Physical Geographer's Perspective

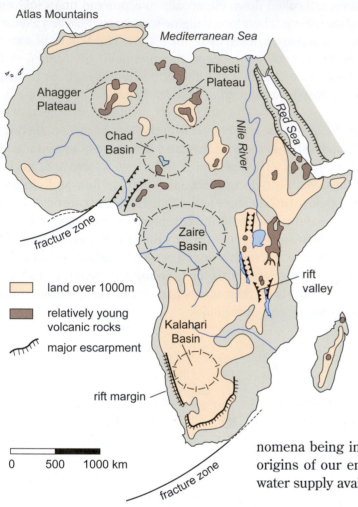

Figure 1.12 The heart of physical geography, like all branches of geography, lies in its maps. Here is a map of the major landforms of Africa. It represents one layer in the larger geographic picture of the continent. Others include climate, drainage, vegetation, and animals, including people.

Figure 1.13 Tens of satellites such as this one monitor the Earth from space, tracking weather, vegetation, snow cover, volcanic activity, and many other geographic phenomena.

Documenting and mapping change in the Earth's surface is central to physical geography. In fact, much of the field is organized around a study of the agents that change the geographic character of the Earth: atmospheric processes, such as hurricanes and thunderstorms; hydrologic processes, such as flood flows and storm waves; biological processes, such as forest growth and the construction of beaver dams; human processes, such as agriculture and urban development; geologic processes, such as volcanoes and earthquakes; and so on. Physical geography seeks to explain the distribution of the various forms and features that constitute the landscapes of our planet by analyzing the processes and systems that create them. Like all branches of geography, physical geography has an abiding interest in the spatial patterns and geographic dimensions of Earth's surface phenomena, like those shown in Figure 1.12, and why they differ from place to place.

Although geography is traditionally associated with global scales of study – and indeed we are today concerned with global distributions as much or more than ever – geographic inquiry is actually carried out at all scales from beyond the global down to areas the size of one's community and smaller. The key is that geographic inquiry must be taken to whatever scale is necessary to provide scientific explanations of the phenomena being investigated. This may take us into the Solar System to explain the origins of our energy supply or into the tiniest particles of the soil to explain the water supply available to vegetation.

New Horizons: In the past quarter century we have witnessed a tremendous explosion of scientific knowledge. You have an opportunity to know more about Earth today than we could possibly have imagined when we were little boys in the 1950s, including insights into the origin of our planet and the geological mechanisms that govern the formation of its continents and ocean basins; the mechanisms governing evolution and the geographic diversity of life; the origins, growth, and development of human societies; the intricate interrelationship of the atmosphere and oceans and its implications in shaping the global climate; and the origins and spread of people, land use, and food-production systems.

In geography, our ability to monitor and map the planet improved immeasurably with advancements in remote sensing and computer data processing. Remote sensors in aircraft and satellites, such as the one depicted in Figure 1.13, allow ready surveillance of the atmosphere, seas, forests, land uses, water systems, and many other geographic features on a day-by-day basis. With these advances has come not just greater factual knowledge but vastly improved understanding about how the planet works, with special insights on how its various parts are interconnected.

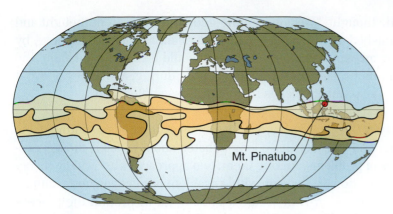

Mt. Pinatubo

Figure 1.14 The airborne ash from the 1991 explosion of Mt. Pinatubo in the Philippines was carried around the world by global wind systems. The same can take place with dust from deserts and smoke from wars and forest fires.

Our Unified Planet: A picture has emerged of Earth as a geographically unified planet, in which distant and often remote parts are linked together by the great systems. These systems are organized in an amazing variety of ways that in many respects we are just beginning to understand. Witness, for example, that changes in atmospheric pressure in the western Pacific can produce a string of changes having profound influences, via an ocean warming event called El Niño, on climate 10,000 miles away in western North America. And deep within the Earth at the base of the mantle, 2000 miles down, pockets of hot, partially melted rock appear to send heated plumes slowly toward the crust where they may fuel clusters of volcanoes. And the global systems of prevailing winds commonly carry the plumes of ash from large volcanic eruptions entirely around the world as the tropical tradewinds did with the debris from the 1991 Mt. Pinatubo eruption in the Philippines, darkening and cooling the atmosphere under them (Figure 1.14). Indeed, discovering the geographic interconnectedness of Earth as a planet may well emerge as one of the great scientific advancements of our times.

1.5 Scientific Thought and its Application in Physical Geography

The story we tell in the pages ahead is part of a great tradition practiced in some form by virtually all peoples, modern and ancient, to explain the nature and origin of the world around them. Our story, however, is different than most Earth stories because it is based on science and most traditional stories are not. So what is it that sets the science story apart from, say, those based on mythology or religion? Both involve considerable thought, both address tough questions, and both provide answers. The difference is the way the thought process is structured.

Science is a way of arriving at understanding by observing, measuring, testing, and reasoning. In the modern world it is often confused with its tools. Science is not, as is commonly thought, the development and application of technology and technical apparatus. Technology and its apparatus, such as microscopes, radars, cameras, telescopes, and computers, are very much a part of science but exist as tools in making observations, measurement, and analyzing data rather than as objects of scientific study themselves.

Realms of Thought: In a general way, the world of thought can be divided into two broad realms: physical and metaphysical. *Metaphysical thought* is based on things such as feelings, beliefs, dreams, and emotions. It often relates to the physical world, as in the way romantic poetry such as Walt Whitman's lines in Figure 1.15 relate to sounds, smells, wind, and morning feelings, but it does not require measuring, testing, or logical discourse. Music, religion, mythology, and much of literature are rooted in metaphysical thought.

Science belongs to the realm of *physical thought*. At its simplest level, physical thought deals with the world we can see, touch, and smell and how to portray what we learn without invoking feelings or beliefs. When people invoke the metaphysical, such as ideas from mythology, to explain natural phenomena like a massive hurricane, they

---

The smoke of my own breath.
Echos, ripples and buzzed whispers....
*   loveroot, silkthread, crotch and vine,*
My respiration and inspiration....the beating
*   of my heart....the passing of blood and*
*   air through my lungs,*
The sniff of green leaves and dry leaves, and
*   of the shore and darkcolored sea-rocks,*
*   and of hay in the barn,*
The sound of the belched words of my
*   voice....words loosed to the eddies of the*
*   wind,*
A few light kisses....a few embraces....a
*   reaching around of arms,*
The play of shine and shade on the trees as
*   the supple boughs wag,*
The delight alone or in the rush of the
*   streets, or along the fields and hillsides,*
The feeling of health....the full-noon trill....
*   the song of me rising from bed and*
*   meeting the sun.*
Have you reckoned a thousand acres much?
*   Have you reckoned the earth much?*
Have you practiced so long to learn to read?
Have you felt so proud to get the meaning of
*   poems?*

                    Walt Whitman (1819–1892)

Figure 1.15 An excerpt from Walt Whitman's book *Leaves of Grass,* first published untitled in the collection in 1855. An example of metaphysical thought that draws on images from the physical world but makes no claims to scientific reasoning.

A huge hurricane sweeping across Florida and the East Coast of the United States.

are not using scientific thought. "We angered the gods with irreverent thought and behavior and they punished us with a destructive hurricane." Scientific thought, by contrast, would pursue an explanation based on what could be discovered about the hurricane from observations and measurements before, during, and after the event followed by a course of reasoning which is built on the assembled facts.

Induction and Deduction: There are several approaches or methods of putting facts and reasoning together and scientists have long debated about which is most appropriate for various types of problems. Among them, two approaches are widely recognized: induction and deduction. Both begin with a problem or question; for example, what causes the formation and growth of hurricanes? With **induction** the process involves gathering copious data on where hurricanes form, their dimensions, air conditions within, above, and around them, their patterns and rates of movement, and so on. These data are then assembled in some fashion, for example, in maps and tables, and, with the help of computers, examined in various ways to see if geographic patterns, trends, and interconnections can be identified.

The data may reveal nothing. On the other hand, maps plotting the geographic locations of hurricane formation, as the map in Figure 1.16 reveals, may show that they always begin and develop over oceans in the tropics. But is this mere coincidence or is there something meaningful about the connection with warm ocean water? Further data are called for and if the tropical-water correlation repeats itself in different oceans over many years, then we may have discovered something meaningful. But strictly speaking we still do not have an answer on what causes hurricanes. We may, however, have narrowed the search.

Deduction, on the other hand, involves putting our efforts first into a line of reasoning rather than into the collection of data. We usually begin with some things we already know from established scientific principles. For example, from physics we can reason that it takes a massive force to grow and drive a weather system the size of a hurricane. With a few calculations, we can even approximate the amount of energy involved. What Earth force can provide that much power? Heat in the atmosphere from the absorption of solar radiation? The tidal force of the solar and lunar gravitational fields acting on the ocean surface? Heat stored in ocean water? Heat from volcanic activity on the ocean floor? A comparison of the

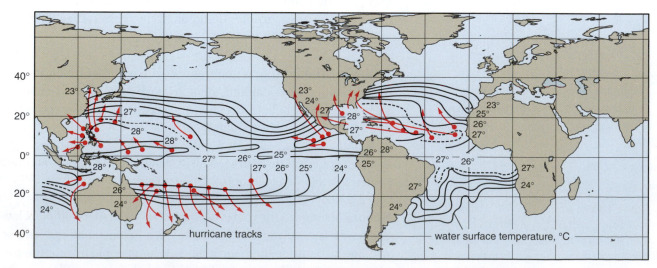

Figure 1.16 Data on the locations of origin of hurricanes (red dots) reveal that all begin over warm tropical seas.

energy required for a hurricane and that available from these sources reveals that all are more or less capable of powering one of the great storms. But we need to know more to advance our thinking and here is where the results of the inductive survey can play a part.

Among other things, the inductive survey revealed that hurricanes consistently originate and develop over warm ocean water. They never develop over land and show no geographic relationship with tides or volcanoes. That leaves us with the heat of air and water. The heat of air is an unlikely source because hot air is also found over land, in fact, the air over tropical landmasses is even hotter than that over the oceans but the landmasses never produce hurricanes. Moreover, hurricanes produce prodigious rainfall and that requires a prodigious supply of water vapor which is not widely available over land. That leaves us with the oceans as a source of heat (and water vapor) and a calculation of the heat available in tropical ocean water (based on temperature data generated in our inductive survey) reveals that there is enough heat energy stored in these great water bodies to spawn and drive scores of hurricanes every year.

Toward an Hypothesis: We now have a concept that with further refinement can lead to a **scientific hypothesis**, that is, to a proposition that can be tested with further measurements. We propose that hurricanes form and are driven by heat energy from tropical seas and this heat is somehow delivered to the atmosphere with water vapor produced from the evaporation of seawater. The quest will reveal that heat energy, called latent heat, is delivered to the atmosphere as part of the water vapor released from the ocean surface in evaporation in the manner shown in Figure 1.5, and that the evaporation process itself is driven by the heat in warm surface water. This begins to explain both the geographic distribution and the seasonal character of hurricanes and it opens the door to a long string of additional questions. How warm does seawater have to be to spawn hurricanes? What processes actually trigger their birth? What controls their patterns of movement?

So which approach, induction or deduction, was most instrumental in our investigation? The answer, of course, is that they both were instrumental. But did we answer the big question on what causes hurricanes and enables them to grow as they do? Strictly speaking we have to say, no. But we narrowed the focus of scientific investigation by defining a candidate source of energy to develop and drive them. And with this finding we have moved closer to a credible hypothesis that, once formulated, can be tested again and again under different circumstances in different places as the diagram in Figure 1.17 suggests. With each test our understanding of this one aspect of hurricanes can be refined and that knowledge will lead to new questions, further investigations, and deeper insights into this astounding phenomenon, and with it more understanding of two of the Earth's great systems, the atmosphere and the hydrosphere.

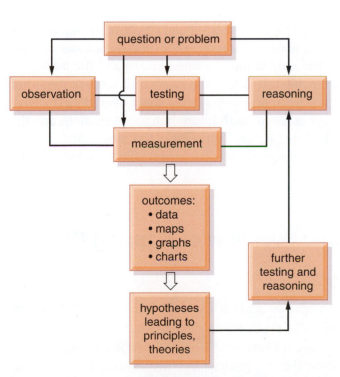

Figure 1.17 The general nature of the scientific method, including both induction (observation and measurement) and deduction (testing and reasoning). A scientific finding is refined by repeated observation, measurement, testing, and reasoning.

1.6 The Concept and Objectives of this Book

Such discoveries have helped shape modern physical geography, and have profoundly influenced the development of this book. In fact, we have given great systems center stage as the framing theme throughout the book. All chapters after this one are devoted to the Earth's great systems and their offspring, beginning with the global energy system and followed by the atmospheric/oceanic system, the biogeographic system, the soil system, the hydrologic system, the geologic system, and the geomorphic (erosional) system. Each chapter weaves a systems perspective into the conventional matter of physical geography in an effort to illuminate the processes and features of both nearby and distant landscapes.

Primary Objective: The book's primary objective is to examine the content and formation of Earth's landscapes so that we may fulfill our intellectual need to understand our planet. This entails a description of the component parts of the terrestrial environment, namely, climate, soils, landforms, water features, and plants and animals, how they are formed, how they interact, and how they are distributed over the Earth. We want to see landscapes as dynamic environments where a host of processes driven by these great systems interact with different results. This must also include the activities of humans because people have become a powerful agent of landscape change as the sculpted hillslopes in Figure 1.18 attest. In fact, it can be argued that humans have in the last several centuries become the most active landscape-change agent on the planet. With world population currently growing by nearly 80 million people per year and technology and consumption surging forward, this role is certain to increase in your lifetime.

Secondary Objective: This brings us to the book's second objective: to improve our understanding of Earth in order to make more responsible decisions about our use of the planet. Earth is the one and only home of human beings and the landscape is the very heart of the human habitat. It is also the habitat for most of the other organisms on the planet. We must learn how to survive without degrading the environment and reducing the quality of life for both future human generations and the other organisms. This is the challenge before humanity in the twenty-first century and it cannot be successfully met without substantial geographic knowledge in the hands of citizens and leaders.

Figure 1.18 Terraced farmland in southern Asia. Once tree-covered, millions of acres of these hills have been sculpted by humans into a landscape of rice paddies in which people rather than nature are the chief controls on water, soil, and vegetation systems.

Chapter Summary and Overview of Mapping Our Course of Study

So we end this first chapter having taken a serious look at a scientific construct called a system and its application to a field of study called physical geography. We argued that in order to understand how things like climate, soils, and landmasses form and change, it is necessary to examine the systems that govern them. Thus, our geographic understanding of the world must go beyond recording where things like deserts and tundra are located and what they are made up of, and must address meaningful questions on how they operate and change over time, for Earth is not a static planet. This is a daunting task that requires scientific thinking – observing, measuring, testing, and reasoning – and as we heap the stress from a rapidly expanding human population upon the ever changing and complex Earth stage already set by nature, the challenge for physical geographers, indeed all of us, is monumental in the least.

▶ **The modern world is pulling us away from our geographic roots.** Yet we long to know the land and nurture our acquaintance with nature. Physical geography advances our understanding of nature and helps us become responsible citizens of planet Earth.

▶ **Earth is constantly changing in response to the work of great systems.** These systems move energy and matter and operate at different rates and geographic scales from the Earth's core to the top of the atmosphere.

▶ **Systems thinking is essential to understanding Earth at all geographic scales.** Most Earth systems involve cycles which are driven by energy and move water vapor, carbon dioxide, ocean water, sediment, rock masses, and scores of other substances among Earth's spheres.

▶ **Geographic systems share several common attributes.** These include the balance between inputs and outputs, organization by geographic scale, and self-adjustment via feedback.

▶ **Feedback in open systems takes two forms.** One produces an amplifying effect on the systems output; the other has a dampening effect on the system's output.

▶ **Great systems are broadly defined by the Earth's global spheres.** They include the atmosphere, hydrosphere, biosphere, and the lithosphere and each contains a whole family of smaller systems.

▶ **The Earth's surface is shaped by the interplay of stabilizing and destabilizing processes.** The relative balance between these forces is defined by thresholds in the landscape, which vary geographically over the Earth.

▶ **Physical geography is concerned with documenting and mapping change in the Earth's surface.** Geographers have an abiding interest in the spatial patterns and geographic dimensions of Earth's surface phenomena and why they differ from place to place.

▶ **Geographic inquiry has benefited from an explosion of scientific knowledge.** Surveillance of Earth's atmospheric and surface conditions has been advanced by remote sensors in aircraft and satellites. The information reveals that Earth is a geographically unified planet tied together by systems that range over the entire globe.

▶ **Science is a way of thinking.** It is based on what can be observed and measured in the world around us. Two approaches are generally recognized in scientific investigations and both begin with a question or problem.

▶ **This book has two main objectives.** The first is to fulfill our intellectual need to understand our planet. The second is to improve our understanding of Earth in order to make more responsible decisions about our use of the planet.

Review Questions

1 What sort of population shift is humanity currently experiencing and what are some of its geographic consequences?

2 On the question of how we come to know nature, do you view it as essential to our existence that we possess a sense of nature and our place in it? Why?

3 How is it that the atmosphere and lithosphere qualify as systems and what makes them great systems?

4 Explain the concept of thresholds in the landscape with respect to stabilizing and destabilizing forces?

5 To become a systems thinker requires looking at the world around us in a particular way. Just what is it that we look for that sets systems thinking apart from non-systems thinking?

6 Name a system with which you are familiar and identify its component parts, its interactions, and its geographic configuration.

7 What evidence is there to justify calling planet Earth an open system?

8 How is it that mountain building and atmospheric carbon dioxide can interact to change atmospheric temperature and where does feedback come into play in this scenario?

9 What is meant by the remark that the Earth is a geographically unified planet? Cite a supporting example or two.

10 Distinguish between the metaphysical and physical realms of thought. Which of the approaches used in the scientific method relies mainly on the generation of extensive data in response to a question or problem?

11 Why should we consider humans a legitimate part of physical geography?

12 Why should we need to understand the geographic character of the Earth in order to be responsible citizens in today's world?

An Overview of Planet Earth:
Some Geographic Observations and Facts

Chapter Overview

Earth is a diverse and colorful planet and there is lot to be learned about it by merely looking down on it from space. Accordingly, we open this chapter with some simple visual impressions of our planet. The first is based on the colors Earth exhibits to the space observer and the second considers the meaning of the patterns we see, particularly the distribution of life and the large-scale motion represented by the great whirls and swirls of the atmosphere and oceans. We then discuss Earth's physical dimensions and how we use a system of north–south and east–west lines, called meridians and parallels, to reference and document locations and other geographic phenomena on the planet's surface. This leads to mapping, the basic properties of maps, and the sorts of maps most commonly employed in geography. And finally, any examination of Earth in physical geography must include time, for the planet's calendar is long, more than four billion years, and filled with events that help explain where we are today geographically.

Introduction

Earth's first great navigators were the Phoenicians, who determined the location of their ships in terms of the length and width of the Mediterranean Sea. The Phoenicians' navigational methods were taught to the Greeks and eventually to the Romans. When translated into Latin, the perpendicular directions of length and width became the longitude (*longus*, or long) and latitude (*latus*, or wide). *Latitude* referred to locations north and south whereas *longitude* referred to locations east and west.

Early sailors used only the latitude in their calculations of geographic position, since all that was required to determine latitude was an instrument capable of measuring angles. One such instrument, shown in Figure 2.1, was the astrolabe, and it is among the earliest known devices for measuring the angle between a star and the horizon. Angular measurements of Polaris (the "North Star") gave sailors the latitude in the northern hemisphere, while other stars were used for southern navigations.

Figure 2.1 A medieval drawing depicting a surveyor instructing a navigator on how to determine latitude by measuring the angle between a celestial body and the horizon.

Sailors did not use longitude because, owing to the constant spin of the Earth, they had no way of measuring it. Simply put, whenever they looked to the east or west, the sky (field of stars) was continually changing. However, they knew there were time differences between locations east and west of each other, such as between Athens and Lisbon, and therefore made the connection between time and longitude. But determining the exact longitude required a comparison of the exact times between two places, specifically, between the local time (such as high noon) and Greenwich Mean Time (GMT) – the high-noon time at the Royal Observatory in Greenwich, England. For this task, a highly accurate clock was required, and in 1714, the English Parliament offered a £20,000 reward for anyone who could find the longitude and demonstrate the method used on a voyage to the West Indies.

The race was on and dozens of inventors took up the challenge. John Harrison, a self-taught watchmaker, developed a chronometer that could withstand the rigors of sea travel – salt, moisture, and shock – and in 1761 a test trip to the West Indies verified its accuracy. Harrison, for all his hard work initially received only half the prize after four years, since the board governing the competition included a number of scientists who still hoped to pocket the money themselves. Only the intervention by King George III got Harrison his full payment, seven years later. With the capability to measure longitude accurately, the inland exploration of continents proceeded rapidly. Maps and descriptions from these journeys often depicted the precise locations of geographical features. In the account of their 1804–1806 expedition, Lewis and Clark, for example, describe and commit to maps such as the one shown in Figure 2.2, the characteristics and coordinates – that is, longitude and latitude – of river intersections, bluffs, mountains, Indian settlements, and scores of other features.

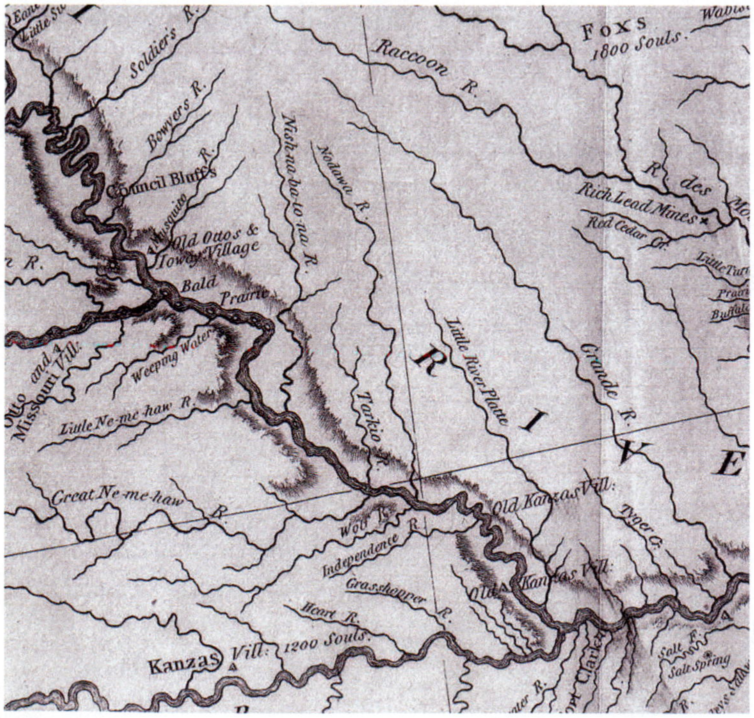

Today maps are made with the assistance of aerial photographs, satellites, and computers. For example, orbiting satellites record the extent of the polar ice packs daily, and from the digital maps made from these data, scientists have traced their rapid decline over the past 40 years. Using computer mapping systems called Geographic Information Systems (or GIS), vast quantities of data about the Earth are collected, analyzed, and displayed by international agencies, national and local governments, universities, and private organizations. The information produced is incredibly diverse, ranging from hurricane damage assessments and watershed conditions to the best locations for a highway or even a new restaurant.

Figure 2.2 A map of a section of the Missouri River from the Lewis and Clark Expedition, 1804–1806, evidence of the remarkable cartographic accuracy possible by the opening of the nineteenth century.

Yet, with all this technological power, the basic process of constructing maps is still the same today as it was in the twelfth century, when the great cartographer Muhammad al-Idrisi was producing maps of the world which would be used by navigators for hundreds of years. To produce a map, you must specify the scale (the area or extent of the Earth's surface the paper map will represent), select the information you want to portray, and then design the map to convey this information clearly to the users. As a result, the ability to observe the Earth effectively, and then organize specific information, is critical to making usable maps.

2.1 Some Geographic Observations about Earth

Earth is one of eight planets in the Solar System. It belongs to an inner set of four small planets called the **terrestrial planets**, so named because they have hard rather than soft (gaseous) surfaces. Earth is the third planet from the Sun, lying between Venus and Mars, at distance of 150,000,000 kilometers from the Sun. Thermally, it is a mild planet, warmer than Mars and much cooler than Mercury and Venus, and it is a wet planet (see the images in Figure 2.3), the only one in the Solar System with a massive supply of liquid water.

Over 97 percent of Earth's water is held in the oceans and these great reservoirs cover 70 percent of the planet's 510 million square-kilometer surface. The remaining 30 percent (148 million square kilometers) is covered by land, represented mainly by six great landmasses. Eurasia, with a land area of nearly 55 million square kilometers, is decidedly the largest landmass on Earth (Table 2.1).

First-level Impressions: From space, one's first impression of Earth is apt to be its distinctive color scheme dominated by blue, white, brown, and green. No other planet in the Solar System is so brilliantly colored, as the comparison with Mercury reveals. The colors are clues to Earth's geographic diversity and organization. Indeed, they are important in today's geographic analysis of Earth because color is one of the attributes detected by the remote sensors mounted in aircraft and satellites and used in building maps of vegetation, land use, storm clouds, and many other phenomena.

The oceans, whose color is produced by blue rays of solar radiation reflected from the sky, and by the scattering of sunlight by the oxygen in the water, are dominated by three great water bodies, the Pacific, Atlantic, and Indian oceans. The oceans are the mainstay of Earth's hydrologic system, supplying the atmosphere with most of its water, which it in turn redistributes around the planet. The oceans are also important in maintaining the Earth's climatic and chemical systems, taking in, storing, and releasing heat, water, carbon dioxide, oxygen, and a host of other elements and minerals.

Although deep blue is the predominant ocean color, near the edges it gives way to other tones and colors, which are indicators of the geographical conditions there. Most are associated with various seas, gulfs, and bays on the margins of the oceans. These are important border environments where the sea and the land intermingle forming some of the richest and most diverse habitats on the planet. And they are often the places where new land is being built such as river deltas, coral reefs, islands, and urban-development fill in shallow waters. As the image in Figure 2.4 shows, these waters may be light blue or tan or even multi-colored depending on climatic conditions, sediment carried by waves and currents, sediment discharged by rivers, the abundance and type of marine life, and pollution emanating from coastal cities.

Clouds and ice caps partially blanket the land and oceans, covering half or more of the planet at any moment. They appear bright white because much of the sunlight striking them is reflected skyward. The Antarctic cap is larger than the Arctic ice cover where much of the ice is sea ice, not glacier ice as in the Antarctic. The polar ice caps are Earth's largest reservoir of freshwater, holding about 75 percent of the

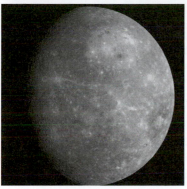

Figure 2.3 Earth and Mercury from space. Mercury is bone dry and scalding hot. Earth is cool and swathed in water.

Table 2.1 Area of Earth's landmasses (km²)

Continent	Area in km²
Eurasia	54,650,000
Africa	30,300,000
North America	24,350,000
South America	17,870,000
Antarctica	13,990,000
Australia	7,680,000

Figure 2.4 Coastal water where the deep blue of the sea changes to lighter tones with shallower depths.

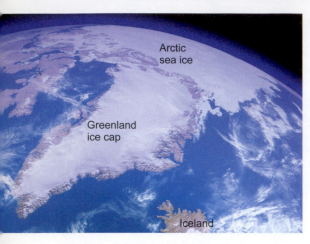

Figure 2.5 The Greenland ice cap, the Earth's second largest reservoir of freshwater. Ice on land and sea covers about 15 percent of the planet.

Figure 2.6 Earth shrouded in clouds. Unlike our neighbor Venus, whose clouds are heavy with sulfuric acid, Earth's clouds are made up of water particles.

world supply. Currently, Earth's ice cover, including the Greenland ice cap shown in Figure 2.5, is shrinking, a trend that is expected to accelerate with continued global warming in this century.

Clouds are composed of water particles that are too small to be pulled out of the atmosphere by Earth's gravity. They are the building blocks of larger precipitation particles leading to rain and snow, the only source of water on the land. Besides their substantial coverage of the planet, as the image in Figure 2.6 bears out, the most striking thing about clouds from our perspective in space is their great, flowing patterns, the products of an active, stormy atmosphere rich in water. Together the clouds and ice give Earth most of its star-like brightness in space. Earth is brighter than the Moon but not as bright as Venus, the only terrestrial planet completely shrouded in clouds.

Excluding the areas covered by ice and snow, the Earth's landmasses are dominated by green and brown, which form great circumglobal swaths trending roughly east–west. There are two green swaths, both apparent in the image in Figure 2.7, one along the Equator and one in the middle-latitude zone of the northern hemisphere, stretching across Eurasia and North America. (There is no comparable middle swath in the southern hemisphere because there is very little land in this zone south of the Equator.) The hearts of both green swaths are occupied by the world's great forests: the temperate and boreal forests in the middle latitudes and the tropical forests along the Equator. On their margins, the forests give way to grasslands and grasslands mixed with trees.

The great belts of green represent the biological core of Earth. The tropical forests are estimated to contain at least half of the world's more than 10 million species of organisms. They are also the most productive biological environments on the planet, where heavy covers of trees and other plants manufacture large quantities of organic matter, through a process called photosynthesis that also helps maintain Earth's chemical balance by recycling gases such as oxygen, water vapor, and carbon

Figure 2.7 Two great swaths of green are visible from space, one in the middle latitudes of the Northern Hemisphere and the other in the tropics. They contrast with great belts of brown where the atmosphere is usually clear of cloud cover.

Figure 2.8 (a) Extensive agricultural development (speckled pattern) around Winnipeg in the North American Great Plains; (b) Massive urban development at the mouth of the Hudson River centered on New York City. This image covers a swath less than 20 miles (32 kilometers) wide and contains five million people or more.

dioxide. One of the most challenging problems in modern geography is documenting the loss of the tropical forests and analyzing the causes and effects of this serious human impact on the global environment.

The middle-latitude belt is less rich biologically, but is singular for the huge number of humans it supports. At least half of the world's 7 billion people live in this zone. It is the world's most prosperous agricultural zone, where vast areas of forest and grassland have been cleared for cultivation (Figure 2.8a). This system of food production is tied to the world's leading industrial economies in North America, Europe, and Asia, which in turn have fostered massive urban development – for example, New York City, which is shown in Figure 2.8b. The resources consumed to support these urban-industrial economies and the high volume of waste they release to the environment in the form of air and water pollution are the main environmental problems of the middle-latitude zone. Once local or regional in scale, these problems now reach deep into the global environment and are changing planetary climate and related geographic systems.

Finally, there are two great swaths of brown visible from space and apparent in Figure 2.7. Each lies between the tropical and middle-latitude green belts, but the swath in the northern hemisphere is much larger than its southern counterpart because of differences in land area. Curiously, the brown swaths are the most visible geographic zones on Earth's surface because the cloud covers there are light or absent. Is there a connection between a cloudless atmosphere and a brown landscape?

Indeed there is, and if we look to the adjacent green swaths, relatively heavy cloud covers are apparent there. Obviously, the Earth's alternating belts of green reflect different climatic zones: one dry, the other wet. But this simple observation raises

Figure 2.9 Swirls of clouds mark storms, one of Earth's most dynamic features and pivotal mechanisms in the planet's water system.

many questions. What mechanisms cause the atmosphere to arrange the distribution of clouds and precipitation in this manner? Should not the driest part of the Earth coincide with what is probably the warmest part, which is located along the Equator? And are these geographic patterns permanent or do they change with the seasons and longer periods of time?

Second-level Impressions: If we watch our planet awhile, we will see that it is constantly changing. Change is most vividly expressed in the motion of the fluids (air and water) that encase the Earth. A clue to the motion of the atmosphere is provided by its cloud cover, especially the great white swirls, like those shown in Figure 2.9, that slide across the planet's face rotating like giant pinwheels. These are the planet's biggest storms and they are products of Earth's atmospheric circulation system, cyclones in the midlatitudes and hurricanes and related storms in the tropics. Like the larger global circulation system, they are driven by solar energy, which we examine in the chapters ahead. Among other things, they bring freshwater to the continents, which in turn gives rise to another system, one of paramount importance to the terrestrial environment. We are referring, of course, to stream systems, the systems chiefly responsible for eroding the land, shaping landforms and ecosystems, and bringing sediment to the seacoast.

Extend our observations by a few months and all this may undergo sweeping changes. Among them are changes in the geographic patterns and the sizes and frequencies of occurrence of storm systems. And with these changes come broad geographic changes in the nature of the land and ocean surfaces. In the midlatitudes, as the belt of circulation and cyclonic storms strengthen with the onset of fall, the whole system shifts equatorward and snow and ice slowly creep down from the poles over both land and water. This is one measure of winter, but it reaches only about half way to the Equator on land and less on the oceans (Figure 2.10). As a geographic boundary, the leading edge of winter is, like most geographic boundaries, seldom stationary because it too is a product of Earth's changing systems.

Figure 2.10 The change of seasons. Snow cover in the northern hemisphere. (a) In September 2005, only Greenland has snow cover. (b) Six months later (March 2006) the snow line reaches well into the midlatitudes.

A few more months of observation reveals another set of geographic changes. The winter storms recede, the circulation system and its storms weaken, and the snow and ice covers of winter retreat. On land, stream systems rise with meltwater, some flood, and virtually all expand their capacity for work, that is, the capacity to shape their valleys and carry sediment to the sea. All this and much more we can observe by merely watching the planet from some vantage point in space. What we cannot see of Earth's dynamics is a much bigger story, of course, but it requires going down to the surface for instrumental observation and measurement. This will give us a glimpse into forces that drive Earth's great systems and the nature of the change that is fundamental to understanding our planet's geography, including the life system, the oceans, and the Earth's interior as well the atmosphere near the ground.

(a)

(b)

Summary on Earth from Space: It takes only a quick glance to conclude that our planet is colorful and geographically diverse, but it takes more than a glance to figure out that the patterns we see are not chaotic, but are part of a basic geographic order. It takes a much longer look to figure out that that patterns are not motionless, but change over different time frames, and that there are correlations among patterns that give us clues about cause-and-effect relations, such as between a cloudless atmosphere and the brown, dry desert landscapes.

2.2 The Shape, Size, and Geographic Organization of Earth

In order to carry our discussion of Earth much further, we need to learn a few hard facts about the planet including its shape, size, and geographic organization. Let us begin with its motion and shape. Our planet **rotates**, or spins on its axis, as it **revolves**, or orbits the Sun. Rotation produces a centrifugal effect, that is, matter in or on the body tends to be thrown to the outside. The faster the rotation, the greater the centrifugal effect. Therefore, at the Equator, where the rotational velocity is highest (greater than 1000 miles per hour), rotation produces a slight bulge which causes the Earth's shape to depart slightly from that of a perfect sphere. This shape, Earth's true shape, is called a **geoid**. If we expand our view to include the atmosphere, Earth's geoid shape is even more pronounced because the centrifugal effect combines with air heating and expansion to produce a thicker atmospheric layer over the equatorial zone and a thinner layer over the poles.

The shape of the solid Earth can be described using the two circles shown in Figure 2.11. One, the **Equator**, is drawn east–west directly around the middle of the planet. The other, which we will call the polar circle, is drawn north–south through the poles. The radius of each circle is the distance from the center of the Earth to a point on the circle at the Earth's surface.

The shape of the Equator is very close to a true circle, with a radius of 6378 kilometers (3963 miles). The polar circle, however, has more the shape of an ellipse rather than a true circle, with a radius of 6357 kilometers (3950 miles) – 21 kilometers less than the equatorial radius. Map makers must take the 21 kilometers difference into account when they construct high-precision maps; however, from space Earth appears as a perfect sphere and it functions as a true sphere in terms of the motion of the atmosphere, oceans, and other essential systems. Therefore, we will consider the Earth a true sphere with a radius of 6371 kilometers (3959 miles) and, as we noted earlier, a surface area of 510 million square kilometers (197 million square miles).

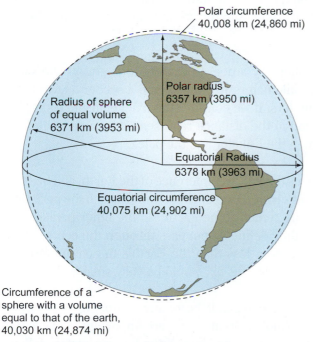

Polar circumference
40,008 km (24,860 mi)

Polar radius
6357 km (3950 mi)

Radius of sphere
of equal volume
6371 km (3953 mi)

Equatorial Radius
6378 km (3963 mi)

Equatorial circumference
40,075 km (24,902 mi)

Circumference of a
sphere with a volume
equal to that of the earth,
40,030 km (24,874 mi)

Figure 2.11 The true shape of the Earth geoid. The polar radius is 21 kilometers less than the equatorial radius, making Earth slightly less than a perfect sphere.

This is the logical point to introduce the geographic reference system used to map and describe locations on the Earth's vast surface. This is the system of crisscrossing lines you see on the globe and other maps, and it is called the **global coordinate system**. It is simply a network, or grid, of lines running east–west and north–south over a geographic area. On a flat surface, such as that used for laying out the streets and property lines in most towns, constructing a coordinate system simply involves building a square or rectangular grid of lines intersecting at right angles. On a curved surface, the job is somewhat more difficult. For the Earth, constructing a coordinate system involves measuring angles on the equatorial and polar circles. We can best describe this system by showing how the lines are drawn.

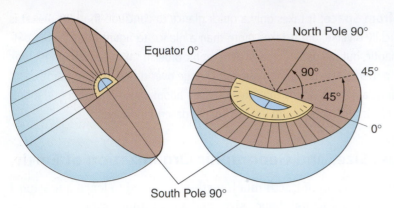

Figure 2.12 Constructing parallels using a protractor to measure the angles between the equatorial plane and the poles.

Parallels and Latitude: The east–west lines are called **parallels**. They are all drawn parallel to the Equator and each is a circle, but no two are the same size because the circumference of the Earth grows smaller from the Equator to the poles. The first step in constructing parallels is to bisect the globe along the polar circle, that is, north–south. Next, a protractor is placed on the plane of bisection, with the base of the protractor aligned with the Equator as is shown in Figure 2.12. Starting at the Equator, angles northward to the pole are ticked off; the procedure is repeated toward the South Pole. The angles are then numbered, beginning with 0 at the Equator and ending with 90 degrees at each of the poles. The point at the edge of the circle (the Earth's surface) marking a 45 degree angle is where the 45 degree parallel is drawn as is illustrated in Figure 2.13. This is the reason why parallels are numbered in degrees – because they represent angles between the Equator and one pole drawn from the center of the Earth.

Parallels are used to measure latitude. The **latitude** of any place on the Earth's surface is a reference to location north or south of the Equator, and it is given in degrees, minutes, and seconds (symbolized °, ′, ″, respectively). There are 360 degrees in a full circle; each degree is divided into 60 minutes and each minute is divided into 60 seconds. In this book, our considerations of angles are generally not precise enough to justify their measurement to the second, so we will generally use just degrees and minutes.

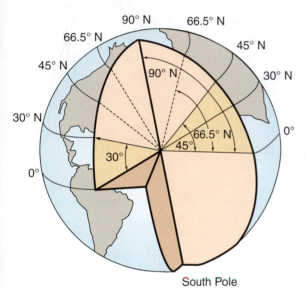

Figure 2.13 Latitude is measured in degrees beginning at the Equator (0 degrees) and ending at the poles (90 degrees north or south).

The distance represented on a sphere by a degree of latitude is always the same. On the Earth, this distance is 111 kilometers (69 miles), and a second (1/3600 of a degree) of latitude is a mere 31 meters. If we consider the true, ellipsoidal shape of the Earth, there is a small variation in the distance represented by a degree of latitude – about one kilometer between the Equator and the poles.

Only one parallel, the Equator, traces the full circumference of the Earth, and because of this feature, it qualifies as a **great circle**. A great circle is defined as the perimeter of any plane that passes through the center of the Earth. All the other parallels are **small circles** because the planes they define do not pass through the center of the Earth; hence, their perimeters represent less than the Earth's full circumference. (One drawn around the pole at latitude 89° 59′ 59″ would be very small indeed.)

The concept of a great circle is not limited to parallels, for any number of great circles can be drawn in any direction on the globe. Great circles are important in navigation because the shortest distance between any two points on the Earth's surface follows a *great-circle route* (Figure 2.14). We should note, however, that although the great-circle principle is employed in the global coordinate system, there are no lines in the grid named great circles per se.

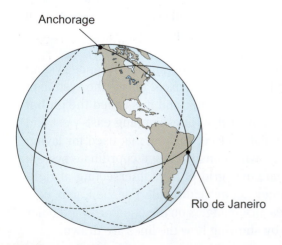

Figure 2.14 Examples of great circles. A great-circle route marks the shortest distance between any two points of the Earth's surface.

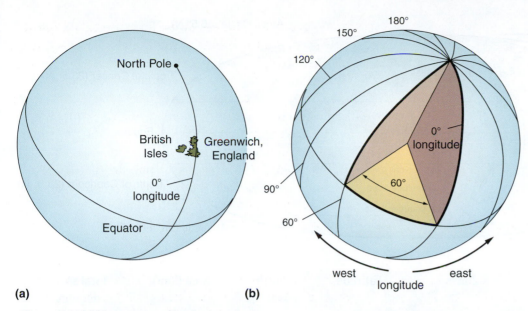

(a)   **(b)**

Figure 2.15 (a) Measurement of longitude begins at the Greenwich, or Prime, Meridian, which is zero, and (b) from that line extends east and west around the globe to 180°.

Meridians and Longitude: The north–south lines of the global coordinate system are called **meridians.** They are constructed in basically the same fashion as the parallels, with one important difference: they do not run parallel to each other but converge at the poles. In constructing meridians, there is a problem as to where to start the system because, unlike the Equator for parallels, there is no geometrically convenient point at which to place zero. According to international agreement, an arbitrary starting point was specified that coincides with the Royal Observatory at Greenwich, England (Figure 2.15a). A north–south line drawn through this point to the North Pole and the South Pole is the **Greenwich (or Prime) Meridian,** and it is labeled 0 degrees longitude. From the Greenwich Meridian all meridians westward to 180 degrees are designated west longitude, and those eastward to 180 degrees are designated east longitude. Thus, every meridian is half a great circle (Figure 2.15b).

Because the 0° meridian does not completely encircle the Earth, but only half of it, the maximum range of longitude is from 0° to 180° W and 0° to 180° E. Both 180° E and 180° W are halfway around the globe, in opposite directions from the Greenwich Meridian. Thus, they are really the same meridian, and for this reason, the direction indication for 180° is omitted. Because meridians converge at the poles, the length (distance) represented by a degree of longitude decreases from the Equator to the poles. One degree of longitude is 111 kilometers at the Equator, 96 kilometers at 30° latitude, 56 kilometers at 60° latitude, and 0 kilometers at 90° latitude.

Zones of Latitude: The Earth's climates and life zones are arranged roughly into several great belts of latitude and much of our discussion in the following chapters uses a framework like that shown in Figure 2.16. Three broad zones of latitude can be defined in both hemispheres: the **high latitudes,** which cover the upper 23.5 degrees of latitude, between 66.5° and 90°; the **middle latitudes,** which extend from 66.5 to 23.5 degrees latitude; and the **low latitudes,** which lie between 23.5° and the Equator. Latitude 23.5° is significant because it marks the highest latitude that receives direct solar radiation, that is, where the Sun's rays hit the Earth's surface perpendicularly (at a 90° angle). (More will be said about sun angles later.) In the northern hemisphere, the parallel at 23.5° is called the **Tropic of Cancer,** and in the southern hemisphere, it is called the **Tropic of Capricorn.** Latitude 66.5° marks the **Arctic** and **Antarctic** circles, above which are the only locations on Earth to experience day-long (24 hours) light and day-long dark each year.

Used correctly, the term **tropics** refers to the Tropic of Capricorn and the Tropic of Cancer. It follows that the zone between 23.5° south latitude and 23.5° north latitude should be the **intertropical** zone, and indeed many scientists do follow this convention. However, the term **intertropical** has declined in usage, and today **tropics** or **tropical zone** seems to be the preferred term for this zone. The **equatorial zone** is the middle belt of the Earth, extending 10° latitude or so north and south of the Equator.

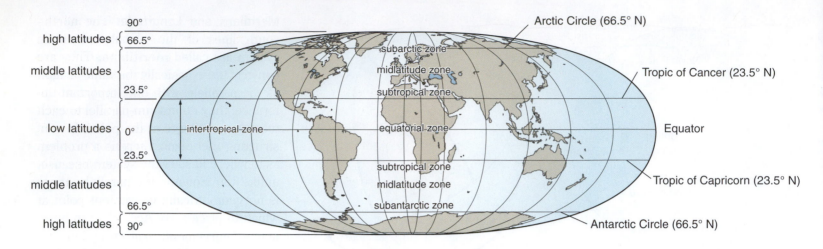

Zones	Latitude	Earth coverage	Area (km²)	Total span, degrees latitude
Low latitudes	0–23.5°	40%	204 million	47 degrees
Middle latitudes	23.5–66.5°	52%	265 million	86 degrees
High latitudes	66.5–90°	8%	40 million	47 degrees

Figure 2.16. Zones of latitude. Each hemisphere is divided into low, middle, and high zones and then into subzones. The low latitudes, or tropical zone, covers the area between the tropics of Cancer and Capricorn.

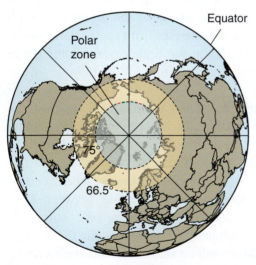

Figure 2.17 Viewing Earth from the top. The Arctic zone lies above 66.5 degrees north latitude and the polar zone occupies the small area (less than five percent Earth's surface area) immediately around the North Pole.

Within the broad belt of the middle latitudes, three additional zones are often designated, though their locations are somewhat arbitrary. Just above the tropics is the **subtropical zone**, which extends from 23.5 to 35° latitude. Much of the American South, the Mediterranean region, Australia, and southern China lie in this zone. The **midlatitude zone** is the center belt and, though its limits are somewhat arbitrary, 35° to 55° is usually given for it. Peking, Tokyo, London, Paris, Berlin, Moscow, New York, Philadelphia, Chicago, Montreal, Buenos Aires, and Sydney lie in this zone. The **subarctic zone** lies between 55° and the Arctic and Antarctic circles with the bulk of the land in this zone occupied by Russia, Canada, and the United States. Beyond 66.5° N and S are the **Arctic** and **Antarctic zones**, the uppermost part of which (generally given as above 75° latitude) is the **polar zone,** the area of which is very small indeed as you can tell from the map in Figure 2.17.

2.3 Mapping Space and Time

Throughout life, each of us thinks and lives within several mental systems or frames of reference that help us make sense of our world. In order to process information about the environment around us, two reference systems are absolutely essential: space and time. It is almost impossible to think about geographical phenomena without a sense of space because our notion of geography is based on distributions. We see the geographical features of the world, including ourselves, situated on the Earth's surface as parts of *various areal or spatial* patterns such as climatic zones, land areas, countries, provinces, states, counties, and so on.

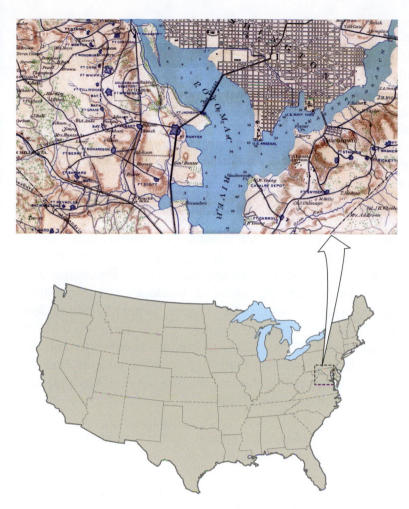

In childhood, our geographic reference system is small and simple. Our memories are tied to local places, rooms, houses, yards, neighborhoods, and the places of events such as birthdays and holidays. As we mature, these places and events become tied together into larger frames of reference and expanded to distant places, such as the locations of geographic features on maps. This requires abstract learning, that is, assimilation of knowledge without the benefit of *empirical*, or first-hand experience. It entails understanding about the world beyond that which you can see and touch, and it is essential to satisfying our mission of becoming responsible citizens in a larger world.

Map Systems: Maps are central to practically everything we examine in physical geography, and they provide a key tool for us to delve into the realm of abstract learning about systems and their processes and work. But in order to read and understand them we need to learn about their key properties, namely, scale, resolution, direction, and symbols.

Scale: **Scale** is defined as the relationship between distance on the map and the corresponding distance on the Earth's surface. Maps of small areas, such as a city, a university campus, or a volcanic island, are called **large-scale maps,** whereas those of large areas, such as a large country or a continent, are called **small-scale maps**. High levels of detail, or resolution, are possible only on large-scale maps and the examples in Figure 2.18 illustrate why.

Figure 2.18 Large-scale maps (top) show small areas in high detail. Small-scale maps (below) show large areas with little detail.

Scale is generally indicated on a map in either a graphic or an arithmetic form and often both are included as part of the map legend. The simplest scale indicator employed is the graphic, or **bar scale**, and some examples are given in Figure 2.19. This consists of an actual line or bar calibrated to indicate a precise map distance and labeled to indicate the corresponding ground distance. Any linear measurement on the map can be compared directly to the bar scale to determine the actual ground distance.

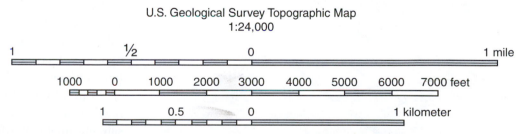

Figure 2.19 A standard map scale from a U.S. Geological Survey map. The representative fraction is 1:24,000. The graduated scales, called bar scales, give the equivalent in standard units of distance.

The arithmetic scale represents a ratio of units on the map to like units on the ground and is called a **representative fraction**. A representative fraction of 1:50,000, or 1/50,000, indicates that 1 unit on the map is equivalent to 50,000 of the same units on the Earth's surface. Since the scale is expressed in terms of a ratio, the proportion between the two distances (map and ground) is constant. Thus, the representative fraction is applicable to all systems and all units of measurement simultaneously. Hence 1:50,000 can be read as "1 map inch to 50,000 ground inches" or "1 map centimeter to 50,000 ground centimeters." Similarly, any other unit of measurement can be substituted, thus obviating the need to convert measurement units such as miles and kilometers (Figure 2.19). The representative fraction also demonstrates why large-scale maps show small areas. A scale of 1/50,000 is much larger than 1/2,000,000 in terms of its representative fraction. But, the large-scale map of 1/50,000 shows less area of the Earth than the map with a smaller scale of 1/2,000,000 (that is, 50,000 ground centimeters versus 2,000,000 ground centimeters).

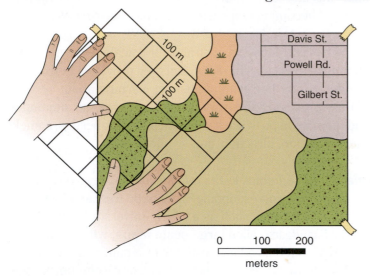

Figure 2.20 Manual techniques for measuring geographic areas, or polygons, of irregular shapes is tedious, slow, and often inaccurate.

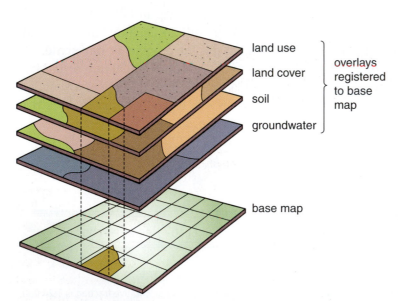

Figure 2.21 Geographic Information Systems not only allow for ready measurement of complex polygons but also for measurement of the spaces formed by the overlap between multiple polygons.

Map scale is necessary to making distance and area measurements. Knowing the area in square kilometers, hectares, or acres covered by a river floodplain, a forest, or a groundwater aquifer is important in geographic analysis. Rectangles and squares such as farm fields and city blocks are easy to measure, but the irregular patterns or polygons formed by natural features like those illustrated in Figure 2.20 are more difficult to measure accurately. This task is made easier by various technologies, including computer mapping systems like the variety mentioned in the introduction, namely, **Geographic Information Systems (GIS)**.

GIS rely on a coordinate system, such as longitude and latitude, which is referenced to real-world locations. This enables the software to compute the distances between the points marking the border of any polygon within the map area, and to derive the area of the polygon. GIS further provides the capacity to overlay polygons representing different features, such as wetlands, land use, and floodplains, and to calculate the areas of overlap among them and then produce a map of the results (Figure 2.21).

Resolution: **Resolution** is closely tied to map scale. It is the level of detail shown on a map. On small-scale maps, such as globes, it is impossible to show much detail, but at larger scales, it is possible to show more detail, that is, more data and information at finer levels of precision. Resolution is critical in interpreting imagery from remote sensors. Some aerial photographs have resolution units less than 1 meter by 1 meter (small enough to identify a person) whereas satellite imagery taken hundreds of kilometers above Earth has much coarser resolution, for example, 1 kilometer by 1 kilometer. Such satellite imagery affords limited detail, but it is excellent for mapping large phenomena such as hurricanes, ocean currents, and snow-cover patterns.

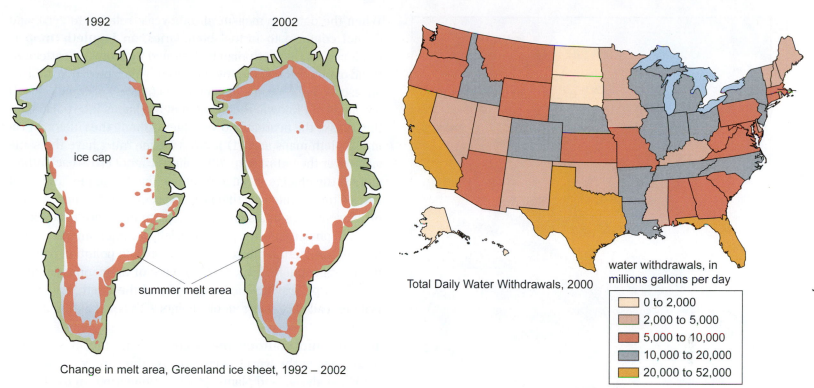

Change in melt area, Greenland ice sheet, 1992 – 2002

Total Daily Water Withdrawals, 2000

water withdrawals, in
millions gallons per day

- 0 to 2,000
- 2,000 to 5,000
- 5,000 to 10,000
- 10,000 to 20,000
- 20,000 to 52,000

Figure 2.22 Example of a thematic map showing the change in summer melt area, Greenland ice sheet.

Figure 2.23 Example of a chloropleth map showing daily water withdrawals by state.

Direction: **Direction** is essential to every map and it is conventional to orient the map with north at the top, south at the bottom and east and west on the right and left respectively. If the convergent point of meridians is used to define north, the map's north arrow will always point directly to the North Pole. However, this alignment will not match the north direction given by a compass for most locations on Earth because *magnetic north* is not at the North Pole. It is located to the south of the pole in northern Canada. Thus, surveyors working in the field must correct their compasses for this difference to align with true north and match with the coordinate lines on maps. Adding to the confusion, some maps include a third north, called *grid north*, which follows one of various grid systems (other than the global coordinate system) used in modern mapping programs.

Map types and symbols: The **symbols** you see on maps depend on the type of map and the phenomena portrayed by the map. The three general types of map produced (thematic, choropleth, and isopleth) are determined by the data and their measurement. Maps produced when data which are simply counted or put into named categories (such as the number of cattle per state, or the climates of each province) are called **thematic maps**, and these are typically represented by either dots or by continuous shadings, as shown in Figure 2.22. Data which are measured on a scale relative to each other or to zero have the capability to be subdivided into classes, and the maps produced from these data are called **choropleth maps**, and one is shown in Figure 2.23. Here, one set of data items, water withdrawal in this case, is typically mapped into a consistent formal boundary, such as a county or a state. The symbolization of this type of map reflects the different classes; e.g. counties or states with low, medium and high values.

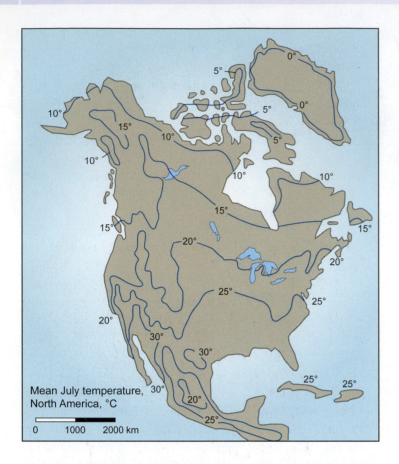

Mean July temperature, North America, °C

0   1000   2000 km

Figure 2.24 An example of an isopleth map showing the pattern of mean July temperatures across North America.

When the data are measured on a scale relative to zero, and do not conform to formal boundaries, an **isopleth map** is produced. An isopleth map is designed to show the pattern or trend of numerical values over an area. This type of map utilizes lines, called *isolines*, to connect points (places) of equal value. If a value is not known, the location of the line is interpolated on the basis of the nearest known values. Among the rules governing isopleth maps are: (1) a given isoline must have the same value over the entire map; (2) isolines cannot cross each other; and (3) the change in value from one line to the next must not exceed the iso-interval; that is, the specified difference in value of adjacent lines in a sequence. Isopleth maps are used extensively in physical geography, especially for phenomena such as air pressure, temperature, precipitation, and pollution patterns in the atmosphere (Figure 2.24). One of their most common uses is in portraying the topography of the Earth, where the isolines, called *contours*, depict points with equal elevations.

Topographic contour maps are perhaps the most widely used maps in the world today because they are so valuable in terrain analysis, land planning, and engineering. In the United States the US Geological Survey is charged with the task of preparing topographic maps for the nation. These maps, called *topographic quadrangles*, are prepared at a variety of scales, for example, 1:24,000, and 1:250,000, and are available to anyone at a relatively low cost. In Canada, topographic maps are prepared by various entities including the Department of Energy, Mines and Resources under a program called the National Topographic System as well as individual provinces.

In addition to contours and elevation data, the US Geological Survey quadrangles provide a great deal of other information about the land. This includes drainage features, forested areas, wetlands, roads, highways, urbanized areas, and even individual structures such as homes and schools in rural areas. Appendix 3 describes the features and use of these maps.

Time Systems: Time is the other reference system needed for recording information about Earth. It is almost impossible to think about nature, for example, without a sense of time because our notion of natural phenomena is one of change over time; we see the natural world around us, and ourselves as well, evolving through time. And, like geographic space, we use different scales to measure time and some are shown in Figure 2.25. We see much of nature changing, including the formation of mountain ranges and the evolution of species, over geologic time, which covers millions of years. We see human societies and civilizations changing over historical time (centuries and millennia), and we see ourselves as individual organisms changing over a life-time covering decades and years. Many Earth processes span even shorter periods, some as little as hours, minutes, and seconds.

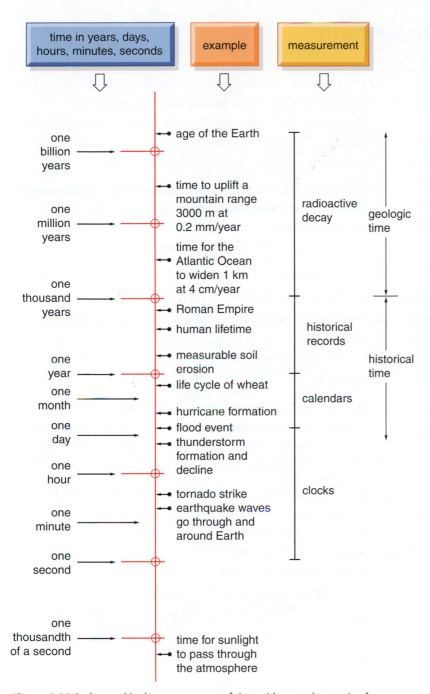

Figure 2.25 Scales used in the measurement of time with examples ranging from the age of the Earth to the travel time of sunlight.

Geologic Time: Earth is about 4.6 billion years old. References to Earth time are based on a scale called the geologic time scale which divides time into epochs, periods, eras, and eons. Each division marks a major segment of time distinguished by certain conditions on Earth, usually in the evolution of life. Some of the breaks between divisions represent transitions when change was gradual. Others, however, represent sharp breaks when change occurred suddenly, the result of a catastrophic event such as an asteroid collision or massive volcanic eruptions.

Evidence of several catastrophic changes or events appear in the geologic column, that is, in the sequence of rocks and fossils exposed in canyons and mountain sides or revealed in rock cores extracted from holes drilled into the Earth's crust. Many of these events were global in scale and often marked a sudden rise in mass extinctions in which half or more of Earth's species died out. Although the record discovered by scientists is far from complete, we do know about several events that have changed Earth dramatically. One of these occurred 65 million years ago when a large asteroid exploded into the Earth. This coincided with a drastic change in climate and hundreds of thousands of species, including the great dinosaurs, were killed off. We will examine this event later in the book.

Chapter Summary and Overview of Planet Earth:
Some Geographic Observations and Facts

In this second chapter of the book, we are concerned mainly with setting the geographic stage. What does Earth look like, how big is it, and how do we measure and map things on its surface? The overview of Earth from space was based on the simple point that unless we have a mental image and some understanding of what we are looking at, a planet with Earth's diversity can appear chaotic, just a random assortment of water, land, clouds, and many other things, and the motions of these things can make it all the more confusing. But science has to be based on more than visual impressions, so the second part of the chapter focused on measurements of Earth's size, shape, and motion, and how locations on its surface can be referenced with geographic precision. Once we know how to document locations using the global coordinate system, we are in a position to build maps. Accordingly, the third part of the chapter examined mapping systems, including modern ones such as Geographic Information Systems that use computers to compile overlays and compare spatial patterns. Finally, geographic change cannot be measured in the absence of time, and this requires a time scale in the same way maps require a scale to measure space.

▶ **Finding our way and referencing locations for mapping purposes has been one of humanity's greatest challenges.** Early sailors learned how to determine latitude using the horizon and certain stars. Determining longitude was more difficult and awaited the invention of an accurate chronometer.

▶ **Earth is one of the terrestrial planets of the Solar System.** It is thermally mild and very different hydrologically from its neighbors Mercury, Venus, and Mars.

▶ **Earth is distinguished among the planets by its distinctive color scheme.** This trait is valuable in mapping the planet with remote sensors and reveals critical information on geographic conditions and their interrelationships.

▶ **Even from a vantage point in distant space it is apparent that Earth is constantly changing.** Change takes place at all time intervals from moments to years to millennia and much longer and at a host of geographic scales.

▶ **Planet Earth is nearly a perfect sphere.** It has two motions in space, one of which causes a slight deviation in its shape

▶ **The global coordinate system is a grid made up of two sets of lines.** One set runs east–west and is used to measure latitude, and the other runs north–south and is used to measure longitude.

▶ **Earth is conventionally divided into three belts of latitude.** These correspond roughly to the major climatic zones in each hemisphere and are defined on the globe by the tropics of Cancer and Capricorn and the Arctic and Antarctic circles.

▶ **Reference systems for space and time are central to physical geography.** Without them it would be impossible to measure change from place to place and over time. Geographers use several types of maps which embody key properties such as scale and direction.

▶ **Our idea of nature involves change over time.** Of the various time scales used in science, the geologic time scale is the largest and includes epochs, periods, eras, and eons.

Review questions

1 Why were early seafarers and explorers so concerned about the development of a durable and highly accurate chronometer (clock)?

2 Why is Earth the most colorful planet among the planets in this part of the Solar System? What do these colors tell us about the geographic character and organization of Earth?

3 What features visible from space reveal that Earth is a highly dynamic planet? What do these features suggest about the systems and processes operating on the planet?

4 What is the difference between revolution and rotation with respect to the Earth's motion in space?

5 Is Earth a true sphere? Elaborate. What is meant by the term geoid?

6 At the global scale, what is the primary difference between the shapes of lines of latitude and longitude? What are these lines called and how are they numbered?

7 Besides mapping, are there any practical reasons for dividing the Earth into distinct belts of latitude and zones of longitude?

8 What is the difference between map scale and resolution? Using your town and state or province, distinguish between large-scale and small-scale maps.

9 Identify the major characteristics of the three general types of maps.

10 Why is it important to select the proper symbols when constructing a map?

11 Topographic maps are an example of what type map? What do topographic maps show and what are they used for?

12 Distinguish between geological and historical time scales. Which of the four main divisions of geologic time best relates to human time?

The Sun–Earth Energy System:

Fuel for a Planet

Chapter Overview

This chapter is about the grandest of all the great systems. It opens with a brief examination of the Sun's nuclear power plant and the nature of the energy it broadcasts into space and to Earth. Is the flow of energy from the Sun a perfectly steady stream, what form does it take, and what happens to that fraction captured by Earth? How does the motion of Earth in space influence the receipt of solar radiation and is this motion important to Earth's geographic character? We will find that, among other things, it governs the huge north–south swings in energy that give us the seasons. We will try to convince you that in order to understand the geographic character of Earth, we need to see it as an energy system, where solar radiation is converted into heat and heat is stored, transferred, and redistributed over Earth. The chapter concludes with a look at Earth's major heat-storage reservoirs, the oceans, landmasses, and atmosphere, and how planet Earth compares with two sister planets, Mercury and Mars, as a thermal body.

Introduction

At three miles above sea level in the remote mountains of the Hindu Kush of Asia, we made camp on a magnificent glacier. It lay in a narrow valley, partly visible in Figure 3.1, framed by great mountain walls on the east and west. These high walls blocked the early morning and late afternoon sunshine, so that only a window of direct radiation fell on the surface of the glacier around midday. From our vantage point near the center of the glacier we could watch the leading edge of the morning Sun creep slowly down the west wall, onto the glacier, and toward our camp. The August Sun shone brilliant white in the thin, cloudless atmosphere, and each morning we patiently awaited its warming rays for relief from the glacier's nighttime chill.

As the hard edge of sunlight crept toward us, the rising sound of rushing water broke the glacier's silence. When the Sun finally crossed over us, our campsite was suddenly transformed into a bleeding mass of meltwater. Rivulets poured around the little rock platforms that served as our sleeping pads, and underneath us the glacier seemed to cough and choke as meltwater rammed its way through cracks and tunnels within the ice.

Figure 3.1 Looking down on the Bandaka Glacier, the site of our first encampment located far below this point near the center of the photograph. Jeff Dozier, Hindu Kush, Afghanistan, 1973.

Figure 3.2 Ice pedestals capped by huge rocks. They mark the approximate level of the ice surface at the time the rocks were deposited on the glacier.

As the Sun rose toward its noon position, we felt its massive energy charge literally destroying the glacier under us. But the drama was not to last very long, for it faded as soon as the Sun fell below the west wall and its afternoon shadow slid across the valley. Within minutes the glacier fell silent. Never had I seen the power of the Sun so strikingly demonstrated. It completely controlled the tempo of the day, including the lives of our party of mountaineers, and shaped the glacial landscape around us. By summer's end artifacts of the Sun's power stood all around us: huge toadstools of ice capped by Volkswagen-sized chunks of rock like those shown in Figure 3.2. The heights of these rocks marked the total depth of the ice loss from the glacier's surface, more than a meter in places.

3.1 Light, Energy, and Work

To virtually all people, nearly everywhere, the Sun is a major part of their immediate geographic environment. This is not surprising; first, because we feel its presence directly in our lives, in the rhythm of the day, the passage of the seasons, the onset and end of life, and much more; and second, because the Sun is only eight minutes away in terms of the time it takes light to get from there to here. That light represents energy, the stuff that drives Earth's great systems, the systems that reach across Earth's vast surface forming a geographic infrastructure that binds land, water, air, and life together into a working whole. They, and the many smaller systems that are embedded within them, are a central theme of this book. But before we ask about their geographic character and how they operate and perform work, we must ask how the Sun drives them. In other words, how does the solar fuel system work?

Solar energy is discharged to Earth in a steady stream of radiation. About 70 percent of the solar radiation reaching Earth is absorbed by the atmosphere, oceans, and land where it is converted into heat. This heat energy warms the planet and drives nearly all the systems, no matter what their size, that operate on or near the Earth's surface, as Figure 3.3 illustrates. Among them are winds, waves, currents, ocean evaporation, rainstorms, rivers, and Earth's life systems. Life systems use only a tiny fraction of the solar radiation reaching earth's surface, but it is enough to support the planet's entire organic network, including more than 6 billion humans, their billions of animals, and nearly all their billions of machines.

Energy is the ability to do work. **Work** is defined as the application of force over distance. Most work involves the displacement of mass (i.e. the movement of a weighted substance) over some distance. When water molecules (objects with mass) are moved from the ocean surface into the atmosphere by the evaporation process, work is being performed. Therefore, when we measure evaporation, we know, based on the laws of physics, that energy is involved because without it the process cannot take place. When currents flow across the ocean and wind moves through the atmosphere, work is performed because mass (kilograms of water and air) is being moved over some distance. In all three of these examples, evaporation, currents, and wind, the work is driven by heat energy derived from the Earth's absorption of solar radiation, which we discuss in the next section.

3.2 The Solar Energy System

Although we think of Earth as a solar-driven planet, and indeed it is, there are actually two other sources that also provide energy to the Earth's surface (see Figure 3.4). One is heat from the Earth's interior and the other is kinetic energy (energy of

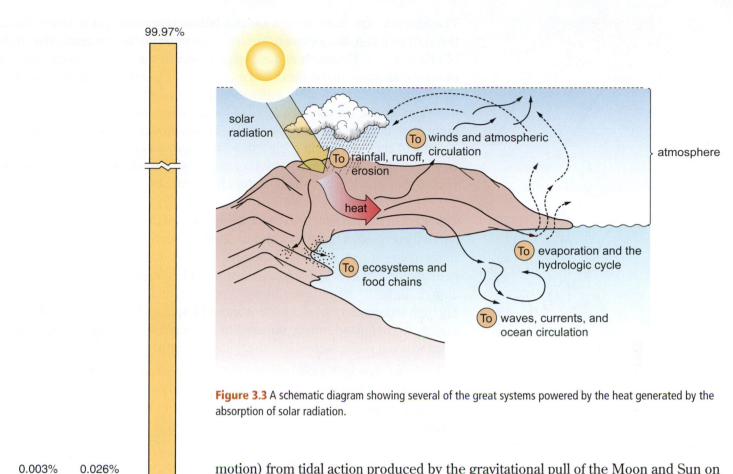

Figure 3.3 A schematic diagram showing several of the great systems powered by the heat generated by the absorption of solar radiation.

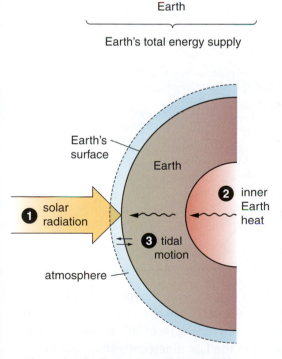

Figure 3.4 The three sources of energy to the Earth's surface and their relative contributions as a percentage of the total energy supply to the Earth's atmosphere, land, oceans, and organisms.

motion) from tidal action produced by the gravitational pull of the Moon and Sun on the Earth's surface. As the graph in Figure 3.4 shows, the relative contribution from the three sources is very lopsided with 99.97 percent coming from the Sun, 0.026 percent coming from the Earth's interior, and 0.003 percent coming from tides. To put this into perspective, for every 1,000,000 units of energy received at the Earth's surface and lower atmosphere, 999,710 come from the Sun, 260 come from interior heat, and only 30 come from tidal motion.

Thus, in the big scheme of things, the Sun provides the lion's share of energy to the Earth. This energy drives virtually all the essential processes on the planet's surface, atmospheric, hydrologic, ecological, and human, with two exceptions. One is tides and the other is geologic processes that originate within or below the crust, such as volcanoes, earthquakes, and the uplift of mountain ranges. Despite its diminutive quantity compared to the Sun, we cannot slight Earth's internal energy system because the geologic processes it drives are extremely important to landscape formation. Indeed, internal energy represents one of Earth's great energy systems, but for the present we will put it aside (until later in the book), and focus on the solar energy system. We will also put tidal energy aside, though it does not hold the rank of a great system.

The Solar Dynamo: The Sun is the dynamo of the Solar System. It is a huge, glowing ball of gas composed mainly of superheated hydrogen and helium. At its center (see Figure 3.5) is a high-pressure caldron of nuclear activity which produces a massive charge of energy that is transmitted outward to the surface of the Sun from which it is broadcast into the Solar System. A tiny fraction of this energy is intercepted by Earth whereupon it enters the atmosphere and illuminates and heats the planet's surface.

The Sun is a star. It originated about 5 billion years ago and is now about half way through its life. Astronomers tell us it is an ordinary star of average size, like billions of others in our Galaxy. It is the largest object in the Solar System, more than 100 times the diameter of the Earth and more than 300,000 times heavier than Earth. The Sun's huge mass produces a gravitational field so strong that it not only holds itself together against the outward force of its considerable nuclear processes, but holds eight planets, scores of small planet-like bodies, and millions of asteroids and smaller chunks of rock and ice in orbit around it. Pluto, one of the outermost of planet-like bodies of the Solar System (which until 2006 was classified by astronomers as a planet), lies nearly 6 billion kilometers from the Sun.

Our objective here is not to examine the physics of the Sun, but rather to learn enough about its processes to advance our understanding of the Earth's energy system. The Sun's main energy generator is the **core**, a high-density zone about 40,000 kilometers in diameter. The core is dominated by nuclear fusion processes (involving the conversion of "light" hydrogen into "heavy" hydrogen and helium) that produce unbelievably high temperatures, calculated at about 14 million degrees Celsius. From the core, energy is conducted rapidly outward to the Sun's surface, called the **photosphere**, from which it is dispatched into the Solar System (Figure 3.5).

Figure 3.5 The solar dynamo, the power plant of the Solar System. This cross-section of the Sun shows the super hot core, the much cooler photosphere, and the corona, a large halo-like zone around the Sun. In the background is Earth which intercepts a tiny fraction of the solar output.

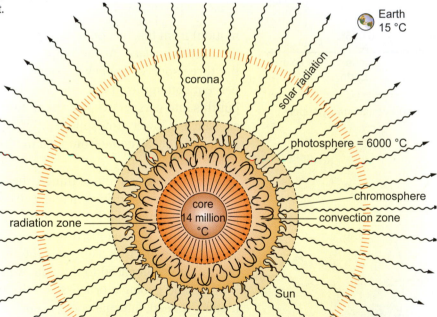

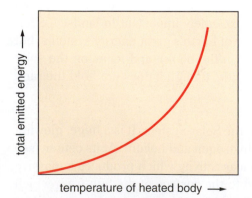

Figure 3.6 The output of radiant energy (solid line) from a heated body such the Sun or Earth increases dramatically with its temperature.

At a temperature of 6000 °C, the photosphere is cool by solar standards, but it is the hottest surface of any body in the Solar System. According to one of the laws of radiation, known as **Stefan–Boltzmann's law** (for physicists Josef Stefan and Ludwig Boltzmann who conducted radiation research in the late nineteenth and early twentieth centuries), the amount of radiant energy emitted by a heated body increases rapidly with its temperature. In other words, the energy emitted increases rapidly as the temperature of the radiating body rises as shown in Figure 3.6. Therefore, the output of radiation from the Sun is astronomical compared, for example, to the Earth whose surface temperature, at 15 °C, is 400 times lower.

The Sun's Energy Output: The flow of radiation from the Sun is equal in all directions from the great sphere. The total radiant energy released per second, called the Sun's **luminosity**, is equivalent to the energy produced by the detonation of about 100 billion 1-megaton nuclear bombs, a number that defies comprehension. This massive flow of radiation is discharged into space, crossing the Solar System at 300,000 kilometers per second, the speed of light. At this rate it takes only 8 minutes to reach the Earth, located 150 million kilometers away, and about 6 hours to reach Pluto, until recently considered the outermost planet of the outer Solar System.

Since the planets, their satellites, and the other objects (such as comets and asteroids) together make up only a minute percentage of the Solar System's total volume, only a minuscule fraction of the Sun's radiation is absorbed by matter in the Solar System itself. The Earth itself intercepts less than one two-billionth's of the Sun's energy output. Like most stars, nearly all the Sun's radiation is dissipated into deep space.

In addition to radiation, the Sun also discharges **ionized particles**. These are highly charged atomic particles (mostly protons and electrons) that are constantly escaping from the Sun's outer atmosphere, the **corona**. They are driven into space by the high temperature of the corona (about 1,000,000 °C), forming a light stream of particles called a solar wind that flows across the Solar System. As the Sun discharges these particles, it is losing mass at a rate of about a million tons per second; thus, the Sun is shrinking. However, the total loss is relatively small; since the origin of the Solar System the Sun has lost less than 0.1 percent of its mass.

To space travelers, the solar wind is imperceptible without the aid of highly sensitive instruments. To Earth residents, the solar wind is intercepted high above the planet by the Earth's magnetic field, but it does create colorful lighting effects above the poles called **auroras** (northern lights and southern lights) and one such display is shown in Figure 3.7. As far as we know, the solar wind has no measurable influence on the Earth's energy system and geographic character.

The Quiet versus Active Sun: Most of the energy output from the Sun is characterized by a steady, predictable flow of radiation. Indeed, it is so steady that the influx of solar radiation at the top of the Earth's atmosphere is called the **solar constant**. This steady-state condition is called the **quiet Sun**, and its behavior contrasts sharply with the more variable, and sometimes explosive, behavior of the **active Sun.** Although the quiet Sun runs the Earth's engine, so to speak, scientists have long been interested in the behavior of the active Sun because it can cause small variations in the output of solar radiation that may affect some Earth systems.

Figure 3.7 When ionized particles discharged by the Sun are intercepted by the Earth's magnetic field, colorful lighting, called auroras, is often created in the skies above the poles. Here is an example of the aurora borealis, or northern lights.

The most common features of the active Sun are **sunspots**, cool (low-energy) regions which appear as dark blemishes on the photosphere. Sunspots measure about 10,000 kilometers across and near their centers are as much as 1500 °C cooler than the surrounding surface. They typically change in size and shape from day to day, last from 1 to 100 days, and often appear in clusters on the photosphere.

Because of their cooler temperatures, the output of radiation from sunspots is somewhat lower than the average for the photosphere. Perhaps the most distinctive aspect of sunspots is the pattern of their frequency of occurrence. After centuries of observation, a fairly regular short-term cycle has been documented in which sunspot frequency varies from nearly zero to 100 or more over a period of 5 to 7 years. The

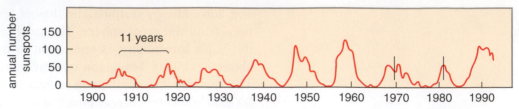

Figure 3.8 The record of sunspot activity in the twentieth century. Sunspots are patches of cooler temperature on the Sun's surface.

peak-to-peak cycle averages 11 years as is illustrated in the diagram in Figure 3.8 for the twentieth century. In addition, the geographic pattern of sunspots changes within each cycle as each new cluster tends to shift toward the solar equator. Sunspots are caused by changes in the Sun's magnetic field, which apparently alter the pattern of convective (mixing) motion and the distribution of energy beneath the photosphere. Despite the apparent short-term regularity of the sunspot cycle, astronomers are quick to point out that the sunspot system can, without warning, shut down every so often as it did between 1645 and 1715, a period known as the Maunder minimum.

The active Sun also produces high-energy events. In the vast areas of photosphere between sunspot clusters, violent eruptions can drive huge sheets of gas off the photosphere. The largest of these are called **prominences**, great looping streams such as the one shown in Figure 3.9 that can rise up to 500,000 kilometers and persist for days or weeks. **Flares** are even more violent eruptions than prominences. They last only minutes, producing temperatures as high as 100,000,000 °C, and releasing massive amounts of energy in one great blast. Unlike prominences which gravity draws back to the Sun's surface, the force of flares is so great that they are catapulted into space where they feed the system of solar wind.

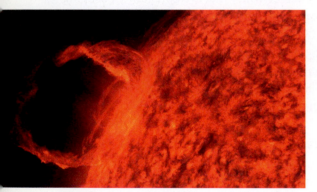

Figure 3.9 An exceptionally large solar prominence detected by sensors on NASA's Skylab.

Variations in Solar Output: The vast majority of the energy discharged from the Sun is produced by nuclear activity deep within the quiet Sun. The result is a great, steady stream of radiation which is intermittently interrupted by the perturbations of the active Sun. These perturbations cause the Earth's total energy supply to vary by a maximum of only about one percent including a variation in ultraviolet radiation that causes ozone concentrations in the upper atmosphere to rise and fall about 1.5 percent. Whether sunspots, prominences, and flares influence Earth's weather and climate remains an open question that scientists continue to investigate.

Some scientists offer evidence of small atmospheric temperature variations related to sunspots. But our focus must be centered on the Sun's steady flow of energy, the solar constant. This energy enters the Earth's atmosphere striking gas molecules, dust particles, clouds, and ultimately the Earth's surface. A large part of it is absorbed and converted into heat. But a large part is also reflected back into space. To understand the behavior of radiation in the atmosphere, we must first recognize that radiant energy is made up of a wide range of radiation forms and types.

3.3 The Measurement and Types of Radiation

All heated objects emit radiation, but the amount and type of radiation emitted vary with an object's temperature. With a surface temperature of 6000 °C, the Sun's radiation output is massive. Only a minute fraction of this radiation is absorbed by Earth, but it is enough to raise Earth's overall temperature to 15 °C. From this heat the

Earth, in turn, emits its own radiation, but the output, consistent with Stefan–Boltzmann's law, is tiny compared to the Sun. In addition there are huge differences in the types of radiation emitted by the Sun and Earth. In order to describe these differences and to understand their significance in the Earth's energy system, we must learn how radiation is measured and described.

All radiant energy, no matter what its source, is called **electromagnetic radiation**. This is energy that travels in the form of waves, much like sound waves or water waves, with one major difference. Electromagnetic radiation does not require the presence of matter, such as air or water, to serve as a medium to transport it. Thus, it can travel through the near vacuum of space. But it can also travel through matter as sunlight can through air, water, glass, and other substances.

Solar Radiation: The Sun's radiation is made up of many types of electromagnetic radiation. Light is a type of electromagnetic radiation; so are radio waves and X-rays. These different types of radiation are distinguished by their **wavelength**, defined as the distance between the crests of two adjacent waves. The principal forms of radiation are portrayed in the scale shown in Figure 3.10, called the **electromagnetic spectrum.**

The wavelengths of electromagnetic radiation vary enormously, from thousands of meters for long radio waves to much less that one-millionth of a meter for X-rays and gamma rays. Between these extremes are found all the forms of radiation known to science, including television waves, microwaves, infrared radiation, visible light, and ultraviolet radiation. Most solar radiation comes to Earth as visible light and infrared radiation. The standard unit of measure used in the study of solar and Earth radiation in physical geography is the micrometer (abbreviated μm). One micrometer is equal to one-millionth of a meter.

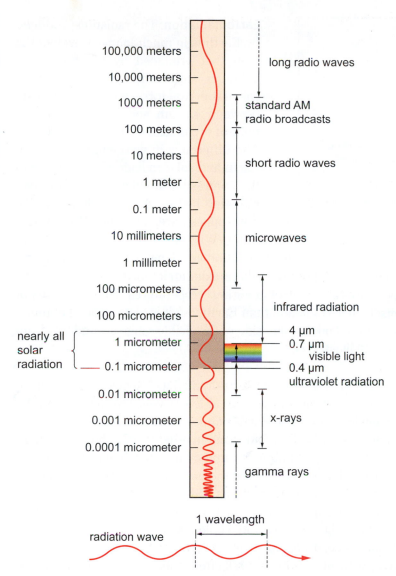

Figure 3.10 The electromagnetic spectrum showing major types of radiation with wavelengths ranging from thousands of meters to small fractions of a micrometer. Nearly all solar radiation falls between 0.1 and 4 micrometers and includes visible light.

The vast majority of the solar radiation, or **insolation** as it is commonly termed (for **in**coming **sol**ar radi**ation**), covers only a small segment or band of the electromagnetic spectrum, between 0.2 and 2.0 μm. This band is made up of three main types of radiation: ultraviolet, visible, and infrared, in the quantities shown in Figure 3.11. At the center is the visible part of the spectrum, the radiation containing the range of colors to which our eyes are sensitive. **Visible radiation** ranges from 0.4 μm (representing the color violet) to 0.7 μm (representing the color red), and constitutes 41 percent of the Sun's radiant energy. The remaining 59 percent is made up of **ultraviolet radiation** (9 percent (and the source of sunburn)), which at 0.4 μm to 0.1 μm is slightly shorter than visible, and **infrared radiation** (50 percent), which at 0.7 μm to 4.0 μm is slightly longer than visible.

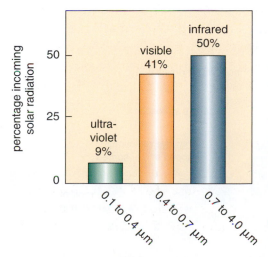

Figure 3.11 The main classes of solar radiation: ultraviolet, visible, and infrared radiation. Over 90 percent is represented by visible and infrared.

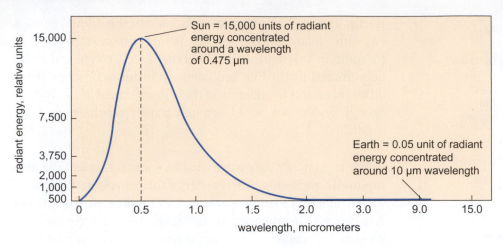

Figure 3.12 The relative intensity (output) and wavelengths of solar and Earth radiation. Because of their huge temperature difference, Earth's radiation is not only in much longer wavelengths, but is a very tiny fraction of the Sun's intensity

Earth Radiation: The radiation produced by Earth is in much longer wavelengths than that produced by the Sun. It is a longer variety of infrared radiation, called **thermal infrared**, which is mostly between 4.0 μm and 20 μm, with the bulk of it concentrated around 10.0 μm. The difference in wavelengths between Earth and solar radiation is attributed to the contrasting temperatures of the two radiating bodies. A second law of radiation applies here. It is called **Wien's law** (after Wilhelm Wien, physicist, 1864–1928) and tells us that the wavelengths of radiation are dependent on the temperature of the radiating body, more precisely, the maximum-intensity radiation grows longer at cooler temperatures and shorter at higher temperatures. Thus, the Sun not only produces immensely more radiation than Earth, but it is concentrated at much shorter wavelengths, centered on the blue side of the visible range around 0.475 μm as the graph in Figure 3.12 illustrates.

The Greenhouse Concept: This difference in the radiation wavelengths produced by the Sun and the Earth is extremely important to Earth's energy balance and the character of our global environment. The Earth's atmosphere is more transparent to shortwave radiation than it is to longwave radiation. In other words, the atmosphere poses less of a barrier to incoming radiation (from the Sun) than it does to outgoing longwave radiation from the Earth. As a result, the energy of longwave radiation is held in by the atmosphere. This phenomenon has come to be known as the **greenhouse effect**. The idea behind this model, which is shown in Figure 3.13, is that the greenhouse glass lets in solar radiation but holds back the outgoing longwave radiation generated from the solar-heated surfaces within the house. The longwave radiation trapped (actually absorbed) by Earth's atmosphere helps heat the atmosphere and the surface environment under it. As a consequence, Earth's climate is warmer and less variable in temperature because of the greenhouse effect. Without it, Earth's equilibrium surface temperature would be –18 °C, 33 °C lower than it presently is! We should add a parenthetical note before leaving this important idea. In reality, greenhouses and the atmosphere function differently as radiation systems, but the analogy has become so widely used over the years that the term has become a standard reference in science and the media, and we shall follow suit.

Figure 3.13 The basic concept of the greenhouse effect. The atmosphere is penetrated by shortwave solar radiation which is absorbed by the Earth and converted into longwave radiation which is trapped (absorbed) by the atmospheric as the greenhouse roof.

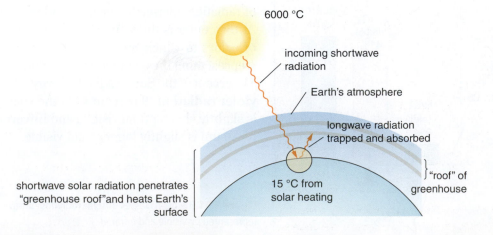

3.4 The Organization and Motion of the Planets

Our next question is how the Solar System operates because the motion of the Earth in space influences the geographic distribution of solar radiation on a daily and seasonal basis. These variations in turn set into motion a wide range of environmental change for most places including that associated with weather and climate, biological productivity, animal migrations, ocean circulation, land-use activity, and many other phenomena.

For as long as we know, humans have had an abiding curiosity about the organization and motion of the Solar System. In the clear skies of ancient Mesopotamia, Egypt, Greece, and Rome, the study of the heavens became a preoccupation for many and led to a host of ideas, some fanciful, some not, on the origin, structure, and motion of the Universe. Among these early thinkers was the astronomer and geographer Ptolemy who laid out a model of the Universe with Earth at its center, and the planets and stars revolving around it. The early Christian church liked this **geocentric** concept, for, among other things, it supported the belief of a Universe centered on Earth, the biblical home God created for Man. The Ptolemaic model became Church doctrine, and while civilization struggled through 1000 years of the Middle Ages, it remained secure under the vestments of Christian dogma.

Heliocentrism: In the 1500s, however, a new idea was introduced by the Polish astronomer, Nicolaus Copernicus (1473–1543). Copernicus revived the old Greek model that the Sun rather than Earth was the center of the heavens and that the Earth and the other planets revolved around the Sun. He placed the orbits of Mercury and Venus next to the Sun, followed by Earth, and then Mars, Jupiter, and Saturn (Uranus, Neptune, and Pluto were at that time unknown). Around this **heliocentric system** he set the rest of the Universe, which he envisioned as a great sphere of fixed stars.

Early in the 1600s, the Italian astronomer, Galileo Galilei (1564–1642), confirmed Copernicus's heliocentric model by using a new instrument, the telescope, to observe the motion of the planets and their moons. He reasoned that the planets moved about the Sun in great circular orbits. In addition to his gifts as a scientist, Galileo was also a brilliant spokesman for **heliocentrism** and soon his teachings and writings threatened traditional scholars who pressed the Church to defend geocentric doctrine and silence Galileo. Unfortunately, much of his later life was spent wrestling with the Church, but he had succeeded in opening the world to a new way of thinking about the Solar System and our place in it.

Around the same time, the German, Johannes Kepler (1571–1630), established that the planets moved about the Sun in elliptical paths, not circular ones as Galileo and Copernicus had thought. He also proposed that the Sun produced some kind of force on the planetary system and that the orbiting motion of the planets was controlled by the Sun's rotation. A few decades later, the most prominent scientist of the time, Isaac Newton (1642–1727), identified that force as gravity, which he called the "universal force of gravity," leading to his model of the Universe as a great mechanical system perfectly balanced among opposing forces.

But the question about the center of the Universe remained. If not the Earth, then was it the Sun as Copernicus thought? The answer did not come until the twentieth century when the prominent American astronomer Edwin Hubble (1889–1953)

An artist's vision of a young Nicolaus Copernicus working with his model of the Sun as the center of the Solar System in the early 1500s.

Galileo debating with church scholars over the concept of heliocentrism.

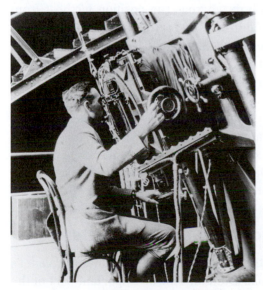

Edward Hubble, 20th century astronomer, offered an answer to the question of the center of the Universe.

established that the Universe has no center because the whole mass – galaxies, stars, and planets – is constantly expanding outward, ballooning into deep space. The idea of a center turns out to be relative to your location in the system. In other words, wherever you happen to be standing in the Universe appears to be its center, because everything around you is moving away.

The Modern Solar System: Our Solar System, which is flying outward with the rest of the star systems, is hardly a microscopic speck in one galaxy, the Milky Way, in one microscopic part of the Universe. But from our standpoint on Earth it is huge. If you envision Earth as a grain of sand, then the Solar System is the size of a baseball field. But how big is the solar ball field? The geographic extent of the Solar System can be defined in two ways: (1) by the motion of the planets around the Sun; or (2) by a much larger entity called the **heliosphere**. The heliosphere is defined by the distance that solar winds reach into space. Its outer limit, termed the **heliopause**, it is estimated at 85 astronomical units (AU), or 12,750 million kilometers, from the Sun. This is about 55 AU beyond Neptune, the outermost planet in the Solar System. An *astronomical unit* is the distance between the Sun and the Earth, 150 million kilometers. We have confirmed the distance of the heliopause with the flight of the spacecraft Voyager I, launched in 1979, which crossed the heliopause in 2004.

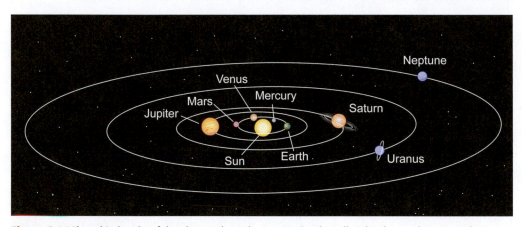

Figure 3.14 The orbital paths of the planets about the Sun. Notice that all eight planets share a nearly common plane of revolution, the plane of the ecliptic.

Table 3.1 Planet distances and year lengths

Planet	Distance to Sun	Year (Earth time)
Mercury	0.39 AU (58 million km)	0.241 (88 days)
Venus	0.72 AU (108 million km)	0.62 (224.5 days)
Earth	**1.00 AU (150 million km)**	**1.0 (365 days)**
Mars	1.52 AU (228 million km)	1.9 (22.6 months)
Jupiter	5.20 AU (778 million km)	11.9 (142 months)
Saturn	9.54 AU (1427 million km)	29.5 (354 months)
Uranus	19.19 AU (2870 million km)	84.0 (1008 months)
Neptune	30.06 AU (4497 million km)	165 (1978 months)

Within the heliosphere are the eight planets of our Solar System (Pluto has lost its status as a planet and is now considered a dwarf planet). All are in orbit or revolution around the Sun and all lie in a nearly common plane of orbit, called the **plane of the ecliptic** as you can see in Figure 3.14. The distance between each planet and the Sun varies over the revolutionary year. This is related to the shape of the orbits. They tend to be slightly elongated or elliptical with the Sun a little off center. Thus, each planet has a near and far position. The position nearest the Sun is called **perihelion** whereas the position farthest from the Sun is called **aphelion**. For Earth, perihelion occurs on January 3, when the Earth–Sun distance is 147 million kilometers, and aphelion on July 5 when the Earth–Sun distance is 152 million kilometers.

The sizes of the planets' orbits increase with distance from the Sun; thus Mercury, at a distance of 0.39 AU, has the smallest, and Neptune, at a distance of 30.06 AU, has the largest. Table 3.1 lists the mean distances between the planets and the Sun and the lengths of the planetary years. All data are in Earth units.

3.5 Earth's Motion in the Solar System

Let us now turn to the Earth itself and examine our planet's motion within the Solar System. This an important part of the story, for understanding the nature of the Earth's movement in space is necessary to explaining the north–south variations

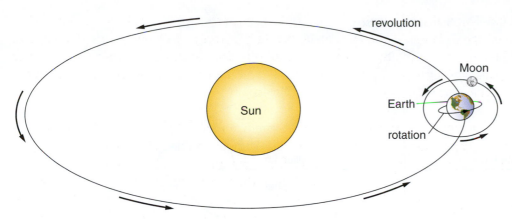

Figure 3.15 The motion of the Earth and Moon in space. All motion is counterclockwise when viewed from above the plane of ecliptic.

in the receipt of solar energy, the change of seasons, and the duration of day and night, when hurricanes form, and many other aspects of weather, climate, and hydrology.

Revolution and Rotation: To begin with, the Earth in space has two principal motions, revolution and rotation, and both are very important in the geographic distribution of solar radiation. The term **revolution** applies to the orbital path about the Sun whereas **rotation** defines the spin of the planet on its axis. If we could observe these motions from a bird's-eye perspective above the Solar System, as Figure 3.15 offers, both would appear counterclockwise. That is, a point on the surface of the Earth rotates from west to east, and the path of the Earth in its orbit around the Sun (and the Moon around the Earth), is in the same direction.

The period of revolution is called the year and is equal to 365.242 solar days. A **solar day** is the length of time it takes a single point on the Earth's surface to make one complete rotation with respect to the Sun. Although we consider it to be a constant, the length of day varies during the year. Recall that the Earth's orbit is slightly elliptical and within this ellipse the Sun is slightly off center. Because of this Earth moves (revolves) a little faster at one end of its orbit and a little slower at the other. Therefore, the length of the solar day is actually an *average* of 24 hours.

In addition, the apparent rotation speed – measured by sighting the Sun – averages slightly slower than the true rotation speed. If we measure Earth's rotation period based on sightings of distant stars rather than the Sun, the length of a day would be slightly shorter. A day measured by the stars is called a **sidereal day**. We must also note that the whole system of Earth revolution varies slightly over periods ranging from about 20,000 to 100,000 years. These variations appear to influence some long-term climatic cycles and we will examine them later in Chapter 8. But for now we are interested in the cause of seasons.

The Seasons: We recognize the seasons as the rhythmical oscillations of climate and landscape that are predictable year after year. In different parts of the world, the seasons have different characteristics. March may be wet or dry; July may be cold or warm, light or dark, or fair or stormy. Throughout most parts of the world, however, the processes of life are closely related to the pattern of the seasons. In turn, the seasons everywhere are related to changes in the amount of energy the Earth is receiving from the Sun. What is the cause of these changes?

One possible explanation is the variation in the Earth–Sun distance related to the Earth's elliptical orbit, which produces aphelion and perihelion. But the variation in distance between Earth and Sun has only a minor effect (about plus or minus 3.4 percent) on the seasonal variations in the receipt of solar energy. What is more, the dates of aphelion and perihelion are opposite the cold and warm seasons in the northern hemisphere.

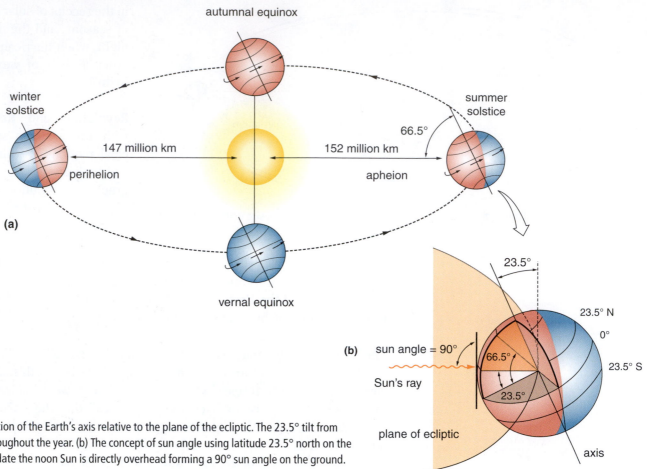

Figure 3.16 (a) The position of the Earth's axis relative to the plane of the ecliptic. The 23.5° tilt from vertical remains fixed throughout the year. (b) The concept of sun angle using latitude 23.5° north on the summer solstice. On this date the noon Sun is directly overhead forming a 90° sun angle on the ground.

The true explanation of the seasons becomes clear when we look at the diagram in Figure 3.16a. As the Earth revolves around the Sun, its axis is not perpendicular to the plane of the elliptic; instead, it inclines at an angle of 66 degrees 33 minutes, or 23° 27' off the vertical. (For our purposes we can round these numbers off to 66.5° and 23.5°.) Because the orientation of the axis remains constant throughout revolution, the tilt does not favor the same hemisphere throughout the year. During one part of the orbit, the southern hemisphere leans toward the sun, and during the opposite part, the northern hemisphere leans toward the Sun. (See the positions marking *summer solstice* and *winter solstice* in Figure 3.16a.) When one hemisphere is leaning toward the Sun, the sun angle in that hemisphere is greater by exactly the same number of degrees that it is reduced at comparable latitudes in the other hemisphere.

The Concept of Sun Angle: The angle at which a ray of solar radiation hits the Earth's atmosphere and surface is called **sun angle**, such as is shown for a 90° angle at 23.5° north latitude in Figure 3.16b. This angle (90°) is the highest sun angle possible on Earth. If you were standing at that location, the noon Sun would be directly overhead. The lowest sun angle is 0°, and this occurs when the Sun's rays travel parallel to the Earth's surface. Standing on the Earth's surface, you would see the Sun on the horizon with only its upper half showing (Figure 3.17). When the Earth is at one "end" or the other of its orbit, and the full 23.5° tilt of the axis is aligned with the Sun, all sun angles over the Earth are either increased or decreased by 23.5°. For example, at a location such as 45° north latitude where the sun angle on March 21 is 45°, the sun angle decreases to 21.5° (45°–23.5°) at one end of the orbit and increases to 68.5° (45° + 23.5°) at the other end. More will be said about sun angles in the next chapter.

Figure 3.17 A photograph from the North Pole showing the position of the Sun in the sky during the equinoxes. The Sun is on the horizon throughout the day and the Sun angle is zero. The situation at the South Pole on these dates is exactly the same.

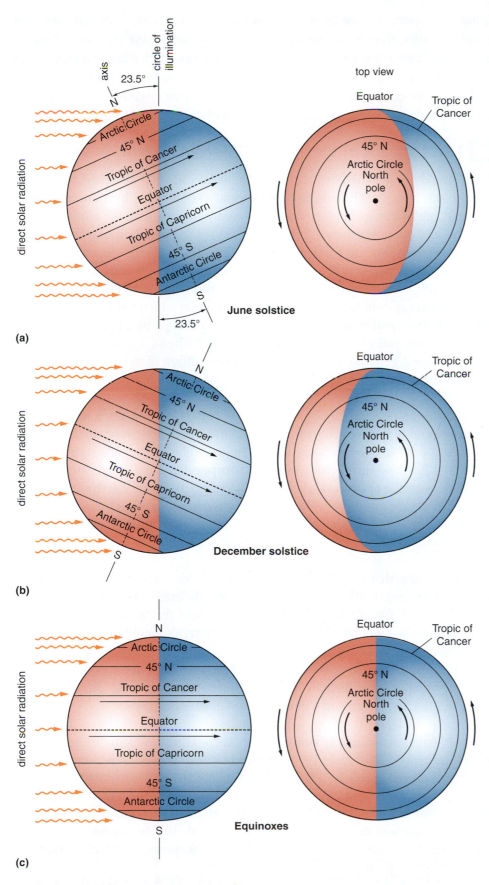

Daylight and Darkness: How many hours per day the Sun shines on a location is also a factor in creating seasons. First, know that exactly half of the planet, about 255 million square kilometers, is illuminated by the Sun at all times. But the geographic position of the illuminated sector changes as the attitude of the Earth's axis changes with revolution about the Sun. On June 20–22, when the axial tilt brings the northern hemisphere toward the Sun by a full 23.5° – what we described above as one "end' of the orbit – the North Pole lies entirely within the illuminated sector and the South Pole entirely within the dark sector as the diagram in Figure 3.18a shows.

The boundary separating the two sectors, called the **circle of illumination**, falls 23.5° of latitude below the poles. This latitude marks the Arctic and Antarctic circles, and this date, June 20–22, is the **summer solstice** for the northern hemisphere. The circle of illumination at this time reaches beyond the North Pole, so that light washes the whole Arctic region from morning to morning, that is, 24 hours each day. The same date in the southern hemisphere is the **winter solstice,** shown in Figure 3.18b. December 20–22 represents the same geometry in reverse and is also called the solstice, though the winter and summer designations are also reversed. Solstice means furthest point, or point of culmination, when the trend toward longer or shorter days reverses.

The word **equinox** is Latin for "equal night." The equinox dates, March 20–22 and September 20–22, are the intermediate points in the Earth's orbit between the solstices. On these days the circle of illumination falls across both poles (see Figure 3.18c). Thus, as the Earth rotates, every geographic location spends exactly half of the day in the dark sector and half in the light sector. Solar radiation is thus broadcast to each hemisphere in equal amounts.

Summary on the Solar Energy System: At a mere eight minutes from Earth, the Sun, from its position at the center of the Solar System, beams a stream of radiation to our little planet that powers almost all our great systems. Most of it comes to us as visible

Figure 3.18 The circle of illumination marks the divides between the light and dark sectors of the Earth. Side views and top views of the Earth during the solstices (a and b) and the equinoxes (c) showing the positions of the circle of illumination.

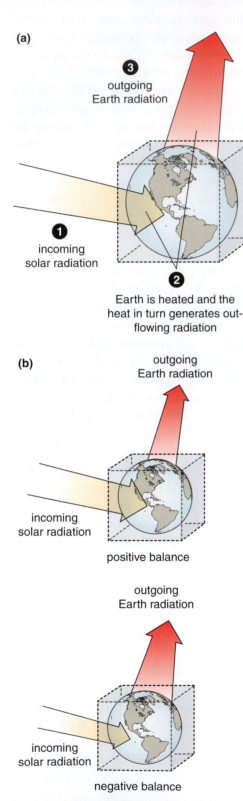

(a)

❸ outgoing Earth radiation

❶ incoming solar radiation

❷ Earth is heated and the heat in turn generates out-flowing radiation

(b)

outgoing Earth radiation

incoming solar radiation

positive balance

outgoing Earth radiation

incoming solar radiation

negative balance

Figure 3.19 (a) A simple model of the Earth as an energy system beginning with incoming solar radiation(1) that generates heat (2), which in turn generates outgoing Earth radiation (3). (b) Positive and negative balances in the Earth energy system.

and infrared radiation. After the systems do their work, the Earth, which itself functions as an input–output system, releases this energy back into space. But because of its modest surface temperature, both the amount and form of Earth's radiation are much different than the Sun's. We should ask why. We should also ask why the reception of energy from the Sun varies geographically over the course of the year. The answer is found in the motion of the Earth in space and how this motion influences sun angle, the duration of daylight, and the change of seasons.

3.6 The Concept of Earth as an Energy System

We have already established that almost all the solar radiation absorbed by Earth is converted into heat, which fuels Earth's thermal system. This heat energy warms the air, land, and water, and in turn drives the hydrologic system, the motion of wind and waves, and other Earth systems and processes. Since all objects that contain heat also emit radiation, the Earth's surface constantly discharges its own radiation. We also know that the rate at which the surface emits radiant energy is governed by its temperature; the higher the temperature, the more radiation emitted. Thus, we have a three-part system (as shown in Figure 3.19a), in which the radiation outflow from Earth is governed by its surface temperature, which in turn is governed by the absorption of solar radiation. The system is in balance when (1) incoming solar radiation, (2) the heat it generates on Earth, and (3) the outflow of radiation from Earth's heat are all equal.

Earth's Energy Balance: If the amount of solar energy received by the Earth's surface over some period of time is exactly the same as the amount of energy discharged by Earth, then the surface temperature should hold steady. This steady-state temperature is called Earth's **equilibrium surface temperature** and it indicates that the energy system for the planet as a whole is in an **energy balance**. Earth's equilibrium surface temperature during most of the twentieth century has been 15 °C. This is a long-term average condition based on many years of temperature readings taken in all seasons at thousands of locations over the entire planet.

But we also know that imbalances of various durations can and do occur when energy inflow exceeds outflow or vice versa. Most imbalances are short-term, lasting only months or a few years, but scientists are convinced that we have entered into a longer period of change that will produce a rise in Earth's equilibrium temperature, popularly known as **global warming**. The prime candidate causing this change is increased atmospheric carbon dioxide, a heat-absorbing gas that is increasing mainly from the burning of fossil fuels by Earth's growing human population. We will say more about global warming and climate change in later chapters.

The energy-balance concept is similar to an accounting problem. To begin with, imagine energy as money and the Earth as a financial account represented by the boxed space in the Figure 3.19. The account is an active one, with money constantly being deposited (as radiation from the Sun) and withdrawn (heat passing from the land to the atmosphere and radiation passing from the atmosphere back into space) from the basic account reserve. When inflows exceed outflows, as we now seem to be experiencing, the reserve grows and the system becomes **positively balanced**, as shown in Figure 3.19b.

When outflows exceed inflows the reserve dwindles, and the system is said to be **negatively balanced**, shown in Figure 3.19b. This occurred about 20,000 years ago when Earth slipped into a period of glaciation and the planet's equilibrium temperature fell to around 12 or 13°C. Precisely between these states is a state of equilibrium, in which the flows are balanced, that is, deposits equal withdrawals. At that point, the system maintains what physicists term a **steady state**. The energy reserve, though continuously undergoing energy loss and replacement, is unchanging in its total amount on account from moment to moment.

Some Governing Principles of Energy Systems: The operation of any energy system, whether as large as the Earth's or as small as your body's, is subject to certain physical principles. First, is the **conservation-of-energy principle**, which tells us that there can be no absolute loss or gain of energy within any system. Energy may change form, as from light (a radiant form) to heat, or it may be stored, but it cannot be created or destroyed and so the system accounting ledger must always balance. In other words, all energy that flows into any energy system, such as an ecosystem or the Earth's atmosphere, must be accounted for in (1) outflows and (2) storage. Any energy used in work must also be accounted for: it must have a source (such as incoming radiation or stored heat), and it cannot, of course, be lost when work is performed, but merely converted into a different form.

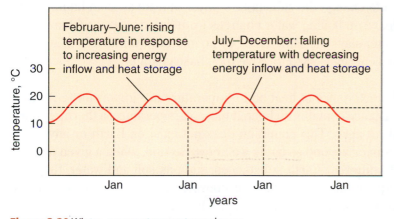

Figure 3.20 Winter–summer temperatures change in the midlatitude zone of the northern hemisphere in response to seasonal changes in solar energy and secondary influences from atmospheric conditions such as cloud cover. The broken horizontal line is meant to illustrate the average of these seasonal changes, that is, the equilibrium surface temperature.

Second, because the energy system is *dynamic* (rather than static), it is capable of undergoing spontaneous adjustments to variations in energy flow. All adjustments represent a trend toward a new state of equilibrium. When the inflow of energy increases and the internal energy reserve grows, the outflow of energy tends to increase correspondingly. When the outflow rate adjusts to a level equal to that of the inflow, the system has attained a new level of equilibrium. In reality, natural systems rarely attain equilibrium, but are usually in a state of continuous adjustment or **dynamic equilibrium** characterized by marked periods of near equilibrium, which accounts for the fact that no two winters or summers are ever exactly alike (Figure 3.20).

Third, when energy is used in **work**, it is always *converted* into another energy form. For example, when the atmosphere is heated by solar radiation and the air expands to create wind, *heat energy* is converted to *kinetic energy*, the energy of motion. When heat on the ocean surface drives evaporation and mountain snowfall eventually results from the water vapor, the deposits of snow represent *potential energy*, defined as mass elevated above sea level. And when solar radiation is converted into organic matter through photosynthesis, *chemical energy* is produced and when it in turn is consumed by an animal, it is converted into kinetic energy (with the animal's locomotion) and heat energy (with respiration). In the end, all Earth's energy is converted into heat, which in turn powers an outflow of radiation from Earth into space. Only through this energy outflow is Earth able to maintain an energy balance, because the energy inflow from the Sun is unending.

Fourth, the Earth is an **open energy system** because it is continuously undergoing spontaneous energy losses and gains. In other words, the Earth is not dependent on a fixed supply of contained energy as would be the case in a closed energy system. With energy constantly entering and leaving the planet, the Earth's energy system does not wind down progressively over time, unless the Sun goes out, of course, and that is not expected for about 5 billion years. Earth's interior (geothermal) energy system, on the other hand, has a more or less fixed energy supply, which is slowly winding down with the loss of heat through the crust.

3.7 Heat Energy and Temperature

We need to understand a few things about heat and temperature before going much further. First, when heat energy is added to any substance, the motion (vibration) of its molecules increases and its temperature rises. **Temperature** is an indicator of the substance's heat content as represented by the average motion of each of its molecules, that is, kinetic energy at the molecular level. **Heat**, on the other hand, is defined as the actual energy content of the substance, defined by the total motion (kinetic energy) of all its molecules, *plus* its potential energy (typically the energy held within the bonds between the molecules). Heat is energy, whereas temperature is an indicator of heat energy and this distinction is illustrated when heat is taken in by a substance.

One of two things can happen when a substance takes on heat: (1) the heat energy can be used to increase the kinetic energy (vibrating motion) of the molecules, which in turn would cause a rise in the substance's temperature; or (2) the heat can be used to increase the potential energy of the molecules by changing the bonds among the molecules. When the bonds change, the heat energy induces a *phase change,* that is, the energy is used to change the substance's physical state as from liquid to gas (vapor). For example, when boiling water changes from liquid to vapor (steam), the temperature of the water stays the same (100 °C) no matter how much heat is added because the additional energy is used to drive the phase change.

A couple of other characteristics of heat and temperature are important to remember. First, when a substance contains no heat, it has no temperature, a thermal condition known as **absolute zero**. This condition does not exist on Earth or anywhere else in the Solar System or beyond as far as we know. Second, when a given amount of energy is added to different Earth substances, such as soil and water, they will develop different temperatures (provided, of course, they are not undergoing a phase change). In other words, even though their heat contents will be exactly the same, their temperatures will not be the same. After a day absorbing summer sunshine, for example, a pond of water will almost always be cooler than an adjacent patch of soil. At a much larger scale, a similar difference will occur between an ocean and a continent.

Measuring Heat: The reason for the variation in different substance's temperature response to heat has to do with a substance's **heat capacity**, a thermal property inherent to all Earth materials. It is defined as the amount of heat energy needed to raise the temperature of one unit of mass (e.g. a gram) or one unit of volume (e.g. a cubic meter) of a substance 1 degree celsius, as illustrated in Figure 3.21. This is, in fact, how a basic unit of energy, the calorie, is defined. A **calorie** is the amount of heat energy applied to 1 gram of water to raise its temperature 1 °C (from 14.5 to 15.5 °C). For a gram of soil, less than a calorie of heat is needed to raise its temperature 1 °C and for air even less heat is needed. Therefore, both soil and air have lower heat capacities than water. (A note on units of measure: Modern science uses the joule (J) instead of the calorie for energy. One calorie is equal to 4.184 joules. In Chapter 4 we will use watts to measure radiant energy. One joule is equal to one watt per second.)

Strictly speaking, then, temperature readings representing different substances such as air, land, and water, can be misleading as measures of the Earth's heat content. Accordingly, data and maps of Earth temperatures must specify whether the

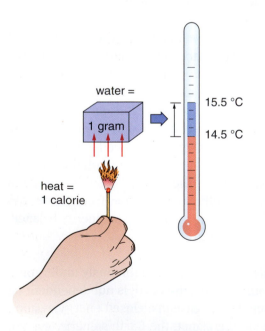

Figure 3.21 The concept of a calorie. The heat energy from the match is transferred to one gram of water producing a temperature rise of 1 °C.

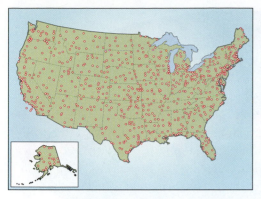

Figure 3.22 The distribution of automated weather stations in the United States where air temperature, precipitation, and other atmospheric data are recorded.

readings represent air, the soil surface, or the surface of the oceans if they are to give an accurate portrayal of distribution of heat on the Earth's surface. And in making geographic comparisons it is necessary to use readings taken from the same substance in different locations. The readings taken at hundreds of automated weather stations (called Automated Surface Observing Sites) in North America record air temperature at a standard 1.5 meters above the ground (Figure 3.22).

Measuring Temperature: The principle of a thermometer is based on the simple fact that when substances take on heat energy, they expand. In a liquid-filled thermometer, for example, a heat-sensitive liquid (usually mercury) is placed in a narrow tube, and when the substance expands and contracts with the addition and loss of heat, it rises and falls in the tube. The tube is scaled numerically so that a temperature can be read from the fluid level.

Several different temperature scales are in use today. The **Fahrenheit scale** sets 32 °F as the freezing temperature of water and 212 °F as its boiling temperature. The **Celsius** (or **Centigrade**) **scale** sets 0 °C and 100 °C for these same two temperatures. The **Kelvin scale** is based on **absolute zero** (0 K), the state at which there is no molecular vibration in a substance and hence no heat, which is equivalent to −273.15 °C. The increment or unit of the Kelvin scale is the same as that of the Celsius.

In the United States, despite efforts to instill the simpler Celsius scale, the media persist in using the more traditional Fahrenheit scale. The rest of the world and the scientific community, however, prefer the Celsius scale. Thus, it is often necessary for Americans to convert from one scale to the other. The chart in Figure 3.23 gives the equivalent readings on the two temperature scales.

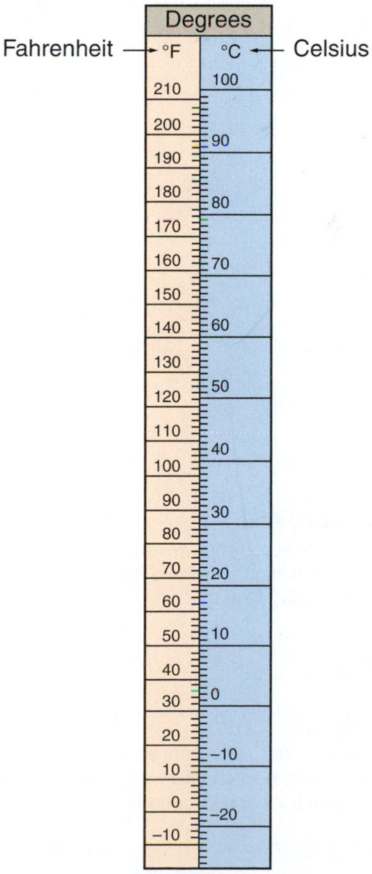

Air is the only component of the Earth's surface environment whose temperature is monitored by virtually all nations of the world on a regular basis. There are several reasons for this. Besides being easy to measure and record, air temperature is one of the two most important elements of climate (the other being precipitation), and humans and their activities are acutely influenced by the temperature of air near the Earth's surface. In addition, air temperature is a good choice to represent the temperature conditions of the planet's surface because air easily reaches thermal equilibrium with the surfaces under it. Therefore, when we examine near-surface air temperatures over broad geographic regions, we are getting a fairly accurate portrayal of thermal conditions at the Earth's surface.

Degrees	
Fahrenheit → °F	°C ← Celsius
210	100
200	
190	90
180	80
170	
160	70
150	
140	60
130	
120	50
110	
100	40
90	
80	30
70	20
60	
50	10
40	
30	0
20	
10	−10
0	
−10	−20

Figure 3.23 A chart showing the equivalent temperatures in Fahrenheit units.

Although the air temperatures at thousands of locations on the continents have been monitored on a regular basis for decades, comparatively few data have been collected over the oceans. However, with the widespread use of airborne and satellite sensing systems in recent years, we have greatly improved our understanding of the thermal conditions of the oceans and the air above them (Figure 3.24). The results are revealing that the oceans are very important controls on global climate, including the climate of the continents. Indeed, some researchers in North America now argue that Pacific Ocean temperature trends, such as El Niño, are important controls on winter weather not only along the American West Coast but across the middle zone of the continent as well.

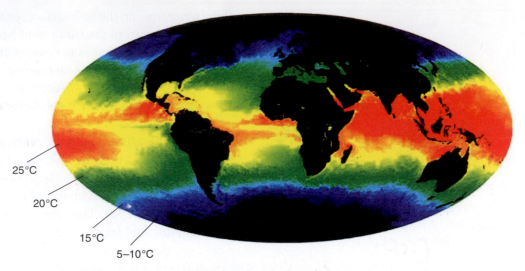

Figure 3.24 A thermal map of the ocean surface prepared from satellite imagery. Thermal conditions of the oceans are now monitored by satellite on a regular basis.

3.8 The Thermal Character of Earth as a Planet

Earth is made up of an amazing variety of thermal environments. They are characterized not only by different temperatures, but by vast differences in the capacity to store and transfer heat. Earth's heat is stored in three main reservoirs, the planet's interior, the oceans, and the atmosphere and, as the diagram in Figure 3.25 reveals, the rates at which each is able to transfer its heat varies dramatically.

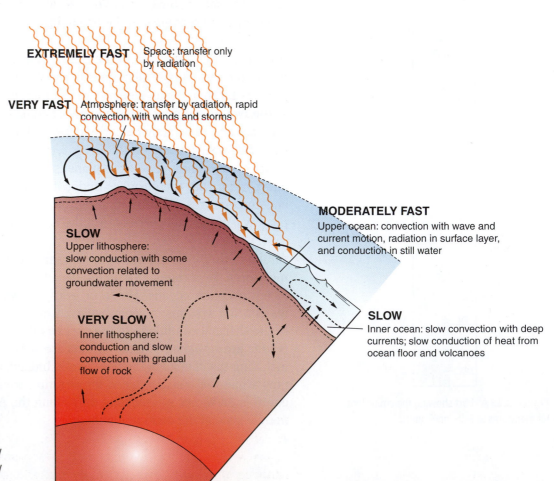

Figure 3.25 Earth's principal heat reservoirs and the relative rates of energy transfer within each. For heat, the inner Earth has the largest supply but transfer rates to the surface are very slow. By contrast, the atmosphere's heat supply is small but transfer rates are very fast.

Earth's Heat Reservoirs: The Earth's interior is unquestionably the largest heat reservoir on our planet. From the Earth's surface downward, temperature increases more or less consistently toward the planet's inner core. At the base of the crust some 50 kilometers below us, the temperature is 800 to 1000°C. The temperature of the inner core, more than 5000 kilometers down, approaches 6000°C, about the same temperature, incidentally, as the surface of the Sun.

With all this heat inside the Earth, why is the planet's surface not hot? The reason is that the Earth is very stingy with its internal (geothermal) heat because rock is a very poor thermal conductor. Therefore, it is very difficult for Earth's heat to get out. In fact, much of the heat deep within the Earth has been there since the planet's origin more than 4 billion years ago. Geothermal heat does seep out of the crust, but it is miniscule compared to the surface heat generated by solar radiation. Only in selected hotspots such as active volcanoes and thermal springs does geothermal heat measurably affect surface environments, but the overall effect on geographic systems is negligible.

Evidence of Earth's massive interior heat reservoir is found in relatively few places which include hot springs, such as this one in Chile.

The oceans are the second-largest heat reservoir on Earth. The oceans receive heat from below with the inflow of geothermal heat from the lithosphere, and from above with the absorption of solar radiation. The geothermal input has little effect on ocean temperature, not only because the input is small but because water has such a high heat-holding capacity. Therefore, as the deep oceans take on heat, water temperature remains cold, about 4°C throughout the world. The surface layer of the oceans is quite different because it is warmed to a depth of 100 m or so by massive inputs of solar energy. All told, the oceans are a great bank of heat energy, and they are not as stingy with their heat as the crustal rock under them. The ocean surface layer gives up huge amounts of energy to the atmosphere, where it drives weather and climatic systems. Indeed, the oceans are considered the atmosphere's major source of energy over much of the world. The deep ocean, at depths of thousands of meters, however, is relatively slow in relinquishing its heat (Figure 3.25).

The atmosphere is Earth's third and smallest heat reservoir. The atmosphere receives heat from the direct absorption of solar radiation, from heat radiated from land and water after these materials have absorbed solar radiation, and from small amounts of geothermal energy released from both land and oceans. The atmosphere has a very small heat-holding capacity, and unlike the oceans and lithosphere, is quick to give up its heat supply. However, it is also quick to take on a new supply. As a result, the heat content of the atmosphere is constantly in a state of flux, fluctuating widely in response to heat input and loss from day to night and summer to winter. Compared to that of neighboring planets in the Solar System, however, this fluctuation is small.

Earth's Surface-temperature Conditions: As we learn more about conditions on neighboring planetary bodies, we can better appreciate the unique thermal conditions on the Earth's surface. Once again, the equilibrium surface temperature for Earth is about 15 °C (58 °F). This temperature is very meaningful for, among other things, it is cool enough to sustain our huge water supply but yet warm enough to drive an active exchange of water, the hydrologic cycle, between the Earth's surface and atmosphere. If Earth were a lot warmer, much of its water would be boiled away and driven off the planet into space. On the other hand, if Earth were only 10°C cooler, it would probably be a much drier place, especially in the middle and high latitudes, with much larger polar ice caps and much lower ocean levels.

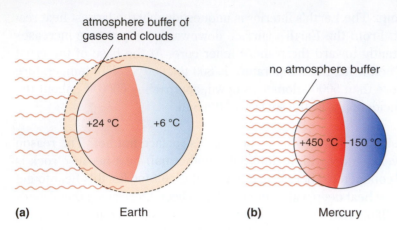

atmosphere buffer of
gases and clouds

no atmosphere buffer

+24 °C   +6 °C

+450 °C –150 °C

(a)   Earth   **(b)**   Mercury

Figure 3.26 (a) With a moderate intensity of solar radiation and an atmosphere to moderate surface temperature, Earth experiences modest day–night temperature differences. (b) Mercury, which is much closer to the Sun and has no atmosphere, has a high solar intensity. Although Mercury's slow rotation contributes to its extreme day–night temperature differences, the absence of an atmosphere helps explain the extreme nighttime cold.

Another important difference between Earth and neighboring planets is the day–night and season-to-season range in temperature. As the Earth rotates, the solar energy system at the planet's surface is literally turned on and off from light to dark. At night, there is effectively no incoming radiation (that from the Moon and distant stars is negligible) and almost every night of the year at every location on Earth the temperature drops. But the temperature drop is relatively modest compared to other planets. Think of a spring day on Earth with a high temperature of 24 °C (75 °F). At night, in the coolest hours before dawn, the lowest temperature typically reaches only 6 °C (42 °F) or so; thus the day–night temperature range is just 18 °C (Figure 3.26a).

Let us take a look at Mercury and Mars for comparison. Mercury, which is an extreme case because it rotates very slowly, has a 600 °C difference in temperature from its light to its dark side (Figure 3.26b). Mars, with a rotation rate nearly the same as Earth, has a daily temperature range of about 100 °C, from around 50 to –50 °C near the Equator. So why is Earth so much less severe thermally? The answer lies mainly in our atmosphere which buffers the planet from a "light-switch" thermal regime. The atmosphere reduces the intensity of incoming solar radiation and at the same time stores heat gained during the daylight hours to offset instantaneous nighttime loss. Mercury has no atmosphere and Mars' atmosphere is very thin; thus, neither planet has much of a buffer to regulate its thermal system.

Chapter Summary and Overview of The Sun–Earth Energy System: Fuel for a Planet

In this chapter we made a case for our overwhelming dependence on the Sun, arguing that because it provides the lion's share of Earth's energy, it is the principal driver behind the great systems. Earth's systems in turn produce work by moving air, water, sediment, and organic matter, rearranging these things, and giving geographic form to the planet's surface. But for most systems to use solar energy, the Sun's radiation must be converted into heat, and Earth is uniquely suited to heat-based systems because, among other things, our atmosphere acts like a greenhouse, storing heat before releasing it back into space. We also argued that all Earth systems are governed by the same principles and they are helpful to understanding how systems function and change. Among them is the concept of an open system, and that for open systems to maintain balance, they must give up as much energy as they take in. We should ask what happens when they do not. And finally we must recognize that Earth is unique among the planets in the Solar System, especially for its capacity to store and move heat energy, a point that is particularly relevant geographically because the distribution and redistribution of stored heat is highly uneven across the planet's surface, thus some places get far more than their fair share of available energy. The following summary diagram may help in visualizing this system.

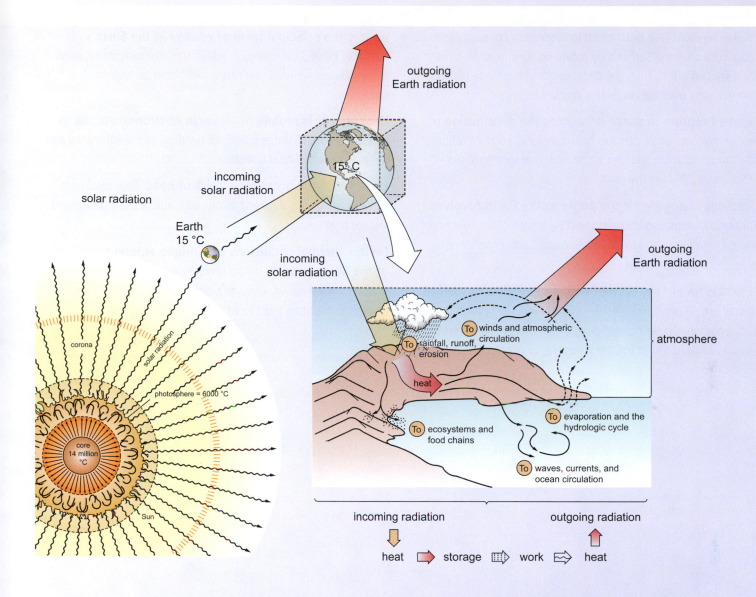

- ► **Energy is the ability to do work.** Energy drives the processes, such as those of air and water, that change the Earth's surface and give our planet its geographic character.

- ► **Earth derives its energy from three sources.** Solar energy is the dominant source, but energy from the inner Earth is also a significant geographic consideration.

- ► **The Sun is the power plant of the Solar System.** Solar energy is discharged from the Sun's photosphere and a tiny fraction of it is intercepted by Earth and the other seven planets of the Solar System.

- ► **The Sun also discharges ionized particles.** The resultant solar wind defines the outer limit of the Solar System and creates Earth's colorful auroras.

- ► **The vast majority of the Sun's energy is produced by the quiet Sun.** The active Sun is defined by more or less irregular behavior with little or no apparent influence on Earth environments.

- ► **The solar energy received by Earth is electromagnetic radiation.** It is measured by its wavelength, is made up of three classes of radiation, and flows to Earth at a steady rate..

- ► **Earth radiation occurs in much longer wavelengths than solar radiation.** This is explained by the temperature difference in the two radiating bodies.

- ► **The greenhouse effect accounts for the slow rate of heat loss from the atmosphere.** It is attributed mainly to gases that absorb thermal infrared radiation.

▶ **The Solar System is a heliocentric system.** It contains eight planets which orbit the Sun as they rotate on their axes. All have elliptical-shaped orbits; thus, the distances between these planets and the Sun vary over the respective years.

▶ **The Earth's motion in space influences the distribution of solar energy.** Sun angle controls the intensity of solar radiation and it changes with the seasons. For midlatitude locations, sun angle varies by 47° over the year.

▶ **The seasons change with sun angle as the Earth revolves.** As sun angle changes, so changes the daily length of daylight and darkness. During the equinox all locations on Earth experience equal hours of day and night.

▶ **The Earth is an energy system with an equilibrium surface temperature.** This system is made up of three interconnected parts. Changes in any part of the system can affect its long-term equilibrium causing a shift toward a positive or negative balance.

▶ **Heat is the principal form of energy at the Earth's surface.** When heat energy is added to Earth substances, they show different temperature responses based on their heat capacities.

▶ **Heat energy is mobile in all Earth environments.** But its rate of movement varies greatly depending on the substance and the mechanism of heat transfer.

▶ **Earth has three great reservoirs of heat.** Their total heat supplies are widely different as are their abilities to transfer and release heat.

▶ **Earth's thermal conditions are unique among the planets of the Solar System.** Unlike Mercury, our planet is cool enough to maintain a huge water supply, but unlike Mars, it is warm enough to maintain most of its water in a liquid state.

Review Questions

1 What are the three sources of energy for the Earth's surface? Which source provides the vast majority of this energy? Name several Earth systems which are powered by this energy.

2 Compare the Sun and the Earth as radiating bodies in terms of surface temperature and principal types of radiation produced.

3 Does the Sun broadcast anything into the Solar System besides radiation and, if so, what is(are) its effect(s) on the Earth's surface environment?

4 Compare the behavior of the quiet and active Sun. What is the solar constant?

5 What is electromagnetic radiation, how does it travel, and what feature of this energy is used to define its various classes?

6 What is the electromagnetic spectrum and what are the three types of radiation that make up the insolation band?

7 Why do the Sun and the Earth produce vastly different types of radiation and why is the type of radiation produced by Earth so important to our planet's thermal conditions?

8 What is the meant by the terms geocentric and heliocentric and who was responsible for developing the concept of the Universe as a perfectly balanced mechanical system?

9 What is the difference between revolution and rotation in the motion of the Earth and what is the attitude of Earth's axis with respect to the plane of the ecliptic?

10 Earth can be described as a three-part energy system. Explain. What is meant by the term equilibrium surface temperature?

11 Briefly explain the difference between heat and temperature. Define the calorie as an energy unit.

12 Why is air near the ground a good medium for monitoring the thermal conditions of the Earth's surface.

13 Name Earth's three primary heat reservoirs and define the sources of heat for each.

14 With respect to the thermal conditions of Earth and Mercury, what is meant by a light-switch thermal regime?

Earth's Radiation and Heat Systems Over Land and Water

Chapter Overview

This chapter is about the concept of balance in the Earth energy system. We deal first with the input side of the system by bringing solar radiation into the Earth's atmosphere to see what happens when it passes through air and strikes the Earth's surface. Since geography is our main concern, we are particularly interested in factors, such as cloud cover and the curvature of the Earth, which influence the global distribution of solar energy. Next is the output side of the system. To keep from overheating, Earth must release back into space the energy it gains from the Sun. But before this can take place, solar radiation must be converted into heat and then converted back into radiation in order to leave the atmosphere. We then examine the balance between Earth's energy inflows and outflows. The chapter ends with a look at global temperature patterns and the geographic controls on those patterns, in particular the world distribution of land and water.

Introduction

Uncle Bill could have been a character out of a Jack London novel. At least that's how I, at the age of eight or nine, saw him. On this occasion he was home on Christmas leave from the Air Force and he thought it might be challenging for us to experience a winter night camping out in the northern Michigan woods. But to make it truly challenging, he suggested that we "rough it" by doing without the standard cold weather gear. "Let's bring just an axe, matches, a few blankets, coffee pot, fry pan, and a little food." So one cold afternoon around the first of January we gathered up our meager supplies and hiked off into the Lake Superior woods. Not until years later did I realize that the whole outing, and indeed our very survival, were an exercise in energy management, in simple terms, how to maintain a favorable balance between energy gains and losses in a system centered on two warm bodies.

We arrived at our destination, a large beaver pond, like the one in Figure 4.1, called Pea Patch Lake, around mid afternoon. "We'd better move quickly," he said, "because it'll be dark in a few hours." First, he chose a camp site sheltered on the north by some large hemlocks. Next, he cut a number of small poles and framed a simple lean-to which we lined, roof and floor, with the green boughs of balsam fir. As he went about these tasks, he explained his reasoning,

Figure 4.1 A beaver pond in winter. Draped in snow and ice, the beaver dam is barely visible.

ending with, "Most important we need a whole lot of firewood." By the time night fell, we had built a pile six or seven feet high, more wood, I thought, than we could possibly burn in one night. To my surprise, by morning only a few pieces remained.

With night came the real lesson. The air grew cold and still and we built a huge blaze that sent flames 10 feet or more into the air. This cauldron acted as a powerful thermal chimney, drawing in cold air near the ground and driving heated air skyward. As a direct heat source the fire gave us little benefit; in fact, it probably had the opposite effect by sucking cold air into our encampment. Our heat supply came instead from another source, the radiation generated by the fire. At a temperature of a thousand degrees Celsius, the fire and hot wood coals discharged generous amounts of radiation which passed through the cold air, against the inflowing drafts being sucked into the flames, and into the lean-to where it was absorbed and converted into heat. Thus our camp functioned as a simple thermal system with radiation providing the primary input and convection of heated air the primary output.

After tending the fire most of the night, Uncle Bill fell asleep a few hours before dawn. As the fire died down, a light wind came up from the north, a chill crept up our backs, and our sense of the energy balance shifted to our own bodies. Because our capacity to generate body heat was diminishing (owing to 12 hours without food), our bodies fell into a negative energy balance, a condition that if extended would lead to depressed core temperatures and hypothermia. It was time to move. We rebuilt the fire and cooked a high calorie breakfast of bacon, eggs, coffee, and cookies. The combination of inputs from the fire, the fat digesting in our bellies, and the new rays of the morning sun quickly shifted the energy balance in our favor and we set off to face the day exploring the frozen beaver pond.

4.1 Solar Radiation and the Composition of the Atmosphere

We begin with the solar constant, the steady flow of solar radiation striking the Earth's atmosphere. Measured at the outer edge of the atmosphere, the solar constant has a strength, or energy equivalent, of 0.14 watt per square centimeter (2.00 calories per square centimeter per minute). Before it reaches the planet's surface, this beam of radiation must pass through the atmosphere where sizeable amounts are reflected, scattered, and absorbed, as shown in Figure 4.2. In other words, the atmosphere acts like a thick screen, buffering the Earth from the full force of the Sun. As a consequence, less solar radiation reaches the surface than enters the atmosphere. How much and what types of solar radiation are not transmitted through the atmosphere depend, among other things, on the composition of the atmosphere.

Earth's atmosphere is a mixture of gases, called air, with a minute quantity of tiny liquid and solid particles at lower levels. Although most air is concentrated within 10 to 12 kilometers of the Earth's surface, the atmosphere forms a layer several hundred kilometers deep that completely envelops the planet. With the exception of two significant constituents, its overall composition near the Earth's surface is almost exactly the same throughout the world. The two exceptions are water vapor and pollutants such as dust particles. The amount of water vapor varies geographically, for example, from regions dominated by warm oceans to those dominated by the great deserts. The presence of pollutants varies with volcanic activity, forest fires, war, and urban development.

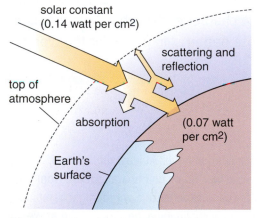

solar constant (0.14 watt per cm2)

scattering and reflection

top of atmosphere

absorption

(0.07 watt per cm2)

Earth's surface

Figure 4.2 The solar constant at the top of the atmosphere has a strength of 0.14 watt per square centimeter, but after passing through the atmosphere it is cut in half because of reflection, scattering, and absorption.

Table 4.1 Composition of dry air

Gas	% volume
Nitrogen (N$_2$)	78.084
Oxygen (O$_2$)	20.946
Argon (Ar)	00.939
Carbon dioxide (CO$_2$)	00.033
Other minor gases	00.003

Table 4.1 gives the composition of dry air. Two elements, nitrogen and oxygen, make up over 99 percent of *dry air* by volume. While they are important to Earth's life system, these two gases are not of central importance to our planetary energy system. The remaining gases are minor constituents, and some of these are very important to Earth's energy system because they are able to significantly alter the stream of radiation coming to and from the Earth's surface. Among the minor gases are argon, an inactive gas of no significance to heat and radiation, and carbon dioxide, an important absorber of infrared radiation, as well as miniscule amounts of a number of very minor gases. One of the very minor gases is ozone (O$_3$), a noxious variation of oxygen (O$_2$), which ranges from 0 to 0.00007 percent, and is important in the absorption of ultraviolet radiation.

Variable and Versatile Water: Water is the most variable and versatile component of the atmosphere. Atmospheric water occurs in three physical forms: vapor (the major form), liquid particles (clouds), and solid particles (snow and various other ice forms). Water vapor is the atmosphere's principal absorber of infrared radiation, outweighing carbon dioxide in this role, and airborne water particles, both liquid and solid, make up clouds, Earth's principal reflectors of solar radiation.

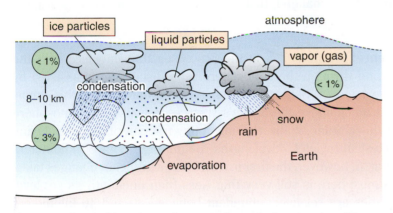

Figure 4.3 The variable and versatile nature of water in the atmosphere. Water vapor content (in green) changes dramatically with altitude in the atmosphere and from place to place across the planet's surface.

Water is versatile. It is a part of every system on the planet, and as the diagram in Figure 4.3 suggests, the atmosphere is pivotal to its wide dissemination. Nevertheless, water is unevenly distributed in the atmosphere, ranging from a small fraction of 1 percent over some deserts to more than 3 percent over tropical seas and rainforests. But in terms of its dynamics, the atmospheric water system is unequalled. The exchange, or recycling, of water between the Earth's surface (mainly the oceans) and the atmosphere is exceptionally rapid, taking only nine days for the atmosphere to give up the equivalent of its entire water content in exchange for a new load of water from the surface!

The Atmospheric Recycling System: In fact, all the constituents of the atmosphere are part of a recycling system linking the atmosphere with the Earth's surface. The atmosphere's compositional balance depends on a complex system of inputs and outputs that can be changed by altering, among other things, conditions and processes on the Earth's surface, including photosynthesis, ocean evaporation, precipitation, volcanic activity, and human air pollution. Those constituents with the largest volumes and slowest recycling times are least subject to significant, short-term change. The speed of recycling is given by the **exchange time**, the time it would take to completely extract and renew the atmospheric reservoir or pool of a gas.

Table 4.2 Atmospheric exchange (recycle) times

Nitrogen	400 million years
Oxygen	6000 years
Carbon dioxide	10 years
Ozone	2–4 months
Water	9 days

The exchange time (see Table 4.2) for nitrogen is estimated at 400 million years; for oxygen it is about 6000 years. The water cycle, the fastest, is driven by mainly by surface heat, and we know that it is possible to change the atmosphere's water-vapor content in a matter of days by suddenly cooling the planet in response to, say, massive volcanic eruptions that darken and cool the Earth's surface and atmosphere. Two other quick cycle gases, carbon dioxide (CO$_2$) and ozone (O$_3$), are presently undergoing appreciable change in response to human activity and atmospheric pollution. Since the Industrial Revolution (1750–1850), atmospheric CO$_2$ has increased

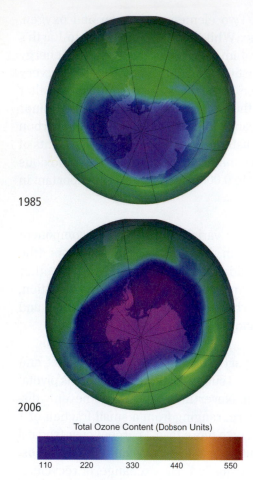

1985

2006

Total Ozone Content (Dobson Units)

110   220   330   440   550

Figure 4.4 The ozone hole over the Antarctic from 1985 to 2006. Ozone is measured in Dobson Units (DU). One DU is equal to 1 part ozone per billion parts of air. Normal ozone concentration is about 300 DU.

at least 30 percent with global population growth and air pollution. Ozone, on the other hand, has decreased with air pollution (from chlorine-based molecules used in spray and refrigerants) and the loss is most concentrated over the Antarctic as Figure 4.4 illustrates. With a world ban on this pollutant, ozone is slowly recovering. CO_2, however, continues to rise worldwide.

4.2 The Effect of the Atmosphere on Solar Radiation

After its eight minute trip from the Sun, the beam of solar radiation enters the atmosphere. As it passes through the atmosphere, its intensity is reduced and its composition is altered as the radiation encounters increasingly denser layers of air and different atmospheric constituents such as ozone, water vapor, and clouds. Two processes are especially important in producing these changes: absorption and reflection. In **absorption**, waves of radiation (electromagnetic energy) are taken up by a substance, such as a gas molecule, and transformed into heat energy. In **reflection**, waves of radiation bounce off a substance, such as water particles, and are sent back into space. **Scattering** is a similar process, but the waves of radiation are sent in all directions, some going back into space, some to Earth. In both reflection and scattering, the energy form is not changed; that is, the radiation maintains its original wavelength.

Atmospheric Interference: We now look at the processes that interfere with the passage of solar radiation through the atmosphere. The discussion is assisted by the diagrams in Figures 4.5 to 4.7. Solar radiation first encounters the Earth's atmosphere several hundred miles above the planet's surface. Because the atmosphere is exceedingly thin at this altitude, it absorbs only a tiny fraction of the solar beam, mostly the very shortest ultraviolet wavelengths. Beginning about 80 kilometers (50 miles) above the Earth, however, the first in a series of significant changes takes place. Although the atmosphere is still very thin (less than one-tenth its sea-level density), ozone molecules are relatively abundant between 80 and 40 kilometers (50 to 25 miles) altitude and they absorb ultraviolet (UV) radiation.

The absorption of UV radiation raises the temperature of the atmosphere (see Figure 4.5), giving rise to a relatively warm zone centered about 50 km (30 miles) above the Earth. This marks the top of the **stratosphere**, a zone of thin air about 40 km deep where UV absorption declines and air temperature falls toward Earth. In the lower stratosphere, temperature is a bitter –60°C and the first clouds appear, thin wispy forms made up of ice crystals, which, as Figure 4.6 suggests, reflect a little solar radiation. At this point, only six miles or so above the Earth, the strength of the incoming beam of solar radiation has been reduced by less than 10 percent.

At 10 to 12 kilometers altitude the beam enters the lowermost zone of the atmosphere, the **troposphere**, the realm of dense air, active weather, and life. Beginning at the top of the troposphere, called the **tropopause**, the atmosphere begins to thicken rapidly and the cloud cover becomes much heavier, blanketing vast areas of the sky (see Figure 4.6). The clouds are highly effective mirrors, reflecting 20 percent of the incoming beam worldwide. This reduction is equivalent to 0.028 watt (0.40 calorie) of energy for each square centimeter of the Earth's surface for every minute the Sun is shining. Without the shade of clouds, Earth would have a much warmer surface.

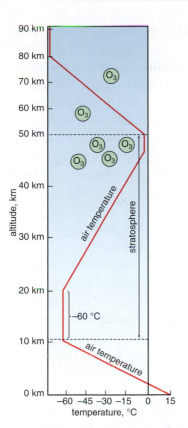

Figure 4.5 The distribution of temperature in the lower 90 kilometers of the atmosphere. The relatively high temperature around 50 kilometers is caused by ozone (O₃) absorption of ultraviolet radiation. At 10 kilometers, temperature rises toward Earth's surface.

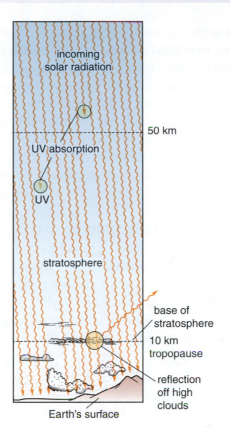

Figure 4.6 The processes of absorption and reflection of solar radiation in the stratosphere. Together they reduce incoming radiation by less than 10 percent.

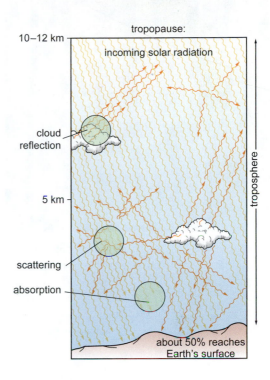

Figure 4.7 The processes affecting solar radiation in the troposphere: cloud reflection, scattering, and absorption.

Scattering and absorption in the troposphere further reduce the solar beam. Gas molecules and particulate matter (including clouds) scatter light rays in all directions across the sky, as the diagram in Figure 4.7 illustrates. Some rays are directed to the ground, where most are absorbed, and some, totaling about 7 percent of the beam, are directed skyward and lost into space. Shorter wavelengths, on the blue side of the visible band, scatter more effectively, and the diffusion of these wavelengths throughout the atmosphere gives the sky its blue coloration.

The other process taking place in the troposphere is absorption, mainly by water vapor, carbon dioxide, and clouds. Absorption in the lower atmosphere accounts for the loss of 21 percent of the incoming radiation, mostly infrared wavelengths. By the time the Sun's radiation has passed through the lower atmosphere, the solar constant has been reduced by about 50 percent for the Earth as a whole. Losses to scattering and cloud reflection total 27 percent and absorption by ozone, water vapor, carbon dioxide, and clouds totals 24 percent.

The remaining half of the solar constant reaches the Earth's surface. Of the three types of solar radiation (ultraviolet, visible, and infrared) that originally entered the atmosphere, almost all the ultraviolet has been absorbed and about one-third of the infrared has been absorbed or reflected and scattered. Almost all of the visible, however, has reached the surface. Thus, the atmosphere is more or less an open window to the visible band of radiation. The time taken for atmospheric passage, including the absorption, reflection, and scattering processes, is less than a millisecond.

Figure 4.8 Shafts of sunlight and areas of shadow illustrate the difference between beam and diffuse radiation.

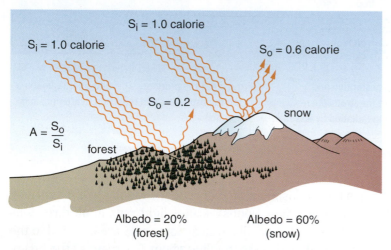

Figure 4.9 The concept of albedo with an illustration of the contrast between mountain forest cover and snow.

Solar radiation actually arrives at the Earth's surface in two forms: beam and diffuse. **Beam radiation** is direct insolation, like that shown in Figure 4.8. It is best characterized by sunlight on a clear, cloudless day. **Diffuse radiation** is indirect insolation, best characterized by the hazy light created by scattering on an overcast day. At the surface, both beam and diffuse radiation are subject to one last round of absorption/reflection. About 94 percent of the radiation hitting the Earth's surface (or 47 percent of the solar constant) is absorbed by land and water and converted into heat. Six percent of the radiation hitting the Earth's surface (or 3 percent of the solar constant) is reflected off water, vegetation, sand, ice, and other materials, and sent back into the atmosphere and thence to space beyond.

Earth's Surface Reflectance: Earth materials vary widely in their capacity to reflect incoming solar radiation. The percentage of incoming solar radiation reflected from a surface is termed **albedo**. It is based on two simple measurements of solar radiation taken immediately over a surface receiving sunlight: one, incoming solar (S_i), and the other outgoing solar (S_o). Dividing S_o by S_i yields the appropriate ratio or percentage (Figure 4.9). The higher the percentage, the greater the reflectance, and the less solar energy absorbed and available to heat the surface. Considering the vast range of materials that coat the Earth's surface, albedo and related surface heating vary appreciably from place to place.

Light-colored, smooth materials such as ice, fresh snow or calm water reflect a high percentage of the radiation that reaches them. They thus have high albedos, typically 60 to 70 percent or more. Dark, rough materials such as a forest canopy or plowed soil reflect light poorly. They have low albedos, typically less than 20 percent. Earth has an overall low surface albedo because relatively dark, rough-textured seascapes and landscapes dominate its surface. But if we pull back from ground level and include atmospheric scattering and cloud reflection, Earth's albedo rises to a total of 30 percent, which is quite high. For comparison, the Moon's albedo is less than 10 percent (Figure 4.10). The difference between Earth's surface albedo and planetary albedo is primarily the result of cloud cover. From space, Earth is one of the brightest objects in the Solar System. We will say more about Earth's albedo and the albedos of other celestial bodies in the next section.

Summary on Solar Radiation in the Atmosphere: The atmosphere is a powerful filter system on solar radiation, absorbing some, scattering some, and reflecting some. These processes take place at different altitudes and involve different components of the atmosphere. Of the three main types or bands of radiation in the incoming beam, visible is the least affected by the atmosphere. Clouds are paramount in the atmosphere's filter system and we might consider what this implies for an Earth with more or fewer clouds than we see today and what might cause such a change. The last step for insolation is the Earth's surface where it is either accepted or rejected depending on the material it encounters. What does this imply for the landscape changes we are making over broad swatches of the planet?

Figure 4.10 The relative brightness (reflectance) of the Earth and the Moon as seen from space.

4.3 Geographic Distribution of Solar Radiation at the Earth's Surface

We now turn to an important geographic question: the distribution of solar radiation reaching the Earth's surface. The atmospheric processes that govern the flow of radiation to the Earth's surface are not evenly distributed over the planet. As a result, some regions, like the great deserts, receive much greater amounts of solar energy than others. Moreover, insolation at all locations varies seasonally, but does so much more at some locations than others. The geographic effects are profound, including dramatic differences over the planet surface in temperature, sunlight on land and water, rates of photosynthesis, and the abundance of life.

Sun Angle and Latitude: We begin with the world map in Figure 4.11 showing average annual solar radiation recorded at ground level. Our first observation is a general one: solar radiation values decrease with latitude. The tropics receive about three times the radiation of polar regions in both hemispheres. This difference is caused by the curvature of the Earth and the resultant change in sun angle. As sun angle decreases toward the poles, the intensity of solar radiation declines correspondingly.

Figure 4.11 Average annual incoming solar radiation (insolation) values at the Earth's surface in watts per square meter per year.

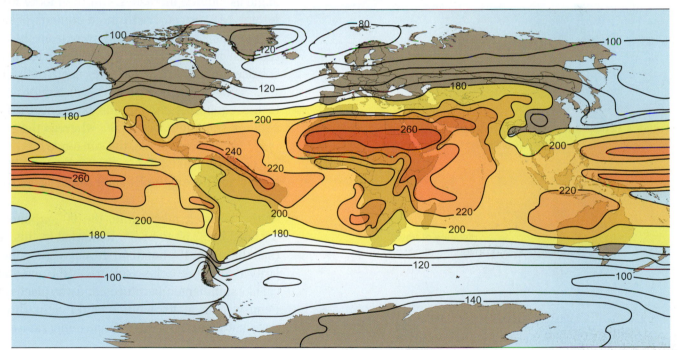

The decrease in solar radiation towards the poles works as follows. When solar radiation arrives at the top of the atmosphere, all of it, from the North Pole to the South Pole, is traveling in lines parallel to the plane of the ecliptic. Consider the case on the equinoxes, March 20–22 or September 20–22, when the Earth's axis is perpendicular to the plane of the ecliptic. On those dates, the radiation striking the Equator hits the Earth's surface perpendicularly, that is, straight on, forming a 90° sun angle (Figure 4.12a). If you imagine an incoming shaft of radiation measuring 1 meter by 2 meters, the area of ground or ocean surface illuminated and heated by it is exactly the same as the area (in cross-section) of the shaft. If we move north to, say, 60° latitude, as Figure 4.12b demonstrates, a similar shaft of radiation strikes the Earth's surface at a much lower sun angle, 30°. The lower sun angle results in two significant differences from the Equator: (1) the shaft of sunlight must illuminate and heat a

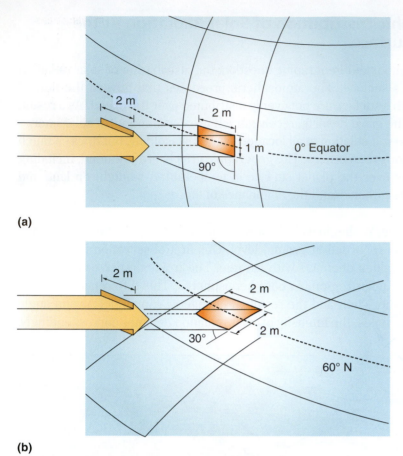

(a)

(b)

Figure 4.12 The effect of the Earth's curvature on solar intensity using (a) the Equator and (b) 60° north latitude on the equinox as examples.

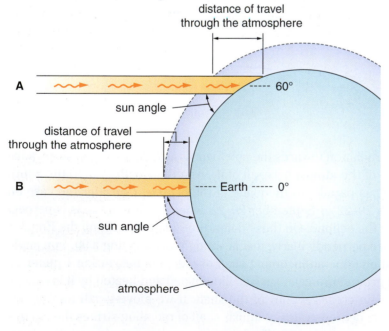

Figure 4.13 The effect of latitude on solar intensity. Not only is the sun angle lower at higher latitudes, but the distance of travel through the atmosphere is greater as well.

surface area twice as large, *which reduces its intensity by half*, and (2) the shaft of sunlight must pass through a much greater distance of atmosphere (see Figure 4.13), *which subjects it to more scattering and reflection*.

If we move to the poles, it appears that neither pole would receive solar radiation at the equinoxes. But if we arrived there on the equinoxes to test this observation, we will actually find a small amount of incoming solar radiation hitting the surface. Some of it is diffuse radiation from atmospheric scattering, and some is beam radiation produced by **refraction**, or bending, of rays as they pass through the atmosphere. As the diagram in Figure 4.14 illustrates, refraction causes radiation to bend a little toward the Earth's surface, making the Sun appear a little higher in the sky than it actually is.

So our first conclusion is that at the broadest scale the intensity of solar radiation at the Earth's surface declines with sun angle as latitude increases. But the model we used, shown in Figure 4.15, placed Earth in an equinox position. Now let us tip the axis 23.5° toward and away from the Sun to represent the solstice positions. For the northern hemisphere's summer solstice, June 20–22, all latitudes north of the Tropic of Cancer (23.5° N) experience a 23.5° sun angle increase compared to the equinoxes. At latitude 66.5° north, for example, the equinox sun angle was 23.5°, whereas the solstice sun angle is 47°. At the North Pole, the sun angle has changed from zero (or nearly so) to 23.5°. In the tropics, the 90° sun angle has shifted north from the Equator to the Tropic of Cancer.

At the Equator, meanwhile, the 90° equinox sun angle has declined by 23.5° to 66.5°. Remarkably, this angle is 2° lower than the one at 45° north on the corresponding date. This similarity in sun angle helps explain why the midlatitudes have tropical-like summer temperatures, and why places like Philadelphia, Pennsylvania, at 40° latitude, and Belem, Brazil, at 1° latitude have just about the same average temperature for July.

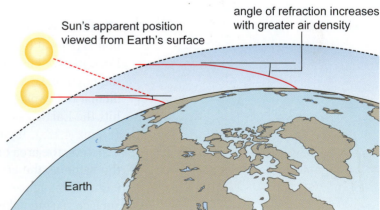

Figure 4.14 The effect of atmospheric refraction on the path of solar radiation through the atmosphere and in turn on sun angle.

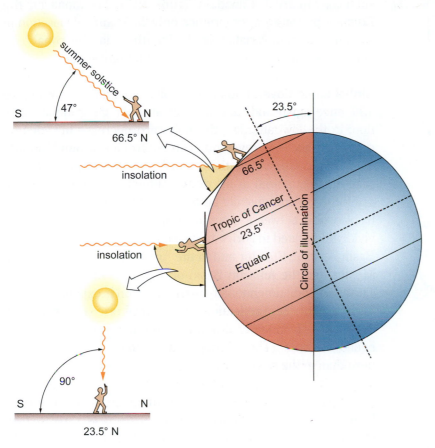

Figure 4.15 Illustrations showing how you would view sun angle from the Tropic of Cancer and the Arctic Circle on June 20–22. On this date the sun angles are 23.5° higher than on the equinoxes.

Table 4.3 Total length of daylight by latitude and date

Latitude	June 15	Sept. 15	Dec. 15
70°N	24 00	13 26	00 00
65°	21 53	13 07	3 42
60°	18 49	12 55	5 56
55°	17 21	12 46	7 13
50°	16 21	12 39	8 06
45°	15 25	12 34	8 46
40°	15 00	12 28	9 21
35°	14 30	12 24	9 50
30°	14 04	12 22	10 14
25°	13 41	12 18	10 36
–	–	–	–
–	–	–	–
–	–	–	–
–	–	–	–
0°	12 00	12 00	12 00

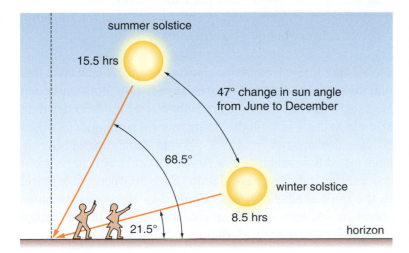

Figure 4.16 The annual range of sun angle and length of day at 45° latitude. On the summer solstice, 65 percent (15.5 hours) of the day is light, whereas on the winter solstice 35 percent (8.5 hours) of the day is light.

Length of Day and Latitude: The approach of summer solstice brings not only higher sun angles but also an increase in the number of hours of sunshine in each day (Figure 4.16). Length of daylight also changes with latitude, but instead of growing smaller toward the pole, as sun angle does, it grows longer. Beyond the Arctic Circle it reaches a full 24 hours on June 20–22. The opposite is true for the winter solstice: the days are shorter toward the poles with total darkness beyond the Arctic Circle. The pattern is exactly the same for the southern hemisphere in opposite seasons. Table 4.3 gives the hours of daylight for latitudes 25–70 degrees on the solstices and equinoxes.

Consider the net result of the changing sun angle and length of day. In the summer, when sun angles are greater and daylight hours longer, latitudes poleward of 23.5° experience an increase in solar radiation receipt, and the lower sun angles at very high latitudes are partially offset by exceptionally long days. At the approach of the winter solstice, the declining sun angle and the shortened daylight hours causes a decrease in solar radiation receipt. Witness the extreme differences from solstice to solstice at 45° north latitude as shown in Figure 4.16. Herein lies the principal reason for the extreme

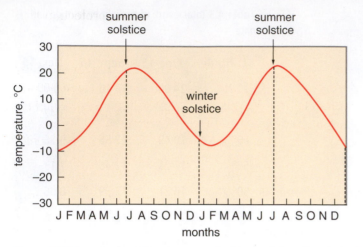

Figure 4.17 Average monthly temperatures for Montreal, Canada, at latitude 45° north. The highs and lows closely follow the solstices.

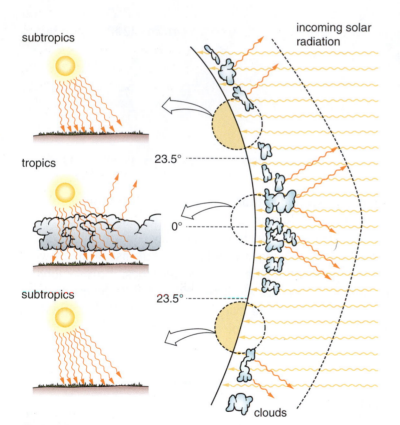

Figure 4.18 Differences in cloud cover between the wet tropics and the dry subtropics are striking and have big influence on the amount of insolation reaching the surface.

thermal contrasts between winter and summer for locations such as Montreal, Canada (Figure 4.17). Locations on the Equator, by contrast, experience only 23.5° annual variation in sun angle and no variation in the length of day. Seasonal temperature variations there are correspondingly small.

Global Cloud Cover: Look at the solar radiation map in Figure 4.11 once again and take note of another geographic variation: in the low latitudes, the poleward decrease in radiation described above is reversed. The subtropics, around latitude 20–30°, receive substantially more radiation at ground level than the equatorial zone does. Annual radiation values at ground level are typically 20 to 30 percent greater in the subtropics, and between some locations it is even more. This is vividly illustrated in Africa between the Sahara and the Congo Basin near the Equator and in the Americas between northern Mexico and the Amazon Basin on the Equator (Figure 4.18). The lower radiation intensity in the Amazon is explained by the high rates of cloud reflection and atmospheric scattering. The atmosphere is much cloudier, more humid, and deeper in the equatorial zone. (The troposphere is 2 to 4 kilometers deeper here than in the subtropics.)

Large parts of the Amazon, for example, are so heavily blanketed with clouds that from an airplane it is rare to ever glimpse the land. On the other hand, desert regions like northern Mexico and Saharan Africa seldom see heavy cloud covers, especially in summer. The drier air and shallower troposphere in these regions produce less absorption and scattering of incoming radiation. Other parts of the world, such as the northern Pacific and northern Atlantic, are also laden with heavy cloud covers and high humidities. Annual insolation values are reduced 50 percent or more, and the radiation received is overwhelmingly diffuse.

Global Albedo: Surface albedo also makes a significant contribution to the global distribution of solar radiation. Both water and land surfaces have higher overall albedos at high latitudes than at low latitudes. For sunlight hitting water at sun angles less than 20°, albedo can be as great as 90 percent. Antarctica and Greenland (as well as much of the Arctic Ocean), with permanent covers of ice and snow, have albedo values of 60 percent or more. In the tropics, albedo of sea is only 5 to 10 percent because most sun angles are above 70°. Albedo values are also low on land in the tropics because of the vegetative cover, especially the dark, green tropical forests.

Albedos in the great deserts may be 30 to 40 percent, and in some areas even higher. The highest values are found in the great sand deserts such as the Empty Quarter of the Arabian Peninsula, a great sea of sand dunes that covers more than 500,000 square kilometers. The clean, tan quartz sand of the Empty Quarter, shown in Figure 4.19, probably reflects more than 50 percent of the incoming solar radiation. The high surface albedo helps explain the traditional dress of the desert nomads,

Figure 4.19 Contrasting landscape albedos: forest cover at 10–20 percent and sand dunes at 40–50 percent.

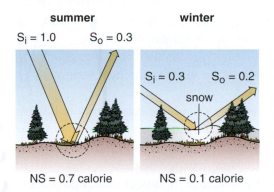

summer   winter

$S_i = 1.0$ $S_o = 0.3$

$S_i = 0.3$ $S_o = 0.2$

snow

NS = 0.7 calorie   NS = 0.1 calorie

Figure 4.20 Net solar radiation is the balance between S_i and S_o. Summer–winter contrasts in NS in the midlatitudes can be as great as sevenfold.

Figure 4.21 Albedo change brought on by urban development in a forested landscape. Manufactured materials are generally brighter than tree canopies.

which typically covers the body from head to toe. One needs solar protection from both above (S_i) and below (S_o).

In the midlatitudes, albedo values rise dramatically in the winter with snow cover. Significant snow cover – that is, an accumulation which lasts two to three months or more – generally begins around 40 to 45° latitude. The first heavy snowfall of the season drives albedo suddenly upward from 20 to 30 percent to 50 to 70 percent. When this happens, temperatures may fall suddenly as the amount of solar radiation absorbed by the landscape is reduced by half or more – a fact that helps explain the sudden onset of winter in some years. Farther poleward and in mountain areas, snow covers last longer, but in the broad belt of the subarctic boreal forests, similar in color to the one shown in Figure 4.19, which stretch across Eurasia and North America, snow is less effective in raising albedo because much of it lies under the dark canopies of the evergreen trees.

Albedo and Net Solar Radiation: In summary, the chief effect of albedo is to reduce the percentage of incoming solar radiation finally absorbed by the Earth's surface. The diagrams in Figure 4.20 illustrate how extreme this can be from summer to winter in a midlatitude landscape. The key question is how much solar radiation is left on the surface, that is, absorbed by the landscape? This value is called **net solar radiation (NS),** and is equal to incoming solar radiation (S_i) minus outgoing (reflected) solar radiation (S_o):

$$NS = S_i - S_o$$

Earth's net solar radiation values not only vary geographically with different landscapes, but over time as well. At the peak of the last major glaciation about 20,000 years ago, glacial ice covered about 30 percent of the Earth's land area, about three times more than today's coverage. This increased albedo values and lowered net solar radiation values over midlatitude North America, for example. Over the past several thousand years, albedo has been changed with the spread of humans, agriculture, and cities (Figure 4.21). Many urban areas like Vancouver, Atlanta, and Shanghai, stand out as brighter landscapes with generally lower solar absorption than the forested landscapes they replaced. On the other hand, desert landscapes converted into agricultural land with the aid of irrigation have much lower albedos and higher rates of solar absorption.

Landforms and Solar Radiation: Finally, we consider the influence of topography and landforms on the distribution of solar radiation. Virtually nowhere does solar radiation strike a truly flat land surface. Even ground that looks flat may be inclined by 2 or 3°, and much of the landscape is inclined in slopes of 5 to 10°. Slopes that face the Sun create steeper sun angles and, as a result, the surface area illuminated and heated is smaller. Therefore, the intensity of solar radiation (also called **incident radiation**) is greater and surface temperatures higher.

On the other hand, slopes that face away from the Sun experience reduced solar intensity and cooler ground temperatures. The amount by which incident radiation is increased or decreased in each case is proportional to the angle of the slope. Look at the diagram in Figure 4.22. At 45° north latitude, a 40° south-facing slope will produce a sun angle of 85° on the equinox. On that same date, a north-facing slope on the other side of the hill will produce a 5° sun angle. These differences can give rise to distinct microclimates, which we will explore a little later.

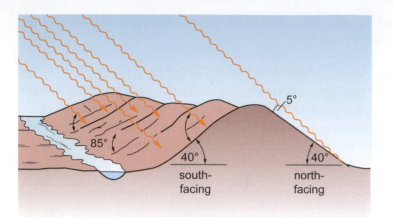

Figure 4.22 The influence of landforms on sun angle. On south-facing mountain slopes in the northern hemisphere, sun angles may be 10 to 20 times higher than those on north-facing slopes.

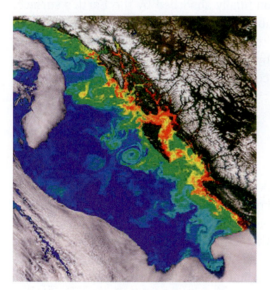

Figure 4.23 Warm ocean currents (in red, orange, and green) such as these along the northern Pacific coast are able to move huge amounts of heat great distances, an essential part of Earth's energy redistribution system.

Summary on the Geography of Solar Radiation: The receipt of solar radiation at ground level varies widely over the Earth. How can this be so if the Earth as a whole is washed over with a steady flow of radiation from the Sun? The answer has mainly to do with sun angles, length of daylight, cloud cover, and albedo. We are led to questions, such as, how do these factors combine to produce a cold December in Minnesota or Manitoba? And following a similar line of reasoning, why are the summer–winter extremes so different in Montana and Venezuela? On balance, things come down to net solar radiation and what accounts for a large NS at some locations and times and small ones at others.

4.4 Heat Transfer in Land, Water, and Air

Now that we have accounted for the receipt of solar radiation at the Earth's surface and its absorption by surface materials, our next question is the disposition of the heat it generates. When air, water, and land are heated by solar radiation, the heat, in turn, is redistributed or dispersed through a larger volume of these substances. The movement of heat from hotter to colder objects or fluids is a consequence of the second law of thermodynamics, which states that over time, differences in temperature, pressure, and density tend to even out in a physical system. We refer to this process as **heat transfer**, and it accounts for the simple fact that heat naturally spreads from its point of origin, such as from a heated soil surface on a sunny day. In general, how quickly heat transfer takes place depends on the nature of the substances holding or receiving the heat (e.g. their phase, temperature, density, molecular bonding, geometry, the type and velocity of the transfer fluid) and the intensity of the heat supply.

Heat Transfer in Earth Environments: Heat is a very dynamic commodity in Earth's surface environment. It is subject to spontaneous movement or transfer often over great distances in short periods of time, as the thermal image in Figure 4.23 suggests. Heat transfer is essential to earth's thermal balance because it ensures that heat is redistributed over the planet, thereby reducing the geographic extremes in hot and cold.

The movement of heat energy takes place by three basic means or mechanisms of heat transfer: conduction, convection, and radiation. Which occurs in the environment depends on the substances through which heat is being transferred, that is, whether they are fluid (air and water) or solid (soil and rock), or transparent (air and water) or opaque (soil and rock). How fast the heat is transferred by each specific mechanism varies with two factors. First, the mechanism involved: radiation is super fast, convection is moderately fast, and conduction is painfully slow. Second, the relative difference in the amount of heat between the two points within a substance. In a teakettle, for example, it is the difference between the heated water at the base near the flame and the cooler water near the surface. This heat differential is called a **thermal gradient**, and it is defined as the rate of temperature (or heat) change over the distance between two temperature points.

Imagine a lake whose surface water warms up to 25 °C on a summer day while the underlying water at a depth of 2 meters (6.6 feet) is fixed at 15 °C (Figure 4.24). The temperature difference between the surface and the 2-meter depth is 10 °C. Thus, the temperature gradient is 5 °C per meter. Heat energy will be transferred downward

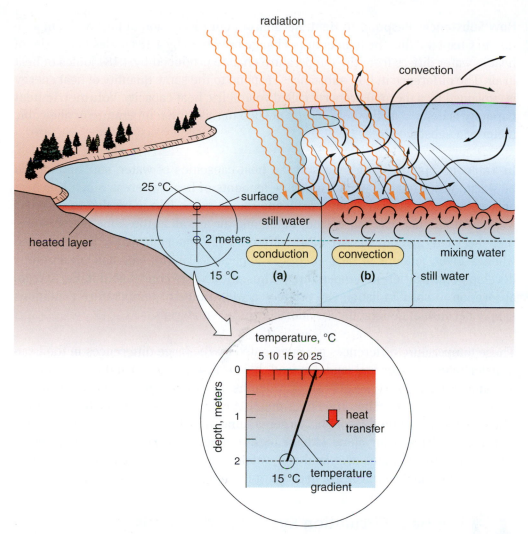

Figure 4.24 Contrasts in heat transfer in lake water by (a) conduction (slow) and (b) convection (fast).

Table 4.4 Relative heat-transfer capacities

Water (still: conduction)	0.6
Soil (still: conduction)	1.0–2.0
Water (moving: convection)	350.0
Air (moving convection)	greater than 3500

* based on the rate of heat passage through one meter of a substance when the temperature gradient is 1 °C per meter.

from the level of high energy to the level of low energy. If the water is motionless (see Figure 4.24a), heat transfer will take place by conduction. **Conduction** is the gradual migration of heat energy from water molecule to water molecule as the molecules vibrate under the influence of the heat.

On the other hand, if the lake is stirred by the mixing action of waves and currents, heat transfer will take place by convection (see Figure 4.24b). In **convection**, the heated molecules themselves are moved, relatively rapidly, with the flow of the fluid. Warm molecules are exchanged with cooler ones, resulting in a rate of heat transfer 500 or more times greater than is possible through conduction. This process can take place in a glass of water or over an entire ocean basin.

Air is even more effective than water in convective heat transfer (Table 4.4). The ability of air to transfer heat rapidly is extremely important to Earth's global climate because atmospheric convection is broadly responsible for redistributing heat across the Earth's surface. In the study of weather and climate, the term "convection" is used to describe mainly the vertical component of atmospheric mixing, whereas the term **advection** is used to describe the horizontal component, that is, wind and air masses moving laterally over the Earth's surface (Figure 4.25). Ocean currents function similarly and satellite imagery can provide vivid illustrations of the mechanism in action (see Figure 4.23).

Heat transfer by **radiation** involves the propagation of energy in electromagnetic waves. When the electromagnetic waves are absorbed by a substance, they are converted into heat. Unlike conduction and convection, this mechanism does not depend on a medium such as air or water to transmit the energy. It can take place in the absence of a connecting medium such as in interplanetary space, or within certain substances, such as air and clean water that are transparent to certain forms of radiation. Substances that are opaque, as soil and dark rocks are to sunlight, do not transmit electromagnetic waves but instead absorb them.

All three mechanisms of heat transfer are active in the atmosphere, but convection and radiation are decidedly the most effective. In the oceans, convection is the principal mechanism, whereas in soil and surface rock, conduction is the principal mechanism of heat transfer. Of the three, conduction is by far the least effective in the geographic redistribution of heat. For example, in a midlatitude landscape, the heat generated after a full summer of absorbing solar radiation typically penetrates only 3 to 5 meters into the ground. On the other hand, once this heat leaves the ground and enters the atmosphere, it typically moves thousands of meters in a matter of minutes or hours.

Figure 4.25 Energy transfer by advection illustrated by a blanket of fog sweeping inland from the ocean.

How Substances Respond to Heat: Recall from our discussion of heat and temperature in Chapter 3 that the temperature change produced by 4.184 joules (1 calorie) of heat in water differs from the temperature change produced by 4.184 joules of heat in air. The difference in how substances respond to the same quantity of heat energy is explained by a thermal property called volumetric heat capacity. **Volumetric heat capacity** is defined as the heat energy needed to raise the temperature of one cubic meter of a substance by 1 °C.

In Table 4.5 note the differences in the heat capacities of water, soil, and air. To be warm, water must contain truly huge amounts of heat, whereas much less heat energy is needed to make land warm, and much, much less is needed to make air warm. Put another way, when land, water, and air are all at the same temperature, their heat energy contents are vastly different. An important consequence is that when a given amount of heat energy is applied to landmasses and oceans they develop greatly different temperatures. Thus, the energy contributed to Earth from absorption of solar radiation can produce a wide range of temperature conditions simply because of geographic variations in surface materials.

Table 4.5 Volumetric heat capacities

Water	4,184,000 joules (1,000,000 calories)
Soil	1,506,240 joules (360,000 calories)
Air	1,213 joules (290 calories)

* *Heat needed to raise the temperature of 1 cubic meter 1 °C*

These temperature differences are augmented by the large differences in the heat-transfer rates in water and land. On land, heat transfer into soil and rock is limited to conduction, and reaches only shallow depths. On water, heat is transferred downward mainly by convection, reaching much greater depths at a much faster rate. Thus, heat is distributed through a larger volume of matter in the ocean than is possible on land. This factor, along with the high volumetric heat capacity of water, helps explain why the climates over land and water are so different and why the oceans are such important reservoirs of heat energy for powering Earth's weather systems.

4.5 The Heat Circulation System in the Landscape

We begin this section with a brief examination of the flows or pathways of the landscape's heat energy system. When the Sun rises in the morning, the relatively cool ground surface receives its first insolation. Heat is generated immediately. Initially it is barely detectable, but as the Sun rises toward its noon zenith, solar intensity increases dramatically, and surface temperature climbs rapidly. In the desert, exposed surfaces may reach a searing 60 °C (150 °F) by midday. Although extreme, such surface temperatures would actually rise much higher were it not for the fact that much of the heat is transferred into the air and deeper into the ground as heat is generated on the surface.

Heat transfer follows the energy gradients that develop above and below the surface. Three **heat fluxes**, or flows, are produced: (1) **ground heat**, which flows downward by conduction into the underlying soil, rock, or water; (2) **sensible heat** (the heat of dry air), which flows upward into the air mainly by convection; and (3) **latent heat** (the heat of water vapor), which also flows upward into the air mainly by convection. Each heat flux, represented by the arrows in Figure 4.26, takes energy away from the surface. Without these flows, the land could not maintain an energy balance; that is, it could

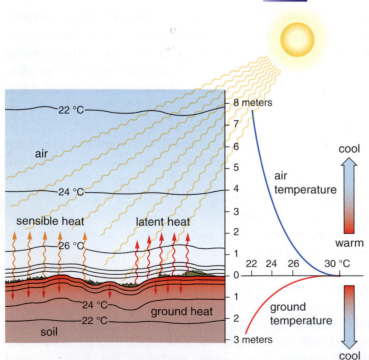

Figure 4.26 Heat fluxes in response to heating of the ground by solar radiation: ground heat, sensible heat, and latent. Each transfers energy away from the surface. (Over water basically the same types of heat flow take place but at different rates, especially downward.)

not achieve equilibrium between energy inflows and outflows. But where does the heat go after it has left the land surface? It goes into storage in the ground and the atmosphere.

Ground Heat: If heat enters the ground after the surface has warmed up during the daylight hours, when does it leave the ground? It leaves at night. After sunset, the surface of the land loses heat and longwave radiation to the atmosphere. As energy is drained away, the surface cools, and the thermal gradient just below it gradually reverses. Now the ground is coolest at the surface and warmer a little farther down. Under these conditions, heat, following the thermal gradient, is transferred upward from the warmer soil beneath as the diagram in Figure 4.27 demonstrates. The amount of heat leaving the ground is just about equal to the amount that went in during the previous day, but it is not quite equal because of day-to-day variations in solar input, ground conditions, and other factors.

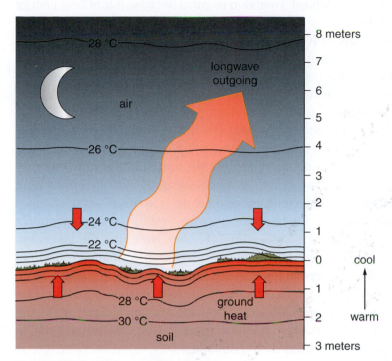

Figure 4.27 Nighttime energy flows from the ground. As the surface cools, heat is transferred to the surface where it feeds an outflow of longwave radiation.

In fact, there are distinct seasonal trends in the ground heat balance that are driven by seasonal changes in sun angle and length of day. The ground heats up in spring and early summer and cools down in fall and early winter. In a year with a perfect energy balance, all heat inflows would equal all heat outflows. Rarely is that the case, however. Measurements reveal that most locations are subject to cooling and warming cycles of several years or decades when average ground temperatures fall and rise a little. As global warming takes place in this century, ground and ocean temperatures are expected to rise over large parts of the Earth.

Sensible Heat: Sensible heat is the energy conducted into the air when air molecules in contact with warm ground take on heat (see Figure 4.26). Sensible heat is the heat you can feel, and it is the type of heat we measure with a thermometer. Most sensible heat is transferred into the atmosphere from the Earth's surface by convection, and since the air near the ground is rarely calm, surface heat is rapidly flushed away and dispersed into the larger atmosphere. We know that such sensible-heat transfer is a highly efficient process because air temperatures 50 to 100 meters above ground closely follow the heating and cooling of the ground itself in response to the daily rise and fall of solar radiation.

Latent Heat: Latent heat is energy stored in the molecules of water as a result of phase changes, that is, changes in water's physical state. From the Earth's surface, latent heat is transferred into the air with the evaporation of water, for example. We know that energy is required to perform work and since evaporation is a form of work, heat energy is required to drive the process. Thus, when water molecules are driven into the atmosphere, they embody, so to speak, heat energy in a stored or latent form. To vaporize 1 gram of water requires about 600 calories of energy. This energy is drawn from the surface of the material containing the water. In the landscape, these materials are soil, vegetation, and water features such as lakes and wetlands. Thus, when evaporation takes place from the ground, heat is lost and the surface is cooled.

Table 4.6 Energy exchange with phase changes in water

Phase change	Process	Rate (cal/gram)
Vapor → Liquid	Condensation	600 released
Liquid → Vapor	Evaporation	600 stored
Solid → Liquid	Melting	80 stored
Liquid → Solid	Freezing	80 released
Vapor → Solid	Sublimation	680 released
Solid → Vapor	Sublimation	680 stored

Latent-heat transfer by evaporation is fundamentally the same process your body uses to cool your skin when you perspire. The heat that drives the process is supplied from a reservoir just below the surface, ground heat in the landscape, body heat in a human. While you work hard, your skin is warm because lots of heat energy is being pumped to the surface by rapid blood circulation. Evaporation of perspiration speeds up and you begin to feel cooler as heat is removed, just as the ground feels cooler with evaporation of soil moisture when it is being heated by the Sun. But in the landscape, ground heat is used up in the evening and evaporation diminishes. At night, the ground grows even cooler and evaporation stops.

Energy Exchange with Phase Changes in Water: The heat energy taken up in evaporation and stored in the molecules of water vapor is released as heat when the vapor returns to the liquid phase upon condensation. Dew, fog, and clouds are all formed by condensation, and during their formation they release heat, termed the **latent heat of condensation**. Latent-heat exchanges are also involved in other phase changes of water: between liquid and solid and between vapor and solid. The amount of heat energy taken in or released in each is given in Table 4.6. Note that the maximum exchange, 680 calories per gram, results from the direct transformation of ice into vapor or vapor into ice. While stored in water vapor, heat energy is often transported great distances over the Earth's surface. Latent heat from distant sources, often thousands of kilometers away, is an extremely important source of energy in driving weather systems, especially storms such as hurricanes, thunderstorms, and the frontal (cyclonic) storms we experience in the midlatitudes.

Since it is surface heat (not air heat) that drives the evaporation process, the more energy that goes into latent heat the less there is available for sensible heat. And since it is only sensible heat that raises the temperature of air, a reduction in the supply of sensible heat results in cooler air temperatures. The ratio between the amount of energy given up as sensible heat and the amount given up as latent heat is called the **Bowen ratio**, and it is an important measure of a landscape's thermal performance.

When we build cities, most moisture sources, such as soil, wetlands, and forests, are eliminated and replaced with asphalt, concrete, and other dry materials. Not surprisingly, air temperatures in cities are often higher than those over the landscape they replaced (as the map in Figure 4.28 illustrates), in part because most of the heat generated by solar radiation is transformed into sensible heat rather than latent heat. Other factors including heat generated from automobiles, factories, and even human bodies also help drive up city temperatures. Building urban parks with lots of vegetation and water features is a way of moderating city heat.

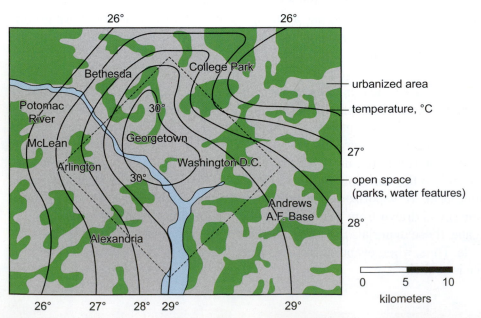

Figure 4.28 Air temperatures over greater Washington, DC, showing distinctly warmer conditions over the drier, built up area in and around the city center.

Over a wet landscape such as a tropical rainforest heat flux favors latent heat. The surface heat taken up in evaporation from vegetation, streams, wetlands, and soil is drawn up into thunderstorms, which carry the heat thousands of meters into the atmosphere. The heat is released as the vapor condenses when the air cools at high altitude. Within hours the system – illustrated in Figure 4.29 – delivers much of this moisture back to the landscape as rain, where, upon evaporation, the water picks up another load of heat. At the same time, tropical forests are drawing up rainwater that seeped into the ground and releasing it in leaf transpiration, also an evaporative process. At any moment many thousands of these systems are operating on the Earth, cooling the landscape while warming the atmosphere aloft.

Thermal Microclimates and Landscape Diversity: The geographic diversity we see in most landscapes – hills and valleys, forests and fields, ponds and rock outcrops, buildings and parks – have a significant influence on the local distribution of radiation and heat. Slopes influence sun angle and solar heating. Snow, vegetation, and soil influence albedo. Water influences albedo, heat storage, and surface temperature. Soil composition influences ground heat flow and storage. Wind influences sensible- and latent-heat flux, because it is a source of convection. As a result, most landscapes develop a maze-like complex of microclimates that vary markedly in geographic scale and pattern from place to place.

Microclimatic patterns are most succinctly defined in hilly and mountainous terrain. In the middle and high latitudes, the differences in incident radiation and heating on north- and south-facing slopes alone can have a profound influence on the landscape (see Figure 4.22). South-facing slopes are not only brighter and warmer, but are less prone to killing frost than their north-facing counterparts. The distribution of many tree, wildflower, and insect species as well as certain land uses such as orchards and vineyards are commonly adjusted to these differences (Figure 4.30).

Where conditions are drier, as in western North America, cooler north-facing slopes often support forests because they have better moisture supplies. They are wetter because, among other reasons, evaporation rates are lower and snow lasts longer in spring. Warmer, south-facing slopes may support only smaller, more drought-tolerant plants, such as grasses and shrubs. The two exposures often develop different soils, runoff rates, and erosion patterns as well.

Valley floors can be cooler than adjacent slopes, despite their lower elevations, first, because water accumulates there, which gives the landscape higher volumetric heat capacity, and second, because cool air collects there at night. When slopes, especially north-facing ones, cool in the evening, the resultant air, because of its higher density, slides downslope. Known as **cold-air drainage**, this phenomenon is often marked by a blanket of fog on valley floors (Figure 4.31).

Summary on Heat Forms and Transfer: Heat on the Earth's surface is held in three great reservoirs, water, land, and air. Besides a capacity for heat storage, each, and especially air and water, has the capacity to redistribute or transfer heat upward, downward, and laterally. In air and water this takes place mainly by convection; in soil mainly by conduction. Variation in the geographic character of the land, particularly landforms, water, soil, vegetation, and land use, can have a pronounced affect on the distribution of heat and in turn the composition and distribution of life.

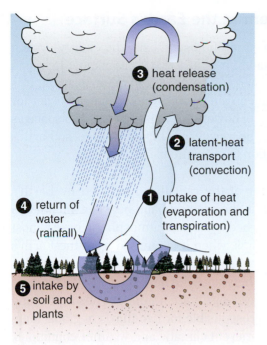

Figure 4.29 The hydrothermal system in the wet tropics involving soil, forest, and atmosphere and five phases of water and heat transfer.

3 heat release (condensation)

2 latent-heat transport (convection)

4 return of water (rainfall)

1 uptake of heat (evaporation and transpiration)

5 intake by soil and plants

Figure 4.30 A section of mountainous, midlatitude terrain illustrating the sharp contrast in vegetative cover on north-facing and south-facing slopes.

Figure 4.31 Valley fog resulting from the downslope drainage of cooler (and denser) air from higher slopes.

Frost on a valley floor is often the result of cold air drainage from side slopes.

4.6 The Energy Balance System at the Earth's Surface

If we ignore the little bit of energy involved in photosynthesis, the energy system that operates at the Earth's surface is made up of just two components: heat and radiation. Solar radiation is the primary energy source and, upon absorption, it is converted into heat, which goes into storage in air, ground, and oceans. In order for the Earth's energy system to stay in balance, however, it must release into space as much energy as it takes in. But heat cannot be transferred through space, so Earth's heat energy must be converted into radiation to leave the planet.

We have already established the character of the radiation produced by Earth's **longwave** or **thermal infrared radiation** in the spectral band between 4.0 μm and 20 μm. Radiation of this wavelength is absorbed by several atmospheric gases, most notably water vapor and carbon dioxide. Thus, when it is emitted from the Earth's surface, longwave radiation is subject to atmospheric absorption, the basic mechanism of Earth's greenhouse. But the atmosphere does not have blanket effectiveness in this process because (1) some infrared wavelengths are immune to absorption and (2) the atmosphere's water vapor, which is responsible for two-thirds of global longwave absorption, varies dramatically with different climatic and weather conditions. In the dry atmosphere of deserts, for example, infrared absorption is relatively low. Much of this radiation escapes into space, which accounts for the rapid nighttime cooling of the desert landscape, as much as 30 °C at times.

When Earth's longwave radiation is absorbed by the atmosphere, heat is generated in the air, and the heated molecules in turn generate their own infrared radiation. This radiation flows both upward and downward (Figure 4.32). The upward component may pass into space or be absorbed at higher levels in the atmosphere. The downward component is absorbed by the Earth's surface, where it adds to the heat being generated by the absorption of solar radiation. If we measure the flows of all forms of radiation at the Earth's surface, we will find there is both an incoming (i) and outgoing (o) component for both shortwave (solar) radiation (S) and longwave (thermal infrared) radiation (L). These four flows of radiation are labeled L_i, L_o, S_i, and S_o. Two bring energy to the Earth's surface (+) and two take it away (−); thus the formula for **net radiation** (NR) is:

$$NR = S_i - S_o + L_i - L_o$$

If the net radiation is positive, then more radiant energy is being taken in by the Earth's surface than is being given up and the surface is heating up. If the net radiation is negative, the surface is giving up more energy than it is taking in and it is cooling down.

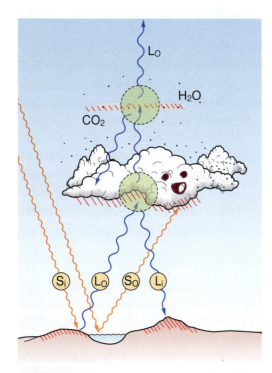

Figure 4.32 Radiation flows in the atmosphere. Unlike those at the Earth's surface, the atmosphere can generate both upward and downward flows.

Day–Night Variations in the Energy Balance: These two conditions occur with day and night at almost every location on Earth. During the day, the net radiation is strongly positive owing to substantial incoming solar radiation (S_i). The heat it generates goes into storage as ground heat (G), sensible heat (S), and latent heat (L) thus:

$$NR = G + S + L$$

At night, S_i falls to zero leaving only longwave radiation (L_i) from the sky to feed the ground with radiation. But the nighttime input of L_i is usually small, especially when compared to its counterpart, outgoing longwave radiation (L_o), which is often quite large. What is the energy source for outgoing longwave radiation at night? The answer is stored heat energy, the G + S + L gained during the previous day when the net radiation was positive.

At night, the net radiation shifts to a negative balance dominated by L_o fueled by stored heat. As this longwave radiation drains energy from the surface (see the big arrow in Figure 4.33), the land cools and the thermal gradients in the ground and air reverse from those of daytime. Heat is transferred to the surface of the land from both above and below (Figure 4.33). Even latent heat contributes as water vapor condenses on the cool ground forming *dew* or, under freezing temperatures, *ground frost* or *hoar*. Although the ground surface is the coolest point in the landscape, it continues to generate longwave radiation because even at freezing temperatures, the air and soil (or water) still contain huge amounts of heat. Remember that absolute zero (the temperature at which a substance contains no heat) is $-273\,°C$; therefore, even substances that are cold to the touch are in fact high in heat energy.

We conclude that the positive net radiation and heat buildup of daytime are counterbalanced by a negative net radiation and associated heat loss at night. By combining radiation and heat flow, we derive total energy balance, or simply **energy balance**. When the inflows and outflows of day and night balance out, the 24-hour energy balance should equal zero, that is, NR ± G ± S ± L = 0. But in reality changes in weather and other transient conditions, such as soil moisture, snow cover, and vegetation, produce variations in the daily energy balance. Exceedingly cloudy, humid weather, for example, can slow the rate of heat loss to longwave radiation coming from the ground. At the same time, humid, cloudy conditions increase the rate of longwave radiation flowing back toward the ground from the sky. Under such conditions air temperatures often fail to cool much from day to night.

As a matter of fact, total incoming longwave radiation for the Earth as a whole typically exceeds total incoming solar radiation. Since longwave radiation is generated from heat supplied by solar radiation, this statement appears to violate the conservation-of-energy principle until we realize that the energy used to generate longwave radiation is recycled several times between the sky and the Earth's surface. Thanks to the Earth's greenhouse effect, most calories in this system are counted more than once. But this is true only when we do our accounting at the bottom of the atmosphere. When we move to the top of the atmosphere, every calorie of incoming solar radiation is matched by a calorie of outgoing radiation made up of longwave (infrared) and reflected shortwave (solar) radiation. If this balance waivers, it means that something within the system, such as global albedo, atmospheric heat storage, or oceanic heat storage, is changing, as summarized in Figure 4.34.

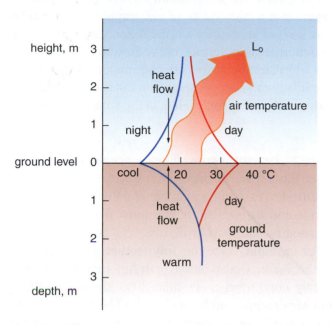

Figure 4.33 The energy balance at night characterized by heat flows to the surface, which in turn feed outflowing longwave radiation.

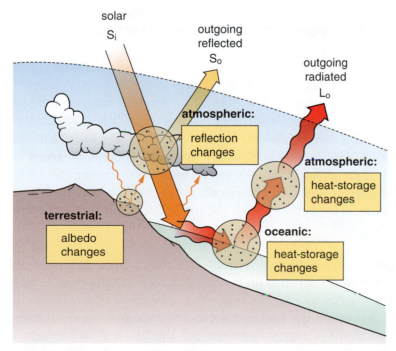

Figure 4.34 Four of many types of changes that can affect the global energy balance.

4.7 Global Temperature Patterns and Controls

We have seen that many factors influence the distribution of heat on the Earth's surface and that these factors operate at a wide range of geographic scales. At the global scale, latitudinal changes in sun angle produce a broadly zonal pattern of temperature ranging from the tropics, where temperatures average around 25 °C throughout the year, to the polar latitudes, where temperatures are continuously below freezing. Between these zones, in the midlatitudes, temperatures vary seasonally with sun angle and length of day, making this zone more tropical-like in the summer and more polar-like in the winter.

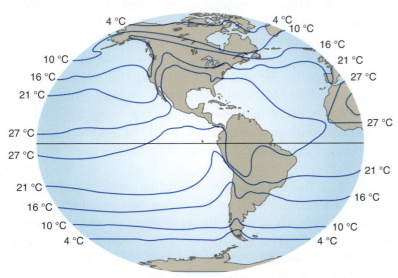

Figure 4.35 The mean July temperature patterns for the western hemisphere, showing the sharp differences between land and water.

But if we trace the **isotherms** (lines marking temperature patterns over the Earth's surface) on a world map, the bands of temperature deviate from a neat zonal pattern and take on curious bends and turns. One of the most distinctive deviations, illustrated by the map in Figure 4.35, is in the middle latitudes. For example, trace the July isotherm representing 21 °C from the Pacific Ocean onto North America. At the coast it jogs sharply northward more than 3000 kilometers (2000 miles) from Mexico into southern Canada. The next isotherm to the north, 16 °C, has an even sharper jog, from southern California to central Alaska. Note that both bends take place where land and water meet.

Because of differences in their thermal characteristics (mainly heat-storage capacities), water bodies and landmasses contrast strongly in their temperature traits. Oceans heat to depths of hundreds of meters and seasonal and daily temperatures stay within moderate ranges, meaning that they are slow to change temperature from winter to summer and day to night. Landmasses, on the other hand, store little heat and have wide ranges in seasonal and daily temperatures, and the larger the landmass the larger the ranges tend to be. Geographers call these temperature systems **thermal regimes**: **maritime** for the oceans and **continental** for the landmasses.

In the midlatitudes, the annual temperature range (based on monthly mean temperature) for a maritime regime is typically less than 10 °C, whereas for a continental regime it is typically 20 °C or more. In North America this is illustrated by comparing San Francisco with St. Louis, Missouri. These cities lie at the same latitude but San Francisco on the Pacific coast is distinctly maritime whereas St. Louis in the interior is distinctly continental. St. Louis's annual temperature range (28 °C) is more than three times greater than San Francisco's (8 °C), but the two cities have almost identical mean annual temperatures (13.5 °C for St. Louis and 13.7 °C for San Francisco).

Earth's Thermal Regimes: Which thermal regime has the greatest influence on climate in any part of the world depends on two main factors: (1) the relative size of land and water surface areas, and (2) the prevailing patterns of atmospheric circulation (wind systems) between the land and water. For example, if a small body of land is located in the middle of a large body of water, the land will be dominated by the maritime regime. This is the case with most ocean islands. Climatically, they are more like the sea around them than like the terrestrial environment of continents.

The same is true for small water bodies within large landmasses. Inland lakes of 50 or 100 acres in area, for example, are thermally dominated by the continental regime around them. The thermal influence of the Great Lakes, on the other hand, is more substantial but still relatively small compared to the continental regime of central North America. The thermal regime of the Great Lakes is limited to a small, but distinct, band around the lake basins. But wind systems can cause sizeable geographic offsets of thermal regimes, blowing them way downwind. Prevailing winds that blow from sea to land, for example, can bring maritime conditions far into adjacent coastal lands. This is the case in northwestern Europe where the prevailing westerly winds drag the maritime regime from the Atlantic well into Germany and Poland. In North America, a similar eastward penetration from the Pacific is blocked by the mountains that parallel the coast.

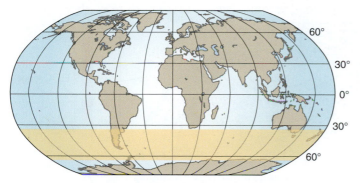

Figure 4.36 The southern hemisphere's midlatitudes contain little land area; therefore, this zone is overwhelmingly maritime.

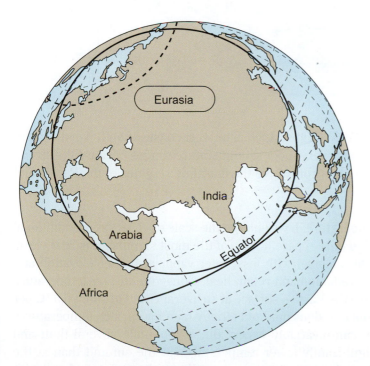

Figure 4.37 Eurasia, Earth's largest landmass is centered in the midlatitudes and produces the strongest continental regime on the planet.

Because of its huge oceans, Earth is predominantly a maritime planet. Water is especially abundant in the southern hemisphere, where it constitutes about 80 percent of the surface area. In the midlatitude and subarctic zone of the southern hemisphere, water forms a continuous belt around the globe that is interrupted only by the southern portions of South America, Australia, and the islands of New Zealand, which together cover less than 15 percent of this zone (Figure 4.36). In sharp contrast, in the northern hemisphere the middle latitude and subarctic zones contain the bulk of the North American and Eurasian landmasses. The largest is Eurasia, which constitutes 37 percent of the Earth's land area (see Figure 4.37) and develops the strongest continental climatic regime on the globe. Another contrast is found in the polar zones: the southern hemisphere's is covered by land (Antarctica) and the northern hemisphere's is covered by water and sea ice (the Arctic Ocean) (Figure 4.38).

On balance, then, we can deduce a great deal about Earth's thermal climate from the distribution of land and water. First, the oceans should have a significant influence on climate for the Earth as a whole. Second, most of the southern hemisphere should favor the maritime regime. Third, the midlatitude zone of the southern hemisphere should be overwhelmingly maritime, whereas the same zone in the northern hemisphere should show a contrasting set of continental and maritime regimes corresponding to the alternating masses of land and water. More particularly, there should be a powerful continental regime over Eurasia. Fourth, the polar zone of the northern hemisphere should be somewhat more moderate in temperature than its counterpart in the southern hemisphere.

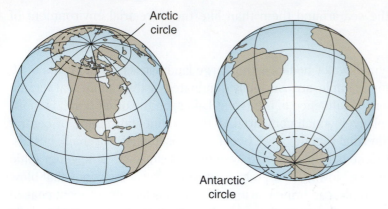

Figure 4.38 The contrasting land and water coverage at the North and South poles.

Indeed, this is what we find in all four cases as the map in Figure 4.39 reveals. This map shows the global distribution of annual temperature ranges based on the extremes of average monthly temperatures (usually July and January). Notice that the southern hemisphere between the Equator and the Antarctic Circle has a modest temperature range, only 5 to 15 °C, reflecting maritime dominance. The same zone in the northern hemisphere shows strong contrasts from oceans to continents, reflecting alternating maritime and continental thermal regimes. For example, trace the change from the Atlantic Ocean to both North America and Asia at the latitude of the Arctic Circle. The north central parts of Asia and North America are ghastly cold in the winter but reasonably mild in the summer. As a result, their annual temperature ranges are

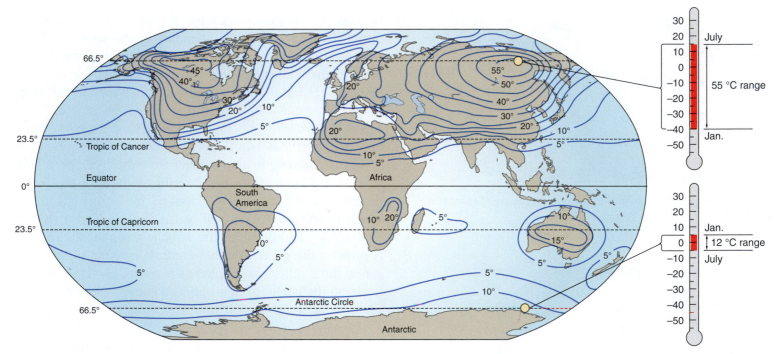

Figure 4.39 Earth's annual extremes in temperature based on average monthly temperature. The difference between maritime and continental thermal regimes is illustrated by Siberia and coastal Antarctica (see the thermometers) (Isotherms in °C).

huge, 55 and 45 °C respectively (Figure 4.39). The west coasts of both North American and Europe, however, have very modest temperature ranges because the strong onshore (westerly) winds in this zone drive the maritime regimes of the Pacific and Atlantic oceans onto land.

Effect of Elevation on Temperature: If we enlarge the scale of observation, additional controls on temperature become apparent. One of the most significant is land elevation. Air grows thinner (less dense) with less water vapor at higher elevation and is thus less able to absorb radiation and retain heat. Consequently, air temperature declines with altitude above the Earth's surface at an average rate of about 6.5 °C per 1000 meters. Although this rate does not hold strictly true for surface temperatures on a mountain slope, we can invariably expect that a mountain several thousand meters high will have significantly lower temperatures at the summit than at the base. Therefore, as we trace isotherms across a continent, they typically bend with major elevation changes in the land, dipping equatorward as elevation increases, and poleward as elevation decreases.

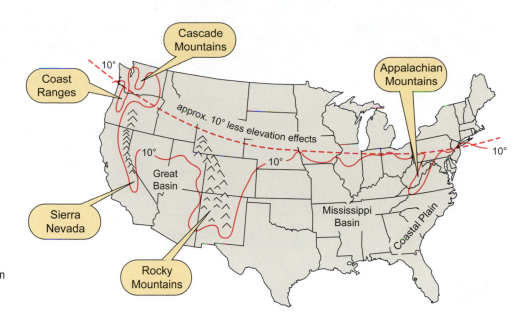

Figure 4.40 The effect of elevation on temperature illustrated by the southward depression of the 10 °C isotherm for average annual temperature. The broken line is the approximate location of this isotherm assuming a flat, lowland surface across the USA.

To illustrate this pattern, we have plotted the isotherm for 10 °C (50 °F) average annual temperature across the United States. This is shown in the map in Figure 4.40. Compare it with the same isotherm's approximate position without the influence of elevation changes, shown by the broken line. Our largest mountain range, the Rockies, depresses the isotherm nearly 2000 km (1200 miles) southward. Similar but less dramatic changes are caused by other large mountain ranges, the Sierra Nevada, the Appalachians, the Coast Ranges, and the Cascades. The isotherm bends northward in the intervening areas where elevations are lower.

Chapter Summary and Overview of Earth's Radiation and Heat Systems Over Land and Water

We explored the idea of Earth's surface as an energy platform where solar energy is received, heat is generated, and longwave radiation is discharged. Of great interest to geographers is that fact that radiation and heat are unevenly distributed at virtually every scale of observation from home town landscapes to regions to continents to the entire planet. This led us to questions about the controls on the radiation balance and to the controls on the generation and storage of heat energy, how they vary from place to place, and how all this relates to the thermal patterns we see in local landscapes and on world maps, including the meaning of the global distribution of land and water. The following summary diagram puts some of these ideas into graphic perspective.

▶ **The Earth's atmosphere is composed of two major gases, a number of minor gases, and small amounts of liquid and solid particles.** Water vapor, along with carbon dioxide and a few other minor gases, govern Earth's greenhouse effect. Air pollution is causing global increases in carbon dioxide.

▶ **All the constituents of the atmosphere are part of a recycling system linking the atmosphere with the Earth's surface.** Exchange times are long for major gases and relatively short for small-volume gases such as carbon dioxide and ozone.

▶ **For the Earth as a whole, incoming solar radiation is reduced by about 50 percent as it passes through the atmosphere.** Three processes are responsible for the reduction and at the Earth's surface an additional radiation is lost to albedo.

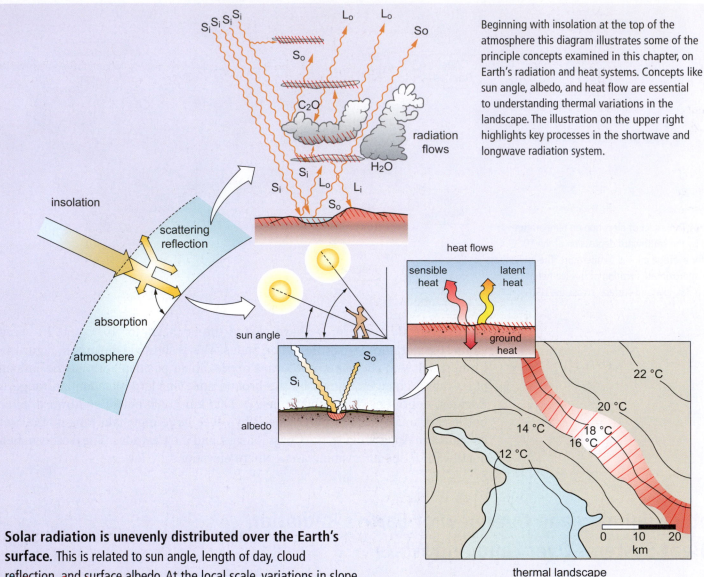

insolation

scattering
reflection

absorption

atmosphere

sun angle

albedo

radiation
flows

heat flows

sensible
heat

latent
heat

ground
heat

thermal landscape

Beginning with insolation at the top of the atmosphere this diagram illustrates some of the principle concepts examined in this chapter, on Earth's radiation and heat systems. Concepts like sun angle, albedo, and heat flow are essential to understanding thermal variations in the landscape. The illustration on the upper right highlights key processes in the shortwave and longwave radiation system.

▶ **Solar radiation is unevenly distributed over the Earth's surface.** This is related to sun angle, length of day, cloud reflection, and surface albedo. At the local scale, variations in slope orientation and steepness affect solar intensities, which in turn may influence landscape character.

▶ **The solar energy absorbed by the Earth's surface generates heat, which is stored in air, ground, and water.** Temperature levels in these materials vary with energy absorption, volumetric heat capacity, and the form in which heat is stored.

▶ **Heat moves in and out of the lower atmosphere, the ground, and the oceans on a daily cycle.** Heat taken in during the day is released at night but the quantities vary with weather conditions, the seasons, and other factors.

▶ **Geographic variations in land form, composition, and cover can produce distinct variations in the thermal microclimate in the landscape.** These differences affect the distribution of plants, animals, and land use as well as the operation of other systems in the surface environment.

▶ **The Earth's energy system is made up overwhelmingly of two components: heat and radiation.** The balance in heat and radiation flows to and from the surface constitutes the Earth's energy balance. Worldwide, the balance is positive because much of the longwave radiation absorbed by the atmosphere is reradiated (recycled) back to the Earth's surface.

▶ **Global temperature patterns are strongly influenced by the distribution of land and water.** These influences can be seen in the patterns of temperature in the middle latitudes of the northern and southern hemispheres.

▶ **Land elevation changes have a pronounced effect on temperature patterns.** Isotherms are depressed equatorward by higher elevations, and for large mountain ranges, the offset may be as much as several thousand kilometers.

Review Questions

1 Name the major factors controlling the amount of solar radiation reaching the Earth's surface?

2 What are the principal roles played by atmospheric ozone, carbon dioxide, and water vapor with respect to incoming or outgoing radiation?

3 Name the three principal processes responsible for the reduction of solar radiation as it passes through the atmosphere and relate these processes to ultraviolet, infrared, and visible forms of radiation.

4 What is albedo and what conditions yield high and low albedos over land and water?

5 Explain the logic of sun angle as a control on the distribution of solar energy on the Earth's surface. How does sun angle change with the seasons and variations in the slope of the land?

6 What would be the equinox and solstice sun angles for the locations on Earth's surface at 40 degrees north latitude?

7 Why do certain areas of the Earth located considerably north and south of the Equator receive more incoming solar radiation at ground level than the Equator itself?

8 Briefly describe the concept of net solar radiation and explain how it would vary in a landscape with and without snow cover.

9 Name and describe the three basic mechanisms of heat transfer on Earth, their relative rates of heat transfer, and the part of the global environment in which each is most active.

10 When air and water are at the same temperature, they contain the same amounts of heat energy. Is this statement true? Why or why not, and what thermal principle applies here?

11 What happens to the radiant energy absorbed by the landscape on a sunny day? Where does it go and what terms are applied to it?

12 Why does evaporation from a surface produce cooling and what is the name and thermal effect of the opposite process, that is, one that warms the surface?

13 In the formula for net radiation, what are the incoming and outgoing components? Which of these operate at night and how is it possible that landscape can register a negative net radiation?

14 What is meant by the term thermal microclimate and what are some of the key factors controlling such phenomena?

15 Can you explain the difference between continental and maritime temperature regimes and how the northern and southern hemispheres differ in this regard?

16 And finally, briefly explain why on a map of North America the pattern of isotherms does not trend directly east–west across the continent.

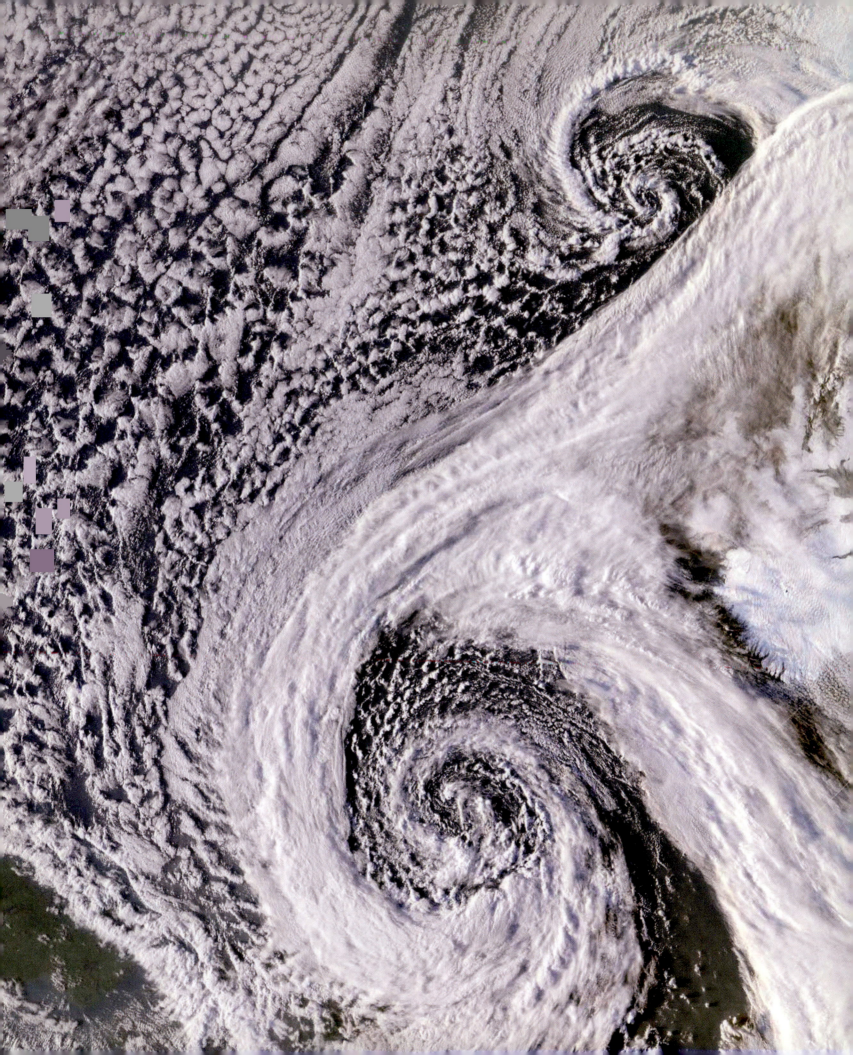

The Great System of Global Air and Ocean Circulation

Chapter Overview

Our goal in this chapter is to sketch a picture of atmospheric and oceanic circulation systems at the global scale. We have some big questions to address, beginning with a very basic one: why air is so mobile and what drives its motion? Once we get a handle on why the atmosphere moves, we can examine the basic structure of the atmosphere's circulation, both at the surface and upper troposphere, and reveal how the ancient mariners used this understanding to navigate the Earth. The next step is to learn about the major wind systems such as the prevailing westerlies in the midlatitudes and the easterly tradewinds in the tropics. In the second half of the chapter we move from wind to waves to ocean currents and the big picture of oceanic circulation, noting this system's important role in shaping global climate and how it feeds into the emerging climate-change scenario.

Introduction

Perhaps none of Earth's great systems is more pivotal to understanding the geographic character of our planet than the atmosphere. Indeed, we might think of the atmosphere as the master system among the great systems: first, because it is the main engine driving other great systems, most notably the hydrologic cycle; and second, because it is the vehicle that ties together systems such as the global water and biochemical cycles. Without the atmosphere, the hydrologic cycle and biochemical cycles (such as oxygen and carbon dioxide exchanges) cannot function. In the case of the hydrologic cycle, the atmosphere takes in water from the Earth's surface (mainly from the oceans), moves it to other, often distant, geographic locations via winds and air masses, and releases it back to the surface. Without the middle step, the transport phase, there is no way to get water onto the land. But in order for the atmosphere to perform this important function, it must itself be mobile with the capacity to circulate over great distances.

Closely tied to the atmosphere is a second great flow system, the oceans. Indeed, so closely linked are the oceanic and atmospheric systems that they can be thought of as two parts of one mother system that is broadly responsible for redistributing energy and matter around the planet. Embedded within the atmosphere/ocean system, beyond the sort of first-level connections we make among, say, the patterns of storms, winds, waves, and ocean currents, are various complex feedback mechanisms which govern, alongside other things, certain weather cycles and longer term global fluctuations in atmospheric cooling and warming. Much of our thinking about the complexities of the atmospheric/oceanic system is in its formative stages, but the basic concept of air and water circulation as a system has a long history, dating back thousands of years. Little did the early mariners like the Phoenicians and the Polynesians, whose lives depended on an intimate understanding of the patterns of winds and ocean currents on the high seas, realize that they were among the first systems thinkers. And this thinking must have given them a view of the world that was remarkably different from that of their land-bound compatriots!

5.1 Atmospheric Mobility, Air Pressure, and Wind

Of all the major substances on Earth with a capacity for flowing motion, which include water, glacial ice, and molten rock, air by far has the greatest mobility. The main reason for this is that among air molecules there is little resistance to movement. Put another way, defined as a fluid, air is not very sticky or viscous. In fact, the viscosity of air is thousands of times less than that of water. So when force is applied to air, it tends to move quickly and efficiently, that is, without much energy loss to internal friction caused by molecules rubbing together. This property applies to air at all temperatures and at all geographic scales from global wind systems to local breezes sliding on and off the seashore.

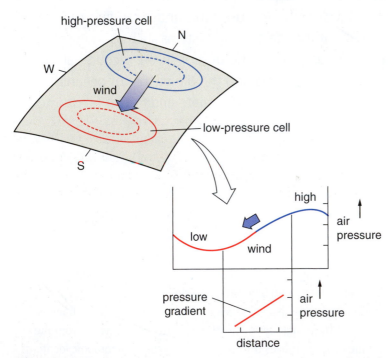

Figure 5.1 Air flow (wind) from a high-pressure cell to a low-pressure cell. Wind velocity is governed by the steepness of the pressure gradient; wind direction by the relative location of the pressure cells.

But what causes motion in air? What drives the atmosphere to produce winds? The answer is mainly the force of air pressure. If pressure is relatively high at one place in the atmosphere and low at another, air will rush from the area of higher pressure to the area of lower pressure as it seeks to equalize its pressure overall. Together, the two areas, or cells, of pressure and the wind linking them constitute a simple circulation system, as shown in Figure 5.1. The system has two key attributes: direction and force. Wind direction is controlled by the locations of the pressure cells, and force, measured by the velocity of the wind, is controlled chiefly by the difference, or *gradient*, in air pressure between the two cells. Big differences (steep pressure gradients) produce fast winds; small ones, slow winds.

So if it were not for geographic differences in air pressure, the atmosphere would be motionless. No wind. And without wind, moisture, carbon dioxide, nitrogen, heat, and many other things would not get moved around the planet. As a result, the pattern and nature of Earth's climates would be much different, probably far simpler than they are. Since air-pressure differences are at the root of atmospheric circulation, we must ask what causes pressure imbalances in the first place? The easy answer is geographic differences in heating of the Earth's surface.

Imagine a body of air, like the one in Figure 5.2, resting over a patch of heated ground (A). According to the processes we discussed in the previous chapter, the air will take on sensible

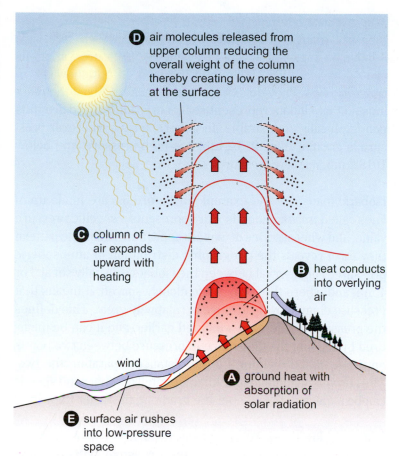

D air molecules released from upper column reducing the overall weight of the column thereby creating low pressure at the surface

C column of air expands upward with heating

B heat conducts into overlying air

wind

A ground heat with absorption of solar radiation

E surface air rushes into low-pressure space

Figure 5.2 A parcel of air over heated ground and the sequence of processes resulting in free convection and the generation of wind. The weight of the air column is reduced and low pressure forms when air molecules are released from the upper column.

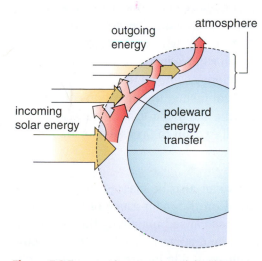

outgoing energy

atmosphere

incoming solar energy

poleward energy transfer

Figure 5.3 To correct its uneven zonal distribution of energy, the atmosphere and oceans transfer massive amounts of energy from the tropics to higher latitudes.

heat from the ground (B) and, as it does, it will expand and push up the column of air above it (C). As the expanding column reaches upward into the atmosphere, it will begin unloading air molecules to the surrounding air, cooler air that has not expanded upward (D). With the loss of molecules, the total weight of the column declines, exerting less force on the Earth's surface. Less weight means lower atmospheric pressure at the base of the column, and lower pressure brings on an inflow of surface air (E) to fill the void. This process is called *free convection* and the product is a small *circulation system* with wind on the surface feeding currents of warm rising air called *thermals*. As the thermals rise, the system carries aloft not only heat, but water vapor and other gases from the surface.

Now think at a global scale, where the heated surface is the belt along the Equator and air is drawn into the low pressure from the north and south. Two great wind systems are created which sweep across vast expanses of both land and ocean and feed the belt of low pressure. Over the ocean, where most of the action takes place, the wind creates waves which roll along with the wind, more or less dragging the surface water along with them. The resultant flows of surface water are ocean currents. Taken together, the air and water circulation systems, winds and currents, produce a huge global energy exchange, as depicted in Figure 5.3, in which the surplus energy of the tropics is redistributed to higher latitudes over the rest of the Earth. Without these systems, Earth would be a far different planet geographically, much hotter in the tropics and much colder everywhere else.

Isobaric Pressure and Pressure Cells: Time for some terms and definitions. **Air pressure** represents the force exerted on a square meter of Earth's surface by the total mass of molecules and particles in the overlying atmosphere. Each air molecule has mass; therefore, the greater the number of molecules in a cubic meter of air, the greater its total mass and pressure. When air is heated, its molecules become more active (that is, vibrate faster), increasing the space between them and causing the air to expand and lose density and pressure. When air cools, it contracts, becomes denser, and its pressure rises.

Air pressure can be expressed in a variety of ways, and a unit called a millibar is one of the most commonly used. A **millibar (mb)** is one thousandth of a bar. A bar is a unit of force. What we need to know is that at sea level, mean atmospheric pressure is 1013.2 mb, or just greater than 1 bar. The reference to sea level is important because above the Earth's surface, air pressure declines rapidly with altitude as the graph in Figure 5.4 shows. At a height of only 5.5 kilometers (3.5 miles), well below the height of tall mountains (Mt. Everest is 8.85 kilometers high), it is half (500 mb) sea-level pressure and at 10 kilometers (6 miles) altitude it is less than 300 mb. This rapid decline in air pressure above the Earth is explained by the fact that air is highly compressible and gravity squeezes the bulk of it down to a thin blanket over the planet's surface. Despite the fact that air is detectable several hundred kilometers above the Earth, over 75 percent of the atmosphere's mass is held tight to the Earth's surface in a thin envelope only 10 to 12 kilometers deep, which defines the **troposphere**.

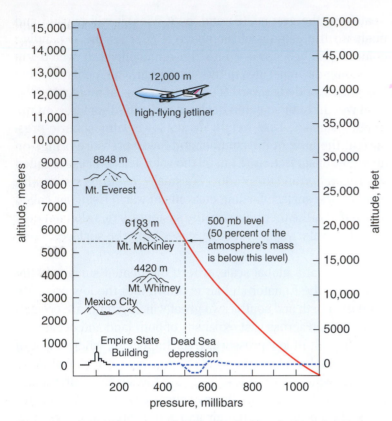

Figure 5.4 Pressure changes rapidly with altitude above the Earth. At 5.5 kilometers, pressure is half that at sea level. In isobaric maps it is important to adjust pressure differences related to elevation to sea-level pressure.

Figure 5.5 An example isobaric map for central North America. The closer the isobar spacing, the faster the resultant wind.

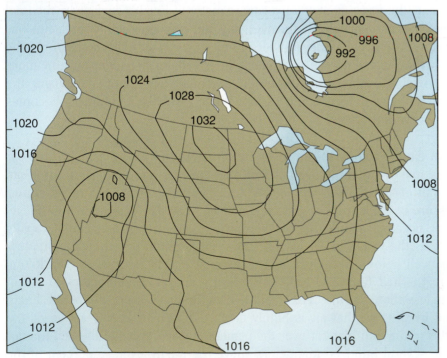

atmospheric pressure in millibars (mb)

The pressure readings we see on weather charts and on air-pressure maps, called **isobaric maps**, represent pressure relative to mean pressure at sea level. Pressure variations caused by elevation changes in the land do not show up on isobaric maps. They do exist, but map makers adjust the readings so that all data represent sea-level readings, and are thus geographically comparable. In general terms, wherever pressure is greater than 1013.2 mb, it is high pressure, and below 1013.2 mb, it is low pressure.

Isobaric maps like the example in Figure 5.5 are made up of lines called *isobars*. Each isobar represents a specific pressure value, and when a series of isobars appear on a map, the pattern they form reveals the geographic distribution of atmospheric pressure. The interval between the isobars is usually set at 3 or 4 mb, and where the isobars are closely spaced, it means that pressure change over distance is relatively great. This defines the pressure gradient, as we noted earlier, and it can be measured by reading the difference in pressure between two points on the map and dividing by the distance separating the two. The answer, in millibars per kilometer, tell us what wind speeds can be expected; the steeper the gradient, the faster the wind.

High- and low-pressure systems occur in all sorts of sizes and shapes. At the largest scale are the **pressure belts**, circumglobal zones of pressure such as the belt of low pressure found near the Equator. Within these belts are geographic centers of pressure, called *cells*, and they vary in size, circulation patterns, and weather characteristics depending on pressure type (high or low), location, and season. We apply the terms **anticyclone** and **cyclone** to high-pressure cells and low-pressure cells, respectively, and each is distinguished by a distinctive circulation system. Cyclones draw in air whereas anticyclones force out air and each rotates in a broad spiral though in opposite directions as shown in Figure 5.6. And to make it more confusing, the directions of rotation change from the northern to the southern hemisphere. We will explain the reasons for this in the pages ahead.

Pressure cells can also be classed as thermal or dynamic based on the main cause of their origin. *Thermal cells* are caused mainly by the heating of air over warm land surfaces such as the free-convection cell illustrated in Figure 5.2 or, conversely, by the cooling of air over cold surfaces such as Antarctica. *Dynamic cells* are caused by mechanisms of fluid circulation, such as two wind systems that collide and drive air upward or by irregularities

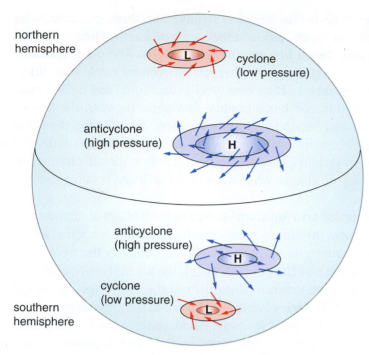

Figure 5.6 Cyclones and anticyclones are the terms given to low- and high-pressure cells. For reasons discussed later, each has a different direction and pattern of circulation and the direction changes from northern to southern hemisphere.

in the flow of fast wind at high altitudes that force air to sink toward the Earth's surface. Some of Earth's largest pressure cells, such as the ones that lie over the world's great deserts, are dynamic and persist so long that they consistently appear on global isobaric maps. These cells are called **semipermanent pressure cells**. At the other extreme are **transient cells**, such as thunderstorms, which may cover less than a square kilometer and last only 2 to 8 hours. Between these scales are the **synoptic systems**, which is what we see on the daily weather charts depicting cells of intermediate sizes (in the range of 1000 kilometers in diameter), and lifetimes usually measured in days. These include midlatitude cyclones (the ones with cold fronts and warm fronts), tropical storms, and hurricanes, all of which we will discuss in detail in the next chapter.

5.2 The Framework of Earth's Pressure and Circulation System

The first questions we must tackle are how does global circulation take place, and what factors control the patterns, scales, and rates of atmospheric and oceanic circulation? To answer these questions, we need to investigate the forces acting on a body of fluid lying on the surface of a rotating sphere. This is a complex system, so let us build it piece by piece beginning with a simplified model of the Earth as a non-rotating planet.

Circulation on a Non-rotating Earth: We start with a planet uniformly covered with one material, say, desert sand, no rotating motion, and an atmosphere like Earth's. The atmosphere would circulate solely in response to latitudinal differences in the heat generated directly from incoming solar radiation. Air would be heated most intensively in the equatorial zone, where it would expand and rise producing a belt of low pressure as shown in Figure 5.7. In contrast, cold air would form over the poles, and because cold air is relatively heavy, high pressure would develop in these regions. With rising air and low pressure at the Equator and high pressure air to the north and south in each hemisphere, we have the basic components of a convectional engine.

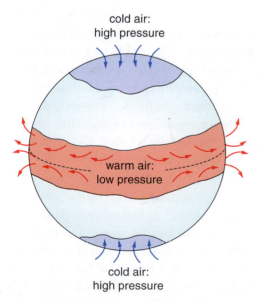

Figure 5.7 Atmospheric heating at the Equator and cooling at the poles and the resultant differences in air pressure and vertical airflow.

How the system works is shown in Figure 5.8. As surface air is heated and driven upward in the equatorial zone, cooler air is drawn in from the polar zones. As this air nears the Equator it, too, is heated and driven aloft. But as it rises, the air cools, fans out at the top of the atmosphere, and then circulates back to the polar zone in the upper atmosphere. Thus, a two-part wind system develops in each hemisphere. For the northern hemisphere, the prevailing surface wind is northerly (because it blows from north to south), but in the upper atmosphere the return flow is southerly as labeled in Figure 5.8. The whole system is driven by solar heating and low pressure along the Equator. It would resemble a pot of boiling water with a flame at the center and cold zones on the perimeter.

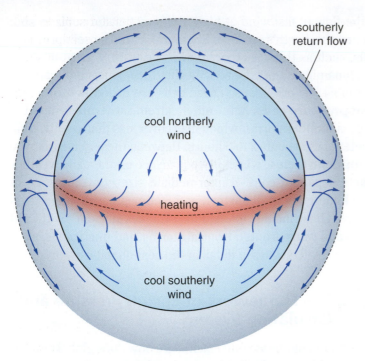

Figure 5.8 Atmospheric circulation between polar high pressure and equatorial low pressure on a non-rotating planet. The system is driven by heating in the equatorial zone with wind systems at the surface and aloft.

How close is this model to reality? The three pressure belts shown in Figure 5.7 do indeed exist on Earth. They are called the **equatorial low** and the **polar highs**. And these pressure belts do originate from thermal differences caused by differences in solar heating between the equatorial and polar zones. In addition, there is a gradual flow of air between the tropics and poles, but the pattern is far more complicated than this model describes. Beyond these similarities, there is little else this model helps explain about Earth's global circulation. Things will change when we set the Earth into rotating motion.

Circulation on a Rotating Earth: The main effect of rotation is to transform the simple north–south circulation described above into a pattern that is more east–west. But since the tropics are still warm and the poles still cold, the air in the upper atmosphere is still drawn poleward from the top of the equatorial low. For a particle of air driven high into the atmosphere above the equatorial zone, it should circulate about the Earth, gradually spiraling toward the pole. As the particle moves toward the pole, however, an important change takes place: its velocity of circulation about the globe increases as its circumglobal path gets smaller. This is explained by a principle from physics called *conservation of angular momentum*, which describes the momentum of a mass – in this case, air – moving in a circular path. The velocity of the air particle circling the Earth will *increase* as the radius of its orbital path (that is, the distance to the Earth's axis) *decreases* with latitude.

When you were a child, you probably applied this principle many times. If, for example, you sat on a swing and "wound yourself" up by twisting the swing ropes and then unwound, you found that you could control your angular velocity by either extending your legs to slow down or drawing them in to speed up. Another example of this principle is the ice skater in Figure 5.9 who folds his or her arms in to spin more rapidly. At the end of a tight spin, the skater can quickly stop by thrusting his/her arms out, that is, by increasing the radius of rotation. In the atmosphere, this principle helps explain the extraordinary wind speeds near the center of a tornado. As the swirling air is drawn toward the center of the tornado (because of the extreme low pressure there) the whirl becomes tighter and tighter. In all of these examples, the rotating mass remains constant, but since the radius of rotation changes, velocity must change.

At the Equator, the Earth rotates at a velocity of approximately 1000 miles per hour (mph) (1600 kilometers per hour), because it travels about 24,000 miles in 24 hours. But at higher latitudes the Earth's rotational velocity is less because its circumference (daily travel distance) is smaller. At the poles it is zero. Thus, as an atmospheric particle is carried poleward – bearing in mind now that it is driven by the rotational momentum of the low latitudes where it started out – the Earth beneath it is moving slower and slower than the particle itself as it gains latitude. To consider an extreme example, at latitude 45° north, a point on the ground would have a velocity of 734 mph. Assuming the atmospheric particle has moved from the Equator to 45 degrees north at 10 kilometers above the Earth, we calculate that its velocity would now be nearly 1500 mph, or more than 700 mph faster than the surface beneath it!

Figure 5.9 The principle of conservation of angular momentum is vividly illustrated by a spinning skater who regulates velocity by changing the radius of rotation with his/her arms and legs.

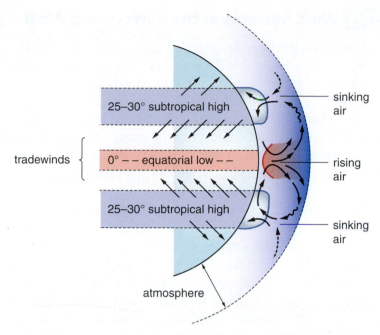

Figure 5.10 Circulation between the equatorial low and the subtropical highs on a rotating planet. This approximates reality and defines a circulation system known as the Hadley cells.

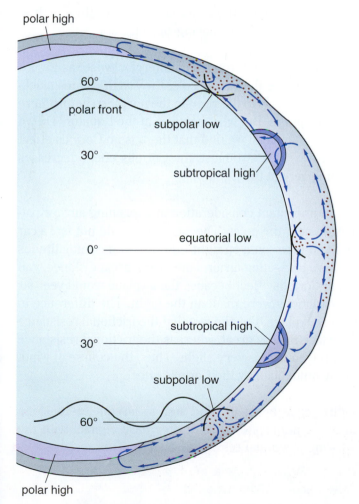

Figure 5.11 Global circulation in cross-section forms two extended figure of eights beginning with the Hadley cells and continuing poleward through the polar highs.

Despite some slowing due to storms and related surface friction, velocities approaching 700 mph are not possible in the Earth's atmosphere, and so the circumglobal circulation system breaks down. This means that the smooth motion of wind in the upper atmosphere becomes irregular, wobbly, and instead of continuing poleward it gradually plunges toward Earth around latitudes 25 to 30 degrees. This descending air produces a buildup of atmospheric mass over the Earth's surface in that area leading to a huge bodies of high pressure, called the **subtropical highs**, in both hemispheres (Figure 5.10). Like the zone of low equatorial pressure, these zones of high pressure more or less encircle the world.

Fed with air forced aloft in the equatorial low, the subtropical high-pressure zones cycle air back to the equatorial low-pressure zone as surface winds. This circulation system is known as the **Hadley cells** (one in each hemisphere), after George C. Hadley (1685–1768), the British physicist who formulated the first theoretical explanation for intertropical surface circulation in 1735. When we compare it to reality, it is very close to the pressure and wind system we find in this broad belt of latitude. There is a system of prevailing surface winds, called the **tradewinds** (shown in Figure 5.10), which blow into the equatorial low from the subtropical highs, though not directly north–south, and there is a return flow of air aloft feeding the subtropical highs.

There are two more zones of pressure and associated winds that need to be added to our sketch of the global pressure system. This is an irregular belt or string of low pressure cells, called the **subpolar low**, that lies in the upper middle latitudes and subarctic zone. It, too, is the result of complications in circulation, but here the cause is cold air and warm air mixing along a major line of contact between polar and tropical air. This contact is called the **polar front**, and in each hemisphere it separates the large mass of cold air that caps the high latitudes from the vast belt of warm air that dominates the middle of the globe. The low-pressure cells along the polar front are powered by latent heat released in condensation as the cold air and warm air mix.

Summary on the Circulation Engine: Expressed in its simplest form, Earth's circulation system in each hemisphere resembles an extended figure of eight, somewhat along the lines of that shown in Figure 5.11. It is made up of four belts of pressure, which alternate low–high–low–high from the Equator poleward, and they are linked together by three wind systems each of which has an upper-level return flow system. If you follow the arrows beginning at the equatorial low, it is possible to trace the path of air from the tropics to the poles. The main driver in this huge circulation system is the belt of equatorial low pressure, which is fueled by a massive supply of solar energy in the low latitudes. We will examine this system in greater detail shortly, but first there is the problem of the broad, swirling wind patterns we see on maps.

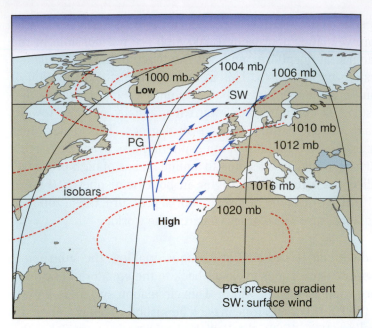

Figure 5.12 Airflow from high pressure to low pressure is driven by the pressure gradient (PG), but surface wind direction (SW) is influenced by the Coriolis effect.

5.3 Wind Systems at the Surface and Aloft

We begin with the force driving it all, air pressure. On isobaric maps, the direction of this force always runs perpendicular to the isobars of surface pressure as shown by the arrow marked PG (pressure gradient) in Figure 5.12. If we inspect a synoptic weather chart, however, we will find that surface winds (SW in Figure 5.12) do not conform to the exact direction of the pressure gradient, for they consistently veer off the line of the pressure gradient, crossing the isobars at an angle.

In the northern hemisphere, the winds veer to the right of the pressure gradient. This shift in wind direction is caused by a factor called the Coriolis effect (named for the French physicist G. G. de Coriolis (1792–1843)). The **Coriolis effect** is created by the rotation of the Earth relative to the path of airborne objects moving over the curve of the Earth's surface. It influences the direction of movement of all airborne objects moving over long distances, including migrating birds, aircraft, and wind as well as water currents in the ocean, and the direction of the deflection is opposite in the northern and southern hemispheres. It is also helpful to know that the Coriolis effect functions independently of the direction of movement of an object, and its magnitude depends only on the rotational speed of the Earth at different latitudes, and the relative speed of the moving object.

When viewed above the North Pole, the Earth rotates counterclockwise, as Figure 5.13 shows. But when viewed from above the South Pole, the Earth rotates clockwise. This explains why the Coriolis effect is to the right in the northern hemisphere and to the left in the southern hemisphere. But what happens at the Equator? Do we find a sudden switch in direction? No, instead we find that there is no Coriolis effect at the Equator. From the poles, where it is strongest, the Coriolis effect decreases with decreasing latitude, becoming zero at the Equator.

Although the Coriolis effect is an important consideration in explaining airflow patterns, it is very weak overall, even near the poles. The fact that you do not feel a car being pulled to the right when you drive on a straight highway in the midlatitudes suggests this. However, if you were navigating an airplane over a great distance, you would notice the influence of the Coriolis effect because the airplane would veer far off course unless directional corrections were made in the flight. The difference in the influence on car and aircraft is explained by the fact that the frictional resistance of the car's tires on the road surface is very great, whereas the frictional resistance of air to changes in direction of the aircraft is very slight. Thus, the course of wind, like that of an aircraft, is strongly influenced by the Coriolis effect.

But there is one significant difference between wind and an aircraft to consider. Surface wind (as opposed to upper atmospheric wind) *does* have contact with the Earth's surface and does feel some frictional resistance from it. This resistance is

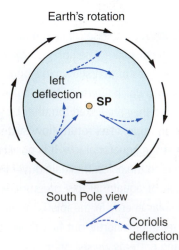

Figure 5.13 The direction of the Coriolis deflection is opposite in the northern and southern hemispheres because of the difference in the direction of the Earth's rotation.

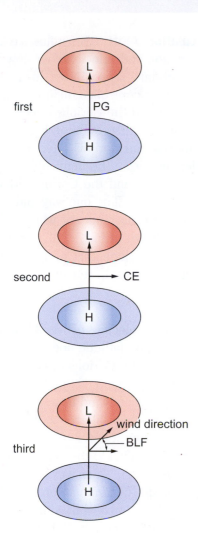

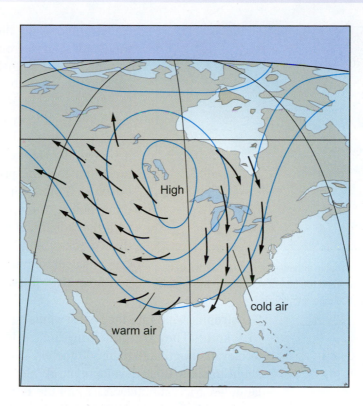

Figure 5.14 The factors affecting the pattern of wind circulation between high- and low-pressure cells illustrated in three sequential steps: first, pressure gradient (PG); second, Coriolis effect (CE); and third, boundary-layer friction (BLF).

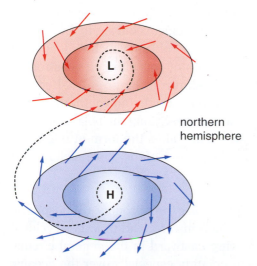

Figure 5.15 The grand pattern of airflow between high (anticyclone) and low pressure (cyclone) cells in the northern hemisphere.

felt mainly in the lower 300 m (1000 feet) layer of the atmosphere, a zone called the **atmospheric boundary layer**, and it tends to offset the full force of the Coriolis effect. Therefore, as a surface wind moves from high pressure to low pressure, it does not veer fully into a Coriolis direction. Instead, it assumes a smaller angle of deflection, which causes it to cross the isobars obliquely as is shown in Figure 5.12. If you picture a pressure cell as a circle then the overall pattern of surface circulation will always be a whirl as is illustrated in Figure 5.6.

In describing the movement of air from a high-pressure cell or into a low-pressure cell, we can simplify it into the steps illustrated in Figure 5.14. Consider *first* the direction of the pressure gradient (PG), *second* the Coriolis effect (CE), and *third*, boundary layer friction (BLF) which partially offsets the Coriolis effect. All high-pressure cells (anticyclones) have outward or divergent airflow, but the direction of deviation from the pressure gradient changes with the hemisphere. In the northern hemisphere, anticyclones have a clockwise whirl, and in the southern hemisphere, a counterclockwise whirl. Low-pressure cells (cyclones) are basically the opposite. They have inward or convergent airflow in both hemispheres, but the direction is counterclockwise in the northern hemisphere, and clockwise in the southern hemisphere. When we combine a cyclone and an anticyclone into a single system, the pattern of airflow forms a great looping path like that shown by the dotted line in Figure 5.15.

This pattern of circulation is an important factor in regional weather patterns and in the zonal transfer of energy. The geographic coverage of both cyclones and anticyclones is often so vast that contrasting types of air are circulated in different sectors of the system. In the case of an anticyclone in the northern hemisphere, for example, the general circulation is northerly along the west side and southerly along the east side, as the map in Figure 5.16 illustrates for North America. The southerly airflow brings warm, humid air northward and the northerly airflow brings cold, dry air southward. The result is a large energy exchange.

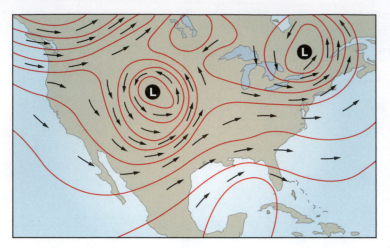

Figure 5.17 A typical pattern of geostrophic airflow over central North America. The altitude is 5.5 kilometers (3.3 miles) and the fastest velocities in this instance are around 160 kilometers per hour (100 miles per hour).

Geostrophic Wind and Circulation Aloft: The influence of the Earth's surface on the direction and velocity of air flow is confined to the atmospheric boundary layer. Above this layer surface friction is absent and wind is controlled mainly by just two factors, the pressure gradient and the Coriolis effect. Air is set into motion by the pressure gradient between high- and low-pressure cells, but as it moves, the Coriolis effect makes it veer to the right or left (depending on the hemisphere). It turns out that the pressure gradient and the Coriolis effect offset each other – think of it as one pulling one way and the other pulling the opposite way – so the resultant wind flows with the isobars, in the manner shown in Figure 5.17, rather than across them. Such winds, or winds that approximate this behavior, dominate the upper atmosphere and are called **geostrophic winds**.

Geostrophic winds are the fastest winds on Earth, reaching velocities of over 325 mph (520 kmh). They are principally limited to zones above 25° latitude and generally to elevations greater than 500 meters. Near the Equator, the Coriolis effect is too slight to allow their development. At elevations lower than 300 to 500 meters at all latitudes, friction between air and the Earth's surface prevents wind from attaining geostrophic velocities and directions of flow. This is the chief difference between surface winds and geostrophic winds. Surface winds stabilize at lower velocities and in directions intermediate to the geostrophic direction (parallel to the isobars) and the pressure-gradient direction (perpendicular to the isobars). Thus, as you ascend from ground level into the atmosphere, wind direction shifts, becoming more parallel to the isobars until it attains geostrophic-type flow. The direction of shift with altitude is, of course, dependent upon the hemisphere. In the northern hemisphere, wind direction shifts clockwise upward through the lower atmosphere. In the southern hemisphere, the shift is counterclockwise.

Thousands of meters aloft in the midlatitudes of each hemisphere the geostrophic winds form a great belt of westerly flow. This belt develops a huge meandering pattern, called **Rossby waves** (after C. G. Rossby, who first described them in mathematical terms in the 1930s), which usually loops equatorward over the continents and poleward over the oceans. To the polar side is the zone of cold air; to the equatorial side is the zone of warm air. The leading edge of the cold air, which we identified earlier as the polar front, is often marked by a stream of concentrated geostrophic flow, called the **polar-front jet stream,** which is shown in Figure 5.18. Because the flanks (east and west sides) of the meanders trend north and south, there is a strong tendency toward zonal exchanges of air and energy along the polar front, that is, between the tropics and the higher latitudes. This takes the form of migrating air masses whose movement generally follows the pattern of the upper airflow as it meanders across the continents and oceans.

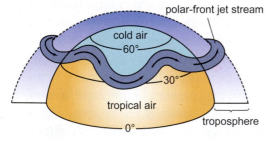

Figure 5.18 A schematic diagram showing the meandering belt of concentrated geostrophic wind called the polar-front jet stream.

The mixing of warm and cold air masses along the polar front produces atmospheric pockets of low pressure, called *disturbances*, which in turn lead to larger cyclonic cells. Although these cells are transitory, moving eastward in the general circulation scheme, they are prominent and frequent enough, especially over the oceans, that semipermanent low pressure is assigned to isobaric maps in the upper middle latitude and subarctic oceanic zones. These are the **subpolar lows,** shown in

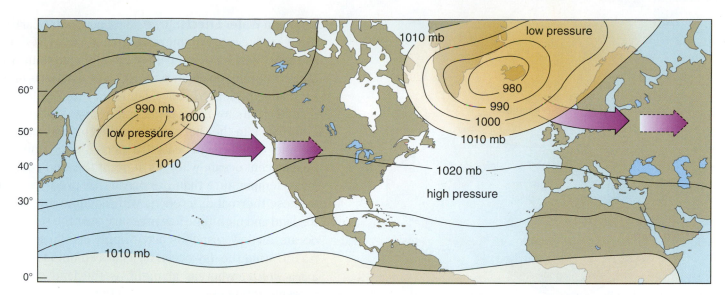

Figure 5.19 The geographic distribution of subpolar low-pressure cells in the northern hemisphere. These cells form with greatest frequency and magnitude over the upper North Pacific and North Atlantic basins.

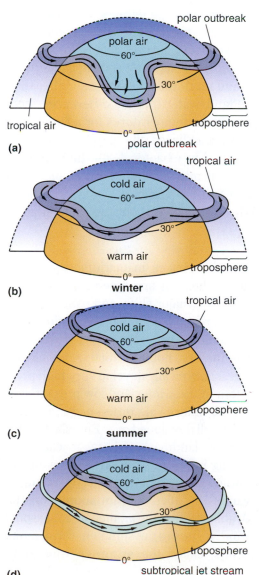

Figure 5.19, and they are a very important ingredient in midlatitude climate, because they migrate eastward with the westerly airflow bringing stormy, wet weather to landmasses downwind. Both North America and Eurasia are dramatically affected by these storms, especially in winter. In the southern hemisphere, these cells are also prominent, but have less influence on terrestrial climate, because there is little land in this zone of latitude.

The polar front and its jet stream are subject to frequent changes in location and cyclonic activity. First, the meander loops typically migrate, sliding gradually eastward, for example. For a given location, such as central North America, this may result in a change from polar to tropical air as the meander moves past the location over the course of only several days. Similarly, the polar front and the jet stream may suddenly shift southward, releasing, for example, frigid air from Canada which may rush deep into the American South as shown in Figure 5.20a. Such *polar outbreaks* are known to cause temperature changes of 30 to 40 °C in a matter of hours. Second, the polar front shifts north and south with the seasons, and as it does so it weakens and strengthens. In winter, it pushes equatorward and strengthens (Figure 5.20b), producing one cyclonic storm after another in rapid sequence (as often as every 3 to 5 days). In summer, the polar front weakens, the meanders grow less distinct and the whole system, including the jet stream, shifts poleward with fewer and weaker cyclonic storms (Figure 5.20c).

In addition to the polar-front jet stream, two other jet streams are found in the upper atmosphere. The **subtropical jet stream** is located around 30 degrees latitude above the subtropical high-pressure cells (Figure 5.20d). It, too, is a westerly flow, but this stream of air is slower and less persistent than the polar-front jet stream. Occasionally, the subtropical and polar jet streams merge, producing a rapid exchange of tropical and polar air. The third jet stream, the **tropical easterly jet stream**, is smaller and weaker yet. It appears only in summer in the northern hemisphere over Africa and Asia and flows westward over the belt of easterly tradewinds.

Figure 5.20 The meandering pattern of the polar-front jet stream is subject to frequent change including (a) outbreaks of polar air that can reach deep into the midlatitudes, and (b and c) seasonal changes in magnitude and location; (d) shows the location subtropical jet stream, a weaker geostrophic stream.

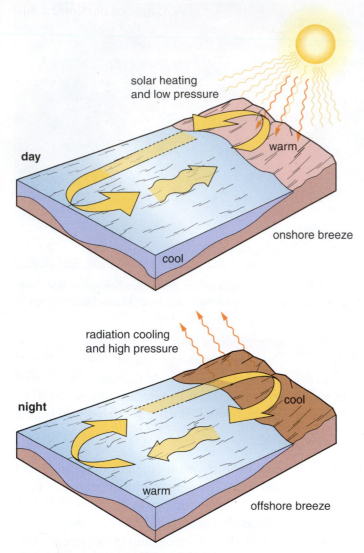

Figure 5.21 Onshore and offshore breezes are a direct response to day/night changes in heating and air pressure over the land relative to adjacent water.

Land and Water Effects: If we float back to the Earth's surface, there are a couple of interesting wind patterns that need our attention, which are not part of the global system. We begin at the local scale, on the seashore, with a pattern of airflow we have all experienced called land–sea breezes. These breezes operate on a day–night cycle and are caused by the dramatic differences in the thermal properties of land and water. On a summer day, the land heats to a much higher temperature than the adjacent ocean, whereas at night the land cools to a much lower temperature than the ocean. Air pressure corresponds to these thermal differences with daytime low pressure over the land and nighttime low pressure over the relatively warmer ocean. Airflow responds accordingly, as illustrated in Figure 5.21. During the day ocean air is drawn onto the land forming a *sea breeze,* whereas at night as the land cools down, the pressure gradient reverses, and a *land breeze* develops.

The land in summer is like a thin hot plate which heats quickly and then equally quickly loses its heat to the atmosphere when the solar flame is turned down at night. In the ocean, by contrast, solar radiation penetrates and heats tens of meters of water, and then the heated water is mixed to even greater depths by wave and current action. In addition, water has a higher heat capacity than land and air, about three times greater than soil and more than 1000 times higher than air (see Table 4.5). The land/water thermal differences show up in a comparison of seasonal temperatures for a land-based city like St. Louis and an ocean-based city like San Francisco, a comparison we mentioned in Chapter 4. As the graphs in Figure 5.22 show, St. Louis is much colder in winter and much warmer in summer than San Francisco, yet both cities lie at nearly the same latitude and receive about the same input of solar energy.

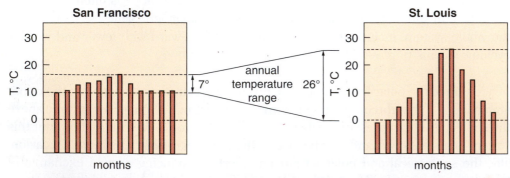

Figure 5.22 Graphs of the monthly mean temperatures for San Francisco and St. Louis reveal the remarkable thermal differences between marine and continental locations.

The differences in maritime and continental heating are essential to explaining some major shifts and patterns in air pressure and seasonal circulation over the continents. The large, midlatitude landmasses, notably North America and Eurasia, typically develop very high pressures in the winter and low pressures in the summer. This difference is particularly well developed over Eurasia, Earth's largest landmass, where it causes a strong inflow of air in one season and an outflow in the other. In Asia, the resultant winds are called the **monsoon** (from the Arabic word *mausim,* meaning "seasonal wind"). During the summer, warm, moist winds, called the **summer monsoon,**

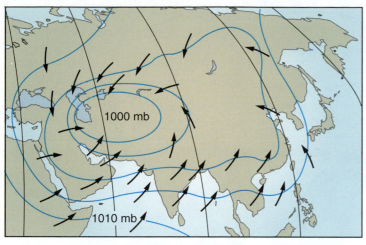

(a) summer monsoon

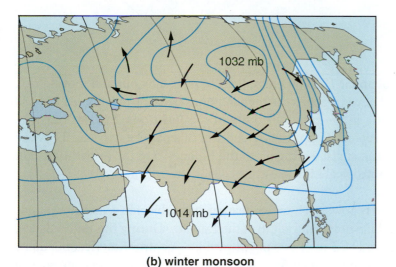

(b) winter monsoon

Figure 5.23 Monsoon circulation over Asia reverses from summer (a) to winter (b) in response to dramatic pressure changes over the great landmass.

blow from the Indian Ocean and Arabian Sea northward into the Asian landmass (Figure 5.23a). In the winter, a strong high-pressure system develops over Asia causing the winds to blow from the land southward to the ocean. This is called the **winter monsoon** and, because of its continental source, it is dry and cool (Figure 5.23b). Monsoon-type circulation can also be measured over central North America, but it is much weaker than that in Asia.

5.4 The General Circulation of the Earth's Atmosphere

We are now in a position to describe the overall global pressure and circulation system of Earth. This includes not only those wind systems and cells that we touched on in our earlier sketch, but the seasonal behavior of these systems and how they shape the basic character of climate in various geographic zones. We begin with the low latitudes.

Equatorial and Tropical Zones: Recall that the equatorial low-pressure system (shown in Figure 5.10) is fed by surface winds from both the north and south. When these winds meet near the Equator, they form a belt of air convergence, called the **intertropical convergence zone** (ITCZ), and within this zone, air is diverted upward. Over land, the upward motion is driven by surface heat, but aloft the power source is latent heat released with water-vapor condensation and cloud formation. Over the ocean, where surface heating is less intense, the upward flow at low altitudes is powered more by the sheer force of the two wind systems ramming into each other. Higher up the continued ascent is also driven by the latent heat of condensation.

This process can be highly variable over the oceans, so much so that the ITCZ is often marked by surface calm with little or no wind movement in any direction. During the days of ocean sailing this was a source of distress, because ships attempting to cross the Equator could float listlessly in the water, sometimes for weeks. As a result, the ITCZ came to be known as the *doldrums*, a word we still use to describe a weary, depressed state of mind.

The winds that feed the ITCZ are the **tradewinds** and those for the western hemisphere are shown in Figure 5.24. North of the Equator they blow from the northeast to the southwest and are called the **northeast trades**. South of the Equator they are called the **southeast trades**. Since these winds originate in the subtropical high-pressure zone several thousand kilometers poleward of the Equator, they are influenced by the Coriolis effect. This explains why the tradewinds do not take a straight north to south route in the northern hemisphere and a straight south to north route in the southern hemisphere. In the days of sailing ships, the northeast trades were a preferred route from Europe and North Africa to the New World. In fact, Columbus's voyages from Spain sailed on the tradewinds, but his return trips sailed on the westerlies to the north. We will examine the westerlies shortly.

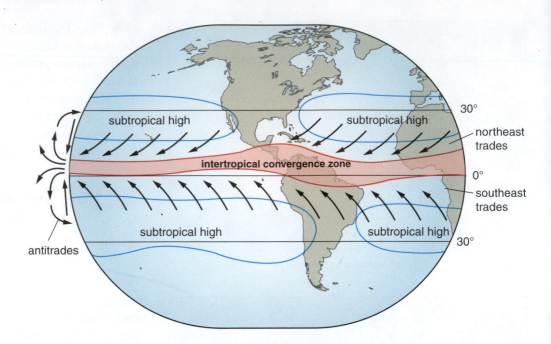

Figure 5.24 The global tradewind system and associated pressure belts. This huge circumglobal wind system covers nearly half the Earth's total surface area.

Aloft, at an elevation of about 12 kilometers, the ITCZ air fans out and moves back toward the poles. But it never reaches the high latitudes because at latitudes 25 to 30 degrees north and south, in the zone called the **horse latitudes**, the air descends, forming the large, subtropical high-pressure cells. In this pattern of cyclical circulation, the tradewinds at the surface bring air to the Equator, and the **antitrades** aloft bring air back to the subtropical high-pressure cells.

Since the ITCZ over land is powered by solar heating, it is subject to seasonal changes in location as the Sun shifts north and south of the Equator. During the June solstice, or shortly thereafter, the ITCZ migrates northward into central America, northern Africa, and south–central and eastern Asia (see Figure 5.25). In the opposite season it lies south of the Equator over central South America, southern Africa, and northern Australia. Over water, on the other hand, where surface temperatures vary by only 2 to 3 °C annually, the ITCZ is less geographically mobile, lying much closer to the Equator year around.

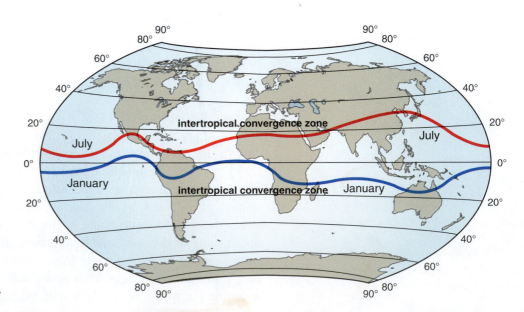

Figure 5.25 The seasonal shifts in the ITCZ are generally greater over land than water. The most extreme shift is between southern Asia and Australia.

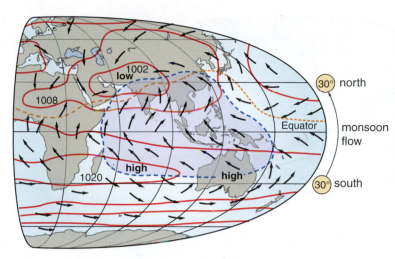

Figure 5.26 Summer low pressure over Asia is so powerful that it draws the southeasterly tradewinds across the Equator where they join the northeasterly tradewinds to form the summer monsoon flow.

Figure 5.27 A traditional sailing boat, called a dhow, on the Arabian Sea, the kind that Vasco da Gama saw when he crossed the Indian Ocean more than 500 years ago.

The one exception to the last statement is the Indian Ocean, where in the northern hemisphere summer the ITCZ shifts far north of the Equator to a region over south central Asia. This shift, which covers a distance of more than 5000 kilometers, has nothing to do with temperature change over the Indian Ocean, but rather with intense heating over the Asian landmass. The low pressure that develops in the Asian segment of the ITCZ is so strong that the northeasterly tradewinds over the Indian Ocean are reversed, as shown in the shaded area of Figure 5.26. Not only that, but the southeast tradewinds, from nearly 30 degrees south latitude, are entrained into this system.

In the sixteenth century, during the period of Portuguese trade with India, ship captains timed their voyages from Europe to India so that the vessels crossed the Arabian Sea from Africa to India during the summer monsoon. They then waited in India until the onset of the winter monsoon for the return trip. The first fleet to sail to India (in 1497–98), commanded by Vasco da Gama, did not do this, but mistakenly started the voyage at the end of August, with the summer monsoon still in progress. It took da Gama three months to cover the 4200 kilometers (2700 miles) distance (a rate of only 30 miles a day) from India to the east coast of Africa, three times longer than the same trip to India during the summer monsoon. Sailing vessels, like the ones in Figure 5.27, are still widely used by Arab and Indian traders and fishermen on the Indian Ocean who follow the same principles of navigation as the early Portuguese.

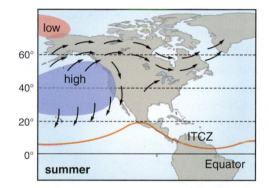

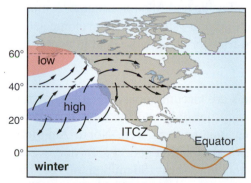

Figure 5.28 As the ITCZ shifts north and strengthens in summer, subtropical high pressure expands and shifts north. In winter the opposite changes takes place as the subpolar low over the North Pacific strengthens and shifts southward.

The subtropical highs also change with the position and strength of the equatorial low. As the ITCZ shifts into the northern hemisphere and intensifies in summer, more air is forced aloft into the antitrades and ultimately into the subtropical highs. The highs in turn expand significantly, as illustrated in Figure 5.28, enveloping areas which, during the winter months, were well within the belt of influence of the midlatitude westerlies. In coastal areas, such as the North American West Coast, this change to subtropical high pressure produces a dry summer climate. This is not surprising since the same pressure system produces the subtropical deserts that lie to the south.

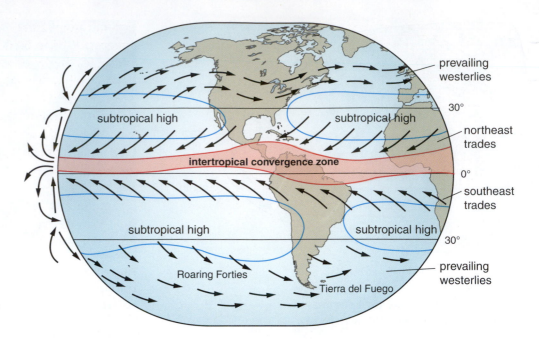

Figure 5.29 The Earth's two main prevailing wind systems. The belts of prevailing westerlies dominate the midlatitudes in both hemispheres.

Midlatitude Zone: Midlatitude circulation begins on the poleward flanks of the subtropical high-pressure system around 30 to 35° latitude. As the air leaves this zone it is deflected by the Coriolis effect into an eastward flow, called the **prevailing westerlies**. The westerlies blow in both hemispheres, forming vast circumpolar belts that dominate the midlatitudes. But there are significant differences in the westerlies between the two hemispheres. Take a look at the map in Figure 5.29.

The major difference is that in the southern hemisphere the westerlies flow almost entirely over the oceans forming a nearly continuous maritime loop. In winter, the system shifts equatorward, strengthens, and brings some of the harshest cyclonic storms on the planet to the southern tips of South America, Australia, and New Zealand. The belt between 40 and 50 degrees south latitude is known to sailors as the Roaring Forties, and the large island, Tierra del Fuego, at the southern tip of South America in the minds of many, including a young Charles Darwin, holds the dubious distinction of being one of the dreariest, wettest, and windiest places on the Earth. The westerly system in the northern hemisphere also shifts equatorward in winter, strengthens, and is associated with stormy weather generated by cyclones that develop along the polar front in the northern Pacific and Atlantic oceans. Compared to the southern hemisphere, there is a much greater rate of winter cyclonic activity here, resulting in four times more poleward transfer of sensible heat in winter than summer.

Since both the tradewinds and the westerlies are circumglobal systems, they function as huge intercontinental beltways moving not only energy, moisture, clouds, and weather systems around the world, but pollutants as well. North America is downwind of China, one of the world's principal pollution sources. In one event in 1997, a dust load from soil erosion in China totaling nearly 150 million tons took less than two weeks to reach North America via upper-level winds. And in 2011, radioactive contaminants from the Fukushima Nuclear Power Plants damaged by the East Japan Tsunami disaster were detected in North America less than three weeks after the event. North America also makes its contributions. The satellite image in Figure 5.30 shows smoke plumes from southern California wildfires being carried westward over the Pacific.

Figure 5.30 Plumes from California wildfires carried over the Pacific by a strong easterly wind called the Santa Ana.

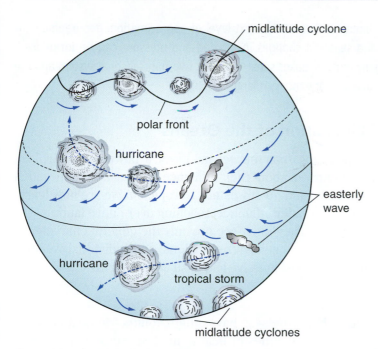

Figure 5.31 Secondary circulation features include midlatitude cyclones, easterly waves, tropical storms, and hurricanes. In the midlatitudes, the cyclones form along the polar front and migrate eastward.

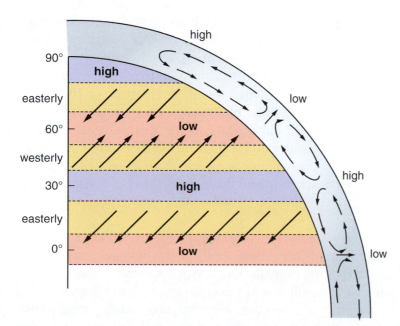

Figure 5.32 A schematic model of the general pressure and circulation system showing the alternating belts of pressure and prevailing wind systems that dominate both hemispheres.

High-latitude Zone: The subpolar low-pressure systems, centered at 55 to 60 degrees latitude, draw air from both the subtropical highs (via the prevailing westerlies) and the polar highs. The flow from the polar highs is deflected by the Coriolis effect to form a westward-flowing wind system called the **polar easterlies**. Although this system is dominant in the zone above (poleward) of the Arctic and Antarctic circles, as a wind system, it is neither strong not persistent in all seasons. Somewhat unexpectedly, the polar easterlies tend to strengthen in summer and weaken in winter. In the northern hemisphere part of this trend is related to the fact that this wind is influenced in summer by the big low-pressure system that dominates south central Asia and is responsible for summer-monsoon circulation.

Secondary Circulation: We have emphasized the general circulation of the atmosphere, because of its critical role in the global redistribution of energy and in shaping the general character of world climates. There is also a second level of circulation influences, called *secondary circulation* features, which are generally smaller in scale and shorter lived than the global features. These features are often characterized as discrete disturbances that result in migratory cells of several days to a few weeks duration. The most common are the *midlatitude cyclones* that originate along the polar front and migrate eastward with the westerlies as shown in Figure 5.31. These storm systems are fed energy from the inflow of tropical air which yields heat with the condensation of moisture. Thousands of such cyclonic storms are produced on Earth every year.

Less common, though often greater in magnitude and duration, are tropical disturbances including easterly waves and hurricanes, also shown in Figure 5.31. *Easterly waves* are modest troughs of low pressure that develop within the belt of tradewinds and migrate westward. They develop over warm ocean surfaces and are fed with moist air that provides a source of latent heat that maintains the low pressure. A much larger and more intensive tropical storm is the **hurricane** (also known as a *typhoon* or *cyclone)* which also develops over warm water and, during its early stages, also migrates westward with the belt of tradewinds. We will have more to say about these storms and their destructive effects in the following chapter.

Summary on the General Circulation of the Atmosphere: At the first level of approximation, the atmospheric circulation system in each hemisphere can be characterized by three great belts of prevailing winds connecting four zones of pressure as illustrated in Figure 5.32. In reality it is a complex, three-dimensional, flow system that takes air on a circuitous ride across the planet, which, in the big picture, produces two important outcomes. First, the zonal transfer energy from the low to high latitudes, and second, the lateral (east–west) transfer of heat and moisture across the oceans and landmasses. Both are important factors in framing Earth's climatic zones. At the next level of approximation, geographic variations in land and water and seasonal changes in heating and pressure can be introduced. Indeed, the circulation in many regions makes sense only as a two-part seasonal system and the monsoon circulation of South Asia is the most striking example.

5.5 The General Circulation of the Oceans

Increasingly, scientists are learning that the oceans exert a powerful control on Earth's climates. We have long known about their influence on coastal regions, such as the North American West Coast and northwestern Europe, but we are now learning that the oceans have a much broader influence, reaching over entire landmasses and indeed across the globe itself. They influence not only the general character of climate, but also long-term trends, short-term oscillations, and individual weather events. It is significant that Earth's most massive and destructive storms, hurricanes, are the exclusive products of the sea.

In sharp contrast to the atmosphere, the seas have a high volumetric heat capacity and, thus, are able to retain great quantities of heat at modest temperature levels. Furthermore, through wave and current motion, the oceans are able to transfer heat to considerable water depths, and redistribute it geographically over broad areas of the ocean surface. Much of the energy redistribution is east–west or west–east. But much is also poleward, resulting in the transfer of energy across the midlatitudes from the tropics to the subarctic and beyond, thereby augmenting the overall poleward heat transfer in the atmosphere. In addition, certain coastal areas are profoundly affected by warm currents or cold currents that give rise to unique coastal climates. Because wind is the principal driving force for waves and currents, we begin our discussion of oceanic circulation with a brief look at water motion and wind.

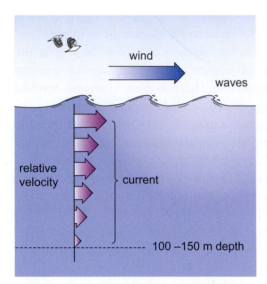

Figure 5.33 Currents result from the drag of strong wind on the ocean surface. Momentum is transferred from wind to the surface layer and then downward with decreasing velocity with depth.

Generation of Currents by Wind: Waves can be generated by a variety of forces including earthquakes and volcanic eruptions, but the waves that dominate the oceans are wind generated. **Wind waves** cover a broad range of sizes from ripples a few centimeters high to great storm waves 20 or more meters high. There are two types of motion associated with wind and waves: (1) mixing beneath the wave, and (2) downwind transfer of water mass. Wave mixing, which is characterized by a circular, churning motion, is an important mechanism in transferring heat from the surface to greater depths in the sea, though usually not more 50 or 100 meters down. The downwind transfer of water mass that results in a current involves the transfer of momentum from wind to the surface layer of the ocean. Once the surface layer of water is set in motion, part of its momentum in turn is transferred downward from one layer to the next with declining force at each level. At a depth of 100 to 150 meters, water velocity reaches nearly zero and this defines the effective base of the current, although very large ones like the Gulf Stream reach much deeper.

Current direction and velocity are governed principally by the direction, velocity, and duration of wind. If brisk winds blow consistently in one direction over a broad reach of ocean, a strong current will develop along the general path of the wind. But as Figure 5.33 shows, the current will flow at velocities much slower than the wind itself. Not only that, but the current's direction, like wind direction, is strongly influenced by the Coriolis effect, causing it to deviate from the wind direction by veering to the right or left. The Coriolis effect also extends to the layers of water beneath the surface layer of the current, with each successive layer veering more to the right or left with increasing depth as illustrated in Figure 5.34.

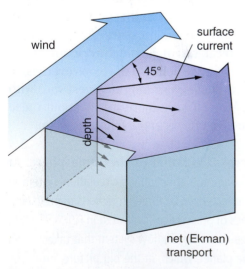

Figure 5.34 The Ekman spiral illustrates the change in current direction with depth under the influence of the Coriolis effect. Ekman transport is the net direction of current flow relative to wind direction.

The Coriolis shift in the surface layer is about 45 degrees from wind direction. The next layer shifts a little more, the next a little more, and so on, until the slow-moving water at the base of the current may be flowing nearly opposite wind direction. This shift in current direction with depth is known as an **Ekman spiral,** named for the Swedish scientist, V. Walfrid Ekman, who first described this phenomenon in the early twentieth century. For our purposes, it is important to define its net effect, called **Ekman transport**, in terms of total water transferred. Ekman transport, represented by the big arrow in Figure 5.34, is 90 degrees to the right (or left) of wind direction. Thus, in mid-ocean, such as in the North Atlantic, where the westerlies trend northeastward and the tradewinds trend southeastward, surface currents tend to travel more directly east or west than the winds that drive them, but still follow the broad path of the prevailing wind systems.

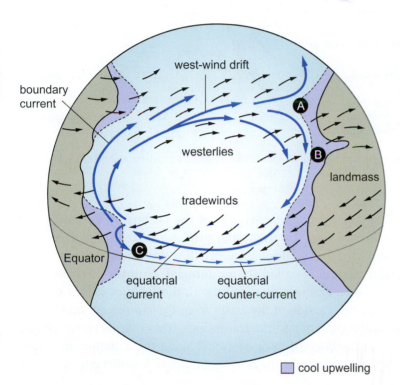

Figure 5.35 A subtropical gyre in the northern hemisphere resembling the one found in the North Atlantic, showing the geographic correlation between the two prevailing wind systems and the two current systems. Two examples of upwelling zones are shown at A and B, while C marks the beginning point of the equatorial counter current.

Ocean Basins and the Formation of Gyres: Because the Earth's water cover is interrupted by landmasses (except for a belt between 45 and 65 degrees in the southern hemisphere), the eastward- and westward-flowing currents are confined to ocean basins. When a current encounters land at the edge of a basin, its flow is deflected, often into two currents, one going northward and one going southward. These currents are called **boundary currents** and they parallel the coast until they enter into the next zone of prevailing wind and merge with the system there. This results in huge circular loops called **gyres** and one is illustrated in Figure 5.35.

In an ocean such as the Atlantic, three gyres form in each hemisphere with each gyre corresponding to a pressure cell and a set of prevailing wind systems. The largest are the **subtropical gyres**. If you look at the illustration of the gyre in Figure 5.35, you will see a westward-flowing **equatorial current** in the tradewind zone which runs into the landmass on the west side of the ocean, where, along with a little influence from the Coriolis effect, it is deflected northward along the coast. The resultant boundary current runs along the coast and then, around 35 degrees latitude, enters the zone of the prevailing westerlies. Here the westerlies set up a strong current called **west-wind drift** which carries warm water eastward to the other side of the ocean, whereupon it splits into two boundary currents, one branching poleward and the other branching equatorward.

As the boundary currents slide along the edge of the ocean basin they pick up cool water from ocean upwellings. **Upwellings** are rises of cool, subsurface water in response to the removal of warmer surface water mainly by surface currents and/or wind. In Figure 5.35, an upwelling is found at (A) where the west-wind drift splits and drags surface water north and south, and at (B) where airflow is onshore but Ekman transport, at 90 degrees right of the wind, drags surface water to the south where the current enters the tradewind zone. In the Atlantic, this occurs off the northeastern coast of Africa where the cool boundary current, called the Canary Current, enters the tropical–equatorial zone, and joins the existing equatorial current, a warm current, thus completing the loop and forming the gyre.

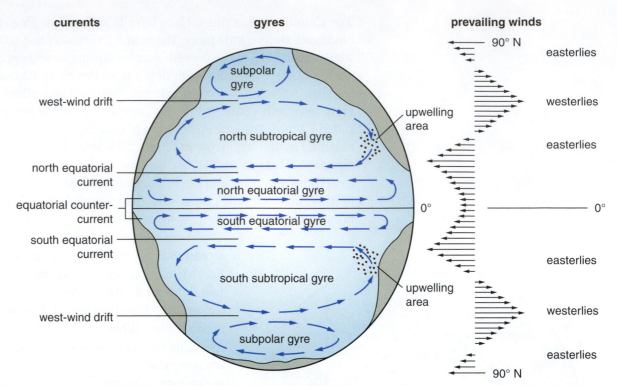

Figure 5.36 A schematic diagram summarizing the correlations among wind systems and ocean circulation systems at a global scale.

Not all the water pushed against the western rim of the oceans by the equatorial current is deflected poleward into the west-wind drift, however. Some (see C in Figure 5.25) is also redirected into a curious reverse flow that runs eastward along the Equator. This return current is called the **equatorial counter-current.** It is driven by an oceanic pressure gradient that forms with the buildup (slight mounding) of water against the western side of the basin from the combined force of the westward flowing equatorial current and the tradewinds. This marine pressure gradient is sufficient to push the current back eastward along the Equator, and since there is no Coriolis effect at this latitude, the flows are more or less directly eastward.

Now turn to the map in Figure 5.36 and note that the equatorial counter-current described above has a counterpart just south of the Equator called the **south equatorial counter-current**. Thus, in the narrow belt between the north equatorial current and the south equatorial current, is a pair of long, skinny gyres know as the **equatorial gyres**. Although they are much smaller and weaker than the subtropical gyres, they are a central player in the well known El Niño system, which we will describe shortly, that involves the periodic build of climate-altering warm water brought to eastern side of the Pacific Basin by the equatorial counter-currents.

On the poleward side of west-wind drift, in the subpolar zone, is another set of gyres, also shown in Figure 5.36, which are smaller than the subtropical gyres. These are **subpolar gyres** and they correspond to the subarctic low-pressure cells. On their western flanks they bring cold water equatorward where it meets the warm waters of the west-wind drift, mixes with the larger current, and is driven eastward. On their eastern flanks, on the other hand, they bring water from the west-wind drift poleward into a zone of colder water.

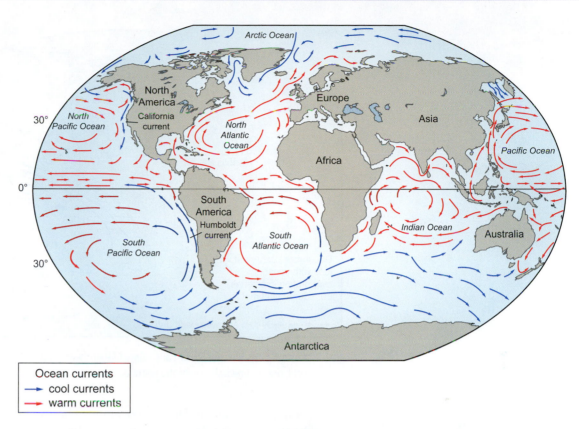

Figure 5.37 The general pattern of global ocean circulation highlighting warm and cold currents.

Although the ocean circulation system outlined above is based largely on the hypothetical model presented in Figure 5.36, it is remarkably similar to the circulation system we find today in the world's oceans, shown in Figure 5.37. There are, of course, some exceptions. The principal one is found in the area around Antarctica, where the ocean basin is not bounded on the east and west by continents, and the west-wind drift circles the entire Earth. Therefore, there are no subpolar gyres in the southern hemisphere. This is extremely important, because without continents to deflect currents northward or southward, not much energy can be exchanged between tropical waters and cooler waters at higher latitudes. It is one of the reasons Antarctica is so deeply frozen. In the northern hemisphere, by contrast, the subpolar zone of the North Atlantic is a region of active north–south energy exchanges that involves a major current called the Gulf Stream.

Heat Transfer by Ocean Currents: The **Gulf Stream** is a warm current that flows from the tip of Florida into the cold North Atlantic as shown in Figure 5.38. By the time it becomes part of the west-wind drift, the Gulf Stream is only several degrees Celsius warmer than the surrounding ocean water. But its water volume is so huge – 100 times greater than the combined flows of *all* the world's rivers – that the additional heat it carries increases by more than twofold the total transfer of heat energy into the atmosphere over the Atlantic. Most heat is transferred into the atmosphere as latent heat produced with the evaporation of surface water.

The heat-transfer process operates at an exceptionally high rate along the North American east coast in winter, where cold, relatively dry continental air is driven over the warm Gulf Stream by the westerlies. This is illustrated in Figure 5.38. The energy transfer from water to air is colossal, well over 12,000 calories per square centimeter

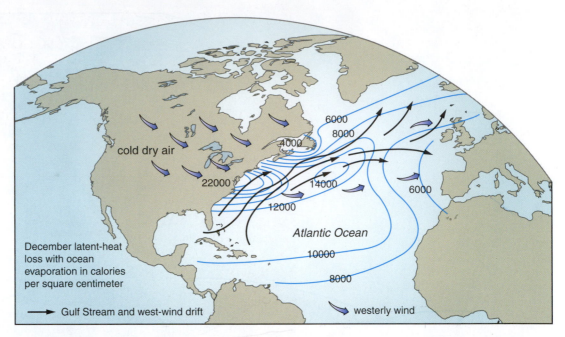

Figure 5.38 The Gulf Stream and related latent-heat flux into the atmosphere in winter related to the flow of cold, dry continental air over the warm ocean.

of ocean surface in December alone. The westerlies then carry the resultant latent heat across the Atlantic and release it with condensation and precipitation on the European side. As a result, the atmosphere over the North Atlantic and the climate of northwest Europe are measurably warmer and wetter than they would otherwise be. Compare, for example, the annual temperature regime of Ireland (around 55 degrees) with a place at the same latitude in the interior of North America.

The opposite tends to be true where cold currents, such as the Humboldt Current along the Pacific coast of South America or the California Current along the California coast, move against a warm coastline (see Figure 5.37). The coastal climate is drier and cooler, because the current cools the air over it while giving up little moisture to evaporation. Since it is the heat in surface water that drives the evaporation process, cool currents tend to dampen latent-heat transfer into the atmosphere. Thus, the coastal air gains little vapor and heat from the ocean.

Figure 5.39 A typical Gulf Stream circulation pattern in the North Atlantic with huge meanders that break off to form large rings that may rotate for months.

Satellites have helped immensely in our understanding of the oceans. Remote sensors are able to monitor the thermal, ecological, and other conditions of the sea on a continuous basis from both orbiting and stationary satellite platforms. Using thermal sensors, the ocean's surface temperatures can be mapped over broad areas, and from the resultant data, changes in circulation patterns can be documented. Satellite monitoring of the Gulf Stream, for example, has revealed (1) a sharp thermal boundary, called a **front**, along the northern side of the current; and (2) great whirls or rings that form when meanders break free of the current. These rings, shown in Figure 5.39, are 100 to 200 kilometers in diameter, and after they become detached from the current, they may rotate for months before dissipating and blending into the greater ocean. Rings are released into waters on both the north and south sides of the current. Those on the north form warm rings. Those on the south form cold rings, and each is identifiable in the satellite imagery not only by its thermal pattern, but also by concentrations of marine life. Marine organisms with a preference for cold or warm water tend to concentrate on the margins and within the rings.

5.6 Thermohaline Circulation: the Hidden System

There is also a second system of circulation operating in oceans. This system operates in the deep parts of the oceans well below the zone of waves and currents. It is much slower than the surface system, but it involves a massive quantity of water. The deep-ocean system is driven by density differences related to water temperature and salinity variations and is called the **thermohaline circulation system**.

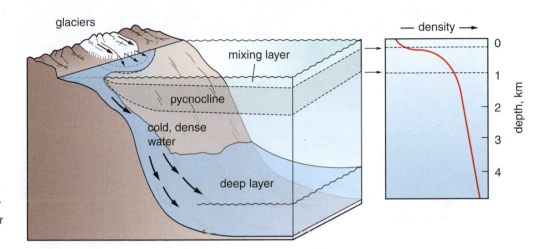

Figure 5.40 The structure of the ocean according to water density. Below the mixing (surface) layer is the pycnocline layer where density increases rapidly with depth. Below it lies the deep layer where water is cold and dense.

Throughout most of the world, the oceans are made up of three layers, shown in Figure 5.40. Uppermost is the **mixing layer,** which is roughly 100 meters deep, where waves and currents work. Next is the **pycnocline layer,** between 100 and 1000 meters, where density changes rapidly with depth. And lowermost is the **deep layer,** below 1000 m, where water is cold (3–4 °C) and very dense. There is little vertical mixing among these layers, especially between the deep layer and the upper two, because, with lower density situated over higher density water, the arrangement is gravitationally stable. In fact, the pycnocline, which also marks the ocean **thermocline,** or thermal-change zone, acts as an effective barrier to vertical movement of water masses between the mixing layer where solar heating takes place and the deep layer which is beyond the reach of solar radiation.

The one exception to this generalization is found in the upper midlatitudes and poleward. Here heat loss to the atmosphere from evaporation off the relatively warm ocean surface, coupled with inflows of cold surface water from glaciers and snowfields (shown on the left side of Figure 5.40), so cools the surface water that no pycnocline develops. Instead, high-density cold water forms on the ocean surface and if cooling is intense enough, this water grows denser than the water beneath it and sinks. If sea ice forms on the ocean surface, water density is increased even further, because of a process called *ice exclusion,* whereby dissolved salts (sea salts) are deleted from the ice as seawater freezes. These salts are left in solution in the surface layer, which increases the seawater's salt concentration and in turn its density.

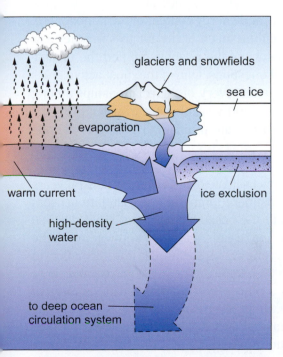

Figure 5.41 The three main sources of high-density water that feed the deep ocean circulation system.

High-density water masses created by these processes, cold-water inflows, evaporative cooling, and ice exclusion (shown in Figure 5.41), form broad areas of **downwelling** which occur only at a few locations in the oceans. The best examples are found at the northern and southern ends of the Atlantic basin, near Iceland and Antarctica. In both locations, high-density water masses sink and enter a deep return flow system that moves equatorward at depths greater than 2000 meters. The North Atlantic

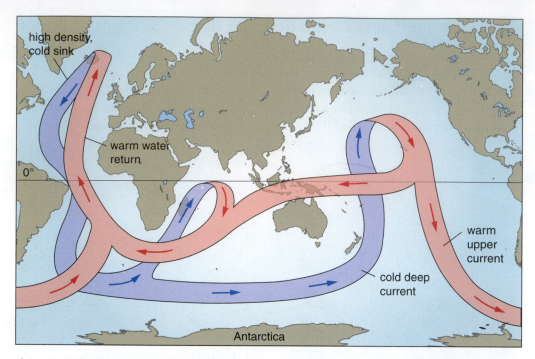

Figure 5.42 The deep ocean circulation system, a great conveyor belt which recycles water from high density sources such as chilled surface currents, glacier meltwater, and ice exclusion.

water flows all the way to the Antarctic where it feeds into a deep-water cold-current system, shown in Figure 5.42, which flows around the Antarctica landmass and in turn feeds into related deep-current systems in the Pacific and Indian oceans. But cold water from the Arctic Ocean, a logical source of high-density water, is excluded from the thermohaline system by topographic barriers (ridges) on the ocean floor that block deep-water circulation with the North Atlantic and North Pacific.

The global deep-water system that is fed from the North Atlantic involves a mass of water equivalent to 20 times the combined discharges of all the world's rivers and it takes about 1000 years to complete the full (global) cycle. This huge current functions as a cold-water return system, a giant ocean conveyor belt that provides a mechanism for recycling surface waters brought north by, for example, the Gulf Stream and the west-wind drift. Once these surface currents discharge their loads of heat into the atmosphere, and undergo a substantial decline in temperature, their water masses are eventually returned to the oceans at lower latitudes by the deep-current system.

5.7 The Influence of the Oceans on the Atmosphere

The oceans occupy nearly 70 percent of the Earth's surface and exert a tremendous influence on the temperature, moisture, and energy content of the atmosphere. Indeed, it appears that the key to understanding much of the Earth's weather and climate is held by the ocean. But on the nature of the relationship and the mechanisms involved, there are many unanswered questions. One of the most active areas of research in this connection centers on the discovery of the relationships among ocean circulation, water temperatures, latent- and sensible-heat transfer from the oceans to the atmosphere, and weather and climate trends and events in various parts of the world.

We have long known that ocean temperatures oscillate over periods of several years in a manner broadly similar to that of atmosphere temperatures. Satellite sensors allow us to monitor temperature changes in the ocean surface, and build detailed maps of weekly, monthly, and seasonal thermal trends. The data show not only the seasonal expansion and contraction of current systems, but also the formation of large bodies of warmer or cooler water, 2000 kilometers or more across and up to 300 m deep, that can last for several years. These bodies, referred to as *thermal anomalies*, can influence the position of the winter jet stream. Over the North Pacific, for example, warm anomalies such as those shown in Figure 5.43 are known to alter the pattern of storm development and movement along the North American West Coast.

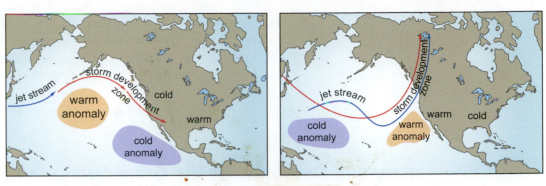

Figure 5.43 The influence of thermal anomalies in the Pacific Ocean on the position of the polar-front jet stream and the zone of storm development along the North American West Coast.

The most celebrated thermal anomaly is **El Niño**, a huge body of warm water that develops in the equatorial zone of the eastern Pacific Ocean. El Niño ("the boy child" in Spanish – an allusion to the Christ Child, because it typically begins around Christmas) forms every five years or so in an area off the coast of Ecuador, and expands northward and southward along the North American and South American coasts. Figure 5.44 plots the general phases of an El Niño event.

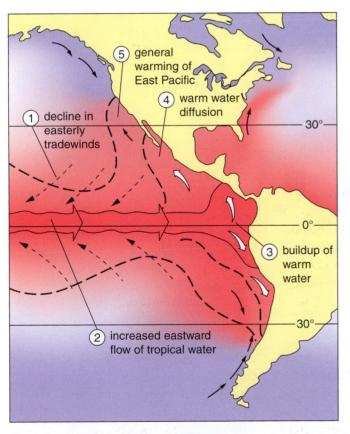

Figure 5.44 The geographic pattern of El Niño ocean warming along the western coasts of the Americas. Temperature increases can reach as high as 6°C and the added energy fuels more and larger midlatitude storms with large increases of winter precipitation.

The El Niño phenomenon is initiated by changes in atmospheric pressure in the eastern Pacific, which result in a decline in the easterly tradewinds. When the tradewinds weaken, the eastward-flowing equatorial counter-current (see Figure 5.37) rises because of reduced wind resistance, pumping thousands of cubic kilometers of warm water into the eastern Pacific. As El Niño builds up, warm water spreads outward into the Pacific and poleward along the west coasts of the Americas, where it displaces the cooler waters of the California and Humboldt currents.

One of the important manifestations of the El Niño-induced weather variation, called the **Southern Oscillation**, is a dislocation of precipitation patterns resulting in dry conditions in some areas that normally receive substantial precipitation, and wet, stormy conditions in some areas that normally receive light precipitation. Another condition related to El Niño is La Niña (girl child), the cold phase of the El Niño cycle. La Niña often appears in the years between El Niño events, and is characterized by cooler than average ocean water, and drier, cooler atmospheric conditions. La Niña is associated with drought conditions in some areas. The severe drought of 1999 in the eastern United Sates, for example, was related to La Niña conditions in the Pacific Ocean along the American west coast. Water temperatures in July and August were as much as 5°C cooler than average, which sharply reduced moisture flux into the atmosphere, thereby limiting the supply of vapor for precipitation. The 1999 event was the most severe drought in the American east in the twentieth century.

Are El Niño/La Niña phenomena limited to the Pacific Ocean? Apparently not, for similar changes have also been documented in the Atlantic and Indian oceans. The one in the Indian Ocean produces shifts in ocean temperature between east Africa and Indonesia, but they do not coincide with the El Niño/La Niña shifts in the Pacific. Like the Pacific, however, the Indian Ocean shifts bring on major weather changes, notable drought with cooling and rainy conditions with warmer temperatures.

Chapter Summary and Overview of The Great System of Global Air and Ocean Circulation

Were it not for atmospheric and oceanic circulation, Earth would be quite a different planet indeed. Because most of the Earth's energy input is concentrated in the tropics, about half the Earth would be ghastly hot and the remainder severely cold. But thanks to the great systems of winds and currents, huge amounts of energy are constantly being redistributed to higher latitudes, as the summary diagram suggests, giving Earth a more gradual climatic transition from Equator to poles. In the course of this zonal redistribution, the wind and current systems also move heat and moisture east and west giving rise to markedly different climatic conditions on different sides of ocean basins and landmasses. The atmospheric and oceanic circulation systems are closely interrelated and the changes we experience in weather and climate commonly involve mechanisms tied to exchanges between the two.

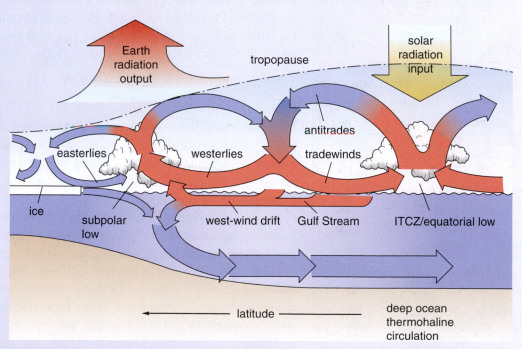

Earth's general energy system with warm and cold flows in both the atmosphere and ocean. The system begins on the right with the input of solar energy and moves via winds systems and ocean currents poleward ending with energy output in the form of Earth radiation.

▶ **There is a large zonal imbalance in energy inputs and outputs to and from Earth's atmosphere.** In order to balance out this difference, it is necessary for the atmosphere and oceans to transfer massive amounts of energy poleward from the tropics.

▶ **Air behaves as a highly mobile fluid which is acutely responsive to pressure differences in the atmosphere.** It flows with the pressure gradient but its direction is influenced by the Coriolis effect at most latitudes.

▶ **Atmospheric circulation is driven by variations in air pressure.** These pressure variations are caused by differences in atmospheric heating and by complications in the fluid flow. At the broadest scale, these factors give rise to belts of pressure that encircle all or part of the globe.

▶ **Atmospheric pressure is measured in millibars and pressure readings on isobaric maps are standardized to sea level.** The circumglobal pressure belts are semi-permanent features that last for months or more whereas small transient pressure cells such as thunderstorms cover only a few square kilometers and last only 2 to 8 hours.

▶ **For surface wind the Coriolis deflection is partially offset by friction with the Earth's surface.** But for winds aloft, where there is no surface friction, the flow is parallel to the isobars.

▶ **As circulation systems, high-pressure cells and low-pressure cells produce two completely opposite patterns of airflow.** In cyclones in the northern hemisphere, surface winds flow in a convergent pattern with a counterclockwise whirl. In the southern hemisphere, because of the Coriolis effect, the whirl direction is the opposite.

▶ **For the Earth as a whole, the atmosphere is composed of seven great belts of pressure systems.** Poleward of the equatorial low each hemisphere has three belts of alternating high and low pressure and each is tied to a prevailing wind system on the surface and another wind system aloft.

▶ **The general system of global pressure and prevailing winds are a major control on Earth's climates.** This includes many of the seasonal characteristics of climate, including the shifting of the intertropical convergence zone (ITCZ) and the migration of the polar front.

▶ **Air pressure and wind are affected by the differences in the heating of land and water.** This is the source of sea and land breezes in coastal areas and, at a much larger scale, causes the monsoonal flows over south and southeast Asia.

▶ **Circulation in the upper atmosphere is characterized mainly by zones of concentrated airflow.** The polar jet stream follows the meandering pattern of Rossby waves corresponding to the polar front on the leading edge of global cold air.

▶ **Secondary circulation systems include midlatitude cyclones and tropical disturbances.** Midlatitude cyclones develop along the polar front and tropical disturbances develop over warm ocean surfaces and may lead to much larger tropical storms.

▶ **The oceans play an important role in determining the weather and climate trends and cycles.** Because of their high heat capacity and geographic circulation, oceans redistribute significant amounts of heat north and south and east and west. Many coastal areas exhibit climates closely related to these heat transfers.

▶ **Over water, wind is the principal force driving waves and currents.** Current systems conform to prevailing wind systems and, like winds, they are influenced by the Coriolis effect.

▶ **Earth's large landmasses deflect the general flow of currents driven by the prevailing winds.** Large circular current systems are formed within the subtropics, equatorial, and subpolar zones that carry heat poleward.

▶ **Deep-water circulation in the ocean is driven by differences in water temperature and salinity.** Thermohaline circulation operates at depths greater than 2000 meters with a very long cycle period.

▶ **Thermal anomalies are part of the ocean–atmospheric system.** They influence weather patterns and the one known as El Niño periodically influences weather well beyond the ocean basins.

Review Questions

1   Is the heat produced from solar radiation evenly distributed over the Earth? If not, where is most of it concentrated? How is this reservoir of heat redistributed and where does it go? Does it make the planet milder or more severe climatically? Why?

2   Describe the change in Earth's rotational velocity from the Equator to the poles. What are the rotational velocities of the Earth and its atmosphere at the Equator and at higher latitudes, and how do these differences in velocity influence the flow of wind from the Equator to the subtropics?

3   Use the following sequence of events to sketch a Hadley cell between the Equator and 30° north latitude: (a) uplift of warm air at the Equator; (b) divergence of this ascending air in the upper troposphere; (c) descent of the air mass at 30° north latitude; and, the return of the surface airflow back to the Equator. What would be the direction of the resulting tradewinds?

4   Referring to Figure 5.7, how does air pressure vary with increasing altitude, and how is air pressure represented on maps?

5   Wind systems occur at several scales from the global to the local. Describe several of the subsystems that operate within the global wind system; specifically, the different types of pressure cells and the geographic scales associated with each type. Use Figure 5.6 for reference.

6   What is the Coriolis effect and what is its influence on wind direction north and south of the Equator? At the Equator? What is the difference in the direction of surface winds and upper-level winds relative to the isobars on a pressure map? What is a jet stream?

7   Examine Figure 5.18. What is the polar-front jet stream and what does it have to do with zonal energy transfer on Earth? How does this jet stream change seasonally?

8   How do you account for the dramatic differences in the annual temperature range of San Francisco and St. Louis? What is it about the thermal characteristics of land and water that produces the day–night changes in local wind on the summer shore?

9   What is the intertropical convergence zone (ITCZ)? Why does it shift with the seasons and what is its relationship to the monsoons in Asia? Where is the ITCZ during the winter monsoon?

10   Referring to the map in Figure 5.29, can you describe the logic of Columbus's routes to and from the New World? What wind systems did he utilize? What current systems?

11   What is meant by long-distance atmospheric transport? What wind systems are involved, what is transported, and how far is it moved? How do you think the atmosphere cleanses itself of these materials?

12   Describe how ocean currents are generated and how it is possible for water beneath the surface layer to move. What is the Ekman spiral and what is meant by Ekman transport?

13   Using Figure 5.37 for reference, trace the gyre in the mid North Atlantic. Identify the sections of this gyre where there is a cold current and a warm current. What are the major wind systems that drive this gyre?

14   Describe the Gulf Stream as an energy system in terms of inputs, transport, and outputs with attention to where geographically these parts of the system are located and what effect the system has on the climate of northwest Europe.

15   Referring to Figure 5.40, identify the primary thermal layers of the ocean, and describe their role in the thermohaline circulation. How do ice exclusion and ocean evaporation contribute to the deep ocean circulation system and how far south does this system carry the water it picks up in the North Atlantic? What is the role of deep ocean circulation in the ocean energy system?

16   What is the equatorial counter-current and what is its connection to El Niño? What changes take place in the ocean that mark an El Niño event. How is the Southern Oscillation related to El Nino?

Atmospheric Moisture, Precipitation, and Weather Systems

Chapter Overview

The atmosphere is a complex system, sometimes described as chaotic in nature. In this chapter we examine one of the principal components of that system, the precipitation system, and find that it is indeed complex, but within that complexity, there is a good deal of order to be found in its processes and patterns. We want to learn how precipitation is produced and how and where it is delivered. We trace our way through the atmospheric moisture system, beginning with a brief examination of water vapor, humidity, and condensation and then go on to the processes and causes of precipitation and their geographic circumstances, including the nature of storm systems such as thunderstorms, tornadoes, and hurricanes. The chapter ends with some insights into the nature of violent storms and some of their consequences.

Introduction

The town was nothing more than a general store and saloon on a side road a few hundred feet off US-21 in northern Nebraska. A half mile or so to the north you could see a few houses and beyond them endless farm fields and plains stretching into the horizon. Thunderstorms with great billowing heads dominated the hot afternoon sky creating shadowy islands in an otherwise sunny landscape. Inside the store we milled around looking at antiques and chatting with the clerk. Suddenly, a siren sounded. Almost immediately the clerk said, "It's a severe storm warning, probably a thunderstorm, we get them all the time." The noise was deafening, but she continued to talk though forced to use a strained, elevated voice. "If it's a short whistle, it's a storm; if it's a long whistle it's a tornado. Never had a long whistle." She kept talking and I pretended to be listening, but I was really wondering about the measure of a short and long whistle.

After two or three minutes, she stopped talking. Her face changed expression and she shouted, "Oh, my God! It's a tornado whistle! What do we do?!" "Do you have a basement?" I asked. "Yes, but it's filled with stuff…and the door's locked…and I don't have a key…the owner's got it." The siren was still screaming and from the saloon across the street, a woman rushed out shouting, "Where do we go, where do we go?" By now gusts of wind were driving dust, leaves, and other debris down the street around her. From the middle of the street she continued her frantic yelling but her words were lost in the rising noise of the air.

As I stepped around the southeast corner of the store, a blast of wind caught me in the face. There it was, off to the southwest, a dark, gray mass hanging to the ground from the base of a huge cloud. No funnel was visible but, tornado or not, the storm was big and rough. From behind me, Alison said, "Will! What are we going to do? There's no tornado shelter around here." By now I was trying to recall the weather map I saw on television that morning, and I told her to head for the car. Inside the car, I tried to explain. "First, the storm looks a lot closer than it is. It's probably 5 or 10 miles away." But the look on her face told me she didn't believe me. "Second, it's probably connected to a front that early today lay along the highway we're traveling on and by now should be well to the south of our route. I know it's risky, but we have no choice but to try driving past it. The highway should take us to the north of the storm." "Are you sure?" she asked. "No, or would you rather sit it out in one of these old, wooden buildings?" She looked at the store's rickety porch and said, "Let's go."

We turned west onto the highway while keeping watch on the storm. The road took a few gentle curves but kept a solid westward bearing for a mile or so and my confidence began to rise. As the wind and rain became more intensive and the road darkened, we slowed down just in time to negotiate a major turn to the south, toward the storm. "Damn Thomas Jefferson and his rectangular land survey," I thought. Things got worse; the winds buffeted the car and the tires slipped on the sheets of water whipping over the pavement. Alison shouted, "Should we turn back?" "No, the road's got to make a turn westward in a mile or so." But I was gambling and a minute or so later I breathed a sigh of relief as the road swung around to the west. After another 5 or 10 intense minutes, we emerged from the storm into sunny skies and were greeted by a line of eastbound cars and trucks, waiting for the storm to clear the highway.

"I don't think we should have done that, do you?" "No," I replied, "but you've got to admit it was some adventure. Nature gave us a light jab, and we're lucky we weren't hit by a knock-out punch. There was probably one waiting for us somewhere in there." But we both knew I was speaking with bravado motivated by a safe escape. We never did learn if there was a tornado churning inside the storm. But after the adrenaline rush, it never really mattered that much.

6.1 The Water-Vapor–Precipitation System

In Section 4.6 of Chapter 4 we talked about the role of water vapor in the exchange of energy between the Earth's surface and the atmosphere. We are now interested in the water-delivery component of this exchange system, which is a subsystem of the great hydrologic system. Figure 6.1 illustrates this system in three basic phases or components, beginning with the vaporization of surface water and ending with that water returning to Earth as precipitation. This model is a good opening for our discussion here.

The entire system of water vapor and precipitation is powered by heat generated by solar radiation. Without this heat, there is no way to drive water off the Earth's surface and into the atmosphere, and without water vapor in the atmosphere, there is no raw material, as it were, for the manufacture of precipitation. But in order for water vapor to become rain or snow, it must first be converted into a liquid or solid form. This process, known as **condensation**, requires cooling and the product of condensation gives us our first visual evidence of the system in action, namely, clouds.

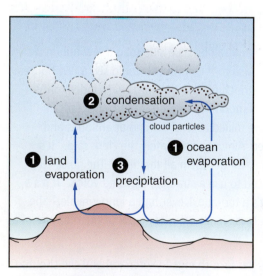

Figure 6.1 The three main processes of the atmosphere's moisture system

Clouds are made up of minute water particles, both liquid and solid. Curiously, most clouds never produce precipitation. First, because the particles in most clouds are too small to fall to Earth, and second, because many clouds simply disappear hours or days after they form as their water particles evaporate into thin air. For the clouds that do yield precipitation, condensation must advance and add enough water to cloud particles that they become too heavy for the atmosphere to support and hence fall to Earth.

Water Vapor: To understand how and why clouds and precipitation form, it is necessary to examine the nature of water vapor. First, recall that water vapor is the most variable gas in the lower atmosphere, ranging from effectively 0 to as much as 4 percent of air. Thus, some parts of the Earth, such as the great desert regions, have practically no vapor in the lower atmosphere whereas others, such as the North Pacific, have so much vapor that clouds are constantly forming and blanketing the sky. Second, vapor is one of three natural phases of water on Earth, each phase contains heat energy and, when water changes phase, energy is exchanged. The two most important phase changes are represented by evaporation and condensation. About 25 percent of the solar energy delivered to the Earth's surface is used in evaporation and the vast majority of the vapor comes from the sea.

Third, it is important to understand that the maximum amount of water vapor air can hold is dependent principally on air temperature. The warmer air is, the greater the amount of water vapor it can hold; conversely, the colder air is, the smaller the amount of water vapor it can hold. This helps explain why – considering the range of temperatures that exist in the atmosphere – the water-vapor content of the atmosphere is so variable. When a body or parcel of air is holding the maximum amount of water vapor for its temperature, it is said to be *saturated*.

Fourth, like all atmospheric gases, water vapor is part of the atmospheric circulation system and thus subject to transfer by wind systems over great reaches of the planet. This is a very efficient system and an important part of the Earth's larger energy system, because as vapor is moved from tropical oceans to midlatitude land areas, for example, huge amounts of energy are also moved. This energy is released as heat when the vapor condenses to form cloud and precipitation particles. To give you an idea of how much energy can be transferred by the water-vapor system, an average-sized hurricane in one day releases an amount of energy equivalent to 200 times the daily worldwide electrical generating capacity.

Measures of Humidity: The water-vapor content of air is referred to as **humidity**. Generally, when we say that air is humid we mean that its vapor content is high. But what this means in terms of the actual moisture content is uncertain unless we specify what measure of humidity we are using. There are several in use and we want to familiarize ourselves with three standard expressions of humidity: absolute humidity, specific humidity, and relative humidity.

Absolute humidity is a measure of the weight (in grams) of the water vapor in a parcel of air with a volume of one cubic meter. Absolute humidity is a useful means of expressing the water content of a large body of air, such as the great air masses that migrate across the midlatitudes, where we are interested in how much moisture is available for precipitation. The maximum amount of vapor that can be held by a

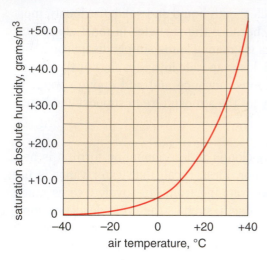

Figure 6.2 The moisture-holding capacity of air is mainly a function of temperature. At 30 °C it can hold six times more vapor than at zero.

cubic meter of air is called saturation absolute humidity, and it increases rapidly with air temperature as the graph in Figure 6.2 illustrates. **Specific humidity** is a measure of the weight of water vapor (in grams) to the weight of the air holding the vapor (including the weight of the vapor). This expression is widely used in weather analysis because the moisture content of air can be evaluated regardless of changes in its volume.

Relative humidity, the most commonly used measure, expresses vapor content in grams relative to the amount of vapor that can be held in the same air when it is fully saturated with vapor. If, for example, a parcel of air contains 5 grams of vapor but its capacity at saturation is 10 grams, then the relative humidity is 50 percent. In other words, the air is 50 percent loaded with water vapor. But since air's vapor-holding capacity changes with its temperature, it is necessary to specify the temperature at which a humidity reading is taken.

Relative humidity can be changed by varying either the temperature or vapor content of air. If we hold air temperature constant, relative humidity rises when vapor is added. If vapor content is held constant, relative humidity goes up when air temperature is lowered and down when air temperature is raised. When air is at 100 percent relative humidity, it is said to be saturated. The temperature at which it reaches saturation is called the **dew point**. Dew point is commonly given in evening weather reports along with the forecasted low nighttime temperature. This can be valuable information for if the night temperature dips below dew point, ground fog and dew can be expected.

Actually, relative humidity can exceed 100 percent under certain conditions. Such air is called *supersaturated* air, and it typically involves dramatic cooling of air, which is resistant to condensation. This is usually very clean air with limited capacity to condense into cloud particles, which we will explain later in Section 6.4.

Air Movement and Moisture Condensation Processes: Condensation is the physical process by which water changes from the vapor to the liquid phase. Under most circumstances, condensation occurs when air is cooled below its dew-point temperature, which is the temperature where the air would be saturated and contains all the water vapor it can hold. The simplest way air can be cooled is for it to pass over a colder surface. This is what happens when you open a freezer and warm air rushes over the cold surface. It cools instantly, briefly forming a light cloud of condensation near the door. In the atmosphere, this process of warm air cooling as it falls into contact with a cool surface is known as **advection**, and it is very common in coastal areas and snow-covered areas crossed by warm winds.

On the west coast of North America, coastal fog often originates when warm, moist air from the Pacific Ocean drifts westward over cold water driven southward by currents along the coast (Figure 6.3). The fog, which is called advection fog, is carried onto the land by the westerly flow, but within a kilometer or so inland, the air is warmed and the fog usually evaporates. Given the reverse arrangement with air passing from warm water onto cold land, condensation and cloud formation are the result. During the winter in the Great Lakes region, for example, air blowing across the relatively warm water surfaces such as Lake Michigan (at about +1 °C), picks up a load of vapor from the lake, then cools downwind as it passes over the frozen land (at about –10 to –15 °C) on the opposite shore, causing condensation and heavy snowfalls on the east side of the lake as shown in Figure 6.4.

Coastal fog being carried onto the land from the Pacific Ocean.

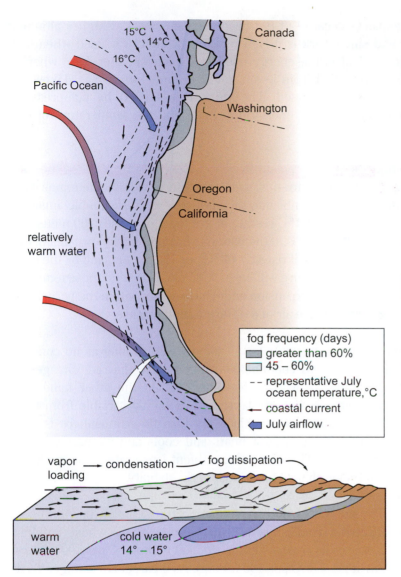

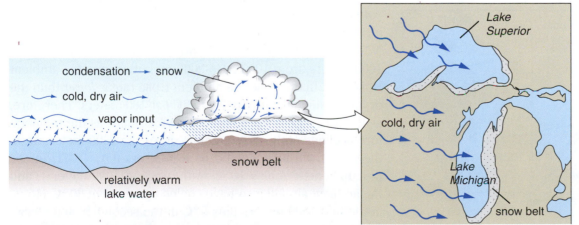

Figure 6.3 The incidence of advection fog along the US West Coast related to airflow over cold coastal water. See adjacent page for a photograph of this type of fog.

But the cooling processes that induce most precipitation involve rising air resulting from some form of **convection** like we see in a thunderstorm. It may rise spontaneously because it is heated over warm ground and like a hot air balloon lifts up because it is lighter than surrounding air; or it may be forced aloft as an air mass passes over a mountain range; or it may be forced aloft when it meets a denser air mass or collides with opposing winds. In all such cases, the rising air undergoes pressure changes that lead to cooling and often to cloud formation and precipitation. The spontaneous ascent of air in the atmosphere is termed **instability** and we want to examine this atmospheric condition in connection with condensation and precipitation.

6.2 Atmospheric Instability

The concept of atmospheric instability can be illustrated using a model called the parcel method. This model takes a large bubble of air, called a parcel, and measures its changes when it moves upward into the atmosphere. The model assumes that no mixing takes place between the parcel and the surrounding air.

The vertical motion of a parcel depends on its mass (weight) relative to that of the of the air around it. Think of the parcel as a mass of fluid within a larger fluid mass, and if it is lighter, it rises, meaning it is **unstable**. If it is equal or heavier than the surrounding fluid, then it either stays still (if equal) or sinks (if heavier), meaning it is **stable**. Two factors determine which of these conditions will prevail: (1) the rate of temperature change inside the moving air parcel, called the **adiabatic lapse rate**, and (2) the rate of temperature change at corresponding elevations in the air through which the parcel is moving, called the **ambient atmospheric lapse rate**. We use temperature instead of density in problems of atmospheric instability for two reasons. First, air temperature changes directly with changes in density, and second, air temperature is so easy to measure compared to density. When the parcel has a higher temperature than the surrounding air, it is lighter and rises; likewise, when the parcel is cooler than the surrounding air, it is heavier and sinks.

Figure 6.4 Advection and snowfall related to Lake Michigan. As air moves across the lake it picks up water vapour, which condenses over the cold eastern shore.

When air rises, it expands because of the decreased pressure at higher altitude. Expansion is work and since work is driven by energy, energy is used whenever expansion takes place. Heat is the energy used in expansion; therefore, when a parcel expands heat is removed from its air, and its temperature falls. Likewise, when air descends, it is compressed. The work represented by compression results in the conversion of kinetic energy into heat, and the temperature of the air rises.

Dry Adiabatic Cooling: The rate of cooling in rising, cloudless air is called the **dry adiabatic lapse rate,** and is equal to –1°C per 100 meters. We use the minus sign to indicate that the air cools as it climbs into the atmosphere. For example, imagine a parcel of air at a temperature of 20°C resting at sea level. If it rises 700 meters, it will cool 7°C to a temperature of 13°C. If the parcel is brought back down again, the adiabatic lapse rate reverses, that is, it rises by +1°C per 100 meters, and by the time it gets back to sea level, its temperature will be 20°C again. Note that these temperature changes are a result solely of expansion and compression within the parcel, *not* of interchange (mixing) with outside air of a different temperature.

The dry adiabatic lapse rate is the *same* for all air parcels, as long as no condensation takes place within them. (If condensation takes place the rate changes, a condition we will discuss shortly.) On the other hand, the ambient atmospheric lapse rate *is* highly variable from time to time and place to place because the atmosphere is constantly changing as it heats and cools and as masses of air shift about. Therefore, the ambient lapse rate turns out to be the major control on atmospheric stability.

To understand this fact, imagine the situation in Figure 6.5 involving a parcel of air surrounded by air whose ambient lapse rate is *greater* (that is, cooler) than –1°C per 100 meters. Say it is –1.5°C per 100 meters. As a parcel rises and cools at the dry adiabatic rate (–1°C/100 m), the air around it is always cooler than the air inside the parcel. Since the parcel of air is warmer, it must be lighter, and must rise. After rising 1000 meters, the parcel will have cooled only 10°C (say, from 30 to 20°C) but the air around it will have cooled 15°C (say, from 30 to 15°C). Being warmer than the air around it, the parcel will continue to rise.

Now let us go to Figure 6.6 and take the parcel up another 1000 meters, from 1000 m to 2000 m, but change the ambient lapse rate. In this case the ambient lapse rate is *lower* than the dry adiabatic rate, say, only –0.5°C per 100 meters. Therefore, as the parcel goes up, it will cool twice as fast as the temperature declines in the surrounding air. At the 2000 meter mark, the parcel will have cooled by a total of 20°C (10°C in the first 1000 meters plus 10°C in the second 1000 meters), and the ambient air will also have cooled by a total of 20°C (15°C in the first 1000 meters plus 5°C in the second 1000 meters). Inasmuch as both the air parcel and the air around it are now at the same temperature, they must also be of the same

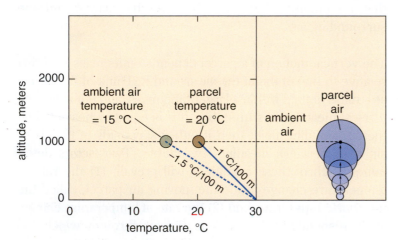

Figure 6.5 Dry adiabatic cooling of unstable air. The rising parcel is unstable at 1000 m altitude because it is warmer than the air around it.

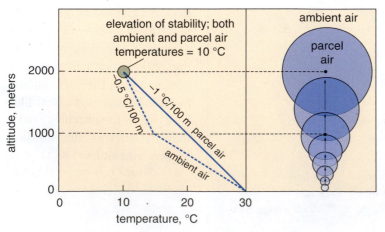

Figure 6.6 Dry adiabatic cooling leading to stability at 2000 m where parcel and ambient air temperatures are the same.

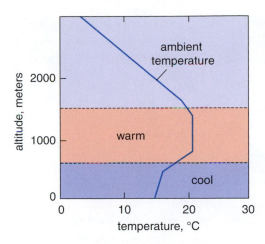

Figure 6.7 Temperature inversion. A common thermal structure that prohibits vertical mixing of the lower atmosphere.

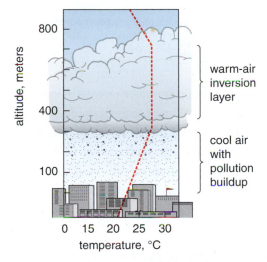

Figure 6.8 Temperature inversion over a city, leading to pollution buildup under the inversion layer.

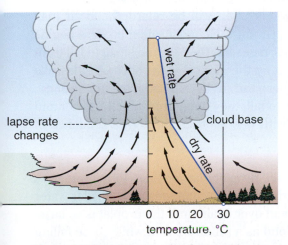

Figure 6.9 Wet adiabatic cooling begins with condensation and is defined by the elevation of cloud formation.

densities. Thus the condition is *stable* and the parcel stops its ascent and blends into the surrounding air.

This illustration of instability and stability in a 2000 meter thick layer of atmosphere is helpful in understanding a form of extreme stability called a **temperature inversion**. In a temperature inversion the ambient atmospheric lapse rate reverses from the surface upward (see Figure 6.7), that is, goes from cool to warmer rather than from warm to cool as is the norm. With cooler (denser) air below and warmer (lighter) air above, no vertical mixing is possible. If an air parcel were to ascend through the lower (colder) layer, it would quickly come to a halt upon entering the warm layer because there the air would be lighter.

Temperature inversions are a common condition in most places and most seasons. They are caused by a number of mechanisms; for example, (a) cool air sliding under a warmer body of air, (b) warm air sliding over a cool surface such as snow, and (c) cooling of ground-level air at night with the outflow of longwave radiation. Since thermal inversions restrict atmospheric mixing, prolonged inversions can be dangerous over urban areas because pollution gets trapped under them and builds up there, as Figure 6.8 shows, creating a health hazard. This was an enormous concern in nineteenth- and twentieth-century coal-burning cities like London, and remains so today over much of the world with some additional considerations. Today's cities are much bigger and the automobile has added significantly to the air-pollution problem.

Wet Adiabatic Cooling: All the above examples of atmospheric instability assume the air parcels are completely dry. What happens if we add moisture to the air and it condenses at some point as the parcel cools? Say it cools to its dew point and condensation sets in? Cloud formation will result, but more important to our discussion, when the water vapor condenses, heat will be given off, causing warming of the air parcel.

The effect of this charge of heat on the behavior of the parcel is to make it more unstable by reducing (that is, slowing) the rate of adiabatic cooling. The saturated air will still cool as it rises, but at a lower rate, called the **wet** (or **saturated**) **adiabatic lapse rate**. The wet adiabatic lapse rate varies depending on the amount of condensation taking place, but a value of −0.6°C/100 m fits most situations. The wet adiabatic lapse rate applies only to air that is saturated, so it is important to remember that even very humid air with a relative humidity as high as 90 percent is considered "dry" until it reaches its dew point. When you spot clouds forming, as shown in Figure 6.9, the cloud base marks the change from the dry to the wet adiabatic rate.

Because much of the water vapor that condenses will fall from the air as precipitation, the wet adiabatic lapse rate, unlike the dry, is generally not reversible. Therefore, air that is lifted beyond its elevation of saturation (1000 meters in the diagram in Figure 6.10) and releases its water vapor will be *warmer* if it is returned to its original elevation. This will happen because the air is dry (non-saturated) on its descent and therefore warms at the dry adiabatic rate (+1°C/100 m), which is nearly a half degree greater per 100 meters than its wet cooling rate on the way up.

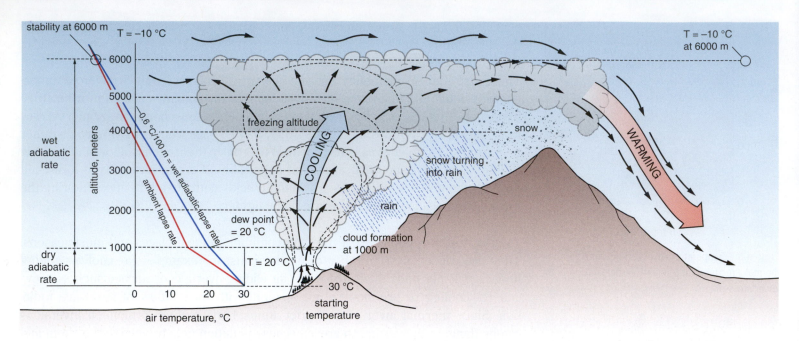

Figure 6.10 An adiabatic cooling and heating system with the movement of air over a mountain range. Cooling and precipitation occur on the windward side. Adiabatic warming and dry conditions occur on the leeward side.

Summary on Humidity, Cooling, and Condensation: The atmosphere receives its moisture from the Earth's surface, holds most of it as water vapor, and returns it to the surface as precipitation. For precipitation to form, the water vapor must be cooled to its condensation temperature. This usually takes place when air rises, and through adiabatic cooling, loses heat energy as it expands. It is important to remember that this process produces more than precipitation; it also releases massive amounts of heat into the atmosphere. We are led now to the questions of what causes air to rise and how do precipitation particles form?

6.3 Condensation and Precipitation Processes

So cooling leads to saturation and condensation, but how does this result in precipitation? In the atmosphere, condensation takes place on **condensation nuclei**, very small particles of dust, salt, or other matter suspended in air. If air is very clean, that is, nuclei particles are largely absent, condensation will be retarded and air may become supersaturated. When nuclei are present the condensation process begins with the formation of minute liquid or ice particles but they are too small to fall from the atmosphere and are held in suspension by the motion of the air molecules. We see this in a fog (a ground-level cloud) where the particles float above the ground. In order for the cloud particles or droplets to fall, they must grow much heavier and they do this by colliding and coalescing with one another, through a sequence of events called the Bergeron process.

This process requires the presence of water droplets larger than 40 micrometers in diameter. As a large droplet falls through a cloud, it combines with smaller water droplets in its path. This is repeated over and over again until the droplet is so large that it is heavy enough to fall out of the cloud as a raindrop. And while these falling drops are typically less than one millimeter in diameter, they often represent a growth in size by a factor of 1,000,000 from the time they began.

Figure 6.11 Photograph of a storm cloud producing precipitation; snow in the upper cloud melting to rain at lower levels.

Recent advances in cloud physics indicate that most precipitation falling to Earth's surface originates as ice (snow) crystals. According to the Bergeron process, named after the Swedish scientist who first suggested it, supercooled liquid water droplets (water that has been cooled below the freezing temperature, but is still in liquid form) collide and coalesce with minute ice crystals. As the ice (snow) crystals grow larger, they eventually become so heavy that they fall out of the cloud. If the air below the cloud is above freezing, the snowflakes melt and fall as raindrops (see Figure 6.11). This process is considered to be the primary mechanism for droplet formation outside of the tropics.

Precipitation Forms: The physical state that precipitation particles assume when they arrive on the Earth's surface is called the **form** of precipitation. Most precipitation falls in the form of rain, meaning that it arrives at the ground in liquid form. Rain may begin as liquid water or as frozen water that melts while falling. However, if the dew point in the condensing air is below freezing, then water vapor is transformed directly into snow crystals, and in cold weather, may fall to the ground as snowflakes little changed from their cloud forms. If small amounts of liquid water condense onto the snow crystals as they fall, the snow is said to be *rimed*. When a snow crystal collides with a raindrop, a small frozen ball, or *graupel*, is formed. *Sleet*, on the other hand, is rain that freezes as it falls. *Hail*, commonly the largest precipitation particle, forms under conditions of strong atmospheric turbulence. A small snow crystal drops into the zone where temperatures are above freezing, collects additional water, is lifted into the frozen zone again, descends once more, is swept aloft yet again, and so on over and over. In this fashion hailstones can grow to large sizes, sometimes several centimeters in diameter, with a structure comprised of concentric rings of ice like those in Figure 6.12.

Precipitation Types: Virtually all precipitation results when air cools by rising through the atmosphere due to some form of instability. It follows that we classify precipitation according to what causes the air to rise and cool. There are four principal causes or mechanisms that drive the precipitation process, and they produce four **types** of precipitation, each of which constitutes a precipitation system:

Figure 6.12 The concentric ring structure of hailstones.

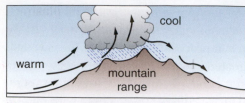

(a)

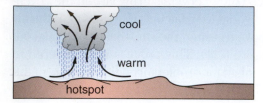

(b)

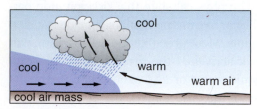

(c)

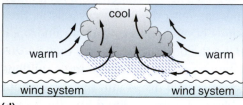

(d)

Figure 6.13 Four precipitation systems:
(a) orographic, (b) cyclonic or frontal,
(c) convectional, and (d) convergent, and the airflow
associated with each.

- **Orographic precipitation**: Caused by airflow over high terrain, usually a mountain range along which air is forced upwards and cooled (Figure 6.13a).
- **Convectional precipitation**: Caused by the free rise and cooling of unstable surface air, usually due to the intensive heating of air near the ground (Figure 6.13b).
- **Cyclonic/frontal precipitation**: Caused by the meeting of air masses of different densities, in which warm air is displaced upward by colder air resulting in cooling (Figure 6.13c).
- **Convergent precipitation**: Caused by barometric or topographic sinks, where air is drawn into large low-pressure areas and then rises and cools (Figure 6.13d).

Each represents a different precipitation system, but all four types involve the release of latent heat with condensation and cloud formation. In fact, it is latent heat that powers these systems, that is, provides the main force driving storm development and precipitation processes in each type of precipitation.

6.4 Orographic Precipitation: Patterns and Processes

Orographic precipitation is the easiest type to describe. When moisture-laden air is forced to pass over a mountain range, inducing cooling and condensation, precipitation is produced. In areas where mountains lie in the paths of moist, prevailing wind systems such as the westerlies or the tradewinds, orographic precipitation rates can be extremely high, the highest in the world. Virtually all areas with rainfall above 500 centimeters (200 inches) per year are orographic situations. Examples of areas of heavy orographic rainfall are the mountains of Hawaii which lie in the tradewind belt, the Himalayan mountain front which lies in the path of the summer monsoon, and the mountainous west coasts of North and South America which lie in the belts of the prevailing westerlies.

Here are a couple of extreme examples. The town of Cherrapunji in the Assam Hills of northeastern India lies on the windward fringe of the Himalayas. Over one year, between October 1, 1860, and September 30, 1861, Cherrapunji recorded an astounding 26.5 meters (87 feet) of rain with 8.9 meters (350 inches) falling in July alone! In Hawaii, on the island of Kauai, northwest of Honolulu, is one of the highest precipitation rates in the world, 1234 centimeters (486 inches) per year (see Figure 6.14). This occurs on the summit of Mt. Waialeale (1569 meters elevation) which, like the entire Hawaiian chain, lies in the path of the northeasterly tradewinds. Only 52 kilometers (32 miles) away on the leeward shore on the island, annual rainfall is less than 50 centimeters (20 inches), comparable to some of the world's great deserts.

Most cases of extremely heavy orographic rainfall result from a combination of the forced rise of air as it passes over the mountain barrier and thermal convection caused by the release of heat with condensation and cloud formation. When massive amounts of latent heat are released, thunderstorms may form, building up thousands of meters over the mountain range. Orographic precipitation in the midlatitudes is often associated with an additional condition; the movement of large air masses from the ocean onto the land. In the Pacific northwest, for example, large bodies of maritime (moist, marine) air, driven by westerly flow over the ocean, sweep against coastal mountain ranges causing episodes of particularly heavy precipitation.

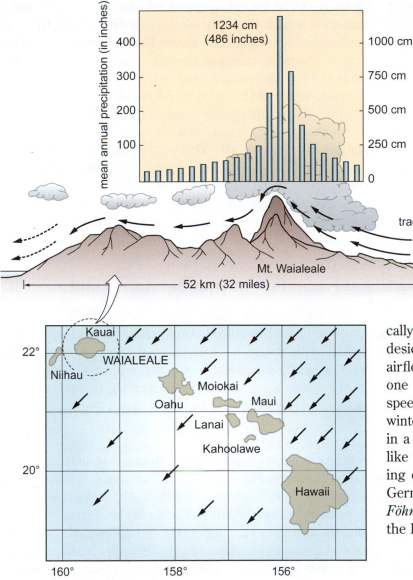

Figure 6.14 Massive orographic precipitation on Mt. Waialeale, Hawaii, one of the heaviest annual rainfall rates in the world.

The diagram in Figure 6.15 illustrates the sequence of precipitation in an orographic system beginning on the windward side with moist air blown against the lower slopes of the mountain range. The air rises, cools, condenses, and is forced even higher by the added charge of latent heat with cloud formation. Heavy precipitation, commonly both rain and snow, fall on the middle and upper slopes as well as across the crest of the mountain range.

The leeward side of the range – that is, the side away from the prevailing wind – is typically quite dry and is termed the **rain shadow**. When the air descends the leeward slopes, it warms adiabatically and can become very hot and dry, resulting in rapid desiccation (drying) of the land. Sometimes the force of this airflow is intensified by strong pressure differences from one side of the mountain to the other, resulting in a high-speed downslope wind. When such winds occur during the winter, they may produce a 20 to 25 °C rise in temperature in a matter of hours. The effects can be dramatic: summer-like weather, rapid melting of snow, avalanches, and sprouting of spring flowers. On the leeward slopes of the Alps in Germany, Switzerland, and Austria such a wind is called a *Föhn*, the German word for hairdryer; on the eastern slope of the Rocky Mountains it is called the *Chinook* or *"snow eater."*

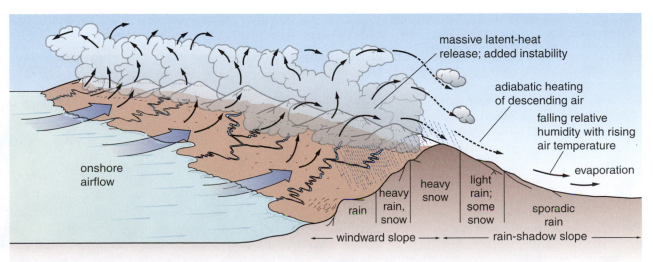

Figure 6.15 Orographic precipitation system along a coastal mountain range such as on the coast of Washington and British Columbia.

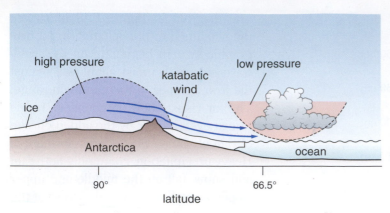

Figure 6.16 Katabatic wind blowing from a zone of high pressure in Antarctica's interior to a low-pressure system over the ocean.

There is also another class of powerful mountain winds, called **katabatic winds**. These are winds caused by cold, high density air draining down mountain slopes like a sheet of water. They may be triggered by strong pressure differences on opposite sides of a mountain range. One of the fiercest katabatic wind systems, depicted in Figure 6.16, is found in Antarctica. It blows from the polar high-pressure cell in the continent's highland interior across the coastal lowlands to the surrounding ocean and into low-pressure cells. This wind is so fierce that it is a main control on the ablation (melting and evaporation) of glaciers, the movement of airborne sediment, and the distribution of life on the continental fringe. Another fierce katabatic wind is the *Santa Ana* of southern California, which dries the landscape and often advances fierce wildfires (see Figure 5.30).

6.5 Convectional Precipitation: Thunderstorms and Tornadoes

Convectional precipitation results from localized **convective storms** that produce gusty surface winds and short, intensive rainfalls that are often accompanied by lightening, thunder, and occasionally tornadoes. Convectional storms can be caused by several mechanisms: the spontaneous heating of surface air, the forced ascent of air along a front or a mountain range, or the ascent of warm, moist air within a hurricane.

Latent heat is the main driver of convectional storm systems. It is delivered to the storm by updrafts, like those shown in Figure 6.17, and then released with condensation in massive quantities and cloud formation. In fact, this charge of energy can be so great that the temperature within the storm cloud may be as much as 10 °C (18 °F) higher than that of the ambient air around it. This produces shattering instability forcing turbulence and cloud development to great heights, 18,000 meters in extreme cases. Such massive development usually produces a thunderstorm, the most violent form of convectional precipitation. Although large thunderstorms are common over much of the world, most convectional precipitation is produced by modest storm systems that rise 5000 to 10,000 meters and produce modest thunder and lightning.

Figure 6.17 The pattern of updrafts in a convectional storm development in the humid tropics.

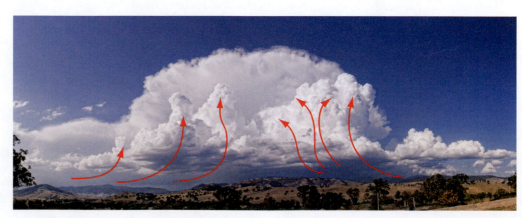

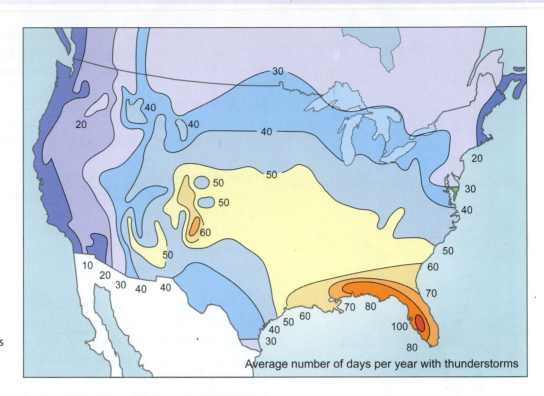

Average number of days per year with thunderstorms

Figure 6.18 The distribution of convectional storms over central North America showing the highest incidence in the American South.

Because convectional systems depend on moist, unstable surface air, they are the most common and largest in humid areas of intensive surface heating. Accordingly, they occur with greatest frequency and stature in the tropics, particularly over land areas with ample moisture supplies such as tropical forests, but with a distinct seasonal rhythm. Beyond the tropics, convectional precipitation is limited mainly to the summer season, and the hotter and closer (more humid) the summer, the more convectional storms. Not surprisingly, convectional activity in the midlatitudes rises sharply in June and subsides in September and, as the map in Figure 6.18 shows, these storms are rare above 60° latitude even in the summer. One more observation: there is a diurnal rhythm to summer convectional activity that follows air temperature with storms rising in the afternoon, subsiding in the evening, and dying away at night.

The size and structure of convectional storms varies enormously. In terms of coverage, individual storms range from 25 to 1000 square kilometers in area. In terms of structure, a storm may be built around a single convectional cell with a central chimney of rising air or around multiple convectional cells with several zones of upward and downward flowing air. Storms may develop as isolated systems, in clusters, or in lines along a weather front and they may be stationary or drift at rates up to 100 kilometers per hour. Thunderstorm and tornado development are possible with any convectional system, although they are more common with large, high-energy systems.

Thunderstorms: A thunderstorm is a convectional storm that emits thunder. Thunder is acoustical energy caused by the explosive expansion of a narrow band of air suddenly heated by a lightning discharge. Most research indicates the power source for lightning is electrical energy that builds up in the cloud from the intensive friction between ice, hail, and graupel, although the actual cause is not yet known. As this electrical field takes shape, positive charges and negative charges develop in different parts of the cloud. The lightning discharge itself is an electrical arc within a cloud, between clouds, or between clouds and the ground.

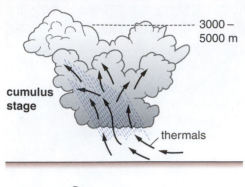

cumulus stage

3000 – 5000 m

thermals

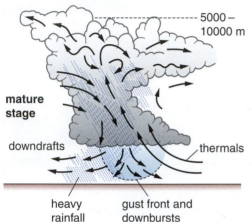

mature stage

5000 – 10000 m

downdrafts

thermals

heavy rainfall

gust front and downbursts

Figure 6.19 The cumulus and mature stages of a convectional storm system.

A great pour of rain discharging from the base of a mature thunderstorm.

Most thunderstorms are classed as **air-mass thunderstorms** because they develop within large, moist high-pressure systems. The storms begin as isolated convectional cells generated by free, or thermal, convection related to local surface heating. The air above is relatively cool which favors instability, and the moisture adds latent heat to drive the storm above cloud-base elevation. These thunderstorms are often described as having three stages of development, cumulus, mature, and dissipating, and most are short lived, lasting only four to six hours.

The **cumulus stage**, shown in Figure 6.19, is characterized by strong vertical development with warm air currents, called *thermals*, sweeping upward into the convective cell at speeds as high as 125 kilometers per hour. Cloud formation begins between 300 and 3000 meters altitude. Above this level, latent heat drives the development of the storm. Precipitation droplets form quickly, but their fall is delayed by powerful updrafts within the cloud.

In the **mature stage** the storm grows rapidly and reaches its full height (Figure 6.19). The cloud billows out to form a huge, anvil-shaped head, typically several kilometers in diameter or even wider in multicellular storms. Internal circulation is complex, with downdrafts as well as updrafts occurring simultaneously in different parts of the storm. Lightning and thunder are usually abundant in this stage. Rainfall is typically intense, often beginning with a downdraft blast of cool air, called a *gust front*, accompanied by hail. Heavy rainfall often lasts only minutes and may fall in a series of short bursts interspersed with lighter rain.

Some mature-stage thunderstorms develop exceptionally powerful downdrafts, called *downbursts* that can reach 200 kilometers (120 miles) per hour. *Microbursts* are highly concentrated downburst streams that can create damage comparable to tornadoes. Microbursts are ranked as the second leading cause of aircraft accidents (after pilot error).

The **dissipating stage** marks the waning hours of the storm. Rainfall is light, thunder is reduced to distant rumbles, and the lower part of the cloud is dominated by downdrafts with occasional light rainfall. By this time, the top of the cloud has often been blown downwind by winds aloft, and the entire storm cell may have drifted miles from its place of origin.

Thunderstorms are also triggered by other mechanisms, including sea breezes rushing onto a heated landscape and warm air forced up along a cold front. Another variety of thunderstorm is a **megastorm** or **mesoscale convective complex**. These are huge thunderstorm clusters that can cover half or more of a midwestern state. They are very powerful and, unlike most thunderstorms, can last through the night and resume activity the next day with a fresh supply of warm surface air.

Tornadoes: Tornadoes are nature's most violent storms. Their form is distinctive: a hard-edged, rapidly rotating funnel cloud hanging from the base of a large storm cloud, usually a thunderstorm. The funnel consists of water droplets and it rotates at speeds in excess of 300 kilometers (180 miles) per hour. The extreme rotational velocity is explained by the conservation of angular momentum principle in which velocity increases as the radius of rotation decreases, that is, as the funnel tightens. When a tornado touches the ground, a debris cloud forms at the base, which usually obscures the end of the funnel. Although most tornado funnels are only about

100 meters in diameter and last only minutes, they are the most feared and locally destructive storms produced by the atmosphere. We will examine the violent effects of tornadoes at the end of this chapter.

All tornadoes are born from parent storms. These include isolated thunderstorms, mesocyclones, cyclonic fronts, and hurricanes. Increasingly, meteorologists find tornadoes associated with massive thunderstorms, called **mesocyclones**. These are thunderstorm systems that cover large areas, 3 to 9 kilometers (2 to 6 miles) in diameter, and develop a broad rotating pattern of circulation. At the heart of a mesocyclone are one or more very powerful convection cells, called supercells, which produce the tornado. The diagram in Figure 6.20 shows the basic structure of a supercell.

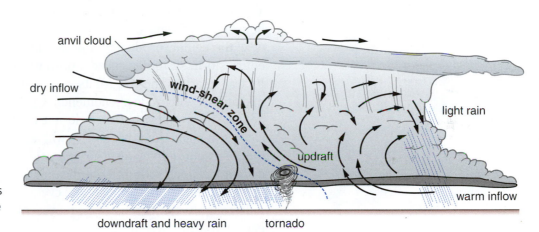

Figure 6.20 The structure of a supercell showing its complex structure and circulation system. Notice the location of a tornado formation.

The stage for the birth of a tornado is set by two large layers of contrasting air separated by a smaller inversion layer. The lower layer is warm, moist surface air whereas the upper layer is cool, dry air. The two are kept from mixing by the inversion, but the balance is very tenuous. The process leading to a tornado begins when a parcel of unstable surface air breaks through the inversion layer and is set into rapid rotating motion by the shearing action of crosswinds. Wind shearing (in the area defined as the wind-shear zone in Figure 6.20) occurs where adjacent wind currents flow in opposing directions, that is, across each other's path. In this case the crosswinds are updrafts and crossdrafts within the storm cell and the result is a wind vortex (spiral) and funnel cloud.

Tornado formation usually takes place on the southwest side of the supercell on the edge of the main zone of updrafts, but not in the zone of downdrafts and rain. As the funnel descends from the cloud base, it intensifies as it lengthens reaching its greatest power when it attains maximum width. At this stage (see the top diagram in Figure 6.21), the tornado's alignment is approximately vertical but within minutes it usually tilts and then takes on a ropelike configuration as it weakens. In the last stage – which is usually only minutes into the event – the rope breaks up leaving a debris cloud on the ground and a truncated funnel cloud hanging from the base of the parent cloud.

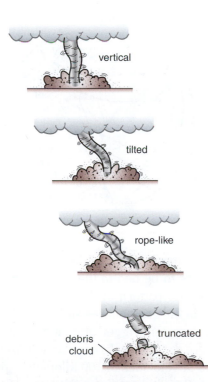

Figure 6.21 Stages in the life of a tornado, a sequence that usually lasts only minutes to hours.

6.6 Cyclonic/Frontal Precipitation: Air Masses and Frontal Waves

Cyclonic/frontal precipitation results from the meeting and mixing of two large bodies of contrasting air, called **air masses**. There are two basic mechanisms involved in this type of precipitation system. The first is the formation of a wave in the line of contact, or **front,** between the two air masses. Normally two fronts form in the wave, as illustrated in Figure 6.22a, each with warm air on one side and cold air on the other. The movement of air is in opposite directions along the front, northerly wind for the cold air and southerly wind for the warm air. As these contrasting types of air move against each other, the warm air is forced to rise and frontal precipitation is initiated.

The second mechanism is the formation of a large system of low pressure, which houses both fronts (Figure 6.22b). Into this system pours surface air which rises near the center, producing a broad interior mass of clouds and precipitation. The entire storm system is called a **midlatitude cyclone** or an **extratropical cyclone**. Its trademark is the two fronts, a cold front and a warm front, and it is uniquely a midlatitude/subarctic phenomenon. Cyclones form in other parts of the world, most notably the tropics, but they do not contain fronts. Midlatitude cyclones develop along the polar front, which you will recall, in the northern hemisphere, separates the atmosphere into warm-air and cold-air sectors or zones. Here air masses from distant locations with different heat and moisture contents meet, mix, and produce these storm systems. Midlatitude cyclones, which operate over a broad band of latitude roughly 40 degrees wide, represent one of Earth's largest and most effective energy transfer systems.

Figure 6.22 (a) Cyclones often begin with the formation of a wave in the polar front. (b) This leads to the formation of a large pressure cell with distinct warm and cold fronts.

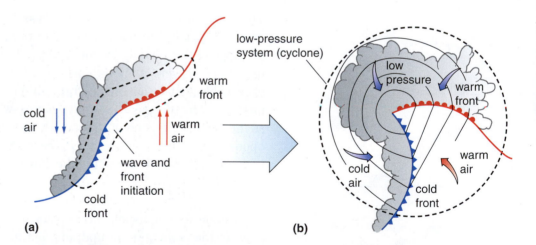

Air-mass Origins and Movements: The atmospheric machinery that drives the movement of air masses across the midlatitudes appears to be tied to the Rossby waves that control the polar front (see Figure 5.18). As the pattern of the Rossby waves shifts, so shift the locations of air masses situated in the great curves and loops along its meandering course. Sudden shifts in this system can trigger dramatic air-mass movements. Unquestionably, the most dramatic are the outbreaks of polar and Arctic air that can rush across the continental interiors at velocities of 100 kilometers per hour or more. In the most extreme cases these outbreaks can cause violent weather and a drop in temperature of 25 to 35 °C in a matter of several hours.

Much of the effort of weather forecasting in the midlatitudes is devoted to analyzing and tracking air masses of different origins. This involves classifying air masses according to the source regions where they originate and then tracing their movement in the broad belt of westerly airflow. Since source regions vary greatly in moisture and thermal characteristics, the air masses they produce also vary greatly in water and heat contents. If the energy content and size of an air mass are known, its potential for storm development can be predicted as it moves along the polar front.

Air-mass classification is based mainly on two attributes of the geographic location where the air originated: temperature and moisture. Air masses that originate over water are wet and are coded **m** (for maritime), whereas air masses that originate over land are dry and are coded **c** (for continental). The temperature characteristic for most air masses is specified as polar (**P**) or tropical (**T**); however, two additional thermal classes are also used, Arctic (**A**) and equatorial (**E**). Arctic is colder than polar and equatorial is warmer than tropical. In addition, once an air mass migrates from its source area, it can be given another designation depending on the temperature of the surface it moves over as it influences the air's stability. If it is cooler (**K**) than the surface under it, it tends to be unstable, whereas if it is warmer (**W**) than the surface under it, it tends to be stable. The latter two designations are less commonly used than the thermal and moisture designations.

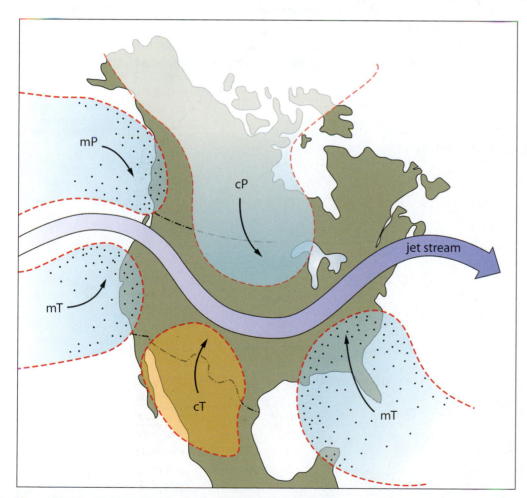

Figure 6.23 The principal air-mass source regions for North America.

The predominant air masses in North America are **mP**, **cP**, and **mT** (Figure 6.23). Most of the polar air masses originate in central and north-central Canada and migrate southeastward. Depending on the location of the polar-front jet stream, occasionally frigid Arctic air masses from above the Arctic Circle penetrate into central and eastern United States (see Figure 5.20a). The maritime polar air masses in North America originate mainly over the North Pacific, whereas the principal source of maritime tropical air masses is the Gulf of Mexico and the Caribbean Sea. The northern Pacific produces frequent maritime air masses of great size and moisture contents. On the other hand, there are few cT air masses of consequence in North America, because the source area – Mexico and Central America – is so small as you can see in Figure 6.23. In Eurasia, on the other hand, the source area for cT air in the continent's dry interior is much larger and the cT air masses it produces are understandably far more substantial.

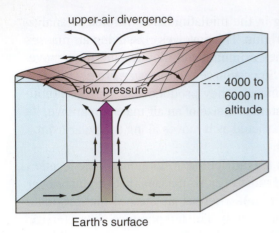

upper-air divergence

low pressure

- - - - 4000 to 6000 m altitude

Earth's surface

Figure 6.24 In a disturbance at the birth of a cyclonic storm, the divergence of air and low pressure aloft draw air up from the surface leading to surface low pressure.

When air masses move from their source areas in the great westerly stream of air flow, they meet with other air masses close to or along the polar front and are driven eastward. This movement appears to be directed by high-altitude winds, principally the jet stream. At some point along the polar front a cyclone is born and it is fed by the heat and moisture provided by the air masses.

Cyclogenesis and Frontal Weather Activity: Most midlatitude cyclones begin as small areas of low pressure, called **disturbances**, along the polar front similar to the wave in Figure 6.22a. The cause of disturbances is debated, but they appear to be related to drops in pressure aloft, above the polar front. This causes surface air to ascend, as shown in Figure 6.24, creating low pressure at ground level which triggers the convergence of surface air. As the air ascends it is diffused by the jet stream in a process termed *upper-air divergence*. This mechanism is critical to the maintenance of the young pressure cell because it prevents the buildup of pressure, which would otherwise quickly obliterate the low pressure.

On the surface, air is drawn into the cell from all directions around the budding storm. Despite its broad coverage, the inflowing air has only two sources depending on which side of the polar front it is drawn from: cold or cool air from the polar side and warm air from the tropical side. The leading edges of these two types of air form a broad, S-shaped wave in the polar front, as shown in Figure 6.22a, the limbs of which form the cold and warm fronts we see on the daily weather maps. In the northern hemisphere, the cold front usually forms on the west or southwest and the warm front, on the east or northeast, as is shown in the map in Figure 6.25.

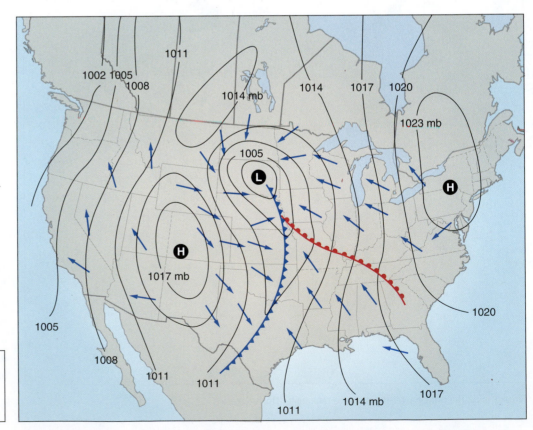

		cold front
▲▲▲		cold front
●●●		warm front
−1014−		isobar, mb
←		wind direction

Figure 6.25 An isobaric map (mb = millibar) showing the position of the cold front and warm front in a cyclonic system.

Three facts about the young cyclone are critical. First, because of the density differences in the two types of air, the warm front and the cold front are going to be structured differently. Second, the inflowing air is influenced by the Coriolis effect; therefore, as the cell develops, it is set into a broad swirling motion. And third, with the convergence of air on the body of low pressure, an upward flow develops in the interior of the cell, consistent with that illustrated in Figure 6.24.

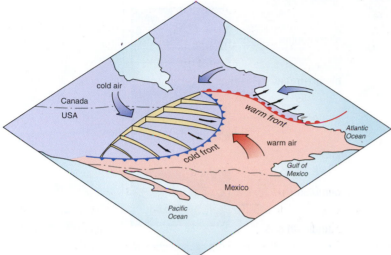

Figure 6.26 The basic structure of a cold front stretching across central North America.

In a cold front, the leading edge of the cold air mass drives under the warm air, forming, as Figure 6.26 illustrates, a relatively steep contact over which the warm air is forced upward. In a warm front, by contrast, the contact angle is reversed and much gentler than the cold front. The warm front has more the form of a long ramp, over which the warm air slides gradually upward. Because cold fronts are steeper and move faster than warm fronts, the rate of uplift along them is usually more dramatic. Consequently, precipitation and turbulence can be very intensive along cold fronts.

Strong cold fronts advancing on warm, moist air often produce a line of rough weather, called a **squall line**. It is characterized by thunderstorms, gusty winds, hail, heavy rainfall, and occasionally tornadoes. Tornado formation is associated with extreme turbulence and thunderstorm development along the squall line. Lateral rolls of rotating air appear to form along the cold air–warm air contact, and as the front advances and instability drives air upward, these rolls tip into a vertical alignment and spin into funnel vortices under the influence of crosswinds.

Cloud development along strong cold fronts typically reaches elevations of 10,000 to 15,000 meters (50,000 feet) and is characterized by great billowing storm clouds, called **cumulonimbus** (or **cumulus** if the cloud is not producing precipitation). (In cloud terminology the suffix *nimbus* or the prefix *nimbo* indicates precipitation.) Because of their great heights, the tops of the cumulonimbus clouds may reach into the zone of fast upper airflow where they are drawn downwind ahead of the front and often ahead of the entire cyclone. These high-altitude clouds, called *cirrus* clouds, have a wispy appearance, and their presence in the sky is often taken as a precursor of bad weather. Cirrus clouds belong to the high-altitude (8 to 12 km) family of clouds made up of tiny ice particles. The three other families of clouds and descriptions of the basic cloud types in each are given in Figure 6.27.

Figure 6.27 Clouds families and common cloud types.

cloud family	cloud types	
high clouds (altitude of cloud base above 6000 m)	cirrus cirrostratus cirrocumulus	 →
middle clouds (altitude of cloud base 2000 m – 5000 m)	altostratus altocumulus	 →
low clouds (altitude of cloud base under 2000 m)	nimbostratus stratocumulus stratus	 →
vertical clouds	cumulus cumulonimbus	 →

Figure 6.28 Landscape damage resulting from an ice storm.

Weather conditions along warm fronts contrast sharply with those along cold fronts. Warm-front cloud formation is mainly horizontal, leading to a broad zone of **stratus** clouds with precipitation (*nimbostratus*). Turbulence is usually modest and precipitation often characterized by prolonged showers or drizzle. If the cold air under the front is below freezing, rain may freeze on contact with surface objects, resulting in the formation of glaze ice which damages vegetation and makes all modes of travel hazardous (Figure 6.28).

With the warm air aloft and cold air below, warm fronts form pronounced temperature inversions that prevent vertical mixing of the surface air. If the frontal system is slow moving, this air may become stalled, or stagnated, over an area for several days. In areas of heavy air-polluting emissions, such events can reduce the flushing capacity of the lower layer of air to negligible levels. When this happens, pollutants will build up, reaching levels dangerous to human health. Some of the worst air-pollution disasters in the midlatitudes have been caused by cyclonic inversions related to warm fronts and low-level cloud masses created by combined warm and cold fronts.

In 1952, an inversion over London led to the formation of *smog* (fog combined with pollutants such as sulfur dioxide and airborne particulates) so thick that 4000 persons died of respiratory difficulties in just a few days (Figure 6.29). Serious air-pollution episodes are also related to inversions of other origins, in particular, adiabatic warming associated with descending air in anticyclones, airflow over mountains, and with advective cooling. If urban areas are located near or within mountain valleys, then the horizontal mixing of the air mass may be inhibited, and this topographic containment can also prolong and exacerbate the air-pollution event.

Cyclonic Development and Movement: Midlatitude cyclones are characterized by three principal motions. The first is that of airflow converging on the center of the cell and then rising into the troposphere. The rate of this flow is dependent mainly on

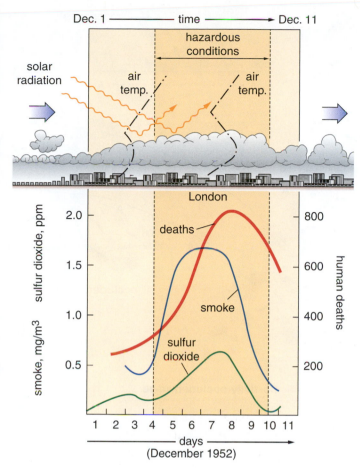

Figure 6.29 Deaths brought on by a stagnated cyclone and heavy air pollution over London during seven days in 1952.

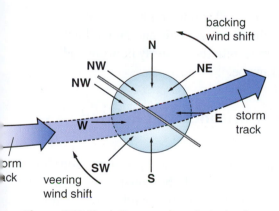

Figure 6.30 Circulation around a cyclonic showing veering and backing wind shifts.

the pressure gradient, which steepens as the system develops. The geographic pattern of airflow around the cell is influenced strongly by the Coriolis effect which causes a spiraling system of wind that is so broad that some air may travel from one side of the storm to the other by the time it takes its ascent. Generally, winds tend to be northeasterly in the northern or northeastern half of the storm and southwesterly in the southern or southwestern half. Therefore, as the storm crosses an area, the types of wind shifts experienced will be different depending on which side of the cell you are located. In a storm moving eastward, two patterns of wind shifts can be experienced called *veering* and *backing* wind shifts (see Figure 6.30). Backing wind shifts occur in the north half of the cell (east to northeast to north to northwest) and veering wind shifts (south to southwest to west to northwest) are common in the southwestern half.

The two other motions of the cyclone are the movement of the fronts themselves and the migration of the entire storm cell along the polar front. We need to understand these motions in order to understand the sequence of weather changes associated with the storm. In the early stage of development, called the **initial wave**, the cold front and warm front are positioned a great distance apart as shown in Figure 6.31. Between them lies a sector of warm air where southerly winds are drawing high-energy mT air into the cell. As the storm develops, the cold front advances rapidly on the warm-air section (usually from the west or northwest), closing the gap between the fronts. When the gap (warm sector) between the fronts approaches at an angle of 90 degrees or so, the cyclone has reached the mature stage of development.

In the **mature stage**, weather conditions contrast sharply among the various sectors of the storm. The warm-air sector is marked by a southerly flow of warm, humid air. Along the warm front this air slides over the cold air, forming a broad belt of stratus clouds and low-intensity rain that dominates the eastern/northeastern sector of the cyclone. The western/southwestern sector, on the other hand, shows rapidly changing conditions behind the cold front as the frontal turbulence gives way to cool, clear cP or cA air from the northwest.

Within a matter of days after formation of the mature wave, the faster moving cold front usually overtakes the warm front. This marks the beginning of the **occluded stage** and the end of the warm-air sector (Figure 6.31). When the fronts reach full occlusion, the remaining warm air has been driven above the cold air, forming a great body of stratus, nimbostratus, and stratocumulus clouds.

The movement of the entire cyclone as a pressure cell appears to be controlled by upper atmospheric circulation along the polar front. Therefore, the movement almost always has a marked eastward component. In addition, cyclonic tracks shift north and south with summer and winter, and this is accompanied by a change in the frequency and magnitude of cyclonic storms. Invariably, cyclones exhibit the

Figure 6.31 The four developmental stages of a midlatitude cyclone. Such storms usually last five to seven days.

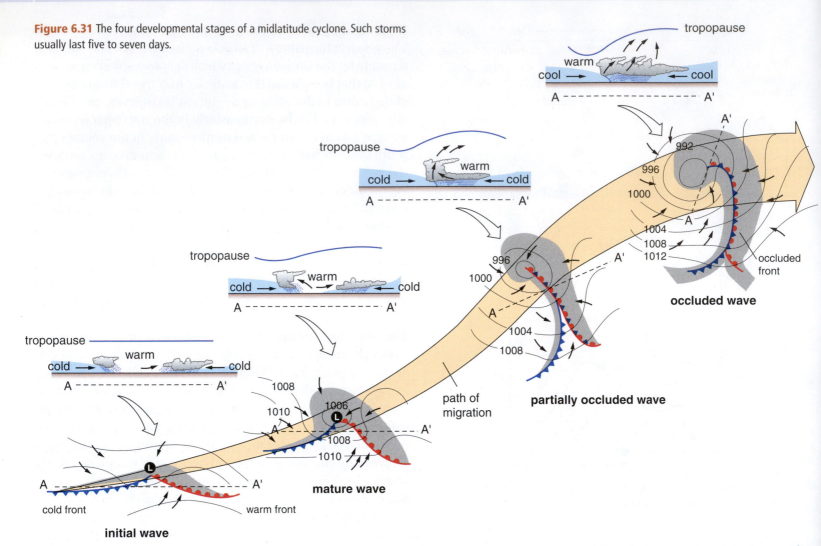

greatest strength and frequency of occurrence in fall and winter and activity is much greater over the oceans than lands. Indeed, the North Atlantic and North Pacific are "breeding grounds" for cyclones, which migrate with the westerly flow onto adjacent land areas downwind, North America and Europe.

6.7 Convergent Precipitation: Hurricanes and Related Tropical Storms

Convergent precipitation results when air moves into a low-pressure trough or enclosed topographic area from which it can escape only by flowing upward. The **ITCZ (intertropical convergence zone)**, which is fed by tradewinds from the north and south, is such a trough of low pressure. Although individual storm cells within the ITCZ may be convectional in origin, especially over land, the net upward flow in this zone is fundamentally convergent like that shown in Figure 6.13d. Over the oceans, convergence is essentially the only source of precipitation because marine surface air is generally not warm enough to trigger widespread thermal instability. In certain areas along the ITCZ, disturbance zones may form areas of weak low pressure related to special circumstances in circulation and surface heating. Thus, weak lows may produce especially heavy convergent rainfall but, typical of most low-pressure cells near the Equator, do not develop into larger cyclonic systems and fade away within days.

On the other hand, some disturbances do become cyclonic storm systems that produce substantial convergent precipitation. Three of these are especially noteworthy. The simplest is a trough of low pressure, called an **easterly wave** that forms within the tradewind belt. Easterly waves always form over the oceans at 5 degrees latitude or more north and south of the Equator and are characterized by a line of showers and thunderstorms. The wave moves slowly (300 to 500 kilometers per day) westward as moist surface air, fed by the tradewinds, converges on the trailing (eastern) side of the trough.

Tropical depressions and **tropical storms** are another source of convergent precipitation. These disturbances appear to be caused by several factors, one of which is a weak cold front associated with an occasional incursion of polar air into the tropics. These polar outbreaks and the weak storms associated with them may reach equatorward to 15 degrees latitude, but they rarely extend to the limits of the tropics. Understandably, their occurrence is limited to regions, such as North America, which produce strong Arctic air masses that can penetrate into the tropical zone, such as the Caribbean.

Hurricanes (or **typhoons**) are the behemoths of tropical storms, and satellite images like the one in Figure 6.32 give us accurate portraits of them. These tropical cyclones are the strongest, most feared, and most destructive storms on Earth. They are distinguished from other tropical storms by their size, power, and wind speed. For a tropical storm to qualify as a hurricane, winds must exceed 65 knots (74 miles per hour), according to international meteorological standards. All hurricanes have distinctive attributes in terms of their origin, distribution, and time of occurrence. Some of the prominent ones are as follows:

- They form only in certain seasons, usually late summer and early fall in both the northern and southern hemispheres.
- They begin as disturbances in the tradewind belt around 5 to 10 degrees latitude.
- They develop only over tropical oceans where surface temperatures exceed 27 °C, which the map in Figure 6.33 approximates.

Figure 6.32 A satellite image of hurricane Katrina over the Atlantic Ocean.

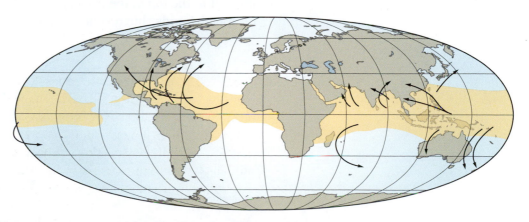

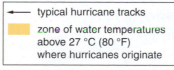

——→ typical hurricane tracks

▨ zone of water temperatures above 27 °C (80 °F) where hurricanes originate

Common hurricane tracks and zones of origin

Figure 6.33 Hurricane spawning grounds in the belt of warm tropical water and typical tracks taken by hurricanes from the tropics to the midlatitudes.

- They do not take on hurricane magnitudes and dimensions until they have moved out of the equatorial zone and well into the tropics and even subtropics, usually between 10 and 30 degrees latitude.
- They move westward and northwestward (in the northern hemisphere) from the tropics into the subtropics and midlatitudes in broad curving tracks.
- They do not form in connection with fronts, nor do they develop fronts once the storm has taken shape.
- They do not occur with any regularity and can develop over any tropical ocean.
- They are many times more intensive than midlatitude cyclones.

Hurricanes usually begin as tropical disturbances such as easterly waves which evolve into tropical storms intensifying as they gain energy from the latent heat supplied by moist equatorial and tropical air. As with midlatitude cyclones, hurricanes are characterized by a great, inward-spiraling circulation pattern that draws in air from all directions. As the air spirals toward the center, wind speed increases tremendously with the conversion of angular momentum, reaching velocities over 250 kilometers per hour (150 miles per hour) near the center of strong storms.

Hurricanes rarely form close to the Equator because the Coriolis effect there is not great enough to produce spiraling circulation. As a result, most hurricanes originate poleward of 10 degrees latitude. Once formed, they move eastward in the belt of easterly tradewinds, veering increasingly poleward under the influence of the Coriolis effect and growing in size and intensity as they move. This track takes many hurricanes into the subtropics and midlatitudes, where they may become caught up in the westerly air flow of the midlatitudes and (in the northern hemisphere) loop back to the northeast or east as is shown by the arrows in Figure 6.33. Those that pass onto landmasses wither quickly (usually in a matter of days) because they lose their supply of moisture-rich maritime air, their primary source of energy.

The detailed structure of hurricanes varies from storm to storm, but the general features are similar and the cross-sectional diagram in Figure 6.34 shows several of them. Most are 350 to 650 kilometers (200 to 400 miles) in diameter and extend upward to the top of the troposphere, 12 to 16 km above Earth. They are made up of **spiral bands**, huge concentric belts of clouds where the heaviest rainfall and

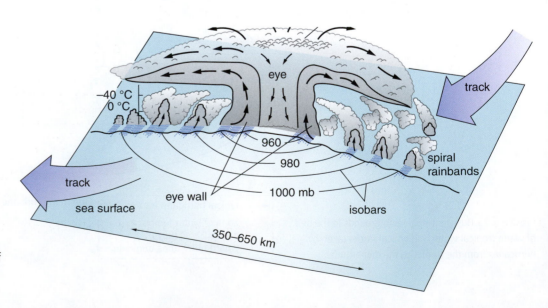

Figure 6.34 The structure, circulation, and major features of a hurricane. The massive cloud cover of the outflow shield is mainly a product of eye-wall circulation.

strongest winds are found. As air moves inward, it speeds up, and is diverted into a strong upward flow around the center of the storm, called the eye. The strong upflows produce a great cloud wall, called the **eye wall**, reaching 8 to 10 kilometers altitude around the eye. At the top of eye wall, flow diminishes and fans out forming a crowning cloud mass called the **outflow shield**, which spreads over much of the hurricane.

The **eye** is a small, cloud-free, corridor that extends upward to the top of the storm. It functions as a return flow conduit where some of the air swept aloft in the adjacent cloud wall returns to the surface. As this air descends, it heats adiabatically, warming the storm's eye and thereby eliminating any chances of condensation, cloud formation, and precipitation there. Thus, the hurricane eye, as Figure 6.34 portrays, represents a small island of peculiarly pleasant weather in the midst of a vast raging storm.

A hurricane is one of nature's most astonishing energy systems. It functions as a massive heat pump converting the latent heat of water vapor into sensible heat and kinetic energy which in turn drive the storm's powerful winds, both laterally and vertically. Since hurricanes do not significantly raise the temperature of the air within them, we conclude that the greatest share of the storm's energy is actually devoted to driving the wind system. The amount of energy available to hurricanes is enormous and appears to be increasing with ocean warming.

As the graph in Figure 6.35 shows, two-thirds of the Earth's precipitable water vapor is found roughly in the intertropical belt, between 30 degrees north and 30 degrees south latitude, precisely where hurricanes develop. A single hurricane can produce up to 20 billion tons of rainwater per day. At an energy conversion rate of close to 600 calories per gram of water condensed, the energy output is obviously massive, and for hurricanes that move distances of hundreds or thousands of kilometers from the tropics into the midlatitudes, the poleward transfer of energy is truly considerable.

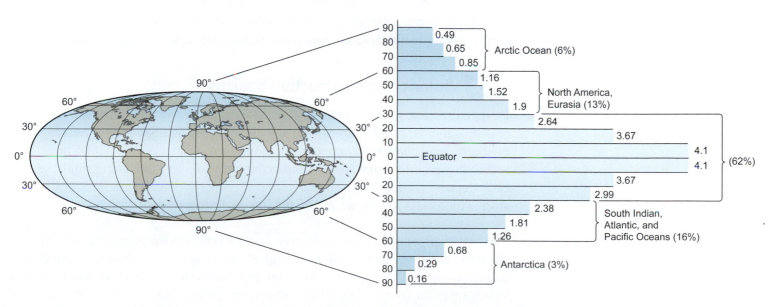

Figure 6.35 The global distribution of precipitable water vapor. Over 60 percent is found between latitudes 30 degrees north and south.

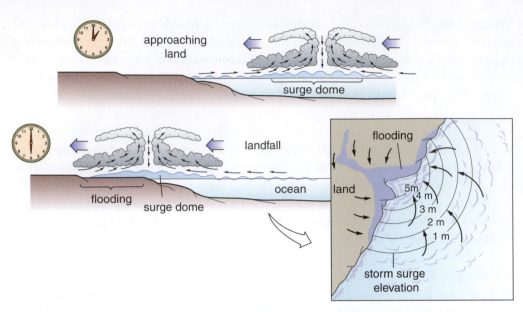

Figure 6.36 Map view of a surge dome with the hurricane advancing onto land from right to left.

Air pressure within a hurricane deepens as the storm develops. At full development, pressure near the storm center is depressed as much as 60 or 70 millibars below normal sea-level pressure, that is, to 940 to 950 millibars. (In the most intense hurricanes, air pressure reaches as low as 900 millibars.) The extreme low pressure not only drives wind to speeds above 150 kilometers per hour (100 miles per hour), but actually elevates the sea surface under the storm creating a broad mound, called a **surge dome**. If you think of the atmosphere as a weighted blanket lying on the ocean, a hurricane represents a hole or a thin spot in the blanket that allows the water surface there to rise up. Surge domes typically reach heights of three to four meters, and grow even higher as large storm waves are heaped on top of them. As the diagram in Figure 6.36 illustrates, the water surface may be pushed even higher if the storm strikes land and the surge dome is buttressed against the shore by the force of onshore wind and waves.

Summary on Precipitation Forms and Types: Precipitation particles form around a nucleus, usually a tiny speck of dust or salt. Most precipitation falls in the form of rain, but probably begins as snow high in the atmosphere. Precipitation types are defined by the mechanisms that cause air to rise and cool. Each type of precipitation represents a system with inputs and outputs of energy (heat) and matter (water). The heaviest annual precipitation rates are produced by orographic systems, but the most intensive rainfalls and the most violent weather are produced by convectional and convergent systems. Cyclonic systems are broad, complex storms that form along the polar front and produce frontal precipitation and variable weather conditions.

6.8 Violent and Destructive Storms

Despite dramatic setbacks from massive storms like hurricane Katrina, the number of people injured or killed by violent storms in the United States and Canada is surprisingly small compared, for example, to highway accidents or violent crimes. Nevertheless, violent storms are a serious concern to society and increasingly so with each decade. During the twentieth century, property damage and loss of life from storms increased worldwide, especially in less developed countries.

The twenty-first century portends even greater setbacks for humanity. Why? Do we not have better scientific understanding of storm systems including sophisticated detection systems such as satellite radar and scanners capable of minute-by-minute monitoring? Of course, but other factors have offset our scientific and technical advances. At the top of the list is global population. Not only has global population increased dramatically in the past century, but its geographic distribution has changed with a greater share of humanity living in coastal lands prone to damaging storms. Today hundreds of millions of people are exposed to hurricanes and the number is rising.

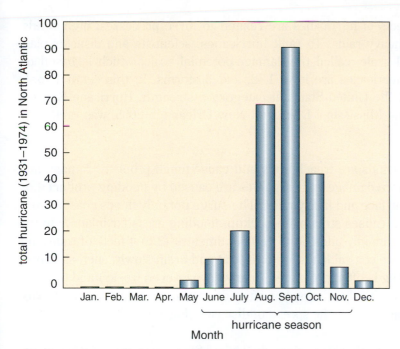

Figure 6.37 The monthly frequency of hurricanes in the North Atlantic, 1931–1974. Hurricane season is June through November.

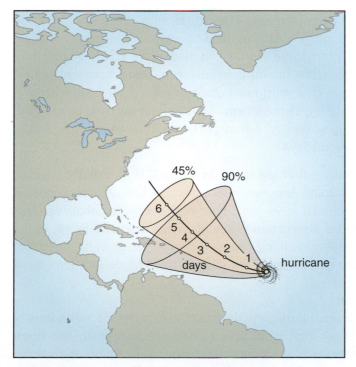

Figure 6.38 The concept of cones of probability for the location of a hurricane landfall.

Second, some governments and societies are unable or unwilling not only to regulate land-use development and settlement but to use available scientific knowledge and detection technology to deal more effectively with storm risks. Third, the planet's climate is warming and the added energy in the oceans and atmosphere appears to be causing an increase in the incidence and magnitude of violent storms in some parts of the world. Documentation of this change is only just beginning, but suffice it to point out that a strong case has already been made for an significant increase in hurricane class storms in the North Atlantic since the early 1990s. Chapter 8 will address this issue in a little more detail.

Hurricanes as Hazards: Hurricanes are decidedly the most destructive storms on Earth. They have greatest impact on the tropical east coasts of islands and continents, as you can see in Figure 6.33, and because their tracks bend poleward, they often travel well beyond the tropics into the midlatitudes. In North America, they frequently strike the Gulf Coast (30 degrees north latitude) and Eastern Seaboard (25 to 40 degrees north latitude) but rarely strike the California coast at the same latitude on the other side of the continent. The Atlantic Ocean produces about ten tropical storms in the average year, of which five to six reach hurricane strength. The Pacific Ocean is much more productive. The northwestern Pacific alone averages 16 hurricanes (typhoons) per year.

The number of hurricane strikes (landfalls) varies widely from year to year depending on the number of storms produced and their tracks of movement. North America can expect one to five hurricane strikes per year. Most occur in September when the oceans have reached their highest surface temperatures (see Figure 6.37). The hurricane "season" for North America, according to the US National Hurricane Center, begins on June 1 and lasts until the end of November. In the southern hemisphere, the hurricane season begins around December and lasts more or less through May.

One of the most important considerations in hurricane forecasting is the geographic location of landfall. Based on the trajectory of a hurricane's track and the behavior of past hurricanes, forecasters are able to define zones or cones of probability – like those shown in Figure 6.38 – for the hurricane's position day by day. The narrower (more specific) the cone, the lower the chances of the forecast's accuracy. Because of the unpredictability of the external factors (such as upper-level wind patterns) influencing hurricanes, some seem to defy landfall forecasting, taking on erratic tracks with sharp turns and even loops that may bring them over some areas twice.

The destructive effects of hurricanes are related to three processes: high winds, storm surges, and heavy rains. To rank hurricanes, scientists and disaster planners use a five-level scale, called the damage-potential scale, which is based on wind speed. Most hurricanes are level 1, 2, and 3 storms. In this century, only two storms have hit the United States at category 5 strength. Hurricane Katrina, which tore across the Mississippi Delta and New Orleans in 2005, was ranked as category 4.

Although hurricane winds are treacherous and cause much property damage and human casualties, more damage and casualties are caused by flooding produced by the elevated ocean surface and heavy rains. The surge not only floods coastal areas with ocean water, but causes streams to backup, flooding areas far inland. On top of that, the storm commonly adds 60 to 180 centimeters (2 to 4 feet) of rainwater to flood levels. Because coastal lands are often flat and drain slowly, such as shown in Figure 6.39, flooding is often prolonged (many days or even weeks) and not only damages houses and other buildings, but often destroys crops, kills farm animals, contaminates wells, and fosters the spread of diseases (Figure 6.39).

Hurricanes killed close to 18,000 people in the United States in the twentieth century. Not surprisingly, fatalities have been higher in the Caribbean where many islands lie within the main belt of North Atlantic hurricane paths. For example, in 1780, 22,000 people were killed in Martinique; in 1930, 8000 were killed in the Dominican Republic; in 1963, 8000 were killed in Haiti and Cuba; and in 1998, 11,000 were killed by hurricane Mitch in Honduras. But these numbers pale beside hurricane deaths in Asia. Several hurricanes in the past century have killed more than 100,000 people each. Perhaps the worst one was the 1970 hurricane in the lowland country of Bangladesh (then East Pakistan) which killed at least 300,000 and perhaps more than 500,000 persons (the real numbers are unknown) (Figure 6.40).

There is also enormous property damage from hurricanes. In 2005, hurricane Katrina laid waste to the city of New Orleans and the Mississippi coast, causing the most damage ever from a hurricane to make landfall in the USA, $60 billion in property damage. In 2008, hurricane Ike hit the Galveston, Texas, area and caused $30 billion in property damage – the third most damaging hurricane in US history, behind Katrina and Andrew (1992). These storms knocked out the local power grid, flooded water-treatment facilities, damaged transportation infrastructure, and thus crippled rescue efforts for several days.

But it is less developed countries that are most threatened. With rapidly growing populations in the countries of south Asia, southeast Asia, and Latin America coupled with the population push into coastal areas and higher sea levels with global warming, hurricane damage and fatalities are certain to rise, perhaps dramatically so, in the twenty-first century. And, if global warming causes the temperature of the oceans to rise, hurricanes may develop farther poleward making midlatitude locations more susceptible to strikes.

Tornadoes as Hazards: The extraordinary destructive capacity of tornadoes is related primarily to their super-high wind speeds, and secondarily to the sharp drop in pressure in the air immediately around the funnel. A common myth holds that the low pressure associated with a tornado causes buildings to "explode" as the tornado passes overhead. In fact, the worst damage is caused by the violent

Figure 6.39 Heavy coastal flooding in Louisiana as a result of hurricane Katrina.

Figure 6.40 Farmstead demolished by the 1970 Bangladesh hurricane strike which killed hundreds of thousands of people.

Figure 6.41 The swath of destruction (from left to right) caused by a tornado that struck the town of La Plata, southern Maryland, USA in April 2002.

Figure 6.42 Doppler radar image from the US National Weather Service showing a line of tornadic thunderstorms (red area) approaching Tallahassee, Florida.

winds and the debris they throw. Analysis shows that the pressure differential between the interior of the funnel and nearby buildings is not sustained long enough to cause the buildings to explode.

Tornado wind speeds can exceed 480 kilometers per hour (300 miles per hour) – the fastest known surface wind on Earth. They are indeed capable of ripping buildings apart. Although most tornado paths are narrow, about 150 meters wide, and only about 8 kilometers (5 miles) long (Figure 6.41), some are 200–300 meters wide and travel more than 160 kilometers (100 miles) on the ground. Flat ground apparently facilitates contact with the landscape. In rough terrain, by contrast, funnel contact with the ground tends to be spotty.

Tornadoes are so small geographically that they do not show up on synoptic weather charts. Although we know what sorts of weather conditions produce them, the actual place and time of tornado formation are difficult to predict. Doppler radars, which can detect storms, wind speed, rain density, and cloud rotation, have proven very effective in detecting tornadic storm systems. Patterns that appear on Doppler radar images, such as "bow echoes," may indicate the presence of cells capable of spawning a tornado as illustrated by the image in Figure 6.42 These images, however, cannot tell us which storms will definitively spawn tornadoes. Therefore, sky watches are an essential part of public protection from tornadoes. In this regard, it is important that most tornadoes occur during daylight hours when they are most visible. Some, however, occur at night, making visual detection impossible and public warning more difficult because many people are asleep.

Tornadoes can occur wherever thunderstorms are found, but they are far more common in some regions than others. Central North America has the highest incidence of tornadoes in the world. On the average more than 800 are produced yearly in the United States, mostly in the prairie and Midwestern states. In fact, the broad belt shown in Figure 6.43, which extends from Texas through Oklahoma, Kansas, Missouri, and on into Ohio, has been branded "Tornado Alley." The reason for the concentration of tornadoes here is simple but the mechanisms involved are complex. This region experiences frequent clashes between cool, dry air from the north and high-energy (moist, warm) tropical air drawn from the Gulf of Mexico. The mechanisms involve cyclonic storms along the contact, the polar front, which spawn thunderstorms, some of which develop into supercells which in turn produce tornadoes.

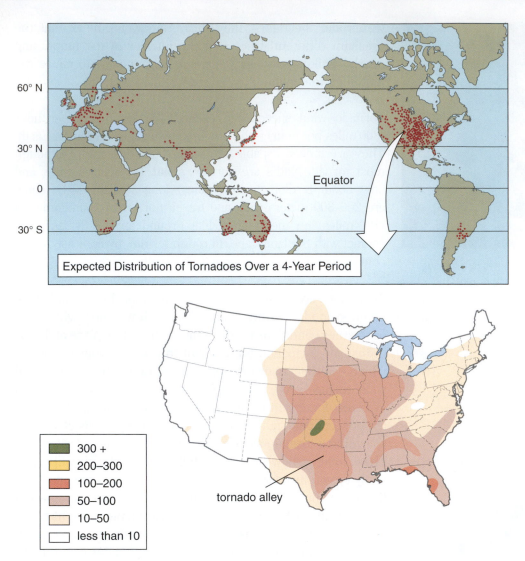

Expected Distribution of Tornadoes Over a 4-Year Period

tornado alley

■	300 +
■	200–300
■	100–200
■	50–100
■	10–50
□	less than 10

Figure 6.43 Expected global distribution of tornadoes over a four-year period based on past occurrences and (below) the distribution of known tornadoes in the United States.

Tornadoes also tend to be seasonal. Although they can occur in any month, there is a recognized "tornado season" beginning in March and extending through August in North America (see Figure 6.44). May is the peak tornado month; however, April is the month of the most violent storms based on the number of deaths. The reason for the April–July concentration is that late spring and early summer are the times in south–central North America when both air mass types, cool and warm, are present and relatively strong. Further, there is also a diurnal pattern to the occurrence of tornadoes. Most occur in the afternoon with 40 percent appearing between 2 and 6 p.m. when thunderstorms are most active.

The destructive effect of a tornado is related to many factors including wind speed, funnel size, time on the ground, and the character of the landscape encountered. A scale has been developed, called the Fujita scale, to classify tornadoes according to their destructiveness. The Fujita scale (named for its founder the meteorologist T. T. Fujita) uses five categories of wind speed and landscape damage. Most tornadoes in the United States are relatively weak. Nearly 80 percent are classed at levels F0 or F1 and only one in a thousand reaches F5, which is able to lift strong houses off their foundations and send them flying to their destruction. It is estimated that since 1990 fewer than 25 tornadoes have reached level F5. In 2007, the Enhanced Fujita Scale was implemented. It has the same basic structure as the original, but also accounts for post-storm damage.

More than 10,000 Americans were killed by tornadoes in the twentieth century, but thanks to improved detection and warning systems, the death rate has been declining despite the growth in the US population. The worst tornado disaster in North America, called the Tri-State Outbreak, occurred in an area overlapping parts of Missouri, Illinois, and Indiana in March 1925 where one large tornado killed 695 people. The second worst disaster was the 2011 Super Outbreak which produced 331 tornadoes over a four-day period in April that killed at least 334 people. But the worst among the known tornado

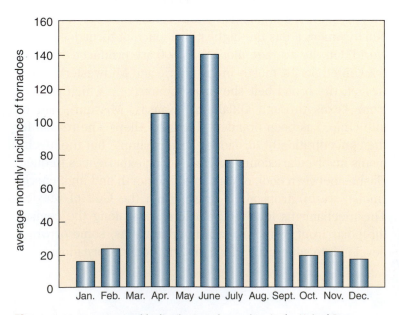

Figure 6.44 Average monthly distribution of tornadoes in the United States showing a four-month peak period, April–July.

disasters in the world occurred in Bangladesh in 1989 where 1109 people were killed, 15,000 injured, and 100,000 left homeless. What are your chances of being struck by a tornado? Based on current data, if you built a house in Tornado Alley, you would have to wait 1400 years to be hit. Thus, the "point probability" is very small, less than one-hundredth of one percent in any year.

Chapter Summary and Overview of Atmospheric Moisture, Precipitation, and Weather Systems

The atmosphere has an enormous capacity to take in and release moisture. And because of its dynamic and extensive circulation system, it also has an enormous capacity to redistribute moisture geographically, a critical factor in the global distribution of energy and in shaping the pattern and character of climates. Where the atmosphere deposits precipitation depends on various geographic conditions on the Earth's surface and in the atmosphere including the location of mountain ranges, air-mass movements, and surface heating. Precipitation systems are also capable of producing violent storms, especially where the atmosphere is rich in heat and moisture. The summary diagram here illustrates three storms of different scales and magnitudes that bring precipitation and rough weather to central Kansas and much of the rest of central North America.

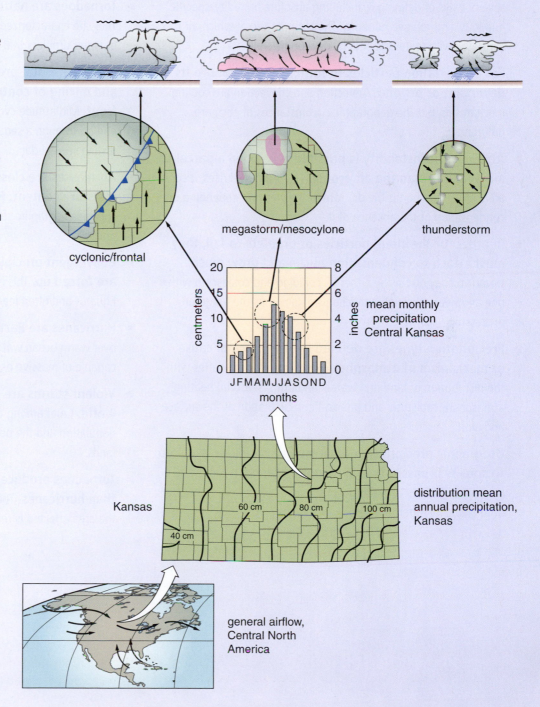

cyclonic/frontal

megastorm/mesocylone

thunderstorm

mean monthly precipitation Central Kansas

centimeters / inches

J F M A M J J A S O N D
months

Kansas

40 cm   60 cm   80 cm   100 cm

distribution mean annual precipitation, Kansas

general airflow, Central North America

- ▶ **Precipitation is one phase in a system of water exchange between the Earth's surface and the atmosphere.** The delivery of water vapor to the atmosphere is driven by heat and the return to the Earth's surface as precipitation is driven by cooling.

- ▶ **Water vapor is the most variable gas in the lower atmosphere.** Cloud formation and precipitation depend not only on the availability of surface moisture, but upon air temperature.

- ▶ **The water-vapor content of air is called humidity.** There are several ways of defining it, including absolute humidity, specific humidity, and relative humidity. The temperature at which air is saturated with water vapor is its dew point.

- ▶ **For precipitation particles to form, air must cool to its dew point or beyond.** Advection and convection are cooling mechanisms, but the geographic circumstances of each are different.

- ▶ **Atmospheric instability is possible only when a parcel of air is lighter than the air around it.** As a parcel rises, it cools adiabatically, either at the dry rate or the wet rate depending on condensation or the absence of it.

- ▶ **In order for the cloud particles or droplets to fall, they must attach to condensation nuclei and grow much heavier.** They do this by either by colliding and coalescing with one another or through a mechanism known as the Bergeron process.

- ▶ **Precipitation types are defined by the principal cause or mechanism of atmospheric cooling.** Cooling varies with the distribution of land and water, landforms, seasonal heating, atmospheric moisture, and atmospheric circulation at the surface and aloft.

- ▶ **Orographic precipitation results when moist air is forced to cool as it passes over a mountain range.** It is most pronounced where prevailing wind systems blow from the ocean onto a coastal mountain range.

- ▶ **Katabatic winds occur when air spills over mountains resulting in compression and heating as they descend.** They are common across the world and can cause extreme environmental conditions including rapid snowmelt, harsh winds, wildfires, and even human behavioral stress.

- ▶ **Convectional precipitation results from ascent of unstable parcels of air.** It may result from local surface heating, the forced ascent of air, or other factors. It is powered principally by latent heat, and commonly results in thunderstorms.

- ▶ **Tornadoes are nature's most intense and violent storms.** They are characterized by a rapidly rotating funnel cloud made up of water droplets that usually forms within large thunderstorms.

- ▶ **Cyclonic/frontal precipitation results from the meeting and mixing of contrasting air masses along the polar front.** Midlatitude cyclones begin as low-pressure disturbances and go through a sequence of developmental stages usually lasting several days.

- ▶ **Air masses are classified according to temperature and moisture content.** For each midlatitude continent, there are specific geographic sources (areas) for different types of air masses.

- ▶ **Convergent precipitation results when winds meet and are forced up.** This is a common process in the ITCZ near the Equator and often leads to convectional storms.

- ▶ **Hurricanes are Earth's largest storms.** They develop only over warm ocean water, move westward as they mature, and are capable of massive destruction in coastal areas.

- ▶ **Violent storms are a serious and rising concern in the world.** Chief among the contributing factors is rising world population and the push to settle high-risk areas such as coastal lands.

- ▶ **Tornadoes produce far less total damage and loss of life than hurricanes.** Their paths are narrow and relatively short so the area affected is relatively small.

Review Questions

1 Draw a diagram of the water-vapor–precipitation exchange system, and identify the exchanges of energy which produce water vapor and precipitation and change their states on the surface and within the atmosphere.

2 What are the three standard measures of humidity and how are they used? How is relative humidity related to the dew point?

3 What is the difference between a stable and unstable parcel of air? How can a stable parcel of air become unstable?

4 Why is the moist adiabatic lapse rate lower than the dry adiabatic rate? What happens to a parcel of air which begins to produce condensation after cooling at the moist adiabatic lapse rate?

5 How do atmospheric temperature inversions occur, and what are the major concerns associated with them?

6 Compare the collision–coalescence and Bergeron processes for precipitation droplet formation, and identify their differences and similarities.

7 Identify the geographical regions where each of the four precipitation types is likely to occur. How can it be said that convection plays a role in all types of precipitation?

8 Describe the three stages of thunderstorm development. What is the relationship between thunderstorms and tornadoes in general and, specifically, what conditions within thunderstorms may give rise to tornadic activity?

9 Using Figure 6.23, identify the primary air masses which influence your local weather. What happens to their relative influence during the changing seasons?

10 Describe the anatomy of a midlatitude cyclone. What are the precipitation patterns associated with its warm and cold fronts?

11 Identify the three primary motions of midlatitude cyclones. Why is their general motion to the east? Why does the cold front move faster than the warm air, and what happens when the cold front catches up with the warm front?

12 Trace the changes a typical Atlantic hurricane undergoes from origin to landfall. Why don't hurricanes form near the Equator? Why do some curve away from North America? What causes their storm surge?

13 What factors determine the amount of damage caused by a hurricane? During hurricane Ike in 2008, 20,000 people ignored the warning to evacuate Galveston and chose to stay on their properties. What factors do you think contribute to this behavior?

14 Why is it so difficult to predict the exact landfall location of a hurricane? What advances in recent years have helped improve the accuracy of these predictions?

15 What are the consequences of recent increases in coastal population concentrations?

16 How do tornadoes and hurricanes differ in their geographical characteristics? When do most tornadoes occur in North America, and what is the key determinant of their damage?

17 Compare and contrast the hazards posed by thunderstorms and blizzards.

Global Climate, Formative Systems, and Human Adaptation

Chapter Overview

Our main objective so far in this book has centered on the systems and processes responsible for the geographic distribution of energy (mainly heat and radiation) and matter (mainly water). Accordingly, this chapter opens with a brief look at the nature of the systems that produce the main ingredients of climate and how they vary in their distributions and behavior, and goes on to examine the two big climate engines, the tropical engine and the midlatitude engine, and how they operate. This is followed by a brief description of a traditional climate classification scheme, which divides the Earth into five main climatic zones. The chapter ends on a practical note: how humans have adapted to climatic conditions through technologies of clothing and shelter.

Introduction

People everywhere brag and whimper about the woes of their early years, but nothing can compare with the Irish version: the poverty; the shiftless loquacious alcoholic father; the pious defeated mother moaning by the fire; pompous priests; bullying schoolmasters; the English and the terrible things they did to us for eight hundred long years.

Above all – we were wet.

Out in the Atlantic Ocean great sheets of rain gathered to drift slowly up the River Shannon and settle forever in Limerick. The rain dampened the city from the Feast of the Circumcision to New Year's Eve. It created a cacophony of hacking coughs, bronchial rattles, asthmatic wheezes, consumptive croaks. It turned noses into fountains, lungs into bacterial sponges. It provoked cures galore; to ease the catarrh you boiled onions in milk blackened with pepper; for the congested passages you made a paste of boiled flour and nettles, wrapped it in a rag, and slapped it, sizzling, on the chest.

From October to April the walls of Limerick glistened with the damp. Clothes never dried; tweed and woolen coats housed living things, sometimes sprouted mysterious vegetation. In pubs, steam rose from damp bodies and garments to be inhaled with cigarette and pipe smoke laced with the stale fumes of spilled stout and whiskey and tinged with the odor of piss wafting in from the outdoor jakes where many a man puked up his week's wages.

The rain drove us into the church – our refuge, our strength, our only dry place. At Mass, Benediction, novenas, we huddled in great damp clumps, dozing through priest drone, while steam rose again from our clothes to mingle with the sweetness of incense, flowers and candles.

Limerick gained a reputation for piety, but we knew it was only the rain.[1]

A view of modern Limerick, Ireland from the River Shannon.

Earth possesses a wide variety of climates, from the perpetually cold to perpetually hot, from the bone dry to the soggy wet such as that of the west of Ireland described above by Frank McCourt. Climate creates impressions and shapes our attitudes about places and perhaps more than any other aspect of the natural environment fixes our notion of the romantic and exotic of Earth. But climate also provides a framing system for understanding Earth environments because it sets the conditions that nurture and limit biota, govern soil development, regulate water supplies, and influence land use and the distribution of people. Perhaps more than any other environmental feature, climate gives Earth its special character as a planet.

What is climate? A traditional classification of climate reads something like this: Climate is the general conditions of the atmosphere at a place on Earth, including not only annual, seasonal, and monthly temperature and precipitation characteristics, but some weather processes and extreme conditions as well. In defining the climate of Miami, for example, it is important to know that it is warm in all seasons, with abundant precipitation throughout the year, and that most rainfall is delivered by convectional storms, and, in most years, a tropical storm or two, including a hurricane, can be expected. But this definition clearly does not fit everyone's needs. A farmer might want to know about summer evaporation rates, the damaging effects of hurricane winds, and the probability of an infrequent killing frost. Air-traffic planners, on the other hand, would want to know about seasonal wind directions, the intensity of heavy downpours, and the occurrence of downbursts from thunderstorms. But these views of climate, though meaningful to the Miami chamber of commerce, a farmer, and an air-traffic planner, only partially meet our objectives. There is also a need to consider climate as a system or, better yet, as a complex of systems, so that we can understand why climate conditions vary as they do in different parts of the planet. This sort of thinking is necessary when we consider how uncertain, in the face of planetary climate change, the geographic pattern of climates is becoming.

7.1 The Nature of Climate Systems

Many systems are responsible for shaping Earth's climates but their roles and order of importance differ geographically. By roles we mean how their influence on climate is actually played out. Some systems operate according to relatively structured and predictable regimes, and are part of all climates throughout the world, whereas others are far less structured, less predictable in their operation, and less universal in geographic coverage. We could frame this concept with two questions. First, to what extent can we expect a particular system to behave in an expected way as a part of climate, that is, does it follow a relatively predictable pattern or regime or does it follow one that is more variable and unexpected? Second, is the geographic coverage of a system regional or global, that is, does it play in all climates across the world or is it exclusive to certain regions and climates?

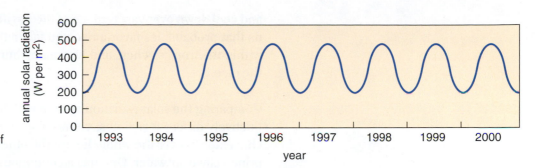

Figure 7.1 Annual fluctuation in available solar radiation for 30 degrees north latitude. The pattern of input in this system is consistent from year to year.

It turns out that geographic coverage and behavior are related. Systems that are less variable in their behavior tend to have greater geographic coverage. The solar-radiation system is a good example. All over the Earth, each year, solar radiation rises and falls according to a predictable regime, one that is marked by the seasons, as shown in Figure 7.1. And year in and year out the geographic distribution of solar radiation follows a distinct global pattern, one that varies broadly with latitude. This is not to say that the solar-radiation system exhibits perfect order, because there are many factors that cause it to vary from year to year and place to place. The most important factor is cloud cover.

Worldwide cloud reflection reduces incoming solar radiation by an average of 24 percent but, as we discussed in Chapter 4, this value is not uniform over the entire planet. Over the great deserts, for example, cloud reflection may be less than 10 percent, whereas over large areas of the wet tropics and subarctic seas, it rises to more than 40 percent. Added to clouds are interruptions in the radiation system from aerosols such as volcanic ash, dust from deserts, and pollutants from land use, which reduce atmospheric transparency, but more at certain places and times. However, as atmospheric systems go, the solar radiation system is remarkably steady, varying in a predictable way with the seasons and latitude. In the midlatitudes we see the solar-radiation system played out every year in every climate, wet and dry, warm and cold, as its input peaks in summer and wanes in winter despite variations in cloud cover and atmospheric transparency.

Not so with many other atmospheric systems that contribute to climate. Let us look at hurricanes as an example. Hurricanes rise and fall seasonally but they are notoriously elusive in terms of their times and places of occurrence. They cover only certain geographic regions and for any location within those regions, as Figure 7.2 shows, the number of storms tends to vary, often radically, from year to year. We know that hurricanes can be expected only over water where and when the ocean surface temperature equals or exceeds 27°C. And when the tropical seas heat up

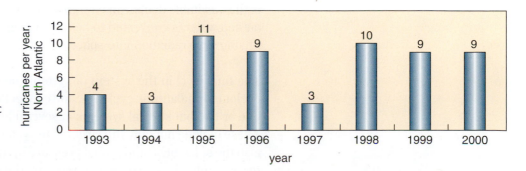

Figure 7.2 Hurricane frequency in the North Atlantic for the period 1993–2000. The development of hurricane storm systems tends to be inconsistent from year to year.

and cool down over vast regions of the Pacific and Atlantic oceans, climatologists tell us that probabilities favor a rise and fall in the occurrence of hurricanes. But where within the tropics, when, and how many hurricanes are produced is largely a role of the dice.

Comparing the solar radiation system with hurricane systems, we find they have distinctively different regimes, yet both systems are central to our concept of climate. One extends over the whole Earth; the other covers only the tropics and subtropics, principally over water. One has a continuous presence over nearly all the globe (the exception is the polar regions in winter) with predictable rises and falls over the year; the other is discontinuous, active in one part of the year, inactive in the other, and highly variable from year to year in frequency, magnitude, and location.

A System of Systems: The topics covered in the chapters previous to this one follow the flow of energy into and through the atmosphere and oceans more or less step by step via a series of systems. These systems are arranged in a hierarchical manner beginning with the Earth's primary energy system, solar radiation, followed by systems such as heat, air pressure, wind, and ocean circulation, all built directly or indirectly upon the solar system. The resultant sequence of systems looks something like the model depicted in Figure 7.3.

At the base of the model is the solar-radiation system. Barring tiny inputs from geothermal (crustal) sources, solar radiation is the sole source of energy driving the Earth's heat system, which is represented by the second level in Figure 7.3. But the amount of energy available to the heat system is less than that in the solar system because in the transfer of energy from one system to the other, some of the solar energy is given up, lost to the Earth. This energy is lost to incoming solar radiation in cloud reflection, backscattering, and albedo. As a result, the bar representing the global heat system is smaller than the one for the solar system.

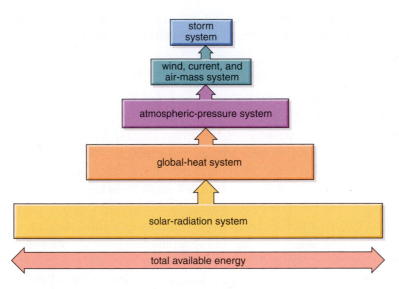

Figure 7.3 A system of systems. Beginning with solar radiation, energy is passed from system to system with large losses at each transfer level.

At the next level, heat (along with other forces) drives the atmospheric-pressure system, producing a global system of pressure belts such as the equatorial low and the polar highs. Once again the transfer of energy from the heat system to the pressure system results in a sizable energy loss, mainly in the release of longwave radiation, from the heat held in the atmosphere. The third level in Figure 7.3 takes us to the global circulation systems, wind systems, ocean circulation, and air masses, the systems responsible for redistributing heat and matter, especially water. Again there is the inevitable decline in energy as the kinetic energy (energy of motion) of moving air and water is converted to heat, which in turn becomes part of the planet's outflow of longwave radiation into space.

The final level in the system of systems takes us to storm systems, such as tropical cyclones, midlatitude cyclones, and convectional storms. They are the products of not one or two but several systems at several levels, working together in complex ways. These include not only the circulation systems (winds, currents, and air masses), but also the solar, heat, and global pressure systems. In the case of a midlatitude cyclone, for example, the storm is generated by the mixing of air masses and the circulation of upper-level and surface-level winds and, for most cyclones, a large part of their energy is provided by the ocean thermal system because it drives the evaporation process that produces the latent heat that powers the storm.

In the case of convectional storm systems, they come in two varieties, more or less, based on how they are generated. One variety is born of intense surface heating by solar radiation and is powered by latent heat released with the condensation of water vapor. The other variety is more complex, for it is spawned by other storm systems, notably, hurricanes, midlatitude cyclones, and orographic storms, which themselves are the products of other systems such as winds and air masses (Figure 7.4).

Uncertainty in Climate Systems: How does this portrayal of a multilevel system of systems fit into our discussion of climate? For one thing, it helps in understanding what we might refer to as the **uncertainty factor** in global climate. Simply put, systems tend to be more uncertain, or more chaotic, at higher levels in the system of systems. This follows our earlier discussion about variability in the behavior of systems, and it is relevant to climate studies because the variability adds the element of risk to the mix of factors that go into a definition of different climates.

Consider the differences between the American South and the Pacific Northwest in terms of storm magnitude, frequency, and types. Both regions have temperate climates with precipitation in all seasons. The South is frequented by hurricanes, thunderstorms, tornadoes, and midlatitude cyclones. In the Pacific Northwest, on the other hand, only one type of storm, the midlatitude cyclone, can be expected year after year in the belt along the coast (Figure 7.5). These storms are often large and they are most frequent in winter, but they are virtually the only storms operating in this climatic zone and, unlike thunderstorms, tornadoes, and hurricanes in the South, they are less uncertain in terms of how and where they operate.

Figure 7.4 Two varieties of convectional storm systems: (above) an isolated cell resulting from local surface heating; and (below) a complex of cells associated with a cyclonic front.

Figure 7.5 Cyclonic storm approaching the Pacific Northwest coast. Cyclones are virtually the only storms to frequent this region.

The uncertainty factor also holds for the geographic coverage of climate systems. Systems of the first three levels, namely, solar, heat, and pressure, are global in coverage, whereas the next two levels are less than global, mostly regional, and in the case of the variety of convectional storms caused by intensive solar heating, they are local. In a general way, variability and risk increase as the geographic scale of the system decreases. Tornadoes, which have very limited geographic coverage, rank among the least predictable storm systems in the atmosphere.

Finally, we are led to ask about the effects of global warming in terms of uncertainty and variability in atmospheric systems. Since circulation systems and storm systems are directly or indirectly products of the global heating system, it follows that they are subject to change with global warming. Put another way, global warming will result in more than a shift in temperature lines on a world map. It will also result in changes in the strength and frequency of storm systems and the precipitation they produce, including changes in their geographic distributions. But how all this atmospheric change in heat, circulation, storms, and precipitation will be played out in different parts of the planet is a tough question, which we will have a serious look at in the next chapter.

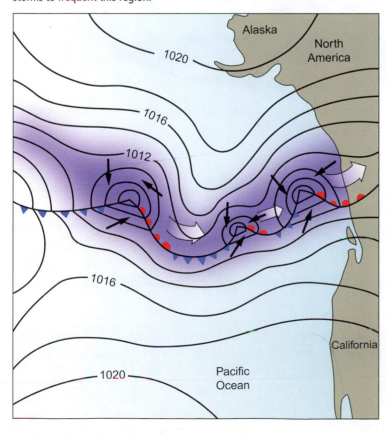

Alaska

North America

1020

1016

1012

1016

California

1020

Pacific Ocean

7.2 The Great Engines of Global Climate

Although many systems participate in shaping the Earth's climates, two big atmospheric engines function as the central drivers, one in the tropics and the other in the midlatitudes. Together these two engines dominate about 90 percent of the planet, but the tropical engine, for reasons we will describe below, is far and away the major player in the larger global climate system.

The Tropical Climate Engine: The broad swath of Earth between the tropics of Cancer and Capricorn, called the tropical zone or intertropical zone, covers a total of 47 degrees latitude and 40 percent of the Earth's surface area. It is decidedly the power center of the planet. Witness that it receives more solar energy than all the rest of the world combined. Its atmosphere stores as much or more latent heat in water vapor than the air over all the rest of the world and is capable of yielding as much precipitation as all the rest of the Earth's atmosphere. Added to the massive reservoir of atmospheric energy is a second massive energy reservoir, the heat stored in the tropical seas, a fact reflected in the high surface-water temperatures shown in Figure 7.6.

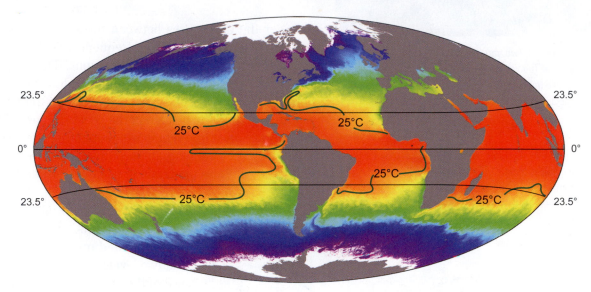

Figure 7.6 Satellite image showing typical ocean surface temperatures, including the large area of tropical water at a temperature of 25 °C or more.

Follow the Sun: The combination of all this heat and moisture makes for a highly dynamic climate system in the tropics. It is driven by solar radiation and the simplest and most direct expression of this on land is the convectional activity produced by intensive solar heating. Indeed, in the literature of this part of the world, whether from South America, Africa, or Asia, it is the daily regime of afternoon rainstorms that is trademark attribute of equatorial and tropical climate. The phrase "follow the Sun" would be an appropriate starting point for our examination of the climates of the tropics.

Recall from our discussion of sun angles that the heart of the equatorial zone (at the Equator) never experiences a sun angle less than 66.5 degrees, and twice a year, on the equinoxes, the sun angle is 90 degrees. Therefore, within the belt along the Equator, there is plenty of energy to drive convectional storms more or less all year.

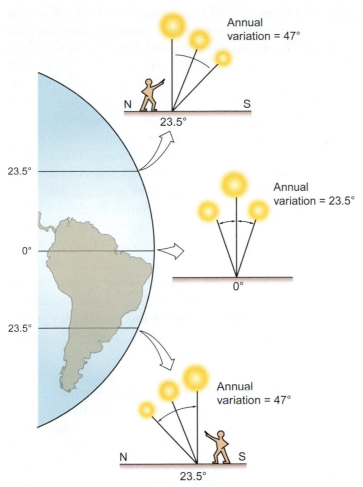

Figure 7.7 Annual variation in sun angle in the tropical (intertropical) zone, only 23.5 degrees in the center, but 47 degrees on the margins

Figure 7.8 A typical African landscape in the tropical wet/dry zone. Summer drought prohibits a full forest cover.

Poleward from the equatorial zone the picture changes, as Figure 7.7 illustrates. For each degree of latitude, the annual range or variation of sun angle increases by one degree. At the very edge of the tropical zone (23.5 degrees latitude), annual sun angle varies from a high of 90 degrees on the summer solstice to a low of 43 degrees on the winter solstice.

This is a big change in sun angle and it produces a big change in solar heating. And if our assertion that climate in the inter-tropical zone follows the Sun, then climate should change from the high-sun season to the low-sun season. And that turns out to be true, especially over land areas where surface heating in response to solar intensity is most pronounced (compared to water). When the Sun is high, thunderstorms abound and it rains; when the Sun is low, thunderstorms are scarce and it is dry. Thus, the equatorial belt in the middle, spanning about 10 to 15 degrees of latitude, experiences substantial precipitation in all seasons of the year, whereas the tropical belts to the north and south in each hemisphere, each spanning about 15 degrees of latitude, experience substantial precipitation only during the summer season.

So at the most elementary level, the broad, 47-degree zone of the tropics can be divided into two climate types, tropical wet and tropical wet/dry. Both are warm in all seasons, but the tropical wet/dry has a distinctive dry season and it is reflected in the landscape. Instead of the heavy forest cover of the wet tropics, including the celebrated rainforest, dominating the landscape, the wet/dry climate is characterized by landscapes of grasses, shrubs, and scattered trees, a response to winter drought. The African savanna, shown in Figure 7.8, is the most familiar landscape of the wet/dry tropical climate.

The Intertropical Convergence Zone: The convectional activity described above that follows the Sun from hemisphere to hemisphere is part of a circumglobal belt of low pressure and converging winds called the intertropical convergence zone (ITCZ). We discussed the ITCZ in some detail in Chapter 5, including its seasonal shifts over water and land and its role in atmospheric circulation in the tropics. You will recall that the ITCZ is fed by the tradewinds, shown in Figure 5.29, a system of prevailing winds, which bring warm, moist air from the oceans into the ITCZ. Here the north-east trades and the southeast trades converge and are driven upward as they meet, resulting in instability and precipitation.

In general the ITCZ can be characterized by a highly active belt of instability, cloud cover, precipitation, and converging tradewinds. These wind systems feed the ITCZ with huge amounts of energy. This energy drives a powerful upward flow that functions, as it were, as the piston of the tropical engine. The air pumped aloft enters the upper atmospheric circulation system, which carries the air poleward to around 20 to 30 degrees latitude where it descends. This descending air produces the subtropical high-pressure cells, huge centers of warm, dry air, the main sources of the world's great deserts.

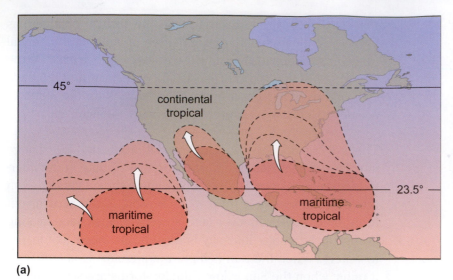

(a)

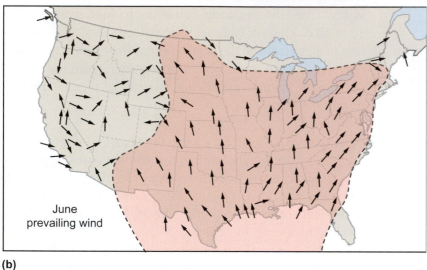

(b)

Figure 7.9 (a) Source areas and general incursion patterns for tropical air masses in North America. (b) Prevailing, monsoon-like, air flow in June over central North America.

The Long Arm of the Tropics: But to think of this tropical climate engine as limited geographically to the tropics would be misleading, for its influence extends way beyond the tropics, well into the midlatitudes. Consider air masses that form within the tropics and migrate poleward into the zone of midlatitude cyclonic mixing. In North America, the most prominent of these air masses, shown in Figure 7.9a, is the maritime tropical air from the Caribbean/Gulf of Mexico region, a major energy source in weather systems throughout the year reaching well into southern Canada. In the summer, as the continent heats up, this poleward flow of tropical air takes on a monsoonal character, even in North America, as the map in Figure 7.9b illustrates. In Asia, the monsoon flow is many times larger, massive by comparison, drawing in its moisture-rich, high-energy air all the way from the heart of the intertropics in the equatorial region of the Indian Ocean (see Figure 5.26).

Another arm of influence from the tropics is the outflow of warm ocean water. We see this in major ocean currents like the Gulf Stream, which carry warm water from the tropics high into the midlatitudes. The energy transfer by the Gulf Stream alone is astounding. With a flow of water exceeding that of all the rivers in the world, the Gulf Stream delivers tremendous amounts of heat energy to the midlatitudes. In fact, scientists estimate this amount to be about 100 times the world's total energy demand, that is, the demand for heat and power from 6.9 billion humans and their machines. As Figure 7.10 illustrates, the Gulf Stream heat is transported northeastward toward Europe where it is given up in evaporation and converted to latent heat of water vapor. So strong is the influence on the climate of the British Isles and Northwest Europe that some climatologists estimate this region would be 4 to 6°C cooler without it.

A third arm of the tropical climate engine is the ITCZ–subtropical high-pressure connection described above. This system is known as Hadley cells, and it is described in Chapter 5 and illustrated in Figure 5.10. The Hadley cells form two great loops of airflow between the ITCZ–equatorial low-pressure system and the subtropical high-pressure systems. When the equatorial/tropical air is driven aloft, it is carried poleward by upper atmospheric winds, called the antitrades. When it reaches around 20 to 30 degrees latitude, it plunges back to Earth and, during its descent, it is heated adiabatically (that is, heated by the force of increasing air pressure with decreasing altitude) so that by the time it reaches the surface it is warm, and thus represents an energy export from the tropics. From the subtropical highs, westerly winds carry this air farther north, as illustrated in Figure 7.10.

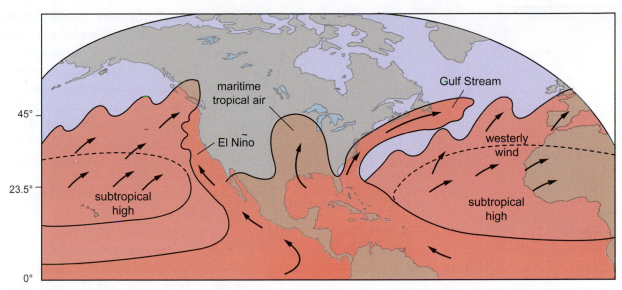

Figure 7.10 The four main sources of tropical energy input to the North American midlatitude zone.

A fourth arm of the tropical climate engine falls under the heading of ocean-driven oscillations and there is no better example than El Niño. Oscillations are regional variations in weather and climate over periods of several years to several decades caused by changes in heat transfer by ocean circulation. El Niño's period is about five years, and you will remember from Chapter 5 that it rises when the Pacific counter-equatorial current system pumps lots of warm water to the west coast of South America resulting in a massive heat reservoir that expands northward and southward along the coasts of both Americas. As Figure 7.10 shows, El Niño water usually reaches high into the midlatitudes where it fuels cyclonic storms, increasing precipitation, often dramatically, over much of western North America.

Summary on the Tropical Climate Engine: The broad belt between the tropics of Cancer and Capricorn is the climate power center of Earth. It receives more solar radiation and stores more heat than all the rest of the planet combined. Within this belt, climate varies mainly with seasonal changes in sun angle and the position of the ITCZ. Beyond this belt, the tropical system's influence extends well into the midlatitudes via both atmospheric and oceanic circulation systems.

The Midlatitude Climate Engine: The climates of the midlatitudes are the most diverse and variable on the planet. Broadly defined, the midlatitudes range from the tropics at latitude 23.5 degrees to the Arctic and Antarctic circles at latitudes 66.5 north and south, and cover 52 percent of Earth's surface. Within this swath in each hemisphere is the atmosphere's main arena of zonal (latitudinal) mixing where the energy contributed by the tropics is transferred poleward to the high latitudes. With polar air on one side and tropical air on the other, the midlatitudes share characteristics of both atmospheric regions, and it is not unrealistic to think of this zone as a huge playing field in which these contrasting types of air vie for position and become entangled in cyclonic storms.

The tropical/polar contrast stands out vividly in the distinct seasonal character of most midlatitude climates. In the continental interiors at 40 or 45 degrees latitude, summers are tropical-like and winters are Arctic- or subarctic-like. This is not surprising considering that the summer high sun angle (68.5 degrees) at 45 degrees north

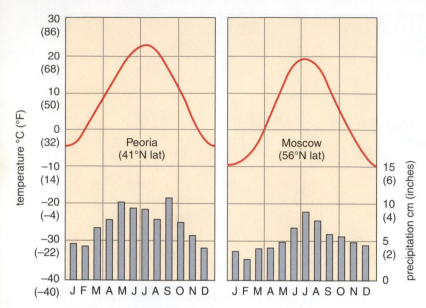

Figure 7.11 Mean monthly temperature and precipitation for Peoria, Illinois, and Moscow, Russia. The tropical-like summers contrast with the Arctic/subarctic-like winters.

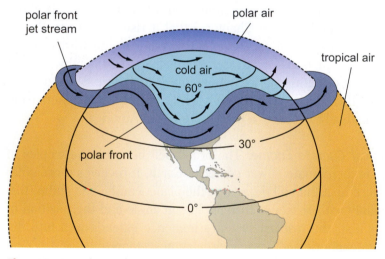

Figure 7.12 Rossby circulation over the northern hemisphere with the polar-front jet stream marking the leading edge of the cold air sector.

latitude is about the same as the sun angle at the Equator on the same date, and that the winter low sun angle (21.5 degrees) is about is about same as the summer high-sun angle at the North Pole. The tropical-like summer conditions produce monthly average temperatures in July around 22 to 24 °C (71 to 75 °F), and the Arctic/subarctic-like conditions produce monthly average temperatures of −10 to 0 °C (14 to 32 °F) in January. For example, Peoria, Illinois, at 41 degrees latitude has a summer high average temperature in July of about 23 °C and a winter low in January of about 6 °C. Moscow, Russia, at 56 degrees latitude, has July and January averages of close to 20 °C and −10 °C (Figure 7.11).

Seasonal contrasts are also found with precipitation, and these are also apparent in the graphs in Figure 7.11. Summer precipitation is about twice that of winter. Much of this is related to the summer heat over the continental interiors, which induces tropical-like precipitation in the form of convectional storms. Unlike the tropics, however, where the winters are marked by little or negligible precipitation, winters in the midlatitudes are decidedly wet, but precipitation comes not from convection but from frontal/cyclonic systems. Not surprisingly, summers see more incursions of tropical air than winters, a major source of moisture for the convectional storms, whereas winters see more incursions of polar and Arctic air, including occasional severe winter freeze.

Rossby Circulation: We highlight these seasonal contrasts to underscore the place of the midlatitude zone in the global climate scheme. Central to understanding these seasonal differences is the system that generates cyclonic storms and drives the mixing process where warm air and cold air exchange energy. The cyclonic engine operates along the polar front, and you will recall from our discussion on atmospheric circulation in Chapter 5 that the polar front is part of an upper atmospheric airflow system called Rossby circulation. Rossby circulation is defined by a wavy band of strong westerly flow that marks the leading (equatorward) edge of the Earth's cold air sector. To illustrate this phenomenon, refer to the diagram in Figure 7.12. As you can see, the polar-front jet stream is the hallmark feature of Rossby circulation.

The behavior of the polar-front jet stream is central to understanding both the weather and climate of the midlatitudes. First, as the meandering waves of the jet stream migrate westward, weather changes more or less from warm to cold and cold to warm depending whether you are on the inside or outside of a wave. Second, the whole system migrates north and south with the seasons. At the same time the system changes strength from strong in winter and to weak in summer. Therefore, in locations such as central North America, the jet stream and the cyclones generated along the polar front are the dominant theme of the winter climate, but in summer the whole system weakens and retreats far to the north leaving this vast region under mostly tropical and subtropical air with most precipitation delivered by convectional storms rather than cyclones.

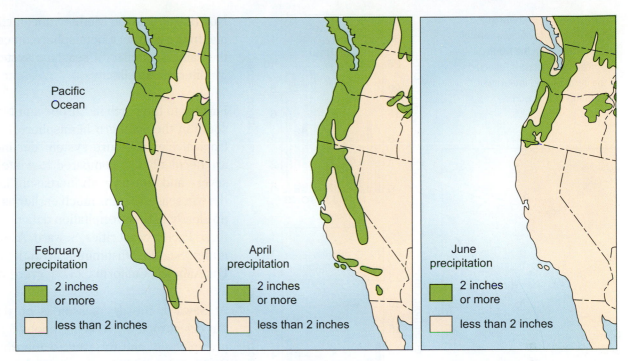

Figure 7.13 The retreating pattern of precipitation on the North American West Coast in response to the decline and retreat of the polar front, February through June.

The Maritime Theater: In addition, there is a significant difference in the jet stream and polar front between the oceans and the continents. Because of the ample supply of moist, high-energy air over the oceans, the winter cyclonic system is much stronger over water than over land. Not only that, but the winter polar front and its jet stream are located farther poleward over the oceans than over the land, as Figure 7.12 indicates. Over the North Atlantic and North Pacific the system is capable of generating a nearly continuous stream of cyclones throughout the winter, which are driven onto the west coasts of North America and Europe. Frontal/cyclonic precipitation is heavy in both places. Toward summer, precipitation declines as the system weakens and retreats and, as Figure 7.13 shows, precipitation follows suit with a steady decline as the system migrates northward.

Why is this decline in frontal/cyclonic precipitation not replaced by convectional precipitation as it is at Peoria and Moscow in the interior of the continent in summer? There are two reasons. First, summer convectional activity is spawned in large part by intensive surface heating, and water masses such as the North Pacific are incapable of generating high enough surface temperatures to trigger instability and convectional processes. This is simply a matter of the difference in volumetric heat capacity and the depth of thermal mixing between land and water, a topic we examined at length in Chapter 4. For a given amount of solar radiation absorbed by water and land surfaces, land produces much higher temperatures than water. Second, as the polar front and jet stream shift poleward, the subtropical high-pressure systems expand and shift poleward behind it bringing stable, dry conditions to the very zone dominated by cyclonic systems in the winter.

This seasonal shift is so pronounced on the southern (equatorward) margin of the coastal midlatitude zone that it gives rise to a distinct climate type, one with a stark summer/winter contrast in precipitation. It is named for the Mediterranean region of Europe whose long, arid, sunny summers are celebrated in Western civilization. In

Figure 7.14 The decreasing influence of the Pacific subtropical high northward on the North American West Coast is reflected in the duration and intensity of the summer dry season.

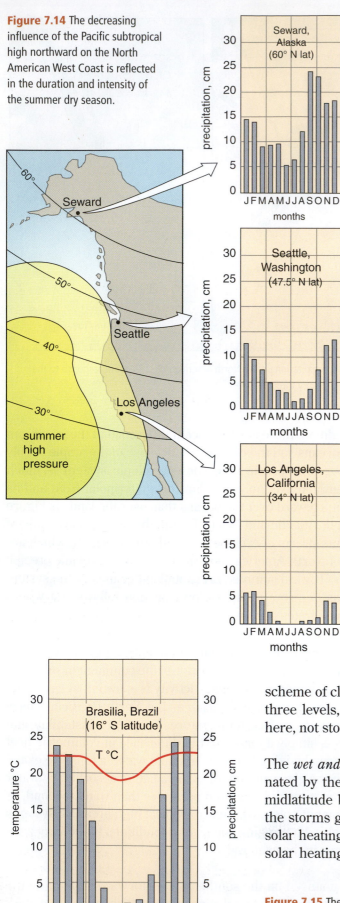

North America, the Mediterranean climate belongs to California. In the southern hemisphere, in central Chile for example, the same sequence of seasonal system shifts also produce a Mediterranean climate.

Farther to the north in Europe and North America (and to the south in the southern hemisphere), the influence of the subtropical high-pressure system declines and, while summers are decidedly dry compared to winter, the dry season is less severe and shorter with increasing latitude. Not surprisingly, the wet season begins much earlier as well. These trends show up clearly in the precipitation data presented in Figure 7.14 for three Pacific-coast cities, Seward, Alaska (60 degrees north latitude), Seattle, Washington (47.5 degrees north latitude), and Los Angeles, California, (34 degrees north latitude).

At Seward, the dry season is only two months long, with the driest month averaging about 6 centimeters (2.3 inches) precipitation, and the wettest month coming in early fall. At Seattle, about 1400 miles (2300 kilometers) down the coast, the dry season is five to six months long, with the driest month averaging less than 2 centimeters (0.6 inches) precipitation, and the wettest month coming three months later than Seward. About 1400 miles further down the North American coast at Los Angeles, the dry season is six to seven months long, with the driest month averaging only a trace (effectively 0) of precipitation.

Overview on the Global Climate System: Broadly speaking, the climates produced by the tropical and midlatitude climate engines tend to fall into two big classes: wet and stormy and dry and stable. Some belong more or less perpetually to one or the other, whereas others fluctuate seasonally from stormy to stable. Good examples of the *dry and stable climates* are the regions of subtropical high pressure, like the Sahara of North Africa, which are dominated year in and year out by clear skies and very light (though variable) precipitation. In the great scheme of climatic systems shown earlier in Figure 7.3, it is the systems of the first three levels, solar radiation, heat, and pressure, that set the character of climate here, not storms, turbulence, and precipitation.

The *wet and stormy climates* are exemplified by those regions in the tropics dominated by the ITCZ–equatorial low, a wet zone of perpetual turbulence, and by the midlatitude belt dominated by the polar front and cyclonic mixing. In the tropics, the storms generated in the ITCZ–equatorial-low zone rise and fall seasonally with solar heating, as illustrated for the capitol of Brazil, Brasilia, in Figure 7.15. When solar heating declines in the winter (March–September) storminess declines and

Figure 7.15 The correlation between seasonal temperature and precipitation in the tropical wet/dry climate of Brazil. Wet and stormy conditions rise with summer heating and development of the ITCZ–equatorial low system.

atmospheric conditions shift toward the dry and clear, that is, toward greater stability and less chaos. And it is not just the convectional storms of the high-Sun season that we are talking about, for hurricanes and related tropical storms also add to the warm-season chaos of the tropics, especially in coastal areas. As you know, these storms rise with the heating of the oceans, which usually reaches its peak near the end of the high-sun season.

In the midlatitudes, wet and stormy conditions rise in winter with the strength of the polar front and jet stream. The magnitude and frequency of midlatitude cyclones increases, and this holds for both the maritime and continental climatic theaters of this zone. With summer, however, the systems change and these two theaters follow markedly different regimes. The continental theater shifts into a wet and stormy regime dominated by convectional systems, whereas the maritime theater shifts toward dry and stable dominated by high-pressure systems.

Continental regions, such as the US South and Midwest, become more chaotic in summer not only because there are more and bigger thunderstorms, but because tornadoes and hurricanes are thrown into the mix. Poleward, summer chaos declines as thunderstorm activity diminishes, tornadoes become rare, and hurricanes are unknown. Over the ocean, the chaos trend is generally the opposite, because with distance poleward the subtropical high fades away and the polar front takes on a greater presence with cyclonic storms playing a role more or less all summer.

A good illustration of the two-part system in the midlatitides is provided by comparing seasonal precipitation in the American southeast (the South) with the coastal southwest (California). The graphs in Figure 7.16 give monthly average precipitation for Chapel Hill, North Carolina (latitiude 36 degrees) and Santa Barbara, California (latitude 35 degrees). Notice the difference in summer precipitation values and the similarity in winter precipitation values. In winter both places are dominated by frontal/cyclonic precipitation, whereas in summer one is stable and dry and the other stormy and wet.

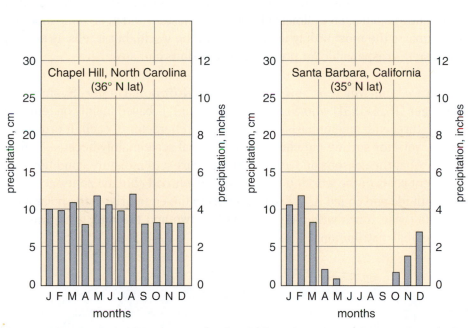

Figure 7.16 Average monthly precipitation for Chapel Hill, North Carolina, and Santa Barbara, California. The summer difference in precipitation reflects the difference in summer climate systems, wet and stormy versus dry and stable.

7.3 Traditional Climate Classification

Early thoughts about classifying the world's climates can be traced back to the ancient Greeks who divided Earth into three climate zones based loosely on temperature: torrid, temperate, and frigid. For them this made sense, considering that their notion of climate was limited to things they could observe in a world that stretched more or less from Egypt to northern Europe and central Asia.

The first attempts at modern climate classification did not come until much later, in the 1800s, when plant geographers and botanists searched for a connection between climate and plant distributions. This was a time of active exploration and observation; however, few data were available on any geographical phenomena including weather and climate. Indeed, for most of the world, average temperature and precipitation values were unknown, so building a map of world climate based on recorded data was virtually impossible and relating it to atmospheric processes and systems was even more remote.

On the other hand, eighteenth-century scientists understood a fair amount about heat and moisture controls on plants. Simple experiments showed that tropical plants, for example, could not survive in the midlatitudes below certain temperature levels and, at a more general level, field studies revealed an apparent relationship between the size and structure of vegetation and different climate types, such as tropical forests and perpetually warm and wet conditions. In the absence of climatic data, plants and vegetation were used as indicators of climate.

The boundaries between climatic zones were drawn according to observable patterns of world vegetation. The resultant climatic zones were then defined according to monthly and annual moisture and temperature conditions suggested by vegetation patterns. In 1900, the distinguished German climatologist, Vladimir Köppen (1846–1940), developed such a climate classification system and it, or some modified form of it, remains the most widely used system in the world, but it tells us nothing about the systems responsible for producing the various climates.

The Köppen system, which is now known as the **Köppen–Geiger System** (for Rudolph Geiger (1894–1981) who worked with Köppen), is based mainly on temperature and moisture, with an emphasis on the seasonal character of each. It provides a simple framework for sorting out climate at the global scale mainly because it gives us an idea of what the seasons and annual temperature and moisture conditions are like in different parts of the world. It begins with five broad climatic zones arranged roughly by latitude and coded A, B, C, D, and E from the equatorial to the polar zone. Table 7.1 lists these zones with a defining trait or two and the map in Figure 7.17 shows their global distribution.

Table 7.1 Main climatic zones of the Köppen–Geiger System

Zone	Defining trait
A	Tropical and equatorial rainy climates
B	Dry climates: the potential exists for evaporation and transpiration (evapotranspiration) to exceed precipitation
C	Temperate, moist climates: long, warm summers, cool winters
D	Cold, snowy climates with forest: short summers, cold winters
E	Polar climates: long, cold winters, brief or no summers (no forests)

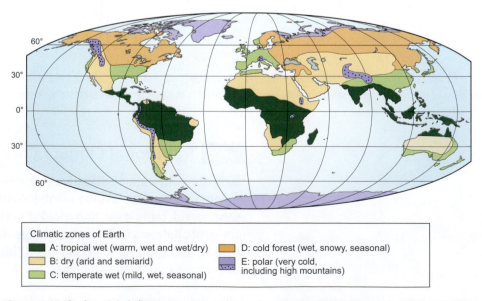

Climatic zones of Earth

A: tropical wet (warm, wet and wet/dry)
B: dry (arid and semiarid)
C: temperate wet (mild, wet, seasonal)
D: cold forest (wet, snowy, seasonal)
E: polar (very cold, including high mountains)

Figure 7.17 The five global climate zones according to the Köppen–Geiger climate classification system.

Earth's Major Climatic Zones: The climatic extremes of the Earth range from the perpetually warm conditions in the tropical climates of the A zone to the perpetually cold conditions in the polar climates of the E zone. Between these zones, the climates can be divided into those that are wet (found in the C and D zones) and those that are dry (found in the B zone). The dry climates are either arid or semiarid and may be warm or cool depending on latitude. The wet climates of the midlatitudes are all distinctly seasonal, characterized either by warm summers and cool winters or cold winters and short summers or, in one case, the Mediterranean climate, by wet/dry seasons.

The A Climates: The tropical wet climates center on the Equator and span the broad intertropical zone from the Tropic of Cancer to the Tropic of Capricorn. This zone is either perpetually warm and rainy year around or seasonally. As we illustrated earlier in this chapter, this is the Earth's main reservoir of heat and moisture, the zone that houses the engine that drives the atmospheric system throughout the tropics over much of the midlatitudes. The A climates are divided into two types, wet in all seasons and seasonally wet/dry. The bulk of the perpetually wet type, called the equatorial wet (or tropical rainforest) climate, is found in a belt about 20 degrees wide centered on the Equator and is represented on land by the three core regions: the Amazon Basin of South America, the Congo Basin of western Africa, and the peninsulas and islands of southern and southeast Asia.

There is one exception to the equatorial location of this equatorial wet climate. It is found on the tropical east coasts, extending poleward to 20 degrees latitude or more in both hemispheres, where the easterly tradewinds hit the coast and provide a steady supply of moist air in all seasons. The result is a coastal arm of the equatorial wet climate, and it is particularly pronounced along mountainous coasts like those in Central America. This is one of the Earth's non-zonal climates, that is, it stretches north–south rather than east–west.

The tropical wet/dry climate lies on the poleward sides of the equatorial wet areas. Often referred to as the tropical savanna climate, because of the mixed grass/tree character of its landscapes, this climate experiences maximum rainfall in summer and drought or drought-like conditions in winter, as illustrated earlier for Brasilia. The most extreme version of the wet/dry tropical climate is found in the monsoon regions of south and southeast Asia where summer rainfall can reach 75 centimeters (30 inches) or more in the wettest month.

The B Climates: The dry climates lie on the shoulders of the A zone and occupy two subzones, one subtropical, and one midlatitude. These are the dry climates. Precipitation is light, typically 20 to 40 centimeters (8 to 16 inches) a year, and evaporation of surface water, when it is available, is so rapid that scarcely any gets into the soil for plants. The subtropical zone includes the world's great deserts, most notably the Sahara, the Arabian, and the Australian, the largest and harshest hot deserts on the planet. Although our image of subtropical deserts is land-based, the same atmospheric conditions also extend over vast areas of adjacent ocean, because the subtropical high-pressure systems that produce the clear, dry desert air also lie over the oceans. Therefore, if we link the oceanic and terrestrial parts together, the belt of B climate can be traced around the world in both hemispheres. There is also a midlatitude subzone of the B climates. It is cooler, smaller in area, except in Asia where it takes up a huge part of continent's interior in the lee of the Himalayas.

The source of most precipitation in the tropics, convectional storms capable of rising high into the troposphere.

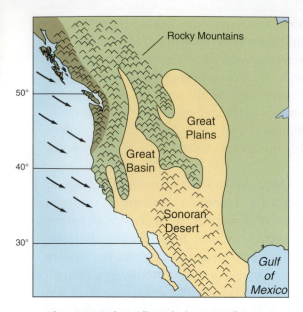

Figure 7.18 The midlatitude dry zones of western North America lie in the lee of two large mountain belts that intercept much of the moisture carried eastward by the prevailing westerly wind system.

The midlatitude variety of dry climate is associated mainly with mountains. In North America, for example, two fingers of arid and semiarid climate extend northward well into the midlatitudes, as shown in Figure 7.18. Both lie in the lee of major mountain chains: the west-coast mountain ranges of California, Oregon, Washington, and British Columbia; and the Rockies to their east. These mountain chains intercept the moisture carried inland by westerly winds and air masses from the Pacific. To a large extent, then, these are rain-shadow deserts and, because they extend north–south, they qualify as non-zonal climates. A geographically "clean" example of the two varieties of B-zone climate (subtropical and midlatitude) is found in southern South America, along the southern Andes (Figure 7.19) The Atacama Desert is a subtropical desert – one of the driest in the world – whereas the desert on the east side of the Andes is a midlatitude rainshadow desert.

The C Climates: The C climates are often described as mesothermal, temperate, and/or subtropical because they exhibit relatively mild temperatures in all seasons. This zone can generally be thought of as transitional between the A climates on one side and the cold-winter D climates on the other. Most of the area covered by C zone climates lies in the low midlatitudes between 25 and 40 degrees latitude. Major blocks of it are found in the southeastern United States and southeastern China where the winters are wet from frequent cyclonic/frontal storms and summers are warm and wet with heavy rainfall from frequent convectional storms. But there is one marked exception to the warm, wet summers.

The exception is found on the west coasts in North America, Europe, South America, Australia, and New Zealand where thermally mild conditions extend poleward, reaching as high as 60 degrees latitude or more where land is available. We touched on the geographic conditions that produce this west coast variant of the C climates earlier in the chapter, though failed to say much about their moderate temperatures related to the influence of the sea and the onshore flow of maritime air. It also bears repeating that the coastal C climates have a distinctly different precipitation regime, one marked by dry, stable summers and wet, stormy winters. The coastal C variety also qualifies as non-zonal, especially along mountainous coasts like Norway, southern Chile, and the Pacific coast of Canada and the United States.

The D Climates: These climates occupy the broad belt of the midlatitudes (excluding the midlatitude dry subzone of the B climates), roughly from the upper subtropics around 40 degrees latitude to the Arctic Circle. This zone covers large blocks of land (including lakes, large bays, and seas) in the northern hemisphere, but in the southern hemisphere it is largely absent for the simple reason that little land is available there above 40 degrees latitude. Two traits characterize the climates of this zone.

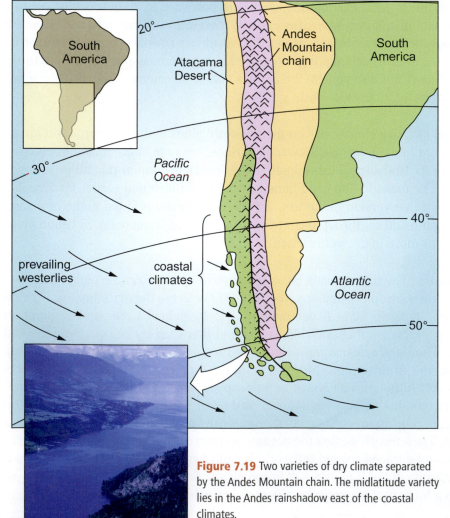

Figure 7.19 Two varieties of dry climate separated by the Andes Mountain chain. The midlatitude variety lies in the Andes rainshadow east of the coastal climates.

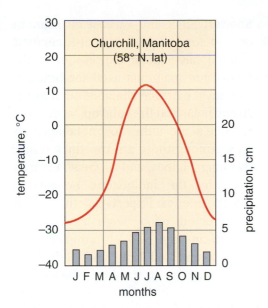

Figure 7.20 The annual temperature and precipitation pattern for Churchill, Manitoba (58 degrees north latitude), an example of a short-summer D-zone climate.

First, they are humid with enough precipitation in all seasons to support forests. Second, they are thermally seasonal with distinct winter and summer seasons. The seasons vary with latitude from modest, snowy winters and warm summers in the south, such as the US Midwest (Peoria in Figure 7.11 is good example) to long, frigid winters and short, cool summers in the north. Churchill on Hudson Bay in northern Manitoba (Figure 7.20) is a good example of the latter.

This cold variety of D climate is most pronounced in the continental interiors of North America and Eurasia where the annual temperature range is commonly 40 to 50 °C (see the map of global temperature ranges in Figure 4.39). So cold is the winter season in these northern interiors that precipitation (snowfall) is limited and in the most extreme region, eastern Siberia, the climate is given a dry-winter designation.

The E Climates: The Earth's coldest climates lie above 66.5 degrees in both hemispheres and cover only about 8 percent of the planet. Winters are long and ghastly cold; summers very short, far less severe, but still below freezing on average. Because of the deep cold, the atmosphere holds little moisture and, therefore, like winters in the northern interiors of the D zone, produces little precipitation. The largest continuous land area belonging exclusively to the E zone is Antarctica. Within the E zone, two climates are recognized, one colder than the other, and a mountain variant, designated ETH, that applies only to very high ranges like the Andes and the Himalayas.

Within each of the five climatic zones, two or three climate types are defined and these are described in Appendix B using a modified version of the Köppen–Geiger System. There you will identify some of the climates we discussed earlier in connection with climate systems, for example, the two tropical climates (tropical wet and tropical wet/dry), the two dry climates (desert and steppe), and the two of the midlatitude climates, the Mediterranean and the humid subtropical. Appendix B also provides a map showing the distribution of these climates in the world.

7.4 Applied Climatology: Shelter and Clothing in a Changing World

The influence of climate on people's lives cannot be underestimated. From rain dances and gods of the harvest to the decision to build a domed football stadium, climate is part of the cultural, spiritual, and economic pursuits of all cultures. To help appreciate the significance of climate to modern humans, consider this. If we were to strip humans of technology, which includes clothing and constructed shelter, and reduced ourselves to lives as biological organisms, we could not survive much beyond the tropical climates.

Technology has dramatically altered our relationship with climate in three fundamental ways. First, with coping mechanisms such as clothing, building materials, and heating systems we can settle in climatic areas that were formerly uninhabitable, or deemed too risky. Second, the chances of being exposed to more environmental hazards such as hurricanes and floods increases as people inhabit areas with higher natural risk, such as coastal zones. Third, the same technology which has enabled us to expand our habitat has, among other things, also created a warmer atmosphere with stronger storms and higher sea levels. This in turn may significantly increase the risks to the millions of people who have located to coastal areas and other high-risk environments, including some cities.

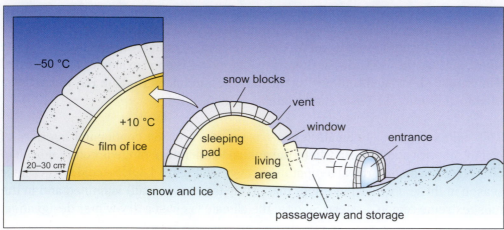

Figure 7.21 The basic design of an igloo where the insulating properties of snow facilitate enough heat retention to create a reasonably comfortable living environment in a polar climate.

Shelter: Since the earliest civilizations and before, adaptations to regional climate have found expression in the forms and materials used for housing. The most extreme climates, such as the Arctic polar and the subtropical deserts, provide some of the most notable examples. Among these are two thick-walled dwellings, igloos, and adobe structures, used by the indigenous peoples of North America. Experts say that a well-constructed igloo, such as the one shown in Figure 7.21, coupled with a very small oil lamp and body heat, can warm the interior of this snow building as much as 40 °C above the outside temperature to 10 °C (50 °F) or higher.

Igloo warming is possible because of the insulating properties of snow and ice. As the ice-crystals in a snow block melt, their rate of melting is slowed by the pockets of cold air around them. Additionally, these cold-air pockets channel some of the heat to the outside of the igloo through convection. As the heat inside melts the surface of the wall it freezes because the outside air is much colder than the inside air. The result is a strong sheet of ice on the inside of the igloo, which strengthens it and adds to the thermal insulation.

Figure 7.22 Heat flow in the roof of an adobe house. The interior is kept relatively cool during the heat of the day by the insulating capacity of the thick, earthen roof.

Though no longer extensively used in the American Southwest, the *adobe*, or mud masonry house is still the principal house type of desert people throughout the world. On first thought, the thick walls and closed design of the adobe house may seem inappropriate for the desert climate. However, on closer examination, as illustrated in Figure 7.22, we find that it is an especially good adaptation to the daily thermal regime of the desert because of the low thermal conductivity of the wood, straw, and soil material used for the roof and walls. Most of the heat absorbed by the roof and walls during daylight hours penetrates only 5 to 7 cm. Then, as air temperatures fall after sundown, the outer surfaces cool down and the temperature gradient in the adobe material reverses. The heat flows back to the surface, where it is released into the outside air. Some heat, however, flows through the roof, which accounts for an early-evening rise in indoor temperature. On balance, in its environment the house is the coolest place during the day and the warmest place at night.

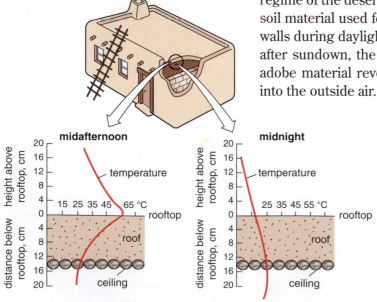

Another form of traditional building adaptation to extreme heat is a device used across the Middle East called a windcatcher. In their simplest form, windcatchers are capped towers with an open face in one or more directions to capture wind high above the ground where it is fast and clean. As shown in Figure 7.23, the wind is directed down the tower shaft and through the interior of the house where it flushes away stale air and improves comfort through air movement. A more sophisticated version of the windcatcher captures the wind outside the house and directs it underground where it is cooled by contact with cool soil and groundwater. The cooled air enters the lower level of the house and is drawn upward into an exhaust tower over the house with its opening facing downwind.

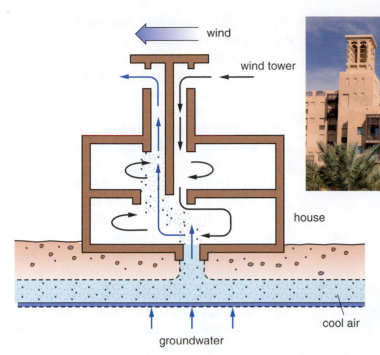

Another ancient building adaptation to climate that is sparking renewed interest in urban areas is the green roof. Green roofs in the form of sod over wood poles have long been used in both hot and cold climates. They were used in the warm climates of Mesopotamia and southern Africa and in the cold climates of Northern Europe. The Vikings, for example, used them in Scandinavia, Iceland, and Greenland. The modern green roof, like the one shown in Figure 7.24, is a bit more complex than sod and poles, of course. It consists of several layers that include structural support, a roofing membrane, a root barrier, insulation, a drainage barrier, a growing (rooting) medium for plants, and the plants themselves. One of the preferred plants is sedum, a small, fleshy, drought-resistant plant requiring little maintenance.

Figure 7.23 The basic design of a windcatcher ventilation system, a system that remains in wide use in the deserts of the Middle East and North Africa.

Green roofs create an indoor environment that is cooler in the summer and warmer in the winter. These attributes make them very attractive options in urban areas, and some cities such as Chicago have, for several reasons, implemented initiatives to increase the use of green roofs. First, they reduce energy-use for both heating and cooling. Second, green roofs naturally absorb excess CO_2 through photosynthesis, and third, roof plants can help remove harmful or unwanted particulate matter from the lower atmosphere.

Clothing: Clothing and the specific materials used to construct clothes are also human adaptations to climate. Different forms of clothing were necessary as humans dispersed beyond our initial biological habitat of the tropics. Most of the early clothing used beyond Africa addressed the colder temperatures associated with midlatitude and Arctic climates. Many cultures developed locally obtained natural fibers for clothing, such as animal hides, feathers, wool, and cotton. For example, the Inuit people living in the Arctic used atiqiks, goose- (or duck-) down parkas housed within animal skins to shield them from the cold. Down is the light, fine feathers that grow under the tougher external feathers. The natural arrangement of these feathers creates pockets of air and provides excellent insulation against the cold.

In the epic race to reach the South Pole in 1911, one reason Roald Amundsen's team beat Robert Scott's team was because the Amundsen team wore down-filled outer garments, while Scott's wore wool. There were other contributing factors, including Amundsen's use of sled dogs and Scott's unsuccessful use of horses, all of which contributed to Scott's tragic end. The Scott party died on the 900-mile return trip from a combination of the effects of extreme cold and hunger. Today, scientific teams at the McMurdo station in Antarctica, the world center of Antarctic research, wear down parkas (Figure 7.25).

The development of wool and cotton for clothing are also related to climate and geography. In the Mediterranean climates of the Middle East, wool was obtained for clothing soon after the domestication of sheep about 10,000 years ago. Wool has good insulating properties and remains a good insulator even when wet because

Figure 7.24 A traditional feature of buildings throughout the world, the sod roof, is now being incorporated into modern buildings as green roofs.

ROALD AMUNDSEN I POLARDRAGT

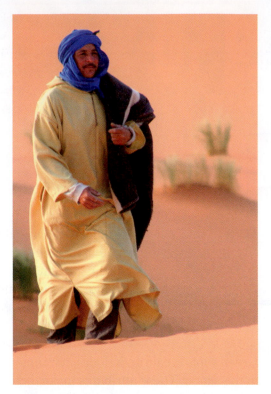

Figure 7.26 Traditional desert clothing of the Sahara favors loose cotton garb, light colors, and head covering.

Figure 7.25 Above, the traditional down parka used by the Amundsen polar explorers a century ago and by the Inuits centuries before Amundsen, remains the standard cold-climate gear for modern scientists in Antarctica, below.

the fibers trap air. Thus, wool was well-suited to the climatic variations that exist over the wide span of mountain elevations used for sheep herding. It was also well suited to the cool, rainy climate of northwest Europe, like that of western Ireland described by Frank McCourt in the opening of this chapter.

By contrast, cotton was adapted for use primarily in warmer and drier climates. Cotton fibers do not insulate well when wetted because the fibers retain water and the air spaces are eliminated. The indigenous people of Mexico grew cotton and used it for clothes almost 8000 years ago, and the Persians of southwest Asia made cotton clothing over 5000 years ago. The outer wraps traditionally worn by nomadic desert people such as the Berbers in the Saharan region of North Africa are also typically made of cotton (Figure 7.26). Here the emphasis is on protection from heat and radiation with attention to fabric selection to help prevent overheating. One layer of light-textured, light-colored clothing is often preferred. Light colors have higher albedos and help reduce the absorption of solar radiation, thereby minimizing surface heating. In addition, skin cooling is facilitated by loose-fitting cotton which allows for air circulation and heat loss by convection.

Modern Technology and Climate: Increasingly we humans are approaching Earth as though it is a fully habitable planet, that no place is beyond our capacity for permanent settlement with all the accoutrements of modern civilization. In reality large parts of the planet are just too cold, too stormy, too unstable, or too dry for long-term human habitation in large numbers. Yet, with modern technology and engineering we have committed ourselves to such zones with a confidence that denies geographic realities.

The false confidence of an invincible humanity has its roots in the Industrial Revolution, but in many respects did not really burgeon until the second half of the twentieth century when advanced technologies and economic power flourished. To overcome the constraints of climate, two approaches were promulgated, which now extend worldwide. The first was modification of the environment to correct the "defects" of climate. We see this vividly expressed in dry regions such as the American Southwest where the natural limitation of water supply was overcome by building great dams, such as Hoover Dam, and piping water large distances to support large cities, such as Phoenix and Los Angeles, where none existed before. The second approach was the invention of "portable" climates where a climatic constraint could be overcome by creating climate-controlled living environments. The invention of air conditioners is such a technology, one that has enabled large increases in population in warm regions such as the American South.

Unsustainable land use. Heated wooded houses in a treeless Arctic landscape where all building materials and fuel must be imported.

Humans have a long tradition of using technology to help them to adapt to challenging environments. The igloo, the adobe house, sod roofs, down clothing, and woolen fabrics are all part of that celebrated tradition. But in all these examples, the technology enabled us to adapt to the environment, and this contrasts with our modern approach to subdue and overcome the environment. But what is so objectionable about the latter? Is it not just a more sophisticated form of adaptation? The problem is that it is not sustainable.

There are reasonable geographic limits to our long-term existence on the planet. When we exceed these *limits to growth*, huge expenditures of resources are needed to (1) build and maintain settlement, and (2) control the environment. After all, dams and aqueducts fill with sediment and wear out in time. But more important are the huge expenditures in energy that are required to maintain portable-climate technology and controls on environmental systems. All this energy use drives carbon dioxide and other pollutants into the atmosphere where they, in turn, alter climate. And here feedback comes into play because the very problems we are treating with technological solutions are exacerbated by the solutions we apply to them, forcing us to throw more technology at the climates which have now become hotter, drier, more stormy, and/or more unstable … unless, of course, we choose a different, more adaptive, strategy.

Unsustainable land use. Suburban development in a hot, dry landscape where most building materials and all water must be imported.

Chapter Summary and Overview of Global Climate, Formative Systems, and Human Adaptation

Climate is the product of a complex set of systems that range from those that are circumglobal in coverage and predictable in their behavior to those like cyclonic storms that are regional in coverage and more or less unpredictable in their behavior. Two big engines run the world's climate systems and produce a number of climate types that can be lumped roughly into two classes: dry and stable and wet and stormy. These engines and the atmospheric circulation they drive are shown in the summary diagram. Conventional climate classification recognizes five climatic zones, however, only one, the A zone, is biologically suited to humans. But with the aid of technology, humans have developed shelters and clothing that permit them to inhabit all Earth's climates, sometimes with impacts to the environment. These impacts often take the form of feedbacks that expand and intensify the climatic constraint.

▶ **Climate is the general conditions of the atmosphere at a place on Earth.** It includes annual, seasonal, and monthly temperature and precipitation characteristics as well as key processes and extreme conditions.

▶ **Many systems are responsible for shaping Earth's climates.** However, their roles and level of importance vary widely across the Earth.

▶ **Climate systems vary in their behavior and geographic coverage.** Behavior ranges from more or less predictable to relatively chaotic; coverage from global to regional or less.

▶ **Climate systems are arranged in a hierarchical pyramid.** Primary energy systems are at the base and storm systems are at the top.

▶ **Two big atmospheric engines are the central drivers of Earth's climates.** The tropical engine is fueled by more energy than is available to all the rest of the world combined.

▶ **The midlatitude climate engine produces the planet's strongest seasonal climates.** Seasonal change includes marked shifts in the types, magnitudes, and frequencies of dominant climate systems, and in summer these vary from the continents to the oceans.

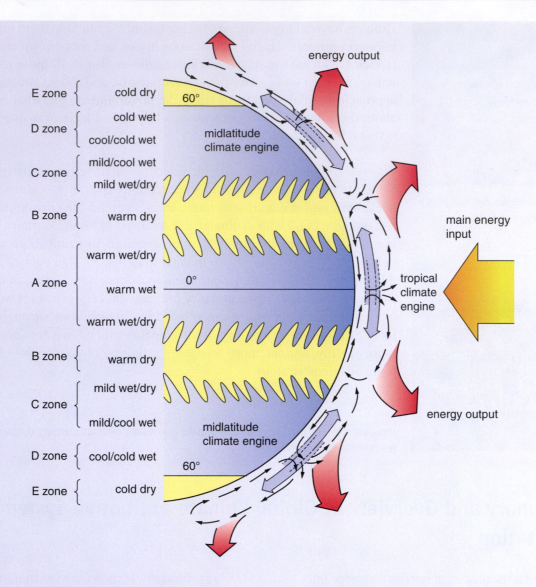

- **Early climatic classification schemes used vegetation to define climatic boundaries.** Many temperature and moisture boundaries were patterned after the size and structure of vegetation in different parts of the world.

- **The Köppen–Geiger Climate Classification System organizes climate into five geographic zones.** The A climates are tropical, the B climates are dry, the C climates are temperate, the D climates are cold and snowy, and the E climates are polar.

- **The tropical climates center on the Equator and span the broad intertropical zone.** The dry climates lie on the shoulders of the A-zone climates except in the midlatitudes where they lie in the lee of mountain ranges.

- **The C climates are transitional between the A and D climates.** They are thermally mild, whereas the D climates are noted for their snowy, cold winter. Beyond the D zone lie the perpetually cold E climates.

- **Climate has profound effects on our lives.** To survive beyond our natural limits in the tropics, humans have had to invent survival technologies, most notably various forms of shelter and clothing.

- **The design of living shelters in both cold and hot climates favor thick walls.** This holds for both igloo and adobe buildings.

- **As technology advances, humans increasingly try to overcome, rather than adapt to climate.** These efforts create huge expenditures and surpluses of energy that may create unwanted feedback into the climate system.

Review Questions

1 When climate is defined as the general condition of the atmosphere at a place on Earth, what is meant by general condition?

2 How are geographic coverage and the behavior of climatic systems related? Explain how this relationship plays out at the global and regional geographic scales, and identify the primary factors involved and their characteristics.

3 Referring to Figure 7.3, identify the energy transfers at these three systems levels: the solar radiation, atmospheric, and storm. How is the climate at a specific location affected by these energy transfers?

4 What uncertainty factors are present at the global, regional, and local scales that influence climates at these scales?

5 What roles do sun angle and the intertropical convergence zone play in the tropical climate engine? How do the circulation systems of the atmosphere and ocean currents expand the influence of this climate engine?

6 Why can the midlatitude climate zone be characterized as "a huge playing field in which contrasting types of air compete for position." How does this game change over landmasses and the oceans?

7 How do the pressure systems differ between the dry and stable and wet and stormy climate types produced by the tropical and midlatitude atmospheric engines? What drives the major changes that produce regional climates within these two broad zones?

8 Using Figure 7.17, identify the five major climatic zones of the Köppen–Geiger classification system, and describe the general characteristics of the temperature and precipitation regimes within each zone.

9 What are the characteristics of North American indigenous shelters that make them good adaptations to the climates where they exist?

10 What climates are particularly well suited for wool and cotton clothing, respectively? What are the factors that originally contributed, and continue to contribute to these climate–fiber associations?

Climate Change:
Past, Present, and Future

Chapter Overview

This chapter is about one of the most pressing issues of our times, climate change. It begins with the question of whether climate change is something intrinsic to Earth history, or something unique to our time on the planet and follows a brief look at how science is able to figure out the nature of climates dating back millions of years. Just how different were ancient climates and, when they changed, what were some of the geographic effects? And then what are the possible causes of climate change and which among them are the leading candidates behind the changes that have helped shape the geographic environment that has nurtured the growth of humans on Earth? Finally, we review modern climate change, beginning at the local and regional scales and ending with global climate change in the twenty-first century and some of its major effects on the environment.

Introduction

The winter wind roared out of the black night sky and sent snow swirling around the little house. Within a few hours it had drifted up to the windows, and where it touched the glass, a patch of frost formed on the inside pane. By bedtime the frost had crept nearly halfway up some windows and it began to feel like the entire skin of the house was slowly freezing around us.

Grandma passed the evening playing solitaire and writing letters. Grandpa smoked his pipe and read. Neither of them took notice when powerful gusts shook the house and caused the woodstove to cough out little puffs of smoke. They'd both been through many Lake Superior blizzards that left home looking like the one in Figure 8.1. At length Gramp looked up, yawned, and said, "Tomorrow we should bank up the house." This meant packing snow along the base of the outer walls to reduce heat loss and wind penetration. I questioned the logic of this practice and Gramp explained that snowbanks work like igloo walls by holding in warm air and, besides, snow itself is usually warmer than winter air.

Figure 8.1 A typical winter scene on the south shore of Lake Superior in the 1950s.

By now Gramp had closed his book and I spotted an opportunity to coax a story or two out of him. "Can you remember a worse winter than this one, Gramp?" "Oh, sure," he replied, "this isn't half as bad as winters in the 20s and 30s when we were commercial fishing through the ice on Lake Superior." He launched into a series of vignettes about the incredible cold, gale-force winds, and shifting ice sheets, about Lake Superior freezing completely over, and about life and death adventures with dog teams and men falling into snow-covered leads (wide cracks) in the ice. It was exactly the fuel a 12-year old boy needed on such a night.

And just when I thought it was over, Grandma added, "Oh, my lands, you can't imagine how bad some of those winters were. When I was about your age, around 1910, we lived in a lumber camp one winter and it seemed to snow every day. By Christmas the cook's shanty was half buried. The camp roofs had to be shoveled off several times, and in the deep woods, the drifts lasted well into May."

They painted a bleak picture of past winters and I firmly believed the weather over Lake Superior was colder, windier, and snowier in those years. The climate of our region, I concluded, must have gotten warmer and life for me must surely be easier than it was for them. As I climbed the stairs to my unheated bedroom, I thought, when I grow up, I must go to the Arctic where the winters are still ferocious and I can test myself against the elements.

Years later, I examined the climatic records for Northern Michigan and found there was actually no evidence of harsher winters in the early decades of the twentieth century. In fact, the Great Lakes Region appeared to have cooled a little between 1940 and 1975. But the pattern of winters then was pretty much as it is now. Each decade had one or two nasty ones and one or two mild ones and, yes, Lake Superior occasionally froze completely over then as it occasionally does now. So what about Gramp and Gram's recollection of past winters? For them, winters really had been much worse years ago because life was lived closer to the snow, cold, and wind. Lumber camp buildings were small and poorly insulated and the contrast between a comfortable winter evening near the woodstove and a winter night in a fisherman's camp on Williams Island in Lake Superior must have been stark in Gramp's memory.

8.1   The Nature and Indicators of Climate Change

We know that climate on Earth has changed many times in the past and that at times it was much different than it is today. We also know that past climate changes covered a wide range of magnitudes. Some were massive, affecting the entire globe and changing Earth's biogeographical character virtually everywhere. Others were relatively modest and limited to sections of the planet or even selected parts of continents with minor biogeographical consequences.

Although our knowledge of Earth's climate history is sketchy, especially for remote time periods, we are virtually certain that climate change has been common to Earth since its origin as a planet. In fact, it would not be incorrect to say that climate change has been the norm for Earth, including the brief time that humans have been around. Thus, it should come as no surprise to us when climate changes in the centuries ahead. But for much of humanity it probably will come as a surprise, especially if it comes suddenly.

Indicators of Climate Change: What do we know about past climates? How do they compare with present climatic conditions on Earth? These are difficult questions to answer with much precision because we have no instrumented temperature

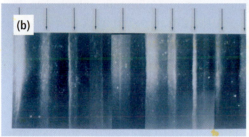

Figure 8.2 (a) An ice core extracted from ancient glacial ice provides clues about past climates. (b) Each layer represents a year and contains bubbles of air that can be analyzed to learn about ancient atmospheric conditions.

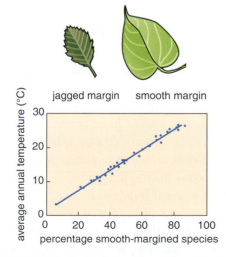

Figure 8.3 A graph showing the relationship between the roughness of leaf margins and air temperature. Smooth margins correspond to higher temperatures.

and precipitation data for Earth much before the nineteenth century. But we do have environmental and historical indicators that provide clues, often referred to as **proxies**, about thermal and moisture conditions, for various places and times. These indicators include chemical and biological markers buried in ancient bedrock, glacial ice, and ocean sediment.

Oxygen isotopes in the shells of ocean organisms and in glacier ice, for example, are a proxy for temperature. This indicator is based on the amount of two isotopes, oxygen-16 (^{16}O) and oxygen-18 (^{18}O), in the water molecules released from the ocean surface in evaporation. As the oceans warm and give up water molecules, the air over them receives more of the light isotope oxygen-16 in the water vapor. When the oceans and the air over them cool, the isotope balance shifts toward the heavier oxygen-18. When precipitation particles form from this vapor, they are marked relatively warm or cold by the ratio of lighter or heavier isotopes. And if the molecules fall in snow and are locked in glacial ice, they can be extracted years later and used as a yardstick of past thermal conditions.

Because they are so old, the Greenland and Antarctica ice sheets are especially good thermal yardsticks. When scientists drill into these thick masses of ice, they find young ice at the top and progressively older ice deeper down. This chronological sequence is related to the way glaciers form. When snow is deposited on their surfaces, it becomes solid ice as it is compacted under the weight of additional snow. Each year's snowfall is thus recorded as a layer of ice which can be plainly seen in the ice core in Figure 8.2. Over time, each layer, including tiny pockets of air trapped within it, sinks deeper and deeper into the glacier as more and more snow is added to the surface. Cores taken at depths of 3.5 kilometers contain ice which originally fell as snow 400,000 years ago. By carefully extracting ice-core samples and analyzing the air for carbon dioxide and the ice for isotopes of oxygen, we are able to construct a long record of temperatures over Greenland and Antarctica, and correlate the results to the average global temperature. It appears that each degree of temperature change revealed by the ice is equivalent to about 0.5 °C of atmospheric change for the Earth as a whole.

The biological markers include the fossil remains of certain organisms, such as reptiles and flowering plants, known to be sensitive to particular temperature levels. When these organisms appear or disappear in sediment deposited over some time span, it can be taken as an indicator of a change in the thermal climate. For example, crocodile bones dating back 34–57 million years have been found in sediments in Colorado, Wyoming, and neighboring states but the same animal today can survive no further north than the Gulf Coast of the United States. We conclude, therefore, that subtropical conditions once extended into central North America.

The fossil leaf imprints from flowering plants are a general indicator of average annual temperatures. As the graph in Figure 8.3 reveals, smooth-margin leaves reflect more tropical conditions, whereas jagged-margin leaves reflect colder climatic conditions. The fossils of these leaves, which are very common over much of the Earth, date back 100 million years or more.

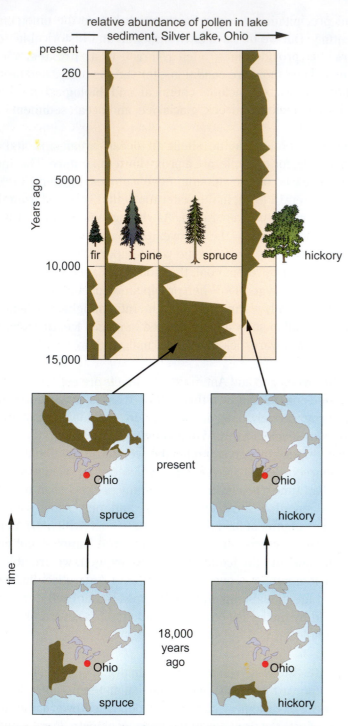

relative abundance of pollen in lake sediment, Silver Lake, Ohio →

Figure 8.4 Above, pollen profiles from a bog in Ohio which show that about 10,000 years ago a major change in vegetation took place in this region. Below, maps showing the change in the distribution of two trees species, spruce and hickory, between 18,000 years ago (when the continental glaciers were at their peak) and the present.

Tree rings. Variations in ring thickness indicate variations in climatic conditions.

Plant pollen buried in bogs and lakes is a reliable indicator of past plant distributions. The presence of pollen from a plant whose range is now distant from a site where its pollen has been found tells us that the plant's distribution has changed and climate change may have been the cause. The pollen counts used for the graphs in Figure 8.4 were taken from a bog in Ohio. They reveal that 10,000 years ago a climate change occurred in central North America, which caused a change in the distribution of tree species. Fir, pine, and spruce declined, while hickory increased.

Another indicator of past climatic conditions is tree rings. Growth rings in both dead and living trees have proven to be useful indicators of climatic conditions for time periods dating back as much as 10,000 years. For example, analysis of trees such as the bristlecone pines in California, which can live for 4000 years, have been helpful in defining periods of heat and moisture change, and these records can be extended by correlating growth-ring patterns with those of even older dead trees.

Historical records and archeological remains are also valuable indicators of past climates. Historical records, for example, telling about the presence or absence of certain agricultural plants in an area provide clues about climate conditions. These records date back 6000 to 7000 years for some locations. Archeological evidence goes back much farther, a hundred thousand years or more, for regions such as the Middle East. Taken together, these various lines of evidence – chemical, biological, historical, and archeological – can reveal a great deal about past climatic conditions, especially in more recent millennia. Beyond a few million years ago the picture tends to be spotty but we do know about some major trends and events.

8.2 Some Climate Changes of the Past

The sediments preserved in ancient rocks reveal that oceans formed on Earth nearly 4 billion years ago. About 3.5 billion years ago, these oceans fostered the first life on the planet. For this to take place the oceans had to have been liquid, so we reason that Earth was warm enough to maintain liquid seas. Relatively warm conditions, with occasional interruption – including several great freeze downs and thaws – seem to have prevailed for the next several billions of years and life remained concentrated in the sea. Animals first appear in the fossil record about 600 million years ago. Around 450 million years ago Earth cooled dramatically, glaciers rose, and ocean cooling triggered a massive die-off. This, of course, was limited exclusively to the sea, for life on land had not yet evolved.

About 400 million years ago, life (beyond simple forms like bacteria and algae) emerged on land and evolved rapidly into new plant and animal forms. Again, Earth had to have been relatively warm for this great change to take place, but it was not without interruptions for there is evidence of climatic perturbations from time to time that caused major extinctions and changed the course of evolution. One of these occurred 250 million years ago, wiping out 90 percent of marine species and 70 percent of terrestrial vertebrate species. Considered Earth's greatest mass extinction, the cause of this event is unknown, but it may have been related to massive volcanic eruptions that darkened the atmosphere and dramatically cooled global climate.

In the period between 200 and 65 million years ago, Earth entered a warm phase in response to an increased greenhouse effect brought on by elevated atmospheric carbon dioxide from volcanic activity. Climate on most continents appears to have been not only warmer but wetter than at present. Shallow seas covered large areas of land, lush tropical and subtropical vegetation was widespread, and dinosaurs, like the ones shown in Figure 8.5, roamed much of the Earth. The remains of the vegetation are now a major source of oil, gas, and coal reserves. This warm period ended abruptly 65 million years ago when climate over most of the globe was plunged into cold. The evidence points to a large asteroid collision as the main cause. Debris from the explosion darkened the atmosphere, blocked out most solar radiation, and snow and ice spread over most of the planet. At least half Earth's species of plants and animals, including the great reptiles, died off. (See Section 17.2 for a more detailed description of this event.)

Figure 8.5 An artist's rendition, based on fossil records, of a typical tropical landscape during the time of the dinosaurs, 200-65 million years ago.

Eventually the atmosphere cleared and, after some additional climatic oscillations, warm conditions returned. But Earth's biogeographical character was irreparably altered. Among other things, mammals replaced reptiles and evolved into the dominant land animals and, in the sea, whales evolved, and corals became the dominant reef-building organism. But then another major change occurred, this time coming from the sea.

About 55 million years ago, the atmosphere suddenly heated up in response to a massive release of methane from the ocean floor. Methane (CH_4) is a carbon-based gas in the same family as carbon dioxide (CO_2), and like CO_2, it also has a high capacity to absorb longwave radiation and increase the atmosphere's greenhouse effect. Geochemists speculate that the methane escaped from sea floor clathrates, methane-trapping ice-crystals that are distributed in sediments on the outer edges of continental margins worldwide. For reasons that remain unknown, the clathrates suddenly began to decompose on a massive scale as the drawing in Figure 8.6 suggests, thereby increasing the amount of methane in the atmosphere and oceans. The rapid release of so much methane, and the methane's oxidation to

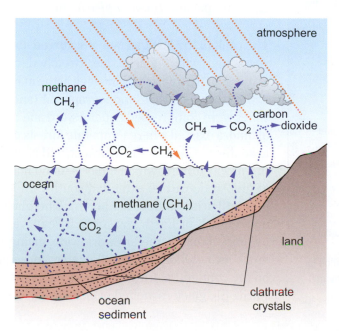

Figure 8.6 The release of methane from the ocean floor followed by its conversion to carbon dioxide.

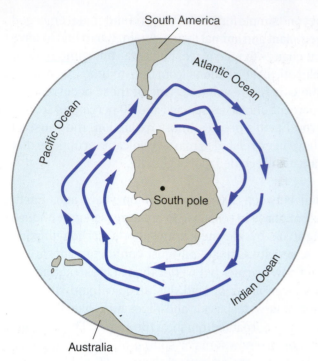

Figure 8.7 The pattern of ocean circulation around Antarctica that developed following its separation from Australia about 34 million years ago.

carbon dioxide, would have significantly altered ocean chemistry, and ultimately the atmosphere and global climate. For a period of 100,000 to 200,000 years, the global temperature was substantially elevated. The temperature of the oceans rose by 5 to 7 °C and vast numbers of marine species perished before methane levels declined and the atmosphere returned to a more modest energy balance.

Plant fossils in North America reveal that over the next 55 million years there were three pulses of global cooling including one around 34 million years ago when Earth entered an ice age. This was brought on by a major shift in the locations of the continents in the southern hemisphere which left Antarctica isolated at the South Pole. Completely encircled by water, a west–east system of ocean circulation, like that shown in Figure 8.7, developed around the continent which limited south–north exchanges with the warmer waters of the tropics. As a result, Antarctica fell into a cold phase which deepened with a global decline in CO_2. Coupled with the change in ocean circulation, the CO_2 reduction led to a massive freeze down and the formation of the continent's great ice cap. This in turn had a cooling effect on the rest of the planet which was followed by a drop in sea level as more water became locked up in ice on the continents.

About 1.8 million years ago, global climate cooled once again, giving way to our time on the planet. During this time the volume of glacial ice expanded to cover as much as 30 percent of the continents then contracted and expanded again. Ocean sediments reveal that this happened many times as global climate warmed and cooled by several degrees. Land deposits, on the other hand, reveal that major **glacials**, or **glaciations**, occurred at least four times in North America and Eurasia. In North America each glaciation began in the subarctic and extended southward beyond the latitude of the Great Lakes. The warm intervals between periods of glaciations, called **interglacials**, lasted for 50,000 to 100,000 years. Sometime during one of the glacials or interglacials about 200,000 years ago, our species (*Homo sapiens*) emerged in Africa.

The peak of the last glaciation occurred about 18,000 years ago when glacial ice stretched completely across North America. If you look at the map in Figure 8.8, you will see that biogeographical (and climatic) zones were compressed equatorward and large areas of tundra existed where temperate forests grow today in southern Canada, the Midwest, and New England. In Europe, tundra, which today is found mainly north of the Arctic Circle, extended southward to the latitude of Paris (49 degrees north). The glaciers took up so much of Earth's water that sea level dropped by more than 100 meters worldwide, giving Earth as much as 3 million square kilometers more land area on the continental fringes. But then climate began to warm, the glaciers melted back, and sea level rose.

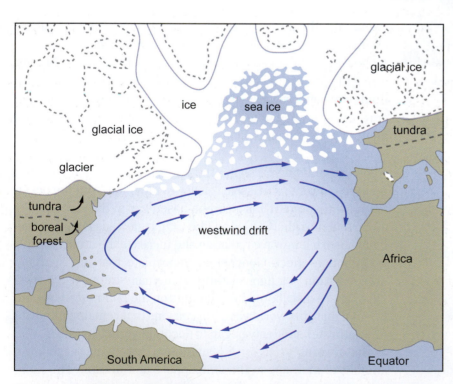

Figure 8.8 The distribution of vegetation, glacial ice, and sea ice, and ocean circulation in an around the North Atlantic at the peak of the last glaciation, 18,000 years ago. Tundra dominates the landscapes where Paris and Washington, DC, are now.

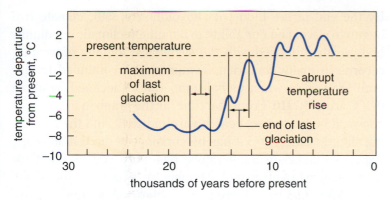

Figure 8.9 The estimated global temperature changes from 24,000 years to 2000 years BP (Before Present). Note the abrupt temperature rise around 10,000 years BP.

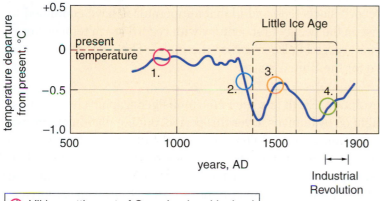

① Viking settlement of Greenland and Iceland
② Bubonic plague in Europe
③ Columbus's voyages to America
④ American Revolution

Figure 8.10 Estimated temperature changes and selected historic events from about AD 700 to the late 1800s. Note the Little Ice Age.

About 12,000 to 14,000 years ago, global temperature rose, then cooled, then rose abruptly 10,000 years ago. Some evidence reveals that this took place in less than a century, perhaps in as little as 20 years. Biogeographical zones shifted, human populations expanded, many large ice-age mammals disappeared, and the first farmers planted crops in the Middle East. As the graph in Figure 8.9 shows, climate continued to warm until about 7000 years ago when global climate was warmer than any time since. In North Africa, the Sahara expanded, became drier, and most of the early Semitic people who lived there were driven out as resources declined. Since then climate has generally cooled, but the cooling trend has been interrupted by oscillations of about 2 °C every 2000 years or so. Scientists generally agree that Earth is still in an ice age, probably an interglacial period, which is periodically interrupted by short cold spells followed by warm intervals.

Our knowledge of climate change over the past 3000 years is fairly good for some regions of the world. For example, we know from historical and archeological records that the great deserts of North Africa and Asia were generally less harsh in the centuries before Julius Caesar (100–44 BC) than they are today. Nomads grazed sheep and goats in parts of deserts where they cannot survive now. It appears that surface water was also more plentiful. There were more oases, and camel caravans crossed sections of desert that similar caravans avoided in the nineteenth and twentieth centuries. But there were also several major droughts each lasting several decades or more which caused famine and population shifts in the Middle East.

Beginning about 2000 years ago, several major temperature trends are apparent in the northern hemisphere. After the fall of Rome around AD 500, a warming trend led to the northward expansion of agriculture in Europe. Grapes grew in England and southern Scandinavia in Medieval times. The Vikings colonized Iceland, Greenland, and North America, but after AD 1200 or so, historical records reveal that the North Atlantic grew stormier and colder, and the Viking colonies in Greenland and North American declined and were abandoned. As the graph in Figure 8.10 shows, this corresponded to a sharp decline in temperature, and by 1400 much of the world had entered a distinct cool phase known as the "Little Ice Age."

The Little Ice Age lasted until about 1850 and history shows that it was a period of significant change in human population and land use. After several centuries of population growth, Europe was hit by a severe outbreak of bubonic plague (Black Death) in the mid 1300s, which ultimately killed more than 20 million people. Though not a direct result of harsher climatic conditions, the colder temperatures did contribute to the plague's spread as people crowded together against the harsh winters. Agriculture declined and famines were frequent. Forests were cut for fuel and building material and by 1700 forests on the island of Great Britain had been completely eradicated. Not surprisingly, mountain glaciers in Europe and North America expanded, pushing far down mountain valleys.

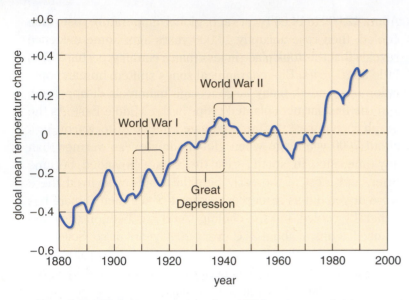

Figure 8.11 Global mean temperature from 1860 based on recorded data according to NASA. Zero is based on the mean temperature for the period 1961–90.

By the end of the Industrial Revolution (1850), climate had begun to warm and our ability to document climatic conditions had greatly improved with the systematic logging of weather records in many countries. Temperatures continued to rise until 1880 or so when global temperature declined by about 0.2 °C around 1910. From that time, as the graph in Figure 8.11 shows, it began to rise, peaking around 1940, and then declining slightly. Since 1975 it has been rising. In the last 35 years, Earth's equilibrium temperature has increased by 0.5 °C or more relative to a mean global temperature based on the period 1961–1990. And this increase is based on air-temperature readings over both the land and oceans. Warming has been especially pronounced in the northern hemisphere and large areas in the Arctic and subarctic of North America and Asia – conspicuous in Figure 8.12 – have registered mean annual temperature increases of 4 °C and higher in recent years. In North America, the 1980s and 1990s were the warmest decades on record since 1880 and global temperatures have continued to rise in the twenty-first century with record warm temperatures consistently recorded for scores of places across the globe. The year 2006 was the warmest year on record for the United States.

Since the Industrial Revolution, Earth has seen enormous geographic change, most of which has been brought on by humans. Over this period, our population has grown sevenfold, from 1 billion to more than 7 billion, and is now expanding by nearly 80 million people per year (Figure 8.13). We have spread across the planet and taken over vast areas of forests, grasslands, and coastal plains. We have seriously altered the global environment by cutting vast amounts of temperate and tropical forests; converting forestlands and grasslands to cropland over about 12 percent of Earth's land area; developing massive urban centers which house nearly half the world's

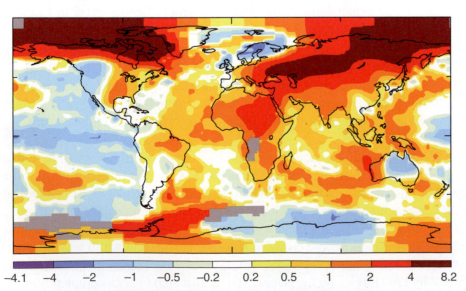

Figure 8.12 The global distribution of change in the annual mean temperature relative to the 1951–1980 mean. Notice the pronounced warming in the northern high latitudes.

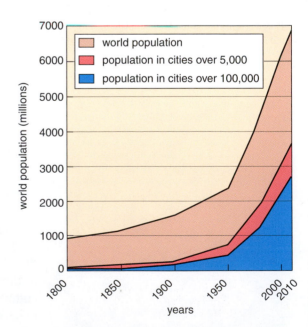

Figure 8.13 The growth of human population across the world and in cities since 1800. Nearly 80 million people are being added per year.

population, most of the industry, and most of the 800 million automobiles; and contaminating the atmosphere with energy-altering pollutants such as carbon dioxide and aerosols which emanate largely from urban centers, their industry, and automobiles.

These changes are altering atmospheric systems and are thus the leading suspect in the search for a cause in the climate shifts we are currently witnessing on Earth. The climate over urban centers is commonly several degrees warmer than the regions that surround them and, as we have noted, global temperature is rising. But changes of this magnitude have happened many times on Earth, especially in the past 200,000 to 300,000 years, and more particularly in the past 20,000 years. So what is the real cause or causes of contemporary climate change? Is it purely humanly driven? Let us try to put the problem into context by examining the list of leading candidates in more detail.

Summary on Past Climate Change: We know with certainty that climate change is common to Earth. Indeed, it is one of Earth's most notable geographic phenomena. And the changes occur at widely different magnitudes and frequencies with a wide range of impacts on the planet. Documenting past changes of climate would challenge Sherlock Holmes for most of it is based on various clues, or proxies, taken from rock, fossils, sediment, ice, tree rings, and archeological ruins, all before the period of recorded history. And since then, only the past century and a half or so has provided instrumented records of temperature, precipitation, and related atmospheric phenomena. If we accept climate change as a planetary norm, the next question is what are the drivers behind the changes, especially during our tenure on Earth?

8.3 The Causes of Climate Change

We know that there are many factors that can bring on climate change. We also know that these factors can act individually or in various combinations and some are interrelated via feedback systems in which the occurrence of one factor may enhance or diminish another. While we can enumerate and describe the factors that lead to climate change, the actual mechanisms by which they change climate are poorly understood for most known factors. Several candidate causes of change have been identified by scientists and we examine seven of them in the paragraphs below.

Variations in Earth's Orbital Geometry: The episodic nature of the Earth's glacial and interglacial periods within the present ice age (the last 1.8 million years) have been caused primarily by cyclical changes in the Earth's circumnavigation of the Sun. These changes comprise three dominant cycles, collectively known as the Milankovitch cycles for Milutin Milankovitch (1879–1958), the Serbian astronomer who is generally credited with calculating their magnitudes. Let us briefly consider these three cycles.

The first cycle is called **eccentricity**. It refers to changes in the shape of Earth's orbit around the Sun, from one that is more elliptical to one that is nearly circular. Figure 8.14 shows extreme versions of these two shapes. The more circular the orbit, the less extreme the seasons, and vice versa. This means that winters in the midlatitudes are harsher when the orbit is more elliptical. The change from more circular to more

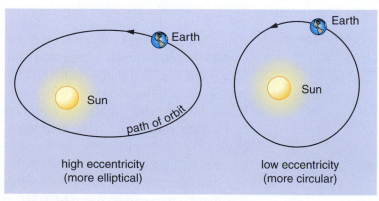

high eccentricity
(more elliptical)

low eccentricity
(more circular)

Figure 8.14 Illustrations of the concept of eccentricity, or change in the shape of the Earth's orbit.

elliptical orbits takes 90,000 to 100,000 years and this interval coincides with the periods of glaciation that have occurred in the past million years. Currently Earth's orbit is nearly circular and the geographic coverage of glacial ice is at a minimum, only about 10 percent of the world's land area, as we noted earlier.

The second cycle is related to variations in the tilt of the Earth's axis, and is called **obliquity**. This angle is currently 23 degrees, 27 minutes (or 23.5 degrees), but astronomers calculate that over a period of 40,000 years it varies from 21.8 degrees to 24.4 degrees. The change does not look like much based on the diagram in Figure 8.15, but it is significant. Larger angles produce greater seasonal extremes. Therefore, winters in the middle and high latitudes would be colder (given that all other variables are held constant). And there is also evidence of the effect of obliquity on climate. Analysis of oxygen isotopes in deep-sea sediment indicates a cycle of global temperature change corresponding to a 41,000-year cycle over the past million years or so. It is noteworthy, parenthetically, that some scientists reason that Earth's axial tilt long ago was more than twice today's. Between 600 and 800 million years ago the axial tilt may have been as much as 54 degrees making the tropics colder than the high latitudes. As the tilt decreased, the seasonal extremes became less severe over the entire Earth and the tropics and the polar regions became uniformly warm and cold respectively.

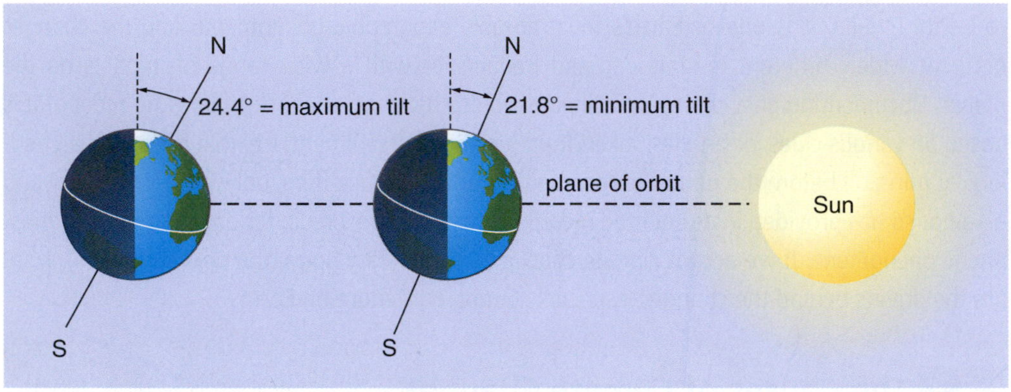

Figure 8.15 The concept of obliquity, or change in the tilt of the axis from 24.4 to 21.8 degrees.

The final orbital cycle, shown in Figure 8.16, occurs with the date of the equinox. This variation is called **precession**, which influences the time in the year when Earth is closest to and farthest from the Sun. The date when the Earth is closest to the Sun (perihelion) occurs on January 5, during the northern-hemisphere winter 2012. As a result, winters should be somewhat milder in the northern hemisphere now than in the past. Precession has a period of 21,000 to 23,000 years and, once again, there is evidence found in ocean sediment of a corresponding period of global temperature change with peaks falling every 21,000 years. When eccentricity, obliquity, and precession are combined, as the diagram in Figure 8.17 attempts to do, the result is a varied pattern of change in solar radiation that correlates with many of Earth's thermal changes over the past 600,000 years.

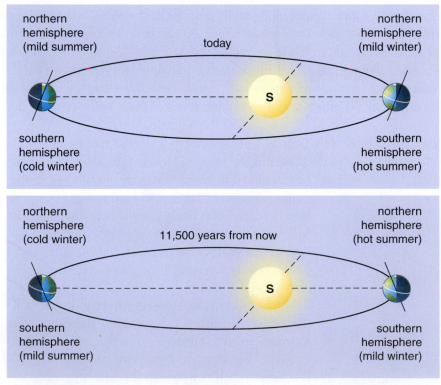

Figure 8.16 Precession or variation in the time of the year when Earth is nearest and farthest from the Sun.

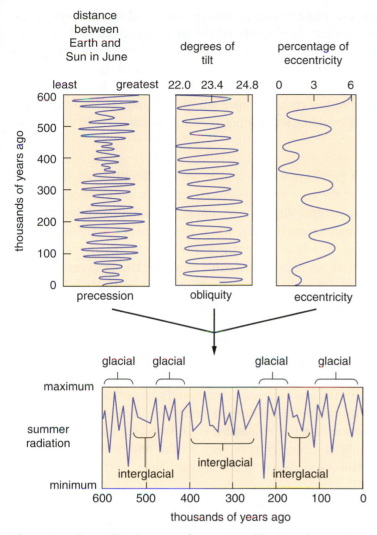

Figure 8.17 The combined patterns of eccentricity, obliquity, and precession over the past 600,000 years and the relationship to the occurrence of glacials (ice ages) and interglacial episodes in the graph below.

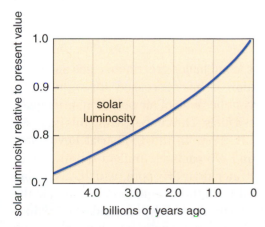

Figure 8.18 The proposed change in solar luminosity since the origin of the Solar System based on the work of astrophysicists and astronomers.

Solar Luminosity: Solar luminosity is the term used to describe the output of radiation from the Sun. According to astronomers, stars go through an evolutionary process during which their energy output changes. They reason that the Sun's output has increased over the past five billion years and it now gives off 30 percent more radiation today than it originally did. A "best-guess" curve is offered in Figure 8.18; however, we have no data to verify this. But we do have data on luminosity related to sunspots going back to the nineteenth century. A link has been discovered between the frequency of sunspots and mean air temperature on Earth. Global air temperatures tend to fall slightly during periods when sunspots occur often, that is, at greater than average frequencies. (The average cycle is 11 years; see the graph in Figure 3.8). The mechanism(s) responsible for the change in Earth's air temperature, however, are unknown, and the change is very small.

Atmospheric Transparency: The transparency of the atmosphere influences the ability of solar radiation to penetrate the atmosphere and heat land, air, and water. Transparency is controlled mainly by particulate matter (aerosols) in the air, minute particles of liquid and solid matter. Simply put, a dirty/hazy atmosphere lets through less sunlight than a clear one.

There are many sources of aerosols including wind erosion of soil, industrial and automobile exhausts, meteor disintegration, asteroid impacts, phytoplankton in the oceans, and volcanic eruptions. In our era, volcanoes are primary sources of aerosols. There is ample evidence that global temperatures often fall during periods of major eruptions as incoming solar radiation is backscattered in a manner similar to that shown in Figure 8.19. For example, the 1991 Mt. Pinatubo eruption in the Philippines spread airborne ash around the world, which so reduced atmospheric transparency that global climate cooled for more than a year. Major volcanic eruptions in the nineteenth century were followed by long harsh winters in the northern hemisphere, and there is evidence of periods in the geologic past when massive volcanic eruptions over large parts of the Earth led to depressed global temperature, climate change, and massive plant and animal extinctions.

Figure 8.19 Dirty air from volcanic ash reduces atmospheric transparency by backscattering incoming solar radiation.

Air pollution from human activities also contributes to decreased atmospheric transparency and atmospheric cooling. Industrial and urban pollution as well as forest and grassland fires discharge large amounts of particulates into the lower atmosphere. Over cities, the concentration is often so heavy that a "dust dome" develops, which can reduce incoming solar radiation by 50 percent or more. If heavy air pollution happens to coincide with a rash of major volcanic eruptions, the combined effect on global temperature could be substantial, perhaps catastrophic to world food production.

Clouds are also made up of particles and they have a powerful influence on atmospheric transparency because they are an efficient reflector of solar radiation. Global cloud cover can increase in several ways. For example, pollution particles and volcanic dust serve as condensation nuclei and when these particles are abundant, cloud formation is advanced. In addition, during warm periods, ocean evaporation can increase and provide more moisture for cloud formation. Under current conditions, clouds are responsible for reflecting nearly 25 percent of Earth's incoming solar radiation back into space.

Atmospheric Absorption: Earth's thermal regime is modulated by the greenhouse effect which is governed by two major gases, water vapor and carbon dioxide, and several secondary gases such as methane, nitrous oxide, and ozone (Figure 8.20). These gases and other manufactured gaseous compounds absorb longwave (infrared) radiation, thereby slowing the loss of energy from the atmosphere. The more abundant these gases, the warmer the atmosphere tends to be, a correlation supported by certain past shifts of global climate such as the one brought on by the great methane burp from the oceans 55 million years ago.

Water vapor is responsible for two-thirds of Earth's greenhouse effect. Most of Earth's water vapor is held in the tropics: first, because that is the largest pool of warm air and the warmer air is the higher its capacity to hold moisture; second, because there is abundant heat in the tropics to drive the evaporation process; and third, because the oceans there provide the largest source area for water vapor on the planet. But despite a warmer surface and atmosphere, there is no firm evidence at this date that water vapor is increasing in the tropics or elsewhere on the Earth and influencing atmospheric heating. However, that may change as the warming trend advances. The evidence surrounding carbon dioxide, on the other hand, presents a more compelling story.

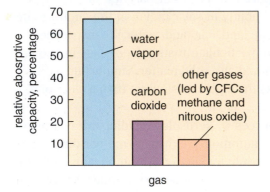

Figure 8.20 A ranking of greenhouse gases according to their effectiveness in absorbing longwave (infrared) radiation.

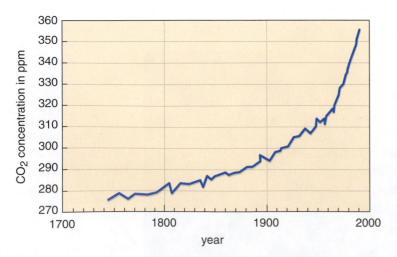

Carbon dioxide is the second major greenhouse gas and since the Industrial Revolution it has increased throughout the atmosphere by 30 percent because of air pollution, principally from combustion of fossil fuels. The curve tracing this increase is shown in Figure 8.21. Prior to the Industrial Revolution, the CO_2 value fell around 275 parts per million (ppm). Today, atmospheric CO_2 is greater than 365 ppm, which, according to air samples extracted from the Greenland ice sheet, is the highest level in 500,000 years. And each year land-use activities

Figure 8.21 The increase in the carbon dioxide content of the atmosphere since the beginning of the Industrial Revolution from about 275 to over 350 ppm.

add about 7 billion tons to the atmosphere. Most of this carbon dioxide comes from the developed world, from industry, urban centers, and automobiles, but it is dispersed around the world and mixed into the atmosphere as a whole by the global circulation system. The increase in atmospheric carbon dioxide correlates with the rise in Earth's surface temperature over the past 100 years or more. In addition, several minor greenhouse gases are also increasing, in particular, methane.

As CO_2 is rising, the capacity of the environment to extract it from the atmosphere is declining in some areas. Tropical-forest eradication is reducing one of the major sinks for CO_2 because trees and other plants remove CO_2 in photosynthesis. At the same time, large areas of cleared forest are being replaced by cattle ranches and cattle digestion and dung are a major source of methane. Another CO_2 sink is soil but it, too, is being reduced as organic-rich topsoil is eroded from cropland and lands subjected to deforestation and overgrazing. Carbon dioxide is a multifaceted system and we will return to it later in Section 8.6.

Distribution of Land and Water: Throughout Earth's history the continents and ocean basins have been changing shape and geographic distributions. These changes are produced by long-term shifts of huge sections of crust, called tectonic plates, which are driven by forces operating in the Earth's mantle. As the plates move, the continents that rest upon them also move. This process is very slow but over millions of years it has produced migrations of landmasses over distances of thousands of kilometers and, in some cases, continents have moved across one or two zones of latitude and from one climatic region to another, often to a contrasting one. An extreme case, India, is shown in Figure 8.22. This landmass was located around 70° south latitude about 300 million years ago, but it is now centered at latitude 20° north. Based on fossils and other evidence left in India's bedrock, over the past several hundred million years it has been host to both glacial and tropical conditions as well as several intermediate climatic conditions.

Figure 8.22 Global geography about 150 million years ago. India is in the process of moving from deep in the southern hemisphere across the tropics and into the northern hemisphere.

The geographic arrangement of oceans also changed with plate movement. The Atlantic Ocean did not exist 200 million years ago, but as North America and South America drifted away from Europe and Africa, the Atlantic Basin slowly unfolded. As the Basin widened a system of gyres eventually developed resembling today's current circulation in the Atlantic but with one major exception. There was a wide gap between North and South America through which poured a westward-flowing tropical current, along the lines shown in Figure 8.23. Because of this current the warm-water input to the Gulf Stream and the west-wind drift of the North Atlantic was much weaker than today, meaning that the marine climate of northwest Europe was likely colder and drier.

About three million years ago, the Panama land bridge formed which blocked the tropical current from entering the Pacific basin and diverted warm water northward. This in turn strengthened and warmed the Gulf Stream and the west-wind drift, sending more warm water poleward toward Iceland and beyond to Europe. Climate no doubt changed as a result. Some scientists have speculated that this warm water may have melted back the sea ice above the Arctic Circle, opened vast areas of ocean to evaporation, resulting in increased snowfall and the growth of continental glaciers in North America and Eurasia.

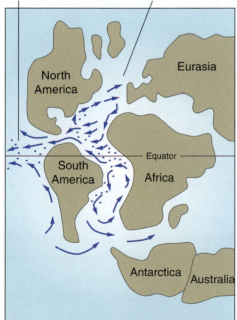

Figure 8.23 The opening of the Atlantic before the Panama land bridge closed. Tropical currents passed into the Pacific through the Panama gap and, as a result, the amount of tropical water diverted northward was much smaller than today.

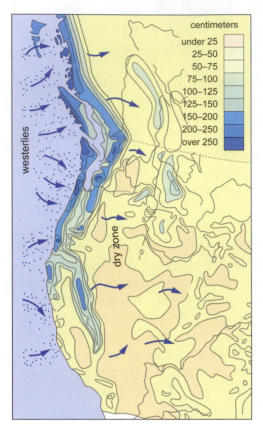

Figure 8.24 Precipitation distribution in western North America shows a strong correlation with the location of mountain belts which lie athwart the prevailing westerlies.

Landforms: As the Atlantic Ocean widened with the westward migration of the Americas, in other parts of the world tectonic plates collided and mountains were thrown up between them, often altering regional wind and pressure systems. The most celebrated collision occurred between India and southern Asia resulting in the formation of the Himalayan mountain chain. This huge mountain range formed a barrier to monsoon circulation blocking the summer (wet) monsoon's penetration into central Asia, leading to reduced rainfall and a change to desert climate there. At the same time rainfall was increased on the windward (Indian) side of the Himalayas making the climate wetter there. (Section 18.7 provides additional discussion on the rise of the Himalayas and related climate change.)

The same sorts of changes have taken place in many other parts of the world. In the midlatitudes, the mountain chains along the west coasts of North and South America formed across the path of the prevailing westerlies. The effects were similar to South Asia. The mountains blocked the maritime air from penetrating into the continental interior. Two distinct climatic zones formed, a wet one on the windward side represented by the marine west-coast climate, and a dry one on the leeward side represented by steppe and desert. Figure 8.24 shows this pattern in western North America where a belt of dry climate stretches northward from Mexico through Arizona and Nevada, into eastern Oregon and Washington, and beyond into central British Columbia.

Land Use and Land Cover: Climate near the ground, in the layer of the atmosphere where we and most other organisms live, is profoundly affected by the character of the landscape. As we illustrated in Chapter 4, landscape materials, for example, affect albedo and when a forested landscape is cleared and lighter materials such as sand are exposed, albedo rises and the absorption of solar radiation is reduced (see Figure 4.19). At the same time, the removal of forest reduces photosynthesis and CO_2 absorption from the atmosphere.

The clearing of vegetated landscapes also influences the balance of heat flows from the ground to the atmosphere. Take a look at the model in Figure 8.25. The vegetated landscape releases much of its heat energy as latent heat with the vaporization of water from plants, soil, and water features. When such landscapes are cleared and replaced by cities, for example, the supply of water is reduced and heat (from the absorption of solar radiation) is released instead as sensible heat. Sensible heat drives air temperature up, whereas latent heat does not. The sensible heat/latent heat ratio (the **Bowen ratio**) is one of the chief factors accounting for the higher air temperatures in cities.

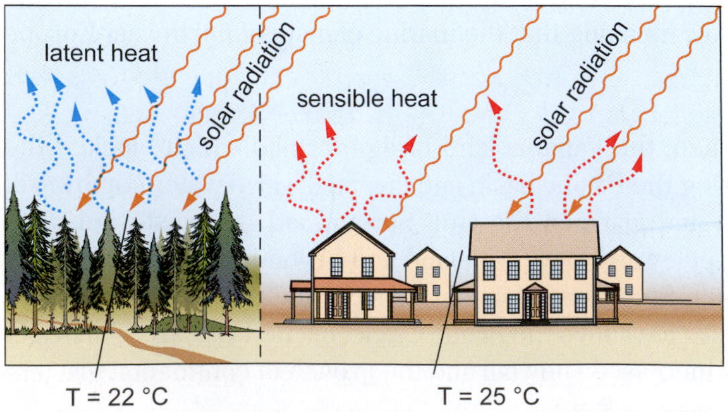

Summary on Causes of Climate Change: The causes of climate change are many, and probably more than we know about. All are driven by systems and fall roughly into three classes: extraterrestrial, Earth surface, and Earth interior. All seem capable of producing a wide range of types and magnitudes of change including those that so alter the Earth's radiation, heat, and water systems that the planet's biological system is thrown off course and nearly arrested. Set into the scale of Earth time, how do we humans stack up as a climate change agent, and at this point, how does the change we are advancing towards appear to rank against past changes in terms of magnitude?

Figure 8.25 The sharp contrasts in heat output between a forested landscape and adjacent developed land.

Figure 8.26 The distribution of urbanized areas, or megacities, in the United States and Canada.

8.4 The Climates of Cities

Cities are the most intensive form of land use yet devised, and it is not surprising that some of the most dramatic alterations of local climate, or microclimate, have traditionally been associated with them. In the past half century urbanized areas have undergone a geographical explosion and now cover such large areas that it is no longer appropriate to call them "localized environments." The greater metropolitan areas of cities such as New York, London, Mexico City, and Los Angeles, for example, occupy areas of 1000 square kilometers or more and in many, such as the Northeastern Seaboard of the United States, strings of cities have coalesced to form megacities (Figure 8.26). We now have massive urban regions throughout the world in both developed and less developed countries, and most are growing rapidly. Some, such as Shanghai, are expected to reach populations of 50–100 million in this century.

In comparing the climate of cities to that of the countryside around them, we find a number of significant differences. Solar radiation is generally less intensive in cities. However, temperature may be as much as 5 to 10 °C higher in cities than in the countryside. Wind at ground level is weaker in cities than in the countryside, and fog is characteristically much more common in urban settings than elsewhere. And for many cities rainfall is greater over the urban region and/or downwind from it.

Alterations in the Flow of Radiation: Cities expel large amounts of gaseous and particulate pollutants into the atmosphere. If the rate of discharge exceeds the rate of removal by airflow through the city, these materials build up in the air forming a pollution (or dust) dome over the city. The air not only smells and tastes bad but is less transparent to solar radiation. Even on sunny days the Sun often has a decided dullness about it. Particulate concentrations are typically ten times greater over cities than over rural lands and, in the lower 1000 meters of the urban atmosphere, the intensity of solar radiation may be reduced by more than 50 percent because of backscattering from the dust dome. Figure 8.27 articulates this phenomenon with a little more detail.

In addition to retarding the entry of incoming shortwave radiation, the dust dome and associated gaseous pollutants decrease the rate of release of longwave radiation from the city. This, of course, is a greenhouse effect, and though it exists in an undisturbed (i.e. natural) atmosphere, it is often greatly increased over cities. The pollutants most responsible for longwave absorption are carbon dioxide, ozone, and aerosols. As a result, heat is retained in the city atmosphere longer than it is in a cleaner atmosphere, thereby inducing higher air temperatures.

The Urban Heat Island: Urbanization may take a variety of forms; closely spaced high-rise buildings in the inner city, sprawling suburban neighborhoods, industrial parks, massive highway systems, and shopping malls, but each has important energy-balance and climatic implications. In some areas albedos may be reduced with the construction of dark roofs and asphalt streets so that more incoming solar radiation is absorbed and converted to heat. More important, the conversion of land from rural to urban surfaces influences the latent-heat flux because hard-surface materials such as asphalt, concrete, and brick nearly preclude the natural moisture flow from soil and plants to the air. As a result, the Bowen ratio is increased significantly. Because a greater proportion of available energy is converted to sensible heat, the ground-level air temperatures in cities are typically higher than those in the neighboring countryside.

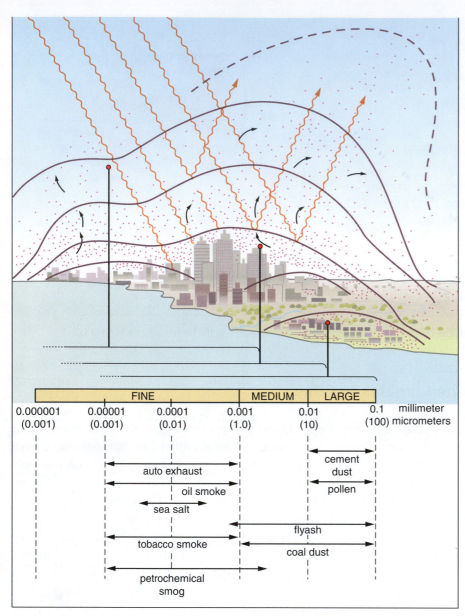

Figure 8.27 The reduction in solar radiation by an urban dust dome with the representative sizes and types of particulates (below).

Another important land-cover factor in the energy balance of cities is the thermal properties of hard-surface materials. Because of the low volumetric heat capacities of asphalt, concrete, and roof materials, the temperatures of these materials tend to rise more rapidly and reach a higher level with the absorption of radiation than do the plants and soil they replaced. Measurements by Rudolph Geiger (1894–1981), a pioneer in the study of microclimate, showed that in early summer the temperature of asphalt would often be double that of the grass. Given both the high Bowen ratio and the low volumetric heat capacities of concrete, asphalt, and brick, it is understandable why the air above hard surfaces tends to heat rapidly. You can verify the remarkable difference in the heating of hard surfaces and vegetated soil surfaces by touching sidewalk and an adjacent lawn during or after a sunny afternoon in any city. Often just walking on them is sufficient to reveal the difference. Little wonder that foot-patrol cops and mail carriers have complained about their hot, swollen feet! From the standpoint of urban planning, this heat differential is a strong argument for incorporating parks, greenbelts, water features, and even "green buildings" with plant-covered roofs into the urban landscape.

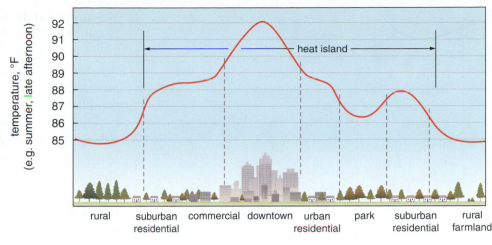

Figure 8.28 The temperature profile of a typical urban heat island on a summer afternoon.

Central Park in New York City, 840 acres of green space in the middle of a most massively built-up urban landscape.

Cities also gain heat from artificial sources, mainly automobiles, factories, homes, office buildings, and institutional and commercial facilities. The heat from these sources escapes into the air around them, and if wind is absent or light, this air can be heated rather quickly. The greatest influence from artificial sources is found in northern cities during winter. In Manhattan in New York City, heat from combustion in midwinter is typically 2.5 times greater than that provided by incoming solar radiation. In summer, on the other hand, this factor drops to about one-sixth (17 percent) of the solar energy input at ground level.

The high emission of heat from hard surfaces plus heat contributions from artificial sources raises the overall air temperature of cities, creating a warm spot, or **heat island**, in the land. Heat islands have a distinctive geographic configuration, which in profile, such as the one represented in Figure 8.28, typically drop off sharply forming a "cliff" where the city gives way to the rural landscape. The magnitude of the heat island – defined as the difference between urban and rural temperatures – increases with city size, based on population.

Figure 8.29 Temperature contrast on one summer day between Golden Gate Park, a large swath of green, and downtown San Francisco, a dense mass of streets and buildings.

But at a finer scale of observation significant differences stand out within the urban heat island. Parks and greenbelts are often distinguishable as cool spots. Here plant transpiration, evaporation from water features, and shade from tree canopies produce much lower Bowen ratios. With less sensible heat available to raise air temperature, these spots are cooler. The map in Figure 8.29 shows the distribution of afternoon temperatures on a summer day in San Francisco. The city is located at the tip of a peninsula with strong thermal influences from the sea. Nevertheless on certain days (usually calm and sunny ones) there are remarkable temperature contrasts between Golden Gate Park, which is overwhelmingly green, and the downtown area near Union Square, which is overwhelmingly pavement and buildings.

But most cities do not have vast areas of parks to cool them down and at times the heat island can be strong enough to affect human health. This takes the form of a disorder known as heat syndrome, which has killed more than 10,000 people in the United States since 1950. Heat syndrome is a clinically recognized disorder in the body's heat-regulation system caused, among other things, by extreme temperatures.

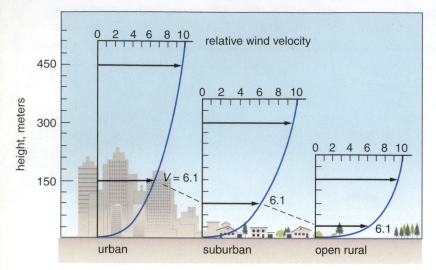

Figure 8.30 Wind-velocity profiles over a city center and surrounding suburban and rural landscapes. The building mass of the city center tends to divert wind upward.

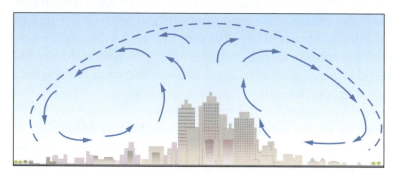

Figure 8.31 Air circulation resembling a convection cell over a city on a calm day in response to the urban heat island.

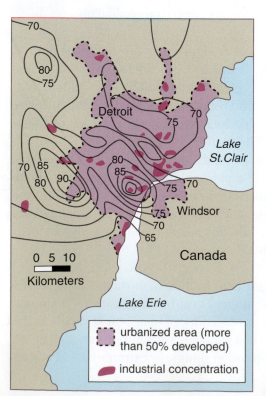

Figure 8.32 Map of annual precipitation over the Detroit–Windsor region in the 1970s and 1980s showing markedly higher values in the heavier built-up areas.

When summer heat waves overtake an urban region, the temperature in inner-city areas may be driven up an additional 5 to 10°C, to as high as 50°C (120°F). For many people, especially the elderly and the infirm, the result is breakdown of the thermoregulatory system which can lead to death. Although air conditioning has reduced the risk of heat syndrome in cities of developed countries, it remains a serious and perhaps growing risk among the poor in the rapidly growing megacities of the less developed world, for example, in China, India, Mexico, and Brazil.

Urban Aerodynamics: Wind is a critical factor in the energy balance and the climate of cities because it is the principal disperser of dirty, heated air. Cities influence wind in two ways. First, and most important, they increase the roughness of landscape, which retards airflow near the ground. As wind moves from a rural area toward the inner city, its velocity profile, such as the one in Figure 8.30, is displaced upward when it encounters taller, more closely spaced buildings. This is a response to the stronger frictional, or drag, resistance posed by the irregular terrain formed by the mass of buildings and other large urban structures. Average wind speed may be 20 to 30 percent lower and extreme gusts, 10 to 20 percent lower over cities than over the nearby countryside. Consequently, the removal of flushing surface air is less efficient in cities than in rural landscapes; surface air tends to linger over urban areas, allowing more time for it to collect pollutants and heat up.

Convection related to the urban heat island is the second influence on wind. It is characterized by a gentle inflow of surface air from the suburbs to the inner city, and is especially likely when regional weather is calm and there is a strong heat island over the city. The heat island produces a weak pocket of unstable low-pressure air, which generates internal upflow and lateral inflow along the surface as the diagram in Figure 8.31 illustrates. Some studies conducted over New York City have recorded not only a marked upward flow over heavily built-up Manhattan Island, but also a downward flow over the nearby Hudson and East rivers. This circulation pattern is consistent with the temperature and pressure regimes that would be expected for surfaces with energy characteristics as different as cool water and warm concrete. The creation of low pressure within a zone of rising warm air, and the subsequent descent of this air into cooler areas gives this microscale urban convection system similar mechanics to the huge Hadley cell circulation driven by surface heating and converging airflow over the tropics.

Precipitation and Heat from Combustion: Do the urban heat island and related conditions influence precipitation? In many areas it appears that convectional rainfall is greater in and around the urban region, but the magnitude of the urban influence is often masked by the broader regional weather and climatic patterns. A study in the Detroit metropolitan region, however, suggests that precipitation (see Figure 8.32)

Table 8.1 Climate changes identified with urbanization

Element	Comparison with rural environs
Temperature	
Annual mean	0.5 to 1.0 °C higher
Summer max.	5 to 8 °C higher
Relative humidity	
Annual mean	6% lower
Winter	2% lower
Summer	8% lower
Dust particles	10 times more
Cloudiness	
Clouds	5–10% more
Fog, winter	100% more
Fog, summer	30% more
Radiation	
Total on horizontal surface	15–20% less
Ultraviolet, winter	30% less
Ultraviolet, summer	5% less
Wind speed	
Annual mean	20–30% lower
Extreme gusts	10–20% lower
Calms	5–20% more
Precipitation	
Amounts	5–10% more
Days with < 0.5 cm	10% more

(From a variety of sources.)

in some urbanized areas can be significantly greater than in the outlying suburban and rural areas. This particular study was conducted in the 1970s when Detroit and its industry were fairly robust. Since then and contrary to the global trend toward megacities, Detroit has lost population and industry. Perhaps the urban heat island and precipitation have also declined?

With respect to cloud cover and fog, there is no doubt that both are more frequent in most urban areas, especially large ones. Although increased cloud cover may be related to instability associated with heat-island circulation, the added clouds and fog are probably related more directly to the abundance of minute airborne particles from air pollution, which serve as condensation nuclei for moisture droplets. The concentration of condensation nuclei over medium-sized cities is typically five to seven times greater than that over rural areas.

Generally, the larger the city and the denser its development, the greater the magnitude of climate change. Since we measure urban climate based on the difference between atmospheric conditions within the cities and those in the landscape surrounding, the nature of the surrounding landscape can be significant in the comparison. A city carved out of a forested landscape may show a stronger heat-island effect, for example, than one set in a desert landscape because the thermal properties of a city should be closer to those of a desert than a forest. Among other things, the Bowen ratio in the desert city should not be extremely different from that of the desert landscape around it.

In fact, the city may exhibit areas where conditions are cooler than the desert. For example, research in the dry Central Valley of California has shown that suburban areas with trees and grass are often cooler than the farmland adjacent to them. But most of what we know about urban climates comes from midlatitude cities in humid zones such as eastern North America and Europe. A summary of the climatic changes associated with urbanization under these conditions is provided in Table 8.1.

Summary and Prospect on Urban Climate: Urbanization is advancing at dramatic rates across the world with profound effects on the energy balance and climate over vast metropolitan areas. Only a few decades ago there were great contrasts in the geographic character of "old-world" cities and the modern industrial cities of developed countries such as Canada, the United States, and Japan with their tall buildings, millions of automobiles, sullied atmospheres, and huge energy outputs. But that difference is rapidly disappearing as cities in less-developed countries, mostly notably China and India, become infested with automobiles and dirty air, and spill over into the surrounding countryside. In fact, many of these urban masses are overtaking cities of developed countries in terms of population, geographic coverage, degraded air quality, and heat-island effects. Mexico City, part of which is shown in Figure 8.33, with a population exceeding 30 million (roughly equal to that of New York and Tokyo combined), is a salient example.

Figure 8.33 Aerial photograph of Mexico City, one of the largest urban masses in the world.

8.5 Climate Change in the Twenty-first Century

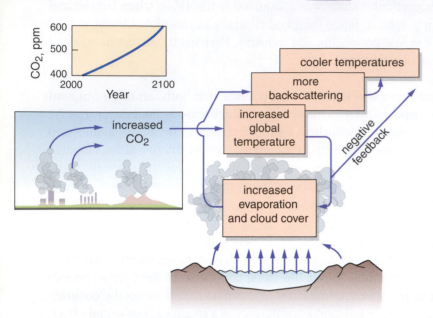

Figure 8.34 Negative feedback in the carbon dioxide cycle. Global warming leads to increased evaporation, cloud cover, and backscattering of solar radiation followed by lower temperatures.

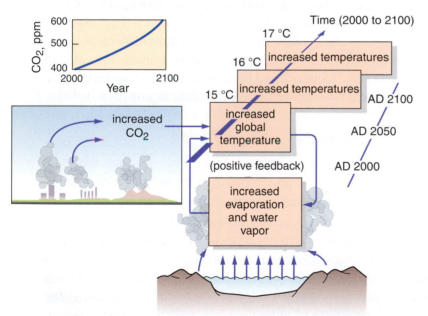

Figure 8.35 Positive feedback in the carbon dioxide cycle. Global warming leads to increased evaporation, higher humidity, and greater atmospheric heat retention.

As we have established, many factors can change Earth's thermal climate. Determining which factors are responsible for a known change begins by looking for correlations between candidate factors and a measured change in climate. This boils down to an examination of trends such as changes in carbon dioxide, sunspot activity, land-use change, or volcanic activity on the one hand and changes in global temperature on the other. And it often requires thinking beyond simple correlations.

We must draw on our knowledge of systems, for the picture is often complicated by complex interactions within the Earth's energy system involving feedbacks and counterbalancing mechanisms. For example, an increase in global temperature caused by rising CO_2 levels (as given in Figure 8.34), can lead to increased evaporation, more atmospheric vapor, and greater global cloud cover. The increased cloud cover could, in turn, lead to higher rates of reflection for incoming solar radiation and, in turn, to cooler atmospheric temperatures. Under such circumstances, a simple interpretation of the relationship between CO_2 and global heating alone would suggest that CO_2 is having little or no effect on temperature, but that would be incorrect.

On the other hand, an increase in atmospheric vapor brought on by global heating related to elevated CO_2 levels could produce the opposite result, that is, further global warming, because increased atmospheric vapor would improve the greenhouse effect. This would be a form of positive feedback and, as illustrated in Figure 8.35, it would repeat itself over and over as CO_2, water vapor, and air temperature rise. Recall that water vapor is the principal greenhouse gas in the atmosphere. Under these circumstances, a simple interpretation might ascribe too much weight to the thermal forcing effect of added CO_2. Thus, the problem of finding causes for climate change is often difficult, especially when it comes to defining the mechanisms of change and their interrelations, especially feedbacks.

The observation that global temperature is rising is widely accepted by the scientific community worldwide. The evidence for warming is compelling and increasingly it is becoming apparent that a major contributor to this trend is the rising CO_2 content of the atmosphere. After assessing decades of climate data recorded everywhere from the depths of the oceans to tens of miles above Earth's surface, leading scientists around the world have reported major advances in our understanding of climate change. In 2007, these findings were examined by the Intergovernmental Panel on Climate Change (IPCC), a group of high-ranking scientists assembled by the World Meteorological Organization and the United Nations

Environment Program and charged with the task of synthesizing current scientific understanding and projecting future climate change using well-established climate-forecasting models.

The two principal findings reported by the Panel were:

It is "very likely" (meaning a probability of 90 percent or more) that emissions of heat-trapping gases from human activities have caused "most of the observed increase in globally averaged temperatures since the mid-20th century." Evidence that human activities are the major cause of recent climate change is even stronger than in prior assessments.

It is "unequivocal" (100 percent certainty) that Earth's climate is warming, "as is now evident from observations of increases in global average air and ocean temperatures, widespread melting of snow and ice, and rising global mean sea level." The report also confirms that the current atmospheric concentration of carbon dioxide and methane, "exceeds by far the natural range over the last 650,000 years." Since the dawn of the industrial era, concentrations of both gases have increased at a rate that is "very likely to have been unprecedented in more than 10,000 years."

The Panel also reported that at continental, regional, and ocean-basin geographic scales, numerous long-term changes in climate have been observed. These include changes in Arctic air temperatures and ice volumes, widespread changes in precipitation amounts, ocean salinity, wind patterns, and aspects of extreme weather including droughts, heavy precipitation, heat waves, and the intensity of tropical cyclones.

The CO_2 System: Since the main driver in all of this appears to be carbon dioxide, Earth's second leading greenhouse gas (after water vapor), our next step is to examine the nature of the CO_2 system on our planet. Most CO_2 in the atmosphere comes from two natural sources: (1) the *biosphere* as a product of biological respiration; and (2) the *oceans* as a product of gas exchanges and biological respiration. The total CO_2 input to the atmosphere from these sources is about 200 billion tons per year. Under natural conditions, however, CO_2 is recycled back to the Earth's surface so that the annual input is counterbalanced by 200 billion tons of CO_2 extraction, mainly from photosynthesis and ocean absorption. Thus, under *natural* conditions atmospheric CO_2 is held in balance by equal levels of input and output.

Now comes industrial-age land use and more than 6 billion humans who add about 7 billion tons of CO_2 to the atmosphere per year but take none out. This represents a surcharge that year by year keeps building the load of CO_2 in the atmosphere. But measurements show, curiously, that the annual buildup rate is actually less than the annual input, closer to 4 billion tons, which means that some part of the environment is extracting the difference, or 3 billion tons a year.

The two leading candidates as sinks for this CO_2 are the oceans and the forests, and it is generally agreed that most CO_2 is probably going into the oceans. This means that CO_2 is building up in the oceans and their ecosystems. But it is uncertain how long this trend can continue. Will these reservoirs reach maximum thresholds and refuse to accept surplus CO_2 at some point, thereby hastening the buildup rate in the atmosphere?

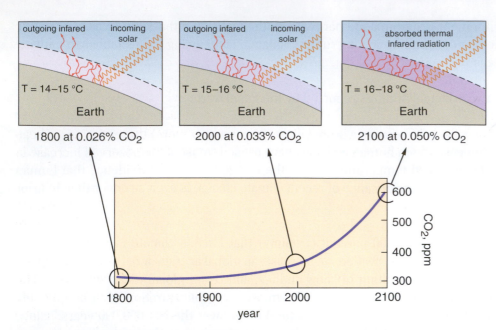

Figure 8.36 Carbon dioxide levels and estimated Earth equilibrium temperatures at three dates: 1800, 2000, and 2100. By 2100, CO_2 will be more than double the 1800 level.

Figure 8.37 An example of a computer-generated map on global warming from the Hadley Centre for Climate Prediction and Research, England. According to this computer model, the projected net change in surface temperature worldwide is 3.2 °C for the period 2070–2100 assuming the world's nations are able to make significant cuts in the use of fossil fuels.

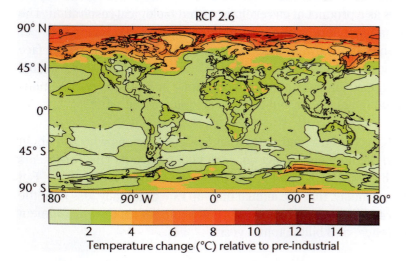

Given the projected loading rates based on world population growth and economic development, atmospheric CO_2 will double pre-industrial levels reaching around 600 ppm by the second half of this century. Figure 8.36 puts this in graphic perspective. If ocean absorption declines in the meantime, the level will be even higher. Although international efforts to curb CO_2 emissions are underway, prospects are not good for significant reductions in the future. Consider this fact. If inputs from human sources were completely stopped tomorrow, it would take 200 years or more for the atmosphere to rebalance itself and return to pre-industrial CO_2 levels.

But the prospects for control and reduction are getting worse with global population growth and economic expansion. Consider, for example, that China's drive toward economic development and a consumer economy is being fueled largely by coal and most is being burned with little or no emission controls. In addition, China plans to increase its use of automobiles. A recently enacted national expressway plan means China will invest 2 trillion yuan (US $240 billion) in an 85,000 kilometer (53,000 mile) expressway network scheduled to be built over the next 30 years. Road building requires making concrete, which is one of the lesser-known activities contributing to CO_2 emissions, being responsible for 7–10 percent of the global total each year (a constituent of concrete is $CaCO_3$, which when heated to make concrete releases CO_2 into the atmosphere). On balance, chances are very high that significant global warming will be well advanced by mid century.

Some Consequences of Global Warming: How much temperature change will produce measurable impacts on the environment? According to various computer models, such as the one used to produce the map in Figure 8.37, we expect an increase in the global temperature of 1.1 to 6.4 °C by the end of the twenty-first century. A change of, say, 1 or 2 °C does not sound like much, but consider that during the Little Ice Age global temperature over much of the Earth declined about 1 °C and history records serious consequences: harsher winters, shorter growing seasons, land-use failures, food shortages, increased health risks, and advances in glaciers worldwide. A 2 or 3 °C change is sure to induce major global changes and place the human population under severe stress in some, if not most, regions of the world.

The early stages of many of these changes have already begun. In fact, scientists recently analyzed the extraordinary weather changes of the United States in the 1990s including severe hurricanes, droughts, and El Niño events and concluded that there is a 90 to 95 percent chance that these are related to greenhouse warming. Especially worrisome is the fact that by the time various climatic changes are fully developed, around 2050, global population will have reached 9 to 10 billion with several billion living in geographic settings such as coastal lands, desert margins, river valleys, and overgrown urban centers, which are especially prone to the effects of changing climate.

Atmospheric Circulation Changes: Again, our thinking must be system-based, for circulation changes in the atmosphere can be understood only by tracing the linkage among energy, pressure cells, and airflow. There is little doubt that large-scale circulation is likely to be altered by global warming. The equatorial low-pressure system, which is driven in large part by heat in the lower troposphere, could intensify, forcing more air into the upper troposphere to feed wind systems aloft. The return flow for this air is the subtropical high-pressure cells on the poleward margins of the tropics. (See Figure 5.10 for an illustration of how this works.) Air enters the high-pressure cells from aloft, and by the time it sinks to the ground it is very dry. This system is what accounts for the arid conditions at subtropical latitudes around the world. The Sahara of Africa and the Sonora Desert of North America are examples. If the flow of air pumped into these cells from the equatorial low is increased, then the intensity and geographic size of the world's arid zones may increase, as well as shift poleward.

Forecasts based on complex computer models of global circulation show poleward shifts in the major climatic zones, including expansion of dry zones and perhaps an increase in the magnitude and frequency of droughts. As some deserts expand, the fringing semiarid landscapes will be transformed into arid ones with drier soils and sparser plant covers. Where lands in these regimes are already under stress from overgrazing, deforestation, and expanding agriculture, **desertification** may accelerate. Desertification is the process of making desert landscapes by the combined action of natural drought and human land use, or excessive human land use.

Global warming is also producing ocean warming throughout the world. And since it is the heat in seawater that drives ocean evaporation, we can expect increased atmospheric loading with water vapor and latent heat. Since the "breeding grounds" for tropical storms is defined by the extent of warm ocean temperatures (27°C minimum), the distribution of these storms is expected to spread poleward as the pool of warm water expands. At the same time, ocean temperatures will be higher within the traditional breeding grounds for storms, giving rise to bigger and perhaps more frequent storms. Recent research (summarized for the Atlantic in the graph in Figure 8.38) reveals there has been around a 60 percent increase in large-category hurricanes over the past 30 years in the Pacific and Atlantic oceans.

In terms of storm distributions, midlatitude areas such as the North American East Coast can expect more hurricanes. These areas may also experience longer growing seasons, more precipitation, and increased plant productivity. On the other hand, they may also experience increased runoff, soil erosion, and flooding. Overall, current scientific forecasts on the location, extent, severity, and effects of regional climatic changes are very uncertain at this time. Other changes, however, are more certain; one is a rise in sea level.

Figure 8.38 The trend in hurricane power over the North Atlantic since 1950 showing a sharp increase beginning in 1993. (Based on the power of hurricane wind speeds and the geographic size and number of hurricanes.)

Hydrological Changes: About 75 percent of the Earth's water supply outside the oceans is held in glacial ice, mostly in the polar ice caps. As the map in Figure 8.12 reveals, this is where global warming is most intensive. In the northern hemisphere, most of the Arctic has warmed by 1 to 3 °C and, in Antarctica, sizeable areas of the coast have warmed by 0.5 to 3 °C. As a result, winters tend to be softer and summers somewhat warmer in these regions with three hydrologic consequences: melting of permafrost, reduction in Arctic sea ice, and decay of the great ice caps.

In Antarctica, pieces of shelf ice (the leading edge of huge ice sheets that extends into the ocean) of unprecedented size have broken away (calved) in the last decade. The satellite images in Figure 8.39 show such a break in 2002 which released 720 billion tons of ice into the sea. And in Greenland, home of the world's second largest ice mass, the ice cap is declining (lowering) by as much as a meter a year in some places. At the same time glaciers in mountain ranges across the planet are melting at accelerated rates. The product of all this melting is the addition of water to the oceans, which is causing a rise in sea level. Meanwhile, ocean water is heating up, which is causing thermal expansion of the oceans, also leading to a rise in sea level. In the extreme, ice melting alone could create massive coastal flooding. Witness that melting of just half the world's current volume of ice would produce a rise in sea level of 40 meters (130 feet).

Given the latest forecasts of atmospheric warming, by the end of this century sea level will be 1 to 2 meters (3.3 to 6.6 feet) higher. Since most of the world's largest cities are located on ocean coastlines, a number of major population centers would be greatly affected. For example, most of New Orleans is at or below sea level (which helps explain the devastating effects of Hurricane Katrina). Most of the 14 million people of the Netherlands live on land below sea level; Alexandria, Egypt, lies only a few meters above sea level; and nearly one-third of the densely populated country of Bangladesh lies within two meters of sea level. Figure 8.40 provides some elevation data for several large coastal cities and most are growing larger each year with development pressing into low-lying coastal lands.

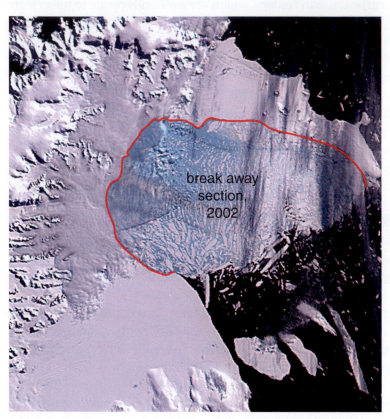

Figure 8.39 The Larsen ice shelf in the Antarctic lost 3250 square kilometers of ice (see red line) in the fall of 2002, an extreme example of a trend affecting the great ice masses worldwide in response to global warming.

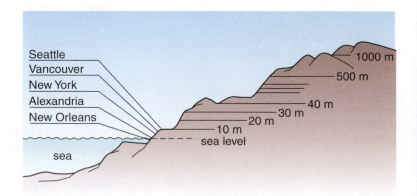

Figure 8.40 Most of the world's major cities are located on the sea coast and most are situated at low elevations where they are vulnerable to disruption by sea-level rise.

Figure 8.41 Massive storm waves striking and washing over a populated coastline, a process that will occur with greater magnitude and frequency with global warming and rising sea level.

Even a modest sea-level rise of 0.5 to 1 meter would be devastating, affecting tens of millions of people. Not only will flooding be a problem, but in many coastal cities, port facilities such as piers and warehouses will be rendered inoperable, dikes/levee systems will be overwashed (as happened with Katrina), and ocean water will back up into sewer lines causing street flooding and health problems. What is more, a rise in sea level will also lead to greater coastal damage from hurricanes and other storms, especially in low areas such as Bangladesh and the American Gulf Coast, because storm waves like those shown in Figure 8.41 will be able to penetrate the shoreline with greater force.

The effect on coastal ecology could also be devastating. It is widely forecast that the rate of rise would be too fast for many intertidal and wetland communities to adjust, that is, shift location with changing water depth. And for many agricultural areas in crowded countries like Bangladesh and India, there are no open lands further inland available for resettlement of displaced people.

Land-use Changes: To support an expanding global population, people are being forced to live in marginal and submarginal lands. These are lands where land use, especially farming, is at risk of failure. In marginal areas, agriculture has a 50:50 chance of success in any year. In submarginal lands chances are less than 50:50. Climatically, marginal lands generally fall into two classes: (1) moisture marginal; and (2) thermal marginal. In thermally marginal lands, the growing season is just long enough to sustain hardy crops, but yields are often poor and crop failures common. If the mean temperature drops by 2 °C, it might well shorten the growing season 30 days or so, causing certain crops to fail. But if temperature rises, the growing season may be lengthened and thermal crop-failure risk reduced. But many cold areas also tend to be dry, so a temperature increase could bring another problem – drought.

Moisture-marginal lands include the desert fringes, steppe, and dry savanna regions where precipitation is light and variable and the balance between a stable landscape and an unstable, desolate one is very tenuous. The stable landscape represents a delicate balance among a number of interrelated factors including soil and its capacity to support vegetation and land use; and forces of the atmosphere such as wind, solar heating, and drought and their capacity to degrade soil, vegetation and land use. Soil is especially critical because of its moisture-holding capacity, which helps buffer the landscape against drought.

Extended periods of drought are inevitable in such climates. To survive a major drought without serious degradation of the moisture-marginal landscape usually requires land-use adjustments such as reduced grazing, population redistribution, and changes in crops. In many parts of Africa and Asia such adjustments are, for many reasons, not possible, and in order to maintain life over the short run, farmers and herders are forced to attempt to squeeze more out of a declining landscape. The result is overgrazing and expansion of croplands, which in turn lead to loss of groundcover vegetation, increased soil erosion, and an overall decline in the landscape's carrying capacity as is suggested by the photograph in Figure 8.42. The downward spiral of land and life is usually irreversible, and the moisture-marginal landscape under the combined stress of drought and land-use pressure falls victim to desertification, even in developed countries.

Figure 8.42 A degraded landscape in Africa where the combined effects of drought and farming have weakened and impoverished the grassland ecosystem.

Ecological Changes: A major international study conducted in 2004 found that global warming may drive a quarter of land animals and plants to the edge of extinction by 2050. According to the researchers' collective results, the predicted range of climate change by 2050 will place 15 to 35 percent of the 1103 species studied at risk of extinction. The numbers are expected to hold up when extrapolated globally, potentially dooming more than a million species.

Sensitive ecosystems such as coral reefs are also threatened by global warming. Although warming of the world's oceans might benefit reefs in some ways, it could add to the environmental stresses already besetting coral reefs. For example, increases in ocean temperature contribute to coral bleaching episodes that cause coral mortality and stress, while future increases in atmospheric carbon dioxide may limit coral growth. A reef bleaches when it loses the tiny algae that live inside the coral polyps. Without the algae, the polyps may starve. In addition to their direct effects, these stresses also act to degrade coral reefs by increasing their susceptibility to pollution, overfishing, predation, and disease. These impacts have significant economic effects, particularly in tropical areas, as coral reefs protect coastal areas from storms, floods, and erosion, provide habitats for thousands of marine species, and attract tourists.

Chapter Summary and Overview of Climate Change:
Past, Present, and Future

Ours is a planet of changing climate, and as the summary diagram illustrates, it comes in a variety of shapes and magnitudes and from a host of sources. It almost always affects life, bringing it to its knees at times, and when it does, the course of evolution has been changed. In short, without climate change, Earth would be a far different planet. If we trace the changes back to their origins, we find that they are all system driven, some Earth bound, some not, and among the Earth-bound systems, we must count ourselves. Today half or more of Earth's 7 billion humans live in and around cities, the most complex and environmentally altering land-use systems we have yet invented and a root cause of global warming.

▶ **Earth's climate has changed many times in the past and the geographic magnitude of change has been highly variable.** Some changes were massive and affected the biogeographical character of the entire planet, altering temperature, precipitation, and biota. Others affected only selected regions of Earth.

▶ **Our knowledge of climate change is drawn from various lines of evidence.** This includes oxygen isotopes in glacial ice, fossil-leaf imprints, buried pollen, tree rings, archeological artifacts, and historical documents.

▶ **Early Earth was warm enough to maintain liquid oceans, which fostered the origin of life.** But there were numerous interruptions in the first several billion years and many since, including one 65 million years ago, which changed the direction of the biological evolution.

▶ **Several periods of cooling occurred in the last 55 million years.** One of these resulted in the formation of the Antarctic ice cap, another in the current ice age beginning about 1.8 million years ago.

▶ **The current ice age is marked by glacial periods and interglacials.** During glacials, global ice coverage expands dramatically and sea level drops. During interglacials, ice coverage shrinks, sea level rises, and there are interludes of warming and cooling.

▶ **In the late twentieth century, Earth began warming in response to CO_2 loading of the atmosphere.** Since 1975, it has been warming rapidly but with more change in some areas than others.

▶ **Over geologic time, many factors have caused climate change, including variations in the geometry of Earth's motion in space.** In the past two million years, three geometric variations have been defined and they correspond to changes in global temperature.

► **Solar luminosity is also subject to change.** It is related to sun-spot activity and atmospheric transparency associated with volcanic activity and air pollution from land use.

► **Earth's greenhouse effect is produced mainly by water vapor and carbon dioxide.** Since the Industrial Revolution, atmospheric carbon dioxide has increased significantly and it corresponds to the current increase in global temperature.

► **Earth's crustal motion can cause climate change as continents move and oceans close and open.** Such changes can alter the pattern and strength of current and wind systems resulting in climate change.

► **Human land use can change climate in many ways.** These include alterations in albedo, surface roughness, wind flushing, and carbon dioxide loading of the atmosphere.

► **Climate change can be a complex process involving interactions of many factors in the Earth's energy system.** Outcomes often hinge on the direction of feedback processes resulting from changes in atmospheric carbon dioxide, ocean evaporation, and global cloud cover.

► **Air pollution from land-use activity is adding billions of tons of carbon dioxide to the atmosphere annually.** Much of the surplus is being taken up by the oceans and global forests leaving a large annual surcharge of 4 billion tons in the atmosphere.

► **Over the twenty-first century global temperature is expected to rise between 1.1 and 6.4 °C.** The geographic effects are expected to be numerous and widespread with impacts on agriculture, cities, coastal lands, and ecosystems.

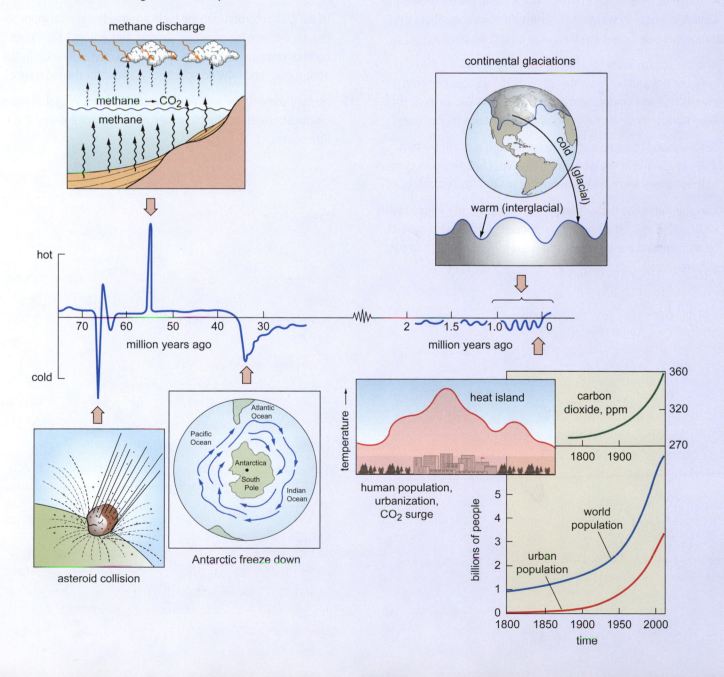

Review Questions

1 Our perception and memory of climatic conditions experienced in life are often different from what the climatic records reflect. What are some of the explanations for this?

2 What are some of the indicators of ancient (prehistoric) climate change and what is the scientific basis for using oxygen-16 and oxygen-18 as a measure of past temperatures? How can land uses of the past (as revealed by archeological and historical evidence) tell us something about past climatic conditions?

3 Can you name at least three major climate-change events of the distant past that affected Earth at a global scale and what were their apparent causes?

4 At different times in Earth's past, volcanoes have contributed to climate change. By what mechanisms do volcanoes affect the atmosphere and what are the major effects on climate?

5 Find the approximate location of Washington, DC on the map in Figure 8.8 and list a number of likely differences between the climate and landscape then and that now. Where in North America are those former geographic conditions found today?

6 Ten thousand years ago and 1400 years ago mark important dates of change in the global climate. What happened to climate at those times and name a number of major effects of each.

7 Can you name the three changes that comprise the Milankovitch cycle and identify the key feature of each? And what is meant by atmospheric transparency and what is its connection to past climate change?

8 What two atmospheric gases are mainly responsible for the greenhouse effect and how have the amounts of these gases in the atmosphere changed over the past 250 years? What accounts for the change?

9 Can you describe in a paragraph or two the major differences in climate between the interior of a large midlatitude city and a forested landscape outside the city on a summer day? What geographic factors are most critical in explaining these differences?

10 What is meant by negative feedback in a system, and in Figure 8.34 explain how it can be argued that an increase in carbon dioxide could lead to atmospheric cooling?

11 In the global carbon dioxide cycle what is the annual input to the atmosphere from human sources and actions? How does this number compare with the actual annual rate of buildup in the atmosphere and how do scientists account for the difference?

12 Identify some of the real and expected impacts of global warming on the atmosphere, the biosphere, the hydrosphere, and the lithosphere.

Earth as an Ecosystem:
Energy, Food, and Life

Chapter Overview

This chapter opens with a brief look at the biosphere as a physical component of the landscape and oceans and then goes on to examine Earth's bio-energy system. Scientists refer to this system as an ecosystem and all life from bacteria and flies to humans and redwoods is inextricably woven into ecosystems. What drives ecosystems and how do plants convert the energy and matter that fuels them into organic energy and how does that energy get passed along to other organisms? And to make things truly geographical, we must examine how the work of ecosystems is distributed over the globe, and how it all relates to geographic conditions of the atmosphere, land, and sea, such as the distribution of heat and moisture. Finally, where do we humans and the agricultural systems we have put together fit into the global ecosystem?

Introduction

Earth's biological system has been developing for over 3 billion years. It began in the sea with simple, single-celled organisms and evolved into a vast network of plants, animals, and microorganisms that today occupies the entire planet. We call this network of life the biosphere, and although it can seem massive when we stand in a forest or dive into the ocean, it is actually quite thin and tenuous compared to the lithosphere, hydrosphere, and atmosphere. Yet the biosphere is central to Earth's geographic character for, among other things, it is instrumental in shaping soil, climate, hydrology, and the physical character of most landscapes. The built environment of humans is itself a product of the biosphere because, like a coral reef or a termite colony, it too is constructed by an organism.

What enabled the biosphere to grow, diversify, and spread over the planet? Was it evolution, the process by which life changes and adapts over time? Part indeed can be attributed to evolution, especially the diversification of life and its occupation of new environments. But evolution has not been the driving force powering the life machine. Evolution, rather, has functioned more as the navigator, charting the direction of change. The driving force behind the life system, and indeed evolution itself, is Earth's unique system of organic energy production.

Every biological process on Earth, from the movement of your eyes to the growth of a giant sequoia tree, runs on a vast system that extracts energy and matter from the geophysical environment and converts it into organic fuel. At the heart of this system is photosynthesis, a process of plants in which sunlight is combined with heat, carbon dioxide, and water to produce a fuel that drives all life, including the plants themselves.

Earth's bio-energy system produces over 200 billion tons of organic fuel every year worldwide. It draws its energy from a distant source, the Sun, which supplies Earth with a never-ending stream of radiation. Solar radiation provides not only light and heat, but powers the geographic systems that make water and carbon dioxide available. The Sun's energy contribution to our planet is huge, yet the biosphere uses very little of it in photosynthesis, less than 1 percent of that reaching the Earth's surface. This is a startling fact, for the entire life system is dependent on a tiny window in the solar radiation system. And this window can change, as it has many times in the past, altering the output of organic fuel by the biosphere and with it the number, type, and distribution of organisms.

9.1 The Form and Function of the Biosphere

Each of the Earth's great spheres (litho-, atmo-, hydro-, and bio-) represents a global system and all four interact by exchanging energy and matter. The biosphere lies at the very heart of these exchanges, centered directly on the Earth's surface. It is made up of several great groups or **kingdoms** of organisms, but biologists are uncertain on how many kingdoms there are, what they should be called, and who belongs to which. The exceptions are the two kingdoms of greatest concern to us, the **Plantae** (plants) and **Animalia** (animals). According to one classification scheme, the rest of living matter, mainly microscale, is placed in three kingdoms: **Monera** (bacteria, very small, one-celled organisms that are neither plants nor animals), **Fungi** (mainly single-celled plant-like organisms), and **Protista** (mostly single-celled organisms that share plant and animal traits). For ease of discussion we shall refer to Monera, Fungi, and Protista as **microorganisms**. Of the three groups (that is, plants, animals, and microorganisms), plants make up the vast majority of the biosphere's mass but, as the graph in Figure 9.1 shows, animals make up the vast majority of its species. Species are the basic unit of biological classification for all organisms. A **species** consists of individuals of the same kind that can breed together but not with other organisms.

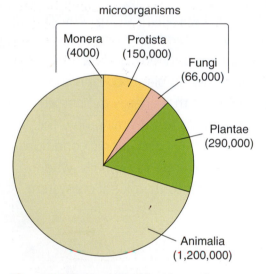

Figure 9.1 The kingdoms of Earth's organisms ranked according to species numbers. The animal kingdom includes far more species than the other kingdoms combined.

The biosphere is the thinnest and most tenuous of Earth's four spheres. The total mass of living matter on Earth is only 8 trillion tons. Compared to the mass of air (5140 trillion tons) and water (1,500,000 trillion tons), living matter is only a miniscule part of Earth. The lithosphere, which represents the rigid outer layer of the Earth, is more than 1,000,000,000 trillion tons.

Set in a global context, the biosphere is little more than a thin membrane draped over the Earth's surface. Although microorganisms can be detected thousands of meters underground, and pollen, spores, and other tiny biological agents can be detected thousands of meters into the atmosphere, and various vertebrates and invertebrates can be detected at great depths in the ocean, the vast majority of all life is concentrated at the Earth's surface within a layer only 10 meters or so deep. Here the life system forms an interlocking web of organisms.

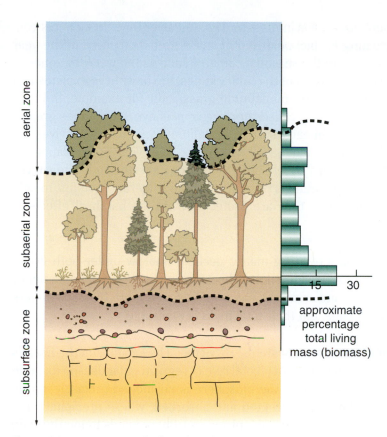

Figure 9.2 A cross-section of a forest biosphere showing the relative distribution of life above and below ground. In forests, the bulk of the living mass is concentrated in the subaerial zone.

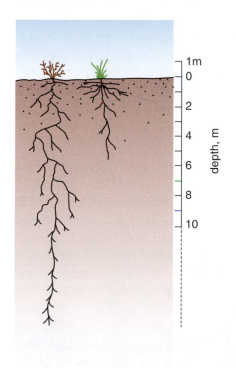

Figure 9.4 As climatic conditions grow harsher, biomass concentrates closer to the ground and underground as illustrated by the deep roots systems of many desert and grassland plants.

The Terrestrial Biosphere: Over most of Earth's land areas the biosphere is made up of essentially three zones as shown in Figure 9.2: (1) the **subaerial**, or **surface zone,** represented by plant bodies, including foliage and root systems, and most animals; (2) a **subsurface zone** represented by the area beneath most plant roots, generally below a depth of 1 to 2 meters where microorganisms are the dominant lifeform; and (3) an **upper** or **aerial zone** represented by non-ground dwelling organisms such as birds, many insects, and wind-blown microorganisms.

The core of the biosphere, defined by the zone containing, say, 75 percent or more its **biomass** (weight of all living matter), is found within the surface zone. On land, this core zone varies in thickness from as much as 50 meters in permissive environments (those with ample supplies of heat, light, nutrients, and water) to as little as several centimeters in harsh environments (see Figure 9.3). Where it reaches its greatest thickness, represented by the large forests of the tropics and midlatitudes, the central mass or heart of the biosphere is roughly 30 meters (100 feet) deep. At its thinnest points, in the harsh deserts and frigid polar lands the surface layer is less than a meter (3.3 feet) deep, and frequently only several centimeters deep.

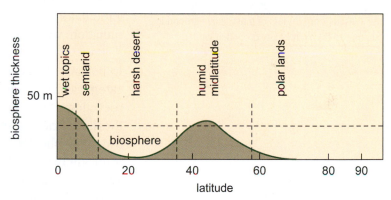

Figure 9.3 The change in the core thickness (depth) of the biosphere with latitude.

In addition, where the center of the biosphere lies relative to ground level generally varies with the harshness of surface conditions, especially ground-level climate. Where conditions are extremely harsh, the center of mass is often situated at or below ground level. For example, some desert shrubs and grasses, like the ones shown in Figure 9.4, which are barely 1 meter tall and make up the bulk of the desert biosphere can send roots 5 to 25 meters into the ground. Arctic plants are also concentrated underground, with only modest subaerial development. Measurements reveal, for instance, that among grasses and shrubs in the tundra, there is two to three times more living matter below ground than above. In temperate and tropical forests, by contrast, where the climate is far milder, only 20 to 40 percent of living matter is located underground and a surprising part of that is often fungi.

Figure 9.5 Relative to ground level, biomass in North American grasslands shifts vertically with the seasons from (a) lots of mass above ground in summer to (b) most below ground in winter.

The terrestrial biosphere also varies in mass and volume over time. In fact, it is constantly changing, fluctuating in thickness and biomass over the course of the year and over longer time frames. In the seasonal climates such as the wet/dry zones of the tropics and the warm/cold zones of the midlatitudes, the biosphere expands and contracts from summer to winter. In midlatitude grasslands, for example, the total mass of living matter above ground, called *standing crop*, is five to ten times greater at the end of summer than it is in early spring. Very little living matter survives the winter cold above ground, as the photographs in Figure 9.5 reveal. Some plants and animals perish in winter, but many others shift their lives underground. The surviving mass of living matter underground is represented mainly by root systems, mostly in the upper meter of soil, where they are protected from the harshest effects of winter.

Below the root zone, the mass of living matter declines sharply under both grasslands and forest. Only microorganisms such as bacteria can be detected much below 25 meters. Some of these minute creatures can even be found more than 500 meters (1650 feet) below the landscape, but their total mass is miniscule.

Biosphere as Mediator of Landscape Systems: By virtue of its location on the Earth's surface, the terrestrial biosphere plays a pivotal role in regulating landscape-forming processes and systems. Consider that the biosphere is positioned precisely where most incoming solar radiation is intercepted and absorbed, where precipitation is received, where runoff and streamflows take place, and where winds touch the solid Earth. The biosphere lies at the interface among these systems. Here it functions as a mediator, mitigating the intensity and altering the interaction and effects of solar radiation, wind, precipitation and many other landscape-shaping agents.

In its capacity as landscape mediator, the biosphere's influences fall into two broad categories: biochemical and geophysical. The *biochemical* influences involve the recycling of energy and matter as a part of biological processes such as photosynthesis, nutrient transport, and respiration. They also involve the maintenance of the atmosphere's balance of gases, soil formation and fertility, and the chemistry of groundwater. The *geophysical* influences involve the role of the biosphere as a complex of architectural forms or physical structures, mitigating the impacts of storms, droughts, and floods, among other things.

The concept of the biosphere as a geophysical mediator is nicely illustrated by a forest such as the one in Figure 9.6a. The forest canopy intercepts rainfall slowing the rate of delivery to the ground, thereby allowing more time for rainwater to soak into the ground. With more water going into the ground, there is less runoff and less soil erosion. The forest canopy also intercepts solar radiation, reducing ground heating, evaporation, and the risk of drought. Wind is subdued by the forest canopy which in turn reduces its drying and erosional effects on the soil. In short, the biosphere creates a surface environment, which is less harsh and less variable than it would be without the complex of living structures.

The influence of the biosphere as mediator of landscape systems varies widely over the Earth. Where the biosphere is deep and dense as in the large forests of the tropics and midlatitudes, its mitigating effects are profound indeed. In places where it is thin and discontinuous, as in arid lands, polar lands, and places where natural events or humans have wiped out most living organisms, like the scene portrayed

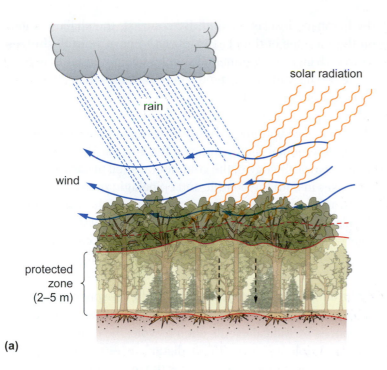

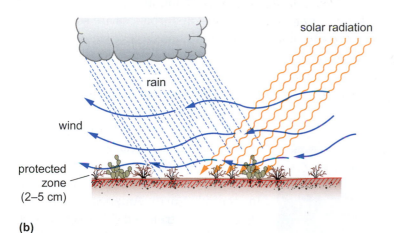

Figure 9.6 Biosphere as mediator of landscape systems: (a) forests provide an effective buffer against the forces of rain, wind, and solar radiation, whereas (b) deserts with poor organic cover do not.

in Figure 9.6b, the mitigating effects of the biosphere can be slight, though rarely insignificant. For example, a thin crust of lichens, barely 2 or 3 centimeters thick, is credited with stabilizing vast areas of barren, wind-eroded farmland on the Great Plains during and after the Dust Bowl of the 1930s.

We should appreciate that the biosphere's role in the terrestrial landscape has also changed radically over geological time. Plants and animals were relatively late in coming to Earth's landmasses. Forest or forest-like covers did not begin to appear on land until about 400 million years ago, several billion years after the landmasses formed. Before this time, Earth's land areas must have been barren, desolate places subject to the drying effects of the direct solar radiation, the unmitigated impacts of rainstorms, the full force of runoff, and the penetrating chill of winter winds. The terrestrial environment had to have been a harsh and challenging place for early life to colonize, somewhat like the one in Figure 9.7.

The Oceanic Biosphere: In the ocean, the biosphere is made up of the three major life zones illustrated in Figure 9.8; benthic, littoral, and pelagic. The **benthic zone** represents the ocean floor, whereas the **littoral zone** represents the belt along the shore where streams discharge and waves and tides are active in mixing the water. The **pelagic zone** is all the rest, the great column of water above the benthic zone and beyond the shore. In the pelagic zone most life is concentrated near the surface in a layer about 100 meters (330 feet) thick, called the **epipelagic zone** or **photic zone**, where there is plenty of solar radiation to support abundant plant life. At greater depths the water is too dark for much plant life, particularly the tiny phytoplankton that make up the bulk of the sea's vegetation, and the biosphere quickly thins out. Life is present to water depths of 1000 meters (3300 feet) and deeper, but it is lightly distributed over a vast volume of water. This deeper zone of life on the faint fringe of light rises and descends with day/night changes in light penetration. At the bottom of the deep ocean, where specialized organisms feed on the debris raining down from the photic zone above, the abundance of life increases somewhat, though it is still very sparse. The one exception is found around vents in the ocean floor where discharges of hot, mineralized water nurture relatively rich concentrations of organisms.

The vast majority of the ocean basins consist of **abyssal plains**, or deep ocean floors, where typical water depths range from 4000 to 5500 meters (13,000–18,000 feet). These vast, dark plains support a surprisingly rich assortment of life, but it has very low density, roughly comparable to biomasses found in

Figure 9.7 An erosion-ridden landscape, similar to those that must have covered huge parts of the continents before the advent of terrestrial vegetation.

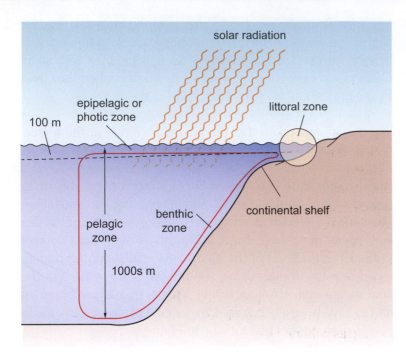

Figure 9.8 The principal zones of marine biosphere: pelagic, epipelagic (photic), benthic, and littoral.

the harshest deserts on land. By contrast, the shallow waters on the margins of the oceans, over the **continental shelves**, teem with life. The continental shelves occupy about 10 percent of the oceans with water depths generally less than 200 meters (650 feet) (Figure 9.9).

Life on the sea bottom, or **benthic zone**, is far more abundant on the continental shelves than on the abyssal plains. From the outer (seaward) edge of the continental shelf toward shore, life increases in both the benthic and epipelagic zones. Near shore the two zones converge forming one of the richest segments in the planet's life web. At the shore itself, the littoral zone is characterized by abundant nutrients delivered by current action, tidal fluxes, and inflowing streams.

Disturbance of the Biosphere: Life in the biosphere is not only subject to daily and seasonal oscillations, but also to sudden disturbances by radical events such as volcanic eruptions, forest fires, and violent storms. These disturbances tear gaping holes in the biosphere and radically alter the geographic distribution of life. Most holes heal over time but, in some areas, such as chronically active volcanic sites, disturbances are so commonplace that the biosphere is continuously in a state of decline and recovery.

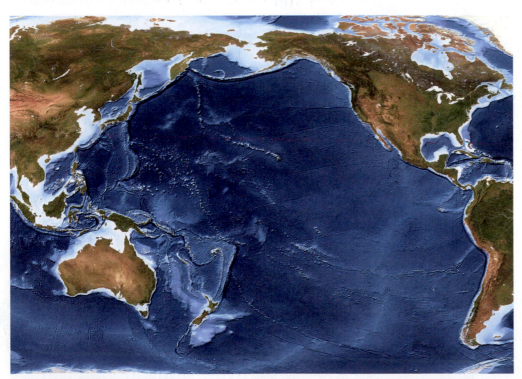

Figure 9.9 In terms of geographic coverage, the ocean biosphere is dominated by the pelagic zone, represented by the deep oceans beyond the continental shelves, identifiable by lighter water tones fringing land. Shallow coastal waters of the inner continental shelves belong to the littoral zone.

The biosphere is also torn open by longer-term and larger-scale geographic changes, such as the spread of continental ice sheets during global ice ages. Only 18,000 years ago, for instance, glacial ice covered 30 percent of Earth's land area. In central North America, a massive sheet of ice stretched from sea to sea, obliterating essentially all the biosphere under it. When the ice sheet melted away, plants and animals recovered lost ground, moving slowly northward in the wake of the retracting ice. That recovery process is still underway and we will touch on it in the next chapter.

Since glaciation, humans have been the biosphere's major change agent. With the advent of agriculture 12,000 years ago, we have spread more than 7 billion people over the Earth, and imposed our land uses over more than half the planet's land area and vast areas of its oceans, which in turn has rendered dramatic changes in the geographic patterns of plants and animals. Today, agriculture, cities, and related land uses are found on all continents except Antarctica. The heaviest concentrations are found in the mid-latitudes, in a belt roughly between 20 and 55 degrees latitude, home to the Chinese, Indians, North Americans, and Europeans (Figure 9.10).

Whenever cropland, cities, and most other land uses are established, the natural biosphere is replaced by a **cultural biosphere**, which is different in form, composition, and function than its natural counterpart. As a whole, the cultural biosphere is ecologically simpler, that is, made up of fewer plant and animal species, and is less biologically stable, that is, less sustainable than the natural biosphere. Where it is dominated by agriculture, the biosphere is heavily dependent on humans, for most crops cannot reproduce without assistance from farmers. Some geographers argue that the human influence on Earth is so pervasive that the planet's entire biosphere should be ranked as cultural biosphere.

The Ecological Context: No organism in the biosphere exists independent of other lifeforms. All life belongs to networks or systems of organisms called **ecosystems**, which extract energy and matter from the physical environment, convert it into organic forms, and redistribute it within the biosphere. Ecosystems are the basic organizational units of the biosphere on both land and water. Ecosystems define not only how different lifeforms, such as various plants and animals are interrelated, but how the biotic (biological) and the abiotic (non-biological) environments are tied together. For example, a female salmon (a biotic component of an ecosystem) requires certain abiotic characteristics of the stream channel (clear water and gravel beds) to spawn and complete its life cycle.

All ecosystems occupy space on the Earth's surface and each ecosystem is dependent on that space for its survival. This space can be thought of as the **habitat** of the ecosystem. Habitat is the particular combination of climate, water, soil, topographic, and biotic conditions within that space. Strictly speaking, the term habitat applies to the living environment of an individual species, but the concept can be broadly applied to the living environment of ecosystems as well, particularly if we think of habitat as the reservoir from which the ecosystem draws its *essential resources*: sunlight, heat, water, nutrients, and shelter (Figure 9.11).

Although we have pinned our concept of ecosystems to those that are fueled by solar energy, Earth has a second type of ecosystem which is based on geothermal energy instead of solar radiation. **Geothermal-based ecosystems** rely on bacteria to capture heat and nutrients emitted from vents in the ocean floor and, through chemosynthesis (i.e. chemical synthesis),

Figure 9.10 The biosphere has a long and eventful history of disturbance. Here a sprawling urban mass has carved a "swatch" in the biosphere.

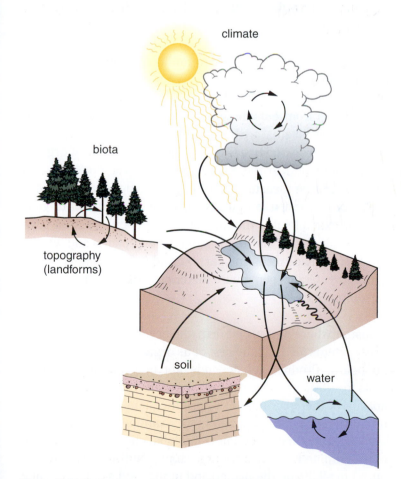

Figure 9.11 The basic components of habitat: climate, topography, shelter, soil, water, and biota. They govern the resources and forces that shape the life environment of an organism.

convert it into organic energy. These vents discharge hot, mineralized water on the dark ocean floor far below the photic zone. (some similar chemosynthesizing bacteria have also been found in soil). Although geothermal-based ecosystems are rich in life, they are limited to a tiny fraction of the ocean floor and, in a global context, appear to be insignificant compared to solar-based systems in terms of biomass, species numbers, energy output, and geographic coverage. Our interests are with solar-based ecosystems.

We can group solar-based ecosystems into three global classes based on the type of habitat they occupy: terrestrial, saltwater, and freshwater. Within these classes there are innumerable types of ecosystems overlapping and interconnecting in multitudes of ways. All are made up of **communities** of organisms, that is, populations of different plants, animals, and microorganisms that live together and interact in various ways. All solar-based ecosystems function similarly, but they vary considerably in size, shape, location, and linkage with other ecosystems.

They also vary as energy systems. Some, such as tropical rainforests, are high-powered systems that extend more or less unbroken over vast areas on the floors of great river basins like the Amazon. They are high-powered because they are able to extract huge amounts of energy from the environment, combine it with raw matter such as soil nutrients and atmospheric gases, and, through photosynthesis, produce organic fuel in the form of plant matter. Others, such as desert badlands, are low-powered systems with irregular, patchy distributions that are barely able to produce any organic matter.

Both ecosystem productivity (plant matter production) and composition (types and numbers of organisms) are dependent on many factors. Where supplies of sunlight, heat, water, and nutrients are abundant and the habitat is not subject to arresting disturbances from outside forces such as hurricanes or land clearing, ecosystems are rich in organisms and are highly productive. **Productivity** is defined as the amount of organic matter plants add to the Earth's surface over the course of a year. It can be measured over any geographic area; a farm field, a belt of forest, a country, or even a continent. In your yard, for example, it is the total output of clippings from lawn grass; petals, roots, and seeds from garden plants and weeds; leaves from shade trees and shrubs; and so on as illustrated in Figure 9.12. Grams per square meter per year are the standard units of measure.

No matter where we are on Earth, the organic matter produced in ecosystems feeds all the rest of us organisms, above ground and below. Productivity depends not only on the ability of plants to extract resources from the environment, but upon the capacity of the environment to produce sunlight, heat, water, and nutrients. In earlier chapters we examined solar radiation, heat, and water in connection with global climates but we have yet to examine Earth's nutrient systems. Nutrient systems are central to all life on the planet, and in the next section we take a look at several of these systems.

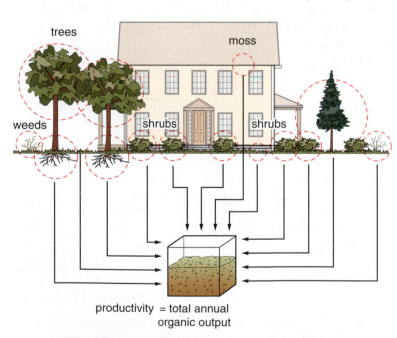

productivity = total annual organic output

Figure 9.12 The concept of productivity applied to a residential lot. The vegetation in the 400 square meter yard area may produce 450 to 600 kilograms of organic matter over a year.

Summary on the Biosphere: The biosphere is a complex web of millions of species and countless billions of organisms spread over the Earth in a thin membrane-like sheet. The biosphere is central to physical geography not only for its role in the greater scheme of Earth systems, but for its role in making Earth's surface a more resilient and livable place. Without life, Earth's landscapes would be far different and harsher places. All life belongs to networks or systems known as ecosystems, which take in energy and matter from the abiotic environment, convert it into organic forms, and redistribute it within the biosphere.

9.2 The Global Nutrient System: Earth's Biogeochemical Cycles

All organic material is made up of molecules of matter. Molecules are comprised of atoms of different elements that are drawn from the environment by organisms. Three elements account for almost all (99.47 percent) living matter: hydrogen, carbon, and oxygen. Most of the remaining half-percent or so is represented by twelve elements (Table 9.1).

The three major elements – plus nitrogen, calcium, potassium, sulfur, and phosphorus – are called **macronutrients** because all of them are needed in relatively large quantities to sustain life. The remaining elements, as well as several others including boron and copper, are called **micronutrients**. They are generally required in much smaller and more variable quantities among different organisms. Both sets of nutrients are drawn from the Earth's spheres (litho-, hydro-, and atmo-), cycled through ecosystems, and ultimately returned to the environment.

Within all organisms the atoms of the nutrient elements are used to build molecules. The molecules can take on many different forms as a result of different chemical and biochemical processes. The atoms themselves, however, are not altered or destroyed by these processes, but remain constant within the different molecular arrangements. Thus, within an ecosystem matter is conserved as atoms are used and reused, or *cycled*, within organisms, among organisms, and between organisms and the environment. As with energy, matter can be neither created nor destroyed, only converted into different forms. Unlike energy, however, which flows from space through the Earth environment and then back into space, matter (beyond a minute quantity of gas molecules that escape from the atmosphere) is not lost to space but is recycled within and between the four major spheres of the Earth and their environmental subsystems.

The cycling of atoms and molecules through ecosystems is termed a **nutrient cycle** or **biogeochemical cycle**. Two types of nutrient cycles are recognized: sedimentary and gaseous. In the **sedimentary cycle** nutrients held in rock are set free by weathering. Weathering involves chemical decomposition of rock, which releases nutrients into the hydrosphere. These nutrients, dissolved in water, are then extracted by plants and animals and transferred through ecosystems, eventually ending up as a sedimentary residue in the soil, in lakes and swamps, or in the ocean.

Huge amounts of organic sediments containing calcium and phosphorous, for example, are deposited in the sea or on the margins of the landmasses, where they are buried, transformed into rock, and ultimately reincorporated into the lithosphere. When the rock is uplifted by geologic forces and once again subjected to weathering, the nutrients are released, and the cycle is completed. The time involved in this cycle

Table 9.1 The 15 most abundant elements of living matter

Element (Nutrient)	Percentage
Hydrogen (M)	49.740
Carbon (M)	24.900
Oxygen (M)	24.830
Nitrogen (M)	0.272
Calcium (M)	0.072
Potassium (M)	0.044
Silicon (m)	0.033
Magnesium (M)	0.031
Sulfur (M)	0.017
Aluminum (m)	0.016
Phosphorus (M)	0.013
Chlorine (m)	0.011
Sodium (m)	0.006
Iron (m)	0.005
Manganese (m)	0.003

(M) = Macronutrient
(m) = Micronutrient

is very long, normally tens or hundreds of millions of years, and the rock segment is decidedly the slowest and largest part of the cycle. Therefore, the lithosphere functions as a great storage bin, or **nutrient sink**, in the system.

In the **gaseous cycle** nutrients are exchanged between the biosphere and the atmosphere without going into the lithosphere. This cycle is much faster, usually tens to thousands of years, and is the chief means of circulation for carbon, hydrogen, oxygen, and nitrogen (see Table 4.1). It involves various processes, but the principal ones are respiration by plants and animals and absorption and release by ocean water. **Respiration** involves the exchange of gases and energy between an organism and the atmosphere. The main gas exchange for plants is the intake of carbon dioxide (with photosynthesis) and the release of oxygen. For animals it is the opposite; oxygen is taken in and carbon dioxide is released. In the oceans, molecules of gases such as carbon dioxide and oxygen are taken in directly from the atmosphere and given up directly to the atmosphere. Within the oceans, these gases are exchanged between the water and aquatic plants and animals.

The Carbon Cycle: Carbon, the principal nutrient in the ecosystems, is moved in both the gaseous and sedimentary cycles as the diagram in Figure 9.13 shows. In the gaseous cycle, carbon occurs in the form of carbon dioxide (CO_2), a free gas in the atmosphere and a dissolved gas in freshwater and saltwater. Although CO_2 represents only 0.033 percent of the atmosphere, the atmosphere is the principal reservoir, or pool, of Earth's carbon dioxide. Between the atmosphere and the Earth's surface a great exchange of CO_2 is continuously in motion, involving billions of tons annually. Every 10 years or so the complete pool of atmospheric CO_2 is exchanged (recycled) with the Earth's surface.

The atmosphere gives up its CO_2 to the Earth's surface by two main processes: (1) intake by land and ocean plants as a part of photosynthesis and respiration; and (2) absorption by seawater. These two processes together take in an estimated 203 billion tons of CO_2 from the atmosphere annually. This mass of CO_2 is stored in ocean water and in the living tissue of terrestrial and aquatic plants.

The atmosphere takes in CO_2 from four sources: living organisms, ocean water, volcanoes, and air pollution. The chief input comes from the respiration of plants and animals, both terrestrial and aquatic. The second input comes as an outflow (**degassing**) from ocean water. Third, volcanic activity releases CO_2 into the atmosphere via a process known as **outgassing**. These three sources release an estimated 200 billion tons of CO_2 into the atmosphere yearly.

Air pollution from land-use activity is the fourth source of atmospheric CO_2. Most CO_2 is produced from burning fossil fuels, but some also comes from burning forests, grasslands, and organic-rich topsoil. The total annual input from air pollution is about 7 billion tons, bringing the grand total CO_2 input to the atmosphere to 207 billion tons a year.

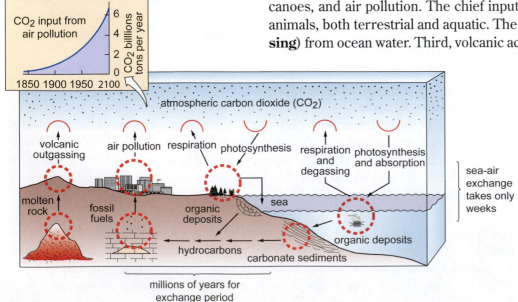

Figure 9.13 The basic components of the global carbon cycle. Input to the atmosphere comes from respiration, seawater, volcanic outgasing, and air pollution. Because of air pollution, inputs exceed outputs by about 3 billion tons a year.

CO_2 input from air pollution

CO_2 billions tons per year

1850 1900 1950 2100

atmospheric carbon dioxide (CO_2)

volcanic outgassing   air pollution   respiration   photosynthesis   respiration and degassing   photosynthesis and absorption

sea-air exchange takes only weeks

molten rock   fossil fuels   organic deposits   sea

hydrocarbons   organic deposits

carbonate sediments

millions of years for exchange period

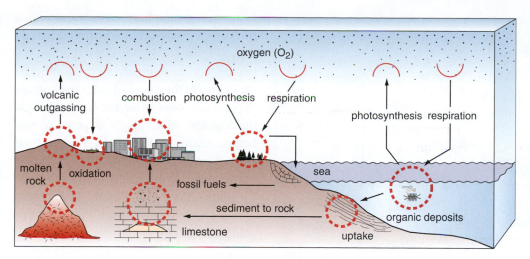

Figure 9.14 The global oxygen cycle generally parallels the CO_2 cycle. Atmospheric oxygen, however, is 700 times more abundant than CO_2, so human activity has little effect on the global O_2 balance.

Thus, CO_2 inputs to the atmosphere (207 billion tons) currently exceed outputs (203 billion tons) by 4 billion tons a year. As we discussed in Chapter 8, CO_2 loading of the atmosphere has serious implications for the global environment, especially for climate. The anticipated impacts are well known and most, such as sea-level rise, drought, and increased storminess are decidedly negative. But there is at least one impact that may be considered positive. As a nutrient, atmospheric carbon dioxide is essential to plant growth. Thus, an increase in available CO_2 could accelerate plant growth rates and the production of organic matter worldwide. Increased plant productivity could also lead to greater global biomass and in turn greater uptake and storage of atmospheric CO_2, thereby helping counterbalance rising CO_2 inputs from air pollution. This is another example of negative feedback in an Earth system.

The Oxygen Cycle: As with carbon, the gaseous cycle for oxygen (O_2) involves massive exchanges between the Earth's surface and the atmosphere. The atmosphere is made up of nearly 21 percent oxygen, and this pool is fed by two sources: a recycle system and a primary supply. Photosynthesis by both terrestrial and marine plants is the recycle source. Volcanic outgassing, which releases oxygen as a part of CO_2 and water (H_2O) molecules, is the primary source of new oxygen (Figure 9.14). However, photosynthesis provides the major input of O_2 to the atmosphere, and it is noteworthy that over the past 3 billion years, and especially over the past half billion years, the oxygen content of the atmosphere has grown significantly with the evolution of plant life on Earth.

Under current biological conditions, it takes about 6000 years or more for atmospheric oxygen to be completely recycled through the Earth's plant cover. Oxygen is removed from the atmosphere by the respiration of animals and microorganisms and by oxidation of minerals on or near the Earth's surface. In addition, oxygen is removed by fires with the burning of fossil fuels and wood, including forest fires. In the oceans some oxygen is also removed in calcium carbonate sediments (such as shells), which are the principal ingredient in limestone.

Although the oxygen content of the atmosphere has increased over geologic time, the last century has seen a slight decline in atmosphere oxygen. The source of the decline is all the burning activity by humans (industrial, domestic, forest fires, and others), the same activity that is causing the increase in atmospheric CO_2. Oxygen, however, is 700 times more abundant than CO_2 in the atmosphere; therefore, the relative change in the two with the exchange of gases in the combustion process is proportionally much greater for CO_2. The decline in oxygen is not a matter of great concern at this date.

The Nitrogen Cycle: The atmosphere is made up of 78 percent nitrogen and is the Earth's principal pool of this gas. The exchange of nitrogen with the Earth's surface is much slower than that of oxygen. Coupled with the fact that it is also more than three times more abundant in the atmosphere than oxygen, it takes a very long time, millions of years, to recycle all the nitrogen in the atmosphere. Nitrogen is extracted

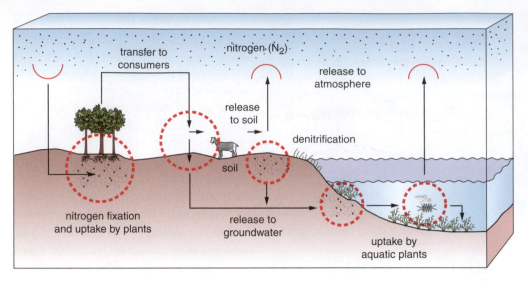

Figure 9.15 The global nitrogen cycle involves both gaseous and sedimentary systems. Nitrogen-fixing crops and manufactured fertilizer are increasing the load of nitrogen in soil and freshwater.

from the atmosphere chiefly by plants through a process called **nitrogen fixation** (Figure 9.15). Nitrogen fixation is a two-part process involving certain bacteria living on the roots of selected species of plants, in particular, most members of the legume family such as peas, clover, beans, alfalfa, and peanuts. The bacteria extract gaseous nitrogen from the atmosphere and convert it into forms usable by the plants, most importantly ammonium, which goes into the building of amino acids and protein.

Through the fixation process, nitrogen enters food chains, such as those of grazing animals and their predators, as well as microflora, insects, and other organisms. On land, the vast majority of the organic nitrogen synthesized by plants and animals ends up in the soil as dead plant and animal tissue. As this organic matter decays, the nitrogen is released, or "unfixed." This process involves bacteria, called **denitrifying bacteria**, which convert the nitrogen back into the gaseous form and release it into the atmosphere.

Large amounts of nitrogen are being added to land and freshwater as a result of human activity; first, by planting nitrogen-fixing crops such as beans, and second, by industrial manufacturing. Using technology, industry is able to convert gaseous nitrogen into solid and liquid forms which are used in a variety of products, but mostly in commercial fertilizer. Total yearly nitrogen production by industrial fixation amounts to about 50 percent of the annual production by natural fixation worldwide. Add to this the amount contributed by nitrogen-fixing crops, such as soybeans, and the number is even higher.

Figure 9.16 The human made phosphorus system. Phosphorus is a major source of nutrient pollution in lakes, streams, and coastal waters where it triggers aquatic plant growth leading to deposits of decaying organic matter and ecological changes.

As a whole, plants are not very efficient in their uptake of nitrogen applied in commercial fertilizer. As a result, most nitrogen is washed through the soil and into groundwater, lakes, and streams. Measurements show that in areas of intensive commercial farming, nitrogen is building up in groundwater aquifers. In the Corn Belt of the United States Midwest, for example, nitrogen has reached such high levels in many aquifers used for drinking water that it poses a serious human health risk.

The Phosphorus Cycle: Phosphorus is also essential to life, but in much smaller amounts than nitrogen. In addition, phosphorus has no gaseous loop in its global cycle and is therefore limited to a sedimentary cycle. This cycle is exceedingly slow and therefore the natural supply of phosphorus to most ecosystems is very limited. As a result, many ecosystems are phosphorus-limited under natural conditions, but this is changing over much of the Earth.

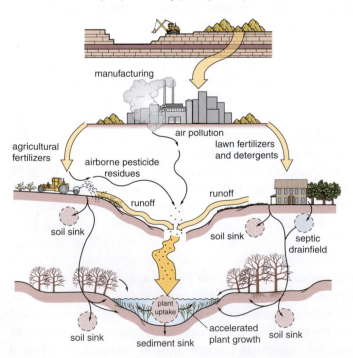

Humans are adding large amounts of phosphorus to ecosystems by processes such as those shown in Figure 9.16. The input comes in three major ways: (1) through atmospheric fallout from air pollution (because phosphorus is contained in fossil fuels such as coal and is released with combustion); (2) through the application of commercial fertilizers to cropland and lawns; and (3) through the release of detergents in wastewater. The phosphorus used in fertilizer and detergent is extracted from phosphate deposits in sedimentary rock. When fugitive phosphorus from these sources is added to lakes and streams, it usually triggers accelerated plant production causing overgrowth and species changes in aquatic ecosystems. Growth comes quickly because nitrogen, the other essential macronutrient, is already there in abundance. As a result, relatively clear waters with limited biomass often become choked with living and dead organic matter. This process, called **cultural eutrophication**, is one of the most apparent changes in the nutrient system related to land use.

Summary on the Global Nutrient Systems: Three macronutrients account for nearly all living matter. These nutrients, along with many micronutrients, are drawn from Earth's water, land, and air systems, cycled through ecosystems, and eventually returned to the environment. Nutrient systems may follow two cycles, sedimentary and gaseous. Carbon utilizes both cycles with reservoirs in the atmosphere and lithosphere. Carbon input to the atmosphere is accelerated by the burning of fossil fuels.

9.3 The Structure and Processes of Ecosystems

Ecosystems can be defined in several ways. First and foremost, they are *energy and matter systems* in which solar energy and nutrients are taken in, converted to chemical energy, and then released as heat into the abiotic environment. Second, they are *biological systems* containing communities of organisms, representing different plant, animal, and microorganism species, living together in an interdependent fashion. Third, they are *food systems* made up of food chains by which organic energy and matter are transferred through the ecosystem. For an ecosystem to thrive, all three systems must work together. Consider that without the biological processes of the community such as reproduction, food chains cannot exist, and without food chains, energy and matter cannot be transferred through ecosystems. And fourth, ecosystems are *geographical systems* because they occupy space on the Earth's surface. That space has form, mass, dynamics, and a location and area of coverage on the Earth.

Food Chains: All ecosystems are made up of food chains. Food chains begin when energy and matter are extracted from the environment by plants and, through photosynthesis, are converted into organic (vegetal) matter. This organic matter is then consumed by other organisms. Plant-eating animals such as deer and buffalo consume plant matter and use it to develop the tissue of their bodies. When these animals are eaten by predators, such as wolves, the tissue provides the predator with nutrition and energy while the waste, such as bones and skin, is consumed by scavengers and decomposers such as bacteria.

We can trace this process by following the transfer of energy from one organism to another. In its simplest form, the route this energy takes defines a **food chain** and several examples are shown in Figure 9.17 along with an integrated set of food chains called a food web. In this system, 1 gram of plant organic matter (dry weight) is equal to 4 kilocalories of energy whereas 1 gram of animal organic (dry weight) matter is equal to 5 kilocalories of energy.

Enriched by phosphorus and nitrogen, this pond is being overgrown with algae and other aquatic plants.

Figure 9.17 Three simple food chains beginning with plants (producers) and followed by three levels of consumers. The food web in (b) is formed by the integration of the three food chains in (a) into a larger and more complex system.

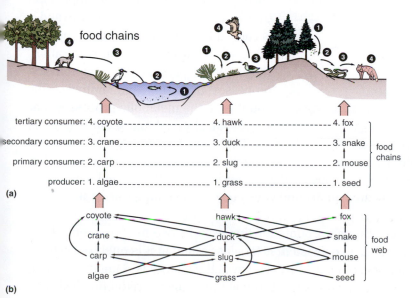

Each food chain is organized into different segments or levels. Each level is defined by the point of energy transfer first from the environment to plants, and then from plants to animals and animals to animals. All food chains are organized in basically the same fashion. Most have four or five levels of energy transfer called **trophic levels** and these levels define the basic structure of all ecosystems. Each trophic level represents a particular function in the ecosystem defined by either production or consumption of organic energy.

For the vast majority of life on Earth, plants are the one and only producers of organic energy in ecosystems. They represent the productivity or **primary production** trophic level of all ecosystems. Primary production is followed by a series of consuming levels. First are herbivores, or **primary consumers**, which make up the second level. Primary consumers include cattle, buffalo, squirrels, and other plant-eating animals. Carnivores, or **secondary consumers**, such as wolves, lions, and weasels, are the third trophic level. These are meat eaters. Specialized carnivores, or **tertiary consumers**, are the fourth trophic level. Hawks, eagles, and sharks are among these, the animals that eat other carnivores. **Omnivores** such as humans, raccoons, and black bears function as both carnivores and herbivores.

In addition to the primary trophic levels, all ecosystems support **detritivores** and they, in turn, are also organized into trophic levels. Detritivores are organisms that live on the refuse of producers and consumers. They include **scavengers** such as vultures and crabs that eat the remains of dead animals and **decomposers** such as fungi and bacteria that break down plant debris, animal droppings, and other dead organic matter.

Figure 9.18 Energy flow in an ecosystem represented by a graphic model called an energy pyramid (center). Most energy is expended in respiration.

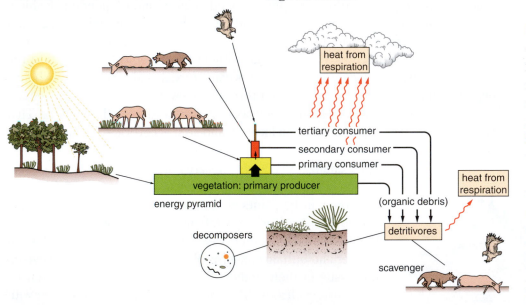

Energy Flow in Ecosystems: Organic energy declines rapidly as it is transferred from one trophic level to the next in an ecosystem. This concept is illustrated in a model, referred to as an **energy pyramid** (shown in Figure 9.18), in which the size of the pyramid's tiers, which represent the trophic levels, shrinks rapidly toward the top. This trend is called **energy attenuation** and it is caused by the huge amount of energy released by organisms in respiration. Respiration represents the maintenance processes of plants and animals. For you it is all the metabolic processes that keep your physical and mental systems running. These processes are fueled by your food intake. This organic energy is converted mainly into heat, which is released with your breath and from your skin into the atmosphere around your body. We can verify this with thermal sensors by mapping an envelope of heated air around your body representing this heat loss.

In most ecosystems, respiration accounts for the loss of close to 90 percent of the energy at each trophic level. So consistent is this loss rate that it is often referred to as the "10 percent rule," meaning only 10 percent of the energy is retained between trophic levels. Thus 1000 calories manufactured at the primary production level can be reduced to only 1 calorie by the time it gets to the tertiary level in a four-tier ecosystem: 1000 > 100 > 10 > 1.

A simple example of an energy pyramid is provided by a common agricultural system that grows alfalfa in support of cows and people. It takes about 65 billion calories of solar radiation to produce 8000 kilograms (18,000 pounds) of alfalfa. This is equivalent to 15 million calories of organic energy. This mass, in turn, is consumed by a cow and converted into 1.2 million calories of body mass, heat, and waste. In this exchange, 92 percent of the energy is given up in heat and bodily waste and 8 percent is synthesized by the cow (that is turned into flesh, bones, etc.). The cow is then eaten by people who turn the beef into 12 kilograms (26 pounds) of body heat, waste, and body mass. This is equivalent to 14,400 calories of energy or a little more than 1 percent of the total cow. The reason this number is much less than the 10 percent attenuation rate generally cited for trophic conversion is that humans only eat a fraction of the cow. Most of the cow is not used at the secondary consumer level by humans but goes to other consumers (dogs, cats, rats, and pigs, for example) and to detritivores as well as to other uses such as shoe leather.

We humans are not exempt from ecosystems but are intricately tied to the governing principles of food chains, dependence on other organisms, and energy attenuation. We are omnivorous consumers and have designed our agricultural systems to produce both plants and meat. As in any ecosystem, however, less total food is available at the carnivore level than at the herbivore level. Therefore, where in the energy pyramid we concentrate our eating has serious implications for the number of people that can be supported by agriculture. Earth's current agricultural system can support a world population of 5.5 to 6.0 billion people at a healthy level if we relied solely on plants. But because we feed huge amounts of our plant energy to farm animals, we fall short of the food supply needed to sustain Earth's current population. Today, about 850 million people live on inadequate food supplies (Figure 9.19).

Figure 9.19 World map of hunger. Nearly a billion people are undernourished, defined as insufficient food and nourishment to sustain proper health and growth.

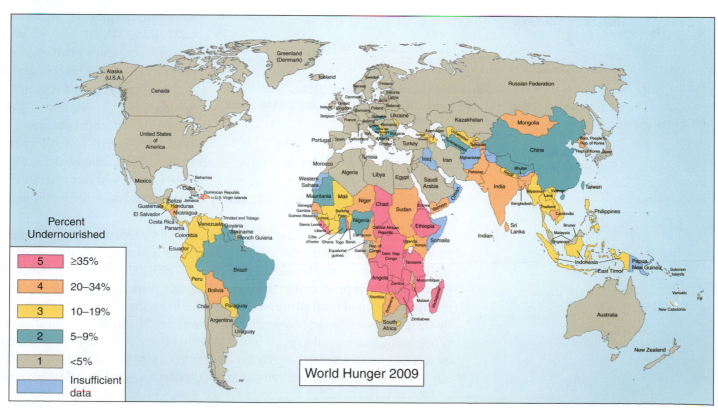

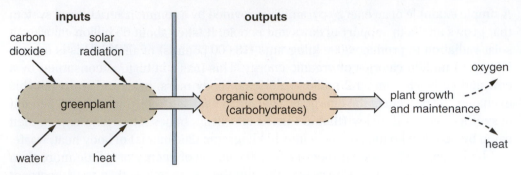

Figure 9.20 A simplified version of the intriguing system known as photosynthesis in which green plants draw in energy and matter from the environment and convert it into organic matter in the form of carbohydrates.

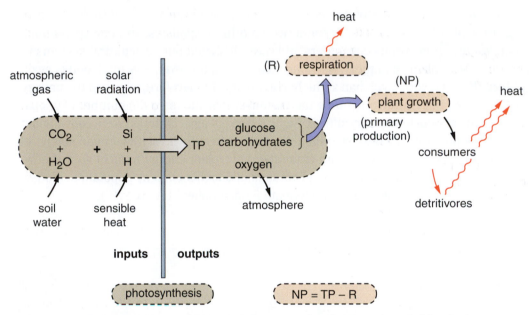

photosynthesis

$$NP = TP - R$$

Figure 9.21 The full productivity system in which the output of total photosynthesis (TP) first feeds respiration and the remaining energy (NP) goes to plant growth. The quantity NP represents the energy that fuels the rest of the biosphere.

Feeding Ecosystems: Living plants are the only organisms on Earth capable of capturing solar energy and converting it into forms usable to themselves and other organisms. As such, plants are the gateway to the Sun's energy for all Earth's creatures (except for those tied to chemosynthesis around hot vents in the ocean floor). At the heart of this great system is a pivotal subsystem called photosynthesis. Like most energy systems, photosynthesis involves energy inputs, outputs, and storage. At the input end, it enables plants to take in solar radiation, combine it with inputs of carbon dioxide, water, nutrients, and heat, and convert it all into organic energy (chemical compounds) in the form of simple carbohydrates, some of which is stored (plant tissue) and some of which is given up as oxygen and heat output (Figure 9.20).

All these ingredients (energy and matter) are extracted from the abiotic environment. For most terrestrial plants, light, carbon dioxide, and heat come from the atmosphere, whereas water and nutrients come from the soil. The organic output from photosynthesis is used to produce cells from which leaves, wood, bark, flowers, and other plant organs are built. A critical attribute of this organic tissue is that once produced it is not subject to spontaneous breakdown (decay) and thus can be stored as biomass. Earth has retained (stored) a small fraction of this biomass in various reservoirs. Some, such as fossil fuels, date back hundreds of millions of years. Shorter-term reservoirs include topsoil and our own food reserves, which, incidentally, are estimated globally at only several months' supply in most years.

Here, in a broad-brush description, is how the system works. We begin with the products of photosynthesis, oxygen, and organic energy (Figure 9.21). Oxygen, which comes from the breakdown of the CO_2 molecule, is released into the atmosphere. Energy is produced in the form of organic compounds and the quantity manufactured is termed **total photosynthesis** (TP in Figure 9.21). This energy is retained by the plant and used in two ways, first to maintain the plant's respiration (metabolic processes), and second to grow new tissue. In Figure 9.21, respiration is represented by R. The energy used in respiration is converted into heat and slowly released to the atmosphere, though the quantity of heat output is too small to affect air temperature.

The amount of energy left after respiration is called **net photosynthesis** and it is labeled NP in Figure 9.21. This is the energy available for building new cells, that

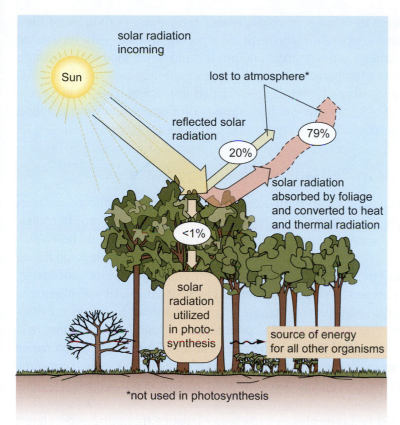

solar radiation
incoming

Sun

lost to atmosphere*

reflected solar
radiation

20%

79%

solar radiation
absorbed by foliage
and converted to heat
and thermal radiation

<1%

solar
radiation
utilized
in photo-
synthesis

source of energy
for all other organisms

*not used in photosynthesis

Figure 9.22 The breakdown of incoming solar radiation within the biosphere using a broadleaf forest as a model. Very little energy goes to photosynthesis. Most goes directly to heat which is released into the atmosphere.

is, for plant growth. These new cells represent the energy that fuels ecosystems above the level of the plants. In other words, it feeds all the various consumers and detritivores and is a measure of the energy balance of an ecosystem. Thus *net photosynthesis* is equal to *total photosynthesis* minus *respiration*, a quantity also called **net primary production**, or just primary production.

Primary production is measured according to the weight (in grams) of organic matter added to a square meter of Earth surface over a year, be it soil, lake bed, or ocean floor. Although most ecosystems are positively balanced as energy systems, that is, they produce more organic energy than they use in respiration, it is possible for ecosystems under stress to decline to zero or even a negative energy balance. Under a negative energy balance, plants would undergo dieback, lose organic mass and ultimately starve to death. This can happen during severe droughts, for example.

Optimizing Conditions for Photosynthesis: The percentage of solar energy that plants are able to convert into organic energy through photosynthesis is termed *efficiency*. Although plants are exposed to a great deal of solar radiation, only about 1 percent of that striking them on average worldwide is actually used in photosynthesis. Where water and nutrients are not seriously limited, efficiency is higher, between 1 percent and 2 percent, and in some agricultural plants it is 6 percent or higher. Thus, for purposes of photosynthesis, Earth receives vastly more solar radiation than plants can use.

What happens to all the rest of the solar radiation that hits plants? Roughly 20 percent is reflected from the foliage. The remainder, about 78–79 percent, is absorbed mainly by the foliage itself and converted into heat. From the foliage, heat is returned directly to the atmosphere as sensible heat, latent heat, and infrared radiation, bypassing the photosynthesis energy path altogether as shown in Figure 9.22.

The conclusion we reach is that on the average there is plenty of solar radiation reaching plants. The problem, of course, is its distribution; it is not consistently available when and where it is needed. Experiments show that plants reach maximum photosynthesis considerably below full sunlight. As photosynthesis rises with increasing light, it peaks at some radiation intensity, call the saturation level. Beyond this level photosynthesis declines, ultimately reaching zero at some high radiation intensity. Aquatic plants commonly reach maximum photosynthesis at 1 to 5 percent of full sunlight but, in a great many terrestrial plants, maximum photosynthesis occurs between one-third and two-thirds full sunlight as you can see in the graph in Figure 9.23. This does not mean that one-third to two-thirds of available sunlight is being used in photosynthesis; efficiency for most plants remains less than 2 percent.

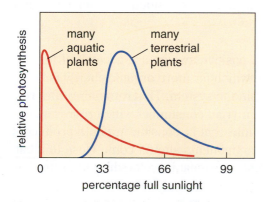

many
aquatic
plants

many
terrestrial
plants

relative photosynthesis

0   33   66   99
percentage full sunlight

Figure 9.23 Relative rates of photosynthesis in aquatic and terrestrial plants as a percentage of full sunlight. Too little and too much insolation inhibit the process.

The same pattern generally holds true for heat, water, and nutrients. That is, there is a middle range where conditions are optimum for photosynthesis. For example, the optimum temperature range for many temperate-zone trees is between 10°C and 25°C. Where these trees grow in cold climates incapable of producing a high summer monthly mean temperature above 10°C or so, the trees tend to be stunted because net photosynthesis is never great enough to yield much food for growth. In the tropics, the optimum temperature range for trees generally falls between 25°C and 45°C. If photosynthesis and growth are retarded in tropical trees, it is usually not for lack of heat but rather because of inadequate water or nutrients, or even light where the sunshine is blocked out by heavy shade from larger trees.

9.4 Ecosystems as Self-adjusting Energy Systems

Ecosystems live in a changing world. Solar radiation, heat, water resources, nutrient supplies and many other conditions upon which they depend and which can affect their performance are constantly changing around them. Therefore, it should be no surprise to learn that ecosystems are themselves constantly changing, rising and falling with oscillations in the environment, especially with the oscillations in input of energy and matter. When input rises, productivity climbs (given, of course, that all other conditions are favorable), biomass increases, and plant and animal life-cycle activities such as reproduction rise. When input falls, productivity, biomass development, and biological activity also fall. If input falls below the basic level needed for maintenance of respiration, the system goes into negative energy balance. That is, it must consume more energy than it is manufacturing, which means it must draw on biomass reserves.

In reality there are just two critical states of energy balance in natural ecosystems: positive and negative. (The intermediate state, equilibrium, represents such a delicate state of balance that it is rarely achieved in nature.) During positive balances, the system is taking in and synthesizing more energy than it is giving up in heat and other losses. During negative balances, the system is giving up more energy than it is taking in and synthesizing. The swings between positive and negative states occur at different periods, ranging from a few days related to weather change, to several months as the seasons change, to many decades or much longer as climate fluctuates with cooling and warming and wetting and drying cycles. The magnitude of the swings varies from small "blips" that are hardly measurable to huge oscillations that are plainly apparent in the landscape based on observable changes in the size (coverage), density (biomass), vigor (health), and composition (species types and numbers) of an ecosystem.

In the long run, when all the negative and positive swings are accounted for, the system should register a positive balance in which far more organic energy is being synthesized than is needed simply to maintain the system. That remaining after respiration is the surplus energy that feeds the myriad consumers which represent the other "half" of the ecosystem. Taken as a whole, that is, considering both producers and consumers, the positive and negative components should come close to a balanced state, or what ecologists call **homeostasis**, meaning "standing the same" or "standing steady."

Feedback: If the geographic environment housing and feeding ecosystems fluctuates in providing resources and producing stresses, these changes can dramatically affect the stability and health of ecosystems. But ecosystems have a way of mitigating such changes through **feedback**. Feedback is classed as either positive or negative depending on whether it has an amplifying or damping effect on the original change. In Figure 9.24, positive feedback is illustrated using the growth system of a red alder

The shrubby desert fringe, an ecosystem that experiences extreme fluctuations in its energy balance seasonally, annually and over longer periods, mainly in response to dramatic variations in water supply.

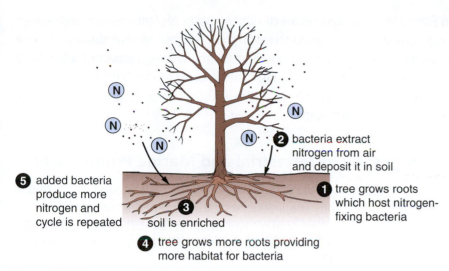

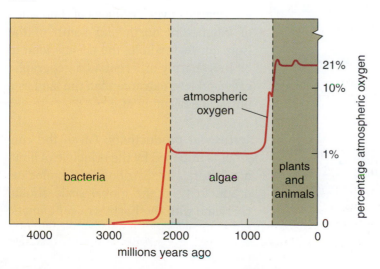

5 added bacteria produce more nitrogen and cycle is repeated

2 bacteria extract nitrogen from air and deposit it in soil

3 soil is enriched

1 tree grows roots which host nitrogen-fixing bacteria

4 tree grows more roots providing more habitat for bacteria

Figure 9.24 The concept of feedback. Positive feedback is illustrated here by the growth of a red alder tree and its root system and the role played by nitrogen-fixing bacteria, which live on the alder roots and enrich the soil, thereby inducing the tree to grow more roots, which in turn promotes more nitrogen input and soil fertility.

The change in atmospheric oxygen since the origin of the Earth and its relationship to the evolution of life forms.

tree. As the young tree develops its root system, the new roots provide a habitat for nitrogen-fixing bacteria. The bacteria draw atmospheric nitrogen into the soil, synthesize it in the form of nodules on the tree roots, thereby enriching the soil and inducing the tree to grow more roots. The added roots draw in more nitrogen, further enriching the soil, promoting even more tree and root growth. This is a positive feedback because it amplifies the primary change, root growth, causing even more tree and root growth via the nitrogen-fixing bacteria. On the other hand, if the increase in soil nitrogen were to lead to such high levels of soil fertility that the alder's tolerance to the nitrogen-saturated soil and related chemical conditions were exceeded and tree growth was impaired, the feedback would be negative. The measure of negative feedback in this case would be a decline in the growth of roots as nitrogen is being added to the soil.

Feedback systems are very complex in nature. For every environmental change, there are a multitude of feedback processes in ecosystems, some with magnifying and some with reducing (damping) effects. Moreover, feedback systems operate at a wide range of geographic scales, from the biocellular to the global. If we consider the Earth itself as one great ecosystem, the question arises as to the capacity of the biosphere to regulate the global environment through feedback processes. Is it possible for the life system to regulate the global environment to its own benefit? In other words, can the life system render broad adjustments in the global environment via, for example, the carbon dioxide cycle or the hydrologic cycle, to serve its own maintenance and survival?

In the past few decades this idea has enjoyed lively debate among scientists. It is called the **Gaia hypothesis** (after the Greek goddess of Earth), and supporters argue that life on Earth does shape the planetary environment in ways that support life. And there is compelling evidence behind the argument. For example, the development of atmospheric oxygen and carbon dioxide levels necessary to nurture life research shows was clearly tied to the evolution and spread of terrestrial vegetation over the past 500 million years.

But there is also contrary evidence including a life-form which seems to be working against the planet's life system, namely, humans. Increasingly, humans modify the global environment, reducing its productivity and ecological balance in many ways. One example is the process called desertification in which grasslands, degraded by land use and drought acting together, are converted into desert. The feedback is positive, that is, the decline in grassland ecosystems brings on increased pressure from land use, as herders, for example, attempt to eke more out of the landscape, which further degrades the ecosystems, increasing their susceptibility to drought. Soil is lost, soil moisture declines, and ecosystem vigor and resilience decline. The effect is downward-spiraling ecosystem productivity. In other words, as a biological agent, humans may be working against rather than for a life-sustaining planetary system.

Summary on Ecosystems: Ecosystems are complex networks of plants, animals, and microorganisms that depend on a resource base centered on photosynthesis for their survival and operation. They are organized in multi-tiered structures in which the energy produced in photosynthesis is transferred from tier to tier but with huge energy losses at each level. Through various feedback mechanisms ecosystems are able not only to adjust to changes in the environment, but render changes to the environment to their advantage.

9.5 The Geography of Terrestrial and Marine Productivity

For the Earth as a whole, the annual average productivity is estimated at 225 billion tons. Of this about 135 billion tons are produced by the continents and 90 billion tons by the oceans. Within both realms productivity is unevenly distributed geographically, but this is not surprising given the limits imposed by the environment over vast areas of the planet.

Limiting Factors: Critical to understanding geographic variations in productivity is the biological **principle of limiting factors**. This principle holds that the rate of productivity at any location is limited by the one resource which is in least supply. This explains the low productivity of deserts, for example, where all resources except water are usually plentiful. In deserts, plants grow and ecosystems come alive, so to speak, only when water is occasionally available. In polar regions, on the other hand, where water is usually plentiful, heat is the limiting factor. In many saltwater and freshwater ecosystems, the limiting factor is often the supply of nutrients such as phosphorus and nitrogen. On balance, we can learn a great deal about the distribution, abundance, and productivity of life by simply examining the global distribution of limiting factors such as those shown in Figure 9.25.

In earlier chapters we described the global distributions of solar radiation, heat, and water in connection with atmospheric systems and world climate. We learned that Earth's landmasses embrace remarkable geographic extremes in surface conditions, from hot, bright, dry conditions to cold, dark, icy conditions, for example. The oceans, on the other hand, are far less extreme geographically. Variations in light are about the same as over land, but average annual water temperatures, on the other hand, typically vary only a few degrees over distances of thousands of kilometers north and south of the Equator. As a whole, the oceans tend to be less sharply differentiated geographically and less extreme than the land as a physical environment.

Figure 9.25 World map of two key limiting factors, water and heat. For mountainous regions they vary at the local scale with elevation changes.

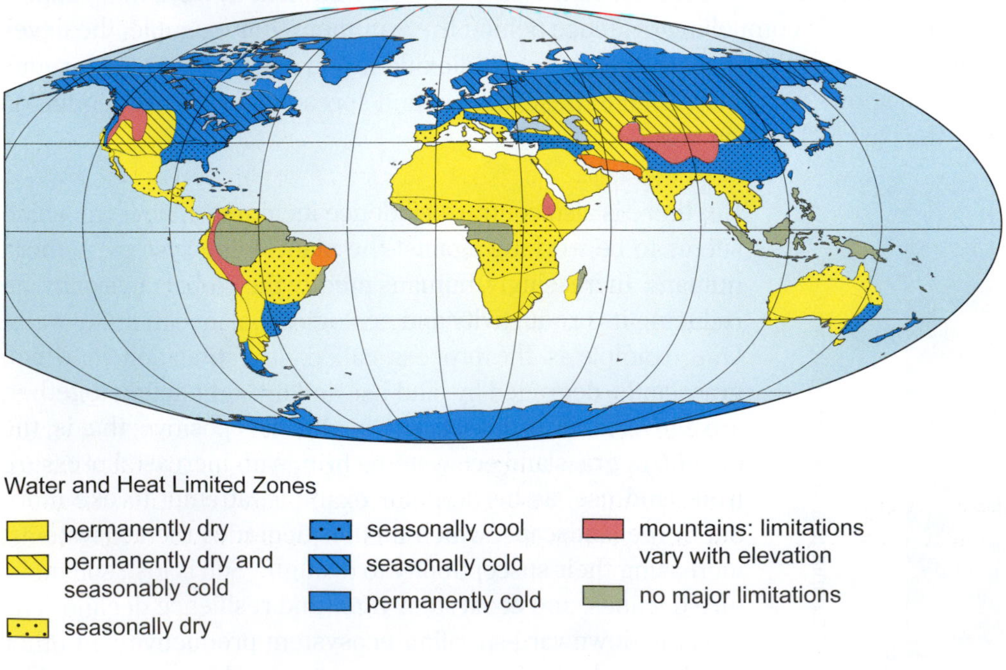

Water and Heat Limited Zones

- permanently dry
- permanently dry and seasonably cold
- seasonally dry
- seasonally cool
- seasonally cold
- permanently cold
- mountains: limitations vary with elevation
- no major limitations

Terrestrial productivity: Given the constraints imposed by the principle of limiting factors, the distribution of biological productivity at the global scale should correlate with the Earth's major climatic

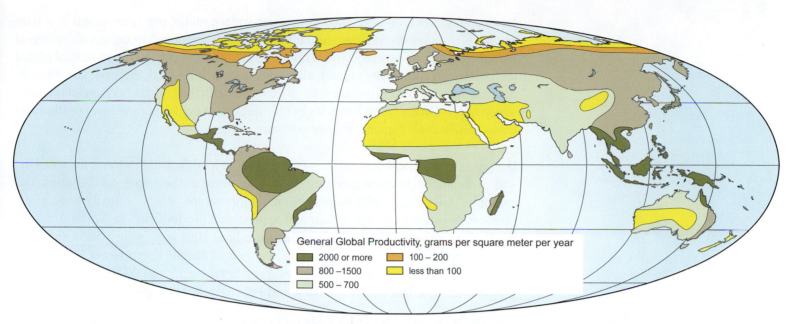

General Global Productivity, grams per square meter per year

- 2000 or more
- 800–1500
- 500–700
- 100–200
- less than 100

Figure 9.26 A general map of primary productivity on the world's landmasses. Notice the patterns are broadly similar to those of the limiting factors shown in the map in Figure 9.25.

zones. Productivity data for terrestrial vegetation show this to be generally so as the map in Figure 9.26 reveals. In the harshest deserts and polar regions, biological productivity, at less than 5 or 10 grams per square meter annually, is barely detectable; whereas in Earth's most permissive regions, the wet tropics, it reaches more than 2000 grams per square meter annually. Not surprisingly, there is a strong correlation at the global scale between mean annual productivity and mean annual temperature and mean annual precipitation. Let us look briefly at these relationships, first temperature and then precipitation, and then ask what effects increased atmospheric CO_2 is expected to have on global productivity.

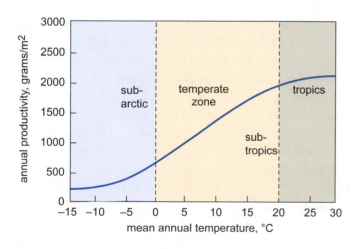

Figure 9.27 The relationship of annual terrestrial productivity and mean annual temperature. Productivity rises fastest between 0 and 20 °C mean annual temperature, corresponding roughly to the midlatitudes.

Temperature and productivity: Examine the graph in Figure 9.27. Starting at the low end of the temperature scale, productivity rises gradually with mean annual temperature (between −15 and −5 °C), and then rises sharply between 0 and 20 °C. The latter temperature range corresponds to the temperate zone of the middle latitudes roughly between the subtropics and the subarctic. The total increase in productivity across this zone is three times or more, from around 500 grams per square meter per year to around 1500 grams or more.

Above a mean annual temperature of 20 °C, productivity begins to level off, or reaches saturation. *Saturation* means that additional heat, the controlling factor, has little or no influence on the dependent factor, productivity. At a mean annual temperature approaching 25 °C, which is found only in the deep tropics, productivity averages over 2000 grams per square meter per year. Given this temperature–productivity relationship, it is interesting to speculate on the influence of global warming on productivity. With a rise of 2 to 3 degrees in mean annual global temperature, productivity should, all other things being equal, increase significantly in the temperate zone but show little or no increase in the tropics because of the saturation factor. On the other hand, higher temperatures could increase evaporation and transpiration rates causing greater moisture stress.

On the face of it, mean annual temperature can be a misleading number for locations with wide annual temperature ranges such as those in the upper midlatitudes. For example, most places with a mean annual temperature of 0 °C actually have a pronounced summer growing season and appreciable productivity. In North America, the 0 °C mean annual isotherm (temperature line) runs east–west across southern Canada through forest lands and the wheat belt of the northern Great Plains. Midsummer monthly temperatures here average between 15 and 20 °C whereas mean temperatures for December, January, and February are below –10 °C.

In the tropics, on the other hand, temperature varies little over the year, allowing continuous plant productivity. In addition, there is plenty of water in the deep tropics, and productivity reaches its highest rate where it is available all year round. These perpetually warm, wet climatic conditions have given rise to the Earth's richest plant cover, the tropical rainforests of South America, Africa, and Asia. As we noted earlier, the only resource that is not present in great quantities over much of the tropics is nutrients through the soil, but this deficiency is made up by recycling nutrients from decaying plant matter to living plants. When productivity is averaged over vast areas such as the Amazon basin, these huge forests produce organic matter at a rate approaching 2500 grams per square meter per year. Within these forest regions, however, there are smaller areas such as river corridors, wetlands, and coastal marshes, where warm, wet conditions combine with abundant nutrients such as phosphorus and nitrogen to yield even higher productivity, as much as 6000 or more grams per year.

Precipitation and Productivity: Terrestrial productivity also rises with mean annual precipitation (see Figure 9.28). However, the rate is not uniform, but rises abruptly at lower values then declines at higher values. The fastest rise comes between zero and 100 centimeters (40 inches) mean annual precipitation, but then beyond 200 centimeters the rate of productivity declines, reaching saturation around 300 cm (120 inches) mean annual precipitation. Only the wet tropics and midlatitude orographic zones are capable of producing such high amounts of precipitation. The tall forests on the rainy west coast of North America, where precipitation exceeds 200 centimeters (80 inches) a year, have the highest rate (close to 2000 grams per year) for any major bioclimatic zone beyond the tropics. The highest precipitation rates on Earth, which exceed 1000 centimeters (400 inches) per year, are found in the tropics, but the productivity rates in these locations are generally no higher than locations receiving half that amount.

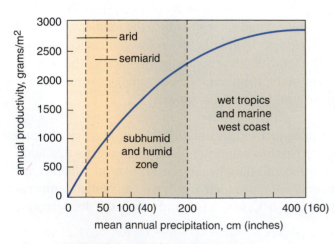

Figure 9.28 The relationship of annual productivity and mean annual precipitation. Notice the rapid rise between 0 and 200 cm mean annual precipitation.

Carbon Dioxide and Productivity: With atmospheric CO_2 on the rise worldwide, scientists have begun to analyze what effects significantly greater CO_2 might have on global productivity in the twenty-first century. Given a doubling of CO_2 by the end of this century, some models estimated

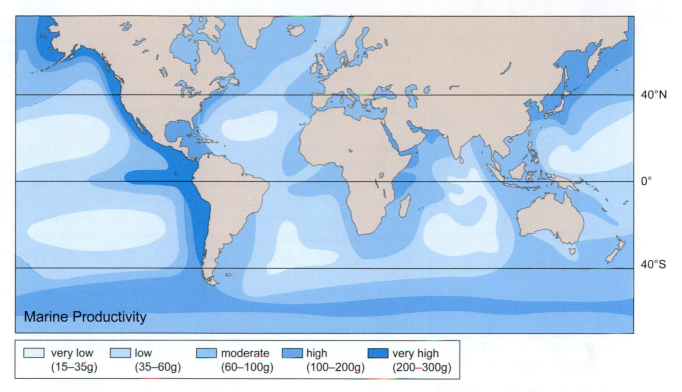

Marine Productivity

very low (15–35g)	low (35–60g)	moderate (60–100g)	high (100–200g)	very high (200–300g)

Figure 9.29 Global marine productivity in grams per square meter per year. The geographic patterns are distinct but markedly different than on land, with very low values over the ocean interiors and high values in coastal areas and higher latitudes.

that global productivity could increase by 35 percent, assuming factors like precipitation remain generally the same. Most of the increase would come from the great forests, especially the tropical forests, and this raises two additional questions. First, will tropical forests continue to decline with the spread of land use, and second, will the increase in productivity help reduce atmospheric CO_2 as more carbon is taken up in plant tissue?

Marine Productivity: Measured in terms of average productivity per square meter of surface area, the oceans are about one-third as productive as the land. Like the landmasses, however, there are strong geographic contrasts in ocean productivity, although the patterns are quite different than on land (Figure 9.29). In the oceans there is only a modest relationship to zonal belts of climate based on temperature and essentially no relationship based on precipitation. Instead, productivity in the oceans varies principally with water depth, circulation, and nutrient availability.

The very lowest productivity, about 1 gram per square meter year, is found in the Arctic Ocean. It is not only cold, but limited in light by the ice cover and long Arctic winters. Beyond the Arctic, the lowest productivity rates are found over the broad expanses of the deep ocean, the area generally referred to as the abyssal plains, which span the tropics and midlatitudes of the Atlantic, Pacific, and Indian oceans. Average annual productivity here is less than 50 grams per square meter, roughly comparable to desert productivity on land. What accounts for such low productivity over such vast areas?

Most of these waters are warm and well-lighted, but the surface layer is deficient in nutrients, especially nitrogen, an essential macronutrient for phytoplankton.

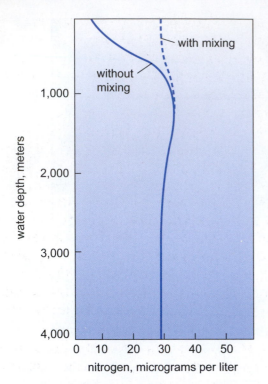

Figure 9.30 Differences in the nitrogen content of ocean water with and without mixing in the upper 1000 meters. Mixing raises nitrogen levels, which induces greater productivity.

However, where ocean mixing processes bring nutrient-rich water into the epipelagic zone (upper 100–200 meters), as the graph in Figure 9.30 reveals, productivity increases significantly. This occurs in upwelling zones in the subtropics (and to a lesser extent along the Equator) and in zones where major currents such as the Gulf Stream flow along the margin of an ocean basin. Productivity in these zones ranges from 100 to 400 grams per square meter per year.

Productivity is also higher on the continental shelves, averaging between 200 and 300 grams per square meter yearly. Water depths are shallow (less than 200 meters) and nutrients are abundant because of contributions from streams and the mixing action of waves and currents. Both the pelagic (water column) and the benthic (bottom) zones contribute to productivity on the continental shelves. In shallow, tropical seas where water temperatures average above 20°C, coral reefs dominate extensive areas of shallow coastal water. Coral reefs can attain one of the highest annual productivities of any major ocean ecosystem, 2000 grams or more per square meter. However, their total coverage is small, less than 0.2 percent of ocean surface area and it is rapidly shrinking as reefs die from the effects of coastal pollution and global warming.

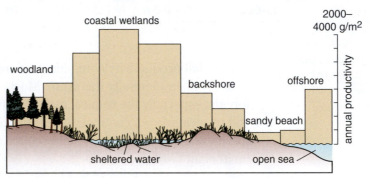

Figure 9.31 Primary productivity varies dramatically in a diverse coastal environments such as between wetlands protected from the forces of the open sea and exposed beaches.

Narrowing our focus to the shore itself, productivity is ranked as variable in the littoral zone. Most of the differences in productivity can be related to the physical character of the coast itself. Coasts with sheltered water, such as lagoons, bays, and wetlands may have very high productivities, as much as 2000 grams per square meter annually, because they are protected from destructive storms and fed with abundant nutrients in runoff from the land (Figure 9.31). River deltas often fall into this class, especially those in temperate and tropical waters. On the other hand, long, straight, sandy shorelines typically exhibit moderate to low productivities because they offer little protection from the forces of the open sea and the substrate (bottom materials) tends to shift about with wave and current action.

9.6 Agriculture, Ecosystems, and Global Productivity

People have been farming for 10,000 to 12,000 years. Before farming we lived by hunting and gathering and Earth was very sparsely populated with humans. Population densities averaged only a few people per square mile in favorable environments and much less in unfavorable ones. As hunter–gatherers, humans had little impact

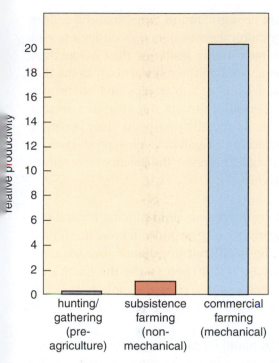

relative productivity

hunting/gathering (pre-agriculture) | subsistence farming (non-mechanical) | commercial farming (mechanical)

Figure 9.32 The relative productivity of three food production systems. Very little hunting–gathering remains on the planet. Subsistence farming is widely practiced in developing countries like China and India. Commercial farming dominates the developed world.

on ecosystems and their productivities. They functioned similarly to many other secondary consumers as a working member of ecosystems, having a role in the food web somewhat like that of bears or wolves. The food production capacity was small, only a tiny fraction of that of modern farming, particularly commercial agriculture (Figure 9.32).

With plant domestication and the advent of crop farming, humans built their own ecosystems. Simpler and far more specialized than natural ecosystems, most agriculture systems were built around two or three main producer crops, such as wheat, corn, beans, and squash. The sole objective was food production for just one consumer and his animals. This system proved so successful that it stimulated rapid growth of human population. Despite setbacks from disease and war, by the beginning of the twentieth century global population had reached 1.5 billion. By the end of the century it stood at 6 billion. To feed this massive population requires a massive system of food production and related resource extraction that reaches deep into the global environment. Of the vast quantity of food consumed by humans, 99 percent currently comes from terrestrial sources and 1 percent from marine sources.

Some observers argue that humans are now utilizing more than 40 percent of Earth's total annual productivity. This includes not only plant material consumed as food, but that consumed as fuel, lumber, and paper, as well as that consumed by livestock on farms and rangeland. Livestock number about 3 billion head worldwide. Humans and livestock, which together total nearly 10 billion, draw on both crop-based and natural ecosystems.

Cropland today occupies nearly 12 percent of the Earth's total land area. On the face of it, this number seems small, but when we consider that roughly half of Earth's land area is fit for agriculture and human habitation, then it appears that crop farming takes up about 25 percent of Earth's productive land area. Add to this the rangeland used for grazing, and it turns out that we are using around 50 percent of Earth's total land area for some form of food production (Figure 9.33). But there is another component to be added: land used for wood production for lumber, fuel, and paper. Wood comes from both natural and farm (plantation) ecosystems, and when added to food production, the combined land area used by humans in wood, food, and feed production now exceeds 65 percent of Earth's total land area.

Figure 9.33 The global distribution of cropland and rangeland. About 50 percent of Earth's land area is used for some sort of food production.

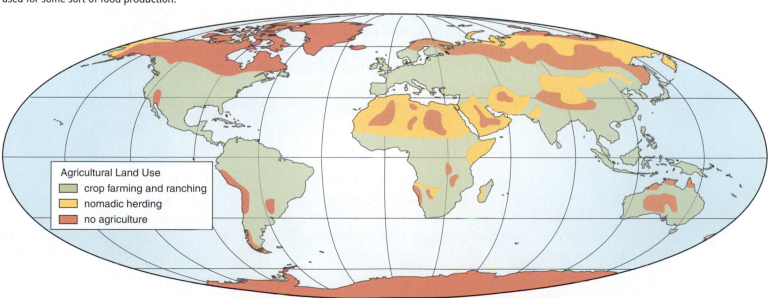

Agricultural Land Use
- crop farming and ranching
- nomadic herding
- no agriculture

The change in total global productivity from pre-agricultural times (natural ecosystems only) to the present (natural plus agricultural ecosystems) is difficult to estimate, but it is a virtual certainty that it is measurably less now than it was then. Most agricultural ecosystems are less productive than the natural ecosystems they replaced. This is particularly so in forested lands of the tropics and subtropics, first, because crops produce for a shorter period of time in the year, and second, because in agriculture there are far fewer plants actually producing than in natural ecosystems. On the other hand, in dry lands where irrigation is used, productivity is raised with introduction of agriculture. As a whole, however, the balance favors lower global production with agriculture than without.

Humans also influence global productivity by degrading and reducing natural ecosystems. This happens as a part of land clearing, overgrazing, soil erosion, wetland eradication, fires, and pollution. This quantity is difficult to estimate, but some scientists place the loss in productivity as high as 30 to 40 percent for the Earth's terrestrial ecosystems. The loss for saltwater ecosystems, related mainly to pollution and sedimentation of coastal waters, is significantly less, probably under 10 percent.

Human population is currently increasing by about 80 million people a year. Demographic projections indicate that global population will continue to grow well into this century, eventually stabilizing at 9 or 10 billion around 2050. This increase will be coupled with increased per capita consumption rates, thereby driving up the demand on ecosystems for additional food, fuel, wood and other plant-based commodities. Cropland will surely expand with much of the expansion going into marginal and submarginal lands, for example, to higher elevations in mountainous terrain as shown in Figure 9.34. Per-acre productivity rates will rise with the development of more efficient crops; extraction of wood, fish, grass, and other ecosystem resources will rise; and ecosystem damage and geographic reduction will also rise as humans press into new lands and intensify global economic activity. The majority of this change will take place in less-developed tropical and subtropical countries, such as Brazil, China, India, and Indonesia, where over 90 percent of Earth's population growth is expected.

If current world population grows by 50 percent to around 10 billion, direct consumption of global productivity will, at the very least, increase by 50 percent over current levels. This will require not only more land under cultivation but higher levels of productivity per acre. Equally important is the question of indirect loss in production in natural ecosystems with damage and geographic reduction. Will losses to terrestrial ecosystems reach 45 to 60 percent? And if so, will such heavy losses reduce the capacity of ecosystems to participate in the larger global systems of water, nutrients, gases, and climate, to say nothing of the health of the ecosystems themselves in terms of species diversity and resilience to disease and other disturbances? What, then, will happen to their feedback mechanisms?

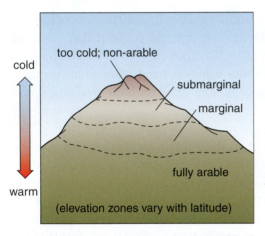

Figure 9.34 The different classes of agricultural land using a mountain as a model. Marginal lands are those with a 50 percent chance of successful farming on the average.

Chapter Summary and Overview of Earth as an Ecosystem:
Energy, Food, and Life

The thin layer of life we know as the biosphere is comprised of a complex network of ecosystems distributed unevenly over the planet. As the summary diagram shows, these systems are one component of a complex of global systems. Both terrestrial and marine ecosystems draw their energy from the Sun and combine it with nutrients from land, water, and air to produce organic energy. Bioproductivity reaches its peak where heat, moisture, light, and nutrients are abundant and dependable. Ecosystem productivity can be disrupted and reduced in a host of ways, and among them are human land uses, principally agriculture, which today ultilizes about half of Earth's land area for food production.

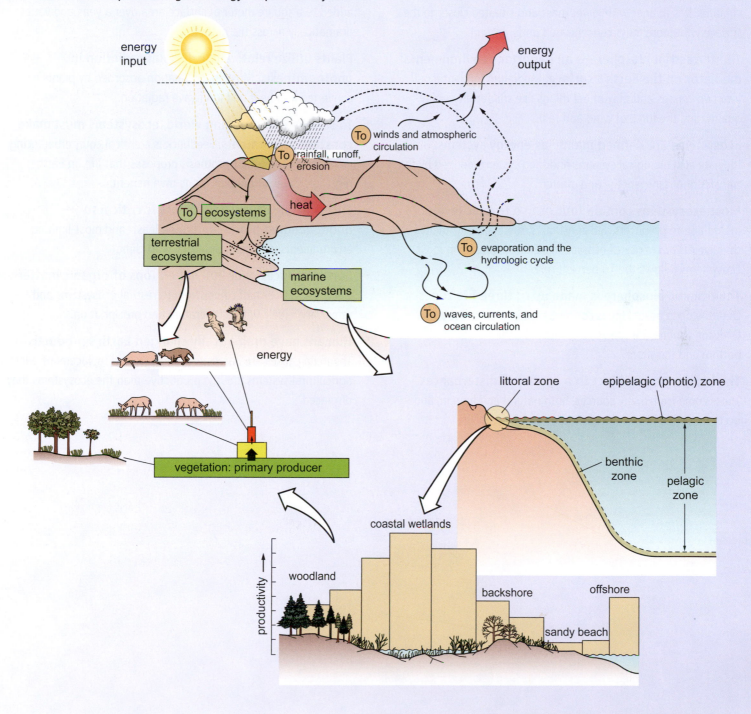

► **The biosphere is the lightest and thinnest of the Earth's four major spheres.** Plants constitute the bulk of its mass but animals and microorganisms account for most of its species.

► **The terrestrial biosphere is made up of three zones.** Most of the biomass is contained within the subaerial zone, which varies dramatically in thickness from place to place on the Earth's surface.

► **The biosphere changes physically with different geographic conditions on the Earth's surface.** In harsh climates, it is thinner with lower mass and situated closer to the surface with more mass concentrated underground.

► **The terrestrial biosphere is an important environmental mediator on the Earth's surface.** Mediating influences fall into two classes and they affect things like soil fertility, water quality, and the force of wind and rain.

► **Ecosystems are defined mainly as energy systems.** But they are also biological systems and food systems defined by the transfer of organic energy and matter.

► **Most ecosystems contain four or five trophic levels.** In addition to producers and consumers, there are detritivores that live off the refuse of ecosystems. Energy transfer in most ecosystems follows the 10 percent rule.

► **The oceanic biosphere is made up of three life zones.** Over the deep ocean, life is concentrated in the surface zone. Over the continental shelves it is most abundant near the bottom and the shore.

► **The biosphere is subject to a variety of disturbances.** These come from many sources, both natural and human, and occur in a wide range of magnitudes and frequencies.

► **All life belongs to ecosystems.** These systems receive inputs of energy and matter from the physical environment, which are converted into organic energy that feeds all life on the planet.

► **All life is made up of elements called nutrients.** Nutrients are cycled among the biosphere and Earth's land, water, and atmospheric systems. Carbon is circulated in both gaseous and sedimentary cycles.

► **Net photosynthesis is a measure of organic energy remaining after respiration.** It is equal to the total tissue added to a square meter of surface area over a year, and varies dramatically across the planet.

► **Plants utilize relatively little solar radiation in photosynthesis.** Most solar radiation absorbed by plants is converted into heat and longwave radiation.

► **To survive in a changing world, ecosystems must make constant adjustments.** Feedback is a critical counterbalancing mechanism. The Gaia hypothesis proposes that life on Earth regulates the environment for its own benefit.

► **The principle of limiting factors is critical to understanding productivity.** The least- and most-limiting environments vary geographically with climate.

► **Earth produces over 200 billion tons of organic matter each year.** Over half comes from terrestrial ecosystems and the rates vary widely with temperature and precipitation.

► **Humans have dramatically changed Earth's productivity.** The main causes are harvesting, damage, and replacement. Most agricultural systems are less productive than the ecosystems they displaced.

Review Questions

1 In the biosphere, what are the five kingdoms of organisms? Which of these groups are considered microorganisms and which of the groups accounts for the bulk of the biosphere in terms of mass and number of species?

2 How would you describe the biosphere's physical character (e.g. mass and thickness) relative to that of the other great spheres of Earth, that is, the hydrosphere, atmosphere, and lithosphere?

3 How would you characterize the distribution of life underground? Under what sort of climatic conditions can we expect to find the bulk of the biosphere concentrated near ground level or underground?

4 What is meant by the statement: "the biosphere acts as a mediator in the formative processes of the landscape?" Provide one example of biochemical and biophysical effects.

5 Name the three major life zones of the oceans and characterize each in terms of the relative abundance of life. How would you characterize the photic zone?

6 Since the last glaciation and the rise of agriculture, humans have become the major change agent of the biosphere. What evidence is there to support this contention?

7 "All life belongs to networks or systems of organisms…" is a reference to what? What is the source of energy that drives these systems? What are geothermal-based ecosystems? Where are they found and how do they rank in terms of relative biomass and geographic coverage?

8 "Carbon dioxide is moved in both gaseous and sedimentary cycles." What does this mean and what are the main inputs and outputs of atmospheric CO_2? Do scientists think that atmospheric CO_2 has changed in the past (before humans) and what might have been the cause?

9 Name four ways of defining ecosystems and to which of these definitions do the following terms belong: *total productivity*, *net photosynthesis*, and *respiration*.

10 Is a food chain a system? What are its component parts? Describe an energy pyramid and relate it to the "10 percent rule."

11 Sketch a diagram illustrating the energy flow to and through a forest beginning with the input of solar radiation. How much energy is taken up in photosynthesis and what form does this energy take?

12 Ecosystems live in a changing world, therefore it should be no surprise to learn that they are capable of self-adjusting their behavior. In this context, what is meant by the terms homeostasis and feedback?

13 Describe several major differences in the geographic patterns of productivity between terrestrial and marine environments at the global scale. Are the oceans more or less extreme than the landmasses? For the terrestrial environment, what climatic zone shows the greatest increase in productivity with mean annual temperature? What accounts for the relatively high productivity rates on the perimeters of the ocean basins?

14 What is the Gaia hypothesis and how does feedback play into this concept? How is it that humans may be having a negative influence on the global ecosystem yet the feedback operating in processes such as desertification is positive?

15 Some observers argue that humans are now utilizing more than 40 percent of Earth's annual productivity. What does this include? How much of the world's land area is now taken up by some form of food production? What the current population of Earth? What is it expected to be by 2050?

Biogeography:
Geographic Distributions of Plant and Animal Types

Chapter Overview

We now examine the types of plants and animals that inhabit Earth and the factors that govern their geographic distributions. These organisms form a great system in which various, and often distant, parts are woven together into complex networks. In terms of sheer mass, the vast majority of these biota is made up of plants whereas most of the species are represented by animals, particularly insects. We will find that many factors govern the distributions of plants and animals, but at the global scale climate and the patterns of land and water, both present and past, are important. On land, smaller scale factors such as regional climate, landforms, drainage patterns, and land use are significant controls. And at an even finer scale, we will see that local factors such as variations in soil and microclimate also play a part.

Introduction

We pushed our way through the brush to the edge of the pond, a simple little basin with the uninspiring name of "Pond 20." It was built to collect stormwater runoff from an equally uninspiring subdivision with the elegant name of Crown Isle, and we had gone there to check its summer water level. We expected little more than an overgrown mud-puddle, but were pleasantly surprised at what unfolded. Beyond the overhang of the canopy, sunlight penetrated the water, warming the surface layer and amplifying the commotion of various bugs and small fish. But closer to shore, within the shade of the alders, a more dramatic scene was being staged. Among the stems of small reeds that lined the shore, hundreds of electric blue damselflies were darting and dancing only inches over the water.

Alison eased herself gently onto the bank while quietly directing me towards the scene. "Will, look, it's a dance of life," she whispered. Pairs of beautiful blue bodies entwined and, one by one, settled onto the reed stems only a few feet from her toes. Once a union was completed, the female crawled down the stem to lay her eggs underwater among the organic debris on the bottom.

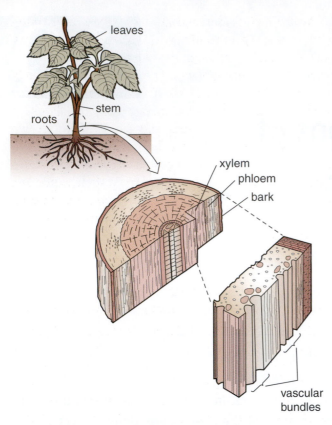

leaves

stem

roots

xylem

phloem

bark

vascular
bundles

Figure 10.1 A representative cross-section showing the basic anatomy of a vascular plant including the vascular bundles responsible for conducting water and nutrients within the three types of tissue: xylem, phloem, and bark.

Figure 10.2 A microscopic photograph of phytoplankton, the dominant plants of the sea and the most widespread vegetation in the world.

We tucked ourselves among the alder boughs and watched transfixed by this elemental dance of life feeling at once like voyeurs and at one with each other and the place. And as we quietly sat more of the play began to unfold. No longer an uninspiring stormwater pond, it became the stage for a much larger drama in which we became willing participants. Birds, snakes, more fish, dragonflies, frogs, and other creatures emerged. The entire pond seemed to be in motion, but each organism seemed to have its own neighborhood. The pond was subdivided into compartments of life, some overlapping, some isolated, and each with a complexity that belied our first glance. In time, interactions began taking place within and among compartments and a faint outline of the pond as a system, an ecosystem, began to unfold.

What controlled these spatial patterns of activity? Certainly the distributions of light and shade were important because they influenced water temperature and it in turn regulated the temperature of the thin envelope of air that blanketed the pond. Water depth was undoubtedly a major influence as well. The distribution of plants was clearly affected by both water depth and light, and the plants in turn affected the distribution of damselflies, frogs, small fish, dragonflies, and other organisms. At another level of organization were the pond's larger predators.

Our union with the scene was rewarded when a small hawk flew over our heads. His interests were also fixed on distributions, for his survival depended on knowing who lived where. From above all, he trained his aerial sensing system on the community below and gave us a demonstration in the finesse of an artful hunter.

10.1 Major Plant Groups

The mass of plants that covers the land and the surfaces and shallow waters of oceans, lakes, and wetlands is called **vegetation**. Broadly speaking, vegetation is made of two size classes of plants: macrophytes and microphytes. **Macrophytes** are large plants that include, among others, trees, shrubs, ferns, and grasses. They may be woody or non-woody (herbaceous), but all macrophytes are **vascular plants**. That is, their bodies and foliage are composed of special conducting tissue, called *vascular tissue*, that forms a pipe-like system of cells arranged in bundles through which water and nutrients are conducted as shown in Figure 10.1. **Microphytes**, on the other hand, are different in almost every respect. Not only are they tiny (most are microscopic) but also they lack vascular tissue, and are exclusively non-woody in composition. Fungi and algae are examples of microphytes.

Macrophytes are the foundation plants of the terrestrial biosphere. They dominate most vegetative covers, giving them not only their basic physical structure but accounting for the vast majority of the productivity in terrestrial ecosystems. By contrast, microphytes are the dominant plants in the marine biosphere. They make up **phytoplankton**, tiny floating plants, such as diatoms and blue–green algae shown in Figure 10.2, which live in the *epipelagic*, or *photic*, zone of all the Earth's oceans and produce most of the organic matter upon which the vast marine ecosystem thrives. Based on total geographic coverage, phytoplankton is Earth's dominant vegetation, covering roughly 70 percent of the planet. But in physical geography we are interested mainly in large, terrestrial plants.

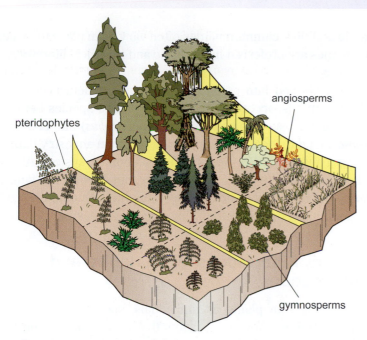

pteridophytes

angiosperms

gymnosperms

Figure 10.3 A schematic illustration of representative examples of pteridophytes, gymnosperms, and angiosperms arranged by size.

Table 10.1 Biological classification, beginning at the class level

Taxon	Size*	Example
Class	Largest	Gymnospermae (cone-bearing plants)
Order		Coniferales (evergreen, needleleaf)
Family		Pinaceae (pine family)
Genus		Pinus (true pine)
Species	Smallest	Pinus strobus (white pine)

* Based on number of members

Terrestrial Macrophytes: Terrestrial macrophytes are represented by three major groups of vascular plants: angiosperms, gymnosperms, and pteridophytes (see Figure 10.3). **Angiosperms** are flowering plants. They occur in both woody and herbaceous forms, range in size from tiny ground plants to huge trees, and are the most abundant vascular plants on Earth. Angiosperms produce seeds and are divided into two major groups according to whether the seeds sprout one leaf, called a cotyledon, or two. The largest group is the **dicotyledons**, which encompasses the broadleaf trees such as maples, birches, and oaks, most shrubs, and familiar small plants such as peas, buttercups, and roses. The **monocotyledons** are mainly non-woody plants with blade-shaped leaves. Familiar examples include grasses, lilies, orchids, and palms.

Gymnosperms are plants that bear naked seeds, meaning the seeds are not enclosed by a mature ovary or fruit as they are in angiosperms. Gymnosperms are dominated by the conifers, such as pine, redwood, and fir, and thus are commonly referred to as cone-bearing plants. They are also distinguished by their needle-shaped foliage and thus are also referred to as needleleaf plants. The **pteridophytes** are also dominated by one group, the ferns. Pteridophytes are small, herbaceous plants which contrast with both angiosperms and gymnosperms by their reproductive systems. Pteridophytes do not reproduce by seeds, but rather by broadcasting tiny particles called *spores*.

Angiosperms not only outnumber both gymnosperms and pteridophyte species, but their geographic coverage and total contribution to Earth's organic productivity is much greater as well. Biogeographers tell us that flowering plants have been growing in species numbers and global coverage during the more than 135 million years or so they have been on Earth, while gymnosperms and pteridophytes, which predate flowering plants in the geologic record, have been declining. Today, angiosperms account for about 80 percent of the major plant groups on Earth. Pteridophytes and their relatives, such as tree ferns, once dominated Earth's landmasses and were a major source of organic matter that was later transformed into the major coal formations that are now found on every continent. Nowadays, pteridophytes are limited to moist shaded habitats on forest floors. Gymnosperms were also more abundant and widespread long ago, but today there are less than 1000 conifer species left on Earth, and they tend to be concentrated in cool climatic zones near the limits of tree growth.

Naming and Classifying Organisms: To keep track of the millions of organisms that inhabit the Earth, we need a standard system of classification and names. The system we use, called the **Universal System of Biological Names**, organizes life into a hierarchical scheme based on genetic and evolutionary relations and gives organisms Latin or Latinized names. The basic unit of the system is the **species**, defined as individuals of the same kind that breed together but not with other organisms. Every species belongs to a **genus**, defined as a group of related species. All genera in turn belong to *families*, and all families belong to *orders* and so on as shown in Table 10.1.

The system is applied to all plants, animals, and microorganisms. An organism is always referred to by two names, genus and species. For example, white pine is *Pinus strobus*. Many organisms also have common names, such as white pine, but most do not. In addition, common names often vary from place to place; hence, Latin names are preferred in scientific and technical literature.

Organisms are classified into genera, families, and orders according to their ancestry, that is, according to genetic lineage. Species belonging to the same genus, such as white pine (*Pinus strobus*) and red pine (*Pinus resinosa*), tend to look alike and have a common ancestry dating back as little as a million years or less in the evolutionary system. At the family level, most members share certain traits, but beyond the family level it becomes increasingly difficult to *see* similarities among different species and genera, even though they share genetic ancestry.

10.2 Floristic Associations, Plant Habits, and Distributions

Our interests are less in plants as individuals, species, and genera and more in the total assemblage of plants that covers an area on the Earth's surface. This assemblage is known as **vegetation**. Virtually all terrestrial vegetation is composed of different mixes or systems of plant species called **floristic associations**. Most floristic associations are dominated by angiosperm species mixed with selected gymnosperms and/or pteridophyte species. What controls the makeup of floristic associations such as the total number of species, the relative balance of species among the three major plant groups, and how they all fit together to give structure and form to the vegetative cover?

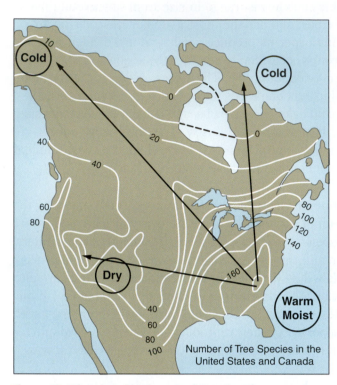

Figure 10.4 The relationship between the number of tree species and climate in North America. The warm, moist climate of the southeast supports eight times more tree species than central Canada and four times more than the arid west and southwest.

The Climatic Factor: Many factors play a part, but climate must be given first consideration because, in general, it sets the limits on the number of species available to form floristic associations. From the standpoint of plants, Earth's climates fall into a broad spectrum from very harsh systems to very permissive or hospitable systems. Harsh climatic systems pose such extremes of cold, drought, heat, and light that few plants can survive in them, and those that can do so only because they have evolved special means, or *habits*, that improve their resilience to harsh conditions.

Permissive climatic systems, on the other hand, pose so few limitations in terms of moisture, heat, and light that multitudes of plants can thrive in them. Between these extremes are climates, such as those of the midlatitudes that tend to combine permissive and harsh traits, usually in a seasonal climatic regime with contrasting wet and dry or cold and warm seasons. North America spans a wide range from fairly permissive conditions in the US southeast and Caribbean to harsh conditions in the dry west and cold north and this range is reflected by the abundance of species as the map in Figure 10.4 illustrates.

Earth's most permissive climate, the tropical wet, nurtures the planet's most varied flora with tens of thousands of plant species mixed together in complex associations. With so many species, the total population representing an individual species over, say, a hectare of land, is not great. As we move toward harsher climates, by contrast, the number (diversity) of species declines while the population of individuals belonging to a given species rises dramatically. This trend is shown in Figure 10.5 and it defines one of the significant differences between tropical and midlatitude vegetation, one that is manifested in the geographic patterns of the forest associations in

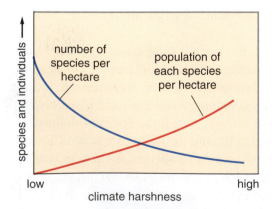

Figure 10.5 Compared to harsh climates, permissive climates support many species but relatively small populations (per hectare) of each.

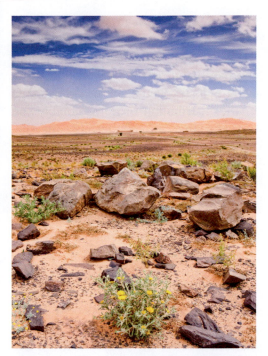

Three habits used by desert plants to contend with drought: (top) ephemeralism, (middle) succulence, (bottom) phreaticism.

the upper midlatitudes. In the boreal forest of the northern midlatitudes (subarctic), for example, vast areas are dominated by one or two tree species, such as spruce and tamarack, whereas in the tropical forest it is rare to find even small patches where just one or two tree species are dominant. Instead, tropical forests are characterized by such diverse plant assemblages that members of individual species tend to be lightly peppered among scores of other species.

Plant Habits: All plants inherit a set of habits that govern their relationship with the environment and in turn influence the makeup of floristic associations. **Habits** are adaptive mechanisms or ways of dealing with the environment that have evolved in a species over long periods of time. They include, among many examples, ways of dealing with cold, drought, fire, flooding, and short growing seasons. In fact, we often group plants according to habit. For example, drought-tolerant plants are called **xerophytes**, water-tolerant plants are called **hydrophytes**, and fire-tolerant plants are called **pyrophytes**.

Two of the most common plant habits are dormancy and annualism. In order to survive a harsh season, such as the cold midlatitude winters, plants must possess the habit of **dormancy**; that is, they must have the capacity to shut down their essential metabolic and reproductive processes to avoid the damaging effects of frost. Cold-season dormancy is common to both deciduous trees, such as maples, oaks, and poplars, and evergreen trees such as pine, cedar, and spruce. Herbaceous plants, such as many wildflowers, in the same climatic zone employ another cold-season survival habit called **annualism**, in which plants complete a full life cycle before the annual harsh season begins. Survival from year to year depends on leaving seeds or root stock over the winter to establish the next generation.

To survive in arid zones, plants must be able to avoid or resist drought. One habit of drought avoidance is **ephemeralism**, the capacity to wait out extended dry periods and then spring to action and complete an entire life cycle when water is occasionally available, which may be only days or weeks every several years. Ephemeralism is common to many desert wildflowers and is a trait well suited to the variable nature of the arid climatic system with its short, unpredictable spurts of water from the sky. Another drought-survival habit, for which cacti are famous, is **succulence**, the capacity to store large amounts of water within plant tissue. And yet another drought-survival habit is **phreaticism**, the capacity to send a long root deep into the ground to draw on the groundwater reservoir (see Figure 9.3b). Both of these habits enable plants to avoid the vagaries of water supply in the desert climate.

These habits allow only drought-tolerant plants (xerophytes) to occupy dry lands where most other plants cannot survive. In other words, *xerophytic habits* control who can live there and who cannot. In traveling from east to west across central North America, for example, climate grows progressively drier. As available moisture declines, hundreds of forest species such as maple, oak, and fir trees are eliminated and replaced by drought-tolerant species such as sage, cactus, and grass. Traveling in the other direction across North America (see Figure 10.6), most dry land species such as prairie grasses are eliminated not because the climate is too wet in the east but because there is too little sunlight for them to survive under the shade of forest canopies.

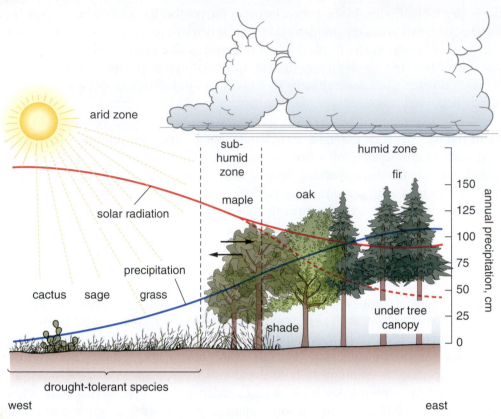

Figure 10.6 A schematic diagram illustrating the change in plant-habit types in response to moisture and light on an east–west transect across central North America.

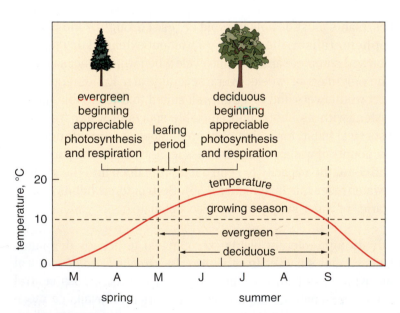

Figure 10.7 A schematic diagram illustrating the advantage of the evergreen habit over the deciduous in getting an early start in the spring.

The Evergreen Advantage: The **evergreen habit** enables plants to keep their foliage year round, through both the harsh and permissive seasons. This habit helps explain some broad patterns in the distribution of flora, notably the concentration of coniferous (needle-leaf) tree species in the relatively cold climates, particularly in subarctic and alpine zones. These zones are geographically marginal for tree species because of their short, cool growing season and long, harsh winters. The evergreen habit gives conifers a distinct advantage here, because they can begin photosynthesis immediately with the onset of warm weather in the spring, thereby maximizing the length of the short growing season. Deciduous trees, on the other hand, need to wait for leaves to emerge in spring before photosynthesis can begin, a process that takes two weeks or more in a growing season only 16 weeks or so long. This difference in the evergreen/deciduous habit in illustrated in Figure 10.7.

Trees with a **deciduous habit** shed their foliage in response to a harsh season. This process is part of the dormancy habit and takes place when environmental stress is too great for the tree to maintain photosynthesis and related processes. Virtually all deciduous trees are broadleaf, but not all broadleaf trees are deciduous, such as some species of oak. The broadleaf forests of the midlatitudes are deciduous, principally in response to winter cold. Extensive tropical woodlands are also deciduous but the stress in this case is seasonal drought. Only in the wet tropics, where there is no prolonged season of stress, are broadleaf forests evergreen.

Geographic Barriers and Disturbance: Other factors also influence floristic associations. Geographic barriers such as oceans and mountain ranges are important in this regard. They limit access that certain plants have to geographical zones where environmental conditions are suited to their growth and reproduction, if only they could get there. For example, the midlatitude forests of New Zealand have a distinctly different flora than forests in North America in the same climatic zone. Prominent North American species such as white pine, red oak, and Douglas fir are not found in temperate New Zealand and prominent New Zealand species such as southern beech and podocarpus (a needleleaf tree) are not found in temperate North

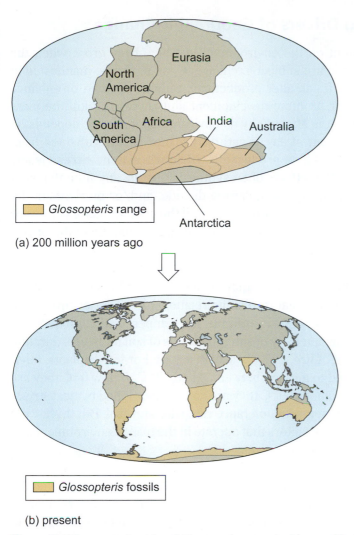

(a) 200 million years ago

(b) present

Figure 10.8 The range of species of *Glossopteris*, a tree-sized fern, and its fossils (a) before and (b) after the breakup of the Pangaea landmass.

Figure 10.9 Grassland fires and forest fires, both common throughout the world, are a major control on the composition and distribution of vegetation.

America. The Pacific is simply too imposing a geographic barrier to allow natural mixing among the two flora. Over the past several centuries, however, humans have broken down this huge barrier and transported scores of plants back and forth across the ocean. Today, New Zealand is reported to have more alien plant species (1700) than native species (1200).

It is noteworthy in passing that ocean barriers have not always been as we see them today. Between 200 and 300 million years ago, the world's landmasses were tightly clustered around Africa in one great continental mass known as Pangaea. Many groups of plants, especially among the ferns and conifers, extended in unbroken distributions, or *ranges*, over vast parts of Pangaea and one is shown in Figure 10.8a. But as the Earth's mantle shifted under this great landmass, Pangaea broke up into smaller pieces that eventually became today's continents – a process called **plate tectonics**. As the map in Figure 10.8b documents, each continent in turn carried with it a large piece of the original stock of many plant families. This accounts for some of the curious similarities among plant types and associations we now see in widely separated land areas of the world. We will say more about plate tectonics and the biogeographical significance of long-distance shifts in Earth's landmasses later in this chapter and again in Chapter 18.

Floristic composition is also influenced by environmental disturbances. **Disturbance** is defined as a force, such as fires, floods, and hurricanes, that damages plants and reduces their capacity to regenerate or eradicates them forthwith from an area. Fires, such as the one shown in Figure 10.9, are common to forests and grasslands in many parts of the world and they restrict certain species from inhabiting those areas. Some plants such as redwood are fire resistant, and others such as jack pine are actually fire dependent. The redwood's thick bark protects it from serious fire damage. Jack pine, which exemplifies the pyrophyte habit, needs extreme heat to crack open its cones and release seeds. Both these habits, fire tolerance and fire dependence, are thus important controls on the makeup of floristic associations in fire-prone areas. When a fire sweeps through a forest, these species will likely survive while other species are wiped out.

But fire is just one of many disturbances affecting the plant cover. Most regions are visited by a variety of disturbances that strike at various magnitudes and frequencies over areas ranging from a few acres to thousands of square miles. The result is a mosaic of overlapping patches, or polygons, of many different sizes and distributions, the net effect of which is floristic diversification of the plant cover. We will examine disturbances further near the end of the chapter.

10.3 Processes and Drivers of Geographic Change

Because plants are dependent on the environment in which they grow, when the environment changes, they too are subject to change. Change can be manifested in different ways among plants. It can alter productivity; reduce, destroy, or enhance reproductive capacity; limit or promote development and stature; or kill the plant itself. Whether and how a plant is affected by an environmental change depends on its **tolerance**, that is, the amount of stress and disturbance it can withstand from various extremes (both high and low) produced by the environment. **Stress** is any limitation in the environment that affects photosynthesis, that is, anything that seriously affects the flow of light, heat, water, carbon dioxide, and/or nutrients in the plant's immediate environment. **Disturbance** covers all those other factors or forces – such as those mentioned above as well as scores of others – that affect the plant's development and survival.

The Concept of Tolerance: For every plant there is an upper and lower limit of tolerances, called **tolerance ranges**, that govern its relationship to the environment. Tolerances are related to many factors; for example, low temperature, high temperature, length of frost-free season, intensity of sunlight, levels of soil moisture, duration of flooding, force of wind, heat of fire, and a host of diseases. For lots of plants, tolerance to cold and heat is limited to a range between 0 and 50 °C. Below 0 °C they are damaged or killed by frost and above 50 °C they are damaged or killed by heat stress. In other words, 0 to 50 °C represents the tolerance range for survival. But this covers just one of the tens or hundreds of factors that operate in the plant's environment and for each there is a particular set of tolerances.

Tolerance ranges may be narrow or wide for different environmental factors and different plants. Plants with narrow tolerance ranges are limited to a very particular, or even specialized, set of environment conditions and can withstand little change in it. For example, many spring wildflowers, such as trillium and trout lily, which grow on the floors of temperate forests, have very narrow tolerance ranges related to light, heat, and soil moisture. In Figure 10.10, we identify them as specialists. Plants with wide tolerance ranges, on the other hand, can manage under a broad set of environmental conditions, and can handle large vacillations in these conditions. Think of them as generalists in Figure 10.10. Plants classified as weeds, such as dandelions, chicory, and kudzu, fall into this category. They have a very wide **habitat amplitude**, meaning they do very well under a wide variety of habitat conditions including harsh disturbances such as land clearing, fires, and farming. For many weeds, it is often difficult to find a place and condition under which they will not grow and reproduce.

We tend to think of the interplay between a plant and the environment in terms of the adult plant. But that view is unrealistic, for plants, like all organisms, go through many phases in their life cycles and each phase has its own and often unique relationship with the environment. To be sure, survival is threatened when a plant is unable to complete any part or phase of its life cycle. **Life cycle** represents the sequence of development, growth, and reproductive phases in a plant's life. A generalized version of the life cycle of a flowering plant in presented in Figure 10.11.

For each phase (numbered in Figure 10.11) there are particular needs and tolerances that are often different from other phases. For angiosperms, the life cycle includes flowering, pollination, embryo formation, seed development, seed dispersal, seed

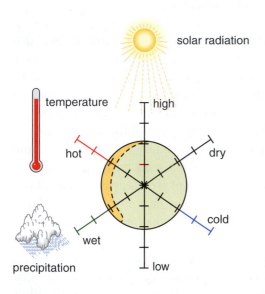

■ narrow habitat amplitude (specialist)
■ wide habitat amplitude (generalist)

Figure 10.10 The concept of habitat amplitude based on solar radiation, temperature, and precipitation. Specialists, such as the species represented by the area in yellow, have very narrow tolerance ranges and specific habitat requirements. Generalists, by contrast, can grow and reproduce under a wide range of habitat conditions.

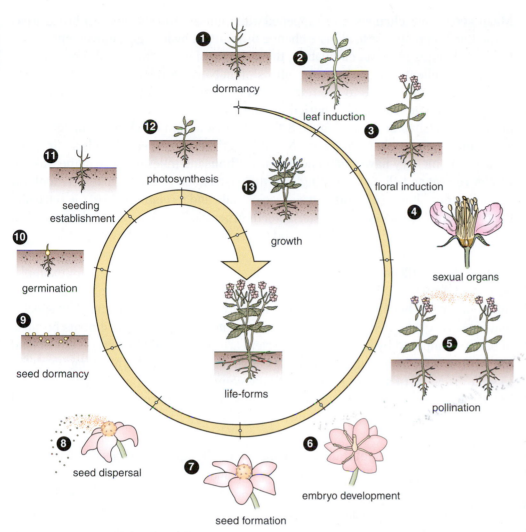

Figure 10.11 A simplified portrayal of the life cycle of a flowering plant. The life cycle of most plants is a complex system in which each phase, such pollination or seed dispersal, depends upon the full completion of each preceding phase.

dormancy, seed germination, and seedling establishment. Disruption of any one of these phases in the life cycle breaks the reproductive sequence, wipes out a generation, and, if repeated, eventually eliminates the plant from that location. And for some phases, such as seedling survival, the odds are not good. To overcome odds against germinating a seed, most plants produce massive numbers of seeds. For example, over a 30-year lifetime, a red alder tree produces about 3 million seeds of which an average of one seedling survives to maturity.

The Nature of Geographic Change in Vegetation:

Elimination (dieback) of vegetation can happen in dozens of ways. It may occur gradually, as with climate change over hundreds of years, or it may occur suddenly, as when plants that require shade are exposed to direct sunlight with the removal of a forest canopy by a tornado, logging operation, or disease. And changes may be bold or subtle. The loss of wildflowers in a prairie fire is bold and dramatic, whereas the decline and loss of wildflowers in response to the die-off of honeybees responsible for flower pollination tends to be subtle. Likewise, the advance of plants into new ground can also be sudden or gradual and subtle. Weeds such as dandelion can populate ground left open after clearing forest or prairie in a matter of months whereas the advance of prairie grasses over sand-dune fields in response to climate change after the retreat of the continental glaciers may take several centuries.

Because the Earth's surface is in a state of perpetual change, the distribution of plants is also perpetually changing. Sometimes the change is wholesale and all plants, an entire vegetative cover such as a wetland or grassland ecosystem, is wiped out. This can happen with severe disturbances such as a massive fire, volcanic explosion, glaciation, or land-use development. Change can also take place piecemeal, species by species, so that in the context of the complex association of plants that make up most vegetative covers, it is hardly noticeable from year to year. Such selective change can often be observed in plants on the edge of their **geographic range** (the area covered by a species, genus, or family) where individual plants have more or less marginal prospects of survival; that is, they have a "toe-hold" type existence. Over time, certain plants drop out and are replaced by others. This often happens when alien species are introduced. The net result is a gradual geographic shift in the composition and distribution of the flora making up the vegetation over an area.

Die-off of trees from acid rain, a vegetation change that may take several decades.

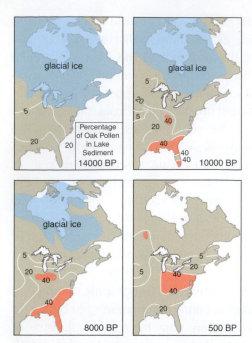

Figure 10.12 Change in the geographic distribution of oak trees in response to climate warming following the retreat of the North American ice sheet since 14,000 years ago.

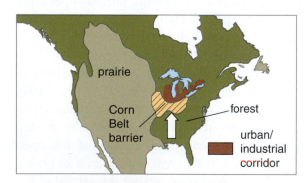

Figure 10.13 The Corn Belt and urban–industrial corridor of central North America together form a major barrier to the northward shift of plant ranges from the American South in response to current global warming. Compare the geographic shift of oak trees in Figure 10.12 to the location of this modern barrier.

Major geographic changes have happened with thousands of species over broad areas of the Earth in response to climate change brought on by ice ages, movement of continents, changes in ocean circulation, the advance of agriculture, and other factors. We have excellent data on the geographic shift in North American flora over the past 20,000 years in response to the retreat of the great ice sheets from the continent. These data come from pollen buried in lake and swamp sediment and some were presented in Chapter 8 (see Figure 8.4) in the discussion on climate change. They show, for example, that prominent North American tree species such as fir, pine, spruce, and hickory have shifted northward a thousand miles or more in the wake of glacier retreat and climate warming over the past 14,000 years. Figure 10.12 looks at the pollen distribution of another tree species, oak, and it relationship to the retreat of the continental ice sheet from eastern North America. It appears that oak has shifted northward from what is now Florida and Georgia to the Ohio Valley region.

Such shifts are still going on, but the game is somewhat different today. For example, we expect changes in the distribution of Earth's flora in response to global warming in this century and beyond. But biogeographers are concerned that some plant species may be severely limited in making the appropriate geographic adjustments because of barriers imposed by humans. The barriers are represented by broad belts of land use such as cropland that stretch across a plant's retreat route, blocking its ability to shift its range to a more environmentally compatible location. The broad region formed by the Corn Belt and the urban–industrial corridor of the US Midwest and southeastern Canada, shown in Figure 10.13, is such a barrier.

Summary on Vegetation and Geographic Change: Angiosperms in various associations dominate the vegetative cover over most of the terrestrial world, but in the oceans microphytes are the dominant plants. At the global scale, climate is the major control in the distribution of vegetation, but many plants have challenged the limits of climate by developing habits that enable them to survive under harsh conditions. Barriers and disturbances also play a part in determining what plants grow where and humans have become a major player in the distribution game by breaking down and erecting barriers and fostering disturbances at many scales. When it is all added up, the overriding theme is one of change, adjustment, and change.

10.4 Major Animal Groups

There are many more animal species on Earth than plant species, but the vast majority of animals are very small compared to most plants. According to biologists, insects are the most abundant animal species on the planet and, among the insects, beetles contain the largest numbers of species. However, the discovery and identification of small species are far from complete. Tens of thousands, and probably more, of unknown insects (and many more microorganisms) await the taxonomist's microscope.

Despite the massive number of animal species, the total biomass represented by animals is small compared to vegetation. Nevertheless, animals, including humans, are important components of the biosphere. Animals are essential in the maintenance and operation of ecosystems besides consuming organic matter. For example, insects are critical in plant pollination and birds are important in the dispersal of seeds, both key processes in plant life cycles. But animals are also important in mechanically shaping landscapes.

Animals and Landscape: Many of the effects of animals on landscape are subtle to the eye, but massive in total impact. For example, the amount of soil moved by earthworms and burrowing mammals such as gophers and shrews in a year exceeds by far that moved by all heavy machinery over the whole Earth in construction, farming, and related activities. Less subtle effects of animals include wholesale damage to vegetation by browsing and grazing animals. As the photograph in Figure 10.14a shows, elephants, for example, are able to strip entire groves of trees to gain access to foliage in the canopies. Taken to the extreme, the distribution and composition of tropical woodlands can be altered significantly by this action.

And let us not forget the work of beavers, also shown in Figure 10.14b. As many as 500 million beavers inhabited the lakes and streams of Canada and the United States before extensive trapping wiped out most of them in the 1700s and 1800s. Today, less than 30 million beavers are at work cutting trees and building dams in the North American landscape. Their dams not only transform woodlands into wetlands, but they also alter drainage systems, modify the flow regimes of lakes and streams, and change soil composition. If we extend our consideration to domesticated animals, the influence of animals on the Earth's landscapes is truly massive. More than 25 percent of Earth's land area is classified as rangeland where herders and ranchers graze more than a billion cattle, sheep, and goats. Overgrazing is common and it often leads to such severe damage to grasslands that prairie landscapes are transformed into desert landscapes.

But without question the most influential animal of all is *Homo sapiens*. Through the practice of agriculture, humans have spread cropland over close to 12 percent of Earth's land area altering entire landscape systems not only by displacing native plants and animals but by modifying landforms, soil, drainage, and ground-level climate. On top of that, humans are long-distance packrats, moving seeds and animals across the planet and mixing aliens with native plant and animal associations. We have already cited one extreme example in New Zealand where aliens outnumber native plant species.

Main Types of Animals: Animals come in a great variety of life forms, from the huge and complex to the tiny and simple. Animals also represent a very complex set of evolutionary patterns and relationships. As a result, classification and description are far more difficult for animals than for plants. Which animals should receive our attention in physical geography is not an easy decision.

On first thought, perhaps it is size that counts most, but many small organisms (microfauna) have an enormous effect on the landscape. Termites are a well-known example. In Africa and Australia, these ant-sized insects build impressive landforms in the shape of mounds and towers like those in Figure 10.15. Ticks, fleas, mosquitoes, and other insects spread diseases such as malaria, sleeping sickness, bubonic plague, cholera, and Lymes disease, which infect humans, who in turn often change the landscape in an effort to control the disease. To control malaria, for instance, people have drained millions of acres of wetland, the habitat of a species of the *Anopheles* mosquito that carries this disease to humans.

Figure 10.14 Some of the effects of animals on landscape, plant habitat, and distributions: (a) trees broken down and killed by elephants in Central Africa; and (b) trees eradicated by beaver cutting, dam construction, and flooding in North America.

Figure 10.15 African termite mounds, an example of an insect's capacity to alter the landscape by creating landforms, altering soil, and changing vegetation.

A simple and convenient way of organizing animals is according to whether they have a backbone or not. **Invertebrates** are animals without backbones and spinal cords. They include more than 90 percent of all animal species and are represented by such familiar organisms as mollusks, sponges, worms, corals, and arthropods. Arthropods include spiders, crabs, insects, and shellfish (crustaceans) and are widely considered the most prominent group of invertebrates.

Vertebrates are animals with backbones that house a spinal cord made up of nerves. They have a long and interesting history on Earth but account for fewer than 10 percent of Earth's species, which includes most of the largest animals. Vertebrates are grouped into five main classes: *fish*, the only fully aquatic vertebrates, which inhabit both saltwater and freshwater; *amphibians*, which are semi-aquatic animals represented mainly by frogs and salamanders; *reptiles*, which are principally terrestrial animals represented overwhelmingly by snakes and lizards; *birds*, which occupy air, land, and water and are composed of nearly 10,000 species, all of which are clothed in feathers and reproduce via eggs; and *mammals*, which are primarily terrestrial and secondarily aquatic animals. Whales and manatees are examples of aquatic mammals. Mammals are the only animals that grow hair and suckle their young. Most mammals are *placental*, which means that the fetus grows within the mother and is nourished though an organ called a *placenta*. A small number of mammals, the *marsupials*, are non-placental in which the young develop in an external pouch.

10.5 Some Animal Habits, Adaptations, and Distributions

Geographic Implications of Thermal Systems: Both birds and mammals are "warm-blooded" or **endothermic** animals (also called **poikilotherms**) meaning they maintain a nearly constant body temperature irrespective of external climatic conditions. In other words, they generate their own internal heat systems, which allow them to survive in thermally hostile climates. This trait has had an enormous influence on the geographic distribution of mammals and birds, enabling them to range over most of the Earth and inhabit environments that many "cold-blooded," or ectothermic, animals like reptiles and amphibians cannot inhabit.

Ectothermic animals (also called **homeotherms**), such as lizards, snakes, and frogs, gain heat from the external environment. Their metabolisms are much slower than endothermic animals and, unless the environment supplies them with heat, they, like the lizards in Figure 10.16, cannot function. Thus, ectothermic animals are limited to perpetually warm climates and seasonally warm climates where the cold season is not so severe that they cannot find thermal refugia (protected spots) such as swamps for winter hibernation.

Birds and mammals have improved on their ability to inhabit harsh environments by evolving special habits and physiological traits. To conserve internal heat in cold ocean water, for example, whales have developed heavy layers of insulating fat beneath their skin. Round body forms and heavy coats of fur or feathers have enabled polar bears, penguins, and other animals to occupy frigid polar environments. At the other extreme are thermoregulatory traits designed for rapid heat loss in hot climates. These include slender body forms, smooth (hairless) skin, and a capacity for evaporative cooling via perspiration. Specialized hot-climate traits include large ears where blood circulates close to the surface of the skin and quickly exchanges heat with the atmosphere.

Figure 10.16 Ectothermic animals such as these sunning lizards gain much of their heat supply from the external environment.

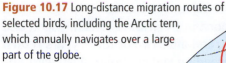

Figure 10.17 Long-distance migration routes of selected birds, including the Arctic tern, which annually navigates over a large part of the globe.

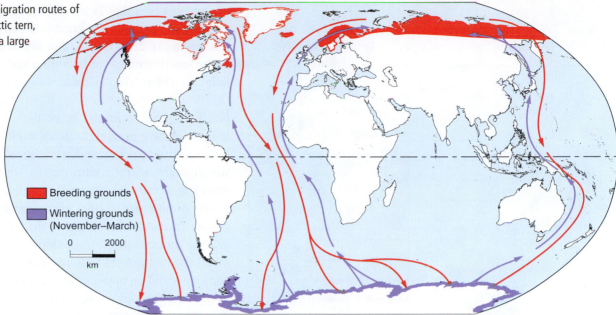

Humans have taken adaptation one step further with technological developments as a part of culture. Among them are two inventions that help us cope with harsh environments, clothing and shelter. Through the clever use of these and a few other inventions we have expanded our geographic zone of occupation on the planet far beyond our biological limits. Without the artifacts of culture, our geographic range would not extend much beyond the tropics. In other words, without clothes and constructed shelters, we would probably not survive a single winter in the midlatitudes, to say nothing of colder climes.

Migration Systems: One of the most successful and widely used habits for dealing with changes in the availability of resources and harsh environmental conditions is migration. **Migration** involves the wholesale movement of animals to more favorable environments. Most migration is seasonal and tied to phases in the migrant's annual life cycles; usually mating, gestation, rearing young, and feeding. The distances involved are related to a species' evolutionary history and its geographic mobility. Birds and aquatic animals commonly employ the longest migration routes and, of the two groups, birds are the champions.

Many birds travel across whole continents. In North America, the migration of ducks and geese from northern Canada to central and southern United States is well known. Most travel in preferred geographic corridors, called *flyways*, which afford many species with feeding opportunities along the way. Many other birds are intercontinental travelers going between North America and South America or between Europe and Africa and back, for example. And a surprising number of birds travel very long routes between the northern and southern hemispheres. The prize for distance goes to the Arctic tern, shown in Figure 10.17, some groups of which migrate annually from the Arctic Ocean to Antarctica and back, a round trip of 70,000 kilometers (43,000 miles).

Most migrations by aquatic animals are shorter than birds but still substantial. Whales have developed some of the longest annual migration circuits. For example, Pacific gray whales every year travel between the Bering Sea, where they feed in the summer, and the west coast of Mexico, where they breed and calve in winter (Figure 10.18). The round trip is 12,000 kilometers (8000 miles) long. Many other long-distance ocean migrants, such as eel and salmon, are organized around a life cycle

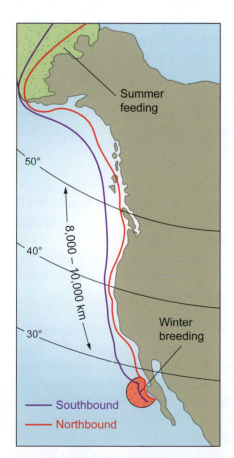

Figure 10.18 Annual migration route of the eastern Pacific gray whale. The round trip can cover 20,000 kilometers.

of several years that is not related to seasonal change over a single year. Salmon are born in freshwater streams, feed and mature in the deep ocean, and after two or three years return to the same streams for breeding and spawning. Fish with this habit are referred to as *anadromous* fish.

For terrestrial animals, migrations are much shorter. The longest involve grazing animals such as caribou in the tundra, bison in the prairies, and wildebeest in the tropical savanna, which migrate from summer to winter rangeland over distances generally less than 1500 kilometers (1000 miles). The shortest migration routes are actually more vertical than horizontal. In mountainous areas throughout most of the world, grazing animals migrate upslope in summer to feed in alpine meadows and retreat to lower elevations in winter as cold and snow descend on upper slopes. Humans also follow this pattern in mountainous areas. Among traditional mountain herders in Asia, Europe, Africa, and South America this practice is known as *transhumance*.

10.6 The Global Dispersal of Terrestrial Animals

There are striking geographic differences, as well as many similarities, in the types and abundance of animals found in different parts of the world. As with plants, many factors play a part and climate is usually the chief control. Warm, moist climates, such as the wet tropical climate, support both the largest numbers of animal species and the largest animal biomass, mostly insects and other small creatures. Beyond the wet tropics, animal life declines as a function of heat and moisture. For example, species numbers decline with latitude in response to temperature as the map in Figure 10.19 illustrates for swallowtail butterfly species.

The same generally holds true for the transition from humid to arid environments. Deserts are able to support a small fraction of the animal species that humid lands can at any latitude. Consider the differences between the American Southeast around Florida and Louisiana and the American Southwest around New Mexico and Arizona. After climate, three other factors are of paramount significance in understanding animal distribution at the global scale: creature mobility, geographic barriers, and the changing distribution of landmasses and oceans.

Unlike plants, animals have the capacity to walk, crawl, fly, or swim to new locations. Although many animals can move rapidly, most distributional changes take place gradually over thousands of years in response to changes in climate, vegetation, topographic and ocean barriers, and geographic connections among landmasses and oceans. However, some groups of animals have been far more effective than others in spreading over the planet. Among terrestrial mammals, nine families (including shrews, rabbits, deer, bears, dogs, and cats) have spread to all continents except Antarctica and Australia. These families and two other mammal families are referred to as "*the wanderers*." Among the continents, Africa – where the ancient evolutionary roots of many animal families are lodged – contains the largest number of terrestrial

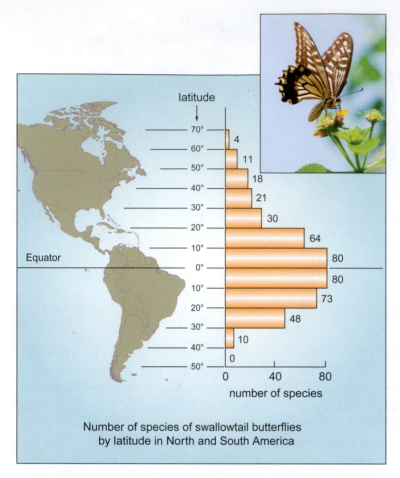

Figure 10.19 Changes in the number of swallowtail butterflies species with latitude, a response to geographic change in temperature with latitude. Between the Amazon (0 degrees latitude) and Florida (25 degrees latitude), species decline by more than half.

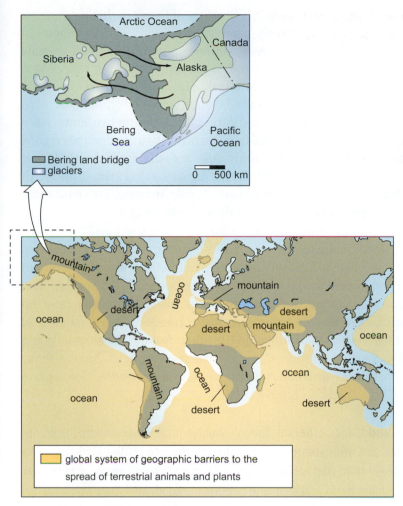

Figure 10.20 The global distribution of major biogeographical barriers: oceans, deserts, and mountain belts. Inset shows the Bering land bridge 15,000–20,000 years age. It provided a land passageway between Asia and North America, which was used by many species including humans.

mammal families (44), and Australia and Antarctica the fewest with 11 and 0 each.

The Role of Geographic Barriers: There are three main classes of geographic barriers to the global spread of animals: deserts, oceans, and mountain belts (see Figure 10.20). Of the three, oceans have been the most invincible barriers. For tens of millions of years large belts of sea have isolated Australia from other landmasses and, as a result, virtually all its terrestrial families are **endemic**, that is, found nowhere else. Africa was also sea-isolated but, about 23 million years ago, it came into contact with Arabia, which linked it to Eurasia enabling ancestors of animals like giraffes, rhinos, and cats to enter the continent.

The narrow ribbon of ocean between Asia and North America, the Bering Strait, on the other hand, has been an on-again/off-again barrier to animals. This has come with the rise and fall of sea level bought on by glacial and interglacial periods over the past 2 million years. During the last major glaciation, which peaked about 18,000 years ago, sea level was much lower, and a dry land bridge – shown in the inset of Figure 10.20 – formed between these continents, which allowed exchanges of large mammals. These included mammoths, bison, musk ox, mountain goats, mountain sheep, and humans. As the glaciers receded and sea level rose, the land bridge closed about 13,000 years ago and the barrier was re-established.

In Africa, the Sahara and connecting deserts have been an imposing barrier to animal migration to and from Eurasia. Although these deserts have existed for several million years, they have not always been as impenetrable as they are today. In the past 2 million years or more, the Sahara has expanded and contracted with climate changes associated with the rise and fall of major glaciations (ice ages). During contracted (cooler) phases it was permeable to some animals such as elephants and apes, which moved into Asia, but impermeable to others such as aardvarks and elephant shrews, which today are not found outside Africa. Our own species, *Homo sapiens*, had its origin in Africa, and we too were influenced by the Saharan barrier as we spread into Asia. This story is taken up in the next chapter.

Plate tectonics has also dramatically affected the distribution of animals on Earth, but over much longer periods of time. For example, around the time mammals first appeared on Earth about 200 million years ago, Pangaea, as map (a) in Figure 10.21 shows, was just beginning to break apart. By about 100 million years ago, the distances of open water separating the continents were great enough to prohibit exchanges of terrestrial animals and thousands of species subsequently evolved as geographically isolated populations. Antarctica shifted southward into deep-freeze isolation at the South Pole, which led to extinction of virtually all of its animals. South America shifted westward, North America shifted northwestward, and the Atlantic Ocean formed between the Americas and Africa. Despite the fact that South America and Africa remained in the same climatic zone and shared much common animal stock, evolution took different directions on the two continents. Excluding the wandering families, Africa and South America today have no mammal families in common.

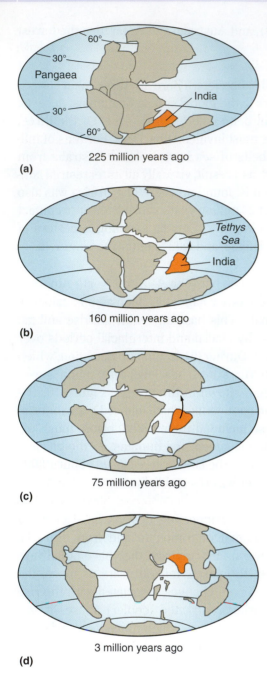

Figure 10.21 Around 225 million years the continents began separating from a common landmass, called Pangaea. India shifted northward across the Equator to join Asia, bringing new animal and plant species to the great continent, and later North and South America joined at Panama.

For more than 150 million years after the initial breakup of Pangaea, India remained isolated far out to sea while shifting slowly northward (Figure 10.21b). About 20 million years ago, it began colliding with Asia, setting up an exchange of animals between the two landmasses (Figure 10.21c and d). At the same time, a large sea, called the Tethys Sea, that separated Africa and Arabia from Asia, closed and, since the Sahara and connecting deserts had not yet become impassable, animals were also exchanged between tropical South Asia and Africa. Today, these two landmasses have the highest level of similarity in mammal families of any two continental regions on Earth. Zoogeographers give Africa and South Asia a *similarity factor* of 70 (not counting the wandering families) compared to 50 for Africa and Eurasia, 46 for South America and North America, and 38 for Eurasia and North America. Two land areas that share all the same families would have a similarity factor of 100.

On the other side of the Atlantic, North and South America remained separated from each other with dissimilar animal species until 3 million years when a string of volcanic islands emerged to weld the two continents together at the Panama Isthmus (Figure 10.21d). The new land bridge led to a massive exchange of animals, referred to by zoogeographers as the **Great American Interchange**. Sixteen families of land mammals dispersed from each continent to the other. Among the animals going from north to south were bears, rodents, horses, llamas, and deer. Among these going from south to north were armadillos, porcupines, and opossums. See Section 18.6 for more on the Great American Interchange.

Summary on Animals and their Dispersal: Though small in total mass compared to plants, animals, both vertebrates and invertebrates, exert a strong influence on the landscape. Like plants, animals have evolved traits and habits, such as endothermism and seasonal migration, that enable them to live under harsh and changing geographic conditions. They also have the capacity to move in response to changes in climate, habitat, and physical barriers, and spread to new locations. Among the efficient animal migrants are a group of mammal families, known as the wanderers, who have spread to all but two continents.

10.7 The Geographic Patterns and Character of Earth's Biota

Describing the composite character of Earth's vegetation and animals can be a formidable undertaking and space limits us in how much we can accomplish here. Since physical geography is chiefly concerned with landscape, our attention is trained on the terrestrial rather than marine biosphere. In addition, the approach we take must use a broad brush and center more on the form and function of biota as part of the landscape rather than its taxonomy as a part of biology. A taxonomic approach would entail describing individual species or groups of species and where they live, whereas a form and function approach addresses more the structure and processes of vegetation and associated animals as we find them in the landscape.

The Concept of a Biome: We will use a biogeographic unit of classification called a biome. **Biomes** are large-scale assemblages of plants and animals whose geographic distribution corresponds roughly to global and regional patterns of climate and soil. Although biomes are made up of both plant and animal communities, they are shaped overwhelmingly by vegetation. Vegetation constitutes more than 90 percent of the biomass of biomes, and sets their basic structure as well. Indeed, we classify

Figure 10.22 Biomes have undergone remarkable geographic change with changes in climate, the location of landmasses and other factors. In the past 100 years or so deserts have expanded while forest have declined mainly in response to human influences.

Figure 10.23 Graphs showing the relationships between (a) productivity and biome type (notice that forest biomes reach over 2000 grams per year, and (b) biome biomass and mean annual precipitation (notice the sharp increase between 100 and 200 centimeter precipitation).

terrestrial biomes by the structure of their dominant plants. **Structure** is the composite form of many plants living together, that is, the overall form of the vegetative cover. For example, an individual tree is a *life form*; many closely spaced trees make up a forest, which is a structural class of vegetation used to define a forest biome.

Biomes are the products of many geographic systems including climate, soil, landforms, and human land use. Despite their great size, they are not static, or even slow-to-change, features of the Earth's surface. Far from it, for the systems that shape them are geographically dynamic and as these systems change, biomes also change. We know, for example, that climate change has caused many big shifts in the global pattern of biomes over relatively short periods of Earth time. About 20,000 years ago, long after our species had found its way into Eurasia and Australia, global cooling associated with the last glaciation caused a major reduction in the area of forest biomes. Beginning about 16,000 years ago, however, forests began expanding in response to global warming. But this latter trend was cut short by the growth of human population and the spread of agriculture about 10,000 years ago. Forest biomes are still declining over much of the world and, at the same time, desert biomes are expanding as a result of desertification, the combined effects of grassland overgrazing and drought. Modern global warming and the growth of human population are expected to accelerate the desertification process (Figure 10.22).

The Primary Classes of Biomes: Five primary classes of biomes are generally recognized: forest, savanna, grassland, desert, and tundra. Each of these, in turn, may contain several smaller biomes, often transitional forms, and many different ecosystems. And as you would expect, the primary biomes are remarkably different not only in their structures but also in their productivity, biomass, and biodiversity. In general, stature, productivity, biomass, and biodiversity are at their highest where heat and moisture are most abundant and, as we discussed in Chapter 9, each declines sharply as heat and moisture decline. Figure 10.23 provides graphs showing the relationship between (a) biome type and productivity, and (b) annual precipitation and biome biomass.

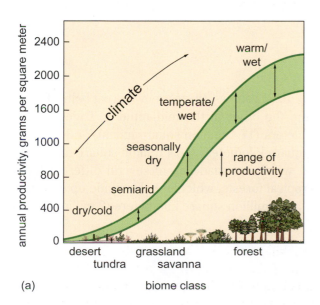

(a)

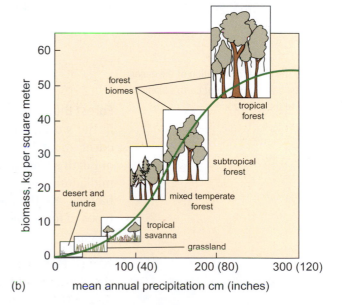

(b)

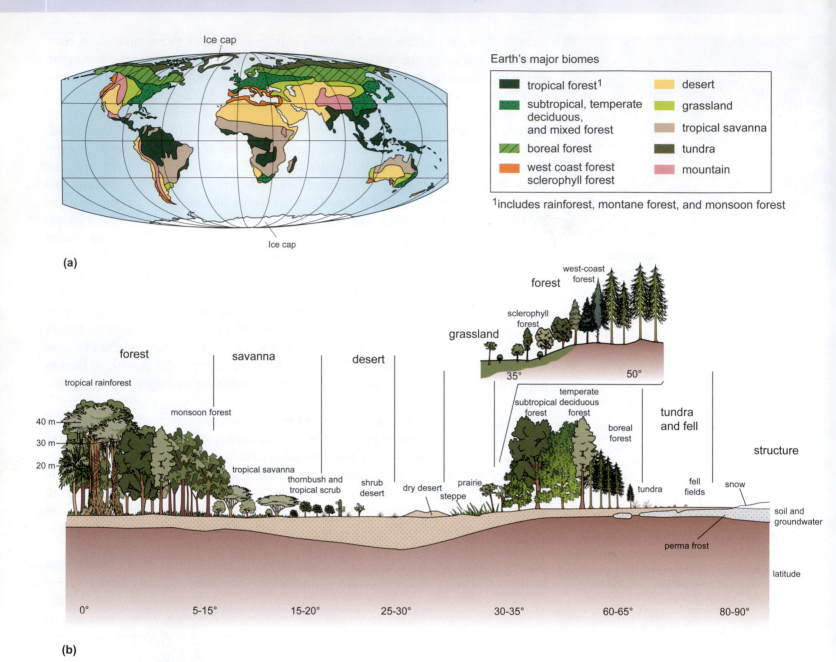

Figure 10.24 (a) The global distribution of the major biomes; and (b) the basic structure of Earth's biomes from the Equator to the polar zone.

Forest Biomes: As the map in Figure 10.24a shows, there are two great belts of forest in the world, one in the tropics and the other in the midlatitudes. Together they cover about 45 million square kilometers (17.4 million square miles), or 30 percent of Earth's land area. At 27 million square kilometers (10.4 million square miles), the midlatitude (or temperate) forests, which are made up of several different forest biomes, are the most extensive. Tropical forests, which are also made up of several types of forest biomes, cover close to 18 million square kilometers (7 million square miles).

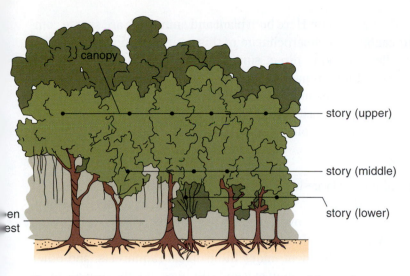

Figure 10.25 The structure of a tropical rainforest consists of three or four stories and an open forest floor.

canopy

story (upper)

story (middle)

story (lower)

en
est

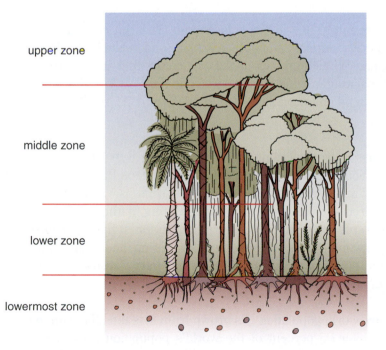

upper zone

middle zone

lower zone

lowermost zone

Figure 10.26 The life zones of a tropical rainforest. The middle and lowermost zones generally display the greatest biodiversity and rank among the richest habitats on the planet.

Tropical Forests: We begin in the hot, wet tropics with what are often described as the deepest, darkest forests in the world, the **tropical rainforests**. Tropical rainforests have come to be our most celebrated forests, not only because they are Earth's richest plant cover, but because they are scientifically intriguing, forming landscapes where much biogeography remains to be discovered, catalogued, and mapped. Rainforests are not dense tangles or jungles, but big tree forests that reach heights of 40 to 50 meters. They are called rainforests because they receive prodigious amounts of rainfall, typically more than 250 centimeters (100 inches) per year, and it comes in all seasons, and in some locations nearly every day of the year. Given an ample, steady supply of moisture, heat, and light, these forests experience no season of dormancy. As a result, tropical-rainforest productivity rates are very high; typically, between 2000 and 3000 grams per square meter annually of wood and foliage is continuously being renewed.

Figure 10.25 shows the canopy of a tropical rainforest consisting of three or four levels, or *stories*, of foliage. In most places this canopy is so dense that the forest floor is continuously shaded. Without direct sunlight, ground cover in the form of herbs and shrubs is very limited, leaving the forest relatively clear of foliage between the ground and the base of the canopy. This structure is referred to as an *open forest*. The bioarchitecture of the open forest determines the vertical distribution of life in the tropical rainforest, giving rise to four distinct life zones, three above ground and one underground, as shown in Figure 10.26.

The **upper life zone** lies at the very top of the forest canopy where the habitat is windy, bright, and often stormy. Though large birds such as hawks and eagles are common here, this zone does not teem with animal life owing largely to its rigorous climatic conditions. The **middle life zone**, represented by the dense central mass of the canopy, does, however, teem with animal life. Indeed, here lies the richest collection of insects, birds, mammals, and reptiles in the world. In a single tree canopy, for example, it is possible to find more than 1000 insect species, most of which have never been catalogued and named. This zone is also rich in plants including large vines, called *lianas* (shown in Figure 10.26), that reach high into the canopy, and plants, called *epiphytes*, that grow on tree branches far above ground without rooting in the soil. Sometimes called "air plants," epiphytes such as ferns and orchids maintain themselves by capturing rain in branch hollows or within their foliage.

The **lower life zone** is the forest floor. Here both plant and animal life are sparse compared to the overlying canopy. The microclimate of this zone is distinctive: shaded, humid, and nearly windless. Lacking direct sunlight, ground plants are relatively sparse, thus forage is limited for animals. In addition, flooding is common in tropical river lowlands such as the Amazon, Congo, and Mekong. In the Amazon Basin, for example, river waters spill over vast areas of forest floor for several months each year. Fish and other aquatic animals follow the floodwaters making seasonal habitat of the forest floor.

Although limited as a habitat for large browsing animals, the warm, moist rainforest floor is an ideal habitat for detritivores that thrive on the organic debris continuously raining down from the canopy above. So active are insects, worms, and related microorganisms that very little organic matter is allowed to accumulate on the soil surface. As a result, topsoil development under the tropical rainforests, and other tropical forests as well, is weak or nearly non-existent. Witness that the organic mass there is roughly comparable to that of the desert, typically less than 8 kilograms per square meter.

The fourth and **lowermost life zone** begins at the soil surface and extends more than 100 meters underground. Within this zone, the upper 5 meters or so supports the richest and densest mass of organisms. Most of this organic mass here is made up of tree roots, as Figure 10.26 shows, but insects and microorganisms make up most of the species. Indeed, soil is considered to be the second great reservoir of species in the tropics and most of these species, especially the microorganisms, have not been catalogued by scientists.

Figure 10.27 (a) Tropical rainforest eradication in progress, Amazon region of Brazil. (b) Conversion of rainforest landscape into commercial agricultural land.

Tropical rainforests are found in two major locations: (1) large core areas on the Equator in South America, Africa, and southeast Asia (and related islands extending into the Pacific); and (2) along mountainous east coasts that are watered by the tradewinds. This latter arrangement is best illustrated in the New World where bands of tropical rainforest extend poleward along the Brazilian and Central American east coasts (Figure 10.24a).

The single largest mass of tropical rainforest, representing 9 to 10 million square kilometers, is found in South America, principally in the Amazon Basin. Prior to 1950 or so, land use here and in other large tropical rainforest regions was generally light, characterized by scattered subsistence farming and hunting/gathering practices. In the second half of the twentieth century, however, the rainforest has been subject to major intrusions by peasant farmers, and ranching, lumbering, and mining interests. Population growth and economic expansion are the primary drivers behind these intrusions and road development has provided access to the forest interior.

Between 1950 and now, world population has grown from less than 3 billion to more than 6 billion. Today over 95 percent of the world's population growth – which amounts to about 80 million persons a year – is occurring in less-developed countries many of which (e.g. Brazil, Malaysia, Nigeria, and Zaire) contain the planet's largest tracts of tropical rainforest. As population and development expand, these forests are being eradicated or damaged (Figure 10.27). Estimates put the rate at 170,000 square kilometers (65,000 square miles) a year worldwide. Only a hundred years ago rainforests covered 14 percent of the Earth's land area; now they cover a mere 6 percent. At this rate the world's tropical rainforests will be virtually gone by 2050.

(a)

(b)

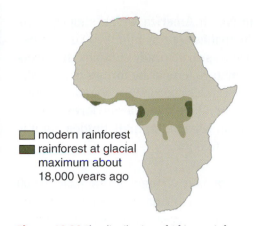

rainforest at glacial maximum about 18,000 years ago

Figure 10.28 The distribution of African rainforest 18,000 years ago at the peak of the last glaciation and today prior to major deforestation.

How does this compare with past reductions, before humans played a part? The global cooling that we spoke of earlier, which reached its maximum about 18,000 years ago, reduced the tropical rainforest in Africa to three relatively small areas, amounting to less than one-third its twentieth century coverage (Figure 10.28). We can assume that Asia and South America experienced roughly similar reductions. Thus, it appears that within each of the world's three main regions of tropical rainforests are core areas of ancient rainforest, ice-age refugia from which the great forest expanded as climate warmed during the interglacial periods. We now live in a time of expanded tropical forest. But that is not Earth's only tropical forest, for beyond these large areas of tropical rainforest in South America, Africa, and Asia, lie other types of tropical forest. These occupy somewhat less friendly tropical environments and are less biologically rich and productive than the rainforest. We will examine two of these, the tropical montane forest and the monsoon forest.

Tropical montane forest is a highland variation of tropical rainforest. It is found on lower mountain slopes such as the windward side of the Andes and the Himalayas between 1000 meters (3300 feet) and 2000 meters (6600 feet) where climate is cooler and often wetter because of orographic influences. Trees are shorter and more widely spaced and secondary vegetation such as mosses, shrubs, ferns, and epiphytes is more abundant. Ground vegetation, especially shrubs, provides ample food for browsers such as the mountain gorillas of the East African Highlands and pandas of south China. Productivity can be very high on lower slopes, comparable to that of lowland rainforest. At higher elevation, these perpetually damp forests, which are also called *cloud forests*, give way to mountain woodlands resembling midlatitude forests with coniferous and broadleaf deciduous trees.

Figure 10.29 Monsoon forest of India, a tropical forest of smaller stature and lower biodiversity than the tropical rainforest.

Most montane forests lie on the wet side of the large tropical rainforest regions. Monsoon forests and other lesser tropical forests, by contrast, lie on the dry side of the tropical rainforests. **Monsoon forests** are deciduous tropical forests, a response to the pronounced winter dry season. They have a different structure than tropical rainforests, with abundant ground vegetation, mainly grass and shrubs, and lower productivity. The monsoon forest structure is explained by the somewhat discontinuous or partially open tree canopy, like that shown in Figure 10.29, which allows direct sunlight to reach ground level. Although the true monsoon climate type is found only in south and southeast Asia, large areas of monsoon-type forests are also found in South America, Africa, and Australia, where the climate is classified as tropical wet–dry but there is enough rain to support a tree cover.

Temperate Forests: In the midlatitudes are five major biomes of temperate forests: subtropical forest, temperate deciduous forest, west-coast forest, boreal forest, and sclerophyll forest. Temperate forests are concentrated overwhelmingly in North America and Eurasia and are geographically separate from tropical forests everywhere except in southeastern China. The gap between the two great belts of forest is mainly attributed to the barrier formed by the dry climates that dominate the subtropics. As the world vegetation map in Figure 10.24a shows, little temperate forest is found in the southern hemisphere because there is so little land available within this latitude range south of the Equator.

In general, temperate forests are smaller, less biologically diverse, and less productive than tropical forests. Average annual productivity ranges from 800 to 1300 grams per square meter, about half that of the tropical rainforest. There is one notable exception and that is the west-coast forest in North America which supports the largest trees on Earth and yields annual productivities approaching 2000 grams per square meter. Temperate forest systems are also less structurally complex than tropical forests. Instead of multiple canopy stories, mature temperate forests are usually made up of one or two stories. In addition, there is far less secondary vegetation such as vines and epiphytes and animal life is far less abundant in temperate forests. That, of course, can be ascribed mainly to the cooler climates of the midlatitudes but it is also related to the relative newness of the temperate flora.

The world's temperate forests have undergone dramatic changes in the past 20,000 years. The continental glaciers in both Eurasia and North America penetrated deep into their midst about 18,000 years ago obliterating vast areas of trees and forcing a southward compression and displacement of forest zones. Think of this as a system's shift. As the great ice sheets melted back over the next 10,000 years, the systems shifted back as forests slowly re-established themselves (see Figure 10.12). Associations of selected species migrated northward from refugia in what is now the zone of subtropical forest. As this was taking place, a new biogeographical system was emerging. Agriculture was spreading from its early hearths in the Middle East, Mexico, and southeast Asia. By the time of the Romans, 2,000 years ago, vast areas of temperate forest in Eurasia had been cleared for crops. Forest clearing was not so advanced in North America until the entry of Europeans who swept across the heart of the continent obliterating the forest there. Let us start with the subtropical forests.

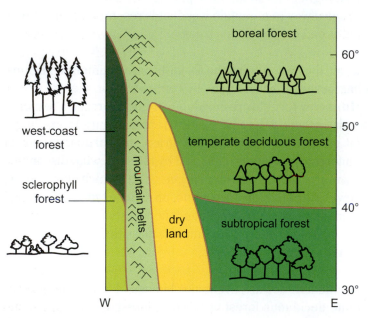

Figure 10.30 A schematic model showing the geographic organization of temperate forests in North America.

Subtropical forests occupy the southern part of the temperate forest zone (Figure 10.30). They are made up of both broadleaf and coniferous (needleleaf) trees. Subtropical broadleaf forests resemble tropical forests not only because they are mostly evergreen, but because they tend to be structurally complex and support a relatively rich mix of plant and animal life. The main regions of subtropical forests in the world correspond to the humid subtropical climatic zones of North America and Asia. In North America, they are found in the southeastern United States and are dominated by coniferous trees, mainly pines. In Asia, they are found principally in southeastern China and southern Japan and are dominated by broadleaf trees such as oaks. In both China and the United States, virtually none of the original subtropical forest remain, most having been cleared for agriculture and other land uses. Only small patches located in areas of rugged topography and land too wet for settlement or farming remain, and most of these remnants are contaminated by non-native or alien plant species introduced with the spread of land use.

North of the subtropical forests are the **temperate deciduous forests** (Figure 10.24). Found mainly in three large tracts, North America, Eurasia, and the Far East (China, Korea, and Japan), temperate deciduous forests are structurally and biologically simpler than subtropical forests with lower productivities (around 1000 grams per square meter yearly). There are fewer tree species, less secondary vegetation, and fewer animals, though many of the mammals such as deer and wolves are larger than similar species in the subtropics. The deciduous habit of this forest cover is a response to the pronounced winter season when frost and snow are dominant for several months.

On the northern margin of all three regions of temperate deciduous forest, the broad-leaf trees combine with coniferous species forming *mixed forests* of pine, maples, birch, balsam fir, and other species. Both temperate deciduous and mixed forests are characterized by patch-like geographic patterns made up of stands of different tree associations and ages. The stands may consist of a single species, such as white pine, or two or three species, such as oaks and hickory. The origins of these stands appear to be tied both to differences in site conditions related, for example, to variations in soil, drainage, slope, and microclimate, and to past disturbances from fires, windstorms, land use, and disease. Northward, species diversity declines further and forest patches tend to grow larger until vast areas may be covered by a single patch of one or two tree species.

Like the subtropical forests, the temperate deciduous and mixed forests have been severely altered throughout Asia, North America, and Europe. In North America, for example, it took only two centuries, from about 1720 to 1920, to cut almost all of these forests for lumber and farms. In Asia, China's temperate forests were all but obliterated as early as AD 1000, and in Europe most of the temperate forests were gone by 1600 or so. Records show that England was practically treeless by 1700.

Beginning mainly in the twentieth century, much reforestation has taken place in Europe and North America as part of conservation and forest management programs. The resultant second-growth forests are usually different in structure, composition, and distributional patterns from the virgin forests they replaced. For example, most planted forests are monocultures, often composed of just one tree species. Virtually all other second-growth forests contain many alien plant species of both advertent and inadvertent introduction. And in most places the forest cover is no longer continuous over large areas but fragmented into patchy patterns characterized by small blocks and short ribbons of trees among larger areas of cleared lands as you can see in Figure 10.31. Not surprisingly, these changes in the phytolandscape are also accompanied by changes in animals. For example, deer, pheasants, crows, and raccoons are more abundant while wolves, bobcats, eagles, elk, and moose are less abundant, and alien species typically abound, especially among birds.

Figure 10.31 A landscape originally covered by temperate forest where the tree cover has been reduced to scattered patches and ribbons along streams and farm fields.

Poleward of about 45 to 50 degrees north latitude, the mixed temperate forests give way to more homogeneous **boreal forests**, where only three or four tree species may dominate extensive forest stands (Figure 10.24). Conifers such as spruce and tamarack are the principal trees, but a few hardy broadleaf trees, such as birch and tag alder, are also common in these great forests. Boreal forests stretch, almost uninterrupted, across Russia, Canada, and Alaska and cover about 12 million square kilometers (4.6 million square miles), representing about 25 percent of the world's forest cover. Boreal forests are also known widely by their Russian name, *taiga*, which in Siberia are dominated by tamarack or larch, a deciduous conifer.

Long, cold winters and short growing seasons limit the growth rates, productivities, and stature of boreal forests. Productivities are typically about two-thirds those of temperate forests, around 800 grams per square meter per year on the average. In addition, extensive areas are swampy, with large patches of wetland interspersed

Figure 10.32 An aerial photograph of the boreal forest of Russia, one of the last great tracts of forest in the world.

with the forest cover (Figure 10.32). Permafrost underlies much of the boreal forest and the cold, damp soils, coupled with subarctic climate, severely limit the number of plant species that can survive there. Animals, such as moose and bear, tend to be large and thickly furred and, as a whole, the fauna and flora of the boreal forest show little mixing with alien species.

On their northern boundaries, the boreal forests grade into treeless tundra. Virtually all borders between major vegetational formations in the world are broad transition zones, and the transition zone here is one in which the characteristic tundra cover is interspersed with patches of trees, many of which exhibit *dwarfism* owing to the stressful Arctic environment. The same type of transition, though much narrower, also appears on midlatitude mountains around an elevation of 3000 meters, where the forest grades into the grassy alpine meadows. Because they lie beyond the limits of most agriculture and urban development, boreal forests are the least disturbed of the major forest biomes in the world today.

Transition from the boreal forest to tundra biome marked by dwarf trees in scattered patches.

Perhaps the most exceptional midlatitude forests are the **west-coast forests** of North America. Taller than tropical rainforests, these giant forests stretch along the Pacific coast from California to Alaska. Here, high humidity, abundant precipitation, and a moderate temperature regime, that precludes a long dormant season, combine to create an environment that nourishes trees that can exceed heights of 100 meters (330 feet). In locations such as the west coast of British Columbia where rainfall is exceptionally heavy (typically more than 300 centimeters (120 inches) per year), these forests commonly harbor dense growths of ground mosses, epiphytes, and related plants. Such forests are often called *midlatitude rainforests*, and although the largest areas of them are found in the marine west-coast climatic zone of North America, smaller areas can also be found in the comparable climatic zone of South America, New Zealand, and Australia.

West-coast forests yield some of the best lumber-bearing trees on the planet. These include the redwoods of California and Douglas fir, Sitka spruce, and hemlock along the coasts and lower mountain slopes in areas farther north. Needless to say, these marvelous trees are highly attractive to the lumber industry and, since the late 1800s, vast areas have been cut for the North American and Asian housing markets. But as with the clearing of tropical rainforests, cutting of midlatitude rainforests has become a source of serious conflict with environmentalists.

Equatorward from the west-coast forests the summers are longer and drier and the trees become much shorter and more widely spaced (Figure 10.33). This is the **sclerophyll forest**, a broadleaf woodland found in the areas of Mediterranean climate in both hemispheres. The sclerophyll forest canopy generally covers only 25 to 60 percent of the terrain, giving it more the character of a savanna than a forest. Productivities are the lowest among the world's forests, similar to those of the tropical savanna.

The sparseness and low productivity of the sclerophyll forest is attributable to two factors. The first is the severe moisture stress of the warm, nearly rainless Mediterranean summer, which has limited the sclerophyll forest to a light cover of drought-resistant trees with shrubs and grasses. The trees include many types of oaks, notably live oaks, cork oak, and white oak, several species of pines, and numerous shrub-sized trees, such as wild lilac and olive. The second factor is the long tenure of human settlement in the Mediterranean regions, especially in the Old World. In areas such as Greece, Italy, and coastal Turkey, where more luxuriant forests once stood, fires, agriculture, grazing, and wood harvesting long ago destroyed them, and the cover we see today is a poor reflection of the past. Much of the landscape is left with a dense shrubby vegetation, which in Europe is called the *maquis* (or *macchia* or *garique*). In California, a roughly comparable vegetation consists of dwarf forests of oaks mixed with shrubs and is called *chaparral*.

Savanna Biome: Between the forests and the open grasslands is a biome of grasses, forbs (broadleaf herbs), and shrubs, mixed with a light cover of trees and shrubs. This is the **savanna**, and although strictly speaking the word *savanna* refers to a tropical vegetation, a similar vegetation formation, called *parkland*, can be found in the midlatitudes where forest has been cleared for farm fields (see Figure 10.31). True savanna biomes or **tropical savannas**, however, are limited to the tropical wet–dry climate of regions of Africa, South America, and Asia (Figure 10.24). They are dominated by a cover of grasses which flourish in the summer rainy season. The dominance of savanna grasses rather than trees is a response to the harsh drought conditions of the winter season, which limit the establishment of a forest cover. But there are likely several other contributing factors as well, in particular fires set by humans and forest clearing for agriculture.

We have come to stereotype this biome by the **tall-grass savanna** of Africa. Here, extensive areas of grass such as elephant grass, which may reach heights of 3 meters (10 feet), are interrupted by scattered individual trees or groves of trees. Many of the trees are umbrella-shaped and, like the grasses, tend to flourish in the wet season, which generally lasts for three to four months a year. Many factors appear to influence the distribution of trees including variations in topography, soils, depth to groundwater, and even the location of termite mounds in some places.

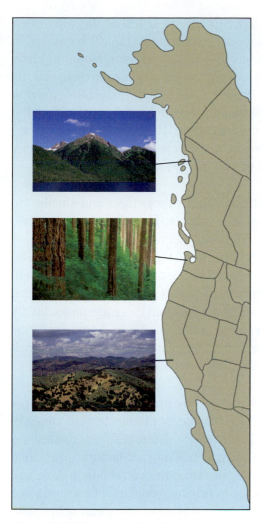

Figure 10.33 Three examples of forests on the west coast of North America. Forest stature, density, and productivity decline southward in response to declining precipitation.

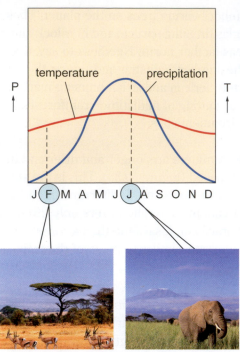

Figure 10.34 A graph of monthly temperature and precipitation representative of savanna landscapes with photographs showing the contrasts between winter (brown) and summer (green) conditions.

The grasses provide excellent pasture for grazing animals, such as the wildebeest and zebra, which are found in great herds across the African savanna. The grazers, in turn, are prey to some of Earth's most active predators, notably lions, leopards, cheetahs, hyenas, and wild dogs. During the dry season the grasses turn brown, and the trees may lose part or all of their foliage (Figure 10.34). As forage declines, grazing animals retreat to wooded areas, waterholes, and diverse rocky knolls to wait out the dry season or migrate to grasslands some distance away. This is also the time when herders may burn large tracts of the savanna in an effort to improve grazing quality and to enlarge the grasslands at the expense of nearby woodland.

In many regions with a wet–dry tropical climate, short, thorny trees and shrubs are found in place of the classical savanna trees and grasses. **Thornbush and scrub savanna** may form a nearly continuous cover, thereby eliminating most grasses, or it may be broken, allowing grasses and other herbs to fill the intervening space. The most impoverished thornbush formations are those where only barren soil is present between the woody plants. Generally, the thornbush and scrub savannas are thought to be responses to longer and more intensive dry seasons, but there are undoubtedly many other contributing factors in various regions, including fire and grazing. Many different regional names are given to this formation; for example, in South America it is called the *pontanal* and *chaco*, and in South Africa it is called the *dornveld*.

Grassland Biomes: Grassland biomes are dominated by grasses of various heights and cover densities. Most grasslands are found in the semiarid climatic zones of the subtropics and midlatitudes (Figure 10.24). In the midlatitudes they also extend well into humid climatic zones; hence, we must conclude that factors other than limited water contribute to the distribution of grassland. In central North America, for example, fire was apparently a major factor in driving woodlands eastwards and allowing the prairie to expand into the humid Midwest prior to European settlement.

The great expanse of prairie grassland (now farmland) interrupted by a corridor of trees following a stream valley.

Trees and shrubs are scarce but not absent in grassland biomes. Trees usually occupy slightly wetter habitats, such as stream valleys, where they form long ribbons trailing through the landscape. Otherwise, the landscape is an endless sea of grass whose color alternates between green and brown from winter and spring to summer and fall.

All major areas of grasslands in the subtropics and midlatitudes border deserts. In the subtropics they are transitional between desert and savanna biomes, whereas in the midlatitudes, they are transitional between desert and forest biomes. Because of this position, grasslands are prone to major geographic, or system, shifts when neighboring deserts expand and contract with large-scale climatic change. Paleobotanists tell us that in the past 30 million years or so Earth's grasslands (including the savannas) have generally been expanding in response to a global cooling and drying trend. This trend has been especially pronounced during the past 2 to 3 million years, the period which produced the current ice age. Most of this expansion has been at the expense of tropical and temperate forests (see Figure 10.28). In the past century or so, however, grasslands have been shrinking as a result of desertification, that is, desert expansion with the damaging effects of land use and drought.

Grassland biomes are distinguished mainly on the basis of grass heights and densities. **Tall-grass prairie** is found on the humid side of the grasslands where the annual moisture budget is moderately positive or about evenly balanced between

Figure 10.35 (a) A remnant patch of original prairie as it looked to the Native Americans and the early settlers. (b) Cultivation of the prairie. Along with cattle grazing, the tractor and plow have eradicated most of the natural prairie.

Figure 10.36 Harsh deserts are not only exceptionally dry and nearly plantless over vast areas but they are prone to erosion by wind and runoff as these gullies reveal.

a winter surplus and a summer deficit. In North America, grasses such as big blue stem, which can reach heights of 2 meters or more, dominated the tall grass prairie in the Midwest before settlement and farming swept over the area. By 1950, virtually all the original tall-grass prairie was replaced by cropland and ranches and today only a few remnant patches of natural prairie, like that in Figure 10.35, remain. With agriculture and settlement of the vast North American grasslands, the dominant native grazing animal, the bison, also declined, falling from a population of tens of millions in the mid 1800s to only a few hundred by 1900. Conversion to agriculture has also been common to other major prairies of the world such as the *Pampa* of Argentina and *Puszta* of Hungary.

From the tall-grass prairie, grass heights and densities decline toward the desert in response to declining moisture. The next major biome is the **short-grass prairie**, which is more widely known by the Russian word **steppe**, taken from the great grassland of the Ukraine and Russia. In North America, short-grass prairie covers most of the Great Plains from Texas to south–central Alberta and Saskatchewan. Common species include buffalo grass and black grama grass and these grasses tend to grow in bunches rather than in carpets like lawn grass. The Great Plains were converted to ranchland in the late 1800s. In the twentieth century irrigated cropland replaced most of the ranchland. Drawing on ample supplies of groundwater stored in shallow aquifers, crop farming continues to push westward across the plains.

Desert Biomes: Desert biomes rank among the lightest and least productive life systems on the planet. Nevertheless, they are places of relatively diverse plant and animal life most of which is related to variations in soil moisture associated with local differences in topography and microclimate. Contrasts are often striking in the desert landscape, as illustrated by the juxtaposition of oases or exotic stream valleys and lifeless sand dunes, where annual productivity may differ by a factor ranging between 5 and 90 grams per square meter per year over as little distance as 10 meters. And where the atmosphere (as opposed to groundwater or runoff) is the sole source of water, precipitation variability is a big factor, because the drier the climate, the lower the dependability of rainfall and the more severe are droughts.

In the harshest deserts, such as the Atacama of northern Chile and the central Sahara of northern Africa, mean annual precipitation is less than 10 centimeters. The landscape is virtually barren of plants. This is the **dry desert** biome. Only in select microenvironments are a few plants able to survive, but they are very small and isolated. Moreover, their survival is dependent on either special physiological traits, such as the capacity for water storage, or on exceptionally deep root systems to draw on sources of water as much as 30 meters or more underground. But even with special adaptations, plant growth and reproduction in the dry desert are exceptionally slow, generally less than 25 grams per square meter annually.

In addition to severe drought, erosional processes can also restrict plant growth in dry deserts. Desert surfaces are typically very active, owing to the absence of a plant cover strong enough to hold the soil in place. Both runoff and wind work and rework the surface. Running water is surprisingly effective and, during infrequent periods of runoff, can move massive amounts of material, including the plants themselves, as the photography in Figure 10.36 suggests. In many parts of the world deserts are expanding with the decline of grasslands related to the desertification process.

(a)

(b)

Figure 10.37 (a) Shrub deserts are common to much of the dry lands of North America. (b) They contrast sharply with the landscapes of the harsher dry deserts such as the Sahara.

At the other extreme are **shrub deserts**, which have considerably heavier plant covers and productivities between 40 and 90 grams per meter annually. Shrub deserts are characterized by diverse plant forms ranging in some places from saguaro cactus of the Sonoran Desert, which reaches heights of 5 to 10 meters, and various shrubs at heights of 1 to 2 meters, to tiny forbs barely 3 to 4 centimeters above the ground. As Figure 10.37 reveals, together they may cover 10 to 20 percent of the ground, and in special locales, such as along dry river beds, where moisture is more plentiful, the coverage may be substantially greater. Although xerophytic vegetation represents one of the most interesting collections of plants in the world, the shrub desert biome is not highly diversified floristically. Compared with forest formations at the same latitude, there are far fewer types of plants in most deserts. Cacti are probably the most famous desert plant, and over time some varieties have increased their geographic range. The prickly pear cactus – once exclusive to North America – is now found in Old World deserts, a result of human introduction.

Tundra Biome: The **tundra biome** is found beyond the thermal limits of tree growth in high latitudes and high mountain zones where the mean annual temperature is well below 0°C. Like the desert biome, the vegetation is generally small in stature, irregularly distributed geographically, and limited to very low growth rates. In addition, frozen ground is a constraint to plants. Most of the tundra biome is underlain by permafrost (which is discussed later in Chapter 23) and much of the land resting on the permafrost layer is poorly drained.

The **Arctic tundra** is a treeless, prairie-like landscape (Figure 10.38a). Here low temperature rather than scant moisture prohibits establishment of a forest cover. Winters are fiercely cold, and summers are short (one to two months) and cool. Short grasses, such as cottongrass and arctic meadow grass and forbs are abundant and productivity, overall, is low, less than 100 grams per meter yearly. In sites protected from the harshest weather and ground conditions, shrubs and dwarf trees can sometimes be found, but they are very slow growing. Because of the geographic diversity and the dynamic character of Arctic environments, grassy tundra tends to have localized distributions, particularly in areas of diverse terrain.

In mountain terrain where elevations exceed the limits of the tree growth is a tundra-like biome called the **alpine meadow**. Mean annual temperatures fall below 0°C, winters are long and fierce and summers short and cool. Vegetation is very similar to the tundra with abundant grasses, forbs, and scattered shrubs and dwarf trees. The main difference between alpine meadow and tundra is geographic distribution. Alpine meadows can be found anywhere there are high mountains, even in the tropics although the altitude where it is found changes with latitude. In addition, the distribution of alpine meadow tends to be patchy, often corresponding to isolated mountain peaks (Figure 10.38b).

Figure 10.38 Biomes of very cold lands. (a) Arctic tundra; (b) alpine meadows.

(a)

(b)

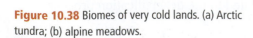

Figure 10.39 Aerial view of an exotic stream, the Nile, marked by a distinctive corridor of green coursing through the desert.

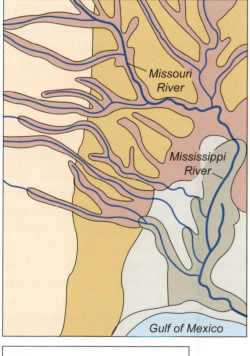

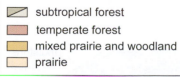

subtropical forest
temperate forest
mixed prairie and woodland
prairie

Figure 10.40 Map showing forest corridors along the Mississippi River and its major tributaries. Notice how the river corridors extend the vegetation of the bioclimatic zones crossed by the river.

Beyond the tundra and alpine meadows are areas with a light, irregular cover of lichens, mosses, and flowering plants that resemble desert formations. These areas are **fell fields**, where the surface is dominated by ice, snow, and rock fragments. Coupled with the harsh polar or high mountain climate, fell is limited to small plants with highly irregular distributions and less than 25 percent ground coverage. Antarctica and Greenland, as well as portions of Alaska, Siberia, and insular Canada are occupied by fell-type landscapes.

Summary on Biome Composition and Distribution: Biomes are large-scale assemblages of plants and animals that provide a convenient framework for describing Earth's biota. They are grouped into five climate-based classes ranging from forest biomes, the most diverse and productive, to desert, tundra, and fell biomes, the least diverse and productive. Since we live in a world of changing systems, particularly climate, hydrology, and land use, we can expect to see changes in biomes, and that is indeed the case, for among other things the desert biomes are expanding and the forest biomes are shrinking over much of the world.

10.8 Vegetation Distributions Related to River and Mountain Systems

The major biomes form broad belts across the continents corresponding to global climatic zones. Within these belts are smaller-scale, or second-order, patterns corresponding to other factors or controls. High on the list of other factors are stream systems and mountain systems.

All continents except Antarctica are drained by several or more great rivers. Great rivers such as the Nile, Ganges, Mississippi, and Amazon are fed by huge watersheds that reach thousands of kilometers into the continents, often cutting across several climatic zones. Within each watershed is a system of streams, which feeds water into a network of trunk channels. The trunk channels form large valley corridors linking the continental interiors to the sea. These valley corridors and the streams within them govern the distribution of the resources such as water and soil, upon which plants and animals depend. In other words, they shape habitat and, in turn, exert a major control on the distribution of biota.

The contrast between the habitat of river corridors and that of the larger climatic zone within which the corridors are situated can be striking. This is best illustrated where streams cross dry climatic zones. In even the harshest deserts, rich ribbons of forest or wetland often inhabit river corridors, stretching for hundreds of miles through the bleak desert landscape. Since the climate within the corridors is basically no different than the desert climate around them, we must of course conclude that river water is the factor enabling the relatively dense biota to thrive there (Figure 10.39).

Beyond the exotic streams of the desert, there are lots of other, though less extreme, examples illustrating the influence of stream corridors on the distribution of biota. The Mississippi River system is a case in point. It crosses several bioclimatic zones beginning with arid and semiarid grasslands in the upper watershed and ending with subtropical forests in the lower watershed. In each zone, the system of stream corridors enables the biota of a lower zone to extend upstream into a neighboring, more limiting, zone. In other words, the stream corridors stretch the distribution of a more permissive habitat beyond the limits of the regional bioclimate. Subtropical forest is extended into the temperate-forest zone, and temperate forest is extended into the semiarid grassland zone as shown in Figure 10.40.

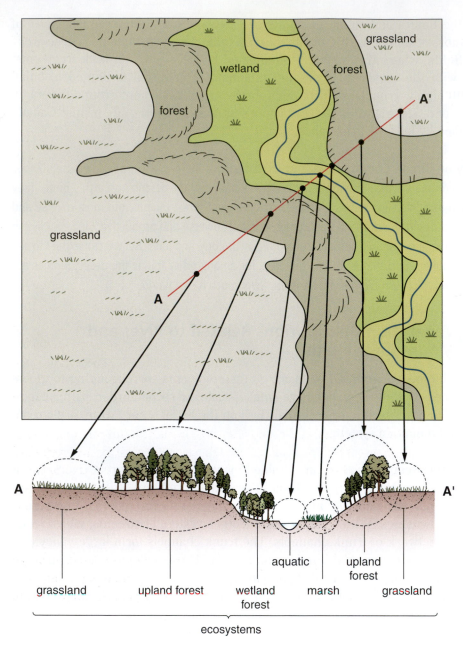

Figure 10.41 A cross-section of a stream valley showing the changes in ecosystems from the stream channel to the uplands beyond the valley.

Mountain systems have the opposite effect. Their high elevations induce cooler temperatures overall and thus less permissive habitat conditions. As a result, colder bioclimatic zones tend to be depressed equatorward along mountain corridors. This is most apparent where mountain systems run north and south as they do in North and South America. The effect is found even in low mountains such as the Appalachians where northern mixed forests extend southward along the mountain chain into the zone of subtropical forests.

If we narrow our focus to individual stream valleys and individual mountains, we will discover another (third) scale of biogeographic distribution. In most river valleys, three or four types of habitat can be defined, each supporting its own ecosystem. The stream channel itself supports an aquatic ecosystem comprised of benthic (bottom) and related communities; the stream banks support a riparian (channel-edge) ecosystem; the valley floor, a flood-plain (wetland) ecosystem; and the valley walls an upland ecosystem. We use ecosystems instead of biomes because within an individual stream valley climate is no longer the controlling factor. Rather, stream-valley habitats are products of many stream-related factors including the availability of water, flooding, channel erosional, and sediment deposition (Figure 10.41).

In mountains, on the other hand, climate changes with elevations and biomes are the appropriate descriptive unit. Significant change in biota is common over short vertical distances, often equivalent to three or four latitudinal zones of bioclimatic change at the global scale. In western North America, for example, the bases of mountains lie around 1500 meters (5000 feet) elevation and the principal biome is short-grass prairie (Figure 10.42). At an elevation between 2000 and 2500 meters, prairie grass gives way to woodland

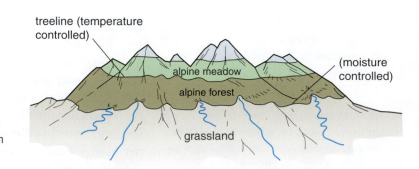

Figure 10.42 Change in biomes with elevation on a mountain, beginning with grassland in the semiarid zone at the mountain base and changing upward with declining temperature and increasing moisture.

or alpine forest. The change from grass to trees is related to increased soil moisture brought on by cooler temperatures (and hence lower evaporative losses) and somewhat heavier precipitation at higher elevations. Around 3000 meters (10,000 feet) elevation, trees in turn give way to grasses. This boundary is called the *treeline* and the grassy zone above is *alpine meadow*. At this elevation moisture is plentiful; thus, the change from forest to meadow is a response to cooler temperatures rather than aridity. The growing season is too short and cool for trees to reach much above dwarf size or to survive at all. Above the alpine meadow, it is too cold for all but small hardy plants and a few seasonal animals.

10.9 Disturbance and the Nature of Change in Terrestrial Biota

In our discussions of the geographic distributions of biomes and ecosystems we have not given much attention to the mechanisms of distributional change. We know that climate, fires, land use, and other controlling factors shift about and these shifts alter the distribution of plants and animals, sometimes wiping out huge areas of biosphere. We know that there have been many episodes of such changes or disturbances in the past, and although most were small in scale, global-scale changes have occurred many times as well. Sixty-five million years ago, it appears that a huge blast from an asteroid collision eradicated most of the organisms on Earth. And only 18,000 years ago, glacial ice covered half of North America and most of Eurasia, world climatic zones were compressed equatorward, tropical rainforests shrank in area, and temperate forests shifted into locations where subtropical forests are found today (see Figure 8.4). In the vast area occupied by the ice, the entire biosphere was more or less wiped out. When the glaciers receded and climatic conditions improved, biomes slowly realigned themselves and plants and animals reinhabited the glaciated landscape.

Figure 10.43 Diagram showing the relationship between the duration of a disturbance event and the geographic area affected, from lightning strikes to continental glaciation.

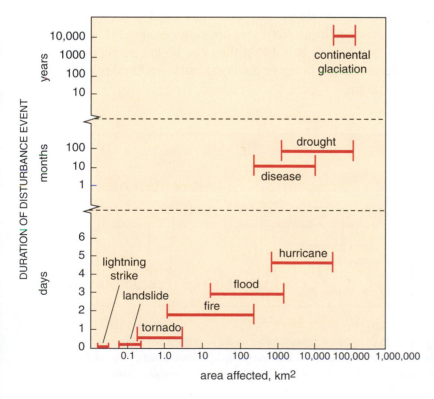

area affected, km2

Disturbances may be instantaneous or thousands of years in duration. Short-duration events such as lightening strikes affect smaller areas and longer-duration events such as droughts generally affect larger areas (Figure 10.43). In addition, the effect of a disturbance on the biosphere can range from near wholesale destruction to selective damage. It is exceedingly rare, perhaps impossible, to destroy all living matter in a disturbance area. Even under great ice masses of the continental glaciers, for example, it appears that some microorganisms, seeds, spores, and other organic matter can survive and remain viable.

An important question investigated by biogeographers is how biota re-establish themselves in an area that has been denuded by a disturbance. A number of concepts or models have been proposed to explain this process and we look at three of them here: the community-succession concept, the individualistic concept, and disturbance theory.

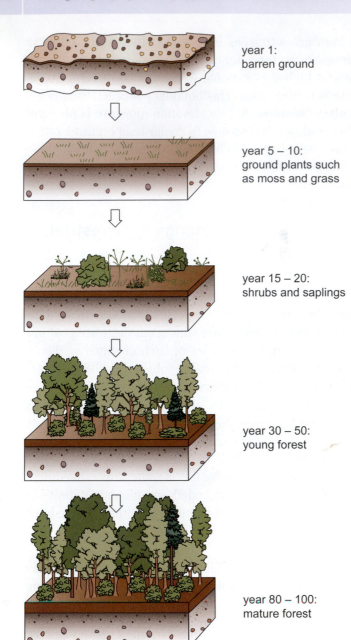

year 1:
barren ground

year 5 – 10:
ground plants such
as moss and grass

year 15 – 20:
shrubs and saplings

year 30 – 50:
young forest

year 80 – 100:
mature forest

Figure 10.44 Diagram illustrating the community-succession concept in which one community of biota follows another until a climax community is reached.

Community-Succession Concept: This concept is based on the notion that organisms reinhabit disturbance sites in successive waves, beginning with selected hardy ground plants, such as the mosses and grasses shown in Figure 10.44, which are called *pioneers*. These organisms help stabilize the site and make it more suitable for other plants and animals. Once this has taken place, the pioneers are replaced, or succeeded, by a second group of organisms which in turn renders further change in the environment. Each group to inhabit the site represents a *community* because it is composed of plants and animals that live together in an ecologically interdependent fashion.

One community succeeds another until eventually one of them, such the forest in the bottom diagram in Figure 10.44, achieves stability, that is, it is able to inhabit the environment on a relatively permanent basis. This group is called the *climax community*, and its presence indicates that a state of equilibrium has been reached in the biota–soil–atmosphere system. In other words, the climax community represents a steady state between a community of organisms and the physical environment in which the community itself functions as a super-organism. Should some outside force change the climate or soil, the balance is interrupted, and a different climax community will probably evolve. Should the climax community be destroyed altogether, succession will begin anew and continue until a climax community is re-established.

Disturbance Theory: According to disturbance theory, the distribution of plants and animals fluctuates in response to the magnitude and frequency of the forces that operate in and around their ranges. We know that every organism has a relatively fixed tolerance to stresses and disturbance such as drought, disease, and fire. When the level of disturbance exceeds the tolerance threshold of an organism, the members living in the affected area are either eradicated or so severely damaged (see Figure 10.45) that they can no longer reproduce. As a result, dieback might occur or the organism loses its edge and gives up part of its range.

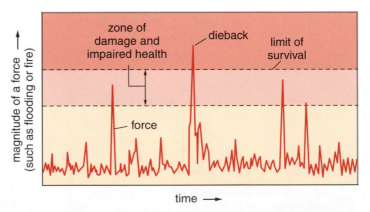

Figure 10.45 Disturbance theory is based on the magnitude and frequency principle in which infrequent but large events such as floods exact change in the distribution and makeup of biota, reversing infilling commonly ascribed to succession.

But events powerful enough to cause significant setback in a community usually occur so infrequently that, in the intervening years, plants and animals may begin to reinhabit the area that was lost in the manner described by the succession concept. According to the disturbance theory, while this is taking place, many smaller, but yet substantial, events are likely to occur that may delay or set back the recovery (or succession) process. In short, if we could compile a time-compressed film of an organism's or a community's distribution, it would appear to be in a state of continuous flux, sometimes expanding and sometimes contracting in response to the rises and falls in the magnitude and geographic coverage of certain forces in the environment. Among those forces are volcanic explosions.

The Mt. St. Helens Story: On May 18, 1980, Mt. St. Helens, a volcano located near the Washington–Oregon border, erupted. The eruption blew out the upper, north flank of the mountain, ripping away 400 meters (1300 feet) of the summit and sending gas, ash, and steam over an area 30 kilometers wide and 20 kilometers long. So powerful was the blast that in the 10 kilometer zone nearest the volcano, no trees were left standing and the entire landscape was buried under thick ash deposits. A new, desolate landscape, shown in Figure 10.46, was created devoid of plants, animals, soils, and streams. Beyond the inner zone damage was less severe but yet considerable. Trees were mowed down and the land was covered with an irregular layer of ash.

Figure 10.46 The ash deposited in the Mt. St. Helens eruption completely buried the forested landscape and created a new playing field for plants and animals.

Within weeks of the eruption, scientists set about monitoring the site, giving particular attention to the processes and patterns by which biota re-established themselves in the denuded land. The results have been instructive. First, there has been no orderly sequence of reinhabitation along the lines of those described in the community-succession concept. Instead, a variety of plants and animals in different combinations have colonized different parts of the site, with trees functioning as pioneers in some places, and small plants like mosses in others.

Second, the types of organisms and the places where they have become established are often related to residuals from the pre-eruption landscape. *Residuals* are, for example, remnant plant populations, seeds, wet spots, and organic matter from dead trees. In other words, the old landscape was not blown completely out of the picture, but is making significant contributions to the rebuilding process. Thus some species have the advantage over others by virtue of the fact they are in a position to get there first or they are favored by site conditions. Finally, it is important to add that the process of reclaiming the 600 square kilometers of denuded landscape is far from complete. Indeed, this is only the first phase of a process that will last centuries and is likely to be disturbed by additional eruptions of different magnitudes and frequencies along the way.

Chapter Summary and Overview of Biogeography:
Geographic Distribution of Plant and Animal Types

Where plants and animals are found on the planet and how they got there are the main questions of biogeography. The search for answers is interesting and challenging and requires understanding not only the environmental opportunities and constraints faced by modern biota, as related to climate and other geographical factors, but also understanding past conditions such as the closing and opening of migration barriers related to plate tectonics and the occurrence of large disturbances like the continental glaciation of North America and Eurasia. The picture we get is one of a changing game board in which plants and animals are continually adjusting their distributions and associations in response to a changing environment at all scales from the local to the global.

▶ **Plants can be classed into two broad groups based on size.** Macrophytes dominate terrestrial vegetation and are represented by three main groups of plants.

▶ **Floristic associations are mixes of plant species that make up vegetation.** At the global scale, climate is the primary control on species numbers and mixes in associations.

▶ **Habits are adaptive means for dealing with obstacles of the environment.** In drought-tolerant plants habits include ephemeralism, succulence, and phreaticism.

▶ **Tolerance is the amount of stress a plant can withstand without damage or death.** Plants with wide tolerance ranges are more geographically versatile than those with narrow ranges.

▶ **To survive and reproduce, a plant must successfully complete all phases of its life cycle.** Failure to do so can eliminate it from part or all its geographic range.

▶ **Animal species are far more abundant than plant species.** But animal biomass is small compared to plant biomass.

▶ **Ectothermic animals gain heat from the external environment.** This trait limits them to warm or seasonally warm climates. Endothermic animals have a geographic advantage.

▶ **There are many geographic differences and similarities in the types and abundance of animals on Earth.** Nine families of terrestrial mammals have spread to all continents except Antarctica and Australia.

▶ **There are three main barriers to the spread of animals.** Until about three million years ago, North and South America were separated by an ocean barrier.

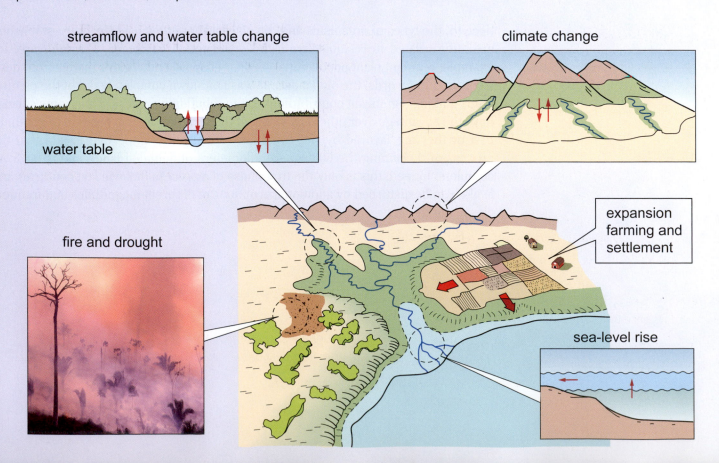

streamflow and water table change

water table

climate change

fire and drought

expansion farming and settlement

sea-level rise

► **Biome is the term used to describe global assemblages of plants and animals.** The distribution of biomes corresponds broadly to the distribution of climate and soils.

► **Five primary classes of biomes are generally recognized.** Forest biomes are found in the least limiting climatic zones.

► **Tropical rain forests are the richest and most productive forest biomes.** Rainforests contain four life zones, each with a different assemblage of plant and animal life.

► **Midlatitude or temperate forests are made up of five principal biomes.** With the exception of the boreal forests, temperate forests have been severely altered by land use.

► **Savanna biomes are found in the tropical wet–dry climatic zone.** There are two main varieties of savanna biomes and one is noted for abundant grazing animals and associated predators.

► **Grasslands are found mainly in the semiarid climatic zones.** Trees and shrubs are scarce and occupy slightly wetter habitats such as stream valleys.

► **Desert biomes rank among the least productive biomes.** Shrub deserts have heavier plant covers and are more productive than dry desert biomes.

► **The tundra biome is found beyond the limits of tree growth.** Beyond the tundra and alpine meadow biomes is fell, which is comparable to dry deserts in productivity and flora diversity.

► **Within the major biomes are secondary distribution patterns.** These occur at small scales and are related to geographic settings like river corridors and mountain ranges.

► **Disturbances can cause major changes in the makeup and distribution of biota.** Disturbances come in all sizes from local to global and all frequencies from days to millions of years.

► **Three major concepts help explain how geographic change takes place in biota.** Disturbance theory argues that the distribution of biota fluctuates in response to the magnitude and frequency of disturbance events.

► **The Mt. St. Helens eruption was major disturbance event.** The patterns of recovery of the disturbance area by plants and animals do not follow the succession-community concept.

Review Questions

1 In terms of geographic coverage, what is the dominant class of vegetation in the world? What are macrophytes and what distinguishes them from microphytes?

2 Of three major groups of vascular plants, which is presently dominant on the planet, how long have they been around, and what is the long-term trend of this group in terms of global geographic coverage?

3 What is meant by the term *floristic association* and what are the major factors that control the makeup of floristic associations?

4 What is meant by the term plant habit? Name several habits, what is accomplished by each, and what sort of environment each is associated with. Is there any advantage to the evergreen habit?

5 Distinguish between plants with narrow and wide tolerance ranges. Is tolerance the same for all phases of a plant's life cycle?

6 Identify an environmental event which is capable of changing the geographic distribution of plants at the global scale. Identify another event which can change plant distributions at a regional scale and another which commonly works at a local scale.

7 What are some of the major environmental impacts of animals? Are their impacts similar, or different to those made by humans? Explain. What factors make soil a key determinant in the principal location for animal impacts?

8 How has plate tectonics influenced the global distribution of plants and animals? What barriers have helped to determine the similarity factors between the continents?

9 Identify the specific associations between the five major biomes (forest, savanna, grassland, desert, and tundra) and climate. How does the presence of water affect their species diversity, and the habits of specific organisms which inhabit them?

10 If there was a poleward or equatorward shift in a biome, what event(s) could possibly have initiated such a geographic change? What landscape features can also induce a biome to shift location?

11 Does the case of Mt. St. Helens invalidate the theory of community-succession? Explain.

Humans as Geographic Agents in a Changing World

Chapter Overview

The Greek playwright Euripides wrote that a bad beginning makes for a bad ending. Although humans had a tough beginning, after thousands of years better times eventually emerged as people spread over much of the world and adapted to different landscapes. But with technological advances and the pressure of growing population, adaptation changed to more and more serious forms of environmental manipulation. What began with ancient peoples as a cooperative relationship with their geographic environment has now evolved into one characterized more by a desire for control. Could the human drama on Earth have a bad ending? This chapter opens with an interesting geographic puzzle – the African origins of *Homo sapiens* and our early migration to Asia, Europe, Australia, and the Americas. It then goes on to consider early agriculture, the rise of population centers, the Industrial Revolution, and their effects on the geographic environment. Measured by the last few centuries, the outcome is a rapidly growing imbalance between human systems and natural systems. The chapter ends by offering a big-picture perspective on human activity on Earth in light of the fact that ours is a planet characterized more by geographic change than by stability.

Introduction

Life has profoundly influenced Earth's geographic character. Of that there is absolutely no doubt among geographers. During its evolution life has directly or indirectly caused many big changes in Earth systems, which have in turn led to major transformations in the biophysical and geographic character of the planet. When the first plants advanced onto land around 400 million years ago they entered a barren, desolate landscape riddled with erosion. As they spread and evolved into larger, rooted forms, their influence on landscape systems took on dramatic proportions. Rainfall no longer battered unprotected soil and runoff and its erosive power declined sharply. More water entered the soil where much of it was taken up by plants and returned to the atmosphere via transpiration. With these changes, the Earth's larger hydrologic, soil, and climatic systems also changed, ultimately transforming the entire global environment.

About 200,000 years ago another Earth-changing biogeographical event began with the emergence of our own species in the heart of Africa. For our first 150,000 years or so, our ancient ancestors were limited to Africa where they eked out a living as minor hunter–gatherers on the fringe of the savanna landscapes and developed tool-making skills and survival strategies to escape large predators. Then sometime between 50,000–80,000 years ago, modern humans moved from Africa into southwest Asia. To do this they had to overcome a huge barrier – the harsh desert climate of North Africa, the region we know as the Sahara.

Over the next 40,000 years, people migrated from southwest Asia eastward into southeast Asia, Australia, east Asia, and westward into Europe. About 15,000 years ago they moved from Asia into North America and rapidly spread across the continent and into Central America and South America. Wherever humans lived their populations were very light and impacts on the landscape were slight. However, about 12,000 years ago the seeds of a geographic revolution were sown with the invention of agriculture.

At its birth, agriculture was carried out with manual labor and rudimentary wood and stone tools. Farm populations were small and impacts on soil and water resources were probably modest. Today, by contrast, farmers throughout the world have the capacity to transform entire landscapes, to radically alter soil, water, and biotic systems to the extent that in some areas as the photographs in Figure 11.1 show, hardly a semblance of the original landscape's form and function remain. Through farming – and later with other land uses – human society has become a formative player in the global environment, a veritable great system onto itself, call it the *anthrosystem*, exercising a hand as powerful as nature's in many respects. Increasingly we have come to see ourselves apart from nature, functioning less as parts of natural systems and more as independent agents, choosing more to manipulate and control than participate in natural systems.

Figure 11.1 (a) A hilly, forested landscape in China before and after its transformation into farmland. (b) A forested landscape in the US Midwest as it may have looked in the 1800s and the way such landscapes look today under modern farming.

11.1 The Physical Geography of Early Human Existence

Homo sapiens is the most recent member of a group of bipedal species, known to scientists as the **hominins**, that emerged in Africa over the past 5 or 6 million years. Except for scanty and often incomplete skeletal remains and finds of the stone tools they manufactured, relatively little is known about the early hominin species. We do know that as many as eight to twelve species (or possibly more) existed before *Homo sapiens*, along with at least two species of hominins (called *Homo erectus* and *Homo neanderthalensis*), which co-existed on Earth for a while with *Homo sapiens*, and we also know that these human ancestors, like most other animal species, survived for periods ranging roughly from 200,000 to 1,000,000 years.

We also know that the earliest hominins evolved and lived in a diverse tropical and subtropical landscape in the highlands of East Africa, a region subject to repeated geographic change related to volcanic eruptions, changing climate, the rise and fall of large lakes and swamps, and the expansion and contraction of forests and grasslands (Figure 11.2). These environmental changes must have been instrumental in the survival and geographic distribution of these hominins. Could they also have been the force that drove human evolution? Or did the hominins, as Darwin hypothesized, evolve through natural selection in the absence of major changes in climate or physical environment? Scientists lean strongly toward the changing geographic environment idea, meaning that alterations in habitat, such as a dramatic decline in forest cover, appear to have driven our early relatives onto new landscapes, such as the grasslands, where they had to adapt socially, culturally, and ultimately biologically, in order to survive.

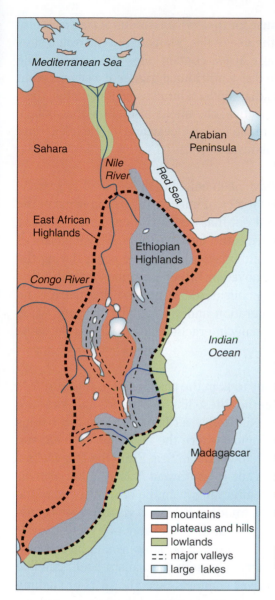

Figure 11.2 The East African Highlands form a broad belt of mountains, plateaus, and lake basins.

Figure 11.3 Primitive bands of hunter–gatherers in Africa and a few other places help us understand what the character and lives of early *Homo sapiens* may have been like.

How much impact these early species had on the African landscape is difficult to ascertain. First, their populations were small, probably extended family bands, and widely scattered. Second, they lived mainly by foraging, a nomadic land use which leaves little imprint on the land. And third, the lands they inhabited were subject to frequent surface alterations, such as volcanic eruptions, which tended to destroy and obscure most signs of habitation. So it is little wonder that the geography of the early hominins is a tough puzzle to unravel.

The picture archeologists piece together is that early humans foraged and scavenged for food in the savanna, forest margins, and along streams and lake shores. Their technologies were very simple and slow to change over time. Tools such as scrapers and hand axes were crafted from stone – there is no evidence that metals were used – and they probably built rudimentary shelters such as lean-tos. Fire may have been used as much as 750,000 years ago, and some scientists suggest that some early hominins may have burned vast areas of savanna and grassland to manage food sources and control predators, but no evidence of this survives. On the other hand, there is ample evidence of our own species using fire 100,000 years ago.

But there is more than scattered archeological remains to help us learn about early humans. We can get a firsthand clue how they lived from primitive hunter–gatherer societies that have survived in remote places, such as the deep rainforest of the Amazon, into the twentieth and twenty-first centuries. All these people live in small groups, typically between 20 and 100 people, with the group size being heavily dependent on local food supply (Figure 11.3). Population density is very light, less than 10 persons per square kilometer, and groups tend to be nomadic or semi-nomadic, shifting about with seasonal changes in the distribution of food and water supplies. The settlement patterns of ancient peoples were probably similar to those of contemporary hunter–gatherers in South America or Africa.

11.2 Leaving Africa: The Geographic Diffusion of Early *Homo sapiens*

The migration of modern humans out of Africa began between 50,000 and 80,000 years ago. Evidence from human fossils suggests the original exodus involved anywhere from 1000 to 50,000 people. Whether the African exodus occurred as one event or multiple events over a longer time period is still unknown. Two ideas have been advanced as the cause of the migration: new technology and climate change.

Archeological evidence in the form of new tools and other artifacts reveals that this was a time of cultural change that may have included increased migratory activity. On the other hand, another line of evidence points toward a sequence of climate changes as the driver behind the migration from Africa. It began with massive droughts that split the native *Homo sapiens* population into small, isolated groups and may even have threatened their extinction, and was followed by improved climatic conditions that enabled the survivors to reunite, multiply, and, in the end, emigrate to Asia. Improvements in technology may have helped some of them set out for new territory, but climate may have been the enabling factor, because the Sahara, which separated *Homo sapiens* from the rest of the world, was a less formidable barrier than it is today.

Figure 11.4 The coast of the Indian Ocean leading to the Red Sea may have been a route of early migration to southwest Asia.

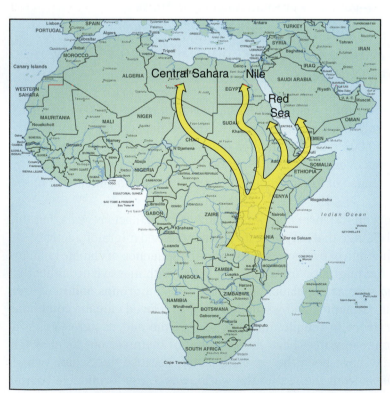

Figure 11.5 The most likely migration routes for early *Homo sapiens* from Africa to southwest Asia.

Even more uncertainty arises about the cause for the African exodus of *Homo sapiens* when we consider new food sources and the use of fire. Many of the excavated sites dating from the initial migrations in South Africa contain piles of seashells. The shells suggest that seafood may have served as a nutritional trigger at a crucial point in human history, providing the fatty acids that modern humans needed to fuel their growing brains. From a geographic perspective, the shells also show that people lived on the coast, and in Africa, the coasts of the Indian Ocean and the Red Sea trend north–south, offering migratory routes northward Figure 11.4.

This sudden increase in brainpower may have been linked to the onset of speech. As for fire, at the Klasies River in South Africa traces of burned vegetation suggest the ancient hunter–gatherers may have figured out that they could encourage quicker growth of edible roots and tubers by clearing the land with fire. These advances in technology, diet, communication skills, and a favorable climate occur during the period known as the dynamic period of innovation, which coincides with the first human migrations out of Africa. All things considered, it may be that there is no single definitive cause for the African emigration and, as so often happens when reconstructing such events, it appears that many interacting factors play a role, some of which no doubt are yet to be revealed.

The Path(s) to Asia: The routes early humans took in the passage from Africa to Asia are unknown, but scholars have pieced together these likely scenarios and some are shown in Figure 11.5. One candidate is the Nile Valley, a ribbon of green that begins in the highlands of East Africa, the very region where the hominins evolved, and flows northward across the desert to the Mediterranean Sea. From the Nile Delta a turn to the right and another 500 km brings one to southwest Asia. Another alternative has early humans migrating from the northern end of the East African Highlands (modern Ethiopia) via the shores of the Red Sea Basin, which at that time was about 50 meters lower than modern sea level. Another possible .route, also shown in Figure 11.5, was north through the Sahara itself. Satellite evidence reveals old river channels (see Figure 11.6) running north from the central Sahara (modern Libya) that may have provided a pathway to the Mediterranean.

By 45,000 years ago, or possibly earlier, humans had moved into south Asia (India) and beyond to Indonesia, Papua New Guinea and Australia (Figure 11.7). We entered Europe around 40,000 years ago, probably moving up the Danube corridor into eastern Europe, and along the Mediterranean coast into southern Europe. By 35,000 years ago, modern humans were firmly established in most of the Old World with populations adapted to a remarkable variety of geographic conditions, ranging from the harsh desert lands of Australia to the humid rainforests of southeast Asia and from the tropics of south Asia to the subarctic lands of central Europe and Asia.

In Europe and Asia, advancing humans ran into another group of hominins, known to us as the Neanderthals, who had preceded *Homo sapiens* in these areas by a hundred thousand years or more. Technologically primitive and probably lacking

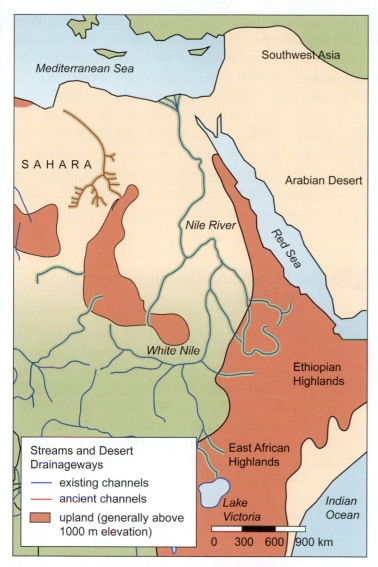

Figure 11.6 Satellite imagery reveals a possible route for early humans across the Sahara (west of the Nile) that may have followed ancient drainage ways produced during a time of wetter climate in this region.

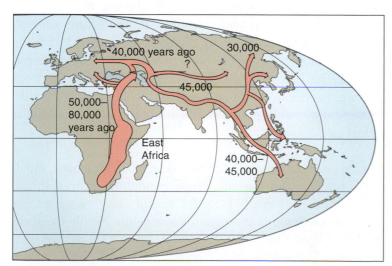

Figure 11.7 Old-world diffusion of humans from southwest Asia into central Asia, south Asia, and Europe.

basic language skills, Neanderthals had adapted to the midlatitude cold, primarily through a biological adaptation that gave them a stout physique. This body form, illustrated in Figure 11.8, gave them a greater ability than slender body forms to conserve body heat, because with rounder forms, there is less surface area per unit volume of body mass. Additional evidence, however, suggests that when the climate started fluctuating between warm and cold periods as the continental glaciers advanced and retreated, Neanderthals were at a disadvantage to modern humans in adapting to the changing climate.

This disadvantage may have resulted from mutations in their cellular-level DNA. Although Neanderthals adapted well to cold temperatures, they had trouble in warm weather because one consequence of these mutations was the production of extra heat in their cells. Whether it was an inferior adaptation to the climate changes, or simply falling victim to the better tools and weapons developed by modern humans, has not been determined. In any case, their numbers slowly declined as *Homo sapiens* numbers rose, and eventually the dwindling Neanderthals fell back into mountain refugia in Europe and southwest Asia (though some scientists argue that they interbred with *Homo sapiens*, and more or less faded away). By 25,000 years ago, they had become extinct.

Finally, around 15,000 years ago, humans crossed from Asia to northern North America and, by 12,000 years BP (before present), had spread into South America (Figure 11.9). Modern humans reached the Americas by walking over the Bering land bridge (see Fig 10.20), or possibly by boat or raft in the northern Pacific. Some of the clearest evidence of early humans in the New World is human DNA extracted from coprolites (fossilized human feces) found in Oregon and recently carbon-dated to 14,300 years ago.

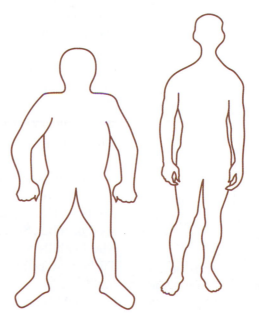

Figure 11.8 The body forms of *Homo sapiens* and Neanderthals. Among other things, Neanderthals were stouter, an advantage in cold-weather survival.

15,000 years ago (ya)
entry via
Bering land bridge

11,000–12,000 ya

12,000 ya

12,000 ya

11,000 ya

Figure 11.9 Geographic diffusion of modern humans from northeast Asia into North America and South America, a process that took about 3000 years.

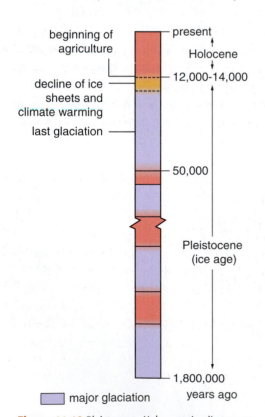

beginning of agriculture

present

Holocene

decline of ice sheets and climate warming

12,000–14,000

last glaciation

50,000

Pleistocene (ice age)

1,800,000 years ago

☐ major glaciation

Figure 11.10 Pleistocene–Holocene timeline covers a time span of about 2.6 million years and includes major shifts in global climate and major glaciations.

Summary on Early Human Existence and Geographic Diffusion: The landscape of early humans in East Africa was geographically diverse and subject to frequent change. Human bands were small and nomadic and left little in the landscape to mark their existence. Scientific evidence suggests that it was a combination of climate change and technical innovation that spurred humans to leave Africa. To cross the desert, early *Homo sapiens* probably followed stream valleys and shorelines toward southwest Asia. From there, they spread across Asia and into southeast Asia, Australia, Europe, and North America, overcoming geographical barriers, adapting to a variety of climates and landscapes, and probably out-competing our hominin cousins the Neanderthals.

11.3 The Origin and Development of Agriculture

We are children of the Ice Age, a period of global cooling and glaciation known as the Pleistocene Epoch, which began about 1.8 million years ago. The last glaciation, when ice sheets covered much of Eurasia and North America, ended 12,000 years ago with the beginning of the present epoch, called the Holocene Epoch. The time scale shown in Figure 11.10 highlights the time of the Pleistocene–Holocene transition, a critical time for humans and many other creatures. It was characterized by large and rapid environmental changes, specifically, reduced climate variability and increases in atmospheric CO_2, global temperature, and rainfall in many regions. The reduction of climatic variability meant that climate was becoming more dependable, and this was critical for the initiation of agriculture because, once the idea of plant domestication was in place, relatively stable conditions were needed, probably over several generations, before a simple system of farming could become embedded in local culture.

Agriculture was a radically new land-use system with serious geographic implications. These included landscape alterations that came with the use of techniques needed to manage water, soil, and the crop itself, including its protection, storage, and distribution. In addition, people's diets probably changed because certain foods like wheat became more abundant while others like nuts traditionally gathered from distant forests declined. But how did humans accomplish the transition from a nomadic hunter–gatherer existence to one of farmer tied to fields, crops, animals, and stationary settlements? And why did agriculture begin where it did in the Middle East? Let us look at this last question first.

During the period of climatic stabilization as the ice sheets were melting back, much of the Earth became subject to long dry seasons. These conditions favored annual plants, such as grasses, which die off in the long dry season, leaving a dormant seed or tuber. Southwest Asia, particularly the region know as the Fertile Crescent (the foothills of Mesopotamia, shown in Figure 11.11, in what is now part of Iraq, Turkey, Syria, and Jordan) was home to a great diversity of annual plants, including 32 of the world's 56 most nutritious grains. These grains grew in abundance and could be efficiently gathered and stored by hunter–gatherers without moving

Figure 11.11 Valleys like this one are representative of the geographic character of the foothills of the Fertile Crescent.

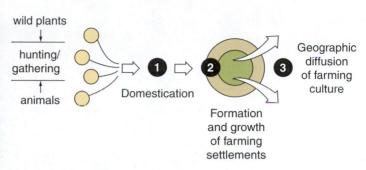

Figure 11.12 The three stages of the land-use transition from plant and animal domestication by hunter–gatherers to the geographic diffusion of farming.

around much, which probably led to a more sedentary life style and eventual domestication of selected grains. The first signs of farming emerged in this region at least 12,000 years ago.

On the first question, we can describe how the transition to farming took place, but not why it took place. We reason that it was accomplished in three major stages (Figure 11.12). Stage one occurred when humans took certain useful animals and plants under their management and exercised the opportunity and ability to *domesticate* them, that is, alter their behaviors, size, genetics, and distribution to better meet their needs. Stage two consisted first of the development of farming settlements and second, of large areas of interconnected settlements of common culture called **cultural hearths**. Settlements were a response to the new land-use system, which required staying in one place year around, and to the need to build facilities for storing harvested crops, moving water, controlling animals, and so on. Cultural hearths grew into major geographic population centers where people shared a common economic system, held similar beliefs, values, and language, and where ideas about new systems of governance and land use emerged. Stage three was marked by the **diffusion**, or spread, of agriculture from the original cultural hearths to other cultures in sometimes distant areas.

Figure 11.13 Two of the earliest domesticated grains, above, wild wheat (einkorn), and below, wild barley. These grains formed the backbone of early agriculture in the Middle East.

Domestication of Plants and Animals: Pre-agricultural humans were familiar with many animals and plants but, for a variety of reasons, relatively few became domesticated. Among what were probably many experiments with plants, two early successes, shown in Figure 11.13, appeared around 12,000 years ago, wild wheat (two varieties, called emmer and einkorn) and wild barley. Both grew in abundance in the Fertile Crescent and were ideally suited to the region's Mediterranean climate. In addition, they were easily harvested and stored and were a good food source by virtue of their relatively high protein contents. The next five plants to be domesticated in the Fertile Crescent (and in the world) also had high protein contents. Four of these plants were legumes (lentil, pea, chickpea, and bitter vetch), and the fifth was flax. Together, these eight plant species are called the **founder crops** and they formed the basis of systematic agriculture in the Middle East, North Africa, India, Persia, and later Europe.

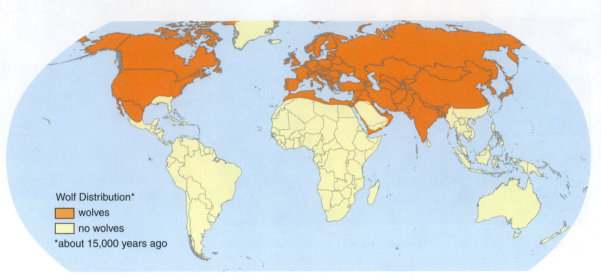

Wolf Distribution*
- wolves
- no wolves
*about 15,000 years ago

Figure 11.14 The original global distribution of the wolf, considered the first domesticated animal and the genetic ancestor of all dogs.

As for animals, we reason that people also selected animals for domestication based on certain criteria. These likely included their behavior, ease of breeding, eating habits, social structure, and growth rates. It turns out that only about 10 percent of the terrestrial mammals eligible for domestication passed the test. To illustrate how these traits apply consider the wolf, whose original distribution is shown in Figure 11.14.

The wolf is the direct ancestor of the dog and was the first animal domesticated (about 15,000 years ago) by Pleistocene hunter–gatherers. First, wolves were not aggressive towards humans nor would they panic in their presence. Second, they bred with each other in the presence of humans, and they grew rapidly from a diet regimen of primarily big game such as elk and moose that were accessible to hunters. Moreover, wolves adhered to organization in packs, and this social structure allowed humans to insert *themselves* as packleaders, thus gaining obedience from the rest of the pack. These traits help explain not only why dogs were the first domesticated animals, but why they soon proved to be invaluable as herders of other animals. When you compare the traits of the wolf with those of, say, a grizzly bear (very ferocious loners) or the antelope (panics when approached) it helps explain why only a few large animals became domesticated.

The majority of meat for human consumption was ultimately supplied by four herbivores: sheep, goats, pigs, and cows. Table 11.1 gives the approximate dates and locations of domestication of 11 large mammal species. All possessed the traits necessary for domestication. For instance, sheep, goats, and cows are non-aggressive pack animals, with easily managed diets of plants common to farmlands and the open range. It is interesting to note that the first five animals domesticated share a common geographic origin in the Fertile Crescent of southwest Asia, a land transitional between the mountains and the dry grasslands of Mesopotamia, which, not incidentally, also possessed a high natural diversity of mammal species.

Table 11.1 Dates and locations of large mammal domestications (from several sources)

Species	Date (years BP)	Location
Dog	15,000	Southwest Asia, China, North America
Sheep	10,000	Southwest Asia
Goat	10,000	Southwest Asia
Pig	10,000	Southwest Asia
Cow	8000	Southwest Asia, India
Horse	6000	Ukraine
Donkey	6000	Egypt
Water buffalo	6000	China
Llama/alpaca	5500	Andes
Bactrian camel	4500	Central Asia
Arabian camel	4500	Arabia

Development of Stable Settlements and Cultural Hearths: As farming expanded and farmers perfected their agricultural techniques, they began to produce food surpluses and, with more food, came change. Population grew, specialized workers and more advanced tools emerged, and villages expanded into towns and small cities. In short, the entire social structure of society underwent radical transformation as the need to administer and distribute the surplus food led to the development of government, distinct social classes, and institutional religion.

Although these changes first took place in Mesopotamia and led to the Sumerian civilization that is credited with the first cities, the same changes also took place in other locations, shown in Figure 11.15, with other cultural hearths. Among the many curious things about the cultural hearths is the fact that, taken as a whole, they represent a remarkable range of geographic settings, including deserts, rainforests, and mountain lands, and underscore once again the remarkable adaptive capacity of humans

Figure 11.15 The cultural hearths of the ancient world occupied a variety of environments, wet and dry, lowland and mountain.

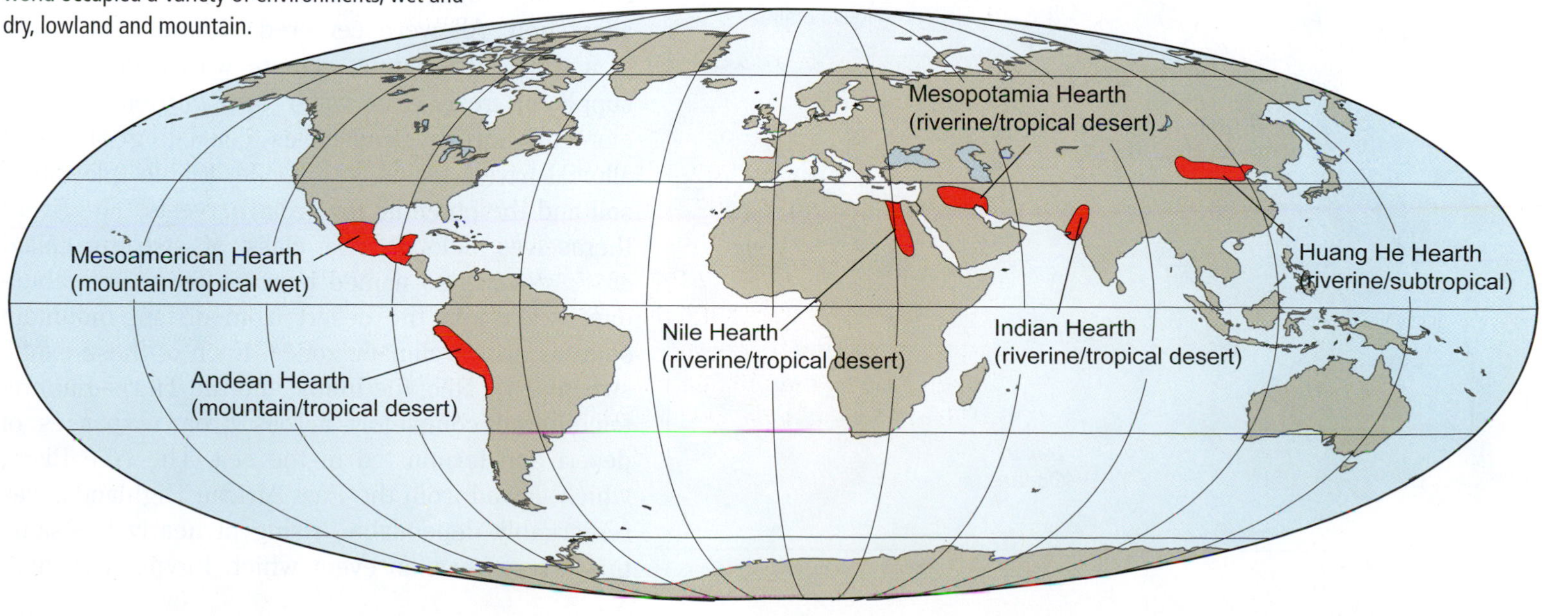

Mesopotamia Hearth (riverine/tropical desert)

Mesoamerican Hearth (mountain/tropical wet)

Huang He Hearth (riverine/subtropical)

Nile Hearth (riverine/tropical desert)

Indian Hearth (riverine/tropical desert)

Andean Hearth (mountain/tropical desert)

Table 11.2 Early cultural hearths

Location	Approximate date founded (years BP)	Climate /primary biome	Physiographic setting
Mesopotamia	6000–7000	Mediterranean / sclerophyll forest	Alluvial plain within a desert landscape
Nile Valley	5000	Desert /desert	Alluvial plain surrounded a desert
Indus/Ganges Valley	4800	Monsoon /deciduous forest	Alluvial plains with desert in the Indus Valley
Huang-He Valley	4500	Humid subtropical/ mixed deciduous and needleleaf forest	Upper reach: mountains with high plateaus; transitioning to an alluvial plain; Middle reach: loess plateau; Lower reach: coastal plain
Mesoamerica	2500	Highland, tropical savanna /broadleaf evergreen forest	Coastal lowlands (Mayans) and mountainous highlands (Olmecs, Toltecs, and Aztecs)
West Africa	2000	Subtropical steppe/ steppe	Hills and low plateaus
Andean America	1500	Highland /subtropical desert, inland tropical rainforest	Mountainous highlands, with transitional plains to coastal-desert lowlands

(Table 11.2). For instance, among those located in major river valleys, climate ranged from arid in the Nile and Indus to humid subtropical in the Huang He of China, while in the mountain hearths, climate ranged from semiarid in Mesoamerica (Mexico) to temperate in the Andes. For several, however, there are distinct geographic similarities that have captured the attention of historians, in particular the great river hearths in deserts of the Old World.

Figure 11.16 Three of the great hearths of ancient civilization located on exotic rivers: the Nile, the Tigris–Euphrates, and the Indus.

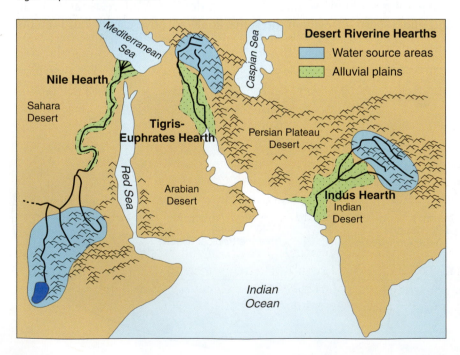

The great river hearths of Egypt, India, and Mesopotamia (see the map in Figure 11.16) share many common geographic traits and forms of human adaptation. All were centered around a big river in a warm, dry, sunny climate with a dependable supply of fresh river water and nutrients. These elements formed landscapes consisting of broad alluvial (river) plains with fertile, friable (plowable) soil and the potential for great harvests. The rivers themselves belong to a class of streams called *exotic streams*, so named because they carry abundant water into the desert from distant mountain sources or wet climatic zones. Each of these exotic streams, the Nile, the Indus, and the Tigris–Euphrates, flowed completely across broad expanses of desert and terminated in the sea. The Nile River, which flowed from the East African Highlands, was remarkably dependable, rising at nearly the same time every year, an event which Egyptian farmers took advantage of by building structures designed to capture its floodwaters and the rich silt they carried. Turn to Figure 10.39 for a satellite image of the lower Nile flowing through the desert.

The cultural hearths provided a new power base for the growing human population. One way this new power was manifested was through the increased manipulation of the landscape. Forests that lined the valley floors were cleared to make way for crop-land and access streams and, as shown in Figure 11.17, stream flows were diverted into canals and rerouted to feed irrigation systems in areas way beyond the natural reach of stream water. Soils were dramatically altered as erosion spiked on cleared land, and land was subjected to intensive grazing by sheep and goats; and the biogeography of plants was changed not only by humans carrying certain plants to new locations, but by modifying the plants' reproduction systems.

Prior to human intervention, for instance, wild wheat dispersed its seeds when the seed head fell off at the end of summer, but after generations of selective breeding by farmers, seed heads stayed put until mechanically removed. Wheat now requires the assistance of humans to complete its life cycle; left to itself, domesticated wheat perishes. As a whole, it is safe to say that the landscape changes in the cultural hearths were probably substantial, but they were fairly limited geographically, by today's standards anyway. Clues, such as vast areas of soils saturated with salt from excessive irrigation by the ancients, remain embedded in the alluvial plains of Iraq today (see Figure 13.2).

Figure 11.17 The remains of ancient irrigation canals dating from the time of early farming in the deserts of the Middle East.

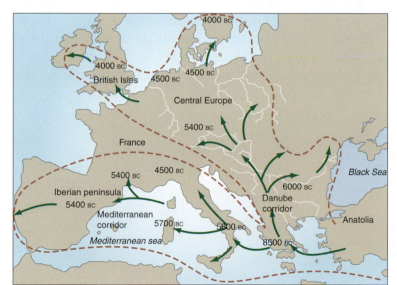

Figure 11.18 The pattern and dates of the diffusion of agriculture from Middle East into Europe following two major corridors.

Agricultural Diffusion beyond the Great Hearths: How did agriculture spread to the rest of the world? The current consensus among geographers is that it was spread by two mechanisms: the migration of people and the diffusion of ideas (word of mouth). Evidence of the first is provided by tracing the spread of language and it reveals that farmers migrated from the Fertile Crescent into Europe, bringing their seeds, tools, and farming methods with them. As for diffusion by ideas, the direct spread of agricultural ideas across cultures was aided by improvements in transportation. After the invention of sails, sea-going vessels helped spread the knowledge of crop domestication outward from the Fertile Crescent to different cultures around the Mediterranean Basin and, as the map in Figure 11.18 suggests, the sea was an important avenue of diffusion.

Over longer distances, ideas were also spread from culture to culture by *stimulus diffusion*. Stimulus diffusion is an idea or innovation sparked by an idea from another culture in which the specific trait, such as a particular plant or a farming technique, may be rejected, but the underlying concept is accepted. An example of this process occurred with the Mesoamerican cultural hearth, where crop domestication began about 5000 years ago. People here had independently domesticated maize, tomatoes, chili peppers, and squash. The *idea* of plant domestication and farming from this hearth spread into northwestern South America where it led to the formation of a secondary agricultural center where the white potato and manioc were domesticated.

Figure 11.19 Primitive methods of crop farming and herding, little changed from ancient times, are still practiced in Asia, Africa, and South America.

11.4 Some Geographic Consequences of Early Agriculture

With the development of agriculture, humans' relationship with the Earth was forever changed. Two types of societies, nomadic herders and sedentary crop farmers, emerged in direct response to this new economic system (Figure 11.19). Both rendered significant change in the Earth's landscapes and the magnitude and geographic extent of these changes increased as agriculture spread and population grew. Perhaps the most notable among the changes was alteration of the biogeographical environment. Farmers and herders found it necessary to clear land of natural vegetation to provide room for crops, expand and improve grazing land, and control predators and vermin, among other things. The vegetation that posed the greatest obstacle was forest and woodland, and the quickest and easiest way to get rid of it was by burning.

Modifying the Biogeographical Environment: Fire was used throughout the world, in climates wet and dry, tropical and temperate. In the savanna of southwest Africa, for example, burning was a common practice among ancient herders. Fires reduced shrub and brush cover and promoted grasses and other herbaceous plants, but it also reduced the richness of herbaceous species. In other parts of Africa, herders repeatedly burned back tropical forest to expand rangeland, a practice still in use (Figure 11.20). After thousands of years, huge areas of tropical forest were lost to herding societies. Far to the north, in the forested landscapes of northern Europe, where farming began around 5000 years BP, fire was the chief means of clearing land for crops and local grazing. In wetlands, however, where the ground was too wet for crop farming, the forests were often exempt, but eventually even they were cleared as the land was drained and converted to agriculture uses. In the tropics of southeast Asia, on the other hand, wetlands were the preferred sites for rice farming, and in coastal areas, mangrove forests were burned and ripped out to make way for rice fields.

Figure 11.20 Burning back forest in Africa for the purpose of expanding grazing land.

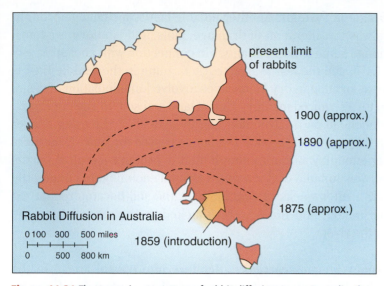

Figure 11.21 The approximate pattern of rabbit diffusion across Australia, the fastest known spread of a colonizing mammal in the world.

Figure 11.22 The diffusion of African bees from Brazil to the American Southwest, a migration that covered 8000 kilometers in less than 35 years.

As agriculture spread, many plants and animals were carried to new locations. By 5000 years ago, for example, wheat, a dry land plant, appeared on farms in central Europe, more than 2500 kilometers from its place of origin in the Fertile Crescent. For a geographic change of this magnitude, centuries of selective breeding were needed to produce new varieties of wheat able to survive in the cool, moist European climate. But other geographic shifts over great distances were much faster and they often involved undesirable (weed) species as well as domesticated plants and animals. One of these shifts was the **Columbian Exchange** which was triggered by Columbus's voyages.

Transoceanic travel brought on rapid exchanges, or jumps, of plants and animals between the Old and New worlds. New foods became staples of human diets and new growing regions were established on both sides of the Atlantic. For example, prior to Columbus, potatoes were not grown outside South America, but by the 1700s they had become a staple in Europe and, by the 1840s, Ireland had become so dependent on them that a major crop failure led to the devastating Irish Potato Famine. Animal exchanges across the oceans sometimes led to changes in the basic lifestyles of native peoples. For example, in the 1600s and 1700s, a European import, the horse, changed the lives and diets of many Native American tribes on the North American Great Plains, allowing them to shift from hunting and gathering on foot to a nomadic lifestyle based on bison hunting on horseback.

And many introduced species had devastating effects on the landscape. A frequently cited example is the rabbit in Australia which was imported by farmers in 1788 first to Tasmania and later (in 1859) to the mainland. Without a natural predator to check their population, rabbits increased by the millions and, by 1900, had spread across about one-third of the continent, destroying grassland habitat and contributing significantly to the decline and extinction of native marsupial species. As Figure 11.21 shows, rabbits now occupy most of Australia.

By far the greatest wave of exchanges has taken place in the past 150 years with increased shipping and commerce. It has included thousands of accidental introductions as plants, insects, and diseases are carried across the seas as parts of ship cargoes, bilge water, dirt, and floating debris, and deposited in ecosystems that are often ill equipped to tolerate them. A simple but blatant example is the African bee, an insect shipped to South America, which escaped from captivity in Brazil in 1957. The bee spread across the continent at an astounding rate, as much as 300–500 kilometers a year (see the map in Figure 11.22), invading native bee habitat and posing a serious threat to humans and animals. It has now spread through Central America and has entered parts of the southern United States.

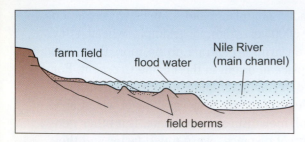

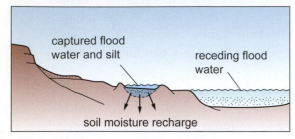

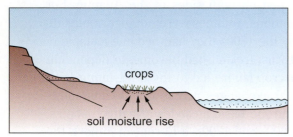

Figure 11.23 Field berms used by ancient Egyptian farmers to capture flood waters and recharge the soil with water and nutrients.

Figure 11.24. Evidence of soil erosion can be found in parts of Iceland where pedestals, called rofabards, mark the elevation of the original soil surface.

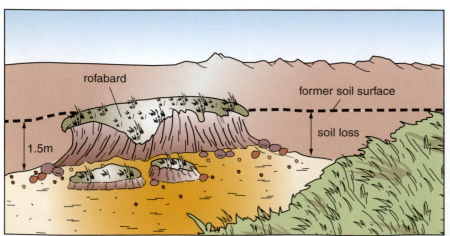

Modifying Land and Drainage Systems: Another notable landscape change brought on by early agriculture was the alteration of drainage, soils, and landform systems. We have already mentioned some of these changes in connection with ancient Mesopotamia (see Figure 11.17), and the lesson from early farming in the Middle East is that almost any attempt at farming in dry landscapes involved manipulation of drainage systems and this in turn required manipulation of land and soil. In ancient Egypt, for example, farmers depended on the Nile flooding their fields and leaving a layer of rich mud onto which they cast their seed. If the floods passed too quickly, farmers found there was neither enough water nor enough mud to produce a crop. A partial solution to this problem was to build earthen dikes around each field to contain the floodwater (Figure 11.23). After weeks of containment, the soil became recharged with water and organic matter, and holes were knocked in the dikes letting the surplus water return to the river.

Beyond the river lowlands, humans carried agriculture into upland areas. In the Mediterranean Basin, herding and crop farming spread up mountain slopes as forests were cut away by the Phoenicians, Greeks, Romans, and others. This led to serious soil erosion and a landscape which could support neither crops, most grazing animals, nor forests. Such landscape degradation was largely inadvertent and it weakened prosperous civilizations, and in time most collapsed. The landscape of the Mediterranean Basin was forever changed. Among other things, the land was drier and warmer with fewer native plants and animals. This pattern of unplanned or uncontrolled deforestation and soil loss was repeated in many places in the world beyond the Mediterranean and always with the same result. Central America and east Asia are notable examples, but there are also many smaller-scale ones such as Iceland, shown in Figure 11.24.

On the other hand, there were impressive examples of *planned* manipulation of the landscape for agricultural purposes and perhaps the most celebrated example is **soil terracing**. Terracing probably began in east and southeast Asia between 9000 and 5000 years ago as a means of extending wet-rice, or *padi*, farming from lowlands, such as floodplains and coastal wetlands, up hillslopes. Wet-rice terrace-farming involves first stripping hillslopes of their tree covers and then building a series of small, step-like reservoirs or terraces along the contour of the slope. These familiar terraces, pictured in Figure 11.25, are arranged in cells and bermed along the perimeter to hold in water and soil. Water is supplied by direct rainfall, local runoff, and various manmade devices. Once the padis are flooded, they function as little ecosystems. Aquatic algae proliferate as the water warms up from solar heating, the algae add nitrogen to the soil and, in combination with decaying plant stubble and doses of manure, tree leaves, and other organic matter, the product is a highly fertile medium for growing rice.

Long before Marco Polo visited the Orient, wet-terrace systems had spread across southern China, Korea, Japan, the Philippines, southeast Asia, Sri Lanka, Madagascar, and parts of India. And here they remain today as the map in Figure 11.26 shows, though with larger geographic footprints, of course. They have proven to be remarkably resilient systems, not because they are uniquely suited to these lands, but more because

Figure 11.25 Elaborate terraces constructed on an Asian hillslope for rice farming. Each padi functions as a micro-ecosystem with measured inputs of water and nutrients.

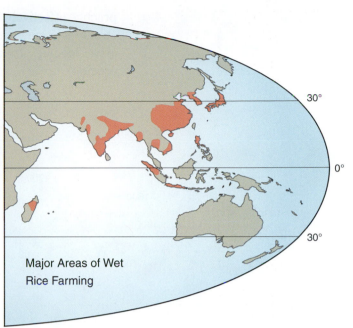

Major Areas of Wet Rice Farming

Figure 11.26 The distribution of wet-rice farming in Asia. This pattern has remained largely unchanged for centuries.

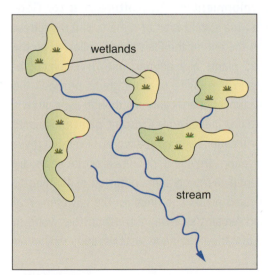

wetlands

stream

(a) before ditching

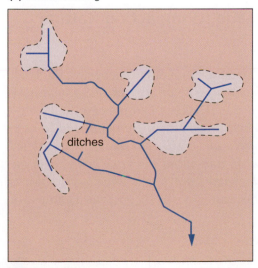

ditches

(b) after ditching

of the huge inputs of human energy and ingenuity they receive. In terms of physical geography, wet-terrace systems are among the best examples of humans as a major driving force on the planet's surface, for their impact on these landscapes over the millennia has been inestimable. Here we see humans designing and simultaneously operating three interrelated systems, water, soil, and vegetation, and doing it with such precision and perseverance that despite social and economic upheavals, setbacks from natural disasters and changing climate, wet-terrace systems have been (and continue to be) able to provide food for hundreds of millions of people.

But besides the sculpted form and managed content of the terraced landscape, how are these systems different than the nature-driven systems they displaced? For one thing, they employ different operational strategies. First, natural systems like watersheds trend over time toward geographic integration. That is, the stream systems that drain the watersheds tend to expand over time forming networks that link together broad geographic areas, routing water toward a common outlet, whereas the wet-terrace systems tend to compartmentalize water and soil into small, manageable-sized geographic units that are more or less disengaged from the larger watershed system. Second, natural systems like hillslope runoff and the watersheds it feeds are highly variable in their behavior. That is, water and sediment move through them more or less in surges separated by long quiet periods, whereas the wet-terrace systems are designed more as steady-state systems with little variability in the movement of water and sediment through them. In Europe and later in North America, by contrast, crop farming often had the opposite effect on drainage and soil systems. Rather than building wetlands such as padis and isolating them from the natural drainage system, the Europeans drained wetlands and forced their integration into the natural drainage system. They accomplished this by digging ditches linking wetlands to stream channels as shown by the example in Figure 11.27. This

Figure 11.27 (a) Before ditching and wetland drainage. (b) After ditching and wetland drainage the stream system is expanded with larger flows.

practice effectively created more stream channels (in the form of ditches) and larger stream systems with bigger surges of water and sediment (because the drainage areas feeding streams were now larger and the soft wetland soil was exposed to erosion).

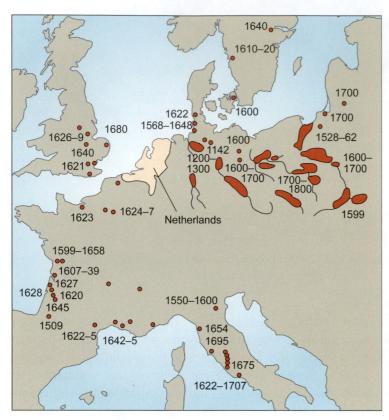

Figure 11.28 Drainage projects conducted by Dutch engineers, the masters of land drainage (or land reclamation), mainly in the seventeenth and eighteenth centuries (from C. T. Smith, 1978).

History records that the Romans drained and cleared countless wetlands to create new farmland (after exhausting most of the better farmland) and stamp out diseases like malaria, but the real champions of wetland drainage were the Dutch. Within their own little country, it is estimated that before 1860 the Dutch moved one billion cubic meters of soil material in drainage projects. So skilled were they at wetland drainage that other countries hired them for large projects. The map in Figure 11.28 shows the major projects planned and managed by the Dutch between the twelfth and eighteenth centuries. The North Americans were also ambitious in eradicating wetlands. In the nineteenth and twentieth centuries, the Americans eliminated as much as 50 percent of the country's vast wetland coverage (less Alaska), estimated at more than 100 million acres, and converted it to farmland, towns, and cities.

Summary on the Development of Agriculture and its Geographic Consequences: The stabilization of climate with the retreat of the great ice sheets at the end of the Pleistocene, followed by the domestication of selected plants and animals, the development of permanent settlements, and the ability to store grains provided the foundation for the establishment of agriculture. Although agriculture is believed to have appeared first in the Fertile Crescent, plant and animal domestications occurred independently at multiple locations around the world. Several of these locations became centers or hearths of early civilization, providing the infrastructure necessary to support large agricultural systems while fostering the geographic diffusion of farming to lands far beyond the ranges of the first domesticated plants and animals. The environmental price exacted by agriculture was often heavy, and remains so today throughout the world in the form of deforestation, soil erosion, alteration of natural drainage systems, and many other impacts.

11.5 The Industrial Revolution and its Impact on Natural Systems

In 1854, Charles Dickens in his novel *Hard Times* wrote about Manchester, England, which he called Coketown: *"It was a town of red brick, or of brick that would have been red if the smoke and ashes had allowed it..."* Dickens had noticed the change in color of the brick houses due to the industrial pollution. He continued to note another change: *"It contained several large streets all very like one another... inhabited by people equally like one another... all went in and out at the same hours, with the same sound upon the same pavements, to do the same work, and to whom every day was the same as yesterday and tomorrow."* The photo in Figure 11.29 gives us a glimpse of these conditions.

Figure 11.29 Early photograph of the British industrial city of Manchester, showing conditions of typical of urban–industrial centers of the Industrial Revolution.

What Dickens saw was a harsh change taking place in the world, one with profound geographic consequences. An Industrial Revolution, which in England had begun 100 years before, had changed the way humans lived and related to their environment. A new force in the form of machines and applied energy had been unleashed, transforming society and land use and – often brutally – reconfiguring the natural landscape. In the countryside, agricultural fields were reorganized and expanded, old-growth forests were cut for lumber, shipbuilding, and fuel, and large patches of land were torn apart in search of coal to fuel the machines. Industry called for more labor and people responded by having more children.

Growth of the Global Population: Perhaps the clearest evidence of the Industrial Revolution's impact on the modern world can be seen in the world-wide human population growth. By the time of the Crusades about 1000 years ago, estimates place the world population at 300 million (a little less than the population of the United States today). For the next seven to eight centuries, the population growth rate in the world averaged about 0.1 per cent per year, reaching 700 million by the dawn of the Industrial Revolution around 1750. By 1800, the world reached one billion people, and then it really took off.

Steam-powered machinery, such as this farm tractor, greatly increased food production in the 1800s.

Improvements in medicine, public health, and living standards during the Industrial Revolution resulted in a population explosion. By 1850, only 100 years after the onset of the Industrial Revolution, the world population grew to 1.26 billion and by 1927 reached 2 billion people. After this, the global population grew exponentially, reaching 6 billion people just before the start of the twenty-first century, a 400 percent population increase in a single century! In the 250 years since the beginnings of the Industrial Revolution, as the graph in Figure 11.30 shows, world population has increased by nearly 6 billion people. Today, it is increasing at nearly 80 million people a year and will continue to grow, it is estimated, until mid century or so when it will stabilize at 9–10 billion.

There are four primary reasons contributing to the accelerated growth in world population: a decline in the death rate, an increase in the birth rate, the virtual elimination of large-scale plagues, and an increase in the availability of food. The latter was brought on by a combination of inventions, including (1) machinery such as tractors and harvesters that reduced manual labor in food production and allowed farming to be expanded into marginal lands; (2) transportation systems such as railroads and highways that greatly improved food distribution and commerce; (3) chemical fertilizers and pesticides such as liquid nitrogen fertilizers and DDT that improved crop yield rates; and (4) new plant varieties such as hybrid corn that added substantially to yield rates. But the costs in terms of environmental degradation, irreversible loss of natural resources, and impacts to landscape systems were considerable.

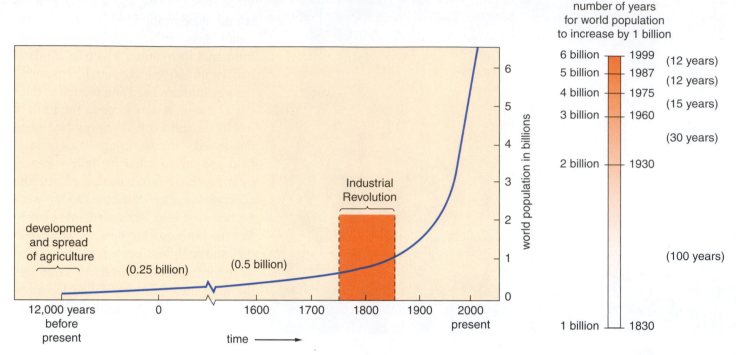

number of years
for world population
to increase by 1 billion

Figure 11.30 The global population growth curve over 12,000 years. Since the middle of the Industrial Revolution, world population has increased about sixfold.

Wetland eradication soared, runoff increased with the construction of drainage ditches, and as a result downstream flooding and water pollution increased. Land clearing, which in the 1800s would take a generation to fully complete for a family farm (Figure 11.31), in the twentieth century, with the aid of tractors, could be completed in a single season. Soil erosion increased as more and more land was taken out of forest and prairie, and along with the sediment and runoff discharged from the exposed fields came pesticides and chemical fertilizers. And since all these farm lands were parts of large runoff systems (watersheds), these inputs were passed on downstream ultimately affecting ecosystems hundreds, even thousands, of miles away in rivers and the ocean. Yet the geographic interconnectedness of the global environment was largely invisible to nations throughout the nineteenth century, and later, in the twentieth century, when science began to reveal it, it was largely ignored.

Figure 11.31 Land clearing in the nineteenth century and before was a slow and arduous process.

Machines and their Impacts on the Landscape: The Industrial Revolution brought on machinery and this machinery dramatically increased the capacity of humans to change the landscape and its systems including the atmosphere. The one invention that really drove the revolution in the eighteenth century was the *steam engine*, first as a power source in coal mining and then as a force in changing locations and sizes of settlements and the economic system.

Before the steam engine, power for mills was derived directly from moving water, thus factories and towns favored locations on streams and rivers with strong, dependable flows. But with steam power that relationship changed and other factors began

Table 11.3 World and US coal production 1800–2007 (in millions of short tons)

Year	World	United States
1800	40	—
1850	130	—
1900	825	240
1950	2500	550
1990	5000	1000
2007	7035	1150

to govern the location of factories, towns, and cities. Coal was now the chief determinant of industrial location and the demand for coal led to huge mining operations and burgeoning industrial cities which burned the coal, producing unparalleled levels of air pollution. This was the opening chapter of the carbon dioxide loading of the atmosphere that has led to the global warming dilemma in which the world is now embroiled. The British, who in 1829 were the world's principal coal producer, mined 15 million tons of coal. Today, coal mining worldwide yields about 7000 million tons a year, with the United States alone producing 1150 million tons annually (Table 11.3).

In addition, coal mining operations themselves led to widespread landscape changes. In the 1800s and early twentieth century, the waste (spoils) from underground (shaft) mining was dumped around mine openings and, after decades of operation, the landscapes there were transformed into a sterile wasteland of odd-looking landforms, stunted vegetation, and altered drainage patterns carrying polluted water. This happened in the United Kingdom, Germany, Russia, Canada, and the United States, among other countries. And then in the 1950s, as the demand for coal continued to rise, the increased use of large-scale earth-moving machines doubled the output from **strip mining**, and this had even greater effects on the landscape.

Strip mining is practiced mainly in the United States where it has affected many thousands of square miles of land, including 2300 square miles in Kentucky alone (Figure 11.32a). It involves the stripping away of soil and surface rock in order to excavate shallow coal seams. The landscape, its vegetation, soil, water features and animal habitats are destroyed, and from the piles of spoils seeps acidic rainwater, which pollutes local streams. A related form of mining for iron ore, gold, copper, oil sands and other minerals, called **open-pit mining**, produced similar effects and it is common throughout the world (Figure 11.32b).

For all forms of mining, the amount of soil and rock excavated exceeds, usually many times, each ton of mineral extracted from the Earth. If we add the earth materials (soil, rock, and minerals) moved by humans in mining operations to that moved in the construction of houses, roads, and dams, the world total is 30 to 40 billion tons a year. This represents more work than the work done by all the world's streams and rivers each year in moving sediment to the sea. And if we add in the earth moved in agriculture by plowing, the annual total goes up to a startling quantity, about 3500 billion tons.

11.6 The Urban Revolution and Modern Land-use Systems

Figure 11.32 (a) Strip mining for coal in the Appalachian region of the United States. (b) Open-pit mining for tar sands (oil) in the northern Great Plains of Canada.

The Industrial Revolution gave rise to the Urban Revolution. As cities grew, they became increasingly chaotic: factory smoke smothered neighborhoods where workers lived; cars demanded an increasing share of the land; and the geographic order of land uses in and around cities became disrupted and fragmented. In the 1920s, the practice of zoning land into specific uses was begun as an attempt to achieve some order among competing land-use systems. What was not anticipated for the decades ahead, however, was the massive rise of the automobile, the frenetic and unplanned growth of suburbs, and the subsequent decentralization of industrial and commercial land use over expanding metropolitan regions. Although these local trends seemed unscripted, they followed a key part of the larger North American trends in land use, particularly the investment in massive road infrastructure fueled by soaring levels of consumption.

Figure 11.33 Expressway development, a trend common across North America and one that promotes sprawl, fragmentation of communities, and air pollution.

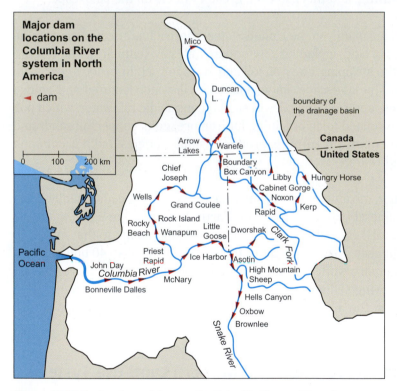

Figure 11.34 The locations of dams on the Columbia River and its major tributaries excluding the upper Snake River.

Roads, Automobiles, and Environment: After World War II the United States began the largest public works project in history, the Interstate Highway System. Over 45,000 miles of new, multiple-lane expressways such as the one in Figure 11.33, were built between 1956 and 1992, linking and encircling cities. Aided by these new routes, urban populations left the cities to live in suburbs where urban sprawl not only ate up more land, but changed the character of the landscape. Land uses became separated. No longer were residential, work, school, commercial, and other activities located more or less in one place. Suburbs became dormitory communities, housing large hordes of urban commuters who daily drove dozens of miles to and from work. Other, often distant, locations supported other land uses, such as new varieties of retail and commercial activities like strip malls, large shopping malls, "big box" retail stores, and corporate parks, all dependent on automobiles for access.

Urban sprawl, brought on by highway development and enabled by the automobile, represented a fundamental change in the geographic relationships between land use and the physical environment. For instance, water resources such as rivers and harbors, once the focal point for locating cities, began to face new forms of land-use development and related demands for water supply, waste disposal, flood control, and recreation areas. These often required radical alteration of streams and their valleys, draining and filling of wetlands, and the construction of dams. In the twentieth century, the United States alone built more than 70,000 dams two meters or more high resulting in flooding of vast areas of forest and farmland. Figure 11.34 shows the major dams on just one large North American river system, the Columbia.

By the 1970s, the massive networks of paved streets, highways, parking lots, and malls covered the land with so much pavement that the soil's capacity to take in rainwater was sharply reduced. The result was increased stormwater runoff which in turn produced flooding in streams valleys and in suburbs, communities, and farmlands downstream. This, in turn, led to more landscape alterations, to building levees along rivers in an attempt to hold in floodwaters, erecting more dams, and deepening and straightening stream channels. On top of that the stormwater generated from the expanded urban areas washed tons of contaminants, such as road salt, oil, auto exhaust residues, and pesticides, off paved land, yards, and other surfaces into streams, lakes, and harbors. (The total area of paved surface in the United States today is 61,000 square miles, an area larger than the state of Georgia. If paved area were made a state, it would be the 24th largest state in the country.)

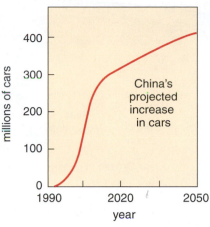

Figure 11.35 Cars and sullied air over a Chinese city with graph of projected increase in China's car population by 2050.

Between 1977 and 2001, the automobile miles driven in the United States rose by 151 percent, about five times faster than the US population growth rate, while energy consumption doubled. In North America, the average person drives more today than 10 years ago, the total number of motor vehicles virtually everywhere in the world is steadily rising, and despite significantly lower pollution emission rates from modern automobile engines, as a whole, air pollution from automobiles is rising world-wide. Of particular concern is the massive contribution this is making to global climate change (Figure 11.35).

Farms, Food, and Fuel: In a parallel timeframe, agriculture began to change as urban areas decentralized after World War II. As more people moved into the sprawling cities, fewer were left to run the farms. Yet food production in the developed world (North America and Europe) continued to rise. This was made possible mainly by new technologies and improved crops. Among the new technologies was larger and more efficient farm machinery that enabled an individual farmer to work huge tracts of land. In combination with more productive crop varieties, pesticides, and chemical fertilizers, a North American farmer could by the end of the twentieth century produce enough food to feed 140 persons. And much of this was facilitated by cheap energy (mainly fossil fuels) used in the manufacture of fertilizer and equipment and in the operation of farm and food-processing machinery. Farms became bigger – the average American farm is now 441 acres – and more specialized, that is, growing only one or two crops year after year.

Traditional farming as practiced by groups like the Amish is no longer able to feed our expanding urban population.

But there were major side effects. The heavy use of fossil fuels contributed to global warming and the heavy use of chemical fertilizers contributed directly to the decline of soil and indirectly to global warming. Chemical fertilizers became an easy substitute for soil conservation because farmers could apply massive amounts of fertilizer to maintain crop yields while largely ignoring topsoil loss to erosion and weathering. Soil organic matter, which in nature serves as a reservoir for storing carbon, declined. As a result more carbon, in the form of carbon dioxide, was available to the atmosphere where it contributes to global warming.

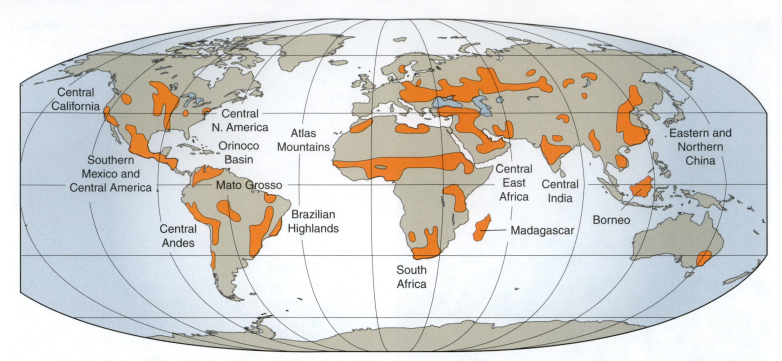

Figure 11.36 Areas identified by the United Nations with seriously degraded soil, mostly from crop farming, grazing, and deforestation.

But soil loss was not just a problem of North America. In fact, as Figure 11.36 demonstrates, it was and is an even greater problem in much of the rest of the world. The United Nations estimates that nearly 12 million square kilometers of the world's soils are seriously degraded because of crop farming, grazing, land clearing, and deforestation. This is an area larger than all of Canada. Erosion rates on farmlands in the world's two most populous countries, India and China, are more than double those in the United States and Canada. And when we put soil loss into an even broader perspective, Earth scientists tell us that the total annual load of sediment (the product of soil erosion) delivered to the oceans by all the world's rivers has more than doubled since the invention of agriculture.

11.7 The Human System in a Global Context

How do we put the human story into context as part of physical geography? One way is to look at our existence on the planet in a much broader context than that provided by history. In Chapter 8 we stated that climate on Earth has changed many times in the past and that these changes have covered a wide range of magnitudes. Of course climate represents only one source of planetary change, and if we add in all the others, the asteroid strikes, glaciations, volcanic eruptions, and so on, we get a picture of Earth as a place characterized more by change than by stability. In this context, it can be argued that the human era fits the global pattern in that we are simply another in a long chain of planetary events, just one more episode of change.

Although the human agency has operated on Earth with a force that rivals and often exceeds that of natural systems, it is appropriate to ask how our system fits into the working whole of the planet. We have learned that all Earth systems, including our own, are wound together in an interdependent fashion such that a change in one system often triggers changes in others, in ways we sometimes understand and sometimes do not. For 10,000 years or more, however, humans have operated in a more-or-less one-dimensional, cause-and-effect manner in which we have pushed on the Earth in various ways and apparently not learned much from the effects, beyond simple and short-term reactions. We have failed to see our actions as parts of more complex networks with broad cause-and-effect relations. A simple example is global warming. It is not the increased air temperature itself that is the main concern, but all the changes in related global systems, such as plant distributions, glacial ice volumes and ocean levels, and the storminess of the atmosphere, that are affected by the warming, and in turn affect it, which are the main concerns.

Why have humans not been able to find a more compatible balance with the planet? Part of the answer has to do with geographic scale and the human perception that actions taken in one place have little to do with the larger whole. In other words, humans have taken a compartmentalized view of their existence, which limits their capacity to perceive networks, connections, and large-scale cause-and-effect relations. Historians tell us that in Medieval times in Europe an average person's concept of the world extended less than 10 miles beyond his/her place of birth. People were flummoxed, for example, by the disappearance of certain birds (migratory species) every fall, and thought that they spent the winter hibernating in mud like frogs. Our view today, of course, is much broader, even global, but many people still have a compartmentalized notion on how the planet works, partly because the scientific concepts of an integrated Earth can be difficult to grasp and partly because it may be inconvenient for them to do so. Witness that many countries ascribe artificial significance to political borders, believing that the nature that operates within their boundaries belongs to them and is functionally exempt from the rest of the planet.

Humans have created an interesting paradox. Through our development and use of technology we have devised remarkably efficient ways to alter Earth's landscapes and the great systems that drive them. Yet, when faced with the consequences of these efforts, we seem only able to apply more technology in an attempt to fix the problems. We have acquired a great faith in ourselves to overcome nature, but have not come to grips with what it takes to fully understand our role as a functioning component within the array of Earth's dynamic systems. Until we do, it remains our monumental challenge to bring the great human system into balance with the rest of the planet.

Chapter Summary and Overview:
Humans as Geographical Agents in a Changing World

The role of humans in changing the face of the Earth began in earnest with the spread of agriculture but, as the summary diagram shows, with the Industrial Revolution it took on massive proportions with accelerated impacts on natural systems. The Industrial Revolution brought on a host of technological advances which launched a global population boom and the concentration of productive forces in factories and urbanized areas. Energy use and coal mining rose dramatically and new modes of transportation flourished. The automobile is most notable here, as it spawned urban sprawl, increased air pollution, and altered the traditional land-use relationships within cities and entire regions. From a geographic perspective, the ability of humans eventually to attain a balance between land-use systems and natural systems will depend on our ability to develop a more integrative outlook in which we see ours as player within Earth's great systems.

► **Early humans evolved and lived in the East African Highlands.** Geographic diversity and environmental change drove them to adapt, culturally, socially, and biologically.

► **Migration from Africa began between 50,000 and 80,000 years ago.** It was probably triggered by some combination of environmental change, technological innovation, and other factors.

► **Human diffusion to other continents proceeded rapidly.** Migration from Africa led to southwest Asia, and then, over thousands of years, to all but one of the world's continents.

► **The first farming appeared during the Pleistocene–Holocene transition period.** It was marked by warmer temperatures, more rainfall, and less-variable climatic conditions.

► **Humans made the transition from a nomadic lifestyle to agriculture in three stages.** Stage three was marked by the geographic diffusion of agriculture from the major centers of farming.

► **The Fertile Crescent was geographically conducive to the domestication of grains.** Nutritious wild grains were abundant there and the climate was well suited to grain farming.

► **The great cultural hearths arose under different geographic circumstances.** All were grounded in agriculture but the plants, animals, geographic settings, and environmental conditions were different.

► **Agriculture spread beyond the great hearths across the continents.** This took place by three processes and led to new centers of agriculture.

► **Agriculture resulted in major alterations of the landscape.** Those affecting the biogeographical environment have been dramatic and extensive.

► **Transoceanic travel accelerated the global spread of plants and animals.** This included weedy plants, insects, and diseases.

► **Soil terracing has been used in farming for thousands of years.** It relies on small, constructed plots for the management of water, soil, and crops.

► **The Industrial Revolution led to population growth and environmental change.** The first was tied to lower death rates, higher birth rates, and increased food production; the second to mining and the burning of coal, among other things.

► **An Urban Revolution accompanied the Industrial Revolution.** Urbanization led to large increases in fuel consumption, infrastructure investment, and urban sprawl.

► **Urban sprawl changed the relationship between land use and the physical environment.** Sprawl was driven by the automobiles and highways with major impacts on the environment.

► **The scale of agriculture changed during the post-war period in North America.** New technology and cheap fossil fuels drove much of this change.

► **Humans have invented powerful ways of altering Earth's landscapes and the great systems that drive them.** But until we understand our position among these systems, balance with the rest of the planet will be tough to achieve.

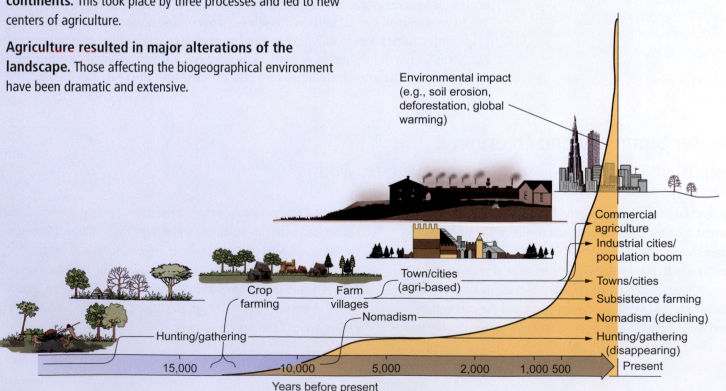

Environmental impact (e.g., soil erosion, deforestation, global warming)

Review Questions

1 What was the nature of the geographic setting in Africa where *Homo sapiens* originated and how would you describe the way they lived in this environment? Are there any humans around today who can give us some insights about how they survived?

2 Most scientists think early human migration out of Africa was spurred either by technological advances or climate change. What do they mean?

3 Using Figures 11.5, 11.6, and 11.9 describe the pattern of human migration and identify the major geographic barriers they crossed en route. If humans entered North America via the Bering land bridge 15,000 years ago and reached the southern tip of South America 12,000 years ago, what was their average annual rate of migration?

4 Why was the Pleistocene–Holocene transition period so critical for human existence?

5 Crop/animal domestication, farming, and diffusion were the three stages of the human transition from a nomadic lifestyle to one based on an agricultural land-use system. What role did the cultural hearths play in this transition?

6 What are the founder crops and what geographic factors are significant to their role in early agriculture? What were the primary requisites for animal domestication?

7 The early cultural hearths arose in a variety of geographic settings. Describe the different geographic conditions in these ancient hearths and identify the particular geographic arrangement common to the Nile, Indus, and Tigris–Euphrates hearths.

8 Agriculture spread beyond the original cultural hearths by human migrations, word of mouth, and stimulus diffusion. What is stimulus diffusion?

9 How did the spreads of the African bee and the rabbit in Australia change those biogeographical environments?

10 Using Figure 11.25 and the related text, describe the features of terracing, and list the major alterations to soil and water brought about by this agricultural method.

11 What major changes to the landscape accompanied the major population growth occurring during the Industrial Revolution?

12 Describe how the increase in automobile use is now affecting the urban landscape worldwide. Outside of urban areas, how has the automobile changed the landscape?

13 What has happened since World War II to the geographical nature of agriculture in the developed nations? What factors have contributed to this change?

14 Should it surprise humans that the Earth is undergoing significant change during our time on the planet? Why? What factors have contributed to humankind's inability to achieve balance with the planet's great systems?

Soil Systems, Processes, and Formation

Chapter Overview

Soil has been described as the excited skin of the Earth and indeed it is, for like our own skin it is constantly changing as it takes in and gives up heat, water, chemicals, and organic matter. And like our skin, soil is a transition medium between two large spheres, and it shares traits of both, including air and water from above and rock and minerals from below. In this chapter we examine soil as a complex of systems, geomorphic, ecological, hydrologic, and biochemical. These systems are responsible for giving soil its basic form and composition, transforming what often begins as a chaotic mix of organic matter, particles of sediment, water, and other substances into an ordered whole. And since these systems are driven by a larger body of geographic systems, such as climate and hydrology, soils tend to vary correspondingly with these systems. That is, prairie soils are different than rainforest soils, which are different than desert soils, and so on.

Introduction

It was a late afternoon call from Jack Goodnoe. "Bill, what do you know about soils on Staten Island?" "Next to nothing," I replied. "Why do you ask?" "Well, we've got a project there planning a new cemetery. It calls for thousands of burials, as well mausoleums, roads, waterlines, and landscaping. We need to know about soils and drainage both on the surface and at depths of 10 feet or so ... some of the burials will be deep with several caskets per grave." Jack Goodnoe, a Michigan-based landscape architect, was leading a team responsible for laying out the cemetery and designing its facilities. The team consisted of landscape architects, architects, scientists, and engineers and, as Jack described the project, I searched my mental files for any geographic information that might shed light on the problem.

"Jack, here are few things that come to mind, but as you know, we'll have to do on-site testing to come up with reliable information. First, Staten Island is near the southern limit of continental glaciation. I seem to recall that during the last glaciation, the ice front reached a little beyond Manhattan Island, so it's likely that Staten Island was ice-covered and left with glacial deposits. They could range from sands and gravels to clays with boulders. Second, the site is near the sea, so I wouldn't be surprised if we found beach and/or marine deposits and they, too, could range from sandy to clayey materials. Third, we know that Staten Island's been

Staten Island, at the mouth of the Hudson River, is geographically diverse including a wide variety of soil materials of different origins.

occupied for a long time. European settlement dates to sometime in the 1600s. Weren't the British garrisoned there during the Revolutionary War? In any case, the Island's been cleared, plowed, and built upon – probably several times – and these activities have certainly had a marked effect on its soils."

"So chances are that this site could be complex with considerable variation in soil makeup and drainage." "That's right, Jack, all the geographic indicators point in that direction. I'd suggest we get on the site with a large backhoe and dig some test pits. There should be plenty of machines available. New York's biggest municipal landfill, called Fresh Kill, is just down the road from the site."

Within a week we were standing on the site studying maps and terrain features. Guided by the distribution of hills, valleys, and drainage features like wetlands and swales, we laid out a sampling plan. The geographic indicators had indeed pointed us in the right direction. Glacial materials dominated most areas, and several holes revealed compact clays of likely marine origin beneath the glacial deposits. In most low spots the soils were wet with thick layers of organic matter over clay. And imprints from all those years of land use were abundant, sometimes bold, sometimes subtle. In one spot, we uncovered a mass of woody debris about 2 meters down that, judging by the trees growing over it, had been buried for decades. In another place, the topsoil had been stripped away, leaving dry sandy patches that were slow in accepting new plants.

The Staten Island cemetery project was a good lesson in the geography of soils. It showed that many factors, including land-use activity, play a major role in soil formation. It also revealed the value of knowing where you are in the big picture because the regional context can provide clues about what to look for in the field. Finally, it showed us that there is a critical scale for sampling and mapping soils, which is governed, among other things, by the size and pattern of landscape features, especially landforms.

Figure 12.1 Soil as a transitional medium between the rock of the Earth's crust and the realm of water, air, and life above.

12.1 Perspectives on Soil

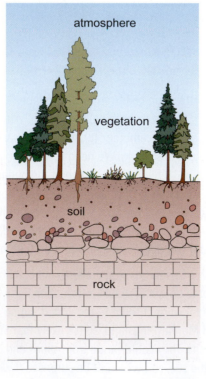

Broadly speaking, soil is a mixture of particles and fluids that blanket the land. It is made up mostly of rock and mineral particles with smaller fractions of air, water, and organic matter. It is not consolidated, meaning its constituents are loose rather than bound together like rock. It is the medium in which most plants grow and upon which humanity and its land-uses reside. It is habitat to a vast system of organisms. Besides the roots of macrophytes such as trees and grass, it contains massive numbers of microphytes, animals, and microorganisms. Witness that an acre of soil can contain as many as a million earthworms and 50 billion nematodes (a microscopic-sized worm), to say nothing of far larger populations of simpler organisms like bacteria.

Think of soil as a transitional medium, part of a larger gradient or continuum from the realm of crustal rock under it to the realms of air, water, and organisms above it, as the graphic in Figure 12.1 suggests. These realms, or systems, represent the

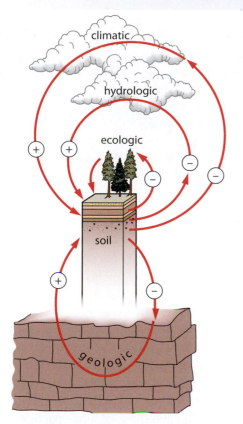

Figure 12.2 The systems that deliver (+) and remove (−) matter and energy to and from the soil.

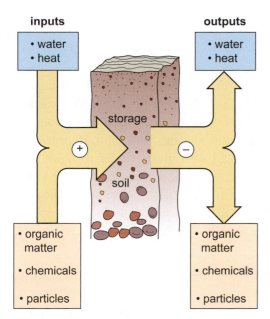

Figure 12.3 Soil as an input–output system. Overall, the relative balance of inputs and outputs determines whether the soil is losing or gaining matter and energy.

lithosphere, atmosphere, hydrosphere, and biosphere and soil is the product of a complex interplay among them. This interplay is driven by both exogenous and endogenous forces, though in most places exogenous forces, which are represented by processes such as runoff, plant growth, and wind, dominate soil formation.

We measure, describe, and classify soil based on its mineral (inorganic), organic, air, and water content. Because these substances – especially mineral particles, organic matter, and water – vary geographically over the Earth, the composition of soil tends to vary correspondingly. This is the first lesson in the geography of soil. Where there is lots of sand, as in areas of sand dunes, for example, soils tend to be sandy. Where little organic matter and water are available, as is typical of deserts, soils tend to be poor in organic matter and water. And where there is lots of rainwater soaking into the ground, soils often tend to be infertile because the water washes or leaches out nutrients as it trickles through them.

Soil as a System: How do substances such as sand particles, water, nutrients, and organic matter get into soil and, once there, are they lodged in place forever? First, the substances found in soil are brought there by larger Earth systems, for example, the hydrologic system brings water with rainfall and runoff, the climatic system brings minerals with gases, dust, and precipitation, and ecosystems bring organic matter from plants and animals. Second, these systems not only deliver matter (and energy as well), but also remove it. Thus, soil is constantly in a state of flux as it receives and gives up matter and energy (Figure 12.2).

In its simplest form, soil functions as an input–output system. Matter and energy (mainly heat) are received at some rate, stored in the soil for some period, altered by processes within the soil, and then released at some rate. The overall difference between the input and output rates (the positive and negative arrows in Figure 12.3) determines whether the major components of the soil system (mineral particles, organic matter, water, air, heat, and various chemicals) over a prescribed area are gaining or losing, that is, whether they are negatively or positively balanced. In the case of organic matter, for example, if more mass is input by plants than is removed by the combined actions of the consumers, detritivores, leaching, and erosional processes, then the soil organic system is positively balanced, meaning that the organic mass is growing over time. Soils used for crop farming provide a simple example of this system in action, though the outcome is commonly negative.

Farming begins with land clearing followed by the replacement of natural vegetation, such as forest, with crops (Figure 12.4). When crops such as corn are harvested, most of the organic mass (ears, leaves, and stocks) is removed from the field, leaving only roots to the soil. Roots provide significantly less input to soil than the original natural vegetation. At the same time, soil erosion is taking place, because the soil, which is barren part of the year, is exposed to wind and runoff. This further reduces the remaining organic matter. Thus, with reduced inputs and increased losses (outputs) the soil organic system falls into negative balance (lower diagram, Figure 12.4). Reserves held in the organic-rich topsoil are gradually (rapidly in some cases) depleted; the topsoil grows thinner, and loses much of its fertility. To put the soil back in balance, the organic system must be augmented in some way. This usually involves mechanical applications of organic matter in the form of manure, plant litter, and other types of fertilizer. But throughout much of the world, augmentation is inadequate or completely neglected, and soils are declining, often dramatically.

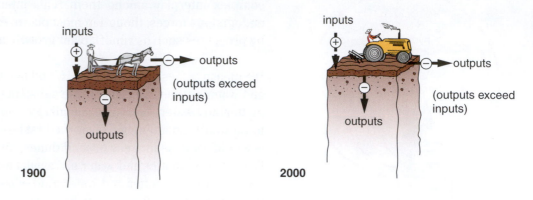

Figure 12.4 Trend in the soil system from pre-agriculture (1800) through pre-industrial farming (1900) to modern industrial farming (2000). Valuable topsoil declines as outputs exceed inputs.

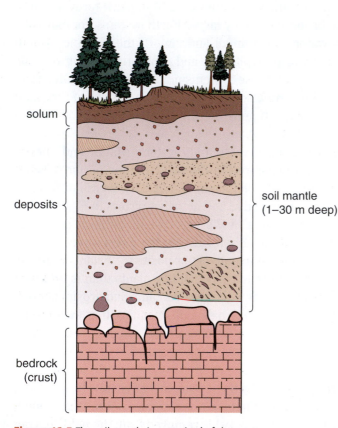

Figure 12.5 The soil mantle is comprised of the loose (unconsolidated) material that overlies the solid crust. The soil mantle includes the solum, various surface and near-surface deposits, and chunks of bedrock.

Figure 12.6 The soil mantle represents a system of particles moving slowly from the continental interior to the coasts and the oceans beyond.

12.2   Soil as a Geomorphic System

Geomorphology is the study of the erosional systems, such as streams and glaciers, that produce sediment and shape the land around us. These systems, or geomorphic agents as they are often called, dislodge material such as sand and clay from the land, move it some distance, and deposit it. These deposits cover most of Earth's 148 million square kilometers of land, forming a collective mass of loose particles that can be thought of as a mantle, the **soil mantle**, which rests on a foundation of bedrock (Figure 12.5). Physically, the soil mantle forms a continuous blanket of uneven thickness and composition, except on steep mountain slopes and parts of rugged uplands and wind-swept deserts where the Earth's crust (or bedrock) is exposed. True soil, or the **solum**, represents the uppermost layer of this mantle where most of the soil biota dwell.

Imagine the soil mantle as a great system of particles slowly creeping from the elevated interiors of the continents toward the lowlands and the sea on the continental perimeter (Figure 12.6). The surface layer, the solum, is moving at the fastest rate, being nudged along by the transport mechanisms of streams, glaciers, and wind. Beneath this layer, chemical processes associated with groundwater dissolve minerals from the soil mantle and deliver them to streams flowing to the sea. Ultimately, these particles reach a final destination where they are deposited. The sand, silt, and clay particles accumulate in thick deposits on the margins of the continents, while the chemicals, such as calcium,

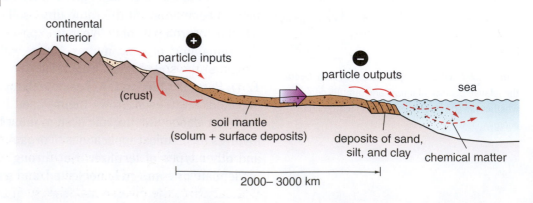

sodium, and iron, are added to the sea. While all this is taking place, the continents are producing more particles and chemical residues, which enter the soil mantle system and begin the slow march to the sea. Viewed this way, the soil mantle functions as a huge input–output system with very long particle-residence (-storage) times.

How fast is the system operating? Prior to the development of agriculture about 12,000 years ago, the rivers of the world (which do the vast majority of the work in moving soil particles) were delivering about 9 billion tons of sediment (mineral particles) to the sea a year. For most of the world this output rate was about equivalent to the input rate of new particles coming from the breakdown of crustal rock, and the ash produced by volcanic eruptions, and various minor sources. In other words, the system was in balance. But with the spread of agriculture and human population, especially in the past 500 years, the system began to change. By mid-twentieth century the output rate had more than doubled. Why? Vast areas of land had been cleared and crop farming and grazing had taken over one-third or more of Earth's land area resulting in accelerated rates of soil erosion, mainly from runoff. Thus, it appears that for the Earth as a whole, the geomorphic system responsible for the mantle of soil particles is currently out of balance and we are losing soil faster than it is being replaced. Let us examine the sources of these particles before going on to the next major system affecting soil.

The Sources of Soil Particles: More than half the total volume of most soil is made up of mineral particles (Figure 12.7). Within the soil these particles are lodged against one another so that together they form a relatively stable skeleton. The particles that make up this skeleton are called **parent material**, and they provide an infrastructure where water, air, organic matter, and creatures reside and operate.

There are two classes of parent material: residual and transported. **Residual parent material** forms from a residue of decayed, or weathered, bedrock. As the rock weathers, a layer of particles forms over it, which, if not eroded away, builds up over time. Parent material formed in this manner often bears a strong similarity to the rock under it; for example, sandy parent material commonly develops over sandstone. But in most places, the weathered residue is not left to accumulate on the surface, because the surface is much too active with wind, runoff, and other processes to allow that. So most is dislodged and carried away. This material collects in bodies or layers, called *surface deposits*, and examination of the soil mantle in most places, such as in Figure 12.5, will usually reveal several different types of deposits representing different transporting agents and different episodes of deposition. Soil material formed in this manner is termed **transported parent material** and it is the principal base material of soil the world over.

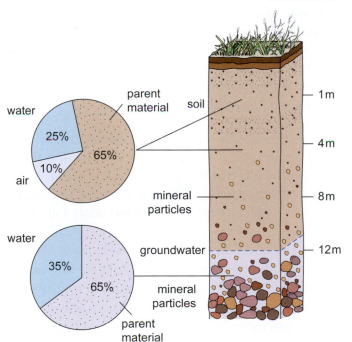

Figure 12.7 The principal components of soil. Mineral particles are the parent material of most soils.

Looking at large geographic regions, we find that the coverage of certain surface deposits is extremely vast. Four examples for eastern United States and southeastern Canada are shown in the map in Figure 12.8. **Loess**, which is wind-deposited silt, covers large tracts of the Midwestern prairies, the Great Plains, and the Lower Mississippi Valley. **Alluvium**, which is river-deposited sediment and includes sand, gravel, and silty material, extends along the floor of every major river valley, and in the Mississippi Valley south of Illinois, it forms a great belt stretching to the Gulf of Mexico. **Glacial drift**, the diverse material, such as sand, clay, and boulders, deposited by

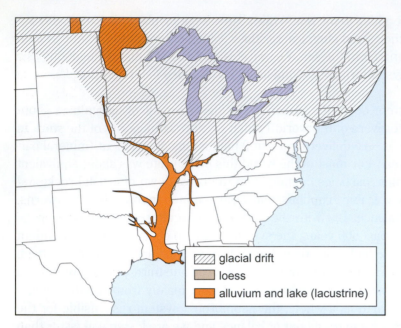

Figure 12.8 The principal surface deposits of the eastern United States and southeastern Canada. These deposits are the parent materials for some of the world's richest agricultural soils.

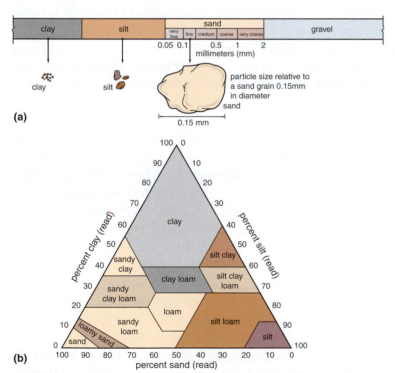

Figure 12.9 (a) The particle sizes of sand, silt, and clay. (b) Soil-texture triangle is used to define textural class when the percentage by weight of sand, silt, and clay particles are known.

glaciers or by the meltwater draining from a glacier, covers a huge region from the Missouri and Ohio rivers on the south to beyond Hudson Bay on the north. **Lacustrine deposits**, which are sediments, usually clayey in composition, deposited by lakes, cover areas around Great Salt Lake, the Great Lakes, and Lake Winnipeg, where glacial lakes stood 5000 to 10,000 years ago.

Soil Texture: Soil particles vary greatly in size, from microscopic clay particles to boulders the size of your head. However, three particle sizes dominate most soils, namely, sand, silt, and clay. The dimensions of each are shown in Figure 12.9a. Sand, silt, and clay each play an important role in soil formation. For instance, sand enhances soil drainage, and silt and clay facilitate the movement and retention of molecular water in the soil, which is used by plants. Small clay particles carry electrical charges that attract ions of dissolved minerals such as potassium and calcium. Attached to the clay particles, the ions are not readily washed away; in this way, clay helps to maintain soil fertility.

Few soils are made up exclusively of particles of one size. Most are a composite of different-sized particles, a property called texture. **Soil texture** is defined as the percentage by weight of the sand, silt, and clay particles in a soil. It is usually expressed by a set of terms, at the center of which is the class *loam*, a mixture of sand, silt, and clay. Loam is composed of 40 percent sand, 40 percent silt, and 20 percent clay. If there is a slightly heavier concentration of sand, say 50 percent, with 10 percent clay and 40 percent silt, the soil is called *sandy loam*. The textural names and related percentages are defined in a three-sided graph, called a **textural triangle** (Figure 12.9b). If you know the percentage by weight of the particle sizes in a sample, you can use this graph to determine the appropriate soil name.

Summary on Soil as a Geomorphic System: So this is where all soils begin, with an accumulation of particles called parent material. More than a pile of dirt, parent material is actually at the heart of a large geomorphic system which is slowly moving particles from the land to the sea. Into parent material creeps water, heat, plants roots, and other creatures. They function as systems, or soil subsystems, and together shape the form and composition of soil.

12.3 Soil as an Ecosystem

Not many years ago we thought that the active organic part of the soil did not extend much more that several meters below the topsoil. **Topsoil** is the organic-rich layer where plant debris and related organic matter make up 20 percent or more of the soil mass and, in most places, gives soil its dark-brown color. We now understand that the life zone is much deeper, with microscopic life detectable tens and hundreds of meters into the soil mantle and bedrock under it.

What kind of life is found at great depths and does it have much to do with soil formation? Only microbes (microorganisms) such as bacteria live deep down and, although their populations are surprisingly large (often millions per cubic centimeter), their total mass is very small indeed, hardly measurable compared to the mass of mineral particles or water. Only near the soil surface does organic matter, both living and dead, represent a large part of the soil mass. In addition, only the upper part of the soil is loose enough and aerated enough to accommodate larger animals, such as earthworms, shrews, and gophers.

Plants are the chief source of soil organic matter with macrophytes such as trees providing the majority of the input. In the soil organic system (Figure 12.10), these large plants deliver most of their organic matter to the soil surface in the form of leaves, branches, stems, and fallen trunks. Organic matter also comes from root systems, which are concentrated overwhelmingly in the upper meter or so of soil. Together, surface organic debris, which is called **litter**, and root systems provide the bulk of the organic energy needed to run the soil ecosystem.

Built on this energy is a complex and highly active system of consumers and detritivores. It is made up of at least four trophic levels (above the producer level) beginning with organisms such as fungi, bacteria, and nematodes, which feed on both roots and litter. They are followed by various secondary consumers including *shredders* such as earthworms that chew up litter as they eat bacteria and fungi. The earthworms in turn are eaten by shrews, among other predators, and shrews in turn are eaten by owls, for example. In the soil, the last we see of the original litter is a dark, more-or-less granular organic material called **humus** (See Figure 12.10), which is made up of decomposed body parts, fragmented and reduced plant debris, and animal droppings.

As with other ecosystems, energy attenuation is rapid in soil ecosystems. As a result, the amount of organic energy available at higher trophic levels is a small fraction of that available to primary consumers. This explains why there are lots of primary consumers and detritivores – bacteria alone may total 20 tons per hectare – but relatively few higher-level predators in soil ecosystems.

Soil Organic Mass Balance: The amount of organic matter in any soil is the product of the rate of plant productivity (total tissue growth) less the rate of destruction of organic matter by consuming organisms and other processes. Other processes include, for example, erosion by runoff and wind. On balance, however, most organic matter is destroyed by decomposer organisms such as fungi and bacteria. What controls the abundance and activity of these organisms?

Fungi, bacteria, and most other soil organisms are very sensitive to heat and moisture conditions. This is underscored by the graph in Figure 12.11, which shows the relative change in bacterial and fungal activity over a year in a midlatitude grassland. In a colder climate, such as the subarctic, populations of these organisms are small and their activity is limited mainly to a few months in summer. As a result, the rate of organic decomposition is very slow and, despite the low rates of plant productivity, organic buildup in the soil may be substantial. This is especially so in wet places because waterlogged soils tend to be cool and poorly aerated, which limits the activity of decomposers.

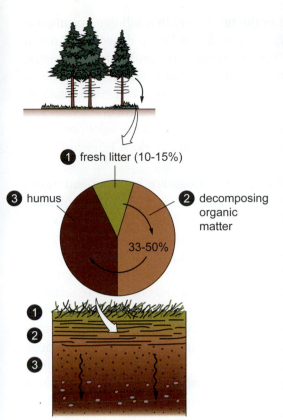

Figure 12.10 The soil organic system begins with fresh litter, which is decomposed and transformed into humus. The humus in turn is lost to weathering, consumption, and erosion.

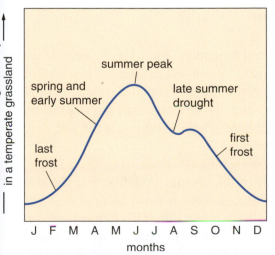

Figure 12.11 Fungi and bacteria in the soil are sensitive to heat and moisture. In temperate climates the peak rate of activity occurs in early summer when the soil is warm and still reasonably moist.

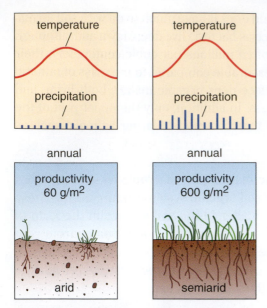

Figure 12.12 Contrasts in soil organic content between an arid and semiarid landscape. Arid lands produce little organic matter; semiarid lands produce much more and lose relatively little to decomposers and consumers.

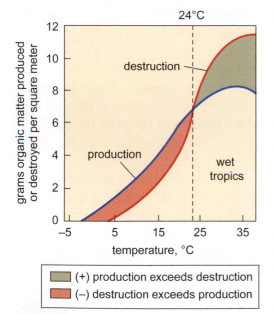

Figure 12.13 Under humid conditions, heat controls the balance between the production and destruction of organic matter. Above a temperature of 24 °C, such as in the wet tropics, destruction exceeds production leaving the soil with little or no organic matter.

In arid regions, it is the lack of moisture in the upper soil that inhibits decompositional activity. But since plant productivity is exceedingly low in deserts, not much organic matter accumulates anyway and that which does accumulate is unprotected by plants and subject to erosion by wind and runoff. Under semiarid conditions, on the other hand, moisture is often adequate to foster appreciable productivity by prairie grasses and other herbs, but inadequate to foster comparable rates of destruction by decomposers and consumers. Hence, organic accumulations in the upper soil can be surprisingly large in grassland bioclimatic regions (Figure 12.12).

In tropical and equatorial areas, where climatic conditions are both hot and moist, plant productivity is very high, but consumer activity is also very high. Consequently, it is impossible for much of an organic layer to develop. Measurements show that under moist conditions at temperatures above 24 °C (see Figure 12.13), organic matter is destroyed nearly as fast as it reaches the ground, giving rise to an apparent paradox in the wet tropics in which heavy biomass and high productivity are coupled with thin rather than thick topsoil. Exceptions are found in swamps, where standing water limits decomposition, but not plant productivity, resulting in heavy accumulations of organic material. Millions of years ago, vast swamps in tropical and subtropical zones produced the massive organic deposits that led to the coal formations found throughout much of the world today. For example, the Powder River Basin in southeast Montana and northeast Wyoming contains the largest reserves of coal in the United States, and was formed when the regional climate was subtropical.

Besides breaking down and consuming organic matter, soil organisms perform many other important functions. These include nutrient cycling, nutrient retention, carbon sequestering, and improved water-holding and water-infiltration capacities. For instance, when organisms such as nematodes consume food, they release the nutrient ammonium in their waste, which, in turn, is taken up by plants. Other examples include nitrogen fixation, and a process in which fungi (molds and mushrooms) break down complex carbon compounds, thereby making the carbon available to other organisms. And earthworms not only decompose organic litter, but mix and aerate soil thereby improving fertility and moisture movement.

12.4 Soil as a Hydrologic System

Water is central to just about every facet of soil formation. Runoff is responsible for most of the deposits that form most parent material and soil moisture is essential to plant growth, organic productivity, and microbial activity, all significant factors in soil fertility. Water is also critical to soil chemistry and the movement and transformation of minerals within the soil mass. Our interest here is with water inside the soil, how it moves, is stored, and is related to soil formation.

Soil Water Types and Processes: Water enters the ground by infiltrating the soil surface. As gravity pulls it into the soil column, it joins a small body of resident water, which coats the surfaces of soil particles in a thin film. This film is made up of **molecular water**, so called because it is controlled by molecular force (bonding among molecules) rather than gravity. The infiltrating water, by contrast, is controlled by gravity and is thus referred to as **gravity water**. The water entering the soil in Figure 12.14 is gravity water. Here we are interested mainly in molecular water.

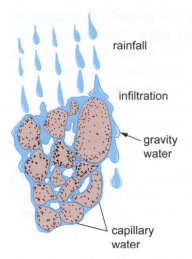

Figure 12.14 Soil particles receiving infiltration water which moves into the upper soil as gravity water. Capillary water resides (over a film of hygroscopic water) on and between the soil particles.

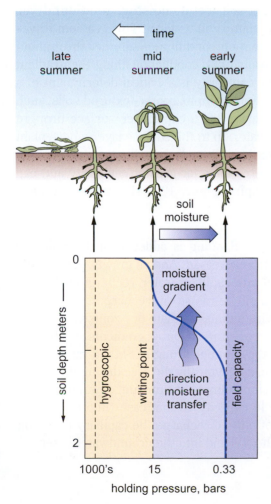

Figure 12.15 The relationship between declining soil water and water uptake by plants. As soil water is drawn down from early to late summer, the remaining capillary water is more and more tightly bound and less available to plants. At 15 bars pressure, most plants wilt.

There are two varieties of molecular water in soil: hygroscopic and capillary. **Hygroscopic water** is represented by a thin capillary of water molecules attached directly to soil-particle surfaces. Often referred to as *tightly-bound water*, it is held under force so great that no natural process is able to move or make use of these molecules. You cannot see it or feel it. Resting on the hygroscopic film is a second type of molecular water, called **capillary water.**

When gravity water enters a dry soil it is taken up by soil particles forming a relatively deep capillary (see Figure 12.14). Because the molecular force holding it is much smaller than that holding hygroscopic water, it is often called *loosely-bound water*. And because it is "loose" rather than "tight," capillary water is capable of movement, not by gravity but by molecular cohesion (molecules sticking to molecules). This phenomenon can be observed, for example, in the "wicking" action of water in a paper towel. In paper, the closely spaced fibers facilitate molecular transfer; in soil, organic fibers or matrices of silt and clay particles do the same thing. The water is drawn or conducted along moisture gradients, that is, from wet spots to dry spots. The capacity of a soil to conduct capillary water is termed **capillarity**. In mineral soils, capillarity is highest with intermediate textures like silt and loam and lowest with coarse textures like sand and gravel. If organic matter is present capillarity is usually increased substantially.

The force that holds capillary water to the hygroscopic film is measured in bars of pressure. One bar is equal to about one atmosphere of pressure, that is, 1013.2 millibars or 14.7 pounds per square inch. Capillary water is held under pressures ranging from 0.33 to 31 bars (between 5 and 456 pounds per square inch). In a capillary film the highest cohesive force is on the inside next to the hygroscopic film and the lowest is on the outside. Thus, the outer water molecules are loosely attached to the soil particle, and can be removed relatively easily.

This is extremely important to vegetation because the amount of capillary water that can be taken up by plants depends largely on the level of cohesive pressure under which the water is held. As capillary water is drawn off by plant roots and the film gets thinner, the remaining fraction becomes increasingly difficult to remove because it is held under greater cohesive force. When the remaining capillary water is held too tightly for plants to draw off amounts needed for transpiration, they begin to wilt as Figure 12.15 illustrates. For most plants, the **wilting point** is reached when capillary water reaches around 15 bars of pressure.

For any soil, there is a maximum quantity of capillary water that it can hold. When this limit is reached the soil is said to be at **field capacity**. Inasmuch as capillary water is held on the surfaces of soil particles, the greater number of particles (and thus surfaces) in a given volume of soil, the higher the field capacity must be. For example, clay can hold nearly six times more water at field capacity than sand can. Once field capacity is reached, any water added to the soil cannot be taken up as capillary water but remains in the liquid form, draining through the soil as gravity water.

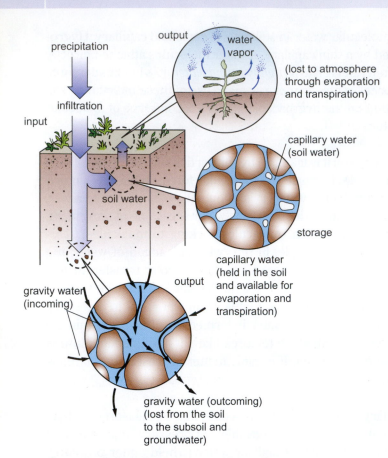

Figure 12.16 The soil-water system with input from infiltration, storage as capillary water, and outputs as gravity water and vapor.

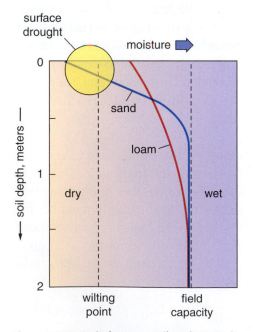

Figure 12.17 Typical summer soil gradients in the upper 2 meters of soil for sand and loam.

Inflow and Outflow of Soil Moisture: A remarkable quantity of water moves into and out of most soils in a year. This system, shown Figure 12.16, has four basic parts. *Input* is provided by precipitation. In the humid regions of North America, for instance, soils commonly take in 50 to 100 centimeters (20 to 40 inches) of infiltration water annually. About half of this water is taken up in *storage* as capillary water and then in the summer returned (*output*) to the atmosphere through evaporation and transpiration, or **evapotranspiration**. The remainder passes through the soil as gravity water, and is released (*output*) to groundwater at some depth. Thus, over the course of a year, there is a considerable amount of water moving up and down in the soil and, as it does, it transports chemicals and particles, thereby altering the internal makeup of the soil.

The loss of soil moisture in evapotranspiration is greatest when air and soil temperatures are high and wind is strong and dry. Losses are even greater when the soil is plant-covered because plants draw moisture rapidly from the upper soil. Uptake rates can be surprisingly large. For example, temperate forests are capable of taking up and transpiring 30 to 40 percent of annual precipitation. In the marine west-coast climatic region of western North America, forests are able to draw 60 to 75 centimeters (25 to 30 inches) of water from the soil annually.

Because moisture is lost first from the upper soil, a **soil–moisture gradient** develops between the dry surface layer and the damp soil under it. This gradient drives an upward transfer of capillary water, a common condition of summer throughout much of the world. The rate of moisture transfer or conduction along the gradient is controlled by the soil's capillarity. Thus, capillarity is a critical determinant of a soil's susceptibility to drying and drought. For example, coarse-textured soils, such as sand, with ample water a meter or so down, often experience surface drought because they lack the capillarity to move water up to the surface layer where plant roots are concentrated. The same situation in a silt or loam soil may produce little or no surface drought because these textures have far greater capillarity and can therefore move water to the surface. The graph in Figure 12.17 shows the difference between sand and loam.

If moisture from precipitation, irrigation, or flooding is added to a dry soil, the uppermost segment of the moisture profile is reversed. And as gravity water is diffused downward, it is quickly transformed into capillary water, recharging the soil-moisture reservoir. Rarely, however, can a single rainstorm recharge the capillary reservoir. Thus, the soil is left stratified with an intermediate dry zone sandwiched between two moist zones as is shown in Figure 12.18a. Double gradients are thereby formed, a common condition in the fall after summer drought. Should rainy conditions prevail for an extended period of time, as it typically the case during the winter months in areas of Mediterranean climate, for example, the amount of water infiltrating the upper soil may greatly exceed field capacity. In that case (Figure 12.18b), enough gravity water can be produced not only to recharge the capillary reservoir but to generate substantial flow through the soil and into the groundwater reservoir below.

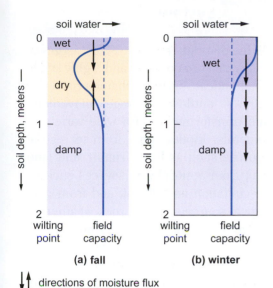

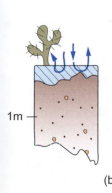

Figure 12.18 Typical soil-moisture gradients (a) during recharge in a dry soil; (b) in a recharged soil where the upper soil exceeds field capacity and releases gravity water to the subsoil.

Figure 12.19 The three principal soil moisture regimes: (a) humid with through-flow of water; (b) arid with little moisture penetration and essentially no through-flow; (c) semiarid with moderate moisture penetration and a little through-flow.

Soil–Climate Moisture Regimes: Outside permanently frozen lands, the climates of the world give us three general soil-water regimes or models. Each is based on the concept of soil as a water system with inputs, outputs, and storage. These models are critical to understanding soil development because the biological, chemical, and physical changes that take place in the soil layer are driven principally by water. An understanding of soil-forming processes thus rests in large measure on an understanding of the directions and rates of soil-water movement.

Humid Regime: In areas of positive soil-moisture balances, that is, where the soil receives more infiltration water than it gives up in vapor losses to evaporation and transpiration, the predominant direction of soil water movement is downward. Field capacity is exceeded for most months of the year and, as a result, gravity water is generated, which drains through the soil to become groundwater. Thus, the net direction of transport of dissolved minerals and small soil particles is downward (Figure 12.19a). Such conditions prevail in the humid, forested bioclimatic regions in the tropics and midlatitudes.

Arid Regime: At the other extreme are dry areas with negative or potentially negative moisture balances. Virtually all infiltrating water in such areas is lost to the atmosphere as vapor, and the potential exists (because of ample surface heat) to vaporize even more if it were available. Precipitation added to the soil surface penetrates only several centimeters before it is absorbed by the soil, taken up by plants, or evaporated (Figure 12.19b). The soil never reaches field capacity and thus generates no gravity water to carry minerals downward. When soil water is present, capillary water moves upward toward the dry surface bringing dissolved minerals with it, which are left on the upper soil when the water evaporates. Such conditions prevail in desert bioclimatic regions such as the Sahara, the Great Australian Desert, and the Atacama.

Semiarid Regime: Zones intermediate between arid and humid, namely, semiarid and wet–dry climatic zones, are characterized more or less by balanced soil-moisture regimes. The soil-water system usually has a strong seasonal component with massive water losses in summer (or winter in the tropical wet–dry climate) and substantial recharge in fall and winter. Winter recharge may exceed field capacity but the gravity water surplus is not great enough to produce much through-flow (Figure 12.19c). Soil moisture is sufficient to support a substantial grass cover, but not great enough to promote high rates of organic decomposition. Therefore, organic matter, its chemical residues, and various minerals tend to remain in the upper 1 to 2 meters of soil.

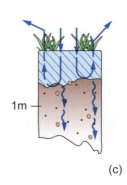

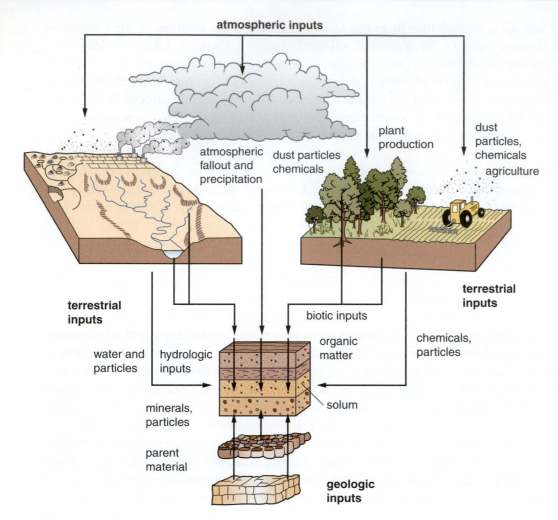

Figure 12.20 The larger system responsible for most chemical inputs to the soil is tied into all Earth's great spheres/systems as well as land use.

12.5 Soil as a Biochemical System

As the diagram in Figure 12.20 suggests, soil is part of a complex and highly active biochemical system characterized by inflows, outflows, storage, movement, and transformation of a wide variety of minerals. Inputs come from terrestrial sources such as biota, runoff, and land use; from atmospheric sources such as precipitation and fallout; and from geological sources such as rock weathering and groundwater. They contribute matter in the form of particles of clay or silt, for example, or in the form of chemical ions (dissolved particles). Once in the soil, chemical matter may be stored, passed through the soil and released, mainly to the water system, or chemically altered and relocated by processes within the soil.

The chemical inputs from biota come from both plants and animals. For example, organic litter falling on the ground from vegetation contains, among other elements, nitrogen, calcium, and phosphorous. These elements are released when organisms such as bacteria break down the litter. This process may involve earthworms, which feed on the bacteria, consuming tiny fragments of litter that are passed through their digestive systems. The worms leave digested bacteria, partially decomposed litter fragments, and ultimately their own bodies in the soil, all of which contribute to the soil's chemical makeup and processes.

A surprisingly large amount of chemical matter is contributed by the atmosphere. It comes not only with precipitation, which may carry both dissolved and particulate matter, but from direct fallout of dust. Among other things, dust may be rich in calcium, silicon, and sodium. Volcanoes and deserts are major contributors. It is estimated, for example, that 240 million tons of dust are blown from the Sahara Desert in the average year. Much ends up in the ocean but much also ends up on land where it is added to the soil. Soil scientists refer to the process of adding minerals to soil through deposition as **enrichment**.

Chemical Processes of the Soil System: Ions are essential components of the soil's chemical system. They are electrically charged atoms or groups of atoms representing different elements such as hydrogen, iron, and potassium. An ion's electrical charge may be positive or negative. Those with positive charges are called **cations**; those with negative charges are called **anions**. Certain substances forming cations,

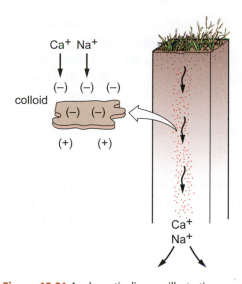

Figure 12.21 A schematic diagram illustrating a soil colloid with abundant negative charges that serve as adsorption sites for cations. In the absence of adsorption, Ca+ and Na+ are washed down with gravity water.

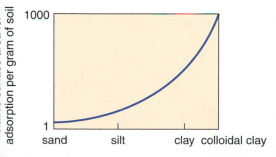

Figure 12.22 A graph showing the difference in cation-adsorption areas (per gram of soil) in sand, silt, and clay.

such as magnesium and potassium, are referred to as **bases** and some are important to soil fertility and hence plant growth.

Equally essential to the soil chemistry system are tiny clay-sized particles known as **colloids**. These minute particles carry small electrical charges on their surfaces, usually positive on one side and negative on the other, somewhat as shown in Figure 12.21. When ions are released into soil water, they are attracted to these charges, giving rise to swarms of ions on each colloid. This process is called **adsorption**, and it operates according to the principle of electromagnetic attraction in which unlike charges attract and like charges repel. Because the negative charge on colloids tends to be appreciably stronger than the positive, colloids adsorb mainly cations. Thus, when colloids loaded with cations are washed from the soil by percolating water, a process termed **translocation**, bases are often lost and soil fertility is reduced (Figure 12.21).

Soil Fertility: Why are some soils so much more fertile than others? Many factors are involved including nutrient supplies, nutrient types, and soil erosion, as well as translocation of nutrients. Supplies can vary, for example, with plant productivity, sources of dust and other sediment, and animal activity. But for soil to be fertile, one thing is certain: it must also be able to retain the nutrients it receives. Nutrient retention takes place through a process called **cation exchange**, which is closely related to the abundance of colloids in soil.

The cation-exchange process involves two groups of ions: nutrient bases and hydrogen. Most hydrogen ions come from water and they are naturally abundant in soil but they do not add to soil fertility as nutrient bases do. When nutrient bases are not present in the soil swarms of hydrogen ions (H^+) occupy adsorption sites on colloids. But when bases such as phosphorus (P^+) and calcium (Ca^+) are introduced, the hydrogen ions are displaced (exchanged) and the bases take up the adsorption sites. Exchange is possible because most nutrient bases have stronger electrical charges than hydrogen ions. In order for plants to absorb these nutrients, however, the bond must not be too tight. Thus, the most fertile soils not only have good base supplies, but bases that can be readily extracted by plant roots.

A soil's **cation-exchange capacity** is controlled mainly by its texture and composition. Clayey soils have much higher cation-exchange capacities than coarse-grained soils simply because clays are loaded with colloids and together these tiny particles contain a huge number of adsorption sites (Figure 12.22). And if the clays are organic, the exchange capacities are even higher. As a result, organic-rich clayey soils are usually good reservoirs of nutrient bases.

The relative abundance of hydrogen ions in soil is used as a measure of the soil pH factor. **Soil pH** is an indicator of the balance between acidity and alkalinity and is expressed in a numerical scale, shown in Figure 12.23, which ranges from 0 to 14. From pH 0 to 7, hydrogen ions are abundant and soil is **acidic**; whereas from pH 7 to 14, hydrogen ions are less abundant and soil is **alkaline** or **basic**. A pH of 7 is neutral. Pure water is neutral. Most soils range between pH 4 and 9, with the soils of dry regions tending to be alkaline and the soils of humid regions tending to be acidic.

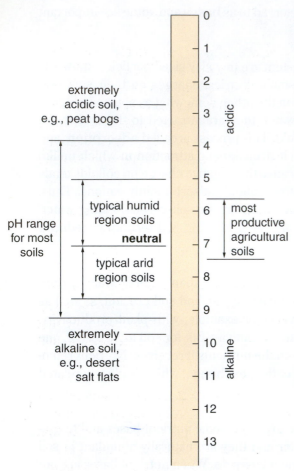

extremely acidic soil, e.g., peat bogs

pH range for most soils

typical humid region soils

neutral

typical arid region soils

extremely alkaline soil, e.g., desert salt flats

acidic

most productive agricultural soils

alkaline

Figure 12.23 The standard pH scale ranges from 0 to 14 with numbers below 7 representing acidic and those above 7 basic. Most soils range between pH 4 and 9 and in geographic terms represent the difference between very wet and very dry climatic regions.

Many plants are pH sensitive and in agriculture it is often necessary to adjust the soil pH for certain crops. If a soil's pH is too low, lime is usually added to "sweeten" the soil. This is often the case with coarse-textured soils in areas of heavy rainfall, for example, where most base nutrients have been washed from the soil. Lime is a finely powdered form of limestone or calcium carbonate ($CaCO_3$) and when added to soil, part of the calcium (Ca) is adsorbed, thereby reducing the number of hydrogen ions. With fewer hydrogen ions, the pH value is raised, making the soil more basic or "sweeter," and thus more fertile for many plants.

Soil Leaching: Most soil water contains an assortment of dissolved minerals, giving it the character of a weak chemical broth or cocktail. When this water moves, it carries these mineral ions with it. The process of translocating or "washing out" mineral ions from a soil is termed **leaching**. Leaching can result in the relocation of minerals to other levels in the soil or in complete removal of minerals from the soil column.

If the washing-out process includes colloids as well as ions, the term **eluviation** is used to describe it, and the layer or zone in the soil losing these materials is called the **zone of eluviation** or zone of removal. From the zone of eluviation, ions and colloids may move up or down depending on the direction of soil-water movement. In dry climates, the movement is usually upward with the transfer of capillary water toward the surface, as illustrated in Figure 12.24a. As this water evaporates, it leaves its load of ions and colloids in the upper soil, a process called **illuviation**. In time, these materials accumulate in the upper soil forming a distinct layer called a **zone of illuviation** or zone of accumulation.

In humid climates, the translocation process is far stronger because a lot of water is passing through the soil. The predominant direction of water movement, and hence ion and colloid movement, is downward with the transfer of gravity water. Some ions, especially calcium, sodium, potassium, and magnesium, are washed completely through the soil and lost to groundwater as the diagram in Figure 12.24b shows. But others, in particular oxides of iron and aluminum, are precipitated out within the soil resulting in a zone of accumulation at some depth, usually between a few tens of centimeters to several meters. In time, this zone can collect such a mass of colloids and mineral oxides that the interparticle spaces become clogged, cementing the parent particles together. The result is a concrete-like layer, commonly referred to as **hardpan**, which can range in thickness from a few centimeters to several meters.

Figure 12.24 (a) The processes of elluviation and illuviation under arid conditions resulting in accumulation of minerals and colloids in the upper soil. (b) The same processes under humid conditions results in downwashing of minerals and colloids producing a zone of accumulation at depth within the soil.

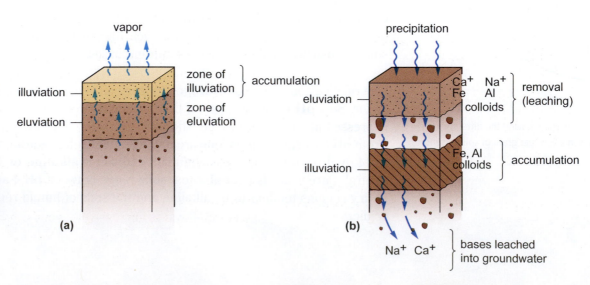

12.6 Soil Horizons, Profiles, and Formative Systems

Working in combination, the various chemical, biological, and physical processes differentiate soil, producing features called horizons. **Horizons** are horizontal or nearly horizontal layers, which are generally distinguishable on the basis of color, texture, and composition. Most horizons form into the upper 1 to 2 meters of soil and virtually all the processes that produce them are driven by or are closely related to water in the soil. And since the availability of moisture is broadly related to climate, it is not surprising that the number and types of soil horizons often reflects the climate of an area.

In the humid climates of the midlatitude environments where the soil surface is stable, well drained, and plant-covered, we can usually expect to see three distinct horizons, an organic-rich upper layer; a zone of eluviation under the organic layer; and a zone of illuviation under the zone of eluviation. In arid regions, on the other hand, only one horizon may be visible, an accumulation zone at or near the surface. In unstable environments, such as sand dunes, beaches, or mountain slopes, no horizons may be found because the parent material may be newly deposited or continually shifting and, therefore, never in place long enough to undergo internal differentiation and horizon formation.

The Soil Profile: The sequence of horizons in a section of soil is called a **soil profile**, and it is the traditional framework used for describing most soils. Since the number and types of horizons vary with geographic conditions, there is no one profile that fits all soils. Not only do the number of horizons vary, but they often come in a wide variety of shapes, colors, textures, and contents, all products of the various systems contributing to their formation. In the field there is the added difficulty of actually defining horizons in some soils because geographic conditions do not lend themselves to neat layering. On top of that the boundaries between soil horizons are often indistinct with one horizon grading almost imperceptibly into another. To address this problem, transitional zones are designated as subhorizons and several can usually be identified in each of the upper two or three horizons. For ease of reference, horizons are usually coded with letters and subhorizons with numbers.

At the most elementary level, just three horizons are recognized and coded, top to bottom, A, B, and C Zone. A is the organic zone; B is the eluviation zone; and C is the illuviation zone. At a slightly more refined level, five horizons are recognized and denoted O, A, E, B, and C as shown in Figure 12.25. They are grouped into an organic set (O and A) and mineral set (E, B, and C). The organic horizons range in composition from concentrated organic matter in the form of humus in the O horizon to mixed organic matter and mineral particles in the A. Generally regarded as topsoil, the A horizon receives inputs of organic matter from the O horizon above it, and also from penetrating roots, fungus, worms and other creatures.

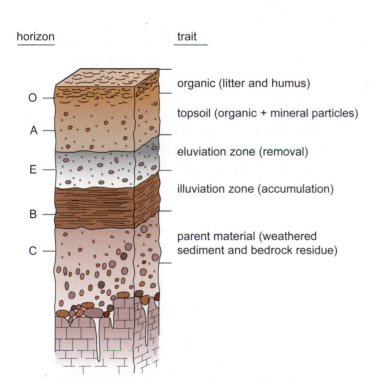

horizon   trait

O —   organic (litter and humus)

           topsoil (organic + mineral particles)

A —

           eluviation zone (removal)

E —

           illuviation zone (accumulation)

B —

           parent material (weathered
           sediment and bedrock residue)

C —

Figure 12.25 A standard soil profile with the O (organic) horizon at the surface and A, E, B, and C horizons below. Different horizons depths and composition reflect different formative conditions.

Below the A horizon is the E horizon, the zone of eluviation, which, in well-drained terrain under a humid climate, may stand out for its washed-out character marked by a grayish color. Next is the B horizon. This layer represents the zone of illuviation where, under the conditions cited above, oxides and colloids accumulate. But if we change climatic conditions to arid, the location and content of the B horizon changes dramatically. Instead of gravity water washing minerals out of the soil and into the groundwater system, under arid conditions soluble minerals are held in the soil, carried upward by capillary water, and, with evaporation, are left to accumulate at or near the surface.

The Systems Context: The point is that climate, as well as other great systems, notably, hydrology, ecology, and geomorphology, drive different responses in the soil-forming system, in the processes that produce horizons, and these differences are reflected in the soil profile in terms of: (a) where horizons are found in the soil column; (b) what they are composed of; and (c) what they look and feel like. Over large parts of deserts the B horizon, for example, is high in the soil column, heavy to salts and often whitish in color with the feel of plaster or weak concrete (see Figure 12.26). In humid, forested regions, it is commonly found a half meter or so down the column, is heavy to oxides of iron and aluminum, and usually rust colored with the consistency of sandstone. And in wet tropical regions, the B horizon is found 1 to 3 meters down, is very heavy to aluminum oxides, reddish in color, almost rock hard, and may extend to depths of 10 meters or more.

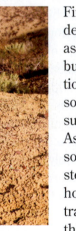

Figure 12.26 Surface exposures of the salt-rich B horizon in a desert soil.

Finally, there is the C horizon. In most places this horizon is made up of surface deposits, that is, beds of sediment originally laid down by a depositing agent such as runoff or wind. It is the parent material in which the overlying soil has formed, but it shares none or few of its horizon features such as the mineral concentrations described above. Nevertheless, the C horizon plays a significant role in the soil system. Most notably, it is an important source of moisture because it is not subject to the severe heating and drying of the surface and near-surface horizons. As a result, in summer the C horizon is a major source of capillary water for the soil root zone (A horizon). In drought-prone climates, many plants, such as the blue stem grasses of the Great Plains, have evolved deep root systems that penetrate this horizon, and deeper, to draw on its moisture reservoir. In addition, with the upward transfer of water by capillary action and plant roots, minerals are transferred from this horizon to the upper soil.

Chapter Summary and Overview of Soil Systems, Processes, and Formation

It is necessary for us to see soil as both a system in itself and the product of a host of larger geographical systems. The larger systems, climate, geomorphology, hydrology, and ecosystems, provide energy and matter to the soil system. They are the drivers of the soil system and the soil system responds by moving chemicals, growing plants, and forming horizons, and if the drivers change when, for example, we move from one location to another, as from the desert of Arizona to the tropical forests of southern Mexico, we expect the soil systems to change accordingly.

The same can be expected when geographic conditions change over time as when the climate of a region changes from wet to dry. But the time it takes the soil system to respond, to reach an equilibrium with new conditions at a given location, may be hundreds or thousands of years. Given that geographic change – not just climate but any of the other external systems as well – is common to our planet, we should expect many soils to bear the imprints of past geographic conditions. And this is where we pick up the story in the next chapter by not only touching on traits inherited from past conditions but also by introducing a new driver ... humans and their plows and animals.

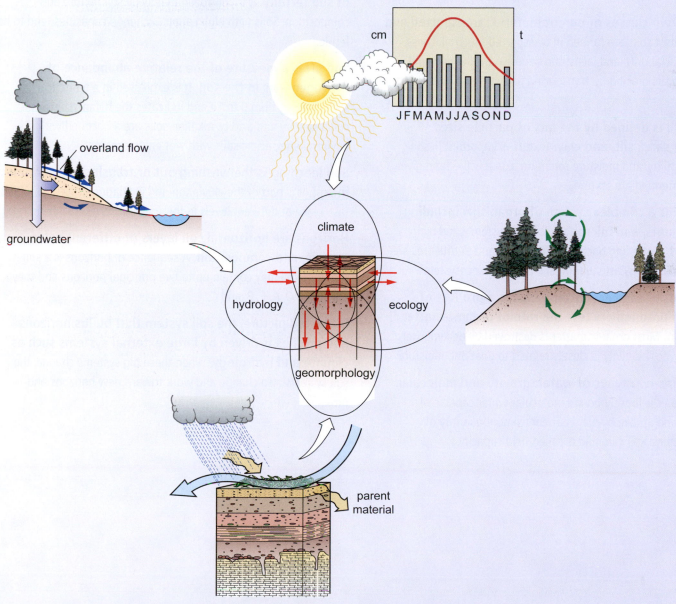

The principal geographic systems of the solum, the systems responsible for most of the inputs and outputs of energy and matter. As these systems change, so changes the soil.

- ▶ **Soil can be thought of as a transitional medium, part of a gradient or continuum between rock below and air, water, and biota above.** It is a mixture of particles and fluids made mostly of rock and mineral particles.

- ▶ **Soil functions as an input–output system.** It receives matter and energy, stores it, alters it, and then releases it. This works for each of the soil's constituents, water, mineral particles, organic matter, air, heat, and various chemicals.

- ▶ **The collective mass of loose particles that covers the land can be called the soil mantle.** This vast blanket is fed with particles weathered from bedrock in the continental interior. As a system, it moves as a system slowly toward the sea where the particles are deposited.

- ▶ **There are two classes of parent material: transported and residual.** Most soils are formed in transported material. Loess, alluvium, glacial drift, and lacustrine deposits are the principal surface deposits making up transported parent material in North America.

- ▶ **Soil texture is defined by the mix of particle sizes, principally sand, silt, and clay.** Texture is important in soil drainage, fertility, and moisture retention. Loam is the term given to soils of intermediate texture.

- ▶ **Soil contains a complex system of organisms including animals, plants, and microbes.** This ecosystem is fed by macrophytes, algae, and bacteria. Consumers and detritivores are abundant and energy attenuation is rapid in soil ecosystems.

- ▶ **The amount of organic matter in soil is a product of the rate of productivity less the rate of organic-matter destruction.** Most organic matter is destroyed by decomposer organisms whose activity is closely related to heat and moisture.

- ▶ **Soil contains two types of water: gravity and molecular.** Capillary water is loosely-bound molecular water capable of movement and uptake by plants. The maximum quantity of capillary water a soil can hold is called field capacity.

- ▶ **Soil moisture balance is the product of moisture inputs minus moisture losses.** For capillary water held in the soil column, evapotranspiration accounts for the outputs. Soil moisture is recharged with infiltration from the surface.

- ▶ **There are three general soil-moisture regimes in the world.** Each is associated with a different climatic zone, humid, dry, and semiarid.

- ▶ **Soil is a complex and highly active chemical system.** It involves inputs, storage, transformation, movement, and output of a variety of chemicals. Ions and colloids are key players in the chemical system.

- ▶ **Cation-exchange capacity is an important determinant of soil fertility.** It is controlled mainly by soil texture and composition. Soils with high cation-exchange capacities tend to be fertile.

- ▶ **Soil pH is a measure of the relative abundance of hydrogen ions in the soil.** It is expressed in a numerical scale ranging from 0 to 14 and indicates the balance between soil alkalinity and acidity. Alkaline soils are associated with dry environments, acidic soils with wet environments.

- ▶ **Soil leaching is the washing out or translocation of ions in soil.** It is part of the eluviation and illuviation processes, which take place at different levels in the soil.

- ▶ **Horizons are horizontal soil layers of different color, texture, and composition.** A sequence of horizons is a soil profile. A soil may contain up to five principal horizons and they are coded O, A, E, B, and C.

- ▶ **In the big picture, the soil system that builds horizons and profiles is driven by large external systems such as climate and hydrology.** When these big systems change, the soil systems also change and work toward new horizons and profiles.

Review Questions

1 Referring to Figures 12.1 and 12.2 and the text, what are some of the forms of energy and/or matter contributed to soil by these four great systems: atmosphere, biosphere, hydrosphere, and lithosphere?

2 If we view soil organic matter with a systems perspective, based on Figure 12.4, can you describe the relative balance of the system before and after agriculture? What would it take to restore equilibrium in the organic system of soils used for crop agriculture?

3 The soil mantle can be thought of as an input–output system. What factors are currently working to increase the outputs from this system?

4 Using the text and the map in Figure 12.8, identify the types of parent material you would expect to find in northern Missouri and southeastern Manitoba? What process is responsible for each of these deposits and what, generally, is the composition of these materials?

5 What is meant by the terms soil texture and loam, and where would you expect to find soils of sandy texture? Does texture have anything to do with moisture or fertility in soils?

6 Refer to Figure 12.10 and consider where the decomposing organic matter in soil originates from. Does soil house ecosystems, and if so, how do the food chains function? Describe one.

7 Using figures 12.12 and 12.13, describe the sort of organic mass balance you would expect in soils under these four general climatic regimes: (a) hot and rainy; (b) hot and dry; (c) cool and moist; (d) cold and dry.

8 Soil is a hydrologic system, which receives inputs of water, stores this water, and then loses (outputs) some or all of this water to other locations. Using the text and Figure 12.16, identify the sources of the soil's water inputs, the types of storage, and the different ways water is released (output) from the soil.

9 Identify the three general soil-moisture regimes in the world. With the assistance of Figure 12.19 and the text, determine the dominant vegetation types present in each regime.

10 What is the role of the atmosphere in soil enrichment?

11 A soil's cation-exchange capacity is an important factor in determining its fertility. Using Figures 12.21 and 12.22, describe the role of the positive and negative ions in cation exchange.

12 Referring to Figure 12.24, how do you account for the differences in the relative locations of the zones of illuviation and eluviation in (a) and (b)?

13 If all of the processes that work within a soil to create horizons are part of the soil system, what external systems drive the soil system and what do they contribute?

Soil Types, Distribution, and Land-use Relations

Chapter Overview

We open this chapter where we left off in Chapter 12, on the nature of the relationship between soil and the geographic environment. The idea here is to provide a simple set-up for classifying soil, beginning at the most elementary level with a two-part scheme and then moving on to classification systems widely used by scientists to describe and map soil. Though sometimes a little challenging to understand, these systems are important learning tools because they enable us build coherent discussions about the character of common soils and what traits make them noteworthy. In addition, without a means of classifying soils, there is no way we can build soil maps and examine the distribution of soils in relationship to other geographic phenomena like climate. The second part of the chapter is concerned with human use and abuse of soil, both past and present. The role of soil and its depletion in the destiny of early civilization is examined, and this is followed by a look at the pressing issue of soil management and food production in the modern world.

Introduction

Besides its foundational role in the terrestrial life system, soil is a tablet upon which are recorded Earth's environmental conditions, that is, the nature of all those things that interact to shape the geographic character of a place on the Earth's surface. Etched into soil, in its composition, its chemistry, its horizons, is an accounting of the systems that have operated there over time. It tells us about climate, moisture, vegetation, erosional processes, and human land use because soil is acutely responsive to these forces. And when we examine soil, the effects of these influences show up in certain diagnostic traits in the soil column.

A diagnostic trait that appears in some soils along the Pacific coast from Alaska to California is a thick layer of organic matter blanketing or packed into dry, sandy deposits along the bank behind the shore. Anywhere else we would associate such concentrations of organic matter with wet places like bogs and swamps, but here there is no evidence of wet conditions, past or present. The explanation has to do with human activities. Over the past several thousand years or more, native North Americans inhabited accessible parts of the coast where they gathered

oysters, clams, fish, and other game, the discarded parts of which ended up in scrap heaps called middens. Old middens, such as the one shown in Figure 13.1, became buried forming deeper organic layers, whereas more recent ones form organic layers closer to the surface.

Only short distances away from midden sites, we are apt to find places where beach sand is being swept inland by the wind. Each year several inches of fresh sand are added to the soil surface just inland from the beach, and over time these sand deposits come to dominate the soil column. No horizons form because this parent material is constantly being renewed. In other words, the soil material is so new that climate, drainage, vegetation, and chemical processes have not had time to do their work and leave their marks.

Further inland, beyond the wind deposits and the native fish camp sites, the landscape is older and more stable and the soils more reflective of regional climate, biota, and drainage conditions. Here the soils formed in parent material deposited long ago and are differentiated into horizons. We can read these horizons as diagnostic indicators of an older soil-forming environment, one that has been operating here at least several thousand years. The point is that soil is principally a product of the environment in which we find it, but in some places that environment may extend back in time to include natural and human events of the past. In the North American Great Plains, for instance, there are vast areas of rich, silty soil (loess parent material) deposited over 10,000 years ago by wind blowing from alluvial plains formed by glacial meltwater (Figure 13.2a) and, in the landscape of Iraq, thousands of square miles of soil still bear the harsh effects (in the form of heavy salt concentrations) from over-irrigation by Sumerian farmers who occupied this area over 5000 years ago (Figure 13.2b).

Fig 13.1 Organic deposits, called middens, left by native North Americans are prominent soil features at some locations along the Pacific coast. The white material is mainly oyster shells.

Fig 13.2 (a) The soils over large areas of the North American Great Plains have formed in parent material of wind-deposited silt (loess) deposited thousands of years ago. (b) The soils of Mesopotamia in Iraq still carry salt deposits such as these from over-irrigation dating back to the ancient Sumerians more than 5000 years ago.

13.1 Soil Formation and the Geographic Environment

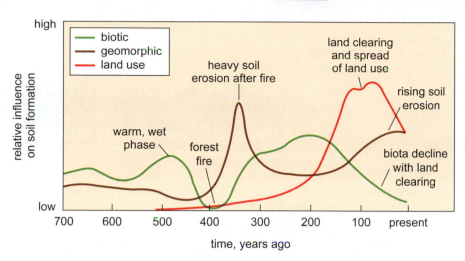

Fig 13.3 The relative influence of three soil-forming factors over seven centuries. Notice how the three are interrelated, for example, when biota decline soil erosion increases and the geomorphic factors take on a more dominant role.

In the big scheme of things it can be argued that soil is the product of a complex interplay among five major systems: climate, biota, drainage (and topography), parent material (including geomorphic factors), and human activities. These systems, along with time, are known as the **soil-forming factors**. In order to understand the character of soil at some location you must consider not only the effects of these factors, but also how long they have been operating on the soil. And it is not merely the passage of years that counts, but the intensity or rate at which the various soil-forming factors operate that is more important, as the graph in Figure 13.3 attempts to show. So we must conclude that not all soil-forming factors share equally in a soil's development. In a great many geographic settings, as the examples in the introduction above illustrate, one or two factors are often dominant and they govern a soil's character.

When we search for a relationship between the distribution of soils and conditions of the environment, we are immediately faced with the problem of geographic scale. At a very broad scale, the general pattern of major soil types correlates fairly well with bioclimatic conditions. At increasingly finer scales, the influence of other factors, such as parent materials, drainage, land use, and vegetation, become more apparent. Recognizing this principle, let us describe the relationship between soils and environment by starting at the broad scale and working our way down to the local scale.

If we ignore the polar and mountainous regions, two global families of soils can be identified corresponding to wet and dry climatic regions. These are called the pedocals and pedalfers. **Pedalfers** are humid-climate soils where leaching is dominant. The *al* and the *fer* in the pedalfer refer to the concentrations of aluminum and iron (ferum) that are found in the zone of illuviation. **Pedocals** are dry-climate soils where evapotranspiration is dominant. The *cal* in pedocal refers to the calcium retained in the soil as a result of the shallow penetration and evaporation of surface water. Pedalfers are generally found where annual average precipitation exceeds 60 centimeters a year, whereas pedocals are found where the annual average precipitation is less than 60 centimeters. In the United States and south–central Canada a north–south line along the 98 degree meridian, as shown in Figure 13.4, can be used as the approximate boundary between these soils, with pedocals found in the drier west and pedalfers in the more humid east.

Bioclimatic Soil-Forming Regimes: Within the pedalfers and pedocal families are several soil-forming regimes corresponding loosely to various bioclimatic zones such as tropical forest of the tropical wet climate. Each regime is characterized by a certain combination of bioclimatic conditions and related soil-forming processes, usually some combination of rainfall, temperature, vegetation, drainage, and biochemical processes. We will briefly examine four such regimes: humid midlatitude, tropical wet, grassland, and desert.

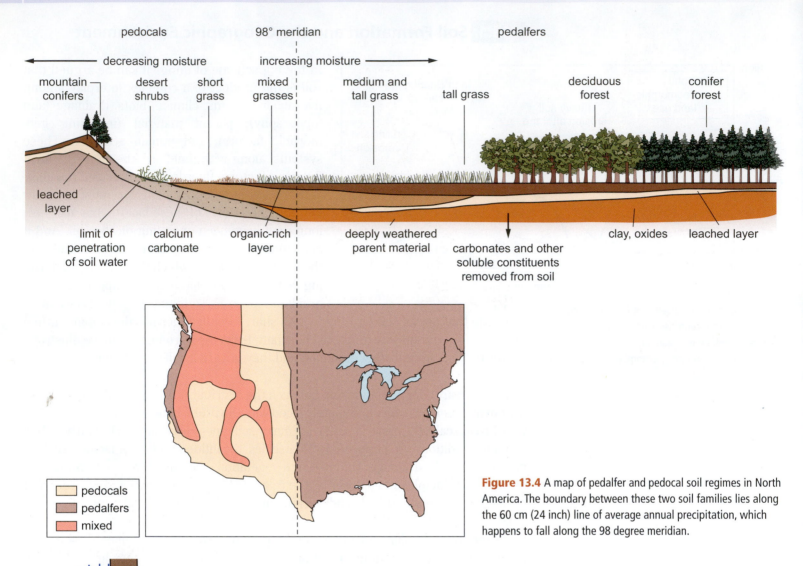

Figure 13.4 A map of pedalfer and pedocal soil regimes in North America. The boundary between these two soil families lies along the 60 cm (24 inch) line of average annual precipitation, which happens to fall along the 98 degree meridian.

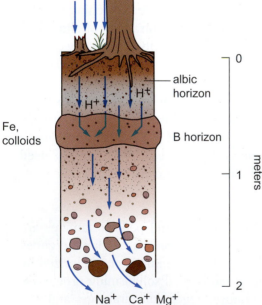

Figure 13.5 The principal features and processes of soils in the humid midlatitudes, such as Ontario or New England. Net water movement is strongly downward and these soils are differentiated into distinct horizons.

The **humid midlatitude regime** is common to forested areas in middle to high latitudes or at high elevations in moist mountainous areas. Climatic conditions in these zones are distinctly seasonal, cold enough in winter to inhibit rapid decomposition of litter, yet warm and wet enough to support substantial forests. Soil-water movement is strongly downward. Bases, colloids, and oxides are eluviated from the upper soil, leaving a grayish horizon just below the organic layer (Figure 13.5). This horizon, called an *albic* horizon, is a distinctive feature of midlatitude forest soils, especially in areas of boreal forest. Another distinctive feature is a dark B horizon where iron oxides, colloids, and clays have accumulated.

The **tropical wet regime** occurs under tropical forests with heavy rainfall and year-round warm temperatures. Vegetative productivity is high, but intensive decomposition rapidly destroys most organic litter so that the topsoil is always thin. Weathering of parent material is also intensive, and most weatherable materials are removed in solution, leaving the soil heavy with clay. As shown in Figure 13.6, water movement is strongly downward and iron and aluminum oxides accumulate in the B horizon forming a hardpan layer, traditionally known as *laterite*, which may become rocklike and massive in some areas. Silica is leached out, and no distinct horizons form other than the laterite layer and the thin layer of organic matter in the O horizon.

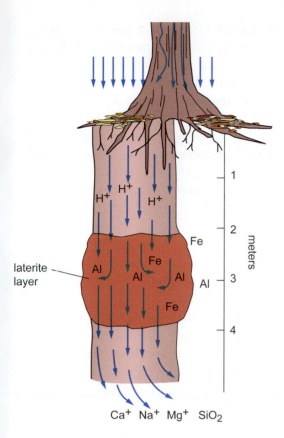

Figure 13.6 Soils formed in the tropical wet regime are deeply weathered with thin topsoil and heavy accumulations of oxides at depth.

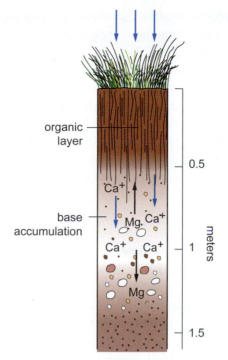

Figure 13.7 Under the semiarid regime, water penetration is modest and abundant grasses play a major role, which helps account for the heavy organic accumulation in the upper soil.

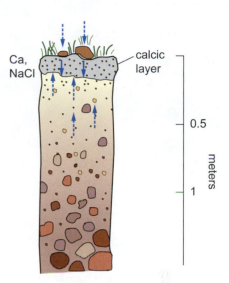

Figure 13.8 Under the arid regime, water penetration is slight and salts tend to accumulate at or near the surface.

Figure 13.9 Wetlands such as these are the result of impaired drainage and saturated soil. They are usually marked by a buildup of wet organic matter and a distinctive plant cover.

The **semiarid regime** is found in grasslands where the moisture balance ranges from even or slightly negative to moderately negative. As there is little leaching by infiltrating water, calcium, magnesium, and other bases remain in the soil. The grasses draw on these bases and produce appreciable organic matter which is recycled back into the soil each year. Limited moisture inhibits destruction of the organic matter, giving rise at length to a fairly heavy organic layer (Figure 13.7). Colloids are not leached out, and they too remain in the soil and may accumulate with calcium carbonate, forming pebble-like nodules below the organic layer.

The **arid regime** is an extreme variety of the semiarid regime. It is limited to desert landscapes where water penetration is slight and soluble minerals are not washed through the soil. As a result, salts accumulate at or near the surface often forming a layer called a *calcic* or *petrocalcic* horizon. Organic accumulation is negligible, of course, because of limited plant numbers and productivity owing to the severe moisture stress and to the heavy accumulation of salts in the rooting zone (Figure 13.8).

Regimes within Bioclimatic Zones: Within each of the bioclimatic regimes described above are smaller areas where soil formation is dominated by factors not directly driven by climate and biota. Among the most prevalent of these is impaired drainage resulting in waterlogged soil conditions. Instead of water moving in and out of the soil column in response, for example, to winter rains and summer drought, for some reason (such as impermeable parent material) water is held in the soil, saturating it throughout. This condition gives rise to wetlands, a common landscape throughout the world, which is often distinguished on the surface by its distinctive plant cover. The wetland-dotted landscape shown in Figure 13.9 is found in the Great Plains of North Dakota, a semiarid climatic zone.

Figure 13.10 Soil formation along the lower slope dominated by sediment brought down by runoff and rock slides along mountain slopes.

(a)

(b)

Figure 13.11 (a) Soil cracks caused by drying and shrinking which, over time, promote mechanical mixing of the solum. (b) Patterned ground related to mechanical mixing of the ground in the active layer over permafrost.

Wetland Regime: Saturated ground conditions can be found in any of the major bioclimatic regimes, including desert. However, they are most common in humid regions where runoff collects in low spots such as stream valleys or where water is prohibited from moving into the ground by some impermeable material such as permafrost or bedrock. This soil-forming regime, call it the *wetland regime*, is usually characterized by heavy accumulations of organic matter. This is explained by two conditions: (1) high plant-productivity rates in wet ground, coupled with (2) slow rates of breakdown in organic litter when it is immersed in water. The outcome is an organic system which is positively balanced. Organic matter builds up over time, reaching thicknesses of many meters in places where the water is fairly deep. The mass of saturated organic matter lacks soil horizons, though a thin clay layer sometimes forms at depth within the mass.

Geomorphic Regime: In rugged terrain, soil formation is often dominated by *geomorphic factors* such as the wind-blown coastal lands described in the chapter introduction or mountain slopes where streams dump their sediment loads (Figure 13.10). The sediment accumulates and shifts about so quickly that plant growth, leaching, and related processes do not have enough time to have much effect on the soil. Such soils commonly lack horizons or horizons are poorly developed. Active sand dunes, shorelines, fresh accumulations of volcanic ash share these soil traits. Soils in these places are called *entisols* or *inceptisols* meaning, respectively, recent soil and young soil.

Geomechanical Regime: In yet other places, *mechanical processes* in the ground dominate soil formation. These processes tend to mix soil, thereby disrupting the formation of horizons related to bioclimatic factors. One version of this regime is found in areas containing clays that shrink and swell with drying and wetting. In the summer, the shrinkage may be so severe that deep cracks form like those shown in Figure 13.11a. From the walls of the cracks, chunks of soil slough off and fall in. Organic matter from the surface, for example, can sink to a depth of a meter or more. In winter, the cracks close as the soil takes on water and expands. After many episodes of cracking, sloughing, and closing, the soil turns over or inverts and the horizons get mixed together. In the American soil classification system, such soils are called *vertisols*, meaning turned soil, and in North America they are found in the semiarid climatic zone where there is a pronounced seasonal wet–dry cycle.

Another version of the geomechanical regime is found in permafrost environments where the upper 1 to 3 meters of ground, called the *active layer*, freezes and thaws from winter to summer. With each freeze–thaw cycle the ground, which is saturated with meltwater, expands and contracts. This action tends to produce a mixing motion in the soil, which drives stones to the surface and separates them from finer particles. In some places the frost action results in patterned ground marked by polygons of various sizes and shapes such as those shown in Figure 13.11b. As the ground is slowly churned by frost action, any tendency toward horizon formation is eliminated in soil development.

Crop farming also produces mechanical mixing action in soil when the ground is plowed. Plowing turns the upper 40–50 centimeters of soil, mixing topsoil and mineral soil while loosening and aerating the soil. This can obliterate the O and A horizons and, considering that nearly 12 percent of Earth's land is subject to crop farming, agriculture must also be ranked as a major force in soil formation.

Figure 13.12 Soil patterns in a farm field. Light and dark patches reflect differences in topsoil development related to subtle variation in drainage and topography.

13.2 Influences on Soil Formation at the Local Scale

If you examine a neighborhood park, a farm field, or a woodlot, you are apt to discover surprising geographic variation in soil over a small area. Various factors are usually responsible, some which may be identifiable in the landscape and some not. One of the most common causes of local soil variations is differences in drainage related to modest changes in topography. Even slight variations in the lay of the land can dramatically affect drainage, which in turn can influence not only the distribution and productivity of vegetation but where water and sediment accumulate. Soils in low spots, like those in Figure 13.12, tend to be wetter with more organic matter and finer textures than those on slightly higher ground.

But the observable surface features may only be part of the picture. Past events such as floods, deforestation, and farming also have marked effects on local soil. For example, a past windstorm that uprooted acres of trees can have striking effects. When a tree's root mass is ripped from the ground, a surprising amount of soil is pulled out with it, leaving a shallow hole or pit. The pit forms a little basin that collects water, organic matter, and sediment. While the pit is filling, the upended root mass decays and the soil particles held among the roots fall to the ground, forming a small mound adjacent to the pit (Figure 13.13). The mound is usually poor in organic matter and dry compared to the pit which is moist and rich in organic matter. Repeat this sequence over a vast area where scores of trees have been blown down. The ground is left checkered with mounds and pits with corresponding patches of organic-rich and organic-poor soils. Such a pattern of soils can remain in the landscape for centuries or longer, long after the upended trees have rotted away.

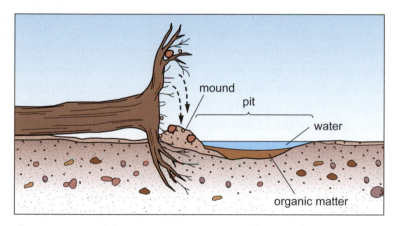

Figure 13.13 Wind-thrown trees create mounds and pits which lead to patchy patterns of soils, especially in organic content.

Figure 13.14 A typical landscape scene in central North America where the influence of soil on the character of the landscape may or may not be apparent.

Summary on Soil Formation and Environment: We end this section with the conclusion that any attempt to relate soil to the environment is limited by the geographic scale of observation. At the broadest scale (barring the polar regions) a simple two-part scheme that divides soils into two great climatic-based families, humid and dry, makes sense. Within these families, 4 bioclimatic soil-forming regimes can be defined, and within these are several soil-forming regimes driven by forces not directly tied to the bioclimatic environment. At the local scale variations in soil are not only related to factors like topography and drainage, but to past conditions including events such as floods, fires, and past land uses. Time erases some of the effects of the past but many remain and contribute measurably to the geographic character of landscapes. Some are apparent at a glance, but others are hidden beneath a mantle of vegetation, sediment, and/or land use (Figure 13.14).

13.3 Soil Classification and Distribution at the Regional Scale

Many soil classification systems are in use today. In general they fall into two groups: (1) those developed by engineers, and (2) those developed by soil scientists. Engineering systems address mainly soil performance related to built facilities such as roads, building foundations, and dams. These systems do not address soil as part of the geographic environment and give no attention to formative processes, diagnostic features such as horizons, and geographic distributions.

Table 13.1 American soil classification system: orders, traits, and landscapes

Soil order	A key trait	Typical landscapes
Entisol	Recent soil	Mountain slopes
Inceptisol	Young soil	Glaciated lands
Andisol	Volcanic ash	Volcanic deposits
Aridisol	Arid soil	Deserts
Mollisol	Soft soil	Grasslands
Spodosol	Ashy soil	Boreal forests
Alfisol	Moist, soft soil	Temperate deciduous forests
Ultisol	Ultimate soil	Subtropical forests
Oxisol	Oxide soil	Tropical forest
Vertisol	Inverted soil	Wet–dry climates
Histosol	Organic soil	Bogs and swamps
Gelisol	Permafrost soil	Arctic tundra

Classification systems developed by soil scientists, on the other hand, *do* address distribution, formative processes, and relations with the geographic environment. For obvious reasons geographers favor these systems. Soil scientists in the United States have developed a system called the **US Comprehensive Soil Classification System** which is based on a concept of diagnostic horizons. **Diagnostic horizons** are distinctive surface and subsurface horizons that can be used to distinguish the soils of one class from another.

The US system is designed to serve as an international system in much the same fashion as the Universal System of Biological Names does for plants and animals. For various reasons, however, it has not been universally adopted. Several countries, including Canada, the United Kingdom, and Australia, have their own systems. Here we shall briefly examine soil classification and related geographic phenomenon as described by the American and Canadian systems.

The US Comprehensive Soil Classification System: The United States system provides six levels of classification. At the broadest (most general) scale it begins with orders, which are subdivided into suborders, great groups, subgroups, families, and series. Each subdivision is more specific than its predecessor with a finer scale of geographic resolution. At the most detailed level, the series class, the system is suitable only for local mapping and application. At this level, thousands of soil polygons (map units) have been defined in the United States.

At the global or national scale, soil orders are the appropriate mapping unit. At this scale, we can examine the relationship between the distribution of soils and the regional patterns of climate, vegetation, and landforms. The following discussion describes the main features of each of the twelve soil orders and identifies significant relationships with the geographic environment. The name for each order ends in "*sol*" which derives from the Latin for soil, solum. Table 13.1 gives a key trait for each order and a representative landscape where it is found. The map in Figure 13.15 shows the geographic distribution of the 12 soil orders in North and South America according to the US Comprehensive Soil Classification System.

Entisols: These are mainly soils of recent origins. Horizons are very weakly developed or non-existent. Entisols are products of the geomorphic regime, that is, of geomorphically active, or recently active, environments such as river floodplains, sand dunes, and mountain slopes like the one shown in Figure 13.16a. Entisols can be found in any bioclimatic region. In the United States, the largest areas of entisols are found in the mountain states, in particular Colorado, Utah, and New Mexico. The agricultural potential of entisols varies with the quality of the deposits in which they form. In the sandy deposits of the deserts and coastlines, they are notoriously poor, whereas in many of the great river deltas, where the deposits are rich in organics, and moist, they are highly productive.

Inceptisols: These are soils in which horizons are just beginning to form. They are found in a wide range of bioclimatic regions but are limited to mainly humid climates with favorable soil-moisture balances. Although one or more horizons are present, visual evidence of horizons is typically weak. Inceptisols are associated principally with young geomorphic surfaces such as those in the mountainous midlatitudes, or subarctic zones, that were deglaciated only several thousand years ago. Beyond these areas, they are widespread in the large river lowlands and mountain valleys (see Figure 13.16b for an example). In the tropics, where many landscapes have been without large-scale geomorphic disturbances for millions of years, inceptisols may be scarce. One exception is selected areas where there are ash deposits from a nearby volcano. A favorable moisture balance generally gives some inceptisols promising agricultural potential, especially where they have formed in rich parent material such as alluvium and volcanic ashes in temperate and warmer regions.

Andisols: These are soils which have formed on recently deposited volcanic ash and similar materials (Figure 13.16c). They are similar to inceptisols and entisols in that soil development has not advanced much. Thus, they tend to show little leaching

Figure 13.15 The distribution of soil orders in North and South America according to the US Comprehensive Soil Classification System.

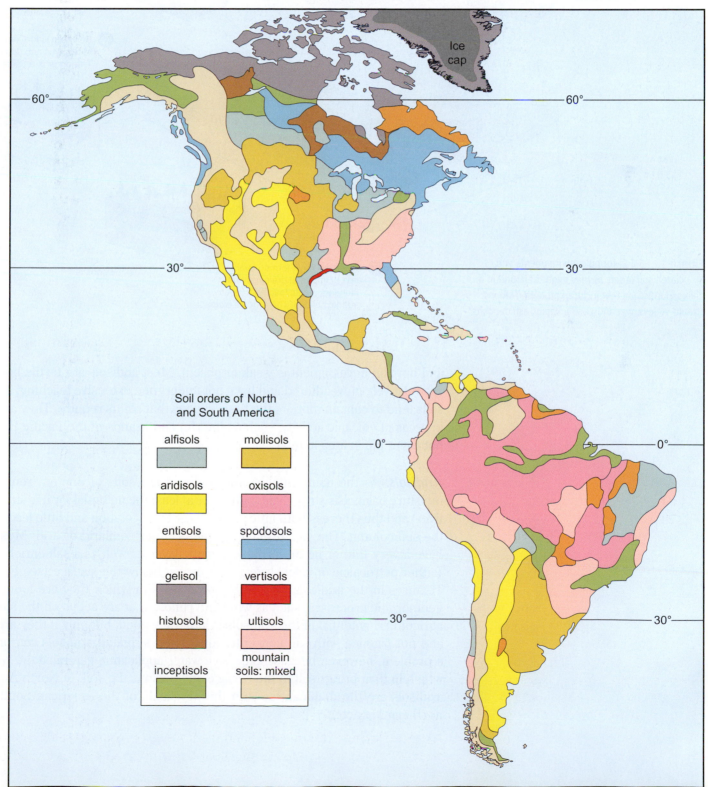

Soil orders of North and South America

alfisols
mollisols
aridisols
oxisols
entisols
spodosols
gelisol
vertisols
histosols
ultisols
inceptisols
mountain soils: mixed

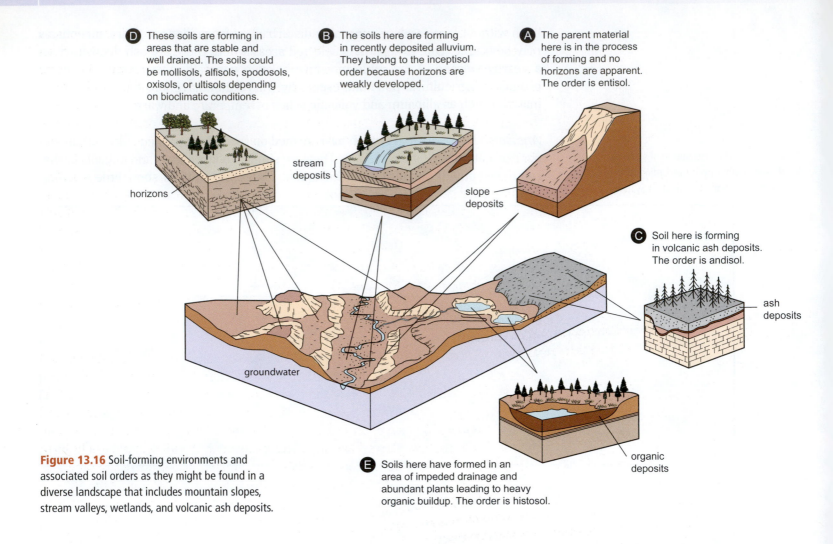

D These soils are forming in areas that are stable and well drained. The soils could be mollisols, alfisols, spodosols, oxisols, or ultisols depending on bioclimatic conditions.

B The soils here are forming in recently deposited alluvium. They belong to the inceptisol order because horizons are weakly developed.

A The parent material here is in the process of forming and no horizons are apparent. The order is entisol.

C Soil here is forming in volcanic ash deposits. The order is andisol.

horizons

stream deposits

slope deposits

ash deposits

groundwater

E Soils here have formed in an area of impeded drainage and abundant plants leading to heavy organic buildup. The order is histosol.

organic deposits

Figure 13.16 Soil-forming environments and associated soil orders as they might be found in a diverse landscape that includes mountain slopes, stream valleys, wetlands, and volcanic ash deposits.

and horizon development is weak or absent. Most andisols are fertile because they are relatively unweathered and have not undergone extensive leaching. Thus, these soils tend to contain more of the basic cations that plants require. They are common in areas of volcanic activity, such as the rim and islands of the Pacific Basin but do not cover much area in total.

Aridisols: As the name implies, the aridisols are soils of dry environments. The moisture balance is strongly negative – or at least the potential for it is strongly negative – and thus there is both little productivity by vegetation and little leaching within the soil column. The soil-forming regime is either semiarid or arid. Most aridisols have a very thin or no discernable organic layer as well as a salt-enriched horizon (either petrocalcic or calcic layer) near or just below the surface (see Figure 13.8). Textures in the aridisols vary with the deposits in which they are found and with geomorphic processes, such as wind and runoff, that are active on the surface. The agricultural potential of the aridisols is generally good, but only if they are irrigated and not plagued with salt concentrations. Even irrigated aridisols can prove to be a problem, however, because excessive water application can raise the water table, which in turn brings salt to the surface and poisons the soil. In North America, the aridisols are the dominant soils in the dry lands of the American Southwest and northern Mexico.

Figure 13.17 A section of a mollisol showing the strong organic horizon.

Figure 13.18 A section of a spodosol showing the distinctive E (whitish zone) B (brown zone) horizons.

Figure 13.19 A rural landscape in Europe, once forest-covered, where alfisols have long provided a reliable base for crop agriculture. The light area is the result of topsoil erosion.

Mollisols: These soils are generally associated with grasslands, and the semiarid soil-forming regime. Geographically, they are found on the humid side of the aridisols in the semiarid and subhumid climatic zones where summer drought is strong enough to induce a substantial upflow of capillary water in the soil column. There is, however, enough moisture to support grasses that in turn recycle nutrient bases in the upper soil and are responsible, together with the slow rate of organic decomposition, for the formation of a thick organic layer, called a *mollic horizon*. This diagnostic horizon, such as the one shown in Figure 13.17, is dark brown to black, generally 25 centimeters (10 inches) or more deep, and is very rich in exchangeable bases. Below the mollic horizon, the B and C horizons are often rich in calcium carbonate and, in the B horizon, this mineral often appears in nodules and white streaks. The mollisols are among the very best grain-farming soils in the world. In North America, they are found in the Great Plains and prairies of the United States and Canada. In addition to inherently high fertility, the mollisols are friable (easily crumbled) and retain soil moisture well because they are often rich in silt (loess) and organic matter. On the other hand, they are easily eroded by runoff and wind, and they are subject to drought by virtue of their climatic location.

Spodosols: These soils are distinguished by a pronounced zone of illuviation called the *spodic horizon*. This is an accumulation of various forms of iron oxides and aluminum oxides together with organic and mineral colloids in a dark (and often hard) B horizon. Immediately above the spodic horizon, many spodosols have a pronounced whitish E horizon (shown in Figure 13.18) called the *albic horizon*, from which minerals and fine particles have been intensively eluviated. Spodosols generally form under forest covers in moist, cool midlatitude climates, but their development is most pronounced under coniferous forests and sandy parent material of the subarctic (Figure 13.15). Spodosol fertility is very poor owing to low base-retention capacity and acidic chemistry. Thus, their agricultural potential is modest and limited principally to crops such as potatoes, which are well suited to low soil pH and short growing seasons. Spodosols are widespread in southern and eastern Canada, the Great Lakes states, the northern Appalachians, the Pacific Northwest, and northeastern United States.

Alfisols: These soils form under forest covers in humid midlatitude climates. Organic matter is appreciable and base retention is fairly high. In the B horizon, clays accumulate, forming a light-colored layer known as an *argillic horizon*. Alfisols have fairly good agricultural potential owing to their favorable moisture balance and good fertility. Indeed, these soils in India and Europe have been successfully farmed for thousands of years giving rise to mature agricultural landscapes like the one shown in Figure 13.19. In North America, the alfisols are abundant on the older glacial deposits in the Midwest, in the loess deposits in and near the Mississippi Embayment, in the inner coastal plain of Texas, and at the northern, wooded, end of the Great Plains in Canada (see the map in Figure 13.15).

Ultisols: These soils are described as being in an advanced state of development, thus, the prefix "ulti" for ultimate. They are found in warm, moist climates, such as the humid subtropical, where the landscape has been free of major geomorphic disturbances for long periods. In many respects ultisols are older, more advanced

Figure 13.20 Unlike the alfisols of the midlatitudes, the ultisols of the humid tropics and subtropics have poor potential for crop farming. Here peasant farmers push back the rainforest (background) to clear cropland.

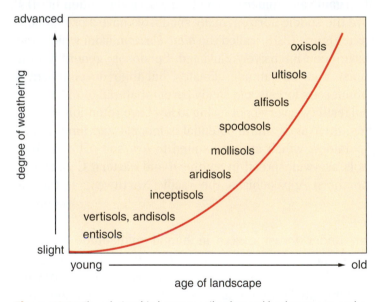

Figure 13.21 The relationship between soil orders and landscape age, or the time a landscape has been without wholesale change. The oxisols are deeply weathered soils that belong to some of the Earth's oldest tropical landscapes.

versions of the alfisols, heavily leached and deeply weathered, with a poor supply of bases that grows ever poorer with depth. Light-colored clays form an argillic horizon (illuvial layer) beneath the E horizon. The C horizon may be deep and underlain by deeply weathered bedrock. Ultisols are the dominant soils in the American South (see Figure 13.15). In China, they are found in a similar geographic location (the southeastern part of the country) and are also associated with the humid subtropical climate. Important areas of ultisols are also found in the wet tropics of South America. The ultisols pose serious problems for agriculture because they are quickly exhausted by crops such as corn, tobacco, and cotton (Figure 13.20). In poor countries of the wet tropics, they are associated with the practice of shifting agriculture in which peasant farmers abandon worn-out land and clear new areas for planting. Where they are used for sedentary agriculture, the ultisols require fertilizer applications to make up for low fertility.

Oxisols: These are soils of warm, humid regions, which have undergone intense weathering under the tropical wet regime. They form mainly in tropical landscapes that have been stable, that is, without major disturbances, for long periods of time, at least a million years. (The graph in Figure 13.21 ranks oxisol landscapes among the oldest soil landscapes known.) Like the ultisols, oxisols are deeply weathered. The organic layer is weak litter because organic matter decomposes rapidly in the wet tropical climates. The underlying soil is rich in kaolinite (a clay) and oxides of iron and aluminum, the residues of weathered parent material. Oxisols do not usually show distinct horizons, though a vague zone of concentrated clays and oxides, called the *oxic horizon,* is usually found within two meters of the surface. At greater depths in some oxisols, a hard, iron-rich layer, called a *plinthic horizon* (or laterite layer) may be present (see Figure 13.6). Like the ultisols, these soils are rapidly exhausted by crops. Agriculture is not productive unless land is allowed to recover after cultivation or fertilizers are used extensively. Farmers who cannot afford fertilizer in areas of oxisols have traditionally resorted to shifting agriculture.

Vertisols: These soils are associated with parent material dominated by clays with high shrink–swell potentials. As we described earlier these soils are subject to a cracking and mixing process associated with seasonal drying and wetting (see Figure 13.11). Vertisols may be streaked vertically, and the sides of cracks and soil blocks (chunks) may show grooves or scratchlike features called slickensides. In addition, these soils may develop a rough microrelief (mini-topography), like that shown in Figure 13.22, characterized by little knobs and hollows called gilgae. Vertisols are commonly found in

Figure 13.22 The bumpy ground associated with soils of the vertisol order.

landscapes of grass and savanna vegetation in subtropical and tropical climates with strong summer drought. In the United States they are most common in Texas. Vertisols generally have low to moderate agricultural potential because the clay makes them difficult to plow when wet and they do not give up water to plants very readily during dry spells. They are, however, quite fertile, being high in exchangeable bases such as calcium and magnesium.

Histosols: These soils are dominated by a thick organic layer. They form under the wetland regime in which impaired drainage is the principal driver (see Figure 13.9). Long-term saturation or shallow standing water, such as is shown in Figure 13.16e, may induce the buildup of organic litter characteristic of the wetland soil-forming regime. The chemistry of the organic mass varies with the type of vegetation and the minerals present in the groundwater.

In the subarctic and arctic zones of North America (principally in Canada and Alaska), histosols have formed extensively under northern forests that yield acidic organic litter low in plant nutrients. The resultant soils are thus low in exchangeable bases and have low pH values. Histosols are used mainly for specialized agriculture depending on their chemistry and drainage; for example, truck (vegetable) farming for those with higher fertility and better drainage and cranberry farming for those with poor drainage and strongly acidic pH values (Figure 13.23). In most cases, histosols must be drained before they are farmed, but this in turn accelerates microfloral action and wind erosion. Thus, the histosols have a limited lifetime under agriculture – as little as 30 to 40 years for the thinner ones.

The Canadian Soil Classification System: In the Canadian system soils are classified according to how they form, based mainly on properties observable in the field such as the thickness and color of the organic layer and the makeup of parent material. The system is structured basically the same as the American system but with five rather than six levels of classification. It, too, begins with soil orders and ends with soil series. However, the Canadians use different soil-order names and classification criteria, which more accurately reflect Canadian soils and soil-forming environments than the American system allows.

The following paragraphs provide a brief description of the nine orders of the Canadian System. We group them into four sets for ease of presentation: forest soils, grassland soils, wetland soils, and cold/active soils. Table 13.2 lists the nine orders of the Canadian System and provides the approximate equivalent order in the American system. Figure 13.24 offers a map showing the distribution of these orders in Canada.

Forest Soils: There are three forest or woodland soils: podzolic, brunisolic, and luvisolic. Together they cover roughly 25 percent of Canada. The **podzolic order**, which is the equivalent of spodosols in the American system, is the most abundant and distinctive of the three. Podzol soils are found in moist areas with sandy parent material and are particularly prevalent in the subarctic region with boreal forest. They are subject to heavy leaching and marked by a signature profile comprised of a thin, dark organic layer, a whitish zone of eluviation, and rust-colored illuviation zone (see Figure 13.18). **Brunisolic soils** are found in drier forest areas of the subarctic in northern British Columbia and the southern Yukon. They form in relatively new

Figure 13.23 A muck farm where the organic mass is partially drained and planted to root crops.

Table 13.2

Canadian order	American equivalent	Landscapes
Podzolic	Spodosol	Boreal forests
Brunisolic	Inceptisol	Mountain conifer forests
Luvisolic	Alfisol	Deciduous and mixed forests
Chernozemic	Mollisol	Semiarid prairies
Solonetzic	Aridisol	Deserts
Organic	Histosol	Wetlands
Gleysolic	Histosol	Wetlands
Regosolic	Entisol	Mountain lands
Cryosolic	Gelisol	Permafrost tundra

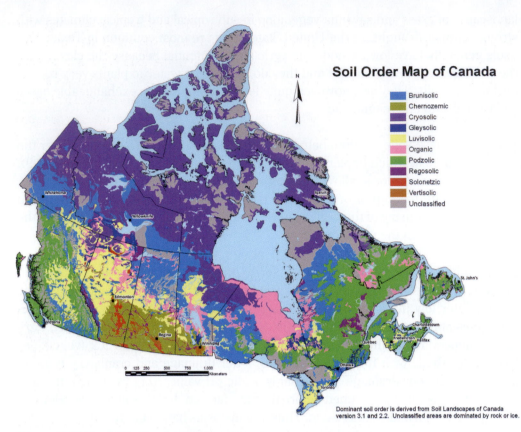

Soil Order Map of Canada

Brunisolic
Chernozemic
Cryosolic
Gleysolic
Luvisolic
Organic
Podzolic
Regosolic
Solonetzic
Vertisolic
Unclassified

Dominant soil order is derived from Soil Landscapes of Canada version 3.1 and 2.2. Unclassified areas are dominated by rock or ice.

Figure 13.24 The distribution of soils in Canada according to the Canadian Soil Classification System.

glacial and mountain valley deposits and are marked by a brownish B horizon. **Luvisols** form in clayey parent material and show less leaching and higher pH values than both the podzols and brunisols. Organic accumulation may be substantial, fertility is generally high, and they are favored soils for farming.

Grassland Soils: There are two grassland soil orders; chernozemic and solonetzic. **Chernozemic** is a prairie soil, which covers less than 10 percent of the country, mainly in the plains of Manitoba, Saskatchewan, and Alberta. It is the equivalent of the mollisol order of the American system with the rich organic and calcium carbonate content (see Figure 13.17). As in the United States, these prairie soils are prized for grain farming. The **solonetzic order** is the desert extreme of grassland soils with heavy salt contents and limited agricultural potential. The order has very small geographic coverage, less than 1 percent of the country.

Wetland Soils: Two orders of wetland soils, gleysolic and organic are defined in the Canadian system. Both are found in wet sites where the ground is saturated most of the year and vegetation is limited to wetland plants. Soils of the **organic order** are equivalent to histosols in the American system. They are characterized by heavy accumulations of organic matter in bogs, swamps, and marshes scattered throughout Canada, but are especially common in the rugged terrain of the boreal forest and tundra (Figure 13.24). **Gleysolic soils** are wetland soils with pronounced mineral content often in the form of a bluish or grey clay horizon. They, too, are common throughout much of the country.

Cold/Active Soils: These soils are found in areas where frost action and erosional and depositional processes are highly active. **Regosols** include recent rocky and sandy deposits in mountain and upland areas. Bedrock may be close to the surface or partially exposed and horizons are absent or very weak. **Cryosolic** soils are dominated by frost and, for most, permafrost lies at a depth of only 1–3 meters (Figure 13.25). As we noted earlier under our discussion of the geomechanical soil-forming regime, the active layer is subject to frost action that causes mechanical mixing of the soil. Cryosols are the most geographically extensive soil order in Canada, covering about 40 percent of the country in the Arctic and subarctic regions (Figure 13.24). They are currently being affected by extensive melting brought on by global warming, which is especially pronounced in the North American Arctic (see Figure 8.12).

Figure 13.25 The Canadian Arctic, a landscape dominated by soils of the cryosolic order where the effects of global warming are especially pronounced.

Summary on Soil Classification: The soil classification schemes used by geographers have been developed by soil scientists. These schemes are based on the concept of diagnostic horizons, the features that reflect the key processes and conditions of a soil's origin and development. These processes and conditions include those contributed by bioclimatic systems as well as by geomorphic, hydrologic, and other systems. The American and Canadian soil classification systems are organized similarly but differ in the number and types of orders, suborders, and other divisions. The difference is mainly a response to the different bioclimatic conditions of the two countries. By including the oxisol order, the US system can be extended to wet tropical regions beyond the American homeland.

13.4 Soil and Civilization in the Ancient World

With the development of agriculture over 10,000 years ago, humans formed a binding and interminable union with soil. Indeed, the story of the rise and fall of early population centers and civilization itself is also the story of the use and misuse of soil, and it is not an overstatement to say that those civilizations which took good care of their soil survived, and those that did not perished. Sumeria, ancient Rome, and the Maya were three soil-dependent civilizations that ultimately disappeared. A quick summary of their legacy with soil offers some insights into the importance of this resource in sustaining civilization, and the methods by which it can be destroyed.

Sumeria: Sumerian civilization (7000–3400 years ago) produced the world's first literate society. The Sumerians arose in a dry area of rich grassland soils between the Tigris and Euphrates rivers, known as Mesopotamia (now in Iraq). Here they farmed wheat and early on learned the advantages of irrigation to dramatically increase crop yields. As food supply increased, the population grew, villages grew into towns, some towns grew into the first cities, society diversified, and complex governance arrangements emerged (Figure 13.26). The result was pressure to produce more and more food, which led to expanded and intensified irrigation, the apparent key to greater wheat yields.

Figure 13.26 Ruins of ancient Sumeria, now part of Iraq, a region once rich with verdant cropland, and laced with irrigation ditches and roads connecting villages and towns.

Initially, the system worked, as more water led to more food, a form of positive feedback in the farming system. But this only encouraged over-application of water, which led to widespread soil saturation. With saturation, salts began to concentrate near the surface leaving a toxic crust in the upper soil as the water evaporated. This process, known as **salinization**, destroyed vast areas of rich farmland, reducing food production and weakening Sumerian civilization. As the situation worsened and the population struggled with declining food supply, the governance structure weakened and the Sumerians fell prey to attacks from invaders, bringing an end to their civilization. Today it is estimated that nearly 50 percent of the soils in Mesopotamian region of modern Iraq have been rendered unusable because of salinization which began with the Sumerians (see Figure 13.2b).

Ancient Rome: The Roman Empire was the largest and longest surviving of many prominent civilizations of the Mediterranean Basin, including the Phoenicians, Greeks, and Hebrews. Yet, despite all we read about the lofty contributions these civilizations made to the arts, philosophy, religion, and government, it is important to remember that all were built on a common, humble foundation, namely, farming. And farming was built on soil and climate systems.

Figure 13.27 Hill slopes throughout much of the Mediterranean region were once forest-covered but wood cutting and soil loss left them denuded, as we find most of them today.

The Mediterranean Basin, with its wet winters and long dry summers, was both a blessing and a curse to agriculture. On the one hand, it was ideally suited for crops such as wheat, olives, and grapes but, on the other hand, it was prone to serious drought in one season and heavy soil erosion where slopes had been deforested on the other. And since every population center needed wood, deforestation and soil erosion were commonplace wherever people settled. For example, more than 500 years before the Romans, the early Phoenicians inhabited what today is Lebanon where forested mountain slopes – featuring the famous cedars of Lebanon – were cut almost entirely away for building huge fleets of ships. Forests were also stripped from the peninsulas and islands of Greece. In their place came herders with sheep and goats, leaving it like that shown in Figure 13.27, which further denuded the landscape, promoting more soil erosion.

The cycle of deforestation swept across the ancient Mediterranean region and the Romans practiced it in the extreme to provide farmland, fuel, and lumber for ships and buildings. Massive structures such as coliseums, bridges, and temples were originally built of wood. The stone architecture we commonly associate with the Romans became dominant only when accessible supplies of quality timber ran low. Almost all space and water heating was done by wood and wood products. The temperature of the public baths was kept at a minimum of 54 °C (130 °F) and even a very small bath required over 100 tons of wood per year. A whole guild of wood-suppliers equipped with 60 ships was created specifically for the purpose of importing bath-heating wood. As for industry, wood was burned to fire the bricks used in building the aqueducts, in making glass windows, in running the lime kilns (for making concrete), and in smelting iron and silver. With the spread of forest cutting and soil loss, the carrying capacity of the Mediterranean Basin declined and people left the land for cities.

As the Empire became more and more urbanized (Rome itself reached over a million people by the time of Christ), the population grew as never before in European history. The enormous population required vast farmlands, and most of the once-forested Italian peninsula was cleared for farming. Widespread grazing by goats, sheep, cows, and pigs made it practically impossible for the forests to recover from this loss. Rich hillside soil washed away and crop yields declined. In an empire that was already struggling to feed its massive population, this depletion was devastating. Since the farms were primarily owned by wealthy absentee landlords and worked by slaves, neither group paid particular attention to the steadily worsening situation. Eventually, Rome came to depend almost entirely on North Africa for food supplies. Food costs soared and, when yields declined because of drought or unrest or when shipments were delayed by bad weather, famine and riots often ensued.

Much has been written about the fall of the Roman Empire and historians tell us that there were many contributing factors. Although wars, politics, and moral disintegration usually get top billing, research reveals that another one of the root causes was actually the breakdown in the food-production system as a result of deforestation and soil depletion. Despite its long history and despite its considerable managerial skills in architecture, engineering, and public administration, Rome fell seriously short at managing its soil resources and building a sustainable agricultural system, and in the fifth century AD, after nearly 1000 years of existence and insurmountable economic, governance, and military problems, it too bit the dust.

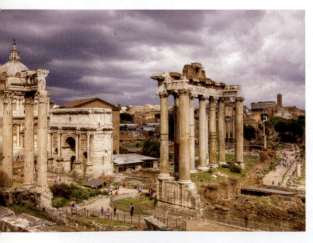

The ruins of the Forum in Rome, the heart of the Empire, considered the first city with a population of more than a million people.

Maya: In 1839, American explorer John Lloyd Stephens and English artist Frederick Catherwood uncovered some curious stone structures in the tropical forests of

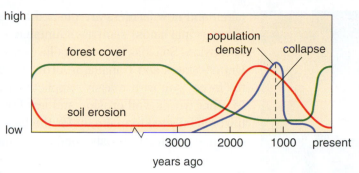

Figure 13.29 Changes leading to the collapse of the Mayan civilization. Forest cover declines and soil erosion rises with the growth in population until the entire system collapses around AD 900.

Figure 13.28 Mayan civilization included hundreds of grand structures such as this and vast areas of farmland cleared from tropical forest.

Honduras, Guatemala, and Mexico that led to the discovery of a once huge civilization comprised of city states, monumental stone architecture, and millions of people (Figure 13.28). Mayan civilization, which dates back 5000 years, is ranked among the most celebrated of ancient civilizations, and its disappearance has long been one of the great mysteries of modern scholarship. Recent research, however, has come up with some clues about what went wrong and, once again, the story is related to forests and soils (see Figure 13.29). From pollen trapped in ancient layers of sediment on the floors of lakes in the Mayan region, scientists have learned that around 1200 years ago, just a few centuries before the civilization's collapse, tree pollen disappeared almost completely and was replaced by the pollen of weeds. In other words, the region had become almost completely deforested.

Without trees, the system responsible for soil maintenance was essentially broken. Topsoil renewal declined, erosion increased, and soil moisture declined with heating of the unshaded ground. (Without the cooling effects of the forest canopy, temperatures over the region, according to NASA scientists, would have been as much as 6 °C warmer.) Despite efforts by the Mayans to remedy the problem, including the extensive use of irrigation, the application of fertilizer (probably human and animal waste), and terracing on slopes, the Mayan civilization – complicated by warring among the city states, administrative problems, and overpopulation – ultimately failed. By AD 900, the Mayan culture had disappeared from an area larger than the state of Florida, an area they once completely dominated. Its people died out and migrated to other locations in Central America.

13.5 Soil and Land Use in the Modern World

History has taught us that healthy agriculture depends on healthy soil and healthy soil depends on sustainable agricultural practices. In order to build sustainable agriculture, we must approach soil as a system with inputs (gains) and outputs (losses) of essential ingredients. A sustainable soil system must be able to form new soil quickly enough to make up for the loss of existing soil, that is, the soil held in reserve. In general, losses cannot exceed more than 5–10 tons per hectare (2–4 tons per acre) in the average year for a soil to be sustainable. Few if any countries in the world today are able to achieve that standard. The annual average soil loss rate on agricultural (cropland) land in the United States is about 30 tons per hectare. Rates in other large developed countries are roughly comparable. Poor countries such as India and China have higher loss rates. We are led to the inexorable conclusion that the world's reservoir of agricultural soil is being depleted.

Why is this so? Lots of factors contribute, but population growth and the need for farmland ranks high. In crowded countries, the best farmland has already been

Figure 13.30 Application of commercial fertilizer to cropland in the North American Corn Belt. An expensive process that is also leading to environmental damage as far away as the Mississippi Delta.

Figure 13.31 Farmland, once fully wooded, exposed to erosion which will carry away tons of sediment with each rainfall.

taken and new farmers are pushed into second- and third-rate lands (marginal and submarginal lands) such as mountainous ground where soils are infertile and prone to erosion. Soil loss and crop failures are inevitable. In developed countries like the United States and Canada, in order to maximize yields commercial farming tends to "force" soil with massive applications of manufactured fertilizer (Figure 13.30). In other words, soil erosion and fertility loss are tolerated because commercial fertilizer is injected to make up for the loss. But this is a short-term solution for as soil declines, production costs, food costs, and environmental damage rise.

Whenever we impose land use on the landscape, soil is affected. When land is cleared for lumbering, for example, the soil-moisture system is altered, runoff often increases, and erosion rates rise. When cities and highways are built, vast areas of topsoil are scraped away and the remaining soil is covered with concrete and asphalt, eliminating water infiltration and plant productivity. These impacts are serious and widespread; however, they are relatively small compared to those related to agriculture.

Crop farming and grazing occupy more than 40 percent of the Earth's land area. They are responsible for many serious soil problems including wetland eradication with land clearing and ditching; salinization from over-irrigation; chemical contamination from pesticide applications; and desiccation and degradation from overgrazing. But without question the most serious problem related to agriculture is soil erosion. Of the 11.5 million square kilometers of land listed by the United Nations as seriously degraded, most is eroded cropland and rangeland. Since the beginning of agriculture about 12,000 years ago, it is estimated that soil erosion has ruined over 4 million square kilometers of farmland, an area equivalent to 42 percent of the total land area of the United States.

Causes and Controls of Soil Erosion: Runoff and wind are the leading causes of soil erosion on agricultural land. Although soil erosion is a natural process, it operates very slowly on undisturbed land (especially forested and prairie lands) mainly because it is held in check by vegetation. Plants not only secure the ground against the force of running water and wind, but they also induce infiltration and absorption or rainfall so that less water is available for runoff. Accordingly, when land is cleared for cultivation, it is necessary to protect the soil from runoff and erosion but farmers generally have not done well in this regard, as Figure 13.31 illustrates. The crops themselves offer poor soil protection for, among other things, they are present only part of the year and plowed ground readily gives way to runoff and wind. To minimize soil erosion, farmers need to select the proper terrain, soil, crops, drainage conditions, and management (or soil-conservation) techniques. These imperatives were demonstrated clearly by the experiences of the three ancient civilizations in the last section.

Simple formulas, called the **Universal Soil Loss Equation** and its updated version called the **Revised Universal Soil Loss Equation**, can be used to forecast soil erosion on farmland. They are based on five factors: (1) plant cover; (2) soil composition; (3) slope; (4) rainfall; and (5) crop management. A worst-case scenario would include the following: fully cleared (denuded) ground; highly erodible soils such as loose sand or silt; steeply sloping ground with a large surface area over which rainfall collects and forms runoff; frequent, heavy rainstorms; and the absence of soil management measures such as the use of cover crops or contour plowing. Such conditions could result in an annual soil loss of hundreds of tons per hectare. A field that loses 250 tons per hectare (2.5 acres) will be lowered roughly 1 inch

(2.5 centimeters) per year, destroying all productive topsoil in a matter of a decade or less. Many places in the world suffer such losses.

Soil-Erosion Processes: There are three main erosional processes associated with runoff: rainsplash, rainwash (also known as sheetwash), and gullying. **Gullying** is the most serious of the three. It is caused by runoff concentrating in small channels that cut into the soil surface. The cuts range in size from rills a few centimeters deep to gullies more than 5 meters deep. Once begun, gullying can "gut" a cleared hillslope in a matter of years, particularly in areas of erodible soils and rainy climate (see Figure 13.32).

Figure 13.32 A landscape being gutted by gullying where the valuable soils are being washed away at alarming rates.

Gullying is most serious in wet climates, especially those subject to intensive downpours, where the plant cover has been destroyed. It was rampant on farmlands in the American South in the decades following the Civil War. The combination of sandy soils, hard rainstorms, avnd the unprotected fields of poor share croppers and tenant farmers was a formula for disaster. Equally bad, if not worse, conditions are found today in many other parts of the world. In the upper watershed of the Huang He (Ho) (China) intensive cultivation of loess- (silt-) based soils is resulting in massive gullying. Not only is valuable farmland being lost, but the silt is clogging the channel of the Huang He, which in turn is increasing the threat of flooding for millions of people. Since the beginning of agriculture in this region several thousand or more years ago, the Huang He's sediment load has increased at least tenfold (Figure 13.33). In Madagascar, gullying is related to tropical-forest clearing in the rainy interior as peasant farmers and herders try to eke out a living on newly formed land. The same holds true in the Philippines, Indonesia, Brazil, Central America, and many other places in the wet tropics where forest is being cleared.

Figure 13.33 The channel of the Huang He is so clogged with sediment that it is forcing the river to higher flood stages that threaten tens of thousands of lives.

Secondary Impacts of Soil Erosion: The billions of tons of soil stripped from the land each year becomes sediment in streams, canals, ditches, wetlands, ponds, lakes, and reservoirs. In streams, it buries benthic (bottom) organisms, degrades spawning habitat, and damages the overall ecology of channels. As channels fill in with sediment, their capacities diminish, and flooding increases. Sediment damages water supply systems that feed communities, industry, and agriculture. Human health is threatened and water-treatment costs are increased. With these losses and the loss in agricultural productivity, soil erosion in the United States is estimated to cost over $44 billion a year. Moreover, pesticides and fertilizer are washed away with sediment. In the 1993 floods on the Mississippi, the largest on record, millions of tons of chemicals were discharged into the great river from farm fields across the Midwest. Much of this ended up in the Mississippi Delta where the nitrogen, a favorite fertilizer of corn farmers, caused massive damage to aquatic ecosystems, an impact that continues year after year.

Soil Erosion and Global Warming: A large part of the carbon synthesized by plants ends up in the soil, mostly as humus and topsoil, where it is stored for various periods of time. In forest ecosystems, for example, it is estimated that soil contains over two-thirds of the carbon stored in the entire landscape (Figure 13.34). Given the enormous amount of organic matter held in soils across the globe, we conclude that soil is a major sink in Earth's carbon cycle. And since carbon dioxide (CO_2) is a major greenhouse gas and the principal driver in the current trend toward global warming, soil therefore plays a significant role in that process. As vegetation such as tropical forest is removed from the planet, not only is the total plant uptake of atmospheric CO_2 reduced, leaving more CO_2 in the atmosphere, but soil erosion also rises and the topsoil is the first to go.

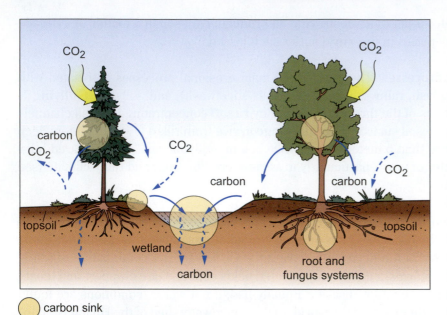

With less topsoil, less carbon is being stored. It is estimated that in the United States, 30 to 70 percent of the carbon in cultivated soils has already been lost. But there is an added twist to this trend. Recall that temperature is a major control on the activity of the organisms that consume organic matter in the soil. Thus, as global warming advances, the rates of organic-matter decomposition may also rise. In the middle latitudes, for example, summers will likely be longer, extending the season of organic decomposition. On the other hand, higher atmospheric temperatures and CO_2 levels may induce increased rates of plant productivity leading to heavier organic contributions to the soil and more carbon sequestration. On balance, the overall outcome is uncertain, but without changes in our forest- and soil-management practices in the face of continued pressure for food and fuel from a growing world population, the picture is not very hopeful for some time to come.

Figure 13.34 The landscape as a sink for carbon dioxide. Soil is a major storage reservoir for carbon which is seriously reduced with land clearing and soil erosion.

Chapter Summary and Overview of Soil Types, Distribution, and Land-use Relations

Soil is the product of a suite of factors, and as these factors vary from place to place and over time, soil form and composition also changes. At the broadest scale, soils vary with bioclimatic regimes but within this framework, many non-bioclimatic factors come into play and these are often the key to understanding soil at the local level. Soil classification systems vary among many countries in the world. Those of the United States and Canada are multilevel systems beginning with soil orders and ending with soil series. As a rule, the story of civilization is the story of soil (see the summary diagram), a lesson that the modern world is struggling to comprehend and apply.

▶ **Soil is the product of the interplay among five major systems.** Known as the soil-forming factors, these have variable influences on soil formation depending on geographic location and time.

▶ **At the broadest geographic scale, soils can be grouped into two families.** Within these are bioclimatic soil-forming regimes, and within the bioclimatic regimes are regimes governed by other factors.

▶ **Geographers use soil classification systems developed by soil scientists.** The US Comprehensive System uses six levels of classification beginning with 12 orders. The Canadian System uses five levels beginning with 9 orders.

▶ **The mollisol/chernozemic soils are among the best agricultural soils on Earth.** Once exposed and plowed, however, they are highly prone to erosion by wind and runoff.

▶ **History clearly demonstrates the importance of the soil resource in civilization.** The Sumerians, ancient Romans, and the Mayans all experienced difficulties managing their soil.

▶ **The fall of Rome may have been rooted in deforestation.** This led to soil erosion and declines in agriculture and food supplies.

▶ **Agriculture has profound impacts on soil.** Erosion is the most serious of these impacts in both affluent and poor countries.

▶ **Runoff and wind are the leading causes of soil erosion.** Gullying is the most insidious erosional process.

▶ **Besides reducing soil productivity, erosion has many serious secondary impacts.** These include damage to stream channels and water supply systems.

▶ **Soil and vegetation are important carbon reservoirs.** Land clearing and soil erosion contribute to global warming.

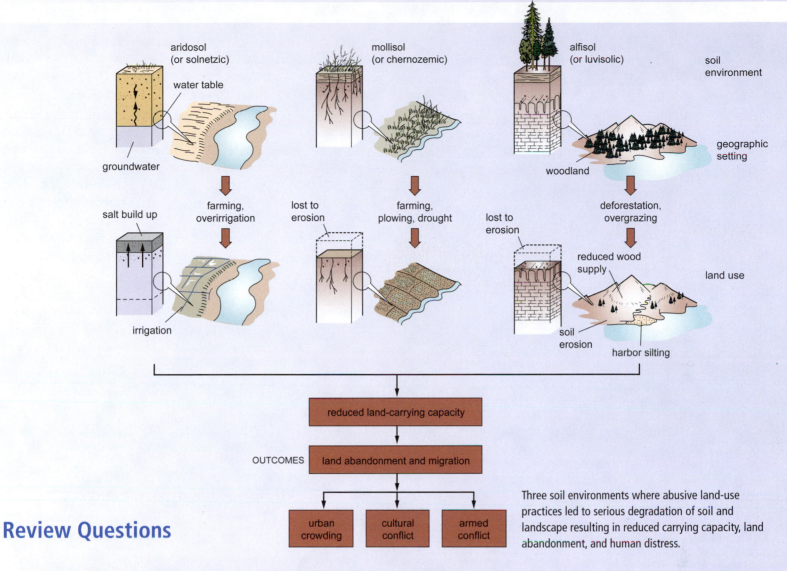

Three soil environments where abusive land-use practices led to serious degradation of soil and landscape resulting in reduced carrying capacity, land abandonment, and human distress.

Review Questions

1 Reflect on the observation that soil is a tablet upon which is recorded the conditions and events at a place on Earth.

2 What are middens and what influence do these features have on soil?

3 What are soil-forming factors and how does time play into the operation of these factors?

4 Where are pedocals and pedalfers found in North America and why?

5 Name four major soil-forming regimes of the bioclimatic variety and relate each to a soil order in the US Comprehensive Soil Classification System.

6 How would you describe the soil-forming regime of a soil whose formation is dominated (a) by the buildup of sand in a dunefield or (b) by the buildup of water in low spots in the land?

7 Soil can vary measurably over areas as small as a park or a farm. Explain how this is possible.

8 What is a diagnostic horizon and what soil classification system is based on this concept?

9 Among the orders in the US and Canadian classification systems, which have good, moderate and poor agricultural potential and why?

10 Why are soils of the cryosolic order so widespread in Canada?

11 What sort of mismanagement and abuse of the soil resource were committed by these ancient civilizations: Roman, Mayan, and Sumerian?

12 Is it natural for land to lose soil to erosion? What is the average annual rate of soil loss from cropland in the United States and what are its effects? How is it that the United States maintains such high agricultural productivity despite its soil losses to erosion?

13 If you had to advise a property owner how to curtail soil erosion on his/her land, what in general would be your first line of advice and why?

14 Describe how gullying erodes soil and how is it that soil erosion is a concern in global warming?

The Global Water System

Chapter Overview

Few things on Earth are as central to physical geography as water or, to put another way, without water Earth's physical geography would be a much simpler matter than it is. So we start this section of the book with an overview of this magnificent system, a glimpse into the big picture of water. The discussion is driven by some compelling questions including how Earth got its water, how much water is held where on the planet, how the water-exchange system, the hydrologic cycle, works, and how on land the system is organized into subsystems. We are led to a model called the water budget that helps us understand how the subsystems like watersheds, glaciers, and soils function. We then give some thought to hydrologic landscapes and what makes humid and arid landscapes different in terms of where water is stored and accessed by plants. Finally, and consistent with one of the book's main themes, Earth is a planet of changing systems, and so it is with the hydrologic system, for no matter where we are this great system is constantly changing around us.

Introduction

Of the many things that make Earth unique as a planet, one of the most striking is its water system. It is not merely the presence of a large water supply that is striking – Mars also has abundant water – but rather the makeup, distribution, and dynamics of Earth's water system. Consider that Earth's thermal state, at an overall surface temperature of 15 °C, is just right for the maintenance of an abundant water supply, with huge amounts of liquid water and smaller, but significant, amounts of ice and water vapor. Mercury, with a surface temperature around 450 °C, is too hot and dissipates its water into vapor, which in turn is driven off into space. Mars is too cold and all its water is frozen. Earth, on the other hand, is able to maintain water simultaneously in all three phases, liquid, solid, and gas.

In addition to its water-holding capacity, there is sufficient thermal energy on the Earth's surface to drive a rapid exchange among the three parts of the system, oceans, atmosphere, and land. At the heart of this three-part system is the atmosphere, which functions as the planet's central water-distribution machine. It takes in huge amounts of water vapor mainly from the oceans and then quickly delivers it back to the surface, but always in a different place and often far away from the input place.

Among Earth's unique qualities as a planet is its dynamic, three-part water system, represented here by atmospheric moisture, glacial ice, and ocean water.

Thus, the behavior of the atmospheric exchange system is very uneven geographically, delivering huge amounts of water to some areas and practically none to others. These differences are central to the geographer's interests, because they contribute to spatial differentiation of the Earth's surface into deserts, forests, ice caps, and so on. These spatial differences are also central to Earth's habitability, both from the standpoint of water as a foundation resource in the planet's life system and as a threat to life in the form of the storms like hurricanes that come with the delivery of water or events like droughts that come with absence of it. It is not unreasonable to think of Earth as a water-driven planet.

14.1 How the Earth Got its Water

For years scientists thought that Earth's water supply formed early in the planet's development mainly by a process called outgassing. **Outgassing** is the release of vapor with volcanic eruptions. Gaseous water released from the crust readily joined the atmosphere where it became entrained into the planet's surface water system, that is, part of clouds, precipitation, the oceans, etc. With increased surveillance of the Solar System in the past few decades, we have gained new appreciation of comets, meteors, and related space debris as a source of Earth's water.

We have long known that the Earth is continuously bombarded by debris from space. Most of the particles are small, essentially stardust, but some are much larger and, as Earth's gravity pulls them through the upper atmosphere, they disintegrate with the heat of friction. Many of these particles are composed of ice and when they hit the outer reaches of the atmosphere, they vaporize and add water molecules to the atmosphere. These vapor processes have been going on throughout Earth's 4.6 billion year history, the products of a more or less continuous pelting of the planet with icy debris ("snowballs"), some as big as a house or larger. And as they approach Earth, they go through a sequence of breakdown and disintegration probably like that shown in Figure 14.1. How much of our water supply can be attributed to snowballs from space is difficult to say, but for the present we are safe in saying that Earth's water supply appears to have been derived from at least two principal sources: (1) outgassing from the Earth's crust, and (2) vapor showers from extraterrestrial ice.

14.2 The Global Water System

No matter where Earth's water comes from, almost all of it ends up in the oceans. The graph in Figure 14.2a certainly bears this out – the oceans hold 97.4 percent of our water, equivalent to 1360 million cubic kilometers. The land holds virtually all the rest, a meager 2.6 percent and, as Figure 14.2b shows, it is unevenly distributed among glaciers, groundwater, and surface water (lakes, ponds, and streams). Ice caps and glaciers hold about 75 percent of all terrestrial water, or about 2 percent of Earth's total water supply (27 million cubic kilometers). The remainder, a little more than 0.6 percent of Earth's total, is the

Figure 14.1 The breakup of an icy comet upon entering the Earth's upper atmosphere. These and millions of smaller ice particles from space are major sources of Earth's water.

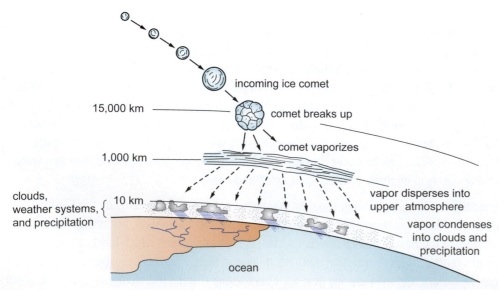

incoming ice comet

15,000 km — comet breaks up

comet vaporizes

1,000 km —

clouds, weather systems, and precipitation { 10 km

vapor disperses into upper atmosphere

vapor condenses into clouds and precipitation

ocean

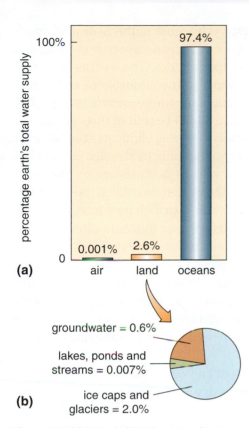

(a)

(b)

Figure14.2 (a) Water held in the atmosphere, on land, and in the ocean as a percentage of Earth's total water supply. (b) The three main sources of terrestrial water as a percentage of Earth's total water supply.

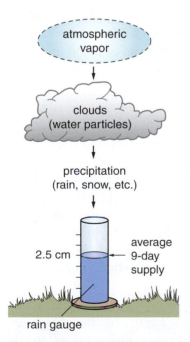

Figure 14.3 Total atmospheric water (vapor) as precipitation is equal to only 2.5 centimeters. This amount is precipitated onto Earth's surface once every nine days.

planet's sole supply of fresh, liquid water. It totals about 8 million cubic kilometers and is the water upon which virtually all terrestrial life depends.

The Atmospheric Phase: The atmosphere is left with only 0.001 percent of the total, by far the smallest amount of water in the global system. This amount is so small that if all of it were precipitated onto Earth at once, it would form a layer only 2.5 centimeters (one inch) deep over the entire planet. Compared to the oceans whose average depth would be more than 2000 meters if spread over the entire Earth, by this measure atmospheric moisture would amount to little more than a micro-thin membrane over Earth's 510 million square kilometer surface.

Paradoxically, the atmosphere is at the center of the engine that drives the Earth's hydrologic system. Once every nine days the atmosphere precipitates onto the Earth a quantity of water equivalent to its entire supply (2.5 cm) and in turn picks up a new supply of water (in the form of vapor) from the oceans and land (Figure 14.3). This happens nearly 40 times a year and, since every nine-day cycle represents 1 inch (2.5 cm) of precipitation, it follows that the Earth's average annual precipitation is about 40 inches (100 cm).

What fuels this powerful engine? Solar energy. Solar radiation heats the oceans, land, and lower atmosphere and this heat directly or indirectly drives the evaporation process. About 25 percent of all the heat from solar energy goes into evaporation. Geographically, most of the evaporation comes from Earth's warmest zone, a broad belt of about 60 degrees latitude centered on the Equator. Wind systems, which are also driven by solar energy, then redistribute the resultant vapor over the Earth and weather systems such as thunderstorms and midlatitude cyclones convert it into precipitation and return it to the Earth's surface. As the diagram in Figure 14.4 shows, in the atmosphere the cycle splits into an ocean branch and a land branch. The land branch cycles its water along two paths: (1) back to the atmosphere as vapor and (2) to the ocean as runoff. The ocean branch cycles its share of the water back to the atmosphere. In its simplest form, this is the **hydrologic cycle** and, among other things, it is important to underscore that the entire system pivots on the atmosphere.

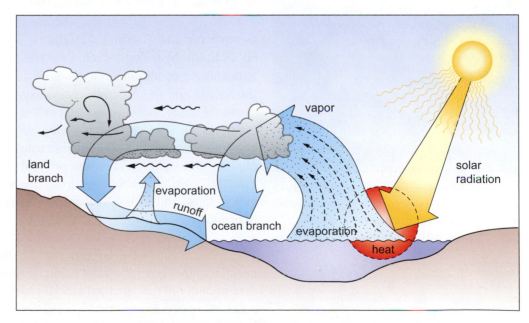

Figure 14.4 A simplified version of the Earth's hydrologic cycle. Powered by solar radiation, the oceans and landmasses release water to the atmosphere, which it then delivers back to the oceans and land as precipitation.

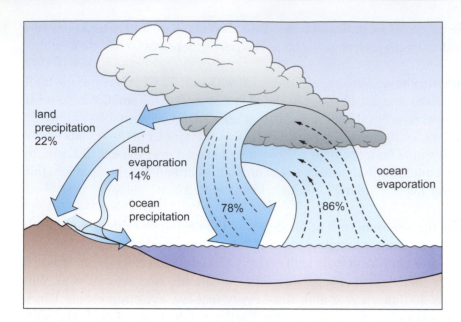

Figure 14.5 Supply central for the hydrologic cycle is the oceans which provide the atmosphere with 86 percent of its water vapor.

The Hydrologic Cycle: Most of the hydrologic cycle is played out over the oceans where it is basically a simple input–output system. The oceans supply 86 percent of the vapor to the atmosphere and in turn receive 78 percent of the atmosphere's precipitation (Figure 14.5). (An added benefit of this exchange is that the evaporation process eliminates the salt from the ocean water so that only freshwater falls back to the Earth.) The world's landmasses, on the other hand, capture 22 percent of global precipitation, which is much greater than their modest contribution of 14 percent water vapor to the atmosphere. In addition, the cycle over land is more complex than over water, involving more phases or subsystems. Three of these systems, streams (runoff), groundwater, and glaciers, are center stage in our rapidly growing global environmental dilemma. We will touch on these terrestrial systems here and in the following chapters.

But first, why the imbalances in the relative input and output of water between continents and oceans? Obviously, land is relatively dry and therefore a much poorer source of water vapor than the oceans. On the other hand, land is highly effective in producing precipitation, much of which comes from vapor blown in from the oceans. Land has more cooling mechanisms, such as airflow over mountains and thermal convection from intensive surface heating, to induce condensation, cloud formation, and precipitation. The resultant precipitation is not found everywhere over the continents, of course, but it does cover large geographic areas, such as the mountainous coastlines of the Pacific coast of North America and the tropical areas fed by moist oceanic air within the Amazon Basin. The annual precipitation in these areas is typically two to four times the global average. The photograph of the Florida Peninsula in Figure 14.6 illustrates the hydrothermal difference between land and water. Clouds from thunderstorms and related convection driven by terrestrial heat and fueled by vapor drawn in from the ocean outline the peninsula.

Figure 14.6 Clouds concentrated over the Florida Peninsula reveal the ability of land to generate large amounts of precipitation, a response to intensive surface heating and convectional processes.

Geographic and Seasonal Variations: There are also enormous geographic differences in the hydrologic cycle over the oceans themselves. It is significant that the atmospheric conditions that produce the world's great subtropical deserts, such as the Sahara and the Australian, also exist over large regions of the oceans. Near the center of these zones, in the belt called the *Horse latitudes* (around 30 degrees latitude), the air is so dry that if ancient sailing ships encountered light wind and were becalmed here, sailors could perish from thirst as water supplies dwindled under this rainless atmosphere. This was probably the setting for Samuel Taylor Coleridge's (1772–1834) memorable lines from the *Rime of the Ancient Mariner*: "Water, water, everywhere, and all the boards did shrink; Water, water, everywhere, nor any drop to drink…"

These vast ocean "deserts," however, are mainstays in the Earth's hydrologic system because they are the world's primary sources of water vapor, giving up 100 to 200 centimeters (40 to 80 inches) of water to the atmosphere per year. The highest evaporation rates occur in winter because the temperature (and vapor pressure) gradients between warm water and the relatively cool air sliding over it are largest at this time.

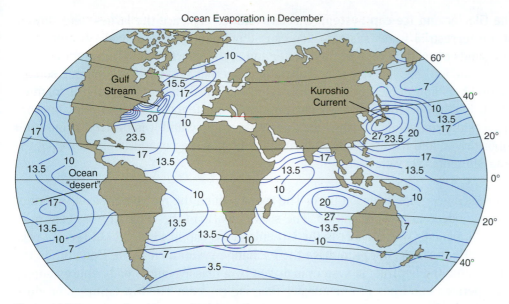

Ocean Evaporation in December

Figure 14.7 Ocean evaporation rates for the month of December (in centimeters of water lost). In June and July rates decline in the northern hemisphere and rise in the Southern Hemisphere, reaching 20 centimeters in a broad belt around 20–30 degrees latitude in the Pacific, Indian, and South Atlantic oceans.

The same holds for the North Atlantic and North Pacific midlatitudes in winter. The seasonal interaction of regional wind and ocean current systems also plays a part. In fact, the Earth's highest monthly rates of ocean evaporation occur where two warm currents, the Gulf Stream and the Kuroshio (shown in Figure 14.7), drive warm water under the cool, dry winds of the prevailing westerlies as they blow from the continents over the western Atlantic and Pacific oceans in December. Charged with this vapor, it is not surprising that the westerlies bring heavy precipitation to northwestern Europe and the west coast of North America at this time of the year. As the air warms with summer, however, the evaporation rates in these areas decline to less than 10 centimeters (4 inches) in June and July.

Since ocean evaporation – and hence the atmosphere's chief input of water vapor – is driven mainly by the heat in ocean surface water, we should ask what effect global warming may have on this system? The oceans are Earth's principal reservoir of heat (beyond that in the core and mantle, of course) and the global-warming trend being measured in the atmosphere is also showing up in the oceans. In the Arctic, for instance, warmer water is reducing the annual period and coverage of sea ice. The outcome of ocean warming is almost certain to lead to higher evaporation rates, more atmospheric moisture, and higher atmospheric energy levels with the input of latent heat, all part of the climate-change scenario.

14.3 The Hydrologic Cycle on the Continents

It is a simple but irrefutable fact that the only source of **terrestrial water** is atmospheric precipitation. All streams, lakes, ponds, glaciers and groundwater are derived from precipitation. Worldwide terrestrial precipitation yields 111,000 cubic kilometers per year, and of this a surprising amount, nearly 65 percent (71,000 cubic kilometers), is lost in evaporation. The remaining water, which collects on the surface or soaks into the ground, makes up three basic storage-flow systems: (1) **ice caps** and **glaciers**; (2) **groundwater** and **soil water**; and (3) **lakes**, **streams**, and related surface water features. These systems are illustrated in Figure 14.8. Each system is distinguished not only by the location and amount of water it holds, but also by the rate at which it recycles or exchanges its water for a new supply. In the end, after evaporation has taken its share, the terrestrial cycle is completed as the remaining water (35 percent of the input) returns to the sea as runoff.

Figure 14.8 The hydrologic cycle featuring the three terrestrial systems: glaciers and ice caps, groundwater, and surface runoff. The recycle times for each are given in Table 14.1.

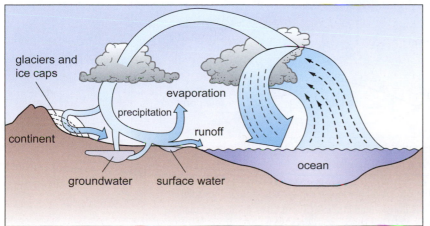

Figure 14.9 The Greenland ice cap, the second largest mass of ice in the world. Near the center it is nearly 3000 meters (10,000 ft) deep.

The Glacier and Ice-cap System: Glaciers and ice caps are the largest and slowest of the terrestrial water systems. Ice caps (continental-scale glacial systems) take thousands of years to recycle their water and in the heart of the great Antarctic and Greenland ice caps lies truly ancient water. Together these two ice masses contain about 11 million cubic kilometers of water, one-third the total supply of freshwater on Earth. Curiously, glaciers also owe their existence to solar energy, because glaciers depend solely on snow for their water supply and it is derived from water vapor driven into the atmosphere through solar-powered evaporation; however, the nourishment is very slow.

There is a geographic lesson here. The rate at which the hydrologic cycle operates generally decreases with latitude. The slowest parts of the cycle, the ice caps at or near the planet's poles, such as the Greenland ice cap in Figure 14.9, receive relatively little thermal energy to drive the exchange system. Near the Equator, on the other hand, the hydrologic system is driven by a massive and steady energy supply and operates hundreds of times faster than ice caps do. The low latitudes produce both the highest rates of evaporation and precipitation on Earth. Table 14.1 gives representative recycle times for five terrestrial water systems.

Despite their slowness, glaciers are fascinating and important hydrologic systems with profound influences on the global environment. As systems they are best understood by examining their water inputs and outputs as illustrated in Figure 14.10 for Greenland. On the input side, water is gained from snowfall which is compressed into ice and stored as it gradually moves downward and outward with the flow of the glacier. On the output side, there are three processes responsible for the release of water: evaporation, melting and runoff, and breaking off (or calving) into the sea.

Since melting and evaporation are driven largely by atmospheric heat, we are led to speculate what would happen if global climate warmed enough to reduce all the world's glaciers and ice caps by, say, 25 percent of their present volume? The increased meltwater would force sea level to rise by about 15 meters (50 feet) causing extensive flooding in coastal areas now occupied by 1–2 billion people including New York City, London, Hong Kong, and Shanghai (see Figure 8.40). Is a change of this magnitude science fiction? Let us recall the forecasts of a sea-level rise of 1 to 2 meters by

Table 14.1 Recycle times*

Ice cap	10,000 years
Groundwater	1000 years
Lakes	10 years
Streams	10 days
Tropical forests*	as little as 1 day

* Representative times: rates vary with the size of individual water features.

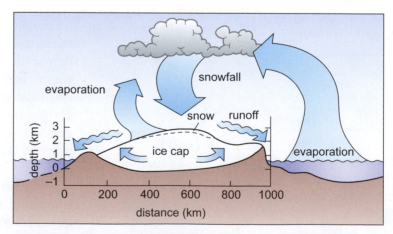

Figure 14.10 The Greenland glacier/ice cap system. Input comes solely from snow which is transformed into ice. Output takes the form of melting, evaporation, and calving (breaking off into the sea).

the end of this century with carbon dioxide-induced global warming of 2–5 °C. While a change of 15 meters would be startling to humanity, it would be small compared to the sea-level changes that our early ancestors witnessed. Only 18,000 years ago, at the peak of the last major glaciation, there was so much ice (three times today's volume) that sea level was more than 100 meters (330 feet) lower than today. In Asia, Europe, Africa, and perhaps North America, Stone Age people lived on extensive coastal plains that are now under water and, over a period of several thousand years, were forced to retreat to higher and higher ground as sea level rose.

The Stream-runoff System: We have no idea how much our Stone Age ancestors understood about Earth's water system but, by the time the Old Testament was written (3000 to 1000 BC), it seems that the idea of a cyclical water system was given serious consideration. Witness this statement from Chapter 1, Verse 7 of Ecclesiastes:

> All the rivers run to the sea, yet the sea is not full; unto the place from whence the rivers come thither they return again.

However, several thousand more years passed before the true nature of the terrestrial phase of the hydrologic cycle was figured out, in particular, the critical relationship between rainfall and runoff. What was the problem? The ancients thought that the amount of rain falling on the land was too small to account for the flow of streams. And if rain was the source of streamflow, then how could streams continue to flow during periods of no rainfall?

In the seventeenth century, a Frenchman named Pierre Perrault (1611–1680) conceived of a way to find the answer. He proposed to measure the rainfall in the watershed of the Seine River in the Province of Burgundy and compare it to the flow of the river. The data revealed that the total volume of rainfall was six times larger than the total volume of streamflow. Around the same time, Edmund Halley (1652–1742), the scientist for whom the comet was named, estimated the evaporation of the Mediterranean Sea. Although his calculations were not quite right, he did establish that the Sea's evaporation was more than adequate to offset the water added by the rivers draining into it. Thus, with evidence that ocean evaporation was sufficient to account for rainfall and rainfall was more than adequate to account for streamflow, the framing concepts for the hydrologic cycle were finally in place.

Table 14.2 Flows of Earth's largest great rivers*

Amazon, South America	180,000
Congo, Africa	40,000
Yangtze, Asia	30,000
Orinoco, South America	28,000
Brahmaputra, Asia	20,000
Yenisey, Asia	18,000
Mississippi, North America	17,500

* Average discharge in cubic meters per second.

Rivers Great and Small: This system maintains the flows of millions of streams that deliver water to the oceans. Most are small with flows of only a few tens of cubic meters per second. On the other hand, the world's largest rivers, which are called **great rivers**, are massive by comparison with flows of thousands of cubic meters per second. In terms of total water delivered, it is noteworthy that the great rivers do the majority of the work. Earth's 70 largest rivers produce more than half the world's runoff and the Amazon stands out head and shoulders among these streams, producing ten times more water than the Mississippi (Table 14.2 lists the seven top great rivers and their average discharges).

For more than a century, the United States and Canada have measured streamflow and precipitation at thousands of locations across North America. The data reveal that for the midsection of the continent the regional distribution of runoff

Figure 14.11 The distribution of precipitation across the United States and southern Canada. The northwest and the southeast are the wettest regions with the highest runoff rates.

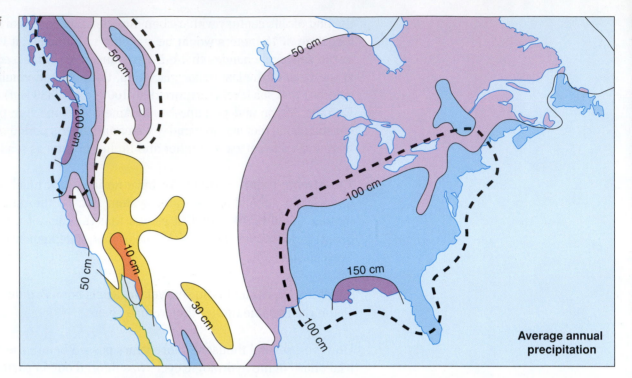

Figure 14.12 The major streams of the United States are located in the northwest and southeast. The Humboldt River is also shown but, unlike the major streams, it never reaches the sea.

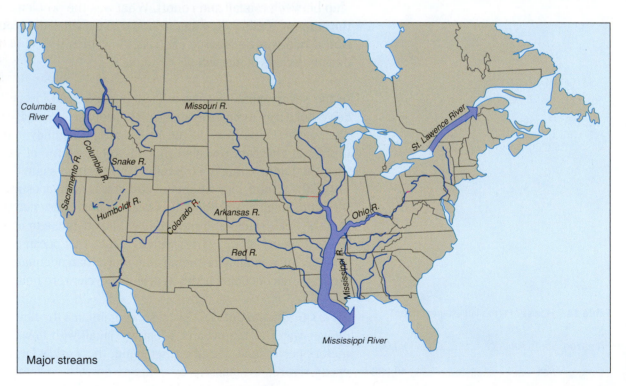

approximates that of precipitation. The maps in Figures 14.11 and 14.12 bear this out. Both precipitation and runoff are heaviest in the northwest and southeast. The runoff in these regions, supplemented by snowmelt and rainfall runoff from the Rocky Mountains, feeds two of the continent's largest rivers, the Mississippi with an average flow of 17,500 cubic meters per second and the Columbia River with an average flow of 7800 cubic meters per second.

In sharp contrast is the Humboldt River in northern Nevada, an area that receives about 25 centimeters (10 inches) annual precipitation. The Humboldt (shown by the broken line in Figure 14.12) gains its flow from mountain runoff, but instead of growing larger downstream as the Mississippi, Columbia, and other wet-climate streams do, the Humboldt declines and finally disappears in the desert as evaporation and the soil take up its water. And with the growth of settlements and irrigated agriculture in Nevada in this century, the Humboldt's flow disappears even sooner than it once did.

The Groundwater System: Groundwater is the third of the land-based systems of the hydrological cycle and in many ways has been the most difficult to understand. Among the reasons is the simple fact that it is hidden underground. Moreover, the ancients (and the not-so-ancients) believed – despite ideas to the contrary advanced by observers like Pierre Perrault – that the Earth's surface was too impermeable to be penetrated by surface water. Therefore, the connection between rainwater and water underground was virtually impossible for most to conceive and, until the nineteenth century, this belief largely stood in the way of thinking of groundwater as a system in the hydrologic cycle.

Like glaciers and ice caps, the groundwater system is both huge and slow. The input of water to this system, a process called **recharge**, does indeed come from surface water, mainly rainwater, infiltrating the soil and trickling downward eventually filling up the spaces between soil particles and in the cracks in rock. Once in the ground it tends to concentrate in porous materials called **aquifers** through which it is slowly transferred laterally under the force of gravity. **Transmission** through an aquifer may take tens or hundreds of years and, at the output end of the system, most groundwater is **discharged** into streams, which carry it to the sea.

There is a vital connection between groundwater and streamflow, which we examine in detail in Chapters 15 and 16. Suffice it to point out here that groundwater is a major source of streamflow and, for large streams, it is the principal source of flow. Unlike surface runoff, which tends to be episodic in nature, groundwater systems provide steady, long-term flows that maintain streams through the often long intervals between rainstorms.

Summary on the Water System Earth is unique among the planets in the Solar System for both the amount and physical states of the water it holds. The oceans hold the vast majority of our water, but the atmosphere is the center of the global water-distribution system. On land, water is cycled through several systems at different rates and, in the course of these cycles, the water takes the form of glaciers, streams, aquifers, and other water features in the landscape around us. Each of these functions as an input–output system, a concept that we take up next.

14.4 The Water-budget Concept

Our effort to understand the Earth's hydrologic system is aided by the use of a concept called the water budget. This concept belongs to a family of scientific models called mass budget models, which are used to describe and analyze open systems. **Mass budget models** resemble simple accounting systems with input (deposits), outputs (withdrawals), storage (account balance), and surplus transfer (to other accounts), but here, of course, the currency is water rather than money (Figure 14.13).

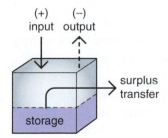

Figure 14.13 A simple input–output model with water storage and surplus transfer.

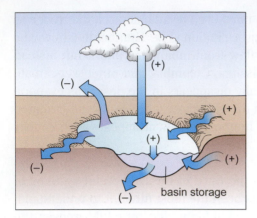

Figure 14.14 Input–output model applied to a lake where input comes from precipitation and stream inflow and output takes the form of stream outflow, evaporation, and leakage to the ground.

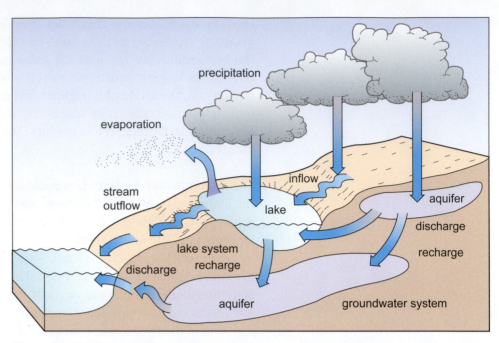

Figure 14.15 Linkages among aquifers and surface water in a groundwater system. Such systems are not only complex but may change their input–output function from season to season.

The Earth's landmasses are laced with water-budget systems. The three terrestrial systems briefly described above – glaciers, streams, and groundwater – are classic examples of water-budget systems. But the water-budget concept can be applied to practically any of the waters we see featured in the landscape. A lake or a reservoir, like the one in Figure 14.14, is a good example. Inputs (+) are provided by inflowing streams, precipitation, and groundwater. The water is stored in a basin, and outputs (−) are provided mainly by evaporation, groundwater loss, and outflowing streams. Wetlands work much the same way as do aquifers: input → storage → output. In a groundwater system, as we illustrated above, water is stored in aquifers (porous rock or soil material), which receive inputs from the surface through recharge, while discharging water to streams and lakes as well as to other aquifers. The size of a lake, wetland, or aquifer can grow or shrink depending on the net balance between total input and total output.

These three systems not only recycle water at different rates, but are usually interconnected in some manner. For example, lakes and wetlands are often tied to complex aquifer systems, as the diagram in Figure 14.15 illustrates. Some provide aquifers with recharge water, others receive aquifer discharge. Some provide recharge in one season, typically winter or spring, and receive discharge in summer. Seasonal cycles also affect the water budgets of glaciers, streams, and soil; and as the diagram suggests, when these systems are linked together they can form very large and complex systems indeed.

Figure 14.16 Application of the water-budget model to a watershed in the Pacific northwest. Stream discharge represents the balance between precipitation and evapotranspiration.

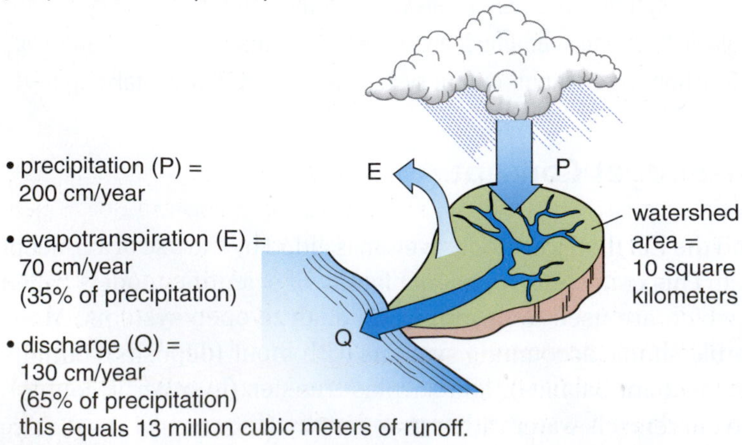

- precipitation (P) = 200 cm/year

- evapotranspiration (E) = 70 cm/year (35% of precipitation)

- discharge (Q) = 130 cm/year (65% of precipitation) this equals 13 million cubic meters of runoff.

watershed area = 10 square kilometers

Watershed Application: The water-budget model can also be applied to much larger geographic entities such as a continent, a country, or a river watershed. A watershed (also called a drainage basin) is an area of land, such as that shown in Figure 14.16, which contributes water to a system of streams via surface runoff and groundwater. Precipitation is its one source of input whereas output is provided by two processes: **evapotranspiration** (*evaporation* from soil and exposed water surfaces plus *transpiration* from the release of water through plants) and **stream discharge** from the mouth of the watershed. Discharge is what remains after evapotranspiration has taken its share.

Evapotranspiration can be huge in warm regions. In Florida, for example, it is 75 percent of annual precipitation but, because of high precipitation rates there, stream-flow is usually substantial. Not so in Nevada, where climate is both warm and dry, and evapotranspiration is 98 percent of annual precipitation. But in western Washington and southwestern British Columbia, where it is much cooler, evapotranspiration is only 35 percent of precipitation and in Alaska it is only 20 percent. The remaining water, 65 percent for Washington and British Columbia (as illustrated in Figure 14.16) and 80 percent for Alaska, is released to streams and discharged into the ocean. Predictably, these two areas have many large streams, especially in the coastal zone where precipitation rates are very high.

The lesson here is that stream discharge can be used as a barometer of a watershed's water budget. As discharge rises and falls the water balance is changing within the watershed reflecting changes in evapotranspiration and/or precipitation. It is not surprising, then, that streamflows, and indeed entire stream systems, not only change geographically in response to differences in climate from place to place, but often change seasonally from summer to winter. And, as we shall see later in this chapter, the water balances of some watersheds are also changed radically by dam and irrigation projects.

The Soil-moisture System: Soil is one of the most important water-budget systems on the planet. Soil water is the sole source of moisture for most plants, including crops, and is a mainstay of terrestrial ecosystems. We discussed the soil moisture system in Chapter 12 but let us refresh our thinking here by applying the water-budget model to a column of soil like the one shown in Figure 14.17.

Water is supplied to the system by precipitation, which seeps into the soil, coating particles and lodging between them. This water is called **capillary water** and it is held in the soil by molecular cohesion in the same way that a film of water is held on and between your fingers after rinsing your hands. Every soil has a maximum amount of capillary water it can store, and this is referred to as its **field capacity**.

When field capacity is exceeded during rainy periods, the excess water entering the soil cannot be held among the particles and must pass downward as **gravity water** (see the right side of Figure 14.17). Named because it responds to gravitational force rather than molecular force, as capillary water does, gravity water represents a surplus in the soil-water system and is discharged from the soil to become part of the groundwater system at greater depths.

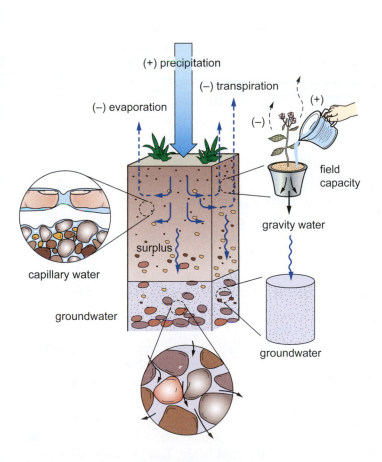

Figure 14.17 The water-budget model applied to a column of soil. Storage is in the form of capillary water, the moisture that most plants draw on. Surplus water passes through the soil as gravity water to recharge groundwater.

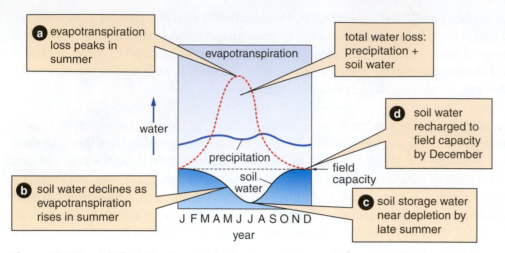

a evapotranspiration loss peaks in summer

total water loss: precipitation + soil water

d soil water recharged to field capacity by December

evapotranspiration

water

precipitation

field capacity

b soil water declines as evapotranspiration rises in summer

soil water

c soil storage water near depletion by late summer

J F M A M J J A S O N D
year

Figure 14.18 A graph illustrating the annual soil-water budget beginning with (a) summer losses to evapotranspiration, (b) soil-water decline and (c) near depletion of storage water, followed by (d) soil-water recharge with cooler conditions in December.

Seasonal Dynamics: The capillary water held in the soil is subject to removal through evaporation and uptake by plants in transpiration. These processes increase and decrease mainly in response to soil and air temperature. When the ground is warm and the air warm and dry (and precipitation is light or absent), capillary-water loss can be rapid, resulting in a negative moisture soil balance, and causing a drawdown on the capillary water stored in the soil (see points a and b in Figure 14.18). If this condition is prolonged, as it often is in summer, then the soil storage water available to plants may be all but depleted, and drought may ensue (point c in Figure 14.18). As climatic conditions cool in fall and winter, evapotranspiration declines and the depleted moisture supply is recharged by infiltrating precipitation water (Figure 14.18d). Within weeks or a few months field capacity is re-established, and any additional infiltration water is converted to surplus and again goes to groundwater.

The soil-moisture budget concept is applicable to nearly all bioclimatic regions of Earth. Because it is a system based on a model that integrates the three essential components of the landscape moisture system – precipitation, evapotranspiration, and soil moisture – it has scientific utility in helping us solve both practical problems, such as how much irrigation water to apply to cropland and when to apply it, and scientific problems concerning, for example, the susceptibility of landscapes to degradation from drought.

14.5 The Hydrologic Landscape

Water has a profound influence on all Earth's landscapes. Its influence on vegetation is apparent nearly everywhere, but here we are also interested in water features such as streams, lakes, and wetlands as parts of the landscape. Broadly speaking, and excluding the polar regions, the world's landscapes can be grouped into three climatic-based hydrologic regimes: humid, semiarid, and arid. Within each of these there are, of course, many variants, but these three will serve to illustrate the concept.

The Humid Landscape Regime: The hydrologic regime in humid landscapes is characterized by a positive moisture balance that yields a water surplus over the entire year or in all but a few months of the year. Annual precipitation rates typically exceed 80 centimeters, except in cold regions, and reach as high as 200 centimeters or more in rainforest climates. Surplus water is discharged downward through the soil column to the groundwater zone below. Groundwater is typically abundant and in many places lies within several meters of the surface where it seeps into stream channels and into low areas forming wetlands and lakes. The groundwater input is remarkably steady, therefore surface water features, as shown in Figure 14.19, tend to be permanent parts of landscapes in humid regions. However, they do fluctuate in size with seasonal changes in groundwater levels, surface runoff, and evapotranspiration. Under the humid climates of eastern North America and western Europe, for instance, the water level in lakes and wetlands may decline a meter or more from spring to late summer and then rise during the winter, but they rarely dry up.

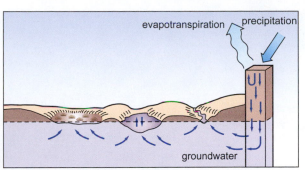

Figure 14.19 A landscape of the humid hydrologic regime. Groundwater is often at or near the surface as is evidenced by this groundwater-fed lake in New England. In the midlatitudes, forests, wetlands, and streams abound in this landscape.

(a)

(b)

The Semiarid Landscape Regime: The semiarid regime is characterized by less precipitation, typically 40 to 80 centimeters a year, and includes not only the semiarid grasslands, but the tropical savanna and Mediterranean landscapes as well. Most water entering the ground goes to recharging soil moisture instead of to groundwater. Recharge takes place in the wet and/or cool season, for example, the winter season in the Mediterranean landscapes of California but most of the water penetrates only 1 to 2 meters into the soil. Surface water is far less abundant in semiarid landscapes, and water features typically appear on a seasonal basis when rainfall and runoff feed stream channels and low spots in the land and then disappear in the dry season (Figure 14.20a). Some groundwater recharge takes place in the wet season, of course, and it causes shallow aquifers to rise and seep into the deeper stream channels and low-lying wetland ponds.

Wildlife ecology is adjusted to this moisture regime, and none is more celebrated than that associated with the water holes in the African savanna as shown in Figure 14.20b. Animal populations tend to be organized geographically according to the distribution and size of these wetland ponds, and when they dry up, as they do every year, animals such as wildebeests, zebras, elephants, and many bird species migrate to areas with better pasturage and water supplies. Indeed, the rhythm of life in savanna, steppe, and Mediterranean landscapes is structured to fit the seasonal moisture regime and the changing distribution of water.

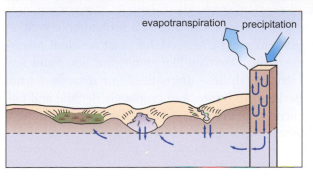

Figure 14.20 Landscapes of the semiarid regime. (a) Stream channels in semiarid landscapes that carry only seasonal flows and occasional stormwater. (b) A watering hole in the African savanna, also a seasonal feature.

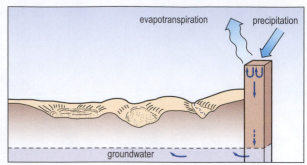

Figure 14.21 In arid landscapes groundwater lies 50 meters or more underground, soil moisture rarely reaches field capacity, and surface water features are rare.

The Arid Landscape Regime: In the arid regime, precipitation is both light and irregular, averaging less than 40 centimeters per year. Water entering the ground penetrates less than 50 centimeters and is inadequate to recharge fully the desiccated soil. Essentially none remains to percolate through the soil to become groundwater and, therefore, aquifers in desert landscapes are in general small and deep as Figure 14.21 illustrates. Where there are more substantial aquifers in this landscape they are either the products of exotic water supplies such as mountain streams fed by snowfields, or past climatic conditions, when things were much wetter.

Most stream channels in arid landscapes carry water only when a cloudburst produces sudden surface runoff. Lakes and wetlands are understandably scarce and, when they do appear, it is in response to the infrequent heavy rainstorm. But most last only hours or a few days during which time many forms of plants, insects, and reptiles may spring to life. The two exceptions to this ephemeral landscape hydrology are, first, large permanent streams such as the Nile and the Colorado, which are called **exotic streams**. We briefly described these streams in Chapter 10 (see Figure 10.39) and you will recall that they bring water into the desert from sources outside the region. The second exception are semipermanent lakes called **playas** which are shallow, salty lakes fed by mountain runoff. As a rule, playas fill and evaporate seasonally leaving dry or mushy salt beds on valley floors but, as we shall see a little later, the large playa basins can maintain a permanent supply of salt water.

14.6 The Nature of Drought

Although deserts are persistently dry, some years are much drier than others. During these years, evapotranspiration greatly exceeds the supply of water available from precipitation and soil moisture. The result is **drought**, a prolonged condition in which available water is inadequate to meet the needs of most resident plants, forcing them to wilt and die, or retreat into dormant-like states. Surface water features such as wetlands and small streams decline dramatically or disappear from the landscape altogether, animals perish or migrate, and croplands decline and fail. In the annals of history, drought represents one of the two most devastating natural disasters; the other, flooding, is also water-driven.

Geographic variations in the incidence of drought are striking on Earth. Droughts can occur anywhere, but are decidedly most severe and frequent in areas that are already dry, the arid and semiarid lands. The reason is related to *precipitation variability*, which follows a simple rule: the drier the area, the more variable the annual precipitation, and vice versa. Look at the maps in Figure 14.22 and compare variability with annual precipitation in a dry region such as the Sahara of North Africa. Now look at a wet area such as the Amazon region of South America. The difference is striking. Not only does the Sahara receive less than 25 centimeters (10 inches) rainfall, but its rate of delivery is highly undependable as well.

In the harshest deserts, total annual precipitation from year to year rarely comes close to the annual mean value. Most years fall way below the mean and a few years fall way above. What is more, the low years tend to occur in clusters so that the effects of several consecutive years of drought are often cumulative. Three such clusters are shown in the graph in Figure 14.23 for Mulka in central Australia. Even worse, clusters may cluster together so that the severest droughts may actually last for decades.

One of the worst droughts in modern times occurred in the Sahel region of North Africa between 1968 and 1974. More than 5 million cattle died and in 1973 alone and an estimated 100,000 people perished of starvation and disease (Figure 14.24). During this period annual rainfall was consistently below average resulting in severe soil desiccation, groundwater decline, and loss of cropland and range land.

When the stress of drought combines with the stress brought on by degrading land use such as overgrazing, the landscape may be damaged beyond recovery. This process, **desertification**, can transform grasslands and savanna into desert in a matter of years. Soil water and groundwater decline irreversibly, the landscape loses its hydrologic resilience, and ensuing droughts may be even more damaging. It is estimated that more than 50,000 square kilometers a year are now subject to desertification worldwide as population pressure, political conflicts, war, and other factors push land use into marginal and submarginal grasslands and woodlands.

Summary on Water Budget and Landscape: All landscapes are water-driven to some extent and the water-budget model helps in understanding how nature allocates its water supply in different landscapes. In most landscapes, the water regime is governed by climate, and since weather and climate vary over time, so vary hydrologic regimes with the dry climates showing the greatest variability, a fact underscored by the magnitude and frequency of drought in arid and semiarid lands.

14.7 Modification of the Hydrologic System by Nature and Humans

Throughout Earth's long history, the hydrologic system has undergone almost continuous modification. We have already touched upon several of these changes in Chapter 8, but let us now take the story a little further by taking a brief look, first, at some of the effects of plate tectonics and mountain building over geologic time, and second, some of the effects of climate change and land use over human time.

Australia has suffered severe drought in the 21st century. Here the remains of a water supply reservoir in Queensland provide vivid testimony.

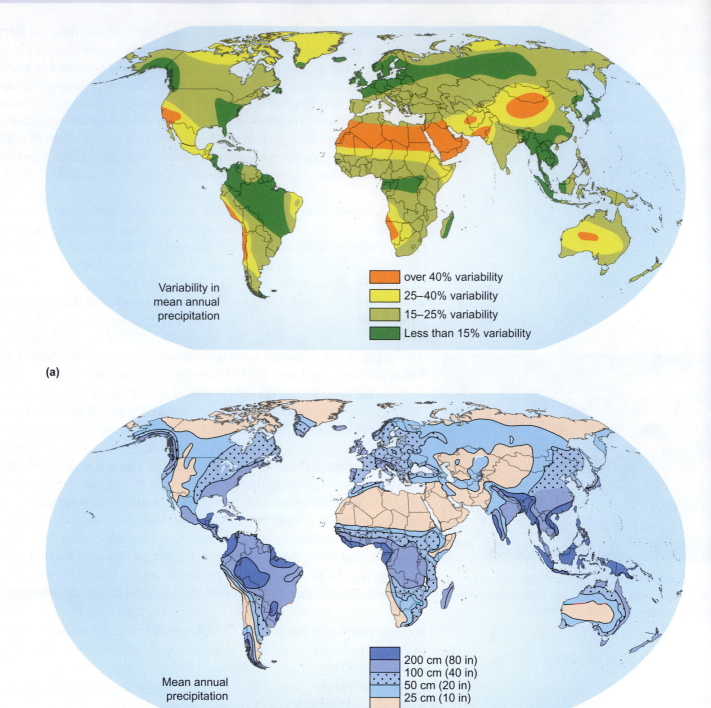

Figure 14.22 (a) Variability in global precipitation and (b) the distribution of global precipitation. Notice the inverse relationship between the two.

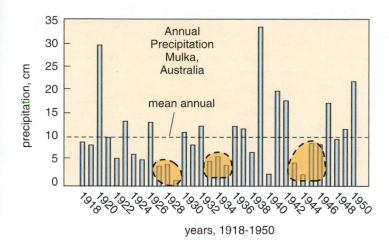

Figure 14.23 Precipitation at Mulka, Australia, for the period 1918–1950, showing the tendency for drought years to occur in clusters.

Figure 14.24 Dying cattle in Sahel during the drought of 1968–74. Goats replaced cattle, and, although hardier than cows, goats are more damaging to vegetation and thus often advance desertification.

Plate Tectonics and Mountain Building: Shifting of great slabs of the Earth's crust (called plate tectonics, which we take up in Chapter 18) has been a major driver of hydrologic change causing, among other things, continents to move thousands of kilometers from one climate zone to another over time spans of millions of years. The rocks and fossils of central North America, for example, reveal changes from wet tropical to desert conditions as the continent changed its geographic position on the planet and as climatic zones shifted over it.

Plate movement has also caused whole oceans to form and fade with the separation and collision of continents, altering, among other things, evaporation and precipitation patterns and amounts. About the time the Atlantic Basin was growing as South America shifted away from Africa (see Figure 18.29), another sea along the southern edge of the Eurasian landmass, known as the Tethys Sea, was shrinking with the northward movement of Africa and the Indian subcontinent. It was a slow process, but over a period of tens of millions of years, the Tethys Sea was reduced in area and finally obliterated as India converged on Asia. The remains of the sea, principally deep layers of sedimentary rock, were crushed and thrown up to form the Himalayan mountain chain, the world's tallest belt of land, which created a barrier separating the Indian subcontinent from the main body of Asia, as depicted in Figure 14.25.

The Himalayan barrier interrupted the seasonal flow of air, the monsoon, to and from the interior of Asia. The summer monsoon, which before the rise of the Himalayas had carried moist air deep into central Asia, now dumped most of its precipitation on India and on the southern flanks of the big mountains. This in turn drove up the flows of three great rivers, the Indus, Brahmaputra, and Ganges while, to the north of the Himalayas in the interior of Asia, conditions grew drier and streamflows declined giving rise to the large deserts which remain there today.

Glaciation and Climate Change: We need not draw only on ancient geologic times for illustrations of nature's alteration of the hydrologic system. The landscape abounds with evidence of more recent examples of alterations and some are striking. For instance, in the American Southwest, only 10,000 to 20,000 years ago, around the time humans were first spreading over the continent, a huge system of lakes, shown in Figure 14.26, existed where only desert valleys and a few remnant salt lakes exist today. The largest of these lakes was Lake Bonneville, which covered 20,000 square miles, a little less than the present size of Lake Michigan (22,400 square miles). Today all that remains of Lake Bonneville is Great Salt Lake with a surface area 2000 square miles.

How could the dry climate we associate with the American Southwest have supported such a system of lakes? This was a period of continental glaciation that produced: (1) cooler climatic conditions with lower evaporation rates; (2) greater precipitation over the region; and (3) large inflows of meltwater from mountain glaciers and snowfields. Acting individually, each of these factors probably would not have produced much change but, as is frequently the case, a combination of factors acting in concert can produce a major shift in the regional water system leading to dramatic changes in the character of the landscape.

Figure 14.25 The closing of the Tethys Sea with the movement of the Indian landmass into Asia. This process led to the uplift of the Himalayan mountain mass, producing a profound change in the climate and hydrology of south and central Asia.

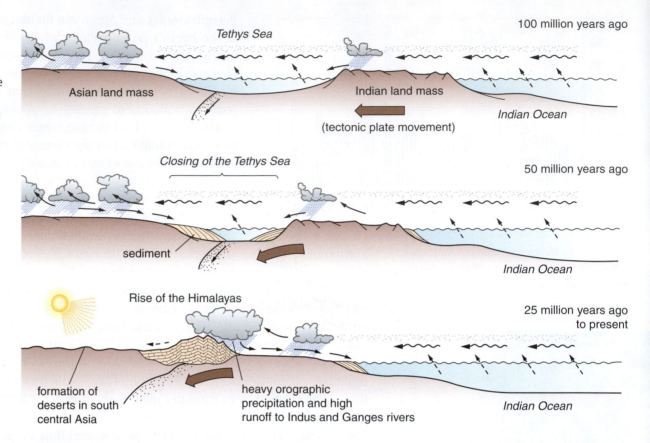

Human Modification of Drainage Systems:
Human modifications of the hydrologic system are so varied and numerous that it is hard to know where to begin. It takes only a quick examination of the historical record to see that the practice of reconfiguring the water system has been one of our greatest preoccupations, first to meet agricultural needs and, later, to meet the needs of cities and related land uses.

The Middle East: For openers we can say that serious modifications of streamflow began with early agriculture in the deserts of the Middle East as much as 10,000 years ago with the diversion of streams for irrigation. Diversions reduced the flow of streams, and for many modest-sized streams on the fringe of the desert, their flows were no doubt depleted early each summer, probably giving rise to the first conflicts among settlements over water rights as well, perhaps, to the first cases of desertification in some locales. As agriculture and population expanded, water-redistribution schemes became more ambitious and elaborate. In many of the great valleys of Asia, centralized irrigation systems provided abundant water, which led to over-irrigation, elevated water tables, and the salt contamination of farmland (described in the previous chapter). In modern Iraq, the cradle of the ancient Assyrian civilization, more than half the once-rich farmland has been lost to salt contamination because of irrigation. Also see Chapter 11 for more on major human impacts on the water system.

But there were also many smaller-scale systems aimed at changing the geography of water. In Asia and North Africa, for example, thousands of long tunnel systems called **qanats** (see Figure 14.27) were dug into mountain slopes to bring groundwater to fields and villages on dry valley floors. It is estimated that in Iran alone there are 22,000 qanat systems. Both of these ancient water-distribution systems – centralized irrigation and qanats – remain in use today, but, in the great river valleys of

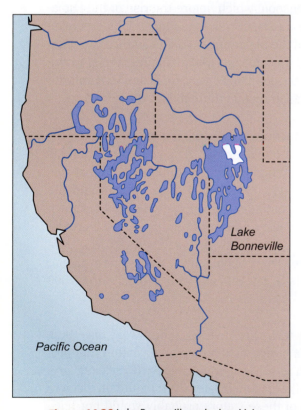

Figure 14.26 Lake Bonneville and related lakes that existed in the American Southwest during the last glaciation of North America. Great Salt Lake (in white) is all that remains of Lake Bonneville.

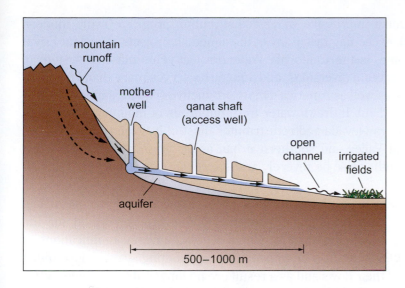

Figure 14.27 Qanats are tunnel systems designed to tap aquifers on mountain slopes and transfer the water to nearby dry valleys. These ancient systems are still widely used in mountainous regions of Asia and North Africa.

antiquity such as in Egypt and China, the centralized systems are bigger and far more ambitious. They now include huge dams and reservoirs such as the Aswan High Dam and Lake Nasser on the Nile River in Egypt, the third largest reservoir on Earth, and dozens of dams on the Tigris–Euphrates river system.

One of the most alarming examples comes from the south–central part of the former Soviet Union (now Kazakhstan and Uzbekistan). It involves the Aral Sea, a large desert lake (originally about the size of Lake Huron) fed by runoff from distant mountains. Between 1926 and 1990, the Soviet Union, driven by an ambitious agricultural expansion program, diverted most of this runoff to feed huge irrigation projects. Deprived of much of its water supply, the sea began to shrink, and by 1990 was only about 60 percent of its natural size with only 35 percent of its original volume. Ports dried up, the water turned saltier, and the sea's ecosystem collapsed. The Aral Sea continues to shrink as the map in Figure 14.28 illustrates.

North America: Some of the world's most elaborate water-modification schemes are now found in North America. Since 1900, the United States has built 75,000 dams 2 meters high or higher for purposes of flood control, water supply, and recreation. On the Colorado River system, America's principal network of desert streams, a massive system of reservoirs has been constructed, which can store an astounding quantity of water – about four years of the river's average total flow (Figure 14.29). The water is diverted to cities (e.g. Phoenix and Los Angeles) and irrigated farmlands, but huge amounts (2 meters or more per year) are lost from reservoir surfaces in evaporation to the arid atmosphere. When the Colorado finally crosses into Mexico only a trickle or less remains. To the north, Canada's system of major dams (defined as those 30 or more meters high) has the capacity to hold the equivalent of two years' discharge from *all* the streams in the country.

Figure 14.28 (a) The decline of the Aral Sea as a result of massive water diversion for agricultural irrigation by the former Soviet Union. (b) Stranded fishing boats on the dry bed of the Aral Sea.

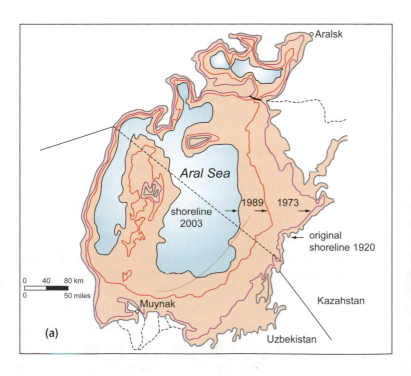

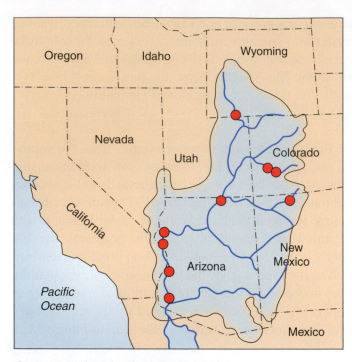

Figure 14.29 The Colorado River system and the location of major dams. Two of these dams, Hoover and Glen Canyon, are ranked among the 50 largest on Earth based on the volume of water stored.

Figure 14.30 One of the main aqueducts in the huge California Water Project, which feeds both cities, such as Los Angeles, and a massive system of irrigated agriculture.

The most impressive water redistribution program in the modern world, exceeding by far any of the aqueduct projects of ancient Egypt or Rome, is the California Water Project. This massive system of dams, stream diversions, and aqueducts serves more than 20 million people as well as America's most productive system of irrigated agricultural land. It draws water from hundreds of mountain streams in the Sierra Nevada and redistributes most of it to lands in the state's Central Valley and southern urban areas around Los Angeles that would otherwise be severely arid (Figure 14.30).

The benefits of the California Water Project are immense. Chief among them is abundant, diverse, and inexpensive food for the United States and Canada. But the environmental ramifications of this system are also immense. The contributing streams lose much or all of their flows and as a result are no longer able to deliver water to the sea. With this has come the loss of spawning habitat for many fish species and the decline of beach sand along parts of the Southern California coast (because streams that naturally transport sediment to the sea have been dammed). In addition, vast areas of wetland have been lost to the massive system of farmland and groundwater has been contaminated with agricultural pesticides in many areas.

Added to these hydro-ecological impacts is the massive amount of energy needed to move the water. Electrical energy is required to lift the water over hills and ridges, in particular over the Tehachapi Mountains, a lift of 2000 feet – the highest lift of any water system in the world. As a result, the California State Water Project is the largest *single* user of energy in California. In the process of delivering water from the San Francisco Bay Delta to Southern California, the project uses 2 to 3 percent of all electricity consumed in the state, a huge amount considering California's massive economy. The energy used to deliver water to residential customers in Southern California is equivalent to approximately one-third of the total average household electric use in the region.

14.8 Global Warming and the Hydrologic Cycle

How the Earth's water system will change with global warming is a major question facing modern science. We briefly explored this question in Chapter 8, but it is worthwhile to review the main line of reasoning behind the thinking.

The Enhanced Hydrologic Cycle: Since the hydrologic cycle is powered by heat stored on the Earth's surface, it stands to reason that global warming should be affecting the cycle's operation. Foremost among the effects are changes in global evaporation and precipitation rates. Simply put, warmer surface temperatures are driving up evaporation rates on both land and water, forcing more water vapor into the atmosphere, which in turn is leading to increases in the amount and intensity of precipitation. Because of this boost in water input and output rates, the global-warming version of this great system is called the **enhanced hydrologic cycle**. Here is how it works.

Roughly 80 percent of the additional heat energy available at the Earth's surface as a result of global warming is used to evaporate water, mostly from the oceans. With each molecule of water evaporated, this heat is transferred to the atmosphere where it is stored with the vapor as latent heat. But this state does not last long for, in a matter of days, the vapor cools and condenses, cloud particles form, and the latent heat is driven into the air, raising air temperature. The remaining 20 percent of the heat from global warming contributes directly to warming of the Earth's surface and the lower atmosphere as ground heat and sensible heat.

The Role of Feedback: Here is where the feedback component of the system comes into play. With higher air temperatures, the atmosphere can now hold more water vapor, and since water vapor is the atmosphere's principal absorber of longwave radiation, the Earth's greenhouse effect is enhanced, meaning that the atmosphere's capacity to trap and retain heat energy is increased. This results in even higher temperatures, which in turn leads to even higher temperatures. Thus, a positive feedback loop is created in which the system spirals towards warmer temperatures with increased evaporation and water-vapor loading of the atmosphere (see Figure 8.36). Not only that, with more water vapor, the atmosphere holds more energy and thus has greater power to drive storms. Storm magnitudes and frequencies, rainfall rates, and runoff rates should be rising.

What do the data show? Are there any trends apparent in the instrumental record that support the concept of an enhanced hydrologic cycle? In a word, yes. Among them are: (1) data from high-altitude weather-balloon measurements indicating that the water-vapor content of the atmosphere is rising; (2) climate records revealing that precipitation amounts over the middle and high latitudes have generally increased in the past 100 years, often in excess of 10 percent; and (3) weather records showing an increase in heavy and extreme rainfall events over the last several decades. Some scientists argue that this is a natural trend, but a great many point to global warming.

Some Hydrologic Consequences of Global Warming: Although a higher global mean annual rainfall and a smaller number of frost days may have some beneficial effects – for marginal mountain agriculture, for example – there are more undesirable consequences. With warmer temperatures, mountain glaciers and snowpacks are shrinking, meaning that less summer meltwater is available to streams, lakes, and groundwater recharge and in turn to agriculture and settlements in regions like south and southwest Asia and the North American West. On the other hand, with larger and more frequent rainstorms, runoff rates can be expected to increase, giving rise to greater soil erosion and flooding, both of which can drive down food production as global population is rising.

Shrinking mountain glaciers and snowfields can lead to reduced water supplies for settlements and farmlands.

Longer, drier summers will place added stress on soil-moisture budgets. Evapotranspiration rates will rise and the period of soil-moisture depletion, shown in Figure 14.18c, will lengthen and deepen, forcing farmers to increase irrigation, change crops, or, lacking these options, abandon the land. History tells us that the latter often leads to landscape degradation, including desertification. Increased irrigation also has a down side, because groundwater recharge rates may also decline with longer summers, thereby hastening aquifer depletion.

Other hydrologic consequences associated with global warming include higher sea levels and increased coastal flooding and erosion, which we examined earlier in Chapter 8. This trend flies in the face of a global trend toward increased population in coastal areas, especially coastal cities. With huge and growing investments in their harbor and urban infrastructures, cities will, against all odds, fight to sustain themselves by building more and even bigger infrastructure to hold the sea back, but the costs and the risks will be extreme.

Chapter Summary and Overview of The Global Water System

A central theme in our discussion of this great Earth system has been the geographic variation in the distribution of terrestrial water. It is curious that in spite of Earth's circumglobal coverage of ocean water and the enormous supply of water vapor made available by the oceans, the availability of water on land is remarkably uneven. The best general explanation for this is that the hydrologic system on land is controlled primarily by the atmosphere, and it is climate, and the various controls on it, including the oceans and mountain barriers, that sets the framework for the Earth's hydrologic system. Within this framework are the various water subsystems and it is these systems that govern the behavior of water in the landscape.

▶ **Earth's generous supply of water makes our planet unique in the Solar System.** This water originated with outgassing from the crust and vapor showers from influxes of ice from space.

▶ **The hydrologic system is driven by solar energy.** It drives a rapid turnover in atmospheric water, most of which is exchanged with the sea.

▶ **The continental branch of the hydrologic cycle is made up of three main systems.** They vary widely in terms of water-storage capacity and exchange (residence) times.

▶ **The water-budget model can be applied to any water system.** In the soil it is driven by precipitation and evapotranspiration; in aquifers by recharge and discharge.

▶ **Hydrologic landscapes are products of Earth's moisture regimes.** The role of groundwater in the landscape generally decreases with total annual precipitation.

▶ **Precipitation variability is linked to mean annual precipitation.** While drought can occur anywhere, the probability of such events increases with declining annual precipitation.

▶ **Earth's hydrologic system has a long record of change.** Change is driven by both natural and human factors. The latter are relatively new to Earth but have resulted in some dramatic alterations in the distribution of water.

▶ **Lakes were far more abundant during the last glaciation in North America.** Over the last 10,000 years the large lakes of the American Southwest have greatly diminished or disappeared with climate change.

▶ **The California Water Project is currently the world's largest aqueduct system.** It serves a massive urban population and North America's most productive agricultural region.

▶ **The Earth's hydrologic system in intricately tied to the global energy system.** One result of global warming is an enhanced hydrologic cycle in which surface heating is promoting higher evaporation rates.

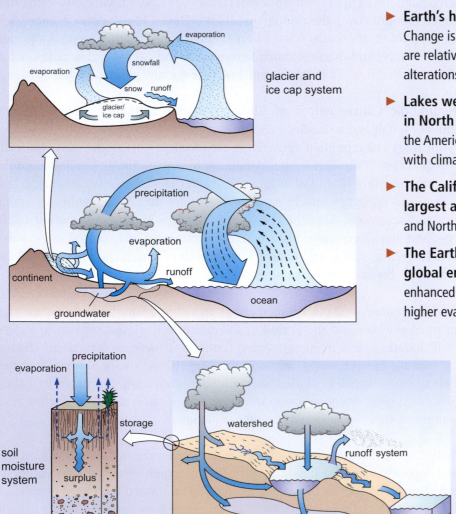

The hydrologic cycle feeding three terrestrial major terrestrial water systems: glaciers and ice caps, groundwater, and stream runoff. Within each of these are various subsystems such as soil-moisture systems and aquifer systems.

Review Questions

1 Can you explain why Earth is so different from Mercury and Mars as a hydrologic environment?

2 What are the two apparent sources of Earth's water and what is the basis of the "snowballs" from space concept?

3 What is the difference in the percentage of Earth's atmospheric, terrestrial, and oceanic water supplies? Which are freshwater and what is Earth's largest supply of liquid freshwater?

4 What is the source of energy that drives the hydrologic cycle, and how is it that the atmosphere with only 0.001 percent of Earth's water is able to produce so much precipitation?

5 How do you account for the fact that Earth's landmasses produce only 14 percent of atmospheric vapor but receive 22 percent of world precipitation in return? What processes enable the continents to "capture" so much precipitation?

6 Nearly 90 percent of the water in the atmosphere comes from ocean evaporation. What geographic regions, conditions, and seasons produce the most vapor?

7 When precipitation falls on the land, how much (percentage) is lost to evaporation and how much goes into storage-flow systems? Name these systems and their respective water recycle times.

8 What is the relationship between ocean level and the world's glaciers and ice caps? How much has sea level risen since the last major world glaciation and why? Is sea level expected to rise in the next century and why?

9 In the development of hydrologic cycle concept what was the chief problem early scientists faced in rounding out the model and how was this problem finally resolved in the eighteenth century?

10 Can you describe the concept of a water-budget model and give several examples of such systems in nature? What are the inputs and outputs in a watershed's budget? In a soil-column water budget? And in this context, how does a soil store water and also give up water to groundwater?

11 Can you describe the principal features of the hydrologic landscapes that develop under these regimes: humid, semiarid, and arid? How does water influence surface conditions in each?

12 What is a drought and how is the incidence of drought related to precipitation variability? How do you define desertification and what are the contributing factors?

13 What hydrologic changes were associated with the closing of the Tethys Sea and the emergence of the Himalayas? And what is the significance of today's Great Salt Lake in terms of North America's changing hydrologic landscape?

14 Have humans had much impact on Earth's hydrologic system? When did human impact begin, what are some of the more glaring examples, and what do you forecast for human alteration of Earth's hydrology in the twenty-first century?

15 Can you explain the effect global warming has on the water-vapor content of the atmosphere and how changes in atmospheric moisture can affect global warming? Why is global warming a concern for countries facing increased population in coastal areas?

Runoff, Streamflow, and Watershed Systems

Chapter Overview

Our story begins with a brief look at the runoff systems that feed streams. We are particularly interested in the role the landscape plays in these systems and, in turn, in streamflow; for example, how trees intercept rainfall and how soil soaks up rain that hits the ground. We are also interested in how the atmosphere delivers rainfall and how rainfalls of different intensities and durations influence streamflow. All this is set into the geographic framework of a watershed where networks of streams of various sizes and patterns form elegant water-moving systems. And no story about streams would be complete without addressing flooding, a phenomenon that has baffled and plagued humans for thousands of years. The chapter concludes on a distinctly geographic note, an overview of the 20 largest watersheds of the world, which include the Amazon, Mississippi, Nile, and Yangtze.

Introduction

Ancient Hindu and Buddhist mythology describe a sacred mountain at the center of the Universe as the source of the Earth's great rivers. Early Christian mythology identifies the Garden of Eden as the place where Earth's waters divide to form the great rivers. Thousands of years later, in the nineteenth century, we were still speculating on the sources of many great rivers. Scores of explorer/geographers driven by the romance of adventure, scientific curiosity, and political directives ventured into the heart of North America, South America, Asia, and Africa to find the few remaining undiscovered river sources.

By the end of the century the geographic record on the sources of the great rivers was largely complete. Two young Americans, Meriwether Lewis and William Clark had documented the sources of the Missouri and Columbia rivers in the Rocky Mountains. To the north, an intrepid Canadian, Alexander Mackenzie, traced a great river that now carries his name more than a thousand miles from its Rocky Mountain headwaters to the Arctic Ocean. About a half century later, after a celebrated search, John Speke, an Englishman, reported to the Royal Geographical Society in London that Lake Victoria was the head of the world's longest river, the Nile (Figure 15.1).

Figure 15.1 A painting of the explorers John Speke and Richard Burton on their expedition in search of the source of the Nile in 1858.

Attention now turned to questions concerning the runoff processes that produced the flows of rivers and streams. While it was clear that precipitation was the root source of streamflow, it was not clear how this water made its way to streams, especially from underground sources. In addition, there were questions about the various controls on runoff rates, and how geographic attributes of stream systems such as watershed size, the density of channels, and land use influenced streamflow. Finally, the age-old question of river flooding, which had perplexed societies for centuries, needed the attention of modern science.

15.1 From Rainfall to Streamflow: The Role of Landscape in Runoff

Streams are nature's most efficient means of draining water from the land. But how does the water that falls as precipitation find its way into stream channels? The easy answer is that it simply runs downhill, and at some point eventually reaches a channel. But if this were their only source of water, then streams would flow only when it rained. Some do just that, but most streams, especially larger ones, flow when it is not raining, even during extended dry periods. Therefore, there must be other sources of streamflow.

In fact there are four systems that feed water to streams: channel precipitation, overland flow, interflow, and groundwater. Two of these, groundwater and interflow, are underground systems and two, channel precipitation and overland flow, are surface systems. Figure 15.2a gives a simple portrayal of how these systems work in and around a small stream channel. As the terms imply, *channel precipitation* is water falling directly on the stream and *overland flow* is rainwater or snowmelt water running over the ground to the stream. *Interflow* and *groundwater*, on the other hand, must first pass through the soil before arriving at the stream. The main difference between these two systems is that interflow moves in the space above the water table whereas groundwater moves in the much larger space below the water table.

Figure 15.2 (a) The four sources of streamflow shown in a small stream-valley setting: channel precipitation, overland flow, interflow, and groundwater.

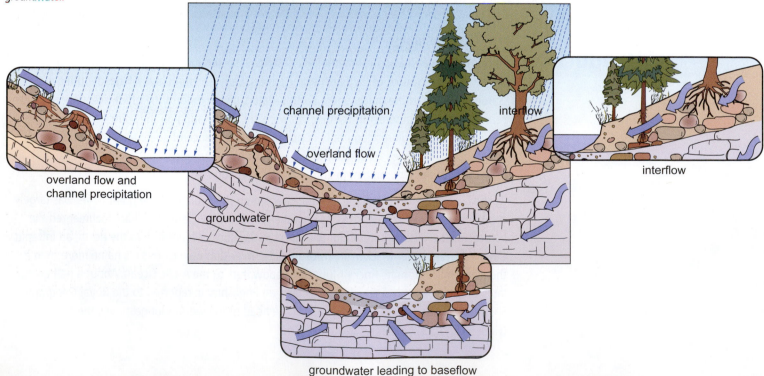

overland flow and channel precipitation

channel precipitation

overland flow

groundwater

interflow

interflow

groundwater leading to baseflow

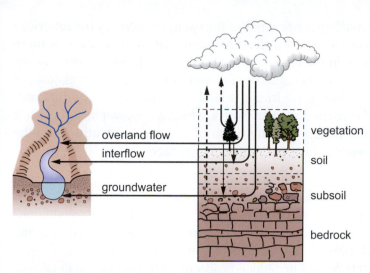

The three terrestrial runoff systems, overland flow, interflow, and groundwater, contribute the vast majority of water to most streams. Each emerges after passing through the landscape and, not surprisingly, the landscape plays a pivotal role in regulating not only how much water is moved by each but at what rate it is delivered to streams (Figure 15.2b). All things considered, the landscape performs four important functions in shaping the nature of streamflow:

- it reduces the variability of runoff thereby making streamflow less sporadic;
- it lengthens the time of the runoff (beyond the period of rainfall) thereby ensuring greater continuity of streamflow;
- it stores water, which is released to streams during dry (rainless) periods; and
- it reduces the total volume of water reaching streams by inducing plant uptake and evapotranspiration.

Figure 15.2 (b) The sources of terrestrial runoff feeding a stream system begin with precipitation passing through the landscape which regulates when and how much water reaches the stream.

In short, the landscape serves as an important hydrologic buffer, or mediator, regulating the disposition of precipitation and the way runoff is delivered to streams and is converted into streamflow. Therefore, as landscape changes from place to place, we can expect runoff and streamflow to change as well.

Sources of Streamflow: Groundwater is a massive reservoir of freshwater extending thousands of meters into the Earth. Its upper surface, called the water table, lies within several meters of the surface in humid landscapes. In arid landscapes, as we noted in the previous chapter, it lies much deeper. As streams erode their channels into the land, they cut into the water table, thereby tapping into the groundwater system. Groundwater gradually seeps into the channel and, since it is such a large reservoir of water, the stream is provided with a steady, long-term source of discharge, called **baseflow**. In humid regions such as the eastern side of the Mississippi watershed, where there is lots of groundwater close to the surface, baseflow makes up the majority, often the vast majority, of a stream's discharge. As much as 80–90 percent of the Mississippi's mean discharge of 17,500 cubic meters per second is baseflow.

Interflow is a smaller source of underground water. It moves closer to the surface in the space between the water table and ground level, flowing laterally among the layers of soil or surface deposits. Some small streams in heavily forested watersheds receive most of their water from interflow fed by rainwater and snowmelt water that has soaked into the ground. When it reaches the channel, interflow combines with two sources of surface water, channel precipitation and overland flow.

Channel precipitation is simply rain and snow falling directly on the stream's surface. For small streams it does not amount to much, but for large streams, 5 centimeters of precipitation, for example, can be appreciable especially as it mounts up downstream with contributions from tributary streams. **Overland flow** is runoff pouring over the ground in direct response to rainfall or snowmelt. It is also called **stormwater** because most is produced by rainstorms. Overland flow is profoundly influenced by conditions of the landscape and, since it is a chief cause of flooding, we need to examine these landscape conditions in some detail.

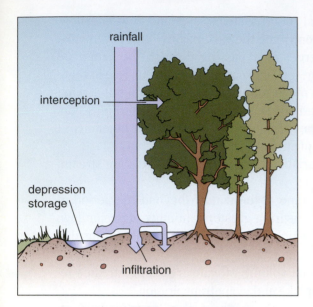

Figure 15.3 Rainfall leading to interception, depression storage, and infiltration. Together, these processes serve as important landscape mediators slowing and reducing runoff.

Landscape Controls on Runoff: In the first half of the twentieth century the American hydrologist Robert Horton (1875–1945) demonstrated that the landscape is more an active than passive player in the terrestrial water system. He and others showed that vegetation, soil, and landforms act as mediators influencing the breakdown of precipitation and regulating its release from the land. We now understand that rainfall is affected by the landscape in three basic ways, which are shown in Figure 15.3: **interception** by vegetation (mainly wetting down of foliage), **infiltration** into the soil, and **depression storage** in little pockets on the surface.

These processes take up water and, for most rainfall events, there is little or no water left over to run off the surface as overland flow. Occasionally, however, conditions are such that the landscape cannot take up all the precipitation delivered to it and the remainder goes to overland flow. Two sets of conditions can result in the production of overland flow: (1) limited or reduced capacity of the landscape to take up rainwater, owing, for example, to thin soil cover or little or no vegetation; and/or (2) a rainfall event of such intensity that the landscape's uptake capacity is exceeded.

How do the landscape mediators (vegetation, soil, and landforms) work? Envision a landscape covered by a dense forest – like the one shown on the opening page of this chapter – with a modest rain falling on it. For the first hour or so most of the rainwater is intercepted by the umbrella of foliage. Should the rainfall last longer or intensify, the tree canopies will become loaded with water and begin giving it up to the ground. This water reaches the ground as drippage, called *throughfall*, or as trickles flowing down branches, stems, and trunks, called *stemflow*. Interception represents a set of processes or a small system that can be significant not only in reducing the total amount of precipitation reaching the ground, but in reducing its rate of delivery as well.

When precipitation does reach the ground, it begins to puddle in low spots as depression storage. It is important to realize that no surface, and especially those with plant covers, is truly level or smooth. Excluding lawns, ball fields, parks, and other mechanically graded ground, most surfaces are very bumpy, making it difficult for water to move across them. Forest floors are especially rough as the graphic in Figure 15.4 reveals. The various pits, pockets, and holes act like little reservoirs that not only hold water but also release it to the ground by infiltration.

Figure 15.4 Some of the factors that roughen the forest floor giving it a high depression-storage capacity.

The **infiltration capacity** of the ground is a primary consideration in all runoff problems. It is controlled mainly by vegetation, soil, and slope. Vegetation loosens the soil, facilitating moisture penetration, and the organic matter in topsoil quickly absorbs water. And if the soil's texture is coarse – for example, sandy, or gravelly – infiltrating water moves quickly downward within the large open spaces among the soil particles. On the other hand, if the soil is composed of compact clay, infiltration is retarded, especially where vegetation is weak or absent. Finally, there is the

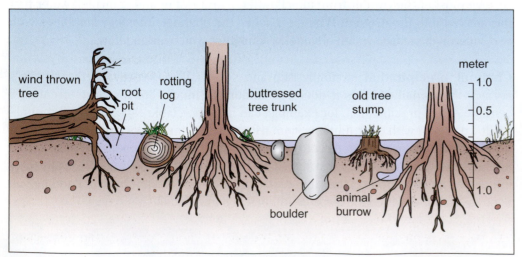

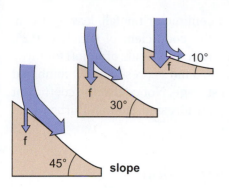

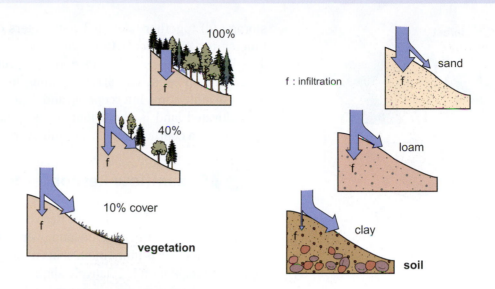

Figure 15.5 The relative influence of slope, vegetation, and soil on infiltration. Maximum infiltration is found on flat or gently sloping ground with heavy forest cover and sandy or gravelly soil.

slope factor, that is, the inclination of the ground. Steep slopes lower the potential for infiltration by encouraging water to run off before it has had time to soak into the ground. At lower slope angles, however, infiltration rises because water tends to linger on the ground. Overland flow, in turn, decreases as infiltration rises as the diagrams in Figure 15.5 suggest.

Transient Conditions and Land Use: Two other sets of factors also influence infiltration capacity: transient conditions and land use. **Transient conditions** include seasonal and shorter-term phenomena such as groundfrost, cloudbursts, and soil saturation by an earlier rainfall. Groundfrost and soil moisture block interparticle spaces, whereas cloudbursts produce such high intensities that they exceed the short-term infiltration capacity of even highly porous ground. Transient factors have contributed significantly to countless major flood events. The runoff that produced the massive Mississippi River flood in the summer of 1993, for example, was caused in part by earlier rainstorms that soaked the soil and cut infiltration capacities to effectively zero before the truly heavy rains came.

When **land use** enters the scene, infiltration, depression storage, and interception almost invariably fall. Vegetation is removed, the ground is graded from rough to smooth, and new materials such as asphalt and concrete are added to the surface. Runoff by overland flow spikes. In agricultural regions, pastures and plowed fields replace woodlands or native prairies, and in urban regions clearing is followed by construction of hard (impervious) ground covers in the form of roads, parking lots, buildings, and related facilities. In the extreme case, when urban development replaces a forested landscape, the ratio of infiltration (plus depression storage and interception) to overland flow will reverse from 9 to 1 to 1 to 9. The result is alarming: surface runoff may increase nine times!

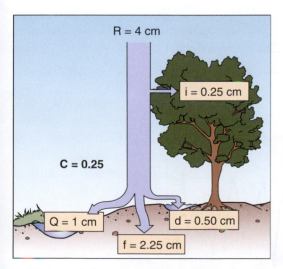

Figure 15.6 The concept of the coefficient of runoff (C). Here 3 centimeters of a 4 centimeter rainfall are taken up by the landscape leaving 1 centimeter, or 25 percent, to surface runoff or overland flow (Q). Hence the coefficient of runoff is 0.25.

For purposes of estimating increases in stormwater runoff resulting from land-use change, it is often necessary to compute the amount of rainwater withheld from the overland flow system by the landscape. Instead of addressing infiltration, interception, and depression storage individually, they can be combined into a single expression called the **coefficient of runoff** (C). This, too, is a ratio indicating the percentage (expressed as a decimal) of a rainfall that is converted into overland flow. In the example presented in Figure 15.6, infiltration (f), interception (i), and depression

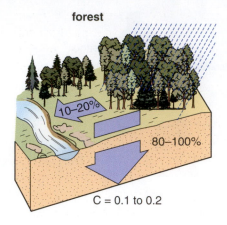

forest

10–20%

80–100%

C = 0.1 to 0.2

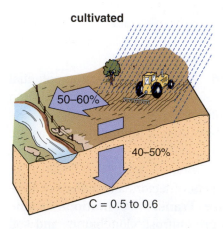

cultivated

50–60%

40–50%

C = 0.5 to 0.6

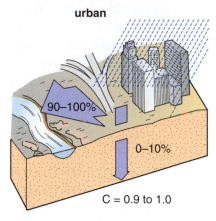

urban

90–100%

0–10%

C = 0.9 to 1.0

Figure 15.7 The influence of land use on the coefficient of runoff. In the conversion of a forested to an urban landscape, the coefficient (and hence overland flow) may increase nine times.

Figure 15.8 The difference in the variability of annual flow between a stream (a) fed preponderantly by a large groundwater system, and one (b) fed mainly by overland flow from rainfall. Notice the dramatic difference between the maximum and minimum years for these streams, such as between the years set off by the broken lines.

storage (d) together take up 3 centimeters of a 4 centimeter rainfall, leaving 1 centimeter (or 25 percent) to overland flow. Thus, the coefficient of runoff is 0.25. The diagrams in Figure 15.7 give some representative coefficients of runoff for three different land uses. The arrow pointing into the ground represents the combined losses to infiltration, interception, and depression storage. Notice that the coefficient for cultivated land is about four times greater than forested land and urban land approaches a coefficient of 1.0, or full overland flow.

15.2 Streamflow Responses to Precipitation

We have established that streams derive their flows from four sources: groundwater, interflow, channel precipitation, and overland flow. Groundwater is slow, steady, and massive; interflow is more modest in volume but somewhat faster than groundwater; channel precipitation is relatively slight and occurs instantaneously when it rains or snows; overland flow also occurs only when it rains, or shortly thereafter; therefore, its contributions to streamflow, like channel precipitation, are highly variable.

Rivers which derive their flows principally from groundwater tend to vary little with short-term changes in precipitation. Those with direct connections to large groundwater systems, such as Silver Springs River in Florida, whose annual discharge is shown in the graph in Figure 15.8a, not only vary little in the short term but tend to maintain relatively steady long-term flows as well. On the other hand, streams with less groundwater support, such as the Brazos River in Texas (shown in Figure 15.8b), which drains a much drier region, produce more radical variations in discharge, both in the short and long term, because they respond more directly to rainfall events. For comparison, in the early 1960s both streams hit very high and very low years in response to changes in annual precipitation in the American South, but the difference between the high and low years was ten times on the Brazos and less than two times on the Silver Springs.

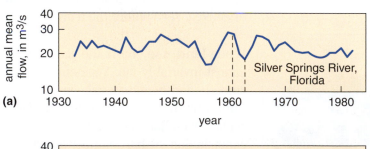

(a)

Silver Springs River, Florida

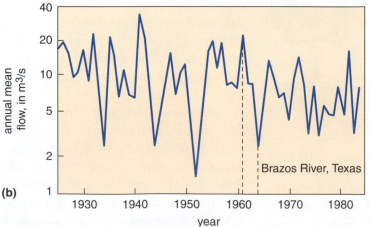

(b)

Brazos River, Texas

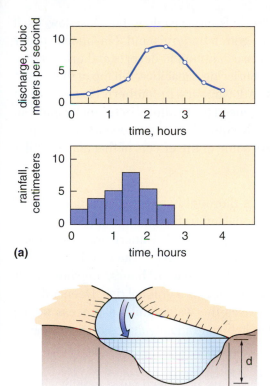

(a)

(b)

Figure 15.9 (a) A streamflow hydrograph shows the response of a stream to a rainfall event. (b) Stream discharge is computed based on measurements of channel width (w) and depth (d) and average flow velocity (v).

The Streamflow Hydrograph: There is a simple model called a **hydrograph**, which is used in streamflow studies to define the relationship between a stream's flow and precipitation. The object of this graph is to show changes in streamflow, or **discharge**, over time in response first to different rainfall events (e.g. light or heavy) and second, to various watershed conditions, such as changes in forest cover, that influence the delivery of runoff to the stream channel. The graph itself is constructed with discharge on the vertical axis and time, usually hours or days, on the horizontal axis. And for ease of comparison, rainfall can be plotted on the same graph or on a separate graph using the same time scale as shown in Figure 15.9a.

Discharge is measured similar to traffic flow on a highway, but instead of vehicles per hour or day, it is cubic feet (or meters) of water per second. Most discharge measurements are made at selected spots along the channel, called gauging stations, where the width and breadth of the stream are known and flow velocity can be measured (Figure 15.9b). Discharge (Q) is calculated by multiplying channel width (w) times average water depth (d) times average velocity (v):

$$Q = w \cdot d \cdot v$$

If velocity is measured in meters per second and channel depth and width in meters, then the product, discharge (Q), is in cubic meters per second (m^3/s). If discharge readings can be made before, during, and after a rainfall event and plotted in a hydrograph, it is possible to determine how the stream responds to things such as a prolonged, heavy rainfall or to urban development in its watershed. The following paragraphs outline four different types of rainfall events and the corresponding changes in stream discharge.

Light Rainfall: Drizzles deliver such small amounts of precipitation at rates so slow that essentially all is absorbed by the landscape in interception, depression storage, and infiltration. Channel precipitation is minuscule, no overland flow results, and the water infiltrating the soil is inadequate to generate interflow or add to groundwater. The stream hydrograph, shown in Figure 5.10, remains unchanged except for the gradual decline in existing baseflow, which would have occurred anyway. In other words, the stream continues to behave as though no rain had fallen.

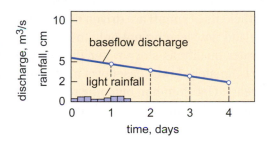

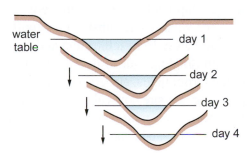

Figure 15.10 Baseflow decline over time when a rainfall is too light to produce overland flow or interflow or recharge groundwater.

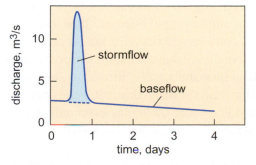

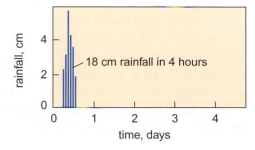

Figure 15.11 Cloudbursts can produce a surge of the overland flow resulting in a sharp spike in the hydrograph. Baseflow, however, remains unchanged because fast-moving stormwater is unable to penetrate the ground.

Short, Hard Rainfall: Cloudbursts deliver so much rain so fast that they often exceed the capacity of the landscape to buffer it. The soil infiltration capacity is overridden and, for a short period of time, usually less than an hour, overland flow pours off the land. The stream rises abruptly with this sudden contribution of stormwater, but falls almost as abruptly with the cessation of the storm. The hydrograph, as Figure 15.11 illustrates,

Figure 15.12 Flash floods can produce massive damage. They can occur in most landscapes but are most common and dramatic in dry lands where the landscape's capacity to slow and reduce runoff is limited. Above, San Antonio, Texas, 1935. Bottom, Toowoomba, Australia, 2011.

shows a large spike superimposed on the existing baseflow, but baseflow itself remains unchanged because little water has gotten underground. Flashfloods occur this way, and it is understandable why they are common in deserts and urban areas because treeless landscapes have such low buffering capacities. On top of that, some dry regions are capable of producing massive downpours. In southern Texas near San Antonio, for instance, a rainstorm in 1935 produced 55 centimeters (22 inches) of rain in 2 hours and 45 minutes, an amount roughly equivalent to the *annual* average precipitation in that region. Figure 15.12 shows some of the effects of that and a similar storm in Australia.

Long, Moderate Rainfall: Long, steady rains thoroughly wet vegetation, gradually fill storage depressions, and generate substantial infiltration, but it is all done at such a slow rate that essentially no overland flow is generated. Underground, however, things are changing as the soil becomes loaded with gravity water and both interflow and groundwater recharge are initiated. Within 10 or 12 hours, interflow water begins to trickle into the channel and the stream begins a gradual rise. Deeper down, the gravity water added to groundwater raises the water table, a process taking several days or more in small valleys, which in turn gradually drives up the stream's baseflow.

Here is the main point. The stream's discharge has increased substantially without stormwater contributions on the surface. And since increases in both interflow and groundwater require precipitation to pass through the various layers of the landscape – via infiltration, percolation, transmission, and discharge – before reaching the stream, the rainstorm's immediate impact has been delayed and drawn out over days (or possibly weeks in the case of groundwater). Additionally, a huge amount of water has been stored in the ground to maintain baseflow until the next rainfall.

Herein lies an important lesson for land-use planning and water management. When we develop land for settlements and agriculture, we invariably change the balance between surface and subsurface runoff. Coefficients of runoff are driven up, water is diverted away from infiltration and into overland flow leading not only to increased stream flooding, but to decreased stream-water quality as stormwater flushes pollutants from fields, streets, parking lots, etc. Ultimately, lower baseflows may result, lowering average flow volumes within the stream. One common outcome of reduced flow volume is an increase in water temperature, which disrupts aquatic ecosystems driving out certain species.

Long, Hard Rainfall: With a heavy rainfall lasting many hours, all pathways leading to streamflow are activated as Figure 15.13 illustrates. The hydrograph may show a small initial rise with channel precipitation at the onset of rainfall, which leads directly into a sharp rise in the curve with the outpouring of overland flow from the lands bordering the channel. The hydrograph continues to rise, reaches its peak, and then subsides gradually over hours as the storm diminishes. While this is occurring, interflow has begun. It is slower than overland flow and smaller in volume, but lasts much longer. And before it has waned – perhaps two or three days after the onset of rainfall – groundwater has begun to come up. The water table may eventually rise a foot or more in the stream valley, thereby establishing a new level of baseflow.

All these processes and the resultant behavior of the stream are dependent on different combinations of geographic conditions in a stream's watershed. These conditions vary enormously from one bioclimatic region to another. For example, the changes

Figure 15.13 Long, hard rainfalls have the capacity to activate all runoff systems producing not only stormwater discharge, but a rise in baseflow.

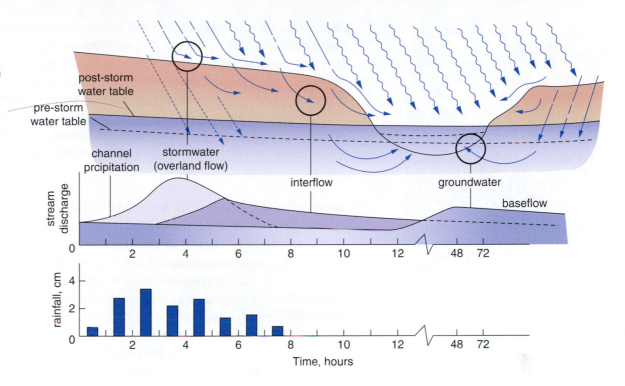

we describe that are brought on by a long, hard rain are limited to humid landscapes where groundwater is in contact with the stream channel. In an arid landscape, by contrast, the baseflow part of the hydrograph would be missing for most streams. Streamflow responses to rainfall also vary with the watershed as a physical system, including its size, shape, and land use as well as with the number of stream channels in it. These factors help determine whether a watershed and its stream system are "fast" or "slow," and whether the system is more or less prone to flooding.

Summary on Streamflow: Streamflow involves more than a simple response to rainwater trickling off the land. It is as much, or more, a response to landscape and its capacity to intercept precipitation, route it underground, reduce its volume, and slow its delivery rate. Therefore, as landscapes change with weather conditions, climate, and land use, streamflows also change. At the same time, the atmosphere's delivery of water to the landscape is not a steady pour of rain, but more an irregular series of spurts and drizzles, which, for a given watershed, yield different sizes and shapes in the streamflow hydrograph, some long and smooth, some high and pointy.

15.3 Watersheds, Channel Networks, and Streamflow

Every stream on Earth, no matter how large or small, receives its discharge from an area of land around it called a **watershed** or **drainage basin**. This is a discrete area of land that is higher on the perimeter and lower near the center (Figure 15.14a). The boundary that marks the perimeter is called the **drainage divide** and it functions precisely as the term implies, separating runoff between two adjacent basins.

Watersheds are master environmental systems. They collect water and drive it downslope delivering it to a central arterial system made up of streams, which link the land to the sea, one of the most critical phases in the hydrologic cycle. As transport systems, watersheds move not only water, but sediment, including organic material, as well. All this matter mounts up as streams merge together, draw in groundwater, and produce bigger and deeper channels with distance toward the sea.

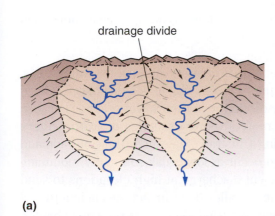

(a)

Figure 15.14 (a) All the land is divided into watersheds (basins) separated by drainage divides.

At the sea, streams pass their fresh water, organic matter, and inorganic sediment on to marine systems, but not the systems of the deep ocean for, at the interface between land and sea, a complex set of coastal systems emerge. Here the freshwater fans out, mixes with saline water, creating brackish water that harbors distinctive and productive aquatic ecosystems. Many of these ecosystems thrive on the organic matter delivered by watersheds, making watersheds part of the great marine food chain. At the same time, the inorganic sediment, the sand and silt, is the principal building material of deltas, beaches, and continental shelves, the setting within which the whole coastal complex operates.

Geographic Organization of Watersheds: Watersheds are organized in a pyramidal fashion in which adjacent small basins combine to form larger basins and these in turn combine to form even larger basins, as the diagram in Figure 15.14b demonstrates. This mode of geographic organization is described as a **nested hierarchy**, because each set of smaller basins is set inside the next larger basin. In the Mississippi watershed, for example, the basins of the Missouri and Ohio rivers are nested within Mississippi's (see Figure 15.14c), and within each of these basins is a whole series of smaller basins ending with those the size of a farm field on the perimeter of the watershed. As each basin adds its runoff to the next, discharge mounts. Therefore, the average discharge produced by most watersheds increases with total drainage area and, within an individual watershed, discharge increases progressively toward the mouth of the basin.

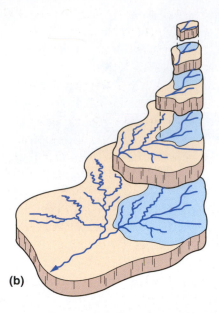

(b)

Mississippi watershed
Missouri basin (watershed)
Ohio basin (watershed)

(c)

Figure 15.14 (b) Watersheds are organized into nested hierarchies in which smaller basins combine to form larger ones and they in turn combine to form still larger ones. (c) Within the Mississippi watershed are nested the basins of the Missouri and Ohio rivers.

But there are important exceptions. First, some modest-sized watersheds produce significantly more discharge than some very large ones because they are located in different climatic zones. Let us compare, for example, the Mekong of southeast Asia with a drainage area of 800,000 square kilometers and the Nile with a drainage area of 3,000,000 square kilometers. The Mekong's mean discharge is about 16,000 cubic meters per second whereas the Nile's is only about 2800 cubic meters per second, nearly six times less despite a drainage area nearly four times larger. Both rivers rise in mountainous regions, but the Mekong watershed is dominated by a wet tropical climate, which averages more than 130 centimeters of rainfall a year, while the Nile watershed is dominated by arid and tropical wet–dry climates that together average less than 50 centimeters of rainfall a year. Moreover, the Nile loses a huge volume of its flow to evaporation in its trip across more than 2000 kilometers of severely dry desert.

A second exception is the downstream trend in discharge. Unlike most watersheds, including the Mississippi, the Rhine, and the Amazon, which gain discharge toward the mouth, some watersheds actually lose discharge in their lower reaches because they flow through deserts. Some of the more striking examples of **exotic rivers** are the Nile, the Indus, the Murray–Darling (Australia) as well as the Colorado and the Tigris–Euphrates. Exotic streams also occur at local scales in many mountainous regions where watersheds gain sufficient discharge at high elevations to send streamflows far into the desert on the adjacent valley floor. In fact the ancient tunnel systems of the Middle East, the qanats that we mentioned in the previous chapter, were designed to reduce evaporation losses in such settings by keeping water underground.

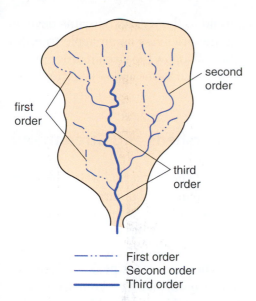

first order

second order

third order

- — · First order
- ——— Second order
- ▬▬▬ Third order

Figure 15.15 According to the principle of stream orders, the branches of a drainage network are ranked by their position in the system. First-order streams have no tributaries and occupy headwater positions.

Figure 15.16 A satellite image of a section of the Martian landscape showing a network of channels carved by running water.

Systems of Channels: At the heart of every watershed is a system of channels, called a **drainage network**. The branches or segments of streams in a network follow the same hierarchical sequence as the basins within a watershed with small streams joining to form larger streams and so on. According to a model called the **principle of stream orders**, the number of streams at each level in the hierarchy, called an **order**, decreases down the network. First-order streams are those without tributaries (that is, branches); second-order streams are those formed by the merger of at least two first-order streams; third-order by the merger of at least two second-order and so on (Figure 15.15). First-order streams are not only the most abundant channels in a network, but usually the smallest and the shortest. In addition, most first-order streams are located in a belt on the perimeter of the watershed where they drain the higher elevation ground. The very head of a first-order stream is the point where runoff changes from overland flow to channel flow. This change marks an abrupt increase in flow velocity and runoff erosive power.

The ratio between the number of streams in one order and the next is called the **bifurcation ratio** (branching ratio), and it gives us an indication of the rate of increase in stream size from one order to the next. We know that when two streams merge to form one channel the size of the resultant stream will be about equal to the sum of the two contributing channels. In general, bifurcation ratios fall close to 3 for most drainage networks, meaning that streams increase in size about threefold with each higher order.

Other networks in nature also follow the stream order model. Some of the best examples are found in the root systems of plants, the branching structure of trees, and the veination system in many leaves. It is interesting that similar branching networks are also found in the Martian landscape. Based on their branching patterns and channel forms, as the satellite image in Figure 15.16 reveals, they were undoubtedly carved by streams, indicating that at least some parts of that planet supported liquid water in the past.

The stream order principle applies to virtually all stream systems irrespective of their geographic pattern. Stream systems take on all sorts of patterns. Some follow geologic structures. Around volcanoes and some plateaus and mountain ranges, for example, streams form **radial** patterns. In areas of parallel-trending folded and faulted geologic structures, such as in the Ridge and Valley Province of the Appalachians, streams flow in **parallel** or **rectangular** patterns (see Figure 15.17). And in

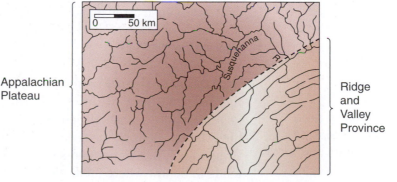

Appalachian Plateau

Ridge and Valley Province

Susquehanna R.

0   50 km

Figure 15.17 Drainage patterns change from dendritic in the Appalachian Plateau to parallel or rectangular in the Ridge and Valley Province, a response to a change in topography.

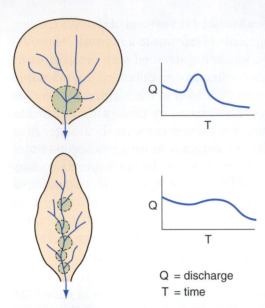

Q

T

Q

T

Q = discharge
T = time

Figure 15.18 The influence of watershed shape on streamflow. Round basins with streams meeting at a central point tend to produce higher hydrographs than elongated ones.

areas where there is no particular directional trend to the terrain, streams develop more or less random or **dendritic** patterns like the branches on an apple tree. No matter how different these patterns may look, they all follow the stream order principle with a progressive sequence of orders from many to fewer and smaller to larger with distance down the network.

15.4 Geographic Influences on Watersheds and Streamflow

Now let us examine some of the variations in drainage networks and watershed conditions that affect streamflow. We have already established that larger watersheds produce larger discharges and that this relationship holds only for basins within a single climatic zone. The slope of watersheds also influences discharge, particularly the magnitude of peak flows. For basins with steeper head-to-mouth gradients, the hydrograph for a given rainstorm will have a higher and faster peak (i.e. shorter response time) than comparable basins with gentler gradients. The same trend, that is, toward higher and faster hydrographs and more and larger floodflows, has also been documented for a number of other geographic conditions and five of these are outlined below.

Circular versus Elongated Basins: There is a tendency in circular-shaped watersheds for streams to join at or near a point in the middle of the basin, sort of like spokes joining at the hub of a wheel. When it rains and each stream mounts a stormflow, the flows meet simultaneously at this point, suddenly driving discharge up in the receiving trunk stream. In elongated or boat-shaped basins, tributaries join at alternate points along the trunk stream. As a result their discharge contributions tend to pass through the basin sequentially producing a longer and lower hydrograph like that shown in Figure 15.18.

Wetland-rich versus Wetland-poor Basins: Wetlands have the capacity to collect and store runoff, reducing both the volume and the rate of streamflow. As a result, watersheds with wetlands (and lakes as well) tend to have more subdued peak flows than similar basins without wetlands. When wetlands are eradicated by agriculture and other forms of development, not only are these moderating influences lost, but the wetland is often replaced by a drainage ditch or pipe that further quickens the release of runoff to streams. The net effect is faster and higher peak discharges (Figure 15.19). In the United States less Alaska, 40 to 50 percent of the original wetlands have been eradicated in the past 200 years, mostly by farming. (In Europe and large parts of China, the number may be even higher.) On the other hand, Americans have constructed millions of small farm ponds and stormwater ponds as well as thousands of reservoirs, which together have at least partially offset the hydrologic effects of wetland loss.

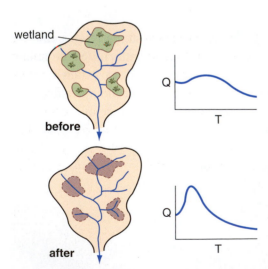

wetland

Q

T

before

Q

T

after

Figure 15.19 Wetlands act as storage basins in watersheds. When they are removed, runoff is conducted downstream faster and in greater volumes.

High Drainage Density versus Low Drainage Density: Drainage density is defined as the total length of all stream channels in a square kilometer of land. Since stream channels are nature's most efficient way of draining water from the land, then the higher the drainage density, the faster runoff is discharged from a basin and, other things being equal, the higher and steeper storm hydrographs will be. Drainage density tends to be low where infiltration is high because water is taken up by the ground before it has a chance to run off and carve surface channels. Land clearing often promotes rapid channel development through an erosional process called gullying. In rainy climates, gullies tied to stream channels can advance into unprotected land at

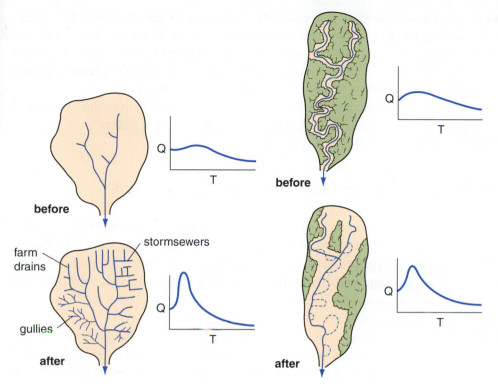

rates of 10 meters or more a year. In time, an entire basin can be etched with a fine network of channels. Drainage density also increases with land-use development. Since the time of the Romans or before, civilization has constructed artificial channels in the form of ditches and pipes (stormsewers) to drain farmland, wetlands, highways, and city streets. The resultant hydrographs are higher and faster, like the one in Figure 15.20.

Engineered-Channel versus Natural-Channel Networks: Once runoff reaches a stream, its rate of passage down the drainage network depends on the flow velocity the receiving channels are able to produce. In natural channels, velocity is regulated by various types of frictional resistance such as obstacles in the form of tree roots and boulders, curves, and bends that cause mixing motion and lengthen the water's travel distance. When streams are engineered to enhance navigation, manage floodflows, or drain cropland, velocity is quickened by removing or reducing these flow-regulating factors. In many small streams, channels are replaced by pipes which drive velocities up from 1–2 meters per second to 3–5 meters per second. Not surprisingly, peak discharges in receiving streams become larger and faster (Figure 15.21).

Figure 15.20 The more channels there are in a watershed, the more efficiently it can conduct runoff. Drainage density is increased with land clearing, farmland erosion, road building, and land-use development.

Figure 15.21 Streams are often too slow and crooked for engineers. In an attempt to reduce flooding and improve navigation, stream channels are often cleared and straightened.

Urbanized Versus Non-urbanized Basins: When watersheds are taken over by urban development, all the changes mentioned above, including alteration of basin shape and size, are invoked. Urban development removes wetlands, eliminates depression storage, and reduces interception and infiltration. To rid streets, parking lots, and yards of nuisance stormwater, various types of stormdrains (gutters, pipes, and ditches) are constructed in yards and along and under streets, parking lots, and buildings, which greatly increases drainage density and often extends the drainage network into areas previously not connected to the channel system. A typical sequence of watershed transformation is shown in Figure 15.22.

Figure 15.22 The change in a drainage system with urban development. Drainage area is enlarged, channels are straightened and piped, and drainage density is increased greatly.

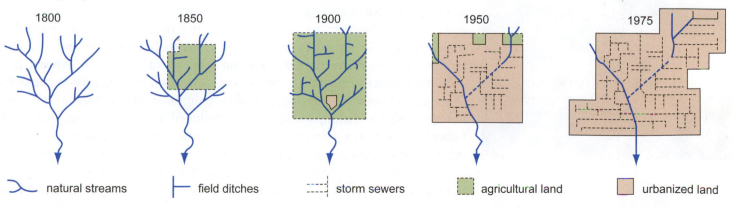

Not only is more stormwater generated, but it is quickly funneled to the new channels that conduct it to streams at rates as much as five times faster than natural processes like overland flow could. Receiving streams become glutted with the excess stormwater, increased flooding ensues, making them "problem" channels, which must then be "improved" (deepened, widened, and straightened) to relieve the overloading. But this only transfers the flooding problem downstream to the margins of the cities where large dams must often be built in an attempt to dampen the massive stormwater surges and the property damage it produces.

15.5 The Causes and Consequences of Flooding

For any stream channel, there is a magnitude of discharge that overtops the stream's banks and sends water onto the low ground along the valley floor, called the **floodplain**. The height a discharge reaches in the channel is termed its **stage** and the critical stage for any stream is the **bankfull stage**, the level reached before it overtops the banks.

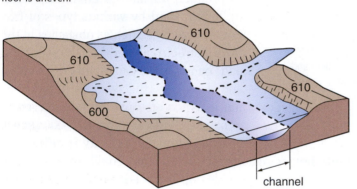

Figure 15.23 Flooding in a stream valley is typically very uneven because the topography of the valley floor is uneven.

Because the banks of most streams vary in elevation, the distribution of the first wave of the floodwater is highly irregular along most streams and very modest in coverage. As the water rises higher and spreads further onto the floodplain the distribution continues to be irregular, owing again to differences in topographic elevation of the valley floor. Natural and human-made obstacles such as levees along channels may prevent the spread of floodwaters into some areas. The ultimate pattern of flooding is always very uneven, particularly along the valley margins as the diagram in Figure 15.23 suggests.

This has not deterred the US Federal Government from mapping the geographic limits of the 100-year flood in all major stream valleys of the United States. This program was instituted as part of the US National Flood Insurance Program in an effort to discourage floodplain development and reduce flood damage. The 100-year flood occurs on the average of once every century and its impacts can be devastating to land uses. Despite efforts to protect land uses in flood-prone areas, flood damage has increased substantially in this century as population and land-use relentlessly push into river valleys.

Figure 15.24 Narrowing of the Mississippi channel near St. Louis, Missouri, has reduced the river's width by more than half since the mid 1800s. To accommodate the river's discharge in a smaller channel, the banks had to be built up with massive levees.

Broadly speaking there are only two general causes of floods: (1) high runoff rates and (2) limited channel capacities, but there are many specific causes of each. High runoff rates can be produced by wet atmospheric conditions, such as extended, heavy rainfall, or by watershed conditions, such as saturated soil, or excessive land clearing and urban development, or by some combination of factors. Limited channel capacity may be caused by natural channel obstructions such as log jams, by flooding downstream such as a hurricane surge forcing a rise in water elevation at the river mouth, or by human-made channel constrictions such as bridges and navigational facilities that pinch channels down and force flows up. At St. Louis, Missouri, for example, the Mississippi's channel width has been reduced from 4300 feet in 1847 to 1900 feet today and the banks have been built up with huge levees (embankments) to contain the higher stages (Figure 15.24).

Let us examine some of the conditions, both natural and human-made, that have produced some major floods. Virtually all are related to a combination of atmospheric, watershed, and channel conditions. We begin with a recent flood event in south Asia and go on to two major floods in North America.

The Indus River Flood of 2010: In July 2010, the Indus River of Pakistan produced a massive and prolonged flood that spread over about one-fifth of the country, displacing millions of people, killing several thousand, and leaving an entire agricultural system in shambles (see Figure 15.25). To understand the causes of this flood it is helpful to return to the discussion on monsoon circulation in Chapter 5 and look at Figure 5.23a, which shows the summer flow of air from the Indian Ocean onto south Asia. The summer monsoon brings a steady supply of moist, tropical air up the Indus Valley where it encounters the high mountains of the Hindu Kush and the Himalayas, as shown in Figure 15.26. Most of the moisture is released in thunderstorms and orographic rainfall, which is discharged into mountain streams and conducted by the Indus back to the ocean. The summer of 2010 was different because the atmosphere over south Asia and the Indian Ocean was much warmer than normal which (a) accelerated ocean evaporation rates thereby increasing the supply of moisture to the monsoon system; and (b) accelerated melting rates on mountain snowfields and glaciers in the upper watershed leading to higher runoff rates.

This was the largest flood on the Indus since 1929. For our purposes let us say that it ranks as a 100-year event, meaning that a flood of this magnitude can, in the long term, be expected here on the average of once a century. Such estimates are based on records of past rainfall and flood events and they are expressed as probabilities as listed Table 15.1. They tell us how often on the average an event of such and such magnitude can be expected but not about *when* it can be expected. In other words, another 100-year flood may occur any time; tomorrow, next month, or 200 years from now. But there is an added problem: Climate change.

Records of past rainfalls and floods may have little meaning today and in the future because the magnitude and frequency of rainstorms are changing with warming of the atmosphere and oceans. A stronger and wetter summer monsoon may now be producing floods like the 2010 event, and bigger ones, more often. And there is yet another problem: Pakistan's growing population, spreading land use, and mountain deforestation. The Indus watershed is less able to store rainfall and regulate runoff rates than in the past. Therefore, flooding would be more frequent and larger even if climate change were not a factor. In 1947, Pakistan's population was 31 million; in 2025, it is expected to reach 225 million, and the vast majority will be crowded into the Indus lowlands, but millions will spill over onto mountain slopes and adjacent valleys.

Lessons from the Red River Flood at Grand Forks, North Dakota: The winter of 1996–97 brought huge amounts of snow to North Dakota, Minnesota, and southern Manitoba. Fargo, located in the Red River Valley along the Minnesota–North Dakota border about 150 miles south of the Canadian border, received a record 288 centimeters (113 inches) of snowfall. But unlike most winters when heavy snows were

Figure 15.25 Massive flooding of the Indus River Valley of Pakistan, summer 2010. More than 700,000 homes were destroyed.

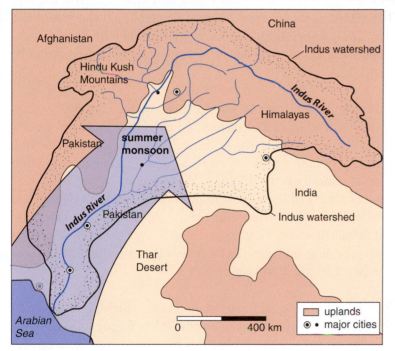

Figure 15.26 The Indus River watershed produces most of its runoff from the monsoon rainfall and meltwater from snowfields and glaciers in the Hindu Kush and the Himalayas.

Table 15.1 Flood probability and average return period

Probability	Return period
1% (0.01)	100 years
2% (0.02)	50 years
10% (0.10)	10 years
20% (0.20)	5 years
50% (0.50)	2 years
100% (1.0)	1 year

Figure 15.27 The dramatic 1997 flood of Grand Forks, North Dakota, which forced evacuation of nearly all the city's 44,000 residents and placed thousands of acres of farmland underwater.

Figure 15.28 The Humber River corridor of Toronto is dedicated to open space as a means of averting flood disasters and providing urban park land.

reduced or eliminated periodically by winter thaws, the snow pack in 1997 tended to build up over the winter. When it began to melt in spring, the wet, clayey soils of the valley floor limited infiltration, forcing large amounts of overland flow into farm drains, streams, and the Red River itself. But the valley floor and river gradient are very flat and the water moved slowly through the system causing it to build up rapidly to flood levels.

The Red River has produced frequent floods over the decades and communities along the river have habitually called on the US Army Corps of Engineers to build facilities to help reduce peak flows and protect low areas from flooding. These facilities have included deepened and straightened channels and large levees along the main channel and its tributaries, all conventional flood-control measures employed by engineers on thousands of other streams across the United States. But in 1997, the decades of engineering and the huge investment of public funds failed and the Red River overtopped the high levees at Grand Forks and spilled across the city forcing almost total evacuation of its 44,000 residents. Once behind the levees the water (shown in Figure 15.27) lingered for weeks, rendering the city useless, contaminating its water supply and flooding its sewer system, streets, houses, schools, and businesses.

Grand Forks is re-evaluating its approach to flood management. Like many other river cities that have experienced similar disasters, Grand Forks now recognizes that a structural (engineering) approach to the problem is fraught with serious limitations. First, no structural program, no matter how elaborate, can control all flood events because the size of natural events such as floods has no upper limit. Just when you think nature has produced its biggest flood there is always a bigger one waiting in the wings. Yet to the north, downstream from Fargo on the Red River, the Canadians have elected to build a massive levee system skirting the city of Winnipeg to protect the city and surrounding land uses. In their view this was the most reasonable alternative to a persistent flooding problem.

Nature works according to probabilities and, based on a rule called the **magnitude and frequency principle**, high-magnitude events occur at low frequencies with low probabilities. But no matter how small the probability of a huge flood may be, the possibility of its occurrence always exists. Second, engineered structures such as high levees along a stream tend to build a false sense of security in communities that the flooding problem has been solved, and may even encourage new development in low areas. Therefore, when a flood occurs, it is met with disbelief and inadequate preparation and people blame public officials rather than nature or themselves for their bad luck.

What is the alternative to engineering? One alternative is to clear vulnerable land uses from flood-prone areas and build new land-use plans for river communities. The huge expenditures that would normally go to engineering can be used for home buy-outs and rebuilding programs and then less vulnerable land uses such as parks and greenbelts can be placed in low areas along the rivers. An example of this type of land-use planning can be found along the Humber River in the Toronto metropolitan area, shown in Figure 15.28. This approach should be combined with public education about the realities of nature in river floodplains, and the need to build realistic risk-management plans. On balance it may be wiser and less expensive to protect the river from the people than the people from the river!

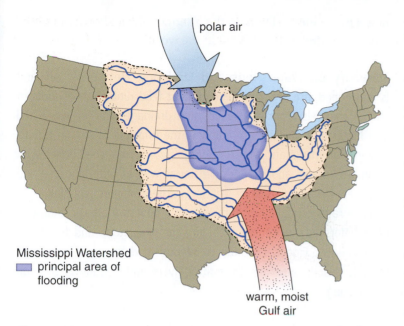

polar air

Mississippi Watershed
■ principal area of
  flooding

warm, moist
Gulf air

Figure 15.29 The area of concentrated rainfall in June and July of 1993 which produced the Mississippi Flood, the largest ever recorded on the great river.

The Great Mississippi Flood of 1993: The 1993 Mississippi River Flood, the largest recorded in over 100 years, was a response to prolonged spring and summer rainfall caused by cyclonic storms over the Midwest fed by a massive flow of warm moist air from the Gulf of Mexico as illustrated in Figure 15.29. Spring rains saturated the soil over more than 780,000 square kilometers (300,000 square miles) of the upper Mississippi watershed, greatly reducing infiltration and leaving soils with little or no storage capacity. As rains continued into June, surface depressions, wetlands, ponds, ditches, and farm fields filled with overland flow and rainwater. With no remaining capacity to hold water, July delivered the final blow with 50 to 75 centimeters (20–30 inches) of rainfall, virtually all of which was forced from the land into tributary channels and thence to the Mississippi.

For more than a month, the total load of water from hundreds of tributaries exceeded the great river's natural channel capacity, causing it to build up within the huge levee system constructed to contain just such an event. But, as the photograph in Figure 15.30 shows, the levee system eventually failed and floodwater spilled onto adjacent floodplains. In places where the levees held, it became apparent after a while that the levees so constricted the flow of floodwaters and forced water levels so high that the Corps of Engineers was forced to break through levees at selected points to release the water onto the floodplain in an effort to lower the channel flow.

The total damage to communities, farms, bridges, levees, and other facilities exceeded $30 billion, and once more government officials and planners were forced to re-examine the traditional approach to flood management. But despite this disaster and many others, including Katrina in 2005 and the Mississippi flood of 2011, and despite the prospects of sea-level rise and related effects of global warming, development continues to push into floodplains and low-lying coastal areas in both developed and less developed countries, and expensive structural measures such as dams and levees continue to be invoked to "protect" it. Although the era of building large dams has largely ended in North America, it remains active in the developing world. China's

Figure 15.30 A portion of the area flooded by the 1993 Mississippi flood. Damage exceeded $30 billion.

Figure 15.31 The huge Three Gorges Dam on the Yangtze River of China, designed in part for flood control, is part of a dam-building tradition that began in North America.

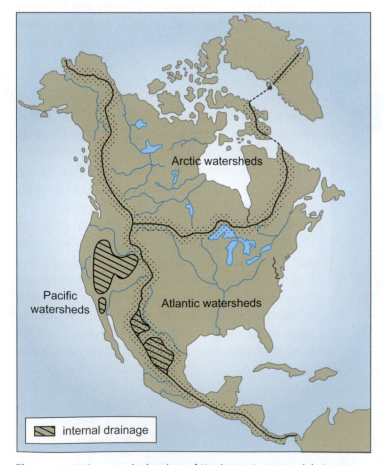

internal drainage

Figure 15.32 The watershed regions of North America. Internal drainage is limited to the dry lands of the mountainous West.

new Three Gorges Dam on the Yangtze River, shown in Figure 15.31, is touted as the largest in the world.

Summary on Watersheds and Drainage Networks: All land is partitioned into drainage systems that feed runoff to networks of stream channels. Both watersheds and drainage nets are organized in a hierarchical fashion according to the principle of stream orders. Within a given climatic zone, discharge from a watershed varies with basin size, shape, and channel density as well as with land use and engineered manipulation of channels. These factors, coupled with variable rates of input from precipitation, govern the basic character of a watershed's discharge including the magnitude and frequency of floodflows. Floods continue to plague land use with increasing impacts despite attempts to control rivers and floodflows through engineering.

15.6 Watersheds of the World

All Earth's land area belongs to a watershed of some size. Watersheds are exceedingly important to the planet's geography because they provide one of the key organizational systems governing the distribution of water resources, soils, ecosystems, and to some extent human population and land use as well. If we think in terms of levels of geographic organization of Earth's environments, climate ranks first because it provides a global framework but, for terrestrial environments, the next level of organization is provided by watersheds. Indeed, it has been suggested by observers no less prominent than John Wesley Powell (1834–1902), noted explorer of the Grand Canyon, that watersheds would be a logical system for geopolitical organization with jurisdictional borders corresponding to major drainage divides, especially in regions of conflict over water resources.

Open and Closed Watersheds: There are many ways to classify and describe the watersheds of the world beginning with the distinction between external and internal basins. **External**, or **open**, **watersheds** are those that drain to the sea. This includes most watersheds, big and small, and represents 80 percent of the planet's land area. The remaining area, about 30 million square kilometers, is drained by **closed** or **internal watersheds** whose discharge remains on land and is evaporated into the atmosphere rather than discharged into the sea.

Internal watersheds form in the interiors of large landmasses that are dominated by arid climate and often surrounded by mountain barriers. For example, in North America, the largest area of internal drainage (see Figure 15.32) lies between the Sierra Nevada Mountains and the Rocky Mountains. Worldwide, the largest area of internal drainage is found in Eurasia

Table 15.2 Internal drainage area by continent

Continent	Total area, million km²	Internal drainage area, million km²	Percentage
Eurasia	54.0	14.5	27%
Africa	30.1	9.6	32%
Australia	8.9	3.9	44%
South America	17.8	1.4	8%
North America	24.2	0.8	3%
Antarctica	14.0	0.0	0%
	149.0	30.2	

Figure 15.33 The Volga Delta on the Caspian Sea, the largest water body of a closed drainage system on the planet.

(14.5 million square kilometers) followed by Africa, Australia, South America, and North America. However, Australia, the driest continent, has the largest percentage of land in internal drainage (Table 15.2).

What is the geographic significance of internal drainage? The fact that the sediment and dissolved minerals carried by these streams are deposited on land and not in the ocean is significant. Each landlocked watershed terminates in an evaporation basin, such as Great Salt Lake, the Aral Sea, and the Caspian Sea, where its load is deposited. The world's largest evaporation basin is the Caspian Sea (surface area 4.5 times Lake Superior's) which is fed with 260 billion cubic meters of water annually by its principal water source, the Volga River. With this inflow each year comes tens of millions of tons of sediment, mostly clay, silt, sand, and smaller but significant amounts of dissolved minerals, which together are slowly filling the sea as the satellite image in Figure 15.33 shows. Although the filling rates of other closed basins are much lower than the Caspian, the trend is consistently toward closure in all landlocked basins. As basins fill with sediment and mineral concentrations increase with evaporation, the water area shrinks and is replaced by salty mudflats or playas. These are among the most desolate and biologically impoverished environments on Earth. See Chapter 14 for more on desert lakes.

Earth's Great Watersheds: About half the world's runoff is carried by only 70 large streams or great rivers. These rivers drain huge watersheds, which together cover more than half the world's land area. The 20 largest watersheds drain more than a million square kilometers each, an area, incidentally, about the size of Egypt. Table 15.3 lists these watersheds by size (area) with their discharges, annual runoff rates, and dominant climate. Notice that this area-based ranking does not match very well with one based on discharge because of differences in the types of climate that dominate these huge basins.

The grandest of the great watersheds is the Amazon with an area of 6.5 million square kilometers and an average discharge of 180,000 cubic meters per second (Figure 15.34). For comparison, the Amazon basin is twice as large as the country of India, and its average discharge is 10 times that of the Mississippi River. If the Amazon watershed were a country, it would be the seventh largest nation on Earth and could boast the heaviest forest cover, the richest biodiversity, the largest supply of freshwater, and one of the lowest human population densities.

Since our interest is in watersheds as hydrologic systems, it is important to ask about differences in runoff rates and what controls them. The graph in Figure 15.35 ranks a number of great watersheds in terms of their average annual runoff per square meter of surface area. The Orinoco and the Amazon head the list, producing an average of nearly one cubic meter of runoff per square meter of watershed. Both are fed by heavy rainfall from large areas of wet tropical climate. The mean annual rainfall over their watersheds is around 2 meters (somewhat less for the Orinoco) which means that roughly half the water received is released in runoff and half in evapotranspiration. In the very center of the Amazon Basin, annual rainfall approaches 3 meters with peaks occurring twice a year. One to two meters go to runoff which moves slowly over the forest floor to large tributaries that flood vast areas of rainforest for 6 to 8 months a year. Up to 500,000 square kilometers of the Amazon watershed may be underwater at these times.

Table 15.3 The great watersheds of the world ranked by drainage area

Watersheds (Principal country)	Area,[1] million km²	Discharge,[2] m³ per second	Annual runoff,[3] m³ per m²	Dominant climate(s)
Amazon (Brazil)	6.5	180,000	0.87	Tropical wet
Congo (Zaire)	4.0	40,000	0.32	Tropical wet
Mississippi (USA)	3.2	17,500	0.17	Mountain, semiarid, humid continental
Nile (several)	3.0	2800	0.03	Arid and wet–dry tropical
Ob (Russia)	3.0	12,300	0.13	Subarctic and semiarid
Yenisey (Russia)	2.6	18,000	0.22	Subarctic
Parana (several)	2.5	18,000	0.23	Tropical wet–dry and humid subtropical
Lena (Russia)	2.5	16,300	0.21	Subarctic
Yangtze (China)	2.0	30,000	0.48	Mountain and humid subtropical
Amur (Russia/China)	1.9	11,000	0.19	Subarctic
Mackenzie (Canada)	1.8	8500	0.13	Subarctic
Volga (Russia)	1.4	8200	0.19	Humid continental
Zambezi (several)	1.3	7000	0.17	Tropical wet–dry and semiarid
Nelson (Canada)	1.1	2900	0.08	Semiarid, subarctic, humid continental
Niger (several)	1.1	6100	0.17	Tropical wet–dry and semiarid
Murray–Darling (Australia)	1.1	570	0.02	Arid and semiarid
St. Lawrence (Canada/USA)	1.0	10,500	0.32	Humid continental
Ganges (India)	1.0	13,000	0.47	Mountain and monsoon subtropical
Indus (Pakistan)	1.0	5000	0.19	Arid and mountain monsoon
Orinoco (Columbia/Venezuela)	1.0	28,000	0.93	Tropical wet and wet–dry
Brahmaputra (India/China)	0.9	20,000	0.50	Mountain and monsoon subtropical
Danube (several countires)	0.8	6400	0.25	Mountain and humid continental
Mekong (several)	0.8	16,000	0.63	Tropical monsoon and mountain
Huang He (China)	0.75	1400	0.06	Mountain and semiarid

[1] *Drainage areas vary among data sources; these figures represent intermediate values. The area of the Amazon watershed and many other great watersheds are not precisely known. Figures for the Amazon vary from 5.78 million km² to 7.19 million km².*

[2] *Average annual discharge, estimated for many rivers.*

[3] *Average amount of runoff produced per square meter of watershed surface.*

The next most productive watersheds are four Asian basins that are dominated by monsoon climates: the Mekong (0.63 m³), Brahmaputra (0.50 m³), Yangtze (0.48 m³) and the Ganges (0.47 m³). Each of these basins has its headwaters in Tibetan Highlands, which not only induce heavy orographic rainfall during the summer monsoon season, as shown in Figure 15.36, but provide additional water from glaciers and snowfields. Not surprisingly, the flow regimes of these and similarly situated Asian rivers follow the seasonal variation of rainfall. Flows decline in winter and then rise

Figure 15.34 The watershed of the Amazon River, the largest river on Earth. This watershed produces nearly one cubic meter of runoff for every square meter of its huge surface area.

dramatically with onset of the summer monsoon (June to September), often producing extensive flooding in the broad, flat lower reaches of these watersheds, which are occupied by tens of millions of peasant farmers. The Indus River flood of 2010 described earlier in this chapter is a prime example of this phenomenon.

The least productive watersheds in terms of runoff output are, expectedly, those in dry climatic zones. The Nile and the Murray–Darling watersheds produce an average of less than 5 centimeters of runoff per square meter. But both support exotic stream systems, so we must appreciate that there are huge geographic differences in runoff rates *within* these big watersheds. The lower half or more of an exotic basin contributes virtually none of its total rainfall to runoff and stream discharge; in fact, through evaporation and infiltration this part of the watershed extracts water from the runoff contributions made upstream. These upstream contributions come from the upper 25 percent or so of the watershed, which not only receives much more precipitation, but loses much less to evaporation and infiltration. Runoff rates there can be as high as 0.50 cubic meters per square meter.

Low runoff rates are also recorded for several Arctic and subarctic watersheds. The Nelson watershed in Canada, with a runoff rate of 0.08 cubic meters, drains both semiarid (Great Plains) and subarctic zones where the mean annual precipitation is less than 50 cm (25 inches). The Mackenzie of Canada and the Ob of Russia, which drain vast interiors of the subarctic climatic region, produce correspondingly low runoff rates, 0.15 and 0.13 cubic meters, respectively. Also relatively low are the Mississippi (0.17 m³), the Niger (0.17 m³), the Zambezi (0.17 m³), the Indus (0.19 m³), and the Amur (0.19 m³). Each of these watersheds contains a sizeable area of dry climate, which in part offsets the runoff contributions from the humid parts of the watershed. The Mississippi watershed, for example, ranges from arid in the southern Great Plains (e.g. the Texas Panhandle) to humid subtropical in Louisiana. Across this climatic spectrum mean annual precipitation ranges from 35 cm to nearly 200 cm. Figure 15.37 shows the locations and coverages of the watersheds discussed above along with some summary remarks on their geographic characteristics.

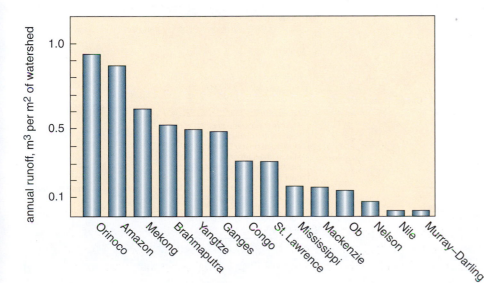

Figure 15.35 Annual average runoff for 13 of the world's great rivers. The runoff figure represents an average value for a square meter of watershed surface. What accounts for the huge differences?

Figure 15.36
The geographic circumstances associated with the summer monsoon of south and southeast Asia, which is responsible for the huge discharges of rivers such as the Mekong, Ganges, and Brahmaputra.

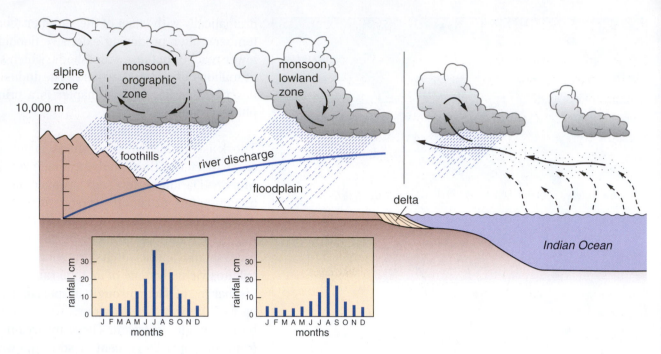

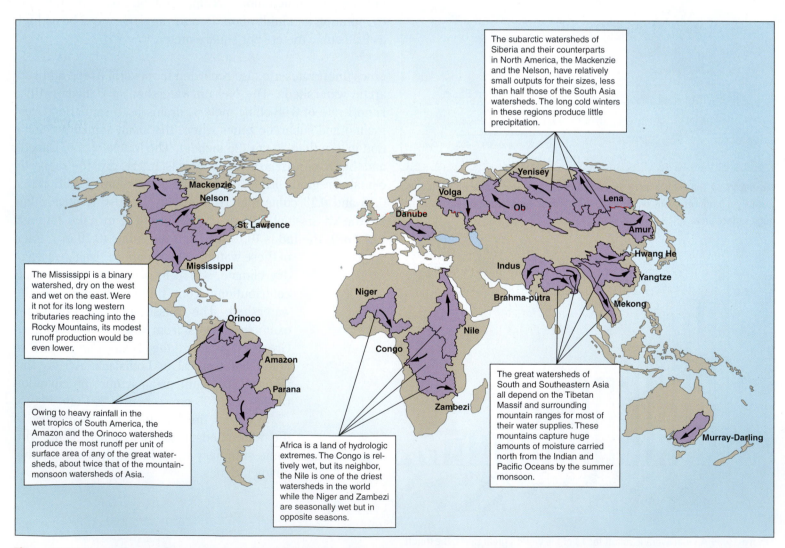

The subarctic watersheds of Siberia and their counterparts in North America, the Mackenzie and the Nelson, have relatively small outputs for their sizes, less than half those of the South Asia watersheds. The long cold winters in these regions produce little precipitation.

The Mississippi is a binary watershed, dry on the west and wet on the east. Were it not for its long western tributaries reaching into the Rocky Mountains, its modest runoff production would be even lower.

Owing to heavy rainfall in the wet tropics of South America, the Amazon and the Orinoco watersheds produce the most runoff per unit of surface area of any of the great watersheds, about twice that of the mountain-monsoon watersheds of Asia.

Africa is a land of hydrologic extremes. The Congo is relatively wet, but its neighbor, the Nile is one of the driest watersheds in the world while the Niger and Zambezi are seasonally wet but in opposite seasons.

The great watersheds of South and Southeastern Asia all depend on the Tibetan Massif and surrounding mountain ranges for most of their water supplies. These mountains capture huge amounts of moisture carried north from the Indian and Pacific Oceans by the summer monsoon.

Figure 15.37 A summary map of river watersheds whose drainage area approaches or exceeds 1 million square kilometers with notes on selected ones. The largest concentration of great watersheds is found in south and southeast Asia, but the two most productive ones in terms of runoff are the Orinoco and Amazon.

Chapter Summary and Overview of Runoff, Streamflow, and Watersheds Systems

Streams and their watersheds are the pivotal hydrologic systems of Earth's landmasses, forming a critical link between the atmosphere and the oceans. Much of our understanding of physical geography hinges on an understanding of watershed systems, for they are broadly integrative, linking landforms, vegetation, soils, climate, and land use together in great working wholes. All watersheds and their streams are fed by precipitation, but in order for rainwater to get to stream channels, it must first pass through the landscape, and here it is slowed, re-routed, stored, and dissipated. Many geographic factors play a part in this, both natural and human-made. The runoff collected by watersheds is directed toward the ocean, the final leg in the terrestrial phase of the hydrologic cycle and, for the large landmasses the huge drainage basins like the Amazon, Nile, and Ganges do a very large share of the work.

▶ **All streamflow is derived from precipitation but the landscape regulates the rates and amounts of water delivered to streams.** The cumulative effect is that less rainwater is delivered to streams than falls from the sky, and it is delivered at a slower rate.

▶ **Streams gain their discharge from four sources.** Each delivers different amounts of water at different rates depending on the character of landscape, including land use, and the nature of the precipitation falling on it.

▶ **The buffering effect of the landscape is reduced by land clearing and development.** In the extreme, overland flow may increase as much as nearly tenfold as a landscape is converted from forest to urban land use.

▶ **A hydrograph can be used to describe and analyze a stream's response to precipitation.** Rainfalls of different durations and intensities activate different discharge sources in the runoff system. Hard, long, soaking rainfalls can produce substantial increases in all sources of discharge.

▶ **A watershed (or drainage basin) is a discrete area of land that feeds runoff to a stream or stream system.** All watersheds share a common hierarchical structure described by the principle of stream orders. In general stream discharge increases with watershed size and stream order.

▶ **Stream discharge is affected by many geographical conditions.** Most of the landscape changes that come with land use induce higher and faster hydrographs and larger and more frequent floods.

▶ **Flooding occurs when streams overtop their banks.** Most flooding is a response to high rates of watershed runoff and/or limited channel capacities. Devastating floods are not limited to the very large rivers.

▶ **Floodplains are favorite places for settlement throughout the world.** The struggle to manage flood hazards has depended overwhelmingly on structural protection. Increasingly, alternatives are being considered.

▶ **All land areas belong to a watershed of some size.** Most of the Earth is drained by external, or open, watershed systems. Eurasia has the largest area of internal drainage.

▶ **Great watersheds approach or exceed a million square kilometers in area.** Their discharges vary radically depending on their size and geographical conditions, especially climate.

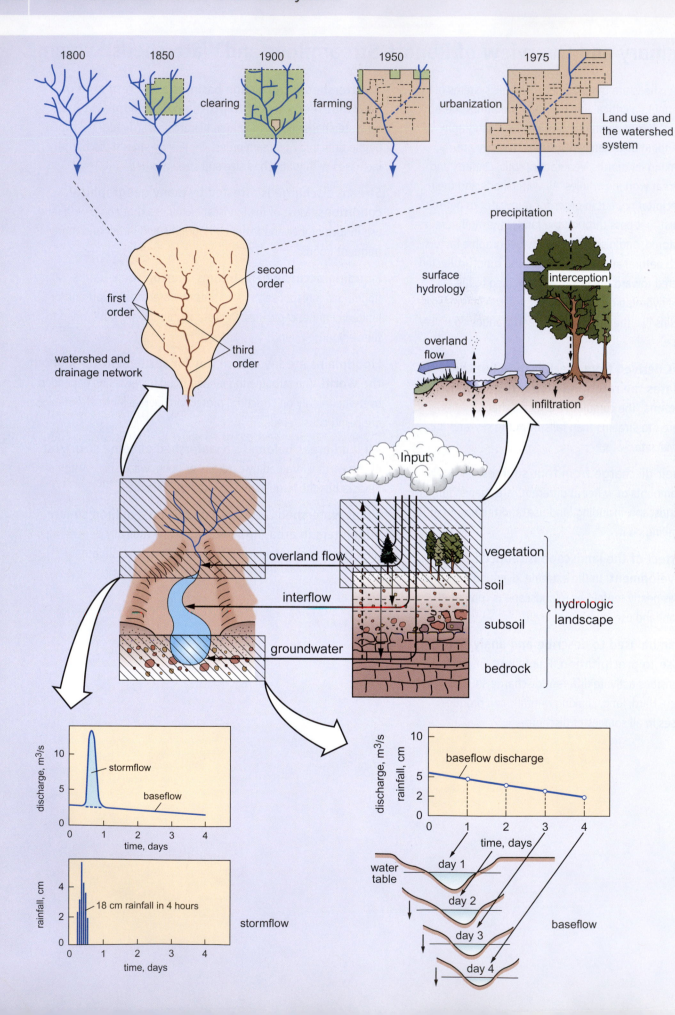

1800   1850   1900   1950   1975

clearing   farming   urbanization

Land use and the watershed system

watershed and drainage network

first order

second order

third order

precipitation

surface hydrology

interception

overland flow

infiltration

Input

overland flow

interflow

groundwater

vegetation

soil

subsoil

bedrock

hydrologic landscape

Summary diagram showing some of the many factors controlling the delivery of runoff to streams. The landscape, including land use, is a critical filter between precipitation and streamflow.

discharge, m³/s

stormflow

baseflow

time, days

rainfall, cm

18 cm rainfall in 4 hours

stormflow

time, days

discharge, m³/s
rainfall, cm

baseflow discharge

time, days

water table

day 1

day 2

day 3

day 4

baseflow

Review Questions

1 Can you explain how the landscape functions as a buffer between terrestrial precipitation and streamflow?

2 What are the four systems that feed water to streams and how would you describe them in terms of relative rates and volumes of water delivery?

3 Can you define interception, depression storage, and infiltration and name several controls on infiltration capacity?

4 What is the coefficient of runoff and why does this value rise with land clearing and development?

5 Long-term records of streamflow show sharp differences between streams dominated by baseflow, and those dominated by overland flow. Can you explain this?

6 What is a hydrograph and how can it be used to analyze a stream's response to a rainfall? Describe the basic hydrographs for a drizzle, a cloudburst, a long, moderate rain, and a long, hard rain.

7 Can you explain why large watersheds are described as nested hierarchies? Does the discharge produced by a watershed always increase with drainage area and distance downstream? Why?

8 What is the principle of stream orders and in a drainage network what is the natural progression in the number of streams from order to order?

9 Is it true that stream networks are the only natural phenomena that follow the stream order model and what controls the geographic pattern of drainage networks as they appear on maps?

10 Can you draft a list of the various features and conditions of drainage basins that influence streamflow and briefly define which tend to increase and decrease stormflows and flooding?

11 What is meant by the term stage and when a stream reaches flood stage does the water always cover the floodplain?

12 Briefly describe the central cause or causes of the Indus River, Grand Forks, and Mississippi floods. Could you write a short essay on the risk management lessons that should be learned from these disasters?

13 What are the external and internal watersheds? How much of the Earth's land area is drained by closed basins and what are the geographic conditions that give rise to such basins? What are some of the significant environmental effects resulting from internal drainage?

14 What is a great watershed and why is the Amazon Basin such a singular geographic feature on Earth?

15 Can you explain why discharge does not increase proportionally with drainage area size in all great watersheds?

16 Can you define the difference in high- and low-runoff basins, give several examples of each, and define the circumstances that account for these differences in runoff production?

Groundwater Systems, Lakes, and Water Resources

Chapter Overview

"The man on the farm over there says he pumps water directly from the ground and the supply is endless … he thinks there must a huge lake down there, at least that's what the old timers tell him." This is a common notion about groundwater, one held by people around the world and one that we hope to dispel in this chapter. There is little doubt indeed that the ground holds immense amounts of water, but how does it get there and how and where is it stored? We will make a case for groundwater as a subsystem in the hydrologic cycle, an open system with inputs from and outputs to the surface water system that we discussed in the previous chapter. We also explore the geographic significance of groundwater and this leads us to the influences it has on the landscape, in particular on streams, lakes, wetlands, and land use. As we will discover, in a world that will see 3 billion additional people in the next 40 years, more than half of whom will be living in dry lands subject to the desiccating effects of global warming, groundwater's role in supplying water to cities, farms, and industry will reach levels of acuity unmatched in history.

Introduction

Groundwater is Earth's single largest supply of fresh, liquid water. Humans have known about it and used it for thousands of years. The first of ancient Rome's aqueducts was fed by groundwater discharged from springs in the hills outside the city. The ancient qanats of the Middle East tapped into groundwater aquifers on mountain slopes, and tens of thousands of public wells have dotted the landscapes of China, India, and Europe since prehistoric times. Yet, by the time the Enlightenment (1700–1800) and the Industrial Revolution (1750–1850) had come and gone, we knew surprisingly little about the nature of this great terrestrial water system. How deep into the Earth does it extend? How did it originate? Does it move?

By the late 1800s, however, things began to change. Drilling techniques were advanced with the discovery of oil (1859) and the portable steam engine enabled drillers to penetrate several hundreds of meters into the ground. In the North American Great Plains the windmill was introduced in the 1880s just when ranchers had overloaded the prairies with cattle and were desperate for a well on every parcel of pasture. Somewhat to their amazement they found a rich supply of groundwater only 25 to 50 meters beneath the parched prairie landscape and,

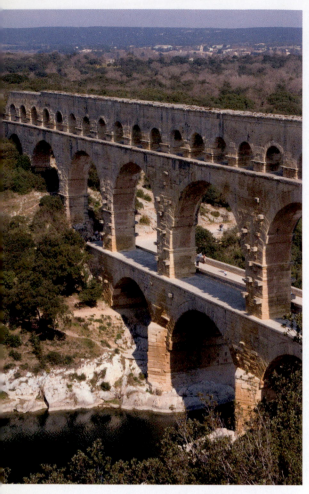

Groundwater technology includes aqueducts dating from Caesar's time to millions of windmills still operating across the world.

even more remarkable, this body of groundwater seemed to stretch all the way from South Dakota to Texas. By 1900, a picture of groundwater distributions and depths was beginning to emerge for some regions.

This new information coupled nicely with another important nineteenth century discovery on groundwater. A French engineer named Henri Darcy demonstrated that the velocity of groundwater flow could be computed based on just two measurements: (1) the inclination of a groundwater body, and (2) the hydro-conductivity of the soil or rock material housing it. This became known as Darcy's law and it enabled scientists to determine not only the rates and patterns of groundwater movement but how long it took to recharge groundwater at different depths.

In this century, our knowledge of groundwater has expanded rapidly with the growth and spread of settlement. Today close to half of all North Americans rely on groundwater for their domestic water supply. Many states and provinces in North America support more than a million groundwater wells and, for the entire continent, the number of wells now approaches 50 million. In most states, well drillers are required to keep records or logs on well depths, locations, water elevations, and underground materials. Thus, as we enter the twenty-first century, our data bank on groundwater is enormous and growing, but there is much to be learned about its dynamics.

16.1 Pathways and Linkages Underground

Let us begin the story of groundwater with a look at the diagram in Figure 16.1. It opens at (a) with water infiltrating the soil and slithering its way downward through the maze of soil particles. As it comes into contact with individual particles, part of the infiltrating water is captured on the particle surfaces and at the contact points between them, forming a molecular film of **capillary water**. As the film of capillary water thickens with the addition of each new molecule, the molecular forces that hold the molecules together weaken. Eventually, the outer water molecules can no longer be held to the film and yield to gravity, flowing down to the next particle.

If enough water percolates into the soil, the soil will soon reach its **field capacity,** which is the maximum amount of capillary water the soil can hold. When additional water is introduced, it must bypass the capillary (or soil water) system and trickle downward as **gravity water** (see (b) in Figure 16.1). In coarse-grained soils like sand and gravel, where the interparticle spaces are large, gravity water moves rapidly. Such materials are said to have high **permeability** or **hydraulic conductivity**, whereas in fine-textured materials like clay, or in tightly cemented or compressed bedrock, the permeability is understandably much lower. Eventually the gravity water reaches a zone at some depth, say, 10 to 20 meters, where all the interparticle spaces are completely filled with water. This zone is called the zone of saturation or **groundwater zone** ((c) in Figure 16.1).

Thus, groundwater is merely an accumulation of gravity water underground. The total amount of groundwater that can be held by any material (soil, surface deposits, or rock) is controlled by its **porosity**, which is a measure of the sum total of the void spaces represented by cracks in rocks, the spaces between particles, and various cavities. Porosity commonly varies from 5 to 50 percent in soils and near-surface bedrock. In general, porosity decreases with depth because the pressure of the massive rock overburden closes out void spaces at great depths. At depths of several thousand meters, it is typically less than 1 percent.

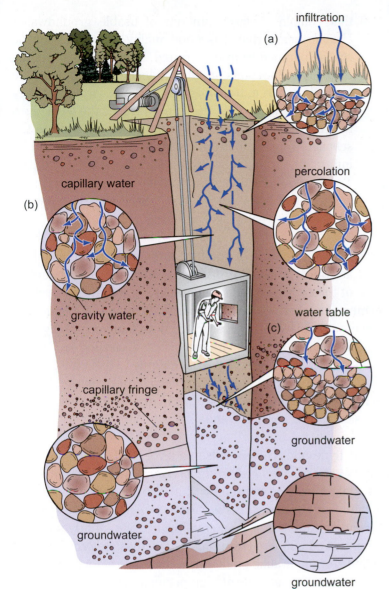

infiltration

(a)

percolation

water table

(c)

groundwater

groundwater

groundwater

Figure 16.1 The processes and zones of water underground: (a) infiltration and percolation; (b) capillary water and gravity water; and (c) water table and groundwater.

capillary water

(b)

gravity water

capillary fringe

groundwater

Figure 16.2 The concept of the hydraulic gradient. Notice the slope on the water table.

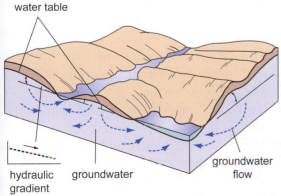

water table

hydraulic gradient   groundwater   groundwater flow

Groundwater Flow: The boundary marking the upper limit of the groundwater zone is the **water table**. In coarse-grained materials, it approximates a boundary line, which you can often observe in the walls of building excavations by the outpouring of water immediately below water-table level. In fine-grained materials, however, the water table has more the aspect of a transition zone, called the **capillary fringe**, which may be several meters wide (see Figure 16.1). Here the water table is not visually apparent in the walls of excavations. The ground will appear wet but there is usually no outpouring of water. Groundwater is there, beneath the capillary fringe, but there is too much resistance imposed by the microscopic-sized spaces among the tiny particles to allow it to pour out.

If we trace the water table across the land, we will find that its elevation rises and falls with broad changes in surface topography in the manner depicted in Figure 16.2. Therefore, the water table almost everywhere is inclined to some degree. These inclines represent gravitational gradients along which groundwater slowly flows and here is where we apply Darcy's law.

Darcy called these gravitational gradients **hydraulic gradients**. If we know the hydraulic gradient in some material (which can be figured out from all those well logs we have been keeping for decades), then the velocity of groundwater flow is governed by the resistance the rock or soil material poses to water movement. Resistance is represented by the material's permeability. The higher the permeability, the lower the resistance, and the faster the water flows. Thus, in its basic form, Darcy's law is: $V = I \cdot k$, where I is the hydraulic gradient and k is permeability.

How fast does groundwater flow? Compared to streamflow, groundwater velocities are miniscule, but they vary widely depending on materials, geographic settings, and depths underground. The fastest flows, around 150 meters a day, are recorded in highly permeable sand and gravel deposits on mountain slopes. But in the broad, flattish interiors of the continents, flows of 15–20 meters *per year* are more typical. These slow rates, however, are offset by the huge volume of groundwater that is in continuous motion. Therefore, it is necessary to think more in terms of the volume of water moved per day or year. In other words, think in terms of groundwater discharge rather than velocity. But in order to address discharge we must define discrete bodies of moving groundwater and this brings us to the concept of a groundwater system.

16.2 Groundwater Systems: Aquifers and Basins

A **groundwater system** is broadly comparable to a watershed in its basic operation. It takes in water, conducts it through a defined space, and then releases it. For groundwater, however, the physical entity that houses the system is an aquifer rather than a watershed. An **aquifer** is defined as any material, such as a rock formation,

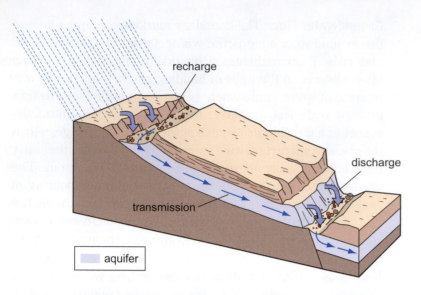

Figure 16.3 A simple aquifer system showing its three basic components: recharge, transmission, and discharge.

Table 16.1

Material	Porosity (%)	Specific yield (%)
Surface deposits		
Gravel	24–38	24
Sand	26–53	25
Silt	34–61	8
Clay	34–60	3
Sedimentary rocks		
Sandstone	5–30	21–27
Siltstone	21–41	15–25
Limestone	0–50	5–45
Shale	0–10	0–3

that contains a large amount of usable groundwater. Usable means that it is good-quality water and can be readily pumped from the ground. In aquifer systems the three basic functions, input, conduction, and output, are termed recharge, transmission, and discharge, and they are illustrated in Figure 16.3. **Recharge** is the term given to gravity-water input from surface sources, such as the seepage from the bottoms of swamps and lakes, and by infiltration through porous soil during and after a rainfall. **Transmission** is the lateral transfer of water through the aquifer, and **discharge** is the release of water, usually outflow to a stream, lake, or spring.

Dynamic Balance in Aquifer Systems: To understand how an aquifer works, it is instructive to examine the interplay of these three basic functions in a groundwater system. If an aquifer of low permeability receives rapid recharge, groundwater builds up because the transmission of water is relatively restricted. With build-up, the hydraulic gradient steepens but, as it does, flow velocity increases, according to Darcy's law. Under such conditions, the hydraulic gradient will continue to rise until the rate of transmission (and discharge, of course) is equal to the rate of recharge. Where recharge is low and permeability is high, of course, the hydraulic gradient will fall until the rates of transmission and recharge are equal.

Under these conditions, the hydraulic gradient falls to a level at which the pressure exerted by the weight of the water (called *hydrostatic pressure*) higher on the gradient is just adequate to sustain a balance between recharge, transmission, and discharge. This is a good example of the *dynamic-equilibrium principle*, whereby a flow system is continuously trending toward a state of equilibrium, or a **steady state**, as it adjusts to changes in inputs and outputs. Owing to the steady-state principle, it is normal in most areas to find steep hydraulic gradients in materials of low permeability and gentle hydraulic gradients in materials of high permeability. An understanding of this concept is important to the management of groundwater supplies where groundwater is pumped for human uses. We will examine the dynamic-equilibrium principle again later in the book in connection with streams as erosional systems.

Aquifer Materials: Many different types of materials may form aquifers, but porous material with good permeability, such as beds of sand and conglomerate formations (rock made up of a mix of sand and stones), are usually the best. An aquifer is evaluated or ranked according to how much water can be pumped from it (without causing a major decline in its overall water level) and, of course, on the quality of its water. Highly mineralized water, such as salt water, which is not suitable for agricultural and domestic uses, is usually not counted among an area's groundwater resources. The amount of water that can be pumped out in a specified period of time is called the **specific yield**, and it is controlled by both porosity and permeability. Notice in Table 16.1 that clay and silt have high porosities but very low specific yields. The low yield rates are related to the limited capacity of groundwater to move in the microspaces among the tiny particles. In addition, some clay particles have plate-like structures and tend to swell when wetted, which further restricts groundwater flow.

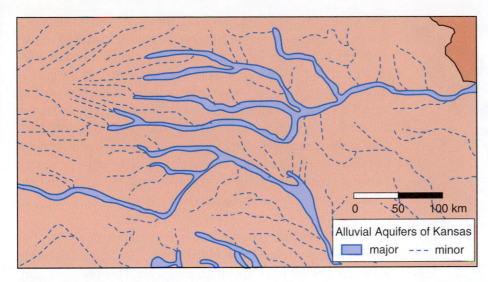

Figure 16.4 Alluvial aquifers form ribbons along rivers. Here they are particularly apparent crossing the plains of Kansas.

Aquifer Types: There are two general classes of aquifers: **consolidated** (mainly bedrock) and **unconsolidated** (mainly surface deposits). In North America the principal unconsolidated materials are glacial deposits, called **glacial drift**, and river deposits, called **alluvium**. Glacial drift covers most of the area north of the Missouri and Ohio rivers to depths typically in the range of 10 to 100 meters (see Figure 12.8). Alluvium is found in the valleys of all large- and moderate-sized streams. The largest single area of alluvium is the lower Mississippi River Valley, which contains huge groundwater reserves extending to depths of 500 meters or more. In the Great Plains, long ribbons of shallow alluvial aquifers, 10 to 50 meters deep, can be traced along thousands of streams, as the map of Kansas in Figure 16.4 suggests. In the mountainous West, alluvial aquifers are found in nearly every valley, even in dry areas, where some reach depths of 1000 meters.

Below the surface deposits are bedrock (consolidated) aquifers, which extend to depths of 5000 meters or more. In the North American interior most bedrock aquifers are found in sedimentary rocks, particularly sandstone and conglomerate. Bedrock aquifers are usually much larger than unconsolidated aquifers, and provide a more dependable water supply. Understandably they are better for large wells (municipal, industrial, and large-scale irrigation), whereas unconsolidated aquifers tend to be favored for residential (domestic) wells.

Recharge Rates: It stands to reason that aquifer recharge rates are broadly related to climate. At the dry end of the spectrum are the subtropical deserts like the Sahara and Atacama where rainfall is so light that water barely penetrates the soil layer before it is taken up as capillary water. Recharge is spotty at best, found only in exotic sites, such as along mountain slopes, where occasional runoff collects and penetrates to a depth beyond plant roots and surface-driven evaporation. At the other end of the spectrum are the wet climates, especially the cool ones like the Marine West Coast climate, where annual recharge rates reach 1 to 2 meters a year along mountain slopes, and may be so great at times that the soil layer is fully saturated. All other climates are more or less intermediate, with the exception of the cold climates where permafrost prohibits the penetration of surface water.

Figure 16.5 Aquifer recharge rates decrease with depth. At depths beyond 1000 meters, it may take thousands of years to renew an aquifer.

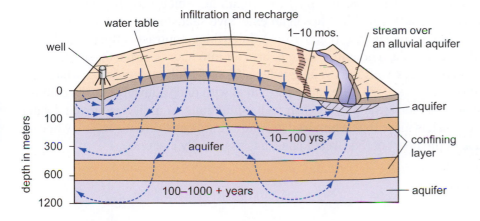

Recharges rates are also affected by surface materials and aquifer depths. Highly permeable soils and surface deposits such as sand and gravel have the capacity to take in water rapidly as does highly fractured bedrock. Some crack systems are so deep and continuous that they serve as recharge conduits. Under such conditions, recharge rates can be as great as climate allows. Aquifer depth is also a factor. Average recharge rates decrease more or less progressively with depth (see Figure 16.5).

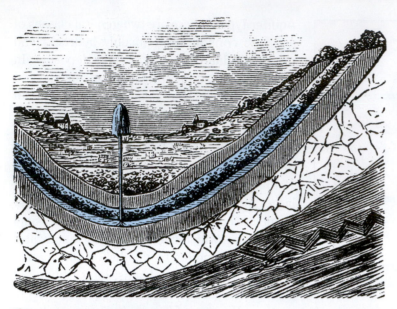

Figure 16.6 An early drawing of an artesian well (center). The water rises to an elevation approximately equal to the head elevation in the aquifer, denoted in blue.

The shallowest aquifers, such as those in alluvial deposits within 50 meters of the surface, recharge in a matter of weeks or months. Aquifers located at depths between 500 and 1000 meters require several hundred years (300-year average) to be completely renewed, whereas those at depths greater than 1000 meters require several thousand years (4600-year average) for complete renewal. This difference is `related not only to the distance recharge water must travel, but also to the fact that percolating water moves slower in the smaller spaces found at great depths and must usually pass through multiple confining (very low permeability) layers.

Artesian Flow: All groundwater exists under pressure, but some aquifers produce such high pressure that their discharge is forced upward to or above ground level. This phenomenon, known as **artesian flow**, results from a geologic condition in which a tilted aquifer, like the one shown in Figure 16.3, is sandwiched between two confining layers. The groundwater in the lower part of the aquifer exists under the hydrostatic pressure of the weight of the water upslope. If the aquifer is tapped by a well, hydrostatic pressure will force the groundwater to flow upward until it reaches a height approximating the head elevation of the water at the upper end of the aquifer. If this elevation happens to lie above that of the land surface, then the water may burst forth into the air forming a spout like the one shown in Figure 16.6. Most artesian flows are less dramatic, and produce a spring or an upwelling of water in a lake or stream channel.

Groundwater Basins: To this point, we have not considered complex groundwater arrangements such as a geologic area housing many interrelated aquifers. A group of aquifers linked together in a large flow system is called a **groundwater basin**. Groundwater basins are typically complex three-dimensional systems characterized by vertical and horizontal flows among the various aquifers and among aquifers and the surface. The size and geographic configuration of a groundwater basin is determined largely by regional geology, that is, by the areal extent and structure of the deposits and rock formations that house the aquifers and related groundwater bodies.

One of the largest groundwater basins in the world is the High Plains Basin in the American Great Plains. This basin extends unbroken 1500 km from South Dakota to north–central Texas and is formed by a large trough of sedimentary rocks. The uppermost aquifer of the basin, called the **Ogallala**, which was discovered in the early decades of cattle ranching, lies in an extensive sandstone formation only 25 to 100 meters below the surface (Figure 16.7). In striking contrast is the groundwater geography of Nevada. The entire state is broken into scores of north–south trending mountain ranges with intervening valleys 25 to 50 kilometers wide, which are discernible in Figure 16.8a. The valleys are filled with deep alluvial deposits containing complex aquifer systems and each valley functions as a discrete groundwater basin. Nevada has more than a hundred groundwater basins (Figure 16.8b), whereas Nebraska, in the heart of the Great Plains, has essentially just one.

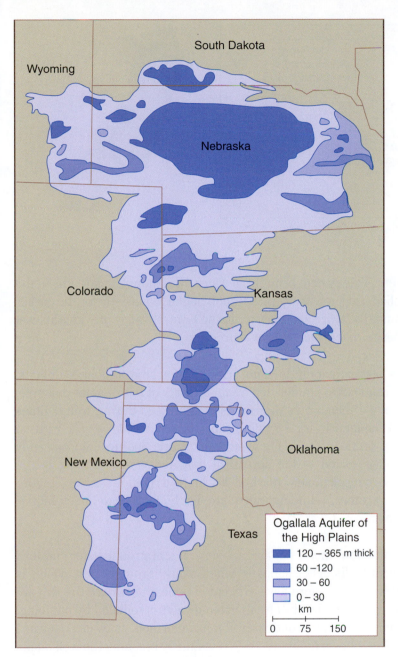

Figure 16.7 The Ogallala Aquifer of the High Plains Basin, the largest aquifer in North America, stretches from South Dakota to north–central Texas.

Ogallala Aquifer of
the High Plains

- 120 – 365 m thick
- 60 –120
- 30 – 60
- 0 – 30
km

0   75   150

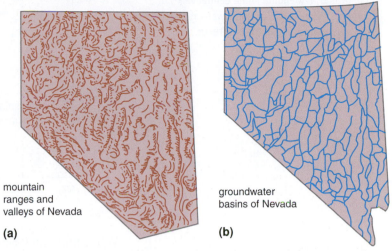

mountain
ranges and
valleys of Nevada

(a)

groundwater
basins of Nevada

(b)

Figure 16.8 (a) The pattern of mountain ranges and valleys in Nevada, and (b) the corresponding pattern of groundwater basins. Contrast the scale of these basins with that of the Ogallala shown in Figure 16.7.

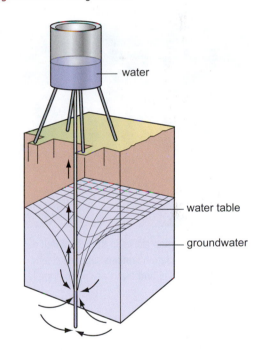

water

water table

groundwater

Figure 16.9 A cone of depression caused by groundwater pumping.

16.3 Human Impacts on Groundwater

Modern land-use activity can profoundly alter both the quantity and quality of water in aquifers. Mechanical pumping tends to draw the groundwater surface down immediately around the well forming a small pocket, called a **cone of depression** (Figure 16.9). Where large municipal, industrial and/or agricultural wells are clustered together, broad sections of aquifers may be depressed up to 100 meters or more. This occurred in the Central Valley of California beginning in the 1930s and 1940s when thousands of large wells drew irrigation water from valley aquifers at rates five times greater than recharge rates. As the graph in Figure 16.10 shows, drawdown continued into the 1960s when the trend was reversed with the development of the Central

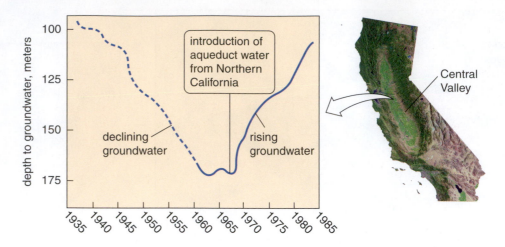

Figure 16.10 The decline and recovery of groundwater at Mendota in the Central Valley of California, 1935–1983. Aqueduct water replaced groundwater pumping in the late 1960s.

Figure 16.11 Irrigation circles in the Wheat Belt of the Great Plains. Irrigation water is supplied chiefly by the Ogallala Aquifer and the consumption rate is enormous.

Figure 16.12 Upconing of salt groundwater in response to the formation of a cone of depression in the overlying fresh groundwater.

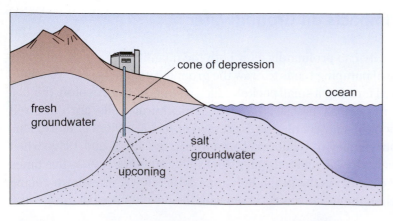

Valley aqueduct system. This system brought in water from mountain streams to the east and farmers applied it at such heavy rates that it began to make up for the earlier groundwater overdrafts. Today irrigation water in the Central Valley provides at least five times more recharge water than natural sources do.

In other dry areas, where irrigation does not make up for large recharge deficits, groundwater decline can lead to aquifer depletion. In parts of Arizona, for example, some aquifers that gained their water thousands of years ago during wetter times (see Section 14.8 in Chapter 14) are declining more than 5 meters a year with no prospects of recovery. Sections of the Ogallala Aquifer are also declining steadily with increased pumping related to the spread of irrigated wheat farming in the Great Plains, which is conspicuously marked by huge irrigation circles (Figure 16.11). With global warming, pumping rates are likely to rise to meet higher evapotranspiration rates.

Groundwater Contamination: In coastal areas, overdrafts of groundwater can lead to saltwater invasion of wells. This happens when saline groundwater, which is denser than freshwater and therefore always situated beneath it where the two are in contact, rises as the overlying layer of fresh groundwater is reduced by pumping. Under large wells where deep cones of depression have formed, saltwater rises in a process called **upconing,** shown in Figure 16.12, eventually reaching the well intake and contaminating the water supply.

Pollution from land use has become an especially serious groundwater problem in most developed countries. The sources of pollution are many, as Figure 16.13 shows. Some are localized and some regional in their distributions, and include buried wastes in landfills, agricultural chemicals, mine debris, accidental spills, leaking fuel tanks, and urban stormwater. Contaminants usually enter aquifers with recharge water. Passage through the soil significantly reduces most and even eliminates some contaminants, but many do reach the water table and enter the groundwater system.

Probably the most disastrous episode of groundwater pollution in human history happened between 1950 and 1990 in the former Soviet Union with the development of its massive industrial economy. It resulted in such severe pollution of soil

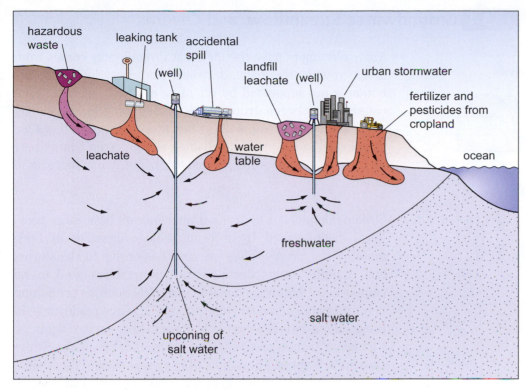

Figure 16.13 Typical sources of groundwater contamination in a coastal urban region.

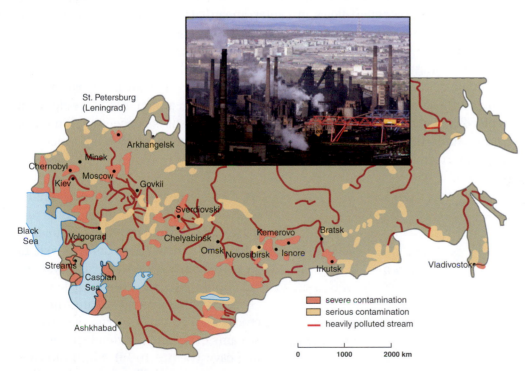

Figure 16.14 Severely polluted lands of the former Soviet Union, the result of indiscriminate storage and dumping of industrial and military waste.

and groundwater that extensive areas – roughly approximated by the map in Figure 16.14 – are now considered beyond remediation and recovery. The effects on human life are not fully documented, and perhaps never will be, but it is likely that millions of people have been affected with illnesses and shortened lives related to contaminated drinking water. Russia's costly experience has served as a warning to other countries, in particular China, which is struggling to regulate its burgeoning industrial economy in the face of serious water management problems.

Beyond the industrial complexes, agricultural fertilizer and pesticide residues are the most serious contaminants in the world. In the United States, for example, intensive fertilizer use in the Corn Belt of the Midwest has led to nitrate contamination of groundwater. In the areas fringing cities, landfills are generally of greatest concern. **Landfills** are buried waste sites and they include sanitary debris (garbage), industrial refuse, and hazardous waste. As buried waste decomposes and mixes with infiltrating water, a contaminated fluid called **leachate** forms, which can seep into aquifers. If it contains hazardous substances such as heavy metals, paint, and petroleum residues, the receiving aquifers may have to be abandoned as water-supply sources. The US Environmental Protection Agency estimates that as many as 50,000 sanitary landfills in the United States contain hazardous waste. But that may be only the tip of the iceberg, for beyond the United States is a rapidly urbanizing world with little or no documentation and control on waste production and disposal. In this century the metropolitan areas of Mexico City and Shanghai are expected to reach populations of 50 million and 100 million respectively.

16.4 Groundwater, Streamflow, and Caverns

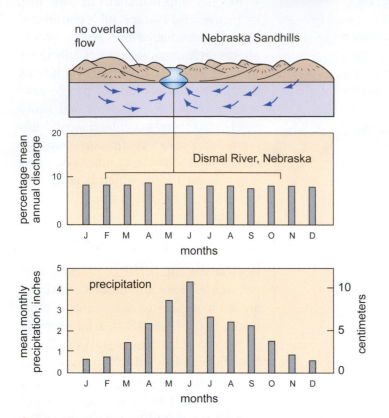

Figure 16.15 The discharge of the Dismal River in the Nebraska Sandhills is controlled overwhelmingly by groundwater. Here each month's average flow is expressed as a percentage of the stream's mean annual discharge.

Figure 16.16 Streamflow in a cavern system. Water enters from the surface through cracks and sinkholes and emerges downstream as springs.

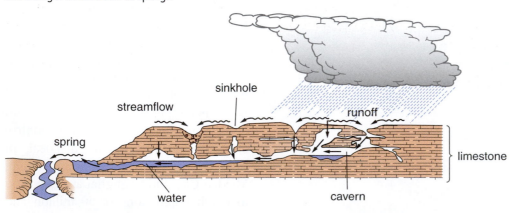

Many old timers have the idea that groundwater comes from underground streams and rivers. For most locations this notion is not supported by scientific evidence, but making a connection between groundwater and streamflow is certainly valid, and some streams do in fact flow underground. As we established in the previous chapter, streams gain a large part of their discharge from groundwater, but some more than others.

Streams dominated by groundwater inflows have large baseflows distinguished by a remarkable uniformity in their monthly discharges. They are less responsive to rainstorms and drought events because their watersheds tend to be composed of highly permeable materials such as sand-dune deposits, porous lava, and glacial gravels, which readily take in precipitation and pass it quickly down to near-surface aquifers. Virtually all runoff takes place underground. This is clearly the case in the Nebraska Sandhills where rivers such as the Dismal (see Figure 16.15) maintain nearly constant flow in all seasons despite large winter/summer variations in precipitation.

Karst Drainage: Surface channels may also be scarce where runoff flows in subterranean caverns. Groundwater has the capacity to dissolve and erode certain types of bedrock, especially limestone, giving rise to a complex terrain of pits, large holes, and caverns called **karst topography**. Water infiltrating from the surface is naturally charged with a light concentration of carbonic acid (carbon dioxide plus water), which reacts with limestone, slowly weathering it away (we will come back to limestone weathering in more detail in Chapter 20, Section 20.5). Where groundwater movement is concentrated in crack systems within the limestone strata, cavities and tunnels are opened up, which can lead to cave formation.

As caves grow they become linked together, forming underground drainage networks that not only carry groundwater, but capture surface runoff as well. In fact streams flowing over cave systems commonly lose all or part of their discharge to underground streams through cracks linked to caverns, or where streams drain into a sinkhole in the roof of a cave (Figure 16.16). **Sinkholes** are deep holes or pit-like depressions that form when part of the roof of a cavern collapses.

In central Texas, where streams cross a belt of weathered limestone called the Edwards Formation, part of their flow is funneled underground into a complex system of deep passageways. The water moves swiftly through the system and is discharged 10 to 15 kilometers downslope in large springs. The discharge from some of these springs is magnificent, typically exceeding a million gallons a day. Upon leaving the spring the groundwater once again becomes streamflow, often rejoining the stream network that originally made the recharge contribution.

Summary on Groundwater: Groundwater occupies the lowest tier in the landscape. It is recharged from the surface, moves by transmission through soil and rock, and at selected sites, such as stream channels and lake basins, is discharged back to the surface. The distribution of groundwater is broadly related to climate but, within this framework, it varies dramatically with landforms, surface and subsurface materials, and depth in the crust. Groundwater has long served as a water-supply source, and in many areas today is being withdrawn at rates exceeding recharge rates. In addition, because it is fed by water passing through the landscape, groundwater is subject to pollution from wastes produced by land-use practices.

16.5 Lake and Wetland Systems

Where the karst processes described above result in large sinkholes or extensive cavern collapses, sizeable basins may be formed and, if they are deep enough, they fill with groundwater, and become lakes. Central Florida has numerous lakes of karst origin, all fed by groundwater (see Figure 16.17). But **karst lakes** are only one variety of at least eight different types of lakes recognized by geographers. The others include desert lakes, riverine lakes, glacial-drift lakes, mountain lakes, shield-margin lakes, permafrost lakes, and tectonic lakes, and all except desert lakes are supported by the groundwater system. The map in Figure 16.18 shows you where each type is found in North America.

All lakes function as simple input–output systems and, for most, groundwater is the principal input. Groundwater's role in a lake basin is basically the same as it is in a stream channel. In a stream channel, groundwater is responsible for baseflow. In a lake, it is responsible for basepool, that is, the relatively permanent body of water produced when groundwater flows into the basin filling it to or near water-table elevation. Added to the basepool are sporadic inputs from precipitation and overland flow and inputs from inflowing streams. On the output side of the ledger are water losses to evaporation, outflowing streams, and in some lakes seasonal losses back into the ground.

Figure 16.17 The Lake District of central Florida, a vast system of karst lakes fed by groundwater.

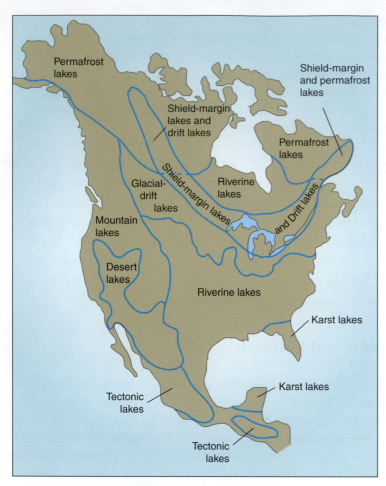

Figure 16.18 The lake regions of North America. The largest lakes are shield-margin lakes, whereas riverine lakes are probably the most widespread, occupying all regions to some extent.

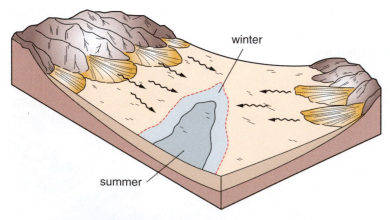

Figure 16.19 The setting, seasonal dimensions, and runoff patterns associated with a playa lake.

Lake Types and Systems: Just as with streams, lakes with substantial groundwater input tend to maintain fairly steady levels and volumes, whereas those supported mainly by surface water sources tend to fluctuate widely in response to precipitation and runoff events. Karst lakes, for example, fluctuate comparatively little, but not so with desert lakes, called playas. **Playas** load up with water from winter and spring runoff and then lose just as much to infiltration and summer evaporation. Consequently, they experience extreme variations in volume and area from summer to winter/spring as the diagram in Figure 16.19 shows. In extreme cases, such as the great Australian playa, Lake Eyre, the lake disappears altogether in the summer only to reappear in the winter spreading over an area (about 40,000 square kilometers) rivaling that of lakes Erie and Ontario combined! But most of the large playas maintain a core pool throughout the summer and then expand by two times or more in the winter. As winter water evaporates, it leaves behind a residue of salt that was washed in with runoff from the desert landscape. Year after year the salt builds up and in time may reduce the lake to a wetland of salty sludge surrounded by a dry salt bed.

Riverine lakes are found in stream valleys and most form in sections of old (detached) stream channel, called oxbows, and in low spots in the floodplain that may or may not be connected to a stream channel (Figure 16.20). Most riverine lakes consist of a basepool of groundwater supplemented by inputs from streamflow and seasonal inputs from floodwaters. In the latter capacity, riverine lakes function as floodwater storage basins. For riverine lakes attached to active stream channels exchange times are usually short, typically several months to a few years depending on lake size and stream discharge. **Exchange time** is the time it takes to completely flush the water from the lake and replace it with a new supply.

Understandably, many riverine lakes, like those in Figure 16.20, fluctuate seasonally in size in response to the flow regime of the feeder stream. Those connected to large rivers with huge spring or rainy season floodflows, vary radically in size. In southeast Asia, for example, the area of Lake Tanle Sap in the lower Mekong watershed, varies more than tenfold from 2700 to 30,000 square kilometers from winter to summer. Similarly large and seasonally variable riverine lakes are found in the central Congo and Yangtze watersheds and one of unknown dimensions forms annually on the forest floor of the Amazon. It may be the largest riverine lake/wetland in the world. Most constructed (human-made) reservoirs also function as riverine lakes and, if the control gates on the dam are lowered or raised, the exchange time can be regulated. This simple concept is the rationale for using reservoirs in flood control and water-supply management.

Figure 16.20 Aerial view of a South American river valley showing riverine lakes of various sizes.

Many **glacial-drift lakes**, on the other hand, may have no inflowing or outflowing streams, but are supported almost exclusively by the groundwater system. These lakes usually form in basins left by large chunks of glacial ice that ranged in size from a few to hundreds of acres. The groundwater is held in the surrounding glacial deposits, often porous and permeable sand and gravels. Infiltration rates in these materials are so high that virtually all precipitation in the area around the lake goes to groundwater recharge. The groundwater in turn flows into the lake basin where it stabilizes at a pool elevation equal to the level of the water table (Figure 16.21). On the output side, the lake gives up water to evaporation and, in some settings, to seepage back to the water table on the downslope side of the lake. Season fluctuations in lake level tend to be modest. Where groundwater inflow to a basin is especially large, glacial-drift lakes often produce outflowing streams that may link with other lakes and wetlands forming larger drainage systems.

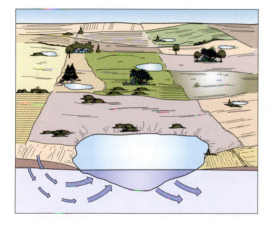

Figure 16.21 Glacial-drift lakes often serve as the headwaters for stream systems by discharging groundwater to outflowing channels.

Mountain lakes are situated in upland valleys where alpine glaciers have scoured basins into mountain sides and eroded deep mountain valleys (Figure 16.22). The most picturesque of these are tiny lakes, called tarns, perched high in alpine valleys. But some mountain lakes are much larger such as those formed where glacial deposits or landslides create natural dams on streams. Although most mountain lakes are supported by groundwater, meltwater from snowfields and/or glaciers is usually a major source of water and one which fluctuates dramatically from summer to winter. Mountain lakes often play a singular role in stream systems in the capacity of headwaters for large watersheds such as those of the Mackenzie, Columbia, and Fraser whose tributaries rise high in the Canadian Rockies.

A lake type of similar origin, but much larger than the mountain glacial lake, is the **shield-margin lake**. These lakes are the largest freshwater lakes in the world based on surface area and include the North American Great Lakes and a series of similar lakes to the northwest of Lake Superior including Great Slave and Great Bear lakes (Figure 16.23). The basins of shield-margin lakes were scoured by the huge ice masses of the continental glaciers as the ice moved from the resistant, interior uplands, called a shield, onto the fringing plains and lowlands. All shield-margin lakes are found in the middle and high latitudes of North America and Eurasia under climates where annual precipitation exceeds lake evaporation. Therefore, these big lakes *could* exist without runoff contributions from their watersheds, but most receive substantial, sometimes huge, inputs from their watersheds. From the Great Lakes watershed comes the St. Lawrence River, the third largest river in North America.

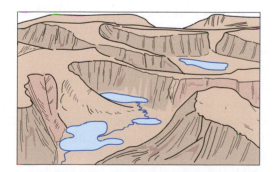

Figure 16.22 Most mountain lakes form in basins sculpted by alpine glaciers. This drawing shows tarns where they are typically found, at the heads of deep glacial valleys.

Figure 16.23 The major shield-margin lakes of North America rank among the largest lakes in the world. Their basins were scoured by continental glaciers during the last ice age.

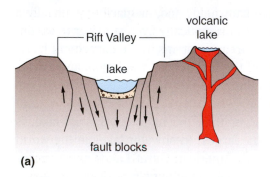

(a)

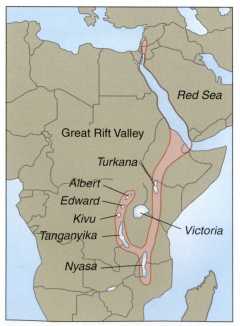

(b)

Figure 16.24 The tundra landscape of northern Canada in summer dotted with permafrost lakes.

Farther north we find permafrost lakes. **Permafrost lakes** form in the thawed layer of ground, called the active layer, which develops each summer on the Arctic tundra, as shown in Figure 16.24. Beneath the active layer at depths of 1–5 meters lies solidly frozen ground, the permafrost, which forms a barrier to the downward transfer of water. As a result, the active layer becomes saturated with shallow groundwater that collects along with rainwater and overland flow in low spots to form seasonal lakes and wetlands. Most are shallow and freeze out with the onset of winter. Some deeper lakes are also found in permafrost regions, but most are tied to large stream systems, such as the Mackenzie and the Yukon, where the permafrost has been thawed to much greater depths. These are groundwater lakes which usually do not freeze out in winter.

Tectonic lakes form in basins created by faulting and volcanism in the Earth's crust. The most common tectonic lake is found in volcanic craters, but the largest is found in rift valleys where faulting has caused large blocks of crust to have dropped down several hundreds of meters or more (Figure 16.25a). The largest tectonic lakes rival shield-margin lakes in surface area and, because of their great depths, exceed them in volume. The largest system of tectonic lakes is located in the Great Rift Valley of the East African Highlands, which contains six major lakes (Figure 16.25b). Most tectonic lakes are very deep; Lake Baikal in central Asia, the deepest lake in the world (1740 meters (5700 feet) deep) lies in a block fault basin.

The Great Lakes of the World: There are close to 20 lakes in the world with surface areas larger than 10,000 square kilometers. This is an area twice the size of the state of Delaware. With one exception, these great lakes are limited to four types, shield-margin, tectonic, desert, and riverine, and they tend to occur in geographic clusters on the world map. Table 16.2 lists the largest lakes in each class. The one exception is Lake Maricaibo in South America which is a coastal lake, or lake-like body, partially controlled by tidal fluxes and ocean water.

Figure 16.25 Tectonic lakes. (a) Rift-valley lakes form in a valley formed by faulting and most volcanic lakes form in craters; (b) the tectonic lakes of Africa lie principally in the Great Rift Valley.

Table 16.2 Great lakes of the world by lake type. (Only five of the largest lakes in each class are listed.)

Shield-margin lakes	Area (km²)	Max. depth (m)	Volume (km³)	Continent
Superior	83,000	406	12,000	North America
Huron	60,000	229	3600	North America
Michigan	58,000	285	4900	North America
Great Bear	32,000	445	1000–2400	North America
Great Slave	29,000	614	1100–2100	North America
Tectonic lakes	**Area (km²)**	**Max. depth (m)**	**Volume (km³)**	**Continent**
Victoria	69,000	92	2700	Africa
Tanganyika	34,000	1470	19,000	Africa
Baikal	32,000	1740	23,000	Asia
Nyasa	31,000	706	7700	Africa
Titicaca	8000	304	830	South America
Desert lakes	**Area (km²)**	**Max. depth (m)**	**Volume (km³)**	**Continent**
Aral Sea	64,100 (38,500)*	68	1000 (360)*	Asia
Balkhash	19,000–(?)	26	110	Asia
Chad	7000–26,000	12	44	Africa
Bangweula	4000–15,000	5	5	Africa
Eyre	0–40,000	0–20	0–23	Australia
Riverine lakes	**Area (km²)**	**Max. depth (m)**	**Volume (km³)**	**Continent**
Dongtinghu	3000–12,000	10	18	Asia
Tanle Sap	2700–30,000	12	40	Asia
Poyanghu	2700–5200	20	–	Asia
Nai Ndombe	2100–8200	6	–	Africa
Taihu	2200– ?	5	4	Asia

* After human-induced reduction in the twentieth century. See Figure 14.28.

Eight of the world's great lakes are shield-margin lakes and seven of these are located in North America (Superior, Huron, Michigan, Great Bear, Great Slave, Erie, and Winnipeg). Four are tectonic lakes: Victoria, Tanganyika, and Nyasa in the Great Rift Valley of Africa, and Baikal in Russia. Lake Baikal has the largest volume of any freshwater lake, whereas Lake Superior has the largest surface area of any lake in the world.

There are six lakes whose areas fluctuate seasonally above and below 10,000 km². These are desert lakes and riverine lakes and their areas are given in the Table 16.2. Although they contrast sharply in terms of climate type, these two classes of lakes have similar and dramatic annual regimes. They expand during the rainy season, reaching two to ten times their dry-season size with depths of only several meters, and then rapidly subside with outflow and evaporation. Not surprisingly, the riverine lakes are extremely important in modulating flooding in connecting river valleys because of the large amounts of excess water they store.

Lake Life Cycles, Climate Change, and Land Use: Although lakes have been a prominent part of Earth's geography for millions of years, most are relatively temporary features in the landscape. Small lakes like oxbows, glacial-drift, and permafrost lakes are prone to infilling with sediment and organic debris, which generally brings them to a close in 15,000 years or less. Life cycles are much longer with big lakes for, among other things, it takes a lot longer to fill them in. The large tectonic lakes appear to last longest, particularly those with deep basins formed in rift valleys. They and the shield-margin lakes probably come to an end by means other than infilling such as renewed volcanic activity or continental glaciation.

All shield-margin lakes date from the last continental glaciation which ended about 10,000 years ago. In all probability it will take another ice age with similarly large continental glaciers to obliterate them. Over the past million years or so there have been episodes of continental-scale glaciation every 50,000 to 100,000 years. Whether these big glaciers scour out the same lake basins with each new advance is not known. In any case, it is probably safe to say that large shield-margin lakes have a life cycle

Lake Baikal, the largest lake in the world by volume, occupies a basin formed in a rift (fault) valley.

Bonneville Salt Flats and Great Salt Lake are all that remain of ancient Lake Bonneville.

Figure 16.26 An overgrown lake in the US Midwest suffering from cultural eutrophication brought on by nutrients from fertilizer, sewage, stormwater, and air-pollution fallout. Global warming will accelerate this process.

in the range of 50,000 to 100,000 years. Barring another ice age, it seems likely, however, that they might last several million years or more before they are filled with sediment.

Many desert lakes, even the very large ones, often have very short life cycles. In Chapter 14 we cited modern Great Salt Lake, the remnant of ancient Lake Bonneville, as an example (see Figure 14.26). Lake Bonneville existed during a wetter time in the American Southwest and covered an area roughly the size of Lake Michigan. As the climate has grown drier over the past 10,000 years, the lake has shrunk to less than one-tenth its original size. The same trend has happened with Africa's largest desert lake, Lake Chad, but over a much shorter time span. In the 1960s, Lake Chad covered 25,000 square kilometers but, because of persistent drought, input from its watershed declined, until by 1990 the lake covered only 2500 square kilometers. Whether lakes like Chad and Bonneville are victims of permanent decline leading to their demise is hard to say. Probably not. The dry climates are notoriously fickle, and in the case of Lake Chad, this extended dry period may be followed by a relatively wet one during which the lake recovers. But then there is also global warming to consider. In the case of Lake Bonneville, the recovery cycle is much longer and recovery may await cooler, wetter conditions with the next period of glaciation, likely many millennia from now.

It appears that global warming is likely to have pronounced effects on many of the world's lakes. Desert lakes may see higher evaporation rates and lower rates of input from rainfall and runoff. As they shrink, large areas of old lake beds will be exposed and the unprotected sediment will be subject to wind erosion, adding to the massive load of dust already blown into the atmosphere from the world's deserts. In the Arctic the effects of global warming are already underway. Both the North American and Eurasian Arctic are now experiencing marked warming (see Figure 8.12) and melting of the upper permafrost has been documented. This will lead to more and larger permafrost lakes and perhaps in turn to expanded drainage systems and higher runoff rates

As warming causes thermal zones to shift poleward, lakes in the midlatitudes will see increased evaporation rates. As a result, some may shift from positive annual water balances to negatives ones. This change, coupled with the inevitable rise in water temperature, could bring on significant changes in the freshwater ecology of lakes, especially if land use is involved. Land use around lakes and in their watersheds promotes the production of pollutants like compounds of nitrogen and phosphorus, which are carried into the lakes with stormwater and groundwater. These pollutants are nutrients for aquatic weeds and, when combined with warmer water temperatures, the plants will flourish and accelerate the lake-infilling process. The natural process of lake infilling with organic debris is called **eutrophication**. The human induced process is called cultural eutrophication (Figure 16.26).

Land use also has major effects on the water balance of lakes, especially those in areas of limited runoff. In Chapter 14 we described the most blatant example yet documented, the Aral Sea in southwestern Asia, one of the world's largest desert lakes (see Figure 14.28). Because its main water supply has been diverted for agricultural irrigation, it has lost 80 percent of its volume resulting in a sixfold increase in salinity and collapse of its once rich fishery.

Figure 16.27 The four basic types of wetlands classified according to the supporting hydrologic system; surficial, groundwater, riparian, and composite.

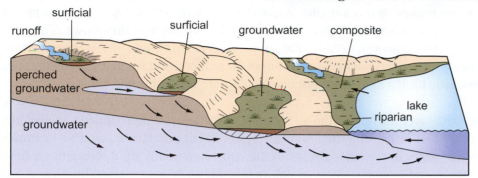

Wetland Systems: Many wetlands are merely shallow versions of lakes. Indeed, it is often difficult to distinguish between the two because most inland lakes are fringed by wetlands and over time lakes are transformed into wetlands as they fill with sediment and organic matter. But wetlands have several attributes that set them apart from lakes: first, most wetland basins are shallow enough to support such extensive vegetative covers that plants occupy more area than open water; and second, wetlands occupy many types of aquatic environments in addition to lakes including rivers, bays, and ocean estuaries. In North America, Canada has the largest area of wetlands, about 18 percent of its total land area. The United States has about 4 percent of its land in wetlands, and Alaska is the state with the most wetland area. A century ago, the area of wetland in the lower 48 states was at least twice that of today. Farming has been responsible for the vast majority of wetland eradication in the United States and the rest of the world.

We can define four general classes of wetland based on the nature of the supporting hydrologic system: surficial, groundwater, riparian, and composite (Figure 16.27). **Surficial wetlands** are dependent on surface sources of water, mainly direct precipitation and local overland flow (stormwater) and with one exception are not supported by groundwater. The exception is those supported by *perched groundwater*, that is, small lenses of groundwater that lie near the surface above the main groundwater system. Where perched lenses intercept the surface, water seeps out saturating the overlying soil giving rise to wetland conditions. This often accounts for the occurrence of wetlands on hillslopes and in the active layer over permafrost. In general, surficial wetlands tend to be small, often isolated, and subject to rather dramatic seasonal changes.

Groundwater wetlands are usually found at lower elevations in the landscape in settings comparable to groundwater lakes such as sinkhole and glacial-drift lakes (Figure 16.28a). These wetland sites lie at or below the water table and therefore receive appreciable groundwater inflows. Unlike surficial wetlands, groundwater wetlands are generally not subject to radical fluctuations with variations in precipitation. The water level in some groundwater wetlands, however, may fluctuate a meter or more from summer to winter/spring or on a longer-term basis with climatic conditions affecting groundwater supplies (Figure 16.28b).

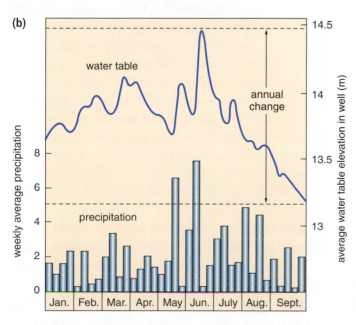

Figure 16.28 (a) Groundwater wetlands in the prairies of North Dakota. These wetland ponds are called prairie potholes. (b) Seasonal variations in water-table elevation with precipitation can be a meter or more.

Riparian wetlands are those in and around major water features such as lakes, streams, and estuaries. These wetlands usually show a strong gradation in habitat conditions with water depth from deep-water aquatic habitat on the wet side to terrestrial habitat on the dry side. The supporting water system governs the wetland's hydrologic regime. If it is a midlatitude stream system, for example, high water can be expected in winter and spring with high discharges and floodflows. This is a two-sided coin, however, because the supporting water system is also the source of destructive processes such as storm waves, floodflows, and ice movements that can cut wetlands back and in some instances obliterate them entirely.

Composite wetlands are those supported by two or more major hydrologic systems. Most large, enduring wetland systems fall into this class. For example, cypress swamps in river floodplains of the American South are dependent on both floodwaters and groundwater. Coastal marshes are often supported by a combination of tidal water, stream discharge, and groundwater. Because each system has a different regime, the principal supply of water to the wetland may change from season to season. In a coastal estuary, streamflow may be the dominant system in the spring, and tidal water the rest of the year. Understandably, the composite class of wetlands is often the most hydrologically complex of the wetland systems.

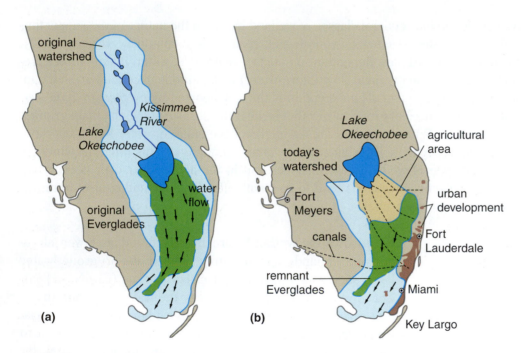

Figure 16.29 The Everglades (a) before and (b) after development of canals, farms, highways, and other land-use facilities. Less than half the original watershed and Everglades ecosystem remains.

Land Use and the Everglades: Land-use alterations such as the construction of roads, canals, navigational facilities, and flood-control structures may have far reaching effects on all wetland systems. Individually, the impacts may initially appear to be minor but their compound effects often lead to substantial impacts in the long run. One of the most controversial management dramas of this sort is currently being played out in the Florida Everglades, one of America's premiere wetland systems, where large drainage canals, highways, levees, agriculture, and urban development have combined to reduce the watershed by more than half, greatly alter the wetland's water balance and ecology, and ultimately reduce its total coverage. Figure 16.29 shows the Everglades before and after development. Water input from Lake Okeechobee was cut off by the construction of a large levee around the lake and by drainage canals designed to enhance agriculture to the south. Cut back by development and cut off from much of its freshwater, the Everglades today are about half their original area. Efforts are now underway to reverse the damages to the great swamp's water system.

Summary on Lake and Wetland Systems: Lakes and wetlands are major features of the hydrologic landscape and, like the streams that often connect them, most are parts of both surface and subsurface drainage systems. Curiously, lakes and wetlands are found in all climatic zones, even arid ones where they rise and fall from winter to summer with changes in runoff contributions and evaporation losses. All lakes and wetlands function as input–output systems and most are tied to both groundwater and stream systems. Shield-margin lakes are the largest freshwater lakes in terms of area, whereas tectonic lakes are the deepest and largest in volume.

16.6 Water Use, Water Supply, and Land Use

The problem with water supply is fundamentally geographical. Humans and freshwater are distributed such that hundreds of millions of people live in places where nature provides too little water. As a result, demands for water among these people far exceed available supplies, especially renewable supplies, and the picture is growing worse day by day. By 2025, nearly 2 billion people, about 25 percent of humanity, will face serious water shortages. Why is it worsening? Two big reasons are population growth (we are adding more than 70 million people to the planet every year) and migration that is leading to major shifts in population toward dry areas and large cities. For example, in the United States since 1960 population in the arid Southwest has increased faster than any other region of country as people leave the Northeast and Midwest for California, Arizona, and Nevada and crowd into sprawling urban centers. Added to all this is the growing prospect of larger and more frequent droughts in some regions related to global warming.

Total water use worldwide is about 4000 cubic kilometers a year. This is a massive quantity of water but, in the great scheme of the Earth's hydrologic system, it is actually quite modest. For example, the total water use in the United States accounts for only 5 percent of the nation's annual water supply from precipitation and of that only 2 percent or so is consumptive use, that is, water lost to evaporation rather than put back into the runoff system after use. If people do not use it, where does the remaining 98 percent go? Refer to the diagram in Figure 16.30. Sixty seven percent goes to evaporation (and plant transpiration), 29 percent goes to stream outflow to the sea, and 2 percent goes to groundwater seepage to the sea. Seems like there should be plenty to go around.

What these comparisons do not put into proper perspective is that in addition to the highly uneven geographical distribution of water, there is the problem of the relative availability of usable, renewable water. Nearly everywhere, even in wet regions, the availability of extractable freshwater is dependent on a host of geographic conditions, including seasonal delivery of rain, aquifer depths, streamflow characteristics, technical ease of water extraction, transport distances, and water quality (lots of water is polluted). Soil water, wetlands, and many small lakes and streams, for example, are not suitable for water supply and many aquifers are also marginal because of the small volume of water held, low yield rates, or poor water quality. On balance, the available supply of useable freshwater is far smaller than the gross precipitation figures suggest no matter where you are.

Water Use in the United States: All the land uses in the United States together extract about 408 billion gallons (1.5 billion cubic meters) of water from the environment on an average day. Power, agriculture, and industry together account for nearly 90 percent of this water (Figure 16.31). The remainder is used in homes and settlements. Of all the freshwater used in the United States, about 76 percent comes from surface sources (streams, lakes, and reservoirs) and 24 percent from groundwater. Fresh groundwater withdrawals have increased by 14 percent between 1985 and 2000, due in large part to the rapid population growth within arid and semiarid regions. A small, but growing part of the increase in groundwater use can be attributed to the use of bottled water in the United States. In 2004, almost 7 billion gallons of bottled water (from primarily groundwater sources) were purchased. Fresh surface-water withdrawals during the same period have varied less than 2 percent.

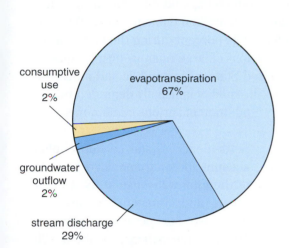

Figure 16.30 The disposition of the precipitation received by the United States in an average year. Only 2 percent is given up to consumptive land uses, but total water use is closer to 5 percent.

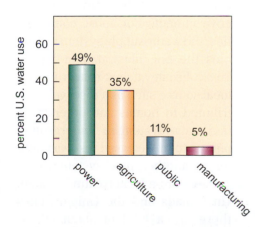

Figure 16.31 The principal water users in the United States ranked by percentage of the national total water use.

Power generation (excluding hydroelectric dams) is the single largest water user at nearly 56 million cubic meters (200 billion gallons) a day. This water, which includes saltwater in some locations, is used for cooling and then released as heated water to streams, lakes, reservoirs, and the ocean. Since 1940 the water used in power generation has increased tenfold in response to increased population and increased per capita energy use. Currently the United States uses about 25 percent of the world's electrical energy.

Agriculture in the United States withdraws an average of more than 541 million cubic meters (143 billion gallons) of freshwater each day. Virtually all this water is used in irrigation, and most is lost to the atmosphere via evaporation and plant transpiration. In California, where water supply has become a crisis issue and the urban population has reached more than 30 million people, about 85 percent of the water is used by agriculture.

The public/domestic sector (settlements and households) is the third-ranked water user at 186 million cubic meters (49 billion gallons) a day and manufacturing is the fourth-ranked water user in the United States, at 76 million cubic meters (20 billion gallons) daily. This figure represents a decline of 24 percent during the period from 1985–2000, with much of this loss related to the loss of large manufacturing capacity in the industrial cities of the Midwestern United States. Most public and industrial water is returned to surface waters, although the quality is usually degraded by the addition of various contaminants.

Figure 16.32 A water-use system for a typical North American community includes both water supply and stormwater systems. Both carry waste, but only the water-supply system is treated.

Settlements and households account for only about 10 percent of the total water use in the United States and Canada. The average daily use per person is between 100 and 150 gallons, or about 500 gallons (2 cubic meters) per home. Although total domestic use is relatively small, the domestic water system is very expensive both in terms of construction, maintenance, and operation, and its impact on the environment. For most communities, the water-supply system resembles the one illustrated in Figure 16.32. Freshwater is drawn from a local source such as a reservoir, distributed to homes and other facilities via a system of underground pipes, then released as wastewater to another pipe system which carries it to a local treatment plant. (In Canada and the United States there are a total of about 60,000 such municipal water systems.) Treatment removes disease-bearing agents and most of the sediment

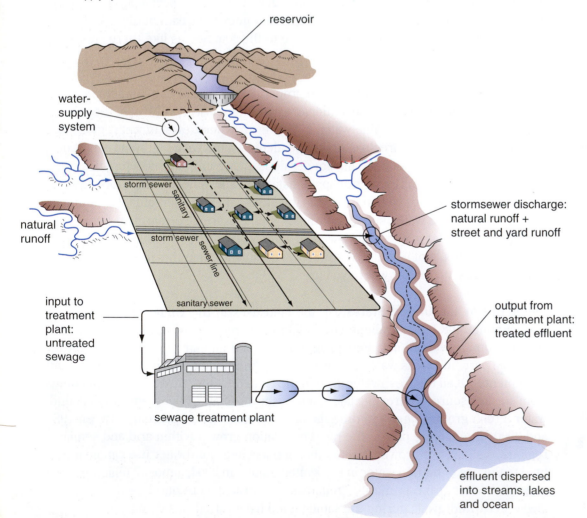

reservoir

water-supply system

natural runoff

storm sewer

sanitary

storm sewer

sewer line

sanitary sewer

input to treatment plant: untreated sewage

sewage treatment plant

stormsewer discharge: natural runoff + street and yard runoff

output from treatment plant: treated effluent

effluent dispersed into streams, lakes and ocean

and biochemical contaminants whereupon the water is released back into the environment, usually a stream, lake, or ocean harbor. Although it is greatly improved by treatment, the wastewater is decidedly poorer in quality than the freshwater originally extracted. However, not all communities are able to attain even the most basic purification standards.

In more than 1000 North American cities, sewage treatment is bypassed during wet weather and raw sewage is discharged directly into a nearby harbor, stream, or lake. The reason for this is that the stormwater system in these cities is tied to the sanitary sewer system and when stormwater runoff is heavy, the sewage treatment plant's capacity can be exceeded and the whole mess, raw sewage and stormwater, passes through untouched. This is a sobering fact to Americans and Canadians, but as a pollution problem it is small compared to that associated with most large cities in the less-developed world, such as Calcutta, Karachi, and Bangkok, which release largely raw sewage directly into rivers, canals, and ocean harbors. As many as 2 billion people worldwide (one-third of humanity) lack access to any sewage treatment.

Figure 16.33 Global water use by continent in cubic kilometers per year. Asia's water consumption is twice that of the rest of the world combined.

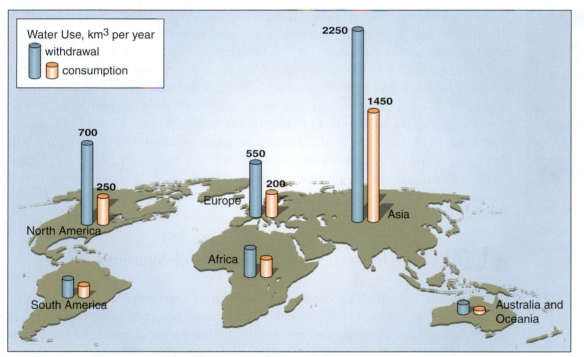

Global Water Supplies and Problems: Figure 16.33 shows global water use by continent in two categories, total withdrawal and total consumption. Consumptive use is water not returned to groundwater or streamflow systems, such as water used in irrigation water, which is lost to the atmosphere in evapotranspiration. Asia stands above all the rest of the world combined in both categories. There are two reasons for this: Asia's huge population and massive agricultural economy. Historically, water use has increased with global population, and the principal water user is agriculture, accounting for 65–70 percent of total water use. Current water use by all countries is twice that of 1950 and 35 times greater than in 1700. World population is projected to reach 9 to 10 billion by 2050 with over 95 percent of the growth in less-developed countries of Asia, Africa, and South America where water supply is already a crisis issue. With 2 or 3 billion additional people in these countries, the crisis will deepen dramatically by the time you expect to retire.

Once again, the principal problem we face is that the supply of usable water is unevenly distributed over the Earth. And despite our efforts to redistribute it with dams, diversions, and aqueducts, in many parts of the world population growth has outpaced water supplies, and we are now facing serious regional shortages on every continent except Antarctica. Among the regions facing unending shortages, the Middle East is one of the most seriously threatened, not only because it is arid and populous, but also because international political disputes threaten regional water planning and redistribution. The various conflicts among Turkey, Iraq, Iran, Kuwait, Syria, Jordan, and Israel illustrate this. An example is Turkey's building of dams and reservoirs on the headwaters of the Tigris and Euphrates, the primary sources of water for Iraq. Other threatened regions include:

- northern China, where freshwater supplies are falling behind demands;
- the United States Southwest, where agriculture and urban development have dangerously stressed existing surface and groundwater supplies;
- parts of Russia and eastern Europe, where water supplies have become so fouled by pollution that they may be unusable for decades or longer;
- south–central Asia, where limited water supplies and undependable delivery systems place several former Soviet republics at risk;
- many parts of Africa, including the Sahara fringe, eastern Africa, and south and southwestern Africa, where water demands of a rapidly growing population are far outpacing supplies; and
- South Asia (India and Pakistan), where a population of 1 billion people places a heavy demand on water for agricultural irrigation.

In Pakistan, for example, it is likely that population will outrun water supplies within the next several decades. Pakistan is mostly arid and semiarid, and its population, which grew from 31 million in 1947 to 126.4 million in 1994, is expected to reach 400 million or more by the year 2035. With the country overwhelmingly dependent on agriculture, and agriculture overwhelmingly dependent on irrigation, serious shortages of both surface water and groundwater are sure to emerge as agricultural expansion attempts to meet rising food demands (Figure 16.34). On top of all this, much of the headwaters of Pakistan's principal surface water source, the Indus River, lies in India, the one country with which it has seemingly unending political conflict. Pakistan is not unique among less-developed countries in dry regions; many African and Middle Eastern countries face similar prospects, including cross-border watershed conflicts.

Figure 16.34 Farm fields and orchards in Pakistan where agriculture is possible only with substantial irrigation.

16.7   Human Impacts on Water-resource Systems

Water use is classified as **consumptive** or **non-consumptive**, depending on whether the water is lost to the atmosphere as vapor or returned to surface waters as liquid. As we noted above agriculture is overwhelmingly consumptive because most irrigation water is lost to evaporation and transpiration. The other major user-classes are only marginally consumptive, consuming in the range of 10 to 20 percent of total use. Most water uses can, however, be defined as **degrading** because they usually pollute the water in some way before being released back into the environment. As we noted earlier, the sewer water from homes and settlements, even when treated, is a source of sediment and biochemical contaminants; agricultural runoff is typically contaminated with sediment, fertilizer residues, and insecticides; water from manufacturing often contains heavy metals and petroleum residues; and the water from power plants is heated (thermal pollution) to a level far above that of receiving waters by the time it is returned to the environment.

Diversions and Dams: Another major impact of water use is the effect of diversion on stream systems. The extraction and re-routing of water from streams often greatly reduces flows. With reduced flows, streams are less able to carry their sediment loads

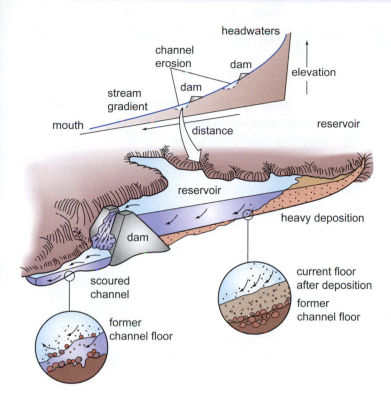

Figure 16.35 Some of the effects of dams on streams. The reservoir catches sediment, which eventually fills the basin, while downstream the stream scours the channel.

and channels often become clogged with sediments. This disrupts stream ecology, such as fish habitat and spawning environments, and reduces the stream's capacity to conduct flood-flows. In addition, lakes and wetlands may be deprived of water by stream diversion. The Aral Sea is perhaps the most dramatic example of diversion in this century (see Figure 14.28). In the nineteenth century, the most dramatic stream diversions were committed by California gold miners who re-routed hundreds of streams and, in one 30-year period, excavated more than a billion cubic meters of sand and gravel from streambeds. After gold mining, these streams were further diverted for irrigation in the Central Valley.

Overall, dams and reservoirs probably have the greatest effect on stream environments. Not only do extensive reservoirs flood entire valleys and obliterate channel environments, but the channel downstream of the dam is severely altered as well. Reservoirs trap the natural flow of sediment, so that the water that emerges from the spillway is without its normal sediment load. As a result, the stream erodes its channel, picking up a new sediment load. This gives rise to deep-cut channels with degraded plant, insect, reptile, and fish habitats (Figure 16.35). Eventually (usually within fifty years for smaller reservoirs), the reservoir basin fills with sediment and the dam is abandoned. In the United States, several thousand dams and reservoirs are classed as retired and typically are looked upon as environmental liabilities. Bigger reservoirs last longer, cover much larger areas, but of course they too have life-cycle limits.

The United States, Canada, China, and the former Soviet Union have built several hundred thousand dams in the twentieth century. Today, about 80 percent of the great rivers and their major tributaries in these countries are interrupted by dams and reservoirs. It bears repeating that the United States alone has built 75,000 dams 2 meters or higher since 1900. The map in Figure 16.36 shows the locations of major reservoirs in part of the American South. If we enlarge a small area on this map dozens of smaller reservoirs will also appear. If we count impoundments as small as farm ponds in our tally, the number of reservoirs in the United States probably totals 5 million or more.

While dam construction in the United States has declined significantly since 1970, in many developing countries it has increased rapidly. China has already built 18,000 dams greater than 50 meters high and is pushing ahead with many more projects. The Three Gorges dam project on the Yangtze River is being touted as the largest on Earth. Brazil is developing massive dams and reservoirs that threaten huge tracts of rainforest. This trend is sure to continue as the demand for power, irrigation water, industrial water, municipal water and flood control rise throughout the world, especially the developing world, in the twenty-first century.

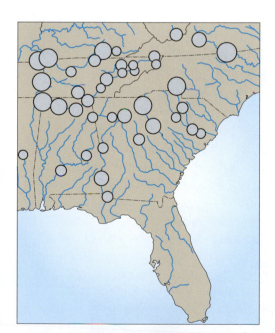

Figure 16.36 Some of the major dams in the American South. Since 1900, the United States has built 75,000 dams 2 meters or higher. However, dam building has largely ceased in the past 25 years.

Chapter Summary and Overview of Groundwater Systems, Lakes, and Water Resources

As the summary diagram suggests, water features such as streams, lakes, and wetlands together with the aquifers that support them, are part and parcel of the larger groundwater system, the Earth's second-largest reservoir of freshwater. Groundwater systems are fed with surface water, which is stored in aquifers for relatively long periods of time, and then discharged back to the surface. Streams and lakes are recipients of this discharge. The input to streams is the source of baseflow, the major part of stream discharge. Humans draw on groundwater and surface water for water supply, but the distribution of available water does not match the distribution of people and economic activity over much of the world.

▶ **Groundwater has been used as a source of freshwater for thousands of years.** But until the nineteenth century relatively little was known about this system and its place in the hydrologic cycle.

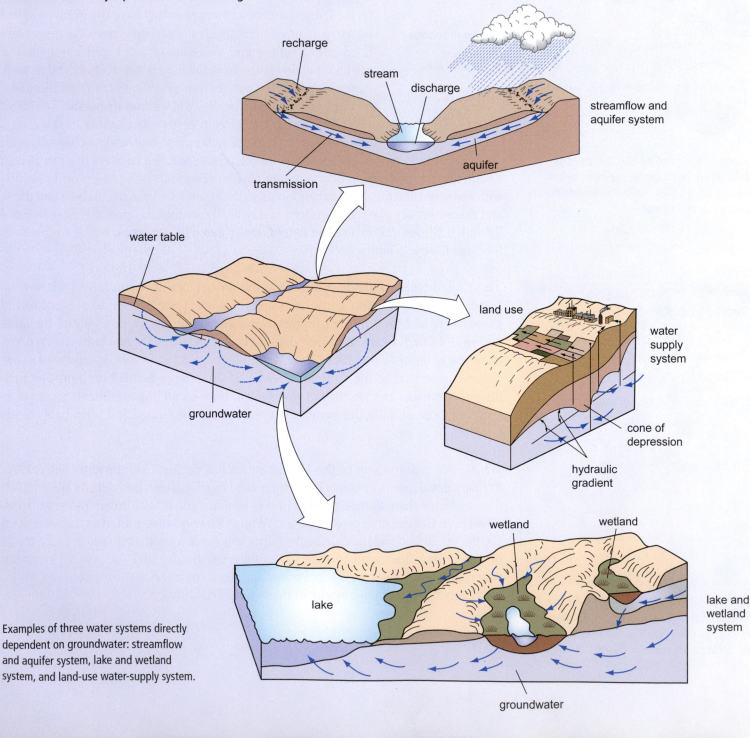

Examples of three water systems directly dependent on groundwater: streamflow and aquifer system, lake and wetland system, and land-use water-supply system.

▶ **Groundwater begins as surface water.** Input rates vary with climate and surface materials. Aquifer storage varies with porosity and flow rates with permeability.

▶ **Aquifers function as input–output systems.** Input is provided by recharge and output by discharge, which may take various forms, including artesian flow.

▶ **Groundwater basins are complex systems of aquifers.** Their size and complexity varies with the size and configuration of geologic formations.

▶ **Land-use activity can reduce and pollute groundwater resources.** The greatest impacts occur in urban and agricultural areas.

▶ **Groundwater is an important source of streamflow.** It provides baseflow for streams and is the source of underground streams in areas of karst topography.

▶ **Most inland lakes are supported by groundwater.** Some depend exclusively on groundwater whereas most other lakes depend on a combination of groundwater and surface runoff.

▶ **Wetlands share many hydrologic characteristics with lakes.** In the United States, wetland area has been reduced dramatically by land use, especially agriculture.

▶ **Worldwide agriculture is the single greatest user of freshwater.** It is also the greatest consumptive user ranking ahead of power generation, the public/domestic sector, and manufacturing in North America.

▶ **Water supply is a serious problem in much of the world.** Available freshwater is sure to become more limited in the twenty-first century, especially in developing countries.

▶ **Most great rivers in the northern hemisphere are interrupted by dams and reservoirs.** The United States and Canada alone have built thousands of dams and millions of small ponds since 1900.

Review Questions

1 Can you name the nineteenth-century advances that led to greatly improved understanding of groundwater?

2 What is the difference between capillary water and gravity water? Between porosity and permeability? Why do some materials have high porosity but low specific yield?

3 What is the water table and how can it be used to illustrate the concept of hydraulic gradient? What are the two main factors in Darcy's law?

4 What are typical groundwater velocities and how is it that groundwater discharge can be high when velocity is low?

5 Can you define the three basic processes of an aquifer system and describe the concept of dynamic balance in an aquifer?

6 Can you name the two basic classes of aquifers and sorts of materials that commonly are found in each?

7 What is an alluvial aquifer, an artesian flow, and a groundwater basin? What controls the size and shape of a groundwater basin?

8 How do cones of depression form and what causes the process known as upconing; what are the causes of the two main trends of groundwater levels in the Central Valley of California?

9 What are the principal sources of groundwater pollution and what is the origin of leachate?

10 Can you define the relationship between groundwater and streamflow and name the sort of ground materials that produce substantial baseflows in streams? What is karst drainage and how can it lead to springs?

11 What are the seven main types of lakes? Which are the largest and which are the most variable in area and what accounts for this variability?

12 How do wetlands differ from lakes and what are the principal sources of water in different types of wetlands?

13 What has caused the vast majority of the loss of wetlands in North America during the past 50 years?

14 What land-use activity is responsible for most water use in the world? In the United States? What is the difference between consumptive and non-consumptive water use and what are the degradational effects of major water users?

15 How have the recent (since 1970) shifts in population within the United States changed the usage patterns of freshwater?

16 Why is there such a strong desire on the part of humans to build dams and divert streams? What are some of the environmental consequences of such actions and what is the expected trend of water use in the twenty-first century?

Earth's Internal System:
Heat, Convection, Rocks, and the Planet's Skin

Chapter Overview

This chapter takes us into the solid Earth and the system of forces that shapes the Earth's crust, the platform of rock upon which the landscape and the oceans rest. This is important to physical geography for crustal processes and features exert a powerful influence on surface systems including oceanic and atmospheric circulation, climate, biogeography, and even the location and operation of watersheds. The story begins in the Earth's interior, the source of energy that powers the crustal processes like volcanoes and earthquakes, and how that energy makes its way toward the planet's surface. To answer this question, it is helpful to take a brief look at Earth's early formation as a planet. This will explain Earth's gross composition and structure and how the internal energy system is constructed. We then go on to examine how the great internal energy system works, how it brings heat to the surface and causes the crust to break apart and move in great sections called tectonic plates.

Introduction

All the great systems that we have discussed so far in this book, including the hydrologic cycle, atmospheric pressure and winds, ocean circulation, and ecosystems, are powered by solar radiation. It is now time to look inside the Earth and examine the system driven by Earth's internal energy system. The term traditionally given to this source of energy is **endogenous**, that is, having origin within the planet. You will recall that the term **exogenous** is used to describe the solar energy source.

The endogenous energy system drives the processes that shape the rock formations of Earth's crust. They include the mountain-building processes of volcanism, faulting, and folding, and the larger-scale processes of plate tectonics that move the continents, change the size and shape of the ocean basins, and control the distribution of most earthquakes and volcanoes. It takes a huge amount of energy to drive these processes, and for decades Earth scientists have investigated both its sources and the mechanisms that move it from the inner Earth to the crust and the Earth's surface.

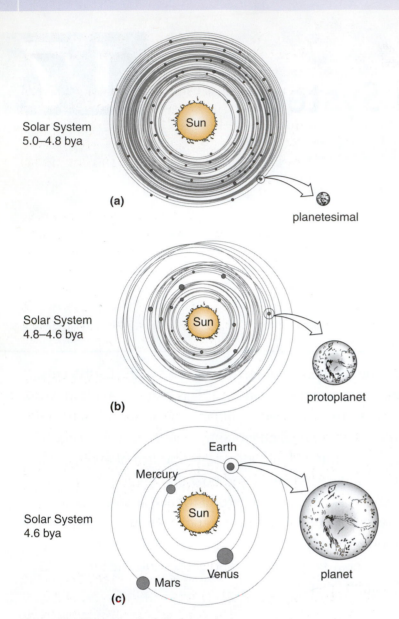

Solar System
5.0–4.8 bya

(a)

planetesimal

Solar System
4.8–4.6 bya

(b)

protoplanet

Earth

Mercury

Solar System
4.6 bya

Mars

Venus

planet

(c)

Figure 17.1 The proposed origin of the Solar System beginning about 5 billion years ago. By 4.6 billion years ago (bya), the terrestrial planets have formed and are in motion around the Sun.

This is no easy task, because it is difficult to measure almost any aspect of the inner Earth. We are able to drill only a short distance into the Earth's rock skin, only 3000 meters or so. Thus, we have never seen the rocks that make up over 99.99 percent of the planet. When a 2008 drilling project in Hawaii, which was looking for sources of geothermal energy, found molten rock pouring out of a 3000-meter- (9,800 feet-) deep drill hole, Earth scientists excitedly described it as a find tantamount to discovering a Jurassic Park dinosaur because no one in the history of science had ever seen fresh molten rock discharged directly from that depth.

How, then, do we learn about the deep Earth and, more importantly, why do we need to know about deep environments anyway? First, we need to know because that is where the forces driving the movement of continents, the opening and closing of ocean basins, and the building of mountains are rooted, and second, since we cannot dig or drill our way to great depths, we must instead rely on various remote-sensing technologies to generate data. The most reliable of these is seismology, which involves sending energy waves into the Earth and, based on the measurable behavior of these waves, differences in rock types, layers, and physical properties can be detected at different depths.

17.1 Origin and Development of the Terrestrial Planets

The story of the Earth's internal energy system actually begins with the formation of the Solar System. You will recall from Chapter 3 that our Solar System is located on the edge of the Milky Way Galaxy, which contains billions of stars arranged in the form of a great spiraling pinwheel. Our Sun is a modest-sized star that formed about 5 billion years ago as the debris from a previously exploded star gathered to form a nuclear mass of some sort. Modern astronomers tell us that the debris lay in a huge plane called a **nebular disc**, which revolved around the young star. The debris nearest the Sun was drawn into it and as the Sun grew, its gravitational field also grew and pulled the disc inward causing it to revolve faster.

This was the early Solar System, the phase that gave birth to the planets. Over the next 500 million years, two groups of planets would take shape: (1) small, rocky or **terrestrial planets**; and (2) large, gaseous or **non-terrestrial planets**. The terrestrial planets, Mercury, Venus, Earth, and Mars, formed close to the Sun by a process broadly described as **accretion**, that is, from the accumulation and concentration of debris.

Accretion and Planet Formation: The debris that went into the terrestrial planets came from the nebular disc and was probably much the same stuff we find in the modern Solar System, but it was much more abundant. It included clouds of gases, streams of dust, and chunks of rock and ice. Around 4.7 to 4.8 billion years ago, a series of changes took place, which are summarized in Figure 17.1, that led to the modern Solar System:

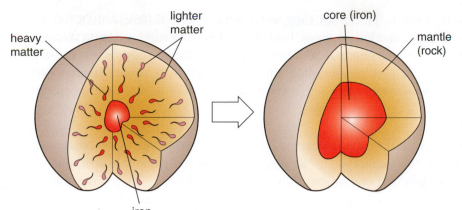

Figure 17.2 A diagram portraying Earth's internal meltdown and reorganization to form the core and mantle about 4.5 billion years ago.

Figure 17.3 A cross-section of Earth illustrating the sorts of processes that dominated the planet in its early days including widespread volcanism, internal convection, and extraterrestrial bombardment. The graph shows the estimated change in Earth's heat output over 5 billion years.

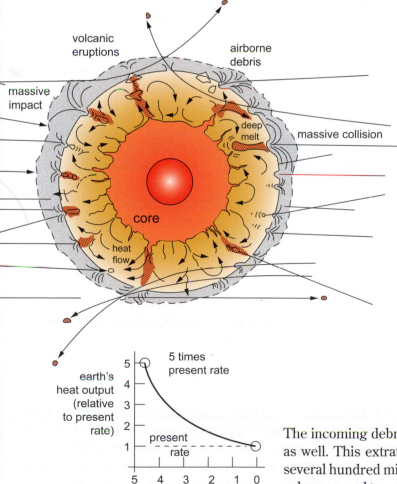

- First, this debris began to coalesce and condense, forming bodies about 10 km in diameter, called **planetesimals** (17.1a).
- Next, the planetisimals combined into still larger bodies called **protoplanets**, which were roughly the size of our Moon (17.1b).
- And finally, the protoplanets were drawn together by gravity to form each of the terrestrial planets (17.1c).

While all this was going on, massive showers of debris rained and crashed onto the new planets, adding both mass and heat to the young planets. But the manner of growth was so disorderly that the planetary compositions were little more than jumbled accumulations of debris. Then, about 4.5 billion years ago, another big event came into play, this time on the insides of the young planets. Internal temperatures got so high, around 7000 °C or more, that the interiors melted and collapsed, resulting in a massive reorganization of matter along the lines illustrated in Figure 17.2.

The heavier elements like iron sank inward driving lighter elements like oxygen upward. This gave rise to a high-density, iron-rich *core* which is still hot, about 6000 °C near the center. Around the core formed a larger sphere of rock, the *mantle*, composed chiefly of silicon and oxygen with densities roughly half those of the iron-heavy core and several thousand degrees cooler. At this stage in their formation the terrestrial planets (after the melt-out and internal reorganization) achieved *gravitationally stable configurations*, that is, they developed roughly stratified structures in which matter was organized according to density from the inside out.

Physical Geography of the Early Terrestrial Planets: The planet surfaces must have been exceedingly violent places, as Figure 17.3 attempts to shows. Not only did debris continue to bombard them, but volcanism driven by heat in the upper mantle was also abundant. Heat output from the Earth's interior was about five times greater than today (see the graph in Figure 17.3). And some of the extraterrestrial crashes and explosions must have been huge, judging from the size of the ancient impact craters that are visible today on Mercury and Mars. One of these was so massive that, according to a concept favored by many scientists, a chunk of Earth was jarred free which fell into Earth's orbit to become our Moon. The Moon's composition (based on rock samples brought back by the Apollo missions 40 years ago) is very similar to Earth's mantle, suggesting that it probably formed after Earth's internal reorganization had taken place, as early as 4.5 billion years ago.

The incoming debris not only brought in rock materials but huge amounts as well. This extraterrestrial water initially was vaporized and driven off several hundred million years of cooling, it combined with other gases relea volcanoes and together they led to an early atmosphere and hydrosphere. B 4 billion years ago, Earth's three main spheres, the lithosphere, hydrosp

Figure 17.4 An ancient lunar volcano. After a long period of early volcanic activity, the Moon cooled and became inactive.

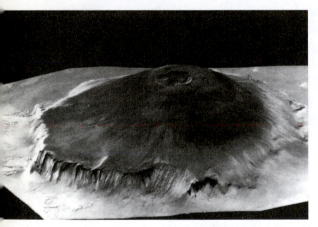

Figure 17.5 A large martian volcano which formed during the planet's active, early stage.

atmosphere, were in place, but they were considerably different than today. The atmosphere contained little oxygen, land areas were small and widely scattered, and the oceans covered as much as 90 percent of the Earth's surface. But it was a lively place with storms, precipitation, lightning, and plenty of volcanic activity. With the exception of ecosystems, all Earth's great systems were in place and operating by this time.

Mercury, Mars, and Venus developed similarly to Earth with cores, mantles, and volcanically active surfaces. The largest two terrestrial planets, Earth and Venus, are still internally hot and volcanically active and, although they have cooled significantly, their internal energy systems remain active today. On the other hand, Mercury, Mars, and our Moon, which are much smaller, cooled much faster in their first 1–2 billion years. As a result, their internal energy systems wound down, volcanic activity subsided, and they became more or less inactive geologically. Volcanoes on Earth are still spewing out millions of tons of lava every day, yet the youngest lavas detectable on the Moon are 3 billion years old (Figure 17.4). We expect that the same generally holds for Mercury and Mars (Figure 17.5).

The planets also developed different atmospheres and hydrospheres. At a distance of only 58 million kilometers from the Sun (Earth is 150 million kilometers), Mercury developed such high surface temperatures that it was unable to retain an atmosphere and hydrosphere. Water vapor and other gases broke down under the intensive heat – which today reaches 450 °C – and were driven off into space. Venus, which lies between Mercury and Earth at 108 million kilometers from the Sun, developed a carbon dioxide-rich atmosphere, which helped push its surface temperature above 400 °C, also too hot for water. Earth, at an overall surface temperature in the range of 10 to 20 °C (recall that today's global equilibrium temperature is 15 °C) retained much of its water with the vast majority held in great liquid oceans.

Mars, located about 80 million kilometers beyond Earth, also retained lots of its water but without much of an atmosphere. Today the Martian atmosphere is cold and thin – less than one hundredth the density of Earth's – and its water is frozen in the ground and in polar ice caps. But in the past, Mars was decidedly warmer at some time, or at various times, perhaps with an active hydrologic cycle including large areas of liquid water, runoff, and streams in some places, as the image in Figure 15.16 reveals. When and for how long and whether it included much evaporation and precipitation, we do not know. Nor do we know why it was warmer, though scientists postulate it was probably related to a denser atmosphere with abundant carbon dioxide.

17.2 Extraterrestrial Sources of Geographic Change

The system of accretion that led to Earth's early growth as a planet still operates in the Solar System, for each day Earth pulls in roughly 200 tons of space debris, mostly dust-sized particles and small meteorites. Although this is a tiny fraction of the tonnage that rained down in Earth's first billion years, it is indisputable evidence that the planet is still growing. Most of the debris today, as in the past, is too small to cause impact explosions. But the net effect of the accretion system, after billions of years of debris collection by all the planets, is to make the Solar System much cleaner overall and the planets a little bigger, of course. However, many large objects remain in the Solar System and, as we will see in the paragraphs below, scientific

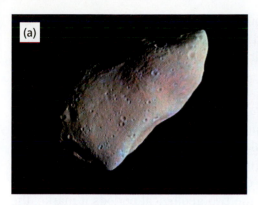

Figure 17.6 Two of the millions of asteroids that inhabit the Solar System as seen from NASA space craft: (a) Gaspra; and (b) Ida and its tiny moon Dactyl (in the circle).

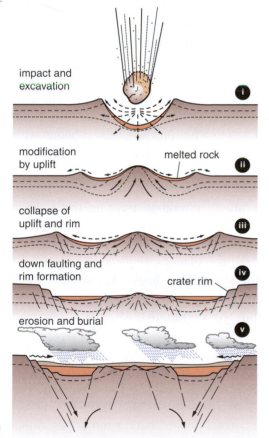

impact and excavation

modification by uplift   melted rock

collapse of uplift and rim

down faulting and rim formation   crater rim

erosion and burial

(a)

(b)

evidence indicates that the incidence of large planetary strikes has *not* declined in the past several billion years.

The Asteroid System: The reason large planetary strikes have not declined is that the Solar System still has a large supply of asteroids. Most lie in a broad, revolving band called the Asteroid Belt located between Mars and Jupiter. It contains millions of large objects like the ones shown in Figure 17.6, which are occasionally thrown out of orbit by the force of Jupiter's huge gravitational field, and some fall into the path of one of the planets. Another supply of asteroids lies beyond the Solar System and the orbits of many of these bodies cut across the Solar System. (Some icy ones form comets.) Telescopic studies reveal there are about 1000 objects larger than 1 kilometer in diameter whose orbits cross Earth's. Such objects weigh millions of tons and each is capable of creating a large impact explosion. The minimum mass needed to create an impact crater is only 350 tons; an object of 1 kilometer in diameter weighs roughly 2 billion tons.

We estimate that an object as large as 10 kilometers in diameter strikes Earth every 40 million years. That means in the past 4 billion years there have been about 100 such strikes, and in the past 400 million years – since *both* terrestrial and marine life have been abundant on Earth – our planet has received around 10 huge strikes. But most have probably hit the oceans which, depending on water depth, may have cushioned the blow. The three or four land strikes, however, would have created an explosion large enough to darken the atmosphere with debris and change climate and living conditions worldwide. So where are the craters from such impacts?

Asteroid Strikes: There is a basic problem in documenting extraterrestrial impacts on this planet: Earth quickly covers up the evidence. When a large asteroid hits Earth, it creates a rapid sequence of changes along the lines illustrated in diagrams i, ii, iii, and iv in Figure 17.7a, ending in a broad crater-like depression with a distinct rim, as shown in the lower two diagrams of 17.7a. Over Earth's long history thousands of these have left their mark. But unlike Mars, Mercury, and our Moon, Earth has few visible impact craters remaining because its surface processes have so been effective in obliterating them by eroding them down and burying them with sediment, as the last diagram in Figure 17.7a shows. It is one of Earth's distinguishing traits among the terrestrial planets of the Solar System. But with the help of satellite imagery, little by little scientists are finding evidence of ancient impacts of magnitudes capable of changing the global environment.

Two of the largest craters – or more accurately fossil craters – found so far are located at Sudbury, Canada, and Vredefort, South Africa. Each crater is about 2 billion years old (therefore, formed before terrestrial life existed) and each measures 140 kilometers (85 miles) in diameter. Another large one, about 100 kilometers (60 miles) in diameter, is the Popigai crater in Siberia. A fourth crater, which also measures 100 km across, and is clearly visible on satellite imagery in Figure 17.7b, is the Manicouagan

Figure 17.7 (a) Diagrams depicting the formation of an impact crater beginning with collision (i), followed by uplift (ii) and collapse of the central peak (iii) and down-faulting (iv) and erosion of the outer rim (v). (b) The outline of the Manicouagan crater in northern Canada. Hidden from view until the advent of satellite imagery, the original crater measured about 100 kilometers in diameter.

crater in Canada. The Popigai and Manicouagan impacts occurred within the past 250 million years, when both the oceans and the continents teemed with life, but little evidence has been uncovered to reveal their impact on the biosphere, though it was probably considerable and global in scale.

The K–T Explosion and Catastrophic Geographic Change: There is one huge impact, however, that we do think had a major impact on the global climate and the biosphere and altered the course of biological evolution. It occurred 65 million years ago and coincides with the eradication of most life on Earth. In popular circles it is known as the event that ended the age of reptiles and eradicated Earth's great dinosaurs. Named the K–T extinction (for the two geological periods it marks, the Cretaceous (K) and Tertiary (T)), this impact first came to light with the discovery of a thin layer of ancient sediment containing a heavy concentration of *iridium*, an element abundant in meteors. This layer has been documented in sedimentary rocks at numerous locations on Earth and the ages of samples from it coincide with the end of the Cretaceous period. Other lines of scientific evidence reveal an impact point for this asteroid in the Caribbean region of North America, and recently researchers located a huge crater, 200 kilometers in diameter, buried under sedimentary rocks in the Yucatan of Mexico, and it also appears to be 65 million years old.

The object responsible for the K–T explosion is estimated to have had a diameter of 10 kilometers or more. As the artist's sketch in Figure 17.8 suggests, the explosion from such a strike must have been enormous and we reason that it set off a chain of catastrophic environmental changes that went something like this. It began with a great cloud of debris blasted into the atmosphere, which plunged Earth into total darkness for at least several months.

Deprived of solar radiation, Earth's temperatures plummet, freezing the surface from the poles to the tropics. Hundreds of thousands of plant, animal, and microorganism species perish immediately. Snow spreads over the continents and ice cover expands on the oceans, which increase the planet's reflectance to sunlight thereby retarding the warming process as the atmosphere begins to clear. For most surviving organisms, food chains are broken, life cycles interrupted, and reproduction stopped. They, too, are eradicated within weeks, months, and years. The cold lasts 10 years or more.

Figure 17.8 An artist's depiction of a large asteroid (called a bolide) slicing through the Earth's atmosphere. One only 5 to 10 kilometers in diameter is capable of creating an explosion large enough to permanently alter the entire global environment.

This is probably followed by changes in atmospheric chemistry leading to acid precipitation from the conversion of nitrogen into nitrous oxide by the explosion. The acidic moisture reacts with limestone in the Earth's crust, releasing large amounts of carbon dioxide that, in turn, causes increased greenhouse heating and drives global climate into a warming trend. Global warming would bring another set of stresses to the planet including drought, causing further biogeographical change and more extinctions. In the end, as much as 75 percent of Earth's species are eradicated and the course of evolution is inexorably changed.

Whether this scenario describes what would happen now with a large extraterrestrial collision, we do not know for certain. We are certain, however, that environmental change would be catastrophic over most or all of the Earth. Would humans survive? Probably not. Millions would perish within hours. Our agricultural systems would collapse in a few days and global food reserves, which are typically less than two to three months, would quickly be depleted. Few, if any of us, would remain by year's end.

Summary on Earth's Origin and Early Change: Earth's beginning was basically the same as the other terrestrial planets up to about 4 billion years ago when it developed a water-rich surface environment. Earth's lithosphere, driven by internal heat, has remained active throughout the planet's life, while the interiors of our Moon, Mercury, and Mars long ago cooled and today these bodies are largely inactive. Asteriod collisions remain a threat on all the planets, with big strikes on Earth every 40 million years. The K–T explosion is widely believed to have caused the end of the age of dinosaurs, extinguished 75 percent of the planet's species, and forever changed the direction of biological evolution on Earth.

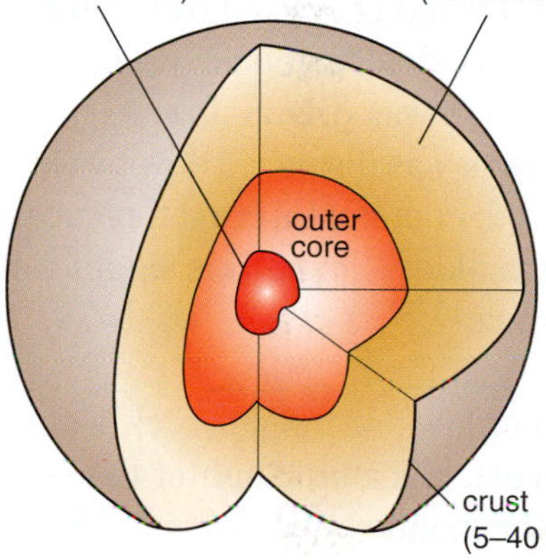

17.3 Earth's Inner Structure and Composition

We return now to Earth's internal system and a more detailed look at the products of differentiation – the sorting and layering of Earth's solid, liquid, and gaseous matter by density. Examine the cut-away diagram in Figure 17.9. The solid surface of planet Earth is traditionally defined by a thin layer of rock, barely visible in the diagram, called the **crust**. The crust forms the outer skin of the mantle where the ocean basins, continents, and mountain belts are nested. The **mantle** is a massive zone of dark, heavy rock that extends nearly 3000 km (1860 miles) into the Earth. Beneath the mantle lies the **core**, the densest part of the Earth, which extends an additional 3500 km (2200 miles) to the Earth's center. The mantle and core contain a massive quantity of heat, geothermal heat. Mantle temperatures reach nearly 4000 °C, and the core is even hotter. This geothermal energy drives the bending, fracturing, thrusting, sliding, and melting processes that deform the crust, cause earthquakes, and raise mountains.

The Crust: A traditional way to describe the crust is to divide it into two zones on the basis of rock types: ocean and continental. The **ocean crust** is composed of a heavy, dark group of rocks called **mafic** rocks which are rich in iron and magnesium. As we noted earlier, iron is a relatively heavy element and it imparts fairly high density to the ocean crust, that is, high mass per unit volume of rock (Table 17.1 provides some comparative density values). The densities of mafic rocks in the ocean floor generally range from 2800 to 3000 kilograms per cubic meter, 2.8 to 3.0 times heavier than the water in the overlying ocean. Throughout most of the ocean basins the crust is relatively thin, only 5 to 10 kilometers (3 to 6 miles) thick and dominated by one suite of mafic rocks, called *basaltic* rock, which are produced by volcanism on the ocean floor. On the edges of the ocean basins the basaltic rocks give way to the continental crust.

Take a look at Figure 17.10. Notice that the continental crust begins not at the coastline but in deep ocean water well beyond the shore. Here a long slope leads several thousand meters up the shoulder of continents and onto the **continental shelf**. From the continental shelf, the crust thickens toward the interior of the landmass and the mafic rocks of the ocean pass beneath the continental mass to form the lower crust. As the land elevation gets higher, the base of the crust gets deeper so that under the highest mountain ranges the crust reaches its maximum thickness, as much as 65 kilometers (40 miles).

The **continental crust** is composed of **felsic** rocks, a group of rocks that are lighter in both color and weight than the rocks of the oceanic crust. Felsic rock densities generally range from 2700 to 2800 kilograms per cubic meter because they are dominated by lighter elements, mainly silicon, oxygen, and aluminum, with smaller amounts of

inner core
(2900–6370 km)

mantle
(40–2900 km)

outer core

crust
(5–40 km)

Figure 17.9 The inner structure of the Earth: core and mantle covered by a thin crust.

Table 17.1 Densities of some common Earth materials*

Substance	Density, kilograms per cubic meter*
Iron ore	5000–5200
Basalt	2800–3000
Granite	2700–2800
Sandstone	2200–2700
Soil	1400–2000
Coal	1100–1400
Water	1000
Ice	919
Snow	150–450
Air	1.3

*The density of Earth materials is measured in kilograms per cubic meter. A cubic meter of water (1000 kilograms) is nearly as heavy as your car. Surface rocks are around 2.5 times heavier than water.

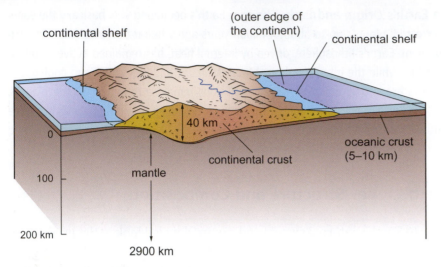

Figure 17.10 The crust is comprised of two main parts, continental and ocean basin, which are different in coverage, thickness, rock types, and density.

heavier elements. Granite is the representative rock type of the continental crust, and it is convenient to refer to rocks of this suite as **granitic** rock.

The Crust as a Rock System: Because both the oceanic and continental sections of the crust are decidedly lighter than the underlying mantle, they "float" on the denser mantle rock. But if the crust is part of the mantle, how do we explain its lower density? It is the result of hundreds of millions of years of differentiation as surface and near-surface rock has been worked and reworked by weathering, erosion, melting, and deformational processes in which low-density rock was driven upward into the crust while heavy rock was displaced downward into the mantle. But this process is far more advanced with continental crust than oceanic crust – partly because the continental crust is much older – and this is significant to understanding the crust's behavior and configuration.

Oceanic crust is formed directly from volcanic outflows of molten rock from the underlying mantle. When this rock solidifies as crust, it is a little lighter than upper mantle rock but still very dense. Thus, it floats on the mantle but, like ice in the sea, it is too heavy to float very high. Continental rock, on the other hand, is lighter because it has gone through many cycles of differentiation, in which it has been broken down, reconstituted, and reformed, thereby reducing its density by eliminating much of the heavier mineral content like iron.

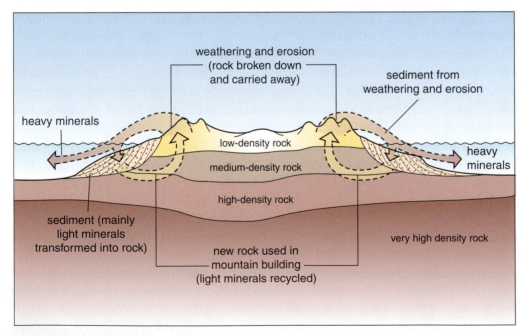

Figure 17.11 The general structure of a continent with lighter rocks on top and heavier ones below, and second, the general system of rock recycling involving mountain building, mountain weathering and erosion, and sediment build up on the perimeter.

This system is not unlike other Earth recycle systems, except that it operates very slowly. With each cycle, the fraction of heavier elements is reduced, as shown in Figure 17.11, and the remaining rock material is proportionally lighter. Because it is lighter, the continental crust is able to float higher on the mantle than the heavier oceanic crust. Evidence of the effectiveness of this recycling system is revealed by the fact that iron makes up only 5 or 6 percent of the crust but 47 percent of Earth as a whole.

Because of these differences in density, when deformational forces drive continental and oceanic sections of the crust against each other, they meet at offset elevations. The lower-riding and denser oceanic crust is inevitably pushed down, back into the mantle, whereas the continental crust is shoved up, more or less, onto the landmass. In the long run, continental rock is conserved by this process while oceanic rock is destroyed, a process we will explore in some detail in the next chapter. Therefore, if ancient rock is to be found anywhere on Earth, it must be within the continents and, indeed, this is the case, for the Earth's oldest known rock, about 4 billion years old, is found there. In startling contrast, the oldest rock in the ocean basins is less than 200 million years old.

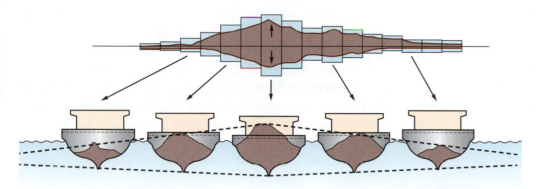

Figure 17.12 The concept of isostasy using a boat with changing loads as a model for a continent (above) with changing thickness from exterior to interior. As the continental rock mass grows thicker, it sinks deeper into the underlying rock.

The Buoyant Crust: The concept that the crust floats on the mantle is at the heart of a principle known as **isostasy**. This principle is based on the idea of buoyancy and is illustrated in Figure 17.12 by the model of a ship floating on the ocean. It floats because its mass is less than the mass of the water it displaces. The continents float on the mantle because their mass is less than the mass of the mantle rock displaced under them. And as a ship settles deeper into the ocean with the loading of cargo, so the continents settle deeper into the upper mantle with loading from the growth of mountain masses or buildup of great sheets of glacial ice, for example. Thus, as the continents gets higher and heavier, they also gets deeper and thicker.

But the reverse is also possible. As the crust is unloaded it rises as mass is removed, a process called **isostatic rebound**. Thus, the crust behaves as an elastic, meaning that it returns to its original elevation when a depressing force is removed. But the process takes place gradually. For example, in the north central part of North America, around Hudson Bay, the crust was depressed under a huge sheet of glacier ice about 20,000 years ago. Around 10,000 years ago, the sheet melted back and the crust began to rebound. In the past 8000 years, it has rebounded as much as 250 meters (800 feet) and it is still coming up. At Churchill on the southwest side of Hudson Bay, the crust is coming up nearly 60 centimeters (2 feet) a century and will continue to rise for thousands of years. This is geographically significant because as the crust rises, the water in Hudson is pushed back; eventually much of the Bay will disappear (see Figure 17.13).

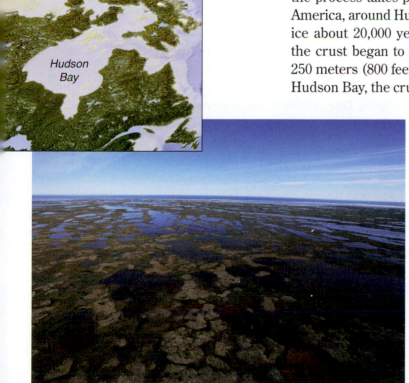

Figure 17.13 Emerging land on the perimeter of Hudson Bay, the result of isostatic rebound (the bay is barely visible). Taken to the extreme, the bay will be reduced to a small arm of the ocean in the distant future.

Probing the Mantle and Core: Most of our knowledge about the composition and structure of the lower crust, mantle, and core has come from analyzing seismic waves transmitted through the Earth. **Seismic waves** are energy waves and their basic character can be illustrated by using the analogy of a struck bell. When a bell is struck, energy waves radiate through the solid part of the bell and, if you are touching the bell, you feel these waves as vibrations. The waves are transmitted within the bell through the bumping together of trillions of atoms. As a wave travels along, successive atoms bump into their neighbors and then return to their original positions. Because the atoms always return to their original patterns after they move, these waves are called *elastic waves*. The elastic waves studied in the Earth are produced by sudden movement of massive quantities of rock in an earthquake or by artificial means such as an explosion from dynamite or a nuclear device. The branch of geophysics that studies Earth waves, or tremors, is called *seismology*.

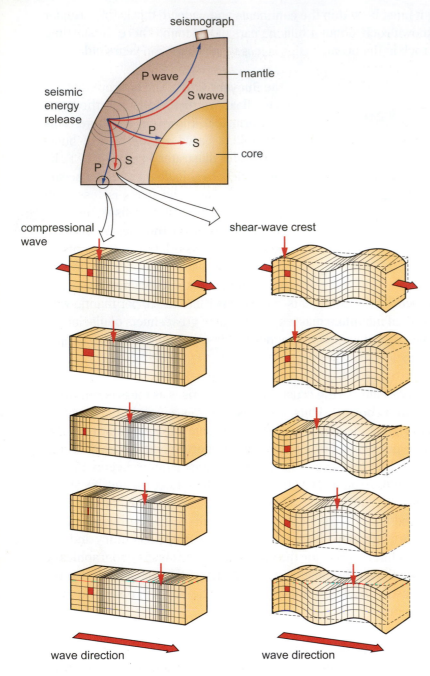

Figure 17.14 The concept of P-wave motion and S-wave motion. P waves create compressional and decompressional motion in the direction of energy propagation whereas S waves create shear motion perpendicular to the direction of energy propagation.

Seismic energy produces several types of waves, but two are particularly important in seismic analysis: **compressional waves** (or P waves) and **shear waves** (or S waves). These waves are distinguished by the pattern of particle motion they create in rock materials. Refer to Figure 17.14 and notice that as compressional waves travel through rock, they generate a push-and-pull or contraction–expansion motion in the direction of wave movement. (Sound waves in air behave similarly and they are also compressional waves.) Shear waves, on the other hand, cause neighboring particles to move in directions perpendicular, or transverse, to the direction of wave movement. Both P-wave and S-wave motions can be detected with devices called *seismometers* and recorded with an associated instrument called a *seismograph*.

By studying the behavior of seismic waves as they pass through a mass of rock, we are able to learn some things about its composition and structure. This includes:

(1) wave velocity (both P and S) generally increases with rock density;

(2) shear waves cannot penetrate molten masses because fluids have no strength against shearing forces (that is, against forces acting at angles crossgrain to each other); and

(3) when waves strike a boundary between two rock layers of different densities, part of the energy is reflected back to the surface and part of the energy is refracted (or bent) as it passes through the second layer.

Seismic Discoveries: The discovery of the base of the crust itself was accomplished through careful analysis of seismic waves. A Croatian geophysicist, named Andrija Mohorovicic (1857–1936) (pronounced Mo-ho-ro-vee-chich), found that wave velocities increased sharply at the contact between the crust and the mantle. This velocity break or discontinuity has come to be known as the M discontinuity or, more commonly, the Moho discontinuity. Originally thought to represent a major change in rock type, the Moho is now interpreted as something more modest, such as a change in crystal organization and packing among iron and magnesium minerals in response to heat and pressure.

Just as the base of the crust is defined by a seismic discontinuity, the remainder of the inner Earth is subdivided on the basis of changes in the velocity of seismic waves. Let us look first at the general picture and then examine some of the details. With one significant exception in the upper mantle, seismic wave velocities increase with depth all the way to the base of the mantle. If you look at the graph in Figure 17.15, you will see that at a depth around 2800 kilometers P-waves reach the highest seismic velocities recorded anywhere in the Earth, more than 13,000 meters per second.

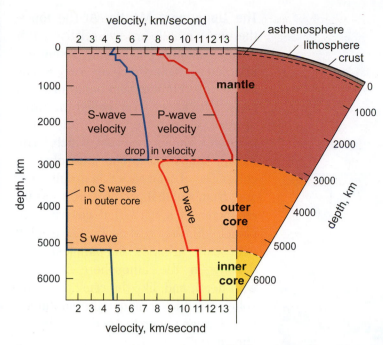

Figure 17.15 P-wave and S-wave velocity profiles from the Earth's surface to the center of the core. Notice the increase through the mantle followed by the abrupt and total loss of S waves and the abrupt decline of P-wave velocity at the outer core.

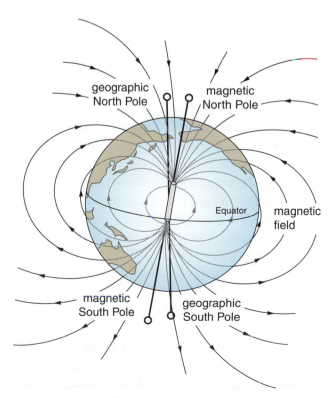

Figure 17.16 A model of the Earth's magnetic field is fundamentally the same as the force field surrounding a simple magnet. Notice the difference in the position of the magnetic and geographic north poles.

But at 2900 kilometers a sudden change takes place that marks the beginning of the core. Compressional- (P-)wave velocities fall by nearly 50 percent and S waves disappear altogether. Since we know that shear waves cannot be transmitted through liquids, the outer part of the core must be liquid rock. At the base of this zone, about 2200 kilometers below the mantle, another abrupt change is recorded. P waves speed up and S waves are again detectable, revealing the lower limit of liquid rock. Both waves continue with modest velocity increases to the center of the Earth.

Structure of the Core and Mantle: Thus, the core is made up of two units: a liquid **outer core** and a solid **inner core**. Together they constitute about 30 percent of the Earth's volume and about 40 percent of its mass. The mass–volume difference is explained by the core's composition of high-density iron, which near the Earth's center exceeds 12,000 kilograms per cubic meter. The Earth's magnetic field is generated in the outer core. The mechanism responsible is not fully understood, but scientists reason that it has to do with the generation of electric currents in the rapidly circulating liquid iron.

This system creates a weak force field in which the Earth functions like a huge magnet with a positive pole at one end and negative pole at the other. The magnetic poles, however, do not coincide with the geographic poles as the diagram in Figure 17.16 shows. The magnetic North Pole is located south of the geographic pole in northern Canada. In addition, for reasons unknown, the polarity of the field reverses every several hundred thousand years or so; that is, the positive and negative poles switch ends. Curiously, this has provided an important line of scientific evidence in documenting long-distance movement of the oceanic crust, which we will touch on in the next chapter.

Returning to the mantle, we find that it, too, is made up of two large divisions, the **upper mantle** and **lower mantle**; however, their differences are less severe than those separating the inner and outer core. In fact, the boundary between the upper and lower mantle is best described as a broad transition between 650 and 900 kilometers depth. Seismic wave velocities here increase in small jumps and geophysicists conclude that this is a response to a change in rock type. Under extreme pressure from billions of tons of rock overburden, iron and magnesium (in the form of two minerals, olivine and pyroxene, both major constituents of the mantle) undergo internal reorganization, taking on tighter, more compact crystal structures. The result is a rock called **perovskite**, which pervades the entire lower mantle. Although it is not found anywhere near the Earth's surface or within the crust, this high-pressure rock is decidedly the most abundant rock on the planet.

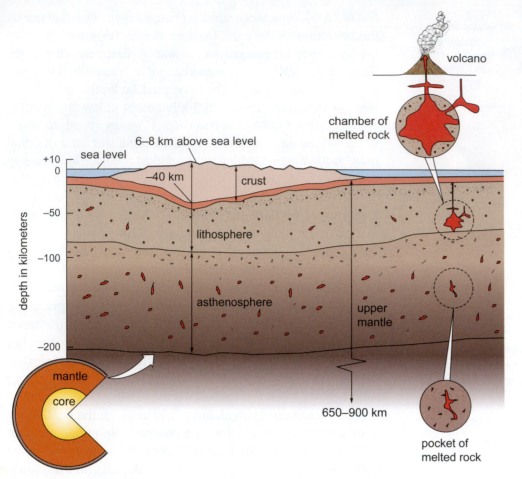

The Uppermost Mantle: At the top of the upper mantle are two relatively thin layers, together averaging about 200 kilometers thick, which are very important to the character and behavior of the crust. The upper layer, called the **lithosphere**, averages about 80 kilometers thick. It includes the crust, which forms its uppermost part. The lithosphere is composed of solid rock that is quite rigid and, from a global perspective, it has the character of a hard rind covering the planet with the crust as its skin (replete with scars, wrinkles, and pimples). Rock density in the lithosphere increases with depth and, like the crust, the lithosphere is thicker under the continents and thinner under the ocean basins, as shown in Figure 17.17.

The lithosphere rests on the second layer, called the **asthenosphere** (also shown in Figure 17.17), a layer of softer rock about 120 kilometers thick. On seismographs, the asthenosphere (from the Greek word *asthenes*, meaning weak) appears as a shadow or low-velocity zone

Figure 17.17 A schematic diagram showing the main components of the upper part of the mantle, namely the lithosphere and asthenosphere.

suggesting that the rock here is at least partially melted. Laboratory experiments simulating seismic waves in such a shadow zone reveal that the asthenosphere contains small amounts of liquid rock, perhaps pockets of mushy rock in a matrix of hot, solid rock.

The asthenosphere is widely recognized as a critical part of Earth's rock-moving system because it is capable of gradual flowing motion. The driving force behind the motion is heat arising from the underlying mantle and, as the asthenosphere moves, it sets the lithosphere above it into motion. The resultant movement is the root cause of crustal deformation as the lithosphere, partitioned into great sections or plates, shifts about on the Earth's surface. The *lithospheric plates* are known as **tectonic plates**, so named because they are the principal components of Earth's *tectonic* or mountain-building system. The question now before us is what do we know about the Earth's internal heat system and its capacity to move mantle rock, which in turn drives the tectonic plates and deforms the crust?

Summary on the Earth's Inner Structure and Composition: The early reorganization of the inner Earth left the planet with a high-density mantle, a higher-density core, and lots of heat. The thin surface layer of the mantle, the crust, is comprised of two major subdivisions, one very old and the other relatively young, both with the capacity to move up and down and recycle rock. We learn about the Earth's interior by analyzing the behavior of seismic waves, which has led to considerable understanding of the structure and composition of the core and mantle as well as insights into the driving force for tectonic plates. This leads to the question of the energy source behind this force.

17.4 Earth's Internal Energy System

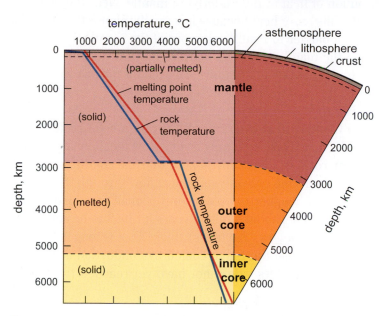

Figure 17.18 The general temperature profile from the Earth's surface to the inner core. The broken line plots the corresponding change in the melting-point temperature of rock. Where the melting-point temperature falls below the rock temperature, melting takes place.

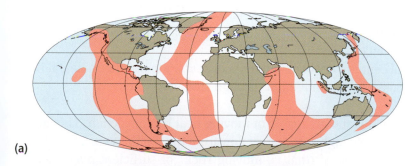

(a)

(b)

Figure 17.19 (a) Global heat flow from the crust showing the higher than average areas in red concentrated mainly along the mid-oceanic ridges (H. N. Pollack et al, 1993). (b) The photograph shows a geothermal field on Iceland, which lies on the Mid-Atlantic Ridge.

We have long known that heat is emitted from the Earth's crust. Hot springs and volcanoes, in particular, are indisputable evidence of this fact. What was not known until this century, however, is that heat is being continually emitted from every square centimeter of the crust's surface. The average rate of emission is about 50 calories per square centimeter per year, which is very small compared to the energy received at the Earth's surface from the Sun (more than 800 calories per square centimeter per day). However, when we multiply this small rate times the surface area of the Earth (510 million square kilometers), the resultant figure is enormous – about ten times the total energy generated and used by all of humanity each year.

Rock Temperatures at Depth: Measurements in deep boreholes reveal that the crust's temperature rises with depth at a rate of 2–3 °C per 100 meters. At the base of the crust, which is beyond the reach of most boreholes, it reaches 800 to 1000 °C. From this point, temperature continues to rise, though at a lower rate, reaching nearly 4000 °C at the base of the mantle. At the core–mantle boundary (see the blue line in Figure 17.18) the temperature rises quickly to about 5000 °C and then continues to increase at a slower rate through the outer core to the Earth's center.

At the Earth's surface the melting point of rock is about 700 °C but, with the exception of the asthenosphere and the outer core, the inner Earth is solid. Given that temperatures in the mantle and core are thousands of degrees, why is the Earth's interior not entirely melted rock? The answer is found in the relationship between the pressure exerted on a rock (by the rock mass around it) and its melting-point temperature. As pressure rises with depth into the Earth, the melting temperature of rock also rises. In the Earth's mantle and inner core, pressure increases slightly faster than rock temperature, thus melting points are always above rock temperature (see the red line in Figure 17.18). Therefore, no melting occurs. Only in the outer core does the rock temperature significantly exceed the melting point temperature, creating full melt-out. In the asthenosphere the two are close to equal which accounts for the semi-melted state of the rock there.

Sources of Earth Heat: What are the sources of Earth's internal heat? One early source was the heat generated with planetary accretion and the impacts of large asteroids. Another was the enormous charge of thermal energy produced by the conversion of gravitational energy into heat when the core formed about 4.5 billion years ago. Added to this reservoir of energy is the heat produced by the radioactive disintegration of the elements uranium, thorium, and potassium, which are contained in the rocks of the mantle, core, and crust. As for the distribution of heat, global heat-flow measurements reveal that the ocean basins actually discharge significantly more heat than the continents (Figure 17.19).

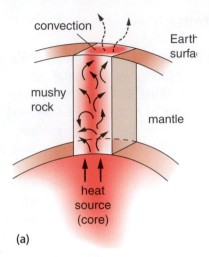

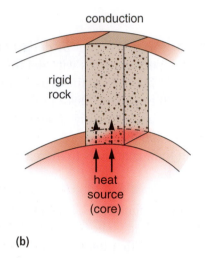

Figure 17.20 The difference in the rate of heat flow through a column of mantle rock by (a) convection, and (b) conduction. Convection requires rock movement, and even very slow movement greatly exceeds transfer through rigid rock by conduction.

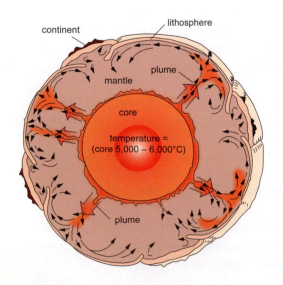

Figure 17.21 A cross-section of the Earth's core and mantle illustrating the convection system that likely operates in the mantle, a system which generates upflows in the form of plumes with enough force to move the lithosphere.

The reason for this is related first to the differences in the thickness of the crust; and second to the distribution of heat in the underlying mantle. Near the center of the ocean basins the crust emits more heat because it is much thinner than the continental crust and thus less a barrier to heat flowing out of the mantle. In addition, the heat-transfer system in the mantle generally delivers more heat to the oceanic regions of the crust than to the continental regions. To account for this we need to understand how heat moves through the lithosphere and mantle.

Heat Transfer Through the Mantle: Heat is transferred through solid rock masses like the lithosphere by conduction. This is not only a very slow mechanism of transferring heat, but the medium, rock, is an extremely poor thermal conductor. It is so poor, in fact, that if conduction were the only means of heat flow through the mantle, the amount of heat reaching the lithosphere would be only a small fraction of what it actually is. It would take, for example, about 5 billion years for heat to be conducted across a slab of rock only 400 kilometers (250 miles) thick. This is longer than the Earth has existed. In other words, had the Earth cooled by conduction alone, heat from depths greater than 400 kilometers would not have yet reached the Earth's surface.

Convection, on the other hand, is a much more efficient means of heat transfer as Figure 17.20 suggests. But convection requires heated rock to flow, and we know from tests conducted on surface rock that solid rock is extremely resistant to deformation by flow. But things are different in the mantle. Recall that the rock of the asthenosphere is partially molten and apparently capable of flow-type movement. Furthermore, tests show that:

(1) at very high temperature and pressure solid rock is capable of flowing motion at *very* slow velocities; and

(2) it takes only several centimeters of movement per *century* to produce a heat flow greater than that possible by conduction through stationary rock.

We thus conclude that the mantle *is* capable of moving heat by convection. This not only accounts for the relatively high *rate* of heat flow from the Earth's crust, but also establishes the presence of a *force* (that is, moving rock) of sufficient magnitude to move the huge tectonic plates. Since the early decades of the twentieth century, the debate over the meaning of a large body of evidence supporting the theory of plate tectonics often stalled on the question of the driving force responsible for plate movement. Because of the size of the lithospheric plates – for example, typically twice the size of the North American continent – it was difficult not only to find a mechanism powerful enough to budge them, but to produce sustained movement over distances of thousands of kilometers.

Mantle Plumes and Plate Movement: But there is now ample evidence that convective flows in the mantle are the driving force behind plate tectonics. Convective currents probably begin at the core–mantle boundary, as the graphic in Figure 17.21 suggests. Here heat concentrates and causes rock to expand and rise through the mantle in some sort of plume configuration. At least 20 of these plumes are currently operating in the mantle. One of these plume systems, called the *Equatorial Plume*

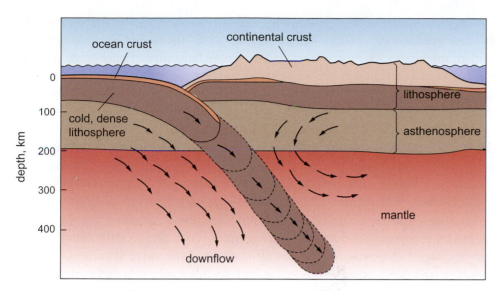

Figure 17.22 A section of lithosphere plunging into the mantle beneath a section of continent. This process is termed subduction.

Group, wells up under the Pacific where it produces volcanic hot spots such as Hawaii and Tahiti. Another, called the *Great African Plume*, rises under Africa and the eastern Atlantic where it feeds volcanic islands such as the Azores and Canary Islands. These and other plume systems not only transfer heat and nurture volcanoes but, as they spread out at the base of the lithosphere, they also split the lithosphere and drive it apart. When they cool with the loss of heat to the overlying rock, the currents of the plumes sink back into the upper mantle.

Just as hot regions in the crust are related to mantle plumes, cold regions in the crust are often related to sinks or downflows of cold rock such as shown in the diagram in Figure 17.22. But in the case of downflows, the lithosphere itself is dragged into the mantle and gravity plays an important role in this process. It is a powerful force that pulls with greatest effect on the heavier parts of the oceanic lithosphere. Thus, the difference in density between cold oceanic lithosphere and hot oceanic lithosphere can determine whether it floats or sinks as part of the mantle convection system.

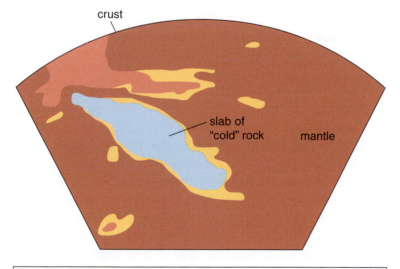

▢	faster P waves, indicating cooler-than-average matter
▢	average-speed P waves, indicating average-temperature matter
▢	slower P waves, indicating warmer-than-average matter

Figure 17.23 An image of a great tongue of subducted lithosphere produce by computer processing of seismic data. The subducted rock is colder than the surrounding mantle rock and therefore produces measurably faster seismic velocities.

As the lithosphere beneath the oceans cools over time with movement and distance from a plume system, it tends to ride lower and lower in the asthenosphere. Eventually its density exceeds that of the underlying asthenosphere and it begins to sink. Where the low-riding lithosphere jams against high-riding lithosphere, such as a continental plate, it breaks and it is driven downward. This process is called **subduction**, and it is capable of driving slabs of lithosphere deep into the mantle as the image in Figure 17.23 reveals. Subduction is one of the key processes in the **theory of plate tectonics** and it is one of several aspects of this intriguing system that we shall examine in the next chapter.

Summary on the Earth's Internal Energy System: The Earth's interior holds an enormous quantity of heat, which is released very slowly through the crust. The system that delivers this heat is facilitated by mantle convection in the form of great plumes that rise up under the lithosphere, melt it in certain places, and propel its movement. The highest rates of heat emission are found in the ocean basins where most of the lithosphere's movement begins.

17.5 The Rocks and Minerals of the Crust

From the question of the Earth's inner structure and the nature of its energy system, we move on to the question of the Earth's rock composition. Our primary concern is with the rocks that are found at or near the surface of the continents because these

Table 17.2 The eight most abundant elements in the crust by percent

Element	Percent
Oxygen	46.6
Silicon	27.7
Aluminum	8.1
Iron	5.0
Calcium	3.6
Sodium	2.8
Potassium	2.6
Magnesium	2.1

are the rocks that influence geographic phenomena such as soil formation, land-forms, and groundwater. Subsurface rocks, especially those that occur at depths of several kilometers or more, are of secondary importance in physical geography, for they are less instrumental in shaping surface environments, though they are often prominently displayed in volcanoes and in the remnants of mountain masses that have been razed by erosion.

Rocks are made up of minerals. Thousands of minerals are known to exist on Earth, however, only about 30 combine to form the rocks of the crust. Most of these rocks are relatively minor in terms of geographic coverage. In fact only eight basic rock types cover more than 90 percent of crust. The ocean basins are covered by essentially one major rock type, basalt. The continents show a greater diversity of rock types, but their coverage is dominated by one major rock group, sedimentary rocks, and among these shale occupies about 50 percent of the surface area of the land-masses. Let us briefly examine the major rock-forming minerals and then describe the principal types of rocks found in the crust.

The Building Blocks of Rocks: A **mineral** is defined as a solid, inorganic substance of natural origin with a specific chemical composition and a predictable crystal structure. Minerals are *chemical compounds* composed of two or more elements such as sodium (Na) and chloride (Cl) which form the mineral halite (or rock salt) or silicon (Si) and oxygen (O) which form the mineral quartz. Only eight elements make up nearly all the minerals found in the rocks of the crust and these are listed in Table 17.2.

Each mineral is unique because of its particular chemical composition and internal structure. Internal structure is determined by the arrangement of groups of atoms and it is defined by a distinctive three-dimensional crystal form. Figure 17.24 shows two crystal forms, quartz with a six-sided (hexagonal) crystal and halite with a cubical crystal.

In magma (molten rock), crystallization begins around 1000 °C and, for a particular mineral, it produces different-sized crystals depending on rate of cooling. Large crystals are the products of slow cooling, which can take place only at depth in the crust or mantle; whereas small crystals are the products of rapid cooling, which commonly occurs at the Earth's surface or under the sea. Some minerals cool so rapidly that they end up with no discernible crystal structure and are referred to as *glassy* or *amorphous* (without form), such as natural glasses formed during volcanic eruptions. Crystallization can also result from the evaporation of a mineral-rich solution such as salt and water.

Rock-Forming Minerals: There are five groups of rock-forming minerals, but only two are significant to our discussion. The primary group is the **silicate minerals**, which form the granitic and basaltic rocks, the principal constituents of the crust. The other group is the **carbonate minerals**, which are noteworthy for limestone, a major surface rock on the continents. Let us briefly describe the silicate minerals and the rocks derived from them.

Figure 17.24 The distinct crystal forms of minerals including the six-sided quartz crystal, and the cubical crystal of halite (salt).

The silicate minerals are built of a basic ion (group of atoms) of silicon and oxygen. To this ion are added atoms of other elements, notably aluminum (Al), potassium (K), sodium (Na), calcium (Ca), iron (Fe), and magnesium (Mg). All these atoms carry a positive electrical charge which enables them to bond with the silicon–oxygen

biotite

olivine

pyroxene

Figure 17.25 Samples of four common mafic minerals: biotite, olivine, pyroxene, and amphibole. Each is dark and relatively heavy. These specimens are 2–4 cm in diameter.

amphibole

ion because it carries a negative charge. Different minerals are formed as different combinations of elements bond with the silicon–oxygen ion and each mineral is distinguishable by its crystal form, color, density, and other traits.

The silicate minerals fall into two subgroups: mafic and felsic. **Mafic minerals** are rich in iron and magnesium and, like the group of rocks of the same name, they are heavy and dark like the samples in Figure 17.25. Four mafic minerals are prominent in the rocks of the crust: olivine, pyroxene, amphibole, and biotite. **Felsic minerals** are poor in iron and magnesium and rich in aluminum, sodium, and calcium, in addition, of course, to the core elements, silicon and oxygen. The four felsic minerals quartz, potassium (or orthoclase) feldspar, plagioclase feldspar, and muscovite mica, are decidedly lighter in appearance and heft than the mafic minerals (Figure 17.26).

The vast majority of the rocks in the crust are composed of these eight silicate minerals, but their distribution, as we noted earlier, shows some distinctive trends. First, the mafic minerals, represented by the basaltic rocks, dominate the oceanic crust whereas the felsic minerals, represented by the granitic rocks, dominate the continental crust. Second, the mafic minerals increase in abundance with depth into the crust, lithosphere, and mantle. A borehole sunk deep into the continental crust would reveal a sequence of change in rock types from those rich in quartz and feldspars in the upper 10 kilometers to those rich in olivine and pyroxene in the lower 10 kilometers. Both trends, of course, are the products of differentiation processes that have worked on rocks and minerals of the crust and mantle for millions of years, sorting them out mainly by density.

Figure 17.26 Samples of three common felsic minerals: potassium feldspar, quartz, and muscovite mica. Each is light-colored and relatively light in weight.

quartz

potassium feldspar

muscovite mica

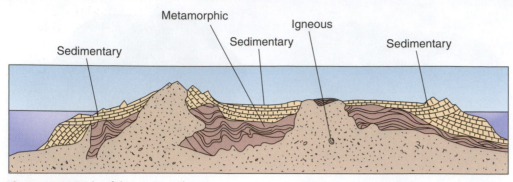

Figure 17.27 Rocks of the continental crust. Sedimentary rocks make up 75 percent of landmass volume, whereas igneous and metamorphic rocks make up 25 percent of the area and 75 percent of the volume.

Rocks of the Crust: All rocks are made up of some sort of mineral particles. They may be large crystals of quartz in solidified magma or minute crystals in volcanic glass. They may be deformed crystals of feldspar, bent and stretched from reheating and pressure associated with deformation of the crust. They may be particles of sediment, like beach sand, eroded from other rocks or they may be tiny chemical particles, such as calcium ions, precipitated from seawater. Each of these descriptions touches on the three basic criteria used to classify rocks: particle origin, particle composition, and particle size (or texture).

All rock belongs to one of three major families: igneous, metamorphic, and sedimentary. With the exception of the upper 1 to 5 kilometers of the continental crust, igneous rocks dominate the entire planet (Figure 17.27). On the landmasses, however, they represent only about 7–8 percent of the surface rock. Sedimentary rocks cover nearly 75 percent of the continents, and the remaining area, about 17–18 percent, is taken up by metamorphic rocks. But if we measure according to the volume in the continental crust it is igneous and metamorphic rocks that make up 75 percent of the continents.

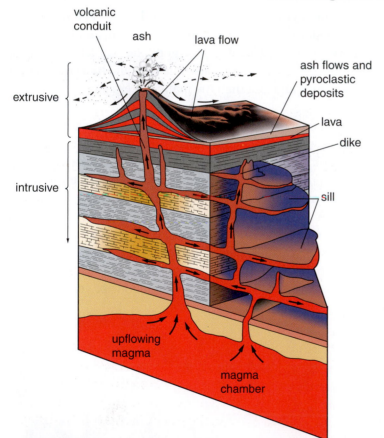

Figure 17.28 Environments of intrusive and extrusive igneous rock formation. Molten rock from deep chambers can be forced into crustal rock forming dikes and sills before extruding onto the surface.

Igneous Rocks: Igneous rocks (from the Latin *ignis*, fire) are composed of mineral crystals formed with the cooling of molten rock. Most igneous rocks form deep underground from magma that originates in vast chambers within the lithosphere and asthenosphere, called **plutons**. From these deep chambers, magma invades surrounding rock by melting through it, dislodging pieces of it, or by forcing open fractures. Figure 17.28 shows such a system. If the magma moves laterally among existing rock layers, the result is a horizontal, tabular feature called a **sill**. If it moves upward through overlying rock, the result is a vertical feature, roughly in the form of a wall, called a **dike**. Magma that has forced its way through pre-existing rock before solidifying below ground is classed as **intrusive** igneous rock. **Extrusive** igneous rocks, by contrast, form from magma that extrudes on the surface of the crust as lava.

If a dike or similar conduit reaches the surface, magma may be discharged onto the ground or ocean floor. The actual nature of extrusive discharges are surprisingly varied, ranging from lava flows that spill silently over the landscape to violent eruptions that blast particles of rock debris high into the atmosphere. All such igneous discharges are referred to as **volcanism** and the materials they produce vary widely in texture, density, form, and composition. The airborne debris from an explosive eruption, called **pyroclastic material**, is a mixture of dust, ash, and rock fragments like that shown in Figure 17.29a. Molten rock discharged as **lava** hardens into a wide variety of textures, densities, and colors ranging from **basalt**, which is dense and dark, to **pumice**, which is so porous (because of its high gas content during cooling) it floats on water. At the surface, the rock may occur in a variety of spectacular physical forms as well, such as the column, rope, and pillow forms shown in Figure 17.29b.

Figure 17.29 (a) Pyroclastic material, commonly referred to as volcanic ash (with a few scattered plants). The large rocks near the middle, called volcanic bombs, are thrown out with the ash.

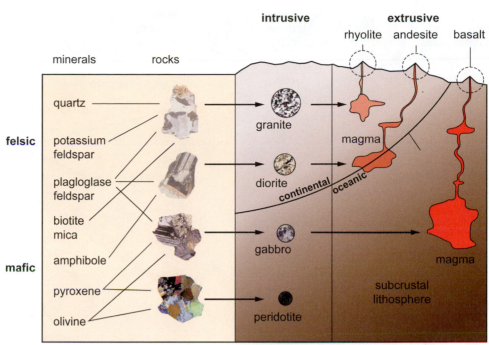

Figure 17.30 A diagram showing common intrusive and extrusive rocks formed from various combinations of felsic and mafic minerals. The diagonal line divides them into continental and oceanic groups.

Figure 17.29 (b) Curious rock forms from the solidification of lava: basalt columns, rope lava, and pillow lava.

The diagram in Figure 17.30 provides an easy way to organize the major igneous rocks. It cross-references them according to mineral content (mafic and felsic) and formational environment (intrusive and extrusive). By combining these two criteria we can identify two important trends: (1) coarse to fine textures; and (2) dark to light (that is, mafic to felsic) mineral composition and relate them both to depth of formation. The intrusive sequence runs from granite, the lightest, to diorite, gabbro, and peridotite, the darkest. **Peridotite** is a deep intrusive rock, darker than gabbro, which is composed exclusively of two mafic minerals, olivine and pyroxene. Peridotite is also described as *ultramafic rock*, and is too deep to have a common extrusive counterpart. However, granite, diorite, and gabbro each have an extrusive counterpart and their locations in the crust are shown in the diagram.

The composition of igneous rock depends mainly on the composition of the stock magma. If magma is 60 to 70 percent mafic minerals and 30 to 40 percent felsic, the intrusive (large crystal) rock type is called **gabbro**. If gabbro is extruded it becomes **basalt**. If the magma is about 60 percent plagioclase feldspar and 35 percent mafic minerals, the intrusive rock **diorite** or the extrusive rock **andesite** will be formed. Basalt and andesite together comprise almost 98 percent of all extrusive rocks. Basalt, as we have seen, dominates the ocean basins whereas andesite is commonly found on the continental margins. Andesite forms in subduction zones where oceanic crust is passing under the continents and is subject to melting and chemical alteration that favors development of feldspar.

If magma is overwhelmingly felsic in composition, the resultant rock is **granite** or its extrusive counterpart, **rhyolite** (see Figure 17.30). Granite is composed of 25–30 percent quartz and 50–60 percent feldspars with the rest in mafic minerals. It is the lightest of all intrusive rocks and makes up more than 90 percent of the igneous rock in the upper continental crust. Rhyolite is the primary extrusive rock in

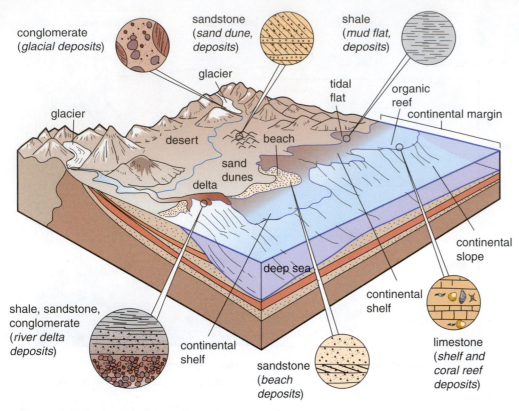

Figure 17.31 A schematic diagram illustrating various types of depositional environments and the sedimentary rock types associated with each. The bulk of the deposits and rock formation belong to the continental shelves.

continental lavas produced from granite-rich magmas. Basalt lavas are rare on land because they have to draw magma from very deep rock sources, near the Moho, or even deeper in the lithosphere or asthenosphere below.

Sedimentary Rocks: Any rock at or near the Earth's surface is subject to weathering and erosion. Think in terms of streams, ice, waves and wind as well as biological and chemical processes weakening and disintegrating rock, dislodging particles of various sizes, shapes, and compositions. From this action, sediment residues (such as clay and sand), biological debris (such as shell and bone fragments), and chemical residues (such as oxides and precipitates) are produced, which accumulate in terrestrial lowlands and basins, on the continental shelves, and on the ocean floor as shown in Figure 17.31. The heaviest accumulations occur on the continental shelves. As layers of sediment grow to thicknesses of thousands of meters, the lower layers are compressed, and in time consolidate and **lithify** (harden) into **sedimentary rock**.

Sedimentary rocks are divided into two main groups: detrital and chemical. **Detrital sedimentary rocks** are formed of particles that have been transported to their resting place by erosional processes, mainly streams. Clay, silt, sand, and pebbles are the most common constituents of detrital rocks. These particles provide the key identifying feature for these abundant rocks, for example, sand in **sandstone**, clay in **shale**, and particles of all sizes in a rock called **conglomerate**. As we noted earlier, shale is the most abundant surface rock on the continents. Sandstone is the second most abundant, occupying 15 percent of the surface area of the landmasses.

Chemical sedimentary rocks are produced mainly by the precipitation of minerals out of water. **Limestone** and **dolomite** are the most abundant chemical rocks, comprising 7 percent of surface rock of the landmasses. Most limestone (calcium carbonate) is formed from (1) animal bodies and shells composed of calcium, which these creatures originally extracted from seawater; and (2) calcium precipitated (a chemical process) directly from seawater. These two modes of limestone formation often occur together. As for dolomite, its origin is disputed, but it is probably an altered form of limestone in which some of the calcium has been chemically replaced by magnesium.

Also included in the chemical group of sedimentary rocks are rocks called **evaporites**. As the name implies, these are the residues left after the evaporation of mineral-rich water. Various kinds of salts, including rock salt, gypsum, and borax, are the most abundant evaporites. There are many places in the world where evaporites are now forming, all in arid climatic zones. In North America, there is no better example than the basin of Great Salt Lake in Utah, and in Asia the Persian Plateau

Figure 17.32 (a) Salt flats in the desert basin of the Persian Plateau of Iran) and (b) mining the massive salt formation under the city of Detroit.

of Iran provides equally good examples (Figure 17.32a). There are also thick formations of salt found at scores of locations deep underground, which date back tens and hundreds of millions of years. One of these, which underlies much of the Lower Peninsula of Michigan, is shown in the photograph in Figure 17.32b.

Another sedimentary rock of the chemical group whose origins are associated with wet environments is coal. **Coal** is a hydrocarbon compound, which forms largely from the buildup and burial of organic matter in ancient swamps and marshes. Coal is classified into three types according to its stage of development in terms of hardness and purity. **Lignite**, a soft brown coal, is least developed, that is, least compact and compositionally pure. It is generally considered to represent an intermediate state between **peat**, the partially decomposed organic accumulations found in bogs and lakes, and true coal. **Bituminous** coal, also called soft coal, is harder than lignite as a result of compaction from deep burial. **Anthracite** is the hardest and purest form of coal. It develops where bituminous coal seams are subject to heat and pressure in areas of mountain building.

Anthracite is the best source of fuel because it burns relatively cleanly with little air pollution. The coal fields of the Appalachian Plateaus and the Midwest, shown in Figure 17.33, comprise the largest body of coal reserves in the world. Appalachian coal, however, is largely bituminous, and is the main energy source for power generation in the eastern United States. The air pollution produced from this coal is the chief cause of acid rain in southeastern Canada and northeastern United States. Because coal deposits are also sinks for the heavy metal mercury, coal combustion is also the major source mercury pollution in the oceans.

Figure 17.33 Major coal fields of North America. Those of the Appalachian Plateaus and US Midwest rank among the largest coal reserves in the world and are the source of power for most electrical plants in the eastern United States.

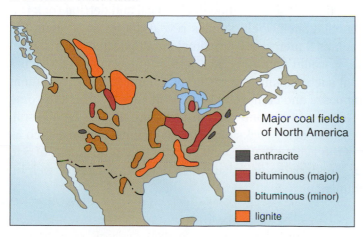

Major coal fields of North America

- ■ anthracite
- ■ bituminous (major)
- ■ bituminous (minor)
- ■ lignite

Building the Continental Shelves: Most sedimentary rocks form under the shallow waters on the continental shelves, where huge quantities of detrital sediment (billions of tons annually) are brought to the sea by rivers. Examine the drawing of the continental shelf in Figure 17.34 and note the rivers entering the ocean carrying sediment. As they slow down they drop their sediment loads, generally according to sediment particle size, largest first. Pebbles and sand are deposited nearest shore. We see them concentrated in beaches throughout the world.

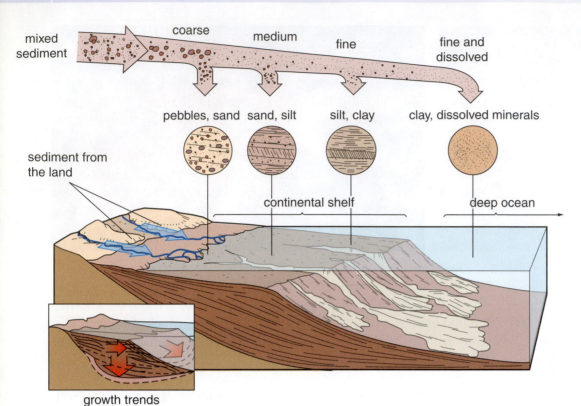

Table 17.3 Chemical composition of seawater

Mineral	Parts per thousand
Chloride	19.3
Sodium	10.6
Sulfate	2.7
Magnesium	1.3
Calcium	0.4
Potassium	0.4
Bicarbonate	0.1
Minor constituents	0.1
Total	**34.9**

Figure 17.34 A diagram of the continental shelf illustrating the general sequence of sediment deposition and the long-term growth trends of the shelf (inset).

Clay particles, on the other hand, which are less than 0.002 millimeters in diameter and can remain suspended in seawater for many days or weeks, are often dispersed by currents far out on the shelf and beyond into the deep ocean. The dissolved minerals discharged with clastic sediment in river water are dispersed even further, mixing into the larger system of deep ocean water. The chemistry of ocean water (see Table 17.3) is a combination of minerals weathered from land and minerals released through volcanic activity (outgassing) on the ocean floor.

The continental shelves are also areas of abundant marine life because light, heat, and various forms of plant and animal nourishment are concentrated there. In fact, the remains of shell creatures and other organisms constitute the primary source of sediment on the continental shelves. Great colonies of calcium-rich corals build up in tropical seas forming limestone reefs and platforms such as the one shown in Figure 17.35. Taken together, the mix of chemical and clastic sediment piling up on the continental shelf can be extremely varied with sand, clay, shell fragments, mushy calcium ooze, and other stuff interbedded in thin and thick layers.

The rate of sedimentation leading to the formation of chemical and detrital rock can be very rapid on the continental shelf. Field studies reveal that in bays along the coast, the rate can be as high as 2 meters a year. Simple multiplication tells us that even at one-tenth this rate, it would take only 5000 years for sediment to reach a thickness of 1 kilometer. But such thicknesses are greatly reduced as the sediment builds up, because it decomposes and compresses under its own weight, and some washes farther out on the shelf and beyond (see inset Figure 17.34). As a result, after millions of years, the sedimentary rocks on large continental shelves, such as that off the Atlantic coast of the United States, may grow to a total thickness of 5 to 6 kilometers. As it does, the underlying crust subsides under the massive weight, making room for more sediment.

Figure 17.35 Part of the Great Barrier Reef of Australia, the largest active limestone (coral) reef in the world.

Metamorphic Rocks: Heat can have a profound effect on the physical and chemical character of any Earth material from clay in the potter's kiln to rock next to a pluton deep in the lithosphere. When rock is subject to high heat and pressure, as well as changes in its chemical environment its mineral and crystalline makeup can be altered even though it does not melt. The result is **metamorphic rock**, or "changed" rock, in both igneous (meta-igneous) and sedimentary (meta-sedimentary) rock.

Because the magma that invades the crust is several hundred degrees hotter than the crustal rock, a great amount of heat is transferred into the resident rock, or *country rock*. Coupled with the pressure exerted on it by the moving magma, the country rock's crystal structure, mineral composition, or both may be altered. In granite, for example, metamorphism produces a realignment of the minerals into bands, and the resultant rock is called **gneiss** (pronounced "nice"). Although meta-igneous rocks are common, most of the metamorphic rocks found on the surface on the continents are derived from sedimentary rock.

When sedimentary rocks are subject to metamorphism they generally become denser and harder. In some cases texture also changes as the rock becomes less granular and sometimes takes on a crystalline character. The metamorphism of sandstone into **quartzite**, for example, results in a welding together of the sand grains to the extent that individual grains are difficult to distinguish. In limestone, metamorphism produces **marble** which is often crystalline, and shale becomes harder when it metamorphoses into **slate**.

Summary on Rocks and Minerals of the Crust: In the beginning all rocks on the Earth's surface were either heavy, dark mantle rock or lava and other forms of volcanic debris. But with the formation of continents, rocks formed that looked nothing like mantle rock, because they had been subjected to weathering and erosion and then reconstituted and recycled. Not once but many times, until new varieties of rocks formed, lighter in weight and color than mantle rocks. Central to this system are the continental shelves where most sedimentary rocks are formed from sediment contributed by rivers, biota, and chemical processes.

17.6 The Rock Recycling System

Rocks, too, are subject to recycling processes but, unlike air and water, the rate with rocks is painfully slow. When viewed in the context of Earth time, however, the work produced by the **rock recycling system** is enormous. It began in the Earth's early years when the crust was dominated by mafic igneous rock and there were no continents. Water probably covered most if not all of the planet. The first sedimentary and metamorphic rocks likely formed on the margins of volcanic island masses. But most of these rocks and the igneous rocks around them were subject to heating and convulsions in the newly forming crust, which forced melting and reincorporation into the lithosphere, the first stage of recycling.

The resultant system produced repeated cycles throughout Earth's early history, and each cycle advanced the separation of mafic and felsic minerals among the igneous rocks. When the Earth was less than a billion years old, it had given rise to small platforms of granitic-like rock or proto-continents. Driven by a mantle convection system that was more active than today's, the young continents were undoubtedly subject to frequent collisions, mountain building, and ongoing rock recycling, which left them structurally and compositionally complex bodies.

Above, a typical granite rock with its salt and pepper mineral pattern. Below, one metamorphic counterpart of granite, gneiss, with its banded pattern.

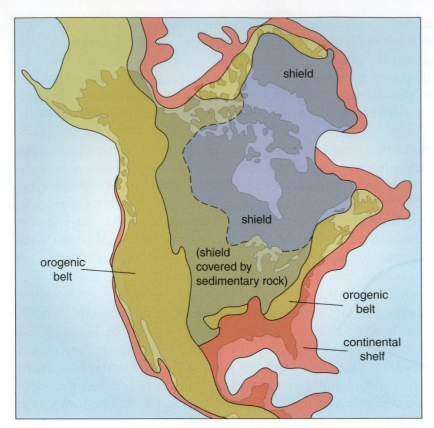

Figure 17.36 North America's three principal geologic regions. The oldest is the shield in the interior and the youngest are the continental shelves on the perimeter.

Geographic Differentiation of the Crust: In Figure 17.36 we see geographic evidence of the rock recycling system in the general distribution of rocks on the North American continent. The interiors or cores of the continents are composed of ancient (2 to 4 billion years old) igneous and metamorphic rocks. These core areas are complex rock masses called **cratons** and they are part of the continents' stable inner platforms or **shields**. The shields are fringed by belts of mountains called **orogenic belts**, composed of younger (100–500 million years old) metamorphic and sedimentary rocks built largely from sediment originally weathered and eroded from the older continental interiors, the shields. The orogenic belts in turn are in the process of breaking down and their sediment is going into the continental shelves, which are less than 100 million years old and still growing. The sediment of the continental shelves is transformed into sedimentary rock, which in time will go into the building of new orogenic belts on the continental margins.

The Mass Balance of Rock in the Crust: Earth's rock recycle system is still very much active with exchanges of rock between the crust and mantle, between the continents and the ocean basins, and within the continents. The crust operates as a two-cycle system, one cycle oceanic and the other continental, as the diagram in Figure 17.37 illustrates. The oceanic cycle is far and away the dominant exchange in terms of tons of rock moved. Input to the ocean basins from the mantle is estimated at roughly 65 billion tons of rock a year while the return flow from the crust to the mantle is estimated at 65–66 billion tons a year. The processes responsible for this exchange are: mantle convection, which drives magma from the

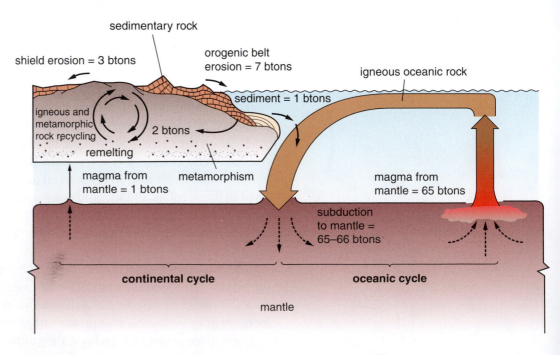

Figure 17.37 The rock recycle system involving exchanges of rock among the continents, ocean basins, and the mantle. The numbers represent best estimates of the amounts of rock moved among the various phases of the system over a year. The net balance between the crust and mantle is zero (btons = billion tons).

mantle to the ocean basins where it pours out building new crust/lithosphere; and subduction, which moves old crust/lithosphere from the ocean basins back into the mantle.

The continental part of the system is a bit more complex with lots of rock shifted around within the continental mass. On the surface, the older parts of the continents are losing huge quantities of rock mass to erosion, mainly by stream and river systems. We estimate that the shields give up 3 billion tons a year to erosion with the resulting sediment going into the formation of sedimentary rock on the continental margins. The orogenic belts give up 7 billion tons of sediment a year most of which goes into building the continental shelves, and eventually sedimentary rock. The continental shelves, in turn, lose about 1 billion tons of sediment to subduction and several billion tons to metamorphism associated mainly with on-going mountain-building in the orogenic belts. The 1 billion tons the continents lose to subduction is offset by 1 billion tons of magma input from the mantle into the continental crust. All the while, billions of tons of igneous and metamorphic rocks each year are continuously being rearranged, melted, and reformed within the continental interiors. But this is an internal recycling process, so the net balance is zero.

All told, the continents do a lot of rock recycling, about half of which involves erosion, sediment production, and the building of new sedimentary rock. Unlike the ocean basins little continental rock is exchanged with the mantle with only 1 billion tons of sediment lost to the mantle in exchange for 1 billion tons of magma input from the mantle. Based on the exchange rate between the crust/lithosphere and the mantle, the oceanic crust is recycled 65 times faster than the continental crust. Moreover, *all* of the ocean crust is subject to recycling with the mantle, whereas most of the continental crust is not. This explains the great difference in the ages of the rocks found in the bulk of the continents and oceans.

Chapter Summary and Overview of The Earth's Internal System: Heat, Convection, Rocks, and the Planet's Skin

Earth's internal energy system dates from the planet's formative period with the development of the mantle and core and their heat reservoir. The questions of the depth and composition of these bodies were answered through deep Earth seismic measurements. On the surface, global measurements of heat emissions from the crust helped point the way to convection as the principal heat-flow mechanism in the mantle. Mantle convection, illustrated in the summary diagram, is central to understanding movement of the lithosphere and deformation of the crust, especially the formation of continents and ocean basins and why these two divisions of the crust are so different in form, age, and rock types.

▶ **Earth's early development paralleled that of the other terrestrial planets in the Solar System.** Each grew by accretion as space debris crashed into and rained down upon the young planets.

▶ **Earth's internal energy system is closely tied to the planet's early formation.** A massive inner reservoir of heat powered an internal reorganization leading to the formation of the core and mantle. Much of that heat remains deep within Earth.

▶ **The K–T impact is associated with the extinction of more than half of Earth's species.** Earth still receives debris from space and large pieces hit Earth at an average of once every 40 million years, which are capable of making massive explosions and altering the global environment.

► **The surface of the solid Earth is defined by the crust.** This thin layer is made up of two divisions which are distinguished by location and rock types.

► **The crust floats on the mantle.** The elevation of the crust varies according to the principle of isostasy, which also helps account for the differences in the elevation of the continents and ocean basins.

► **We learn about the composition and structure of the inner Earth by analyzing seismic waves.** They reveal changes in rock density, phase (liquid–solid), and layering in the crust, mantle, and core.

► **The upper mantle is capped by two important layers.** One is a solid slab that includes the crust and is partitioned into huge sections or plates. It rests on a softer, partially melted layer.

► **Geothermal energy is continuously being discharged from the crust.** It is derived from two sources in the crust and mantle but its emission rate is unevenly distributed geographically over land and seafloor.

► **Heat is transferred through the mantle by convection.** The process is very slow and appears to take the form of great plume systems that originate at the mantle–core boundary.

► **The Earth's lithosphere is subdivided into tectonic plates.** They are capable of movement driven by motion in the mantle, and as they cool they are drawn into the mantle along their leading edge.

► **Relatively few minerals of the crust combine to form rocks.** Two groups of rock-forming minerals are significant in physical geography.

► **All rock belongs to one of three major families.** One group dominates the ocean basins and another dominates the surface of the continents.

► **The mix of felsic and mafic minerals changes from the continents to the ocean basins.** It also changes with depth into the continental crust from granite to basalt to diorite.

► **Igneous rocks are classed as intrusive or extrusive.** Sedimentary rocks are classed as detrital or chemical. Both igneous and sedimentary rocks have metamorphic counterparts.

► **The rocks of the crust are recycled by a system driven by the Earth's internal energy.** The continental part of the system involves very small exchanges of rock with the mantle. The oceanic part involves massive exchanges of rock with the mantle.

► **Among the long-term effects of the rock recycle system is the geographic differentiation of the continents.** The continental core contains the oldest rock, the orogenic belts are significantly younger, and the continental shelves are younger yet.

The Earth's internal energy system operates via mantle plumes to bring new rock to the ocean basins and enough force to drive the tectonic plates. Although the ocean basins are destined to destruction, the continents float high enough to escape subduction.

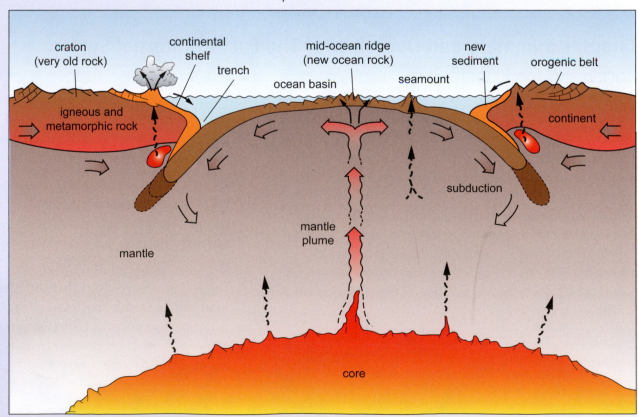

Review Questions

1   What is a nebular disc and what did it have to do with the origin of the terrestrial planets?

2   What was similar and dissimilar about the development of Earth, Mercury, and Mars as planets?

3   What evidence is there for past asteroid strikes on Earth, what were some of the geographic effects of big ones, and do such events remain a threat to the Earth today?

4   What are the key differences between the continental crust and oceanic crust on Earth?

5   What is isostasy, what does it have to do with the crustal elasticity, and the metaphor of a ship floating on the sea?

6   What are seismic waves and how have they helped us understand Earth's internal structure.

7   What role does the asthenosphere play in the mantle with respect to the movement of the tectonic plates?

8   What is the main difference between the transfer of heat through rock via conduction, and that via convection? Why is this difference in the two heat-transfer mechanisms significant in the Earth's internal energy system?

9   What are mantle plumes and, besides transferring heat, what influence do they have on the crust and the movement of continents and ocean basins?

10 What is a mineral and what are the major rock-forming minerals in the Earth's crust?

11 Distinguish between felsic and mafic minerals, give several examples of each, and relate them to the igneous rocks commonly found in the continents and ocean basins.

12 What would be the best way to distinguish between an intrusive and extrusive igneous rock?

13 What do the continental shelves have to do with the formation of sedimentary rocks and what are the principal sources of sediment that go into these rocks?

14 Describe the rock cycles of the continents and oceans. How has the continental rock cycle produced geographic differentiation of the continents?

The Formation and Geographic Organization of the Continents and Ocean Basins

Chapter Overview

Plate tectonics is the window to understanding the geographic arrangement of so many of the things we once took for granted when looking at maps of the world. Why the sizes and shapes of the continents and ocean basins and what about those large islands, chains of islands, and the great belts of mountains hugging the edges of the continents? We begin with a brief review on the development of the theory itself and then go on to describe the gross features of the Earth's crust to provide a geographic framework for the ensuing discussion. This discussion looks into the nature of plate motion, the conditions and features on the plate borders, and the processes that produce earthquakes and volcanoes. We end the chapter briefly examining a few of the many geographic implications of plate tectonics, in particular the distribution of marsupial and placental mammals and the climate and drainage of monsoon Asia.

Introduction

This is the story of plate tectonics, a theory that ranks with evolution as one of the monumental advances of natural science in the past two centuries. For many reasons, some we have already touched on and many that lie in the pages ahead, this concept is foundational to physical geography. Although the theory of plate tectonics has been with us for only several decades, the roots of this revolutionary idea actually go back several centuries or more. The main motivation for early ideas along this line seems to have come from the apparent puzzle-like fit of widely separated landmasses, suggesting that they had once been joined and then "drifted" apart. The remarkable coincidence between the coastlines of Africa and South America is, by itself, compelling.

In the twentieth century, the idea of geographically mobile landmasses became a subject of active debate in scientific circles. But one individual, a German meteorologist named Alfred Wegener (1880–1930), deserves credit for the first major treatise on the idea, which he called **continental drift**. In a book published in 1915, Wegener proposed that the present continents were once united in a single supercontinent that he called *Pangaea*. Pangaea, he reasoned, existed 600 to 225 million years ago and was centered on the African landmass. Over several hundred million years, the continents gradually drifted apart and, as they did, ocean basins formed between them, and the continents, in new locations thousands of kilometers from old

Figure 18.1 The breakup of Pangaea began about 225 million years ago and led to the formation of the Atlantic Ocean and current global distribution of land and water.

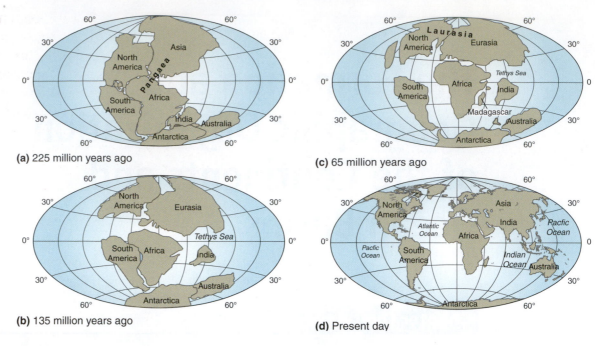

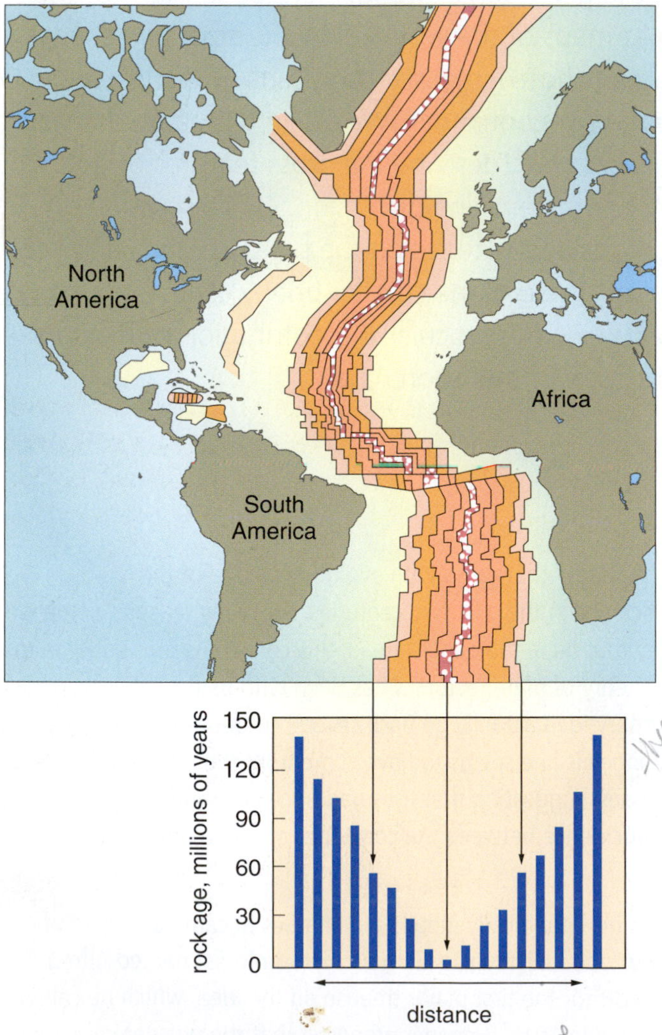

(a) 225 million years ago

(b) 135 million years ago

(c) 65 million years ago

(d) Present day

Pangaea, took on new climatic and biogeographical characteristics. The maps in Figure 18.1 show how modern scientists think this took place.

Although he relied on the puzzle-like geography of the continents as the cornerstone of his argument, he also drew on various lines of scientific evidence to support his case. Among the strongest evidence cited was that of striking similarities in rocks, fossils, and ancient deposits on continents now located in markedly different geographic zones. Similarly, the marks made by ancient glaciers on rocks in South Africa matched with corresponding marks in South America, India, and Australia. And the glaciers that made these marks would have to have formed in a warm climatic zone, unless, of course, the continents moved to their current locations long after the glaciers had disappeared. Then there was the issue on how the continents moved, and he reasoned that if landmasses could move isostatically (up and down) – a fact that was well established by 1915 – then why could they not move horizontally as well?

In the decades following 1915, advocates of continental drift refined and expanded Wegener's ideas. But two key questions remained unanswered. First, what mechanism is powerful enough to move a continent, and second, what firsthand (as opposed to coincidental) evidence is there documenting drift movement? As a result, continental drift remained largely speculative in the minds of most earth scientists. But evidence was mounting, and in the decades following World War II, the problem was cracked open. Seismological surveys, for example, revealed the plastic nature of the rock below the lithosphere, thereby opening the possibility that the rock there could move by slow flowing motion, and surveys of the ocean floors revealed that the rock there was very young, nowhere more than 200 million years old. Moreover, the rocks in the ocean floors were marked by curious, paired sets of magnetic bands made by reversals in the Earth's magnetic field and these bands suggested that the seafloors had spread outward from volcanic ridges in the middles of the ocean basins. And, later, when detailed data on the ages of ocean rocks were available, maps such as the one in Figure 18.2 showed a corresponding pattern of rock chronology across the ocean floors – young in the middle, older on the outsides.

Figure 18.2 Magnetic banding in the volcanic rocks on the Atlantic floor. Bands occur in paired sets and trend from youngest rock in the middle to oldest rock on the outside, suggesting that the ocean basins formed by spreading outward from the middle.

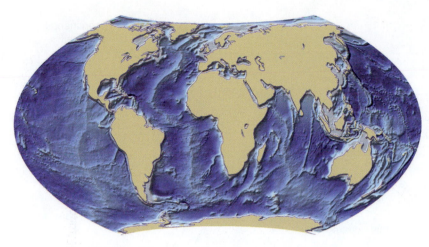

Figure 18.3 Naked Earth. Without its cloak of water, Earth's two main divisions, continents and ocean basins, are conspicuous at a glance.

By 1970, the idea of a geographically mobile crust was firmly in place. However, the idea of drifting continents was discarded because it had become apparent that the continents themselves were not the vehicles of movement. Rather, the continents are being carried around on the tops of larger sections of lithosphere or plate-like slabs in the manner of decks being carried around on a ship or barge. In addition, it became clear that most of the Earth's tectonic activity was taking place on the edges of the plates and the task of mapping the borders of tectonic plates involved little more than tracing the global patterns of earthquakes and volcanoes. What emerged from the twentieth century was a unified theory on a truly great system, one responsible for the formation, growth, and movement of the continents and ocean basins, the distribution of major mountain belts and most earthquakes and volcanoes, and a host of other geographical phenomena including the locations of many great rivers and their drainage basins, the distribution of many plant and animal species, and yes, even climate change.

18.1 Major Features of the Lithosphere

We set the stage with a brief look at the major topographical features of the lithosphere. We do this not only to set a geographic frame of reference for discussions to follow, but because it is virtually impossible to grasp this big idea without some knowledge of the lithosphere's major surface features. Begin with Figure 18.3 by imagining Earth stripped of its ocean water. Two major divisions of the crust are immediately apparent, continents and ocean basins. The continents cover about 40 percent of the Earth and ocean basins about 60 percent.

Besides geographic coverage and shape, the continents and ocean basins are different in other, more fundamental, ways. Continental rocks are not only older and lighter (both in weight and color) than rocks in the ocean basins, but the continental crust and lithosphere are thicker. And because of their lower rock densities, the continents ride much higher on the upper mantle. The graph in Figure 18.4 bears this out. The difference between the mean elevation of the continents, at 840 meters above sea level, and that of the oceans, at 3790 meters below sea level, is nearly 5000 meters. The mean combined elevation of the continents and ocean basins, that is, the average elevation of the entire planetary surface, taking into account the difference in the ocean–continent coverage, is 2610 meters below sea level. Given Earth's present volume of sea water, if the planet were transformed into a smooth sphere, it would be completely covered with a layer of water more than a mile deep.

Earth's area (hundreds of millions of square kilometers)
0
1
2
3
4
5
elevation (kilometers)
12
10
8
6
4
2
0
depth (kilometers)
2
4
6
8
10
12
elevation (miles)
8
6
4
2
0
depth (miles)
2
4
6
highest point =
Mt. Everest 8.85 km (5.5 mi)
mean land elevation
840 m (2760 ft)
sea level
mean planet
elevation 2610 m (8560 ft)
mean depth of sea
3790 m (12,430 ft)
lowest point =
Mariana Trench ~ 11 km
0
20
40
60
80
100
% Earth's area at this elevation or higher

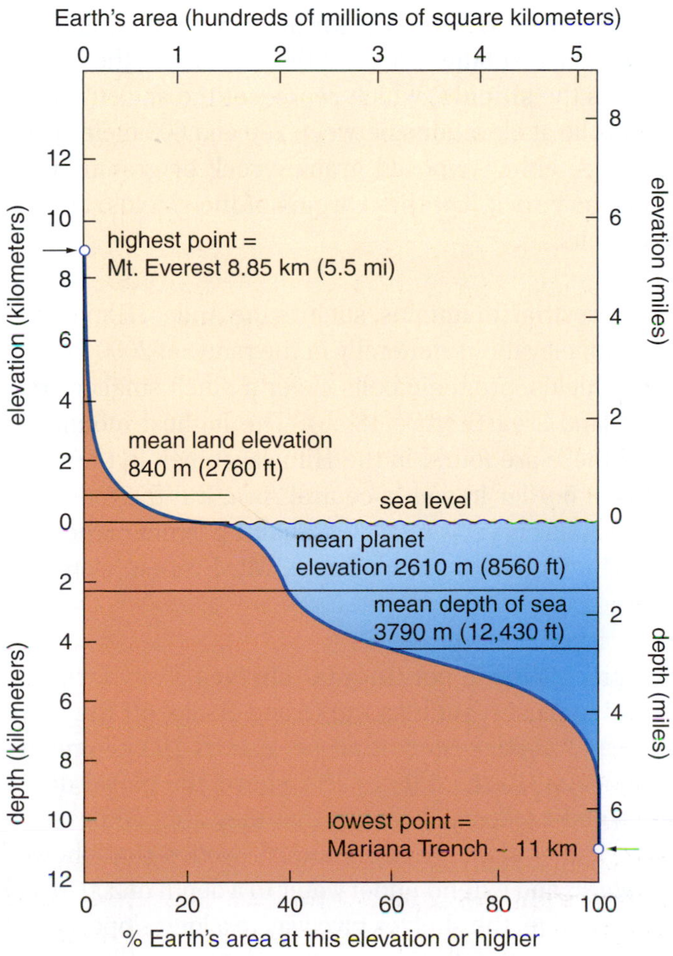

Figure 18.4 A graph showing the distribution of the Earth's surface area, both lands and ocean basins, by elevation. Note that the mean elevation of the Earth's surface is well below sea level.

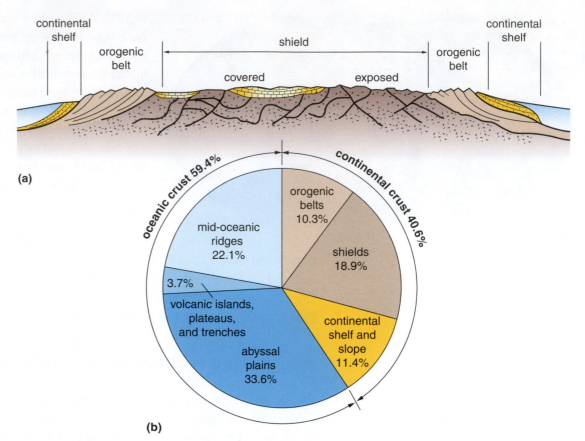

Figure 18.5 (a) A schematic model of a continent showing its three main subdivisions, and (b) the percentage geographic coverage of the Earth's surface by the subdivisions of the continents and ocean basins. Notice that more than half the Earth is covered by mid-oceanic ridges and abyssal plains.

Features of the Continents: Both the continents shown in Figure 18.5a, and the ocean basins are made up of three major subdivisions. On the continents, the largest subdivision in terms of coverage is the **shields**, which represent the ancient interiors of the landmasses. The shields lie at elevations between 200 and 600 meters and have two main surface expressions: either exposed granitic rock or granitic rock covered with a veneer of sedimentary rock. On the margins of the shields, but not surrounding them, are orogenic belts.

Orogenic belts are systems of terrestrial mountains, such as the Andes, Himalayas, and the Rocky Mountain chain, with elevations generally in the range of 2000 to 4000 meters. Much younger than the shields, orogenic belts cover a much smaller area, about 10 percent of the Earth's land area (Figure 18.5b). The highest mountains exceed 6000 meters and most of these are found in the Himalayan belt in the vicinity of the Indian–Nepalese–Tibetan border in south–central Asia. But the total land area in the world at such high elevations is very small, less than 1 percent of the total surface area of the continents. The highest point on Earth, Mt. Everest, stands at 8848 meters (29,028 feet) above sea level.

The third subdivision of the continents is the **continental shelves**. They form the "shoulders" of the continents and in places, such as the Grand Banks off the Newfoundland coast or the North Sea of northwestern Europe, extend 200 kilometers or more seaward from the continental shores. Figure 18.6 shows the global distribution of the continental shelves and Figure 18.7 shows that they are composed of thick masses of sedimentary rock that slope gently seaward. Notice that shelves usually begin on land as *coastal plains* and extend underwater to a depth of 200 to 300 meters below sea level. At the outer edge, the shelves give way to a long slope, called the *continental slope*, that leads down to the deep ocean floor, or in some locations

Q#3

Figure 18.6 The global distribution of the continental shelves. Together with the continental slopes, these fetures cover more than 11 percent of Earth's surface.

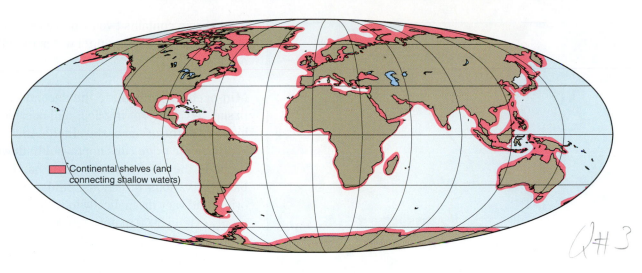

▮ Continental shelves (and connecting shallow waters)

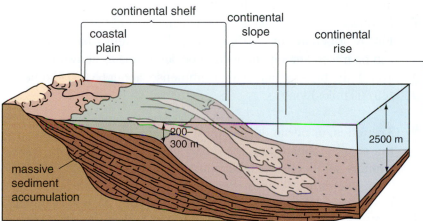

Figure 18.7 The principal features of continental shelves, beginning on land and sloping seaward to the deep ocean floor.

to great chasms in the ocean floor called trenches. Together, the continental slopes and shelves cover a surprising area, over 11 percent of the crust globally. Sediments that build up on the outer continental shelves spill down the continental slopes forming a broad apron on the adjacent ocean floor, called the *continental rise*.

where I hew deep.

Features of the Ocean Basins: The largest of the three subdivisions of the ocean basins is the **abyssal plain**, or deep ocean floor. In terms of area, abyssal plains are Earth's dominant geographic feature, covering over one-third the planet's total area (see Figure 18.6b). They form broad, flat to hilly surfaces lying generally 4000 to 6000 meters below sea level, which are interrupted in selected spots by isolated volcanoes or clusters of volcanoes called *seamounts* and by larger volcanic masses or *plateaus* (refer to Figure 18.8). Each

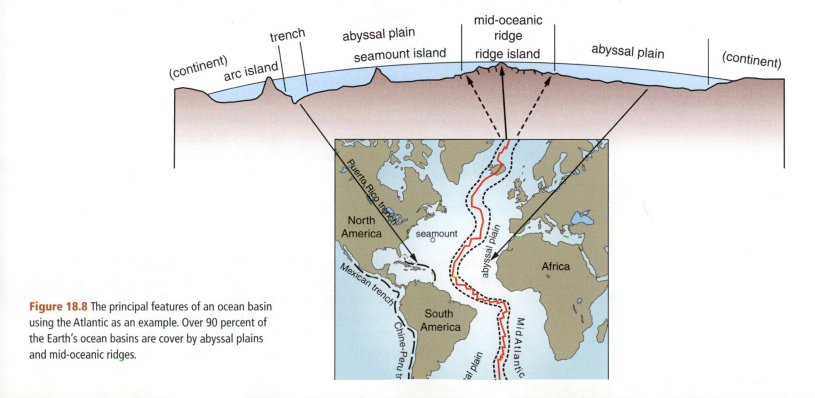

Figure 18.8 The principal features of an ocean basin using the Atlantic as an example. Over 90 percent of the Earth's ocean basins are cover by abyssal plains and mid-oceanic ridges.

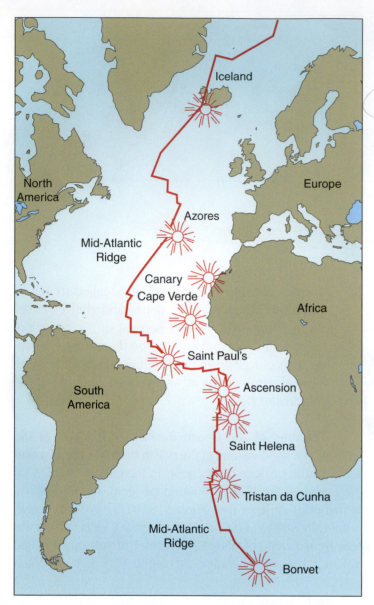

Figure 18.9 The distribution of mantle plumes in the Atlantic Basin. Seven plumes lie on or very near the mid-oceanic ridge and are marked by volcanic islands.

ocean basin contains two or more regions of abyssal plains which are separated by long underwater mountain systems called mid-oceanic ridges.

Mid-oceanic ridges are the second-largest feature of the ocean basins. As the term implies, mid-oceanic ridges are found in the interiors of the ocean basins. In the Atlantic Basin, shown in Figure 18.9, a single ridge runs north–south down the center of the entire basin. In the Pacific, on the other hand, multiple ridges cross the basin in various directions. Mid-oceanic ridges are volcanic in origin and vary dramatically in width and height. The highest commonly extend 6000 meters (20,000 feet) above the ocean floor and, in some places, may form island chains. *Ridge islands* are not to be confused with two other types of volcanic islands; *seamount* islands in the interiors of the abyssal plains, and *arc islands* which are found along trenches near the outer margins of the abyssal plains. Finally, it is noteworthy that as a global system (that is, one that can be traced among connecting ocean basins as shown in Figure 18.10), the mid-oceanic chain is immense, about 65,000 kilometers (40,000 miles) long.

Trenches, or ocean deeps, are the third and smallest division of the ocean basins. They are found on the margins of the basins running parallel to mountain chains. Some such as the Chile–Peru Trench and the Mexican Trench (see Figure 18.8) lie tight against the continental shelf and an adjacent orogenic belt but, as the map in Figure 18.10 reveals, most lie along island arcs (volcanic mountains), such as the Japanese chain. Between the island arcs and the mainland, separate, smaller basins, or marginal seas, are formed such as the Sea of Japan. Most trenches reach depths of 2000 to 3000 meters below the abyssal plains. The deepest known is the Mariana Trench, near Guam in the Pacific Ocean, which reaches about 5000 meters below the nearby ocean floor, giving the ocean here a total depth of about 11,000 meters.

18.2 Essential Processes of Plate Tectonics

A good place to begin our examination of the processes of plate tectonics is on the ocean floor, for this is where most plates are born. In fact, most tectonic plates originate in the deep interior of the ocean basins with the formation of the mid-oceanic ridge. According to conventional thinking in Earth science, mid-oceanic ridges lie over or near mantle plumes, which deliver huge amounts of heat and kinetic force to the lithosphere. It appears that that there are seven such plumes in the Atlantic Basin beginning with Iceland in the north (see Figure 18.9).

Seafloor spreading: As a plume spreads out in the upper mantle, it pulls the lithosphere apart in the manner shown in Figure 18.11. Vertical fractures develop and, as the lithosphere separates, large blocks of rock slip down with jarring motions that create earthquakes. Fractures along which rock has been displaced are called **faults**, and the displacement process is called **faulting**. Most earthquakes produced by faulting in mid-oceanic ridges are relatively small and shallow (less than 20 kilometer

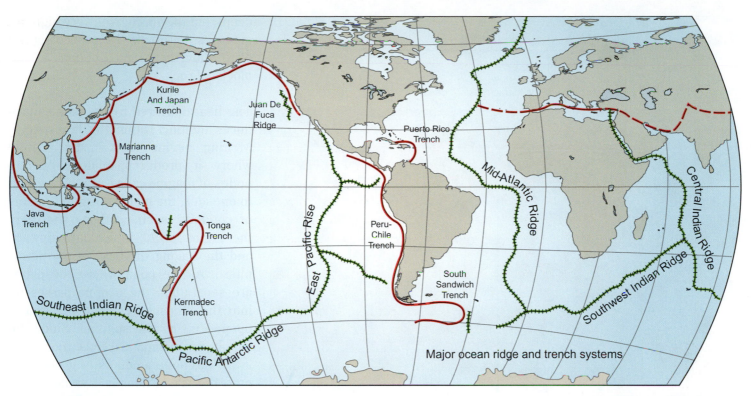

Figure 18.10 The global system of mid-oceanic ridges extends through all the ocean basins and is about 65,000 kilometers long, making it one of the longest, continuous geographic features on the planet. (Major trenches are shown in red.)

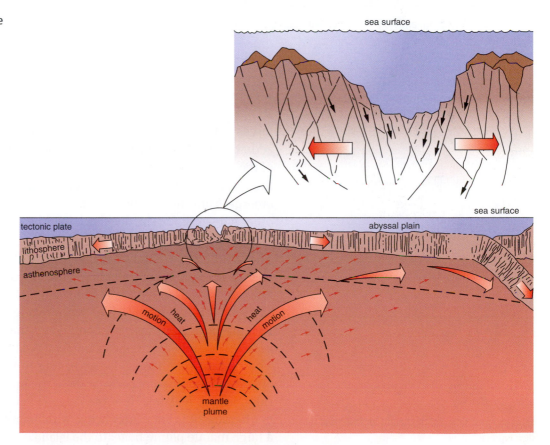

Figure 18.11 Seafloor spreading driven by convection from a mantle plume. As the lithosphere separates, faulting and volcanism are induced and ridge formation ensues.

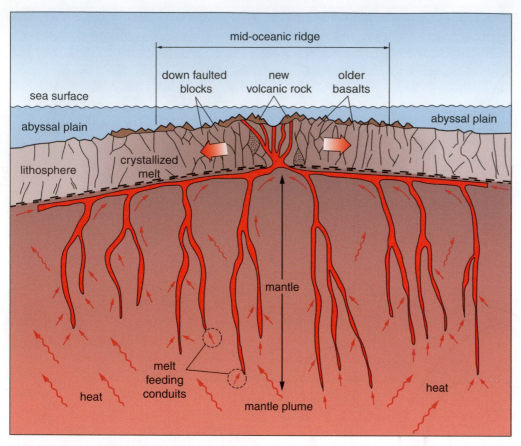

Figure 18.12 Volcanism and ridge formation fed by a vast system of conduits that conduct melt upward from the mantle.

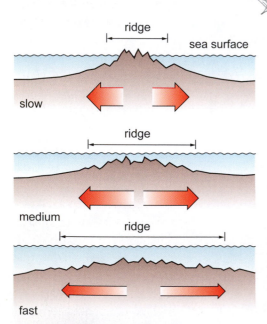

Figure 18.13 The relationship of the rate of seafloor spreading and the height and width of ridge development.

deep), but they occur with very high frequencies, and their occurrence signals plate movement and shifting among blocks of rock.

The heat delivered by the plumes produces melting within the lithosphere. Melting begins deep in the mantle where it produces millions of tiny nodes of magma called *melt*. According to one idea, the melt feeds into systems of conduits, such as those shown in the Figure 18.12, which flow upward and then along the base of the lithosphere. At the point where the plates are separating, the magma passes up into the lithosphere filling the spaces opened up by the splitting. Since the supply of magma exceeds the space available, huge amounts of it pour out onto the ocean floor in lava flows and volcanic eruptions. Eighty-five percent of the Earth's volcanic activity occurs on the mid-oceanic ridges. But as this great outpouring solidifies, it is slowly dragged away as the lithosphere slowly splits apart and spreads laterally.

This process is called **seafloor spreading** and it leads to the formation of new lithosphere as fresh rock emerges onto the ocean floor. Simply put, as the convective forces in the asthenosphere drive the lithosphere apart, new rock is simultaneously added to the departing sections of lithosphere. These sections are the trailing edges of the tectonic plates and the new rock is welded on to each as they slide away from the ocean center. The plates operate somewhat like two great conveyor belts that emerge from the mantle, flow across the ocean floor, and then upon reaching the other side of the basin, dive back into the mantle below.

The height and width of the mid-oceanic ridge depends on the rate of magma discharge and the rate of spreading. Fast spreading and slow discharge produce a wide, low ridge, whereas slow spreading and fast discharge produce a high, narrow ridge as Figure 18.13 suggests. The highest ridges build up far above the ocean surface forming prominent islands. Look down the Mid-Atlantic Ridge in Figure 18.9 and note the islands there, for example, Tristan da Cunha, St. Helena, and Ascension. The current rate of seafloor spreading in this section of the Mid-Atlantic Ridge is only 2 to 3 centimeters per year. Near the north end of the ridge is the world's largest ridge island, Iceland. Here the ridge is very wide, not because it is spreading fast, but because it is extremely active with massive outpourings of magma generated by a huge mantle plume beneath the island.

As the plates move outward from mid-ocean, the ridge is literally pulled apart and carried across the ocean floor. Why, then, is the ocean floor not completely covered with irregular volcanic topography, rather than the more modest topography that makes up the vast abyssal plains? The answer is that the plates grow thicker with distance from the ridge as additional rock is apparently welded on to their underside, and denser with cooling over time. This causes them to ride lower and lower on the ocean floor, as the diagram in Figure 18.11 illustrates.

The Subduction Process: Subduction is the process that consumes tectonic plates. In the seafloor-spreading system it marks the output end, opposite the mid-oceanic ridge at the input end. In other words, subduction is the process that destroys old lithosphere by ramming it back into the mantle. Intuitively you would expect that subduction zones are fraught with geologic activity, and indeed they are. They produce the greatest number of large earthquakes and volcanoes of any place on Earth.

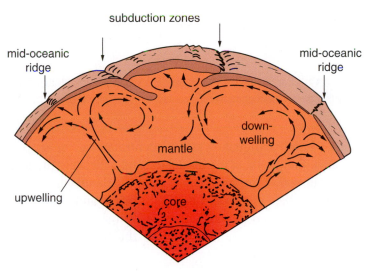

Figure 18.14 One concept of the mantle convection system showing upwelling under ridges and downwelling in subduction zones where the lithosphere descends into the mantle.

Subduction, too, appears to be driven mainly by the convectional system of the mantle. In a thermal convectional cell, heat causes upwelling and this motion is counterbalanced by downwelling as matter cools. In the upper mantle, upwelling takes the form of plumes that rise under the mid-oceanic ridges (among other places), and downwelling takes the form of subduction as material is returned to the mantle as Figure 18.14 shows. But there is another force helping draw the lithosphere back into the mantle, namely, gravity pulling on the plate's huge mass. The latter is based on the observation that as a plate moves across the ocean floor and grows older, it gets thicker and colder. As a result both its mass and density rise with distance from the mid-oceanic ridge, causing the plate to ride lower and lower in the upper mantle. Eventually it is dragged down under the combined forces of gravity and downwelling as Figure 18.15 attempts to show. Some scientists argue that down-dragging also helps drive the movement of the entire plate across the ocean basin.

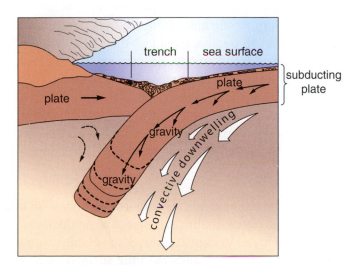

Figure 18.15 In subduction, the forces of convective downwelling and gravity pull on the plate's huge mass driving the plate into the mantle. On the surface, subduction is marked by a trench.

As an oceanic plate subducts under its neighbor, the crust is drawn down to form a **trench** on the ocean floor which becomes a depository for sediment. This is explained by the fact that the plate itself acts as a sediment-collection system. During its long trip across the ocean floor, which takes more than 100 million years for large plates, a sheet of ocean sediment builds up on the plate and, when the plate reaches the trench and subducts, this sediment is scraped off and accumulates in the trench. In time, a wedge-shaped mass of sediment, called an *accretionary wedge*, builds up in the trench as is illustrated in Figure 18.16. And if subduction takes place along the edge of a continent, additional sediment pouring down from the continental shelf is mixed into the wedge. The resultant mass may be compressed, transformed into sedimentary rock, and upheaved to form a *wedge island* and eventually part of the continental platform.

Beneath the trench, the subduction process is characterized by a jerking, downward motion as the plates slip along the angular contact between them. However, it is not along this

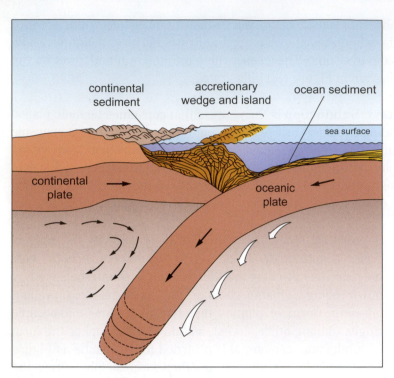

Figure 18.16 In their trip across the ocean floor, tectonic plates collect sediment that is delivered to the trenches where it accumulates to form accretionary wedges, which may be jammed up into islands.

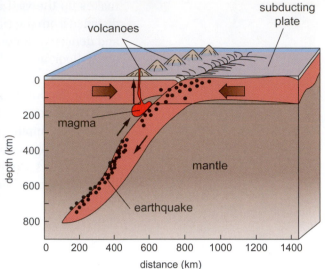

Figure 18.17 The distribution of earthquakes in a subduction reveals that most occur within the plate rather than at the contact with the adjacent plate.

contact where earthquakes are concentrated as Figure 18.17 illustrates. Rather, most earthquakes, and especially the large ones, occur within the subducting plate as it bends downward. Because the plate is cold and rigid, it is brittle and readily fractures as huge rock masses are displaced in sudden jars within a complex system of faults. Each displacement releases seismic energy into the lithosphere which produces an earthquake of some strength.

We will discuss the measurement, magnitude, and effects of earthquakes in the next chapter, but for the present let us note that the strength of an earthquake at the Earth's surface depends on the amount of energy released at the point of displacement, the depth of the fault, and other factors. If the fault is shallow and the rock is rigid, the tremors at the surface may be very strong and potentially very destructive. In subduction zones, earthquakes occur at all depths, from less than 10 kilometers in the lithosphere to 600 kilometers or more in the mantle. Because subducting rock is brittle and thick and under huge amounts of stress, earthquakes can be massive; however, only those in the upper 50 kilometers or so of the lithosphere are capable of producing heavy surface damage. At greater depths their destructiveness is buffered by the surrounding rock and the higher rock temperatures. The massive earthquake near the coast of Japan in March 2011 was centered at a depth of only 16 kilometers below the ocean floor.

Figure 18.18 Fed by deep sources of magma, volcanoes in subduction zones often form in lines or chains parallel to the trench. The Cascade Mountain chain of Oregon and Washington is shown here.

Volcanic activity is also prominent in subduction zones. The source of volcanism is found at the lower contact between the two plates. In the early stages of subduction, a body of magma forms at the spot marked in red in Figure 18.17. Some of this magma melts its way through the overlying plate and discharges onto the surface forming a volcano. According to one interpretation, initially a single line of volcanoes forms such as the one shown in Figure 18.18, but as the plate subducts further, and the magma body is dragged down with the plate, a broader zone of volcanism is created, leading to more or less a second line of volcanoes. As the volcanoes grow in size and number, long *island arcs* such as the Aleutian Islands form parallel to the trench. Eventually, individual volcanoes merge with others and they in turn coalesce to form large islands such as main islands of Japan.

In the advanced stages of subduction, the plunging slab of lithosphere extends through the asthenosphere and deep into the underlying mantle (see Figure 17.23). Because the slab is much colder than the mantle rock, it extends intact as deep as 700 kilometers in some areas, near the base of the upper mantle. But as the slab heats up and grows less brittle, earthquakes subside and eventually (in millions of years) the rock melts away and is reincorporated into the mantle, thus ending where it began hundreds of millions of years before.

18.3 Distribution and Motion of Tectonic Plates

It is a simple but important fact that at any moment tectonic plates cover the entire planet. There are no gaps where we can look down to the underlying asthenosphere. This is remarkable because the whole system of plates is in motion across the entire globe, sliding in different directions at different rates. Yet as soon as a plate or set of adjacent plates vacates an area, new lithosphere immediately forms in its place and, at the same time, in a distant subduction zone an equal amount of lithosphere is destroyed. In other words, the system maintains a zero net balance.

The Principle of Plate Motion: To understand the nature of movement in tectonic plates, it is necessary to appreciate that each plate behaves as a single, rigid unit. No one part of a plate can move without corresponding movement throughout the rest of plate. Further, all parts of a tectonic plate, as the diagram in Figure 18.19 illustrates, must move in the same direction. The result is that each plate moves in a rotating fashion, pivoting about its own pole. This follows a mathematical principle known as **Euler's theorem** (for its author, the Swiss mathematician and physicist Leonhard Euler, 1707–1783) which describes the movement of a curved surface such as a piece of orange peel (or a tectonic plate) on the surface of a sphere such ball (or a planet). For different plates, the pole of rotation, or *Euler pole*, is situated in a different location with respect to the plate itself. Some plates rotate about a centrally located pole within the area of the plate, whereas others, such as the one in Figure 18.19, rotate about a distant, mathematically defined pole.

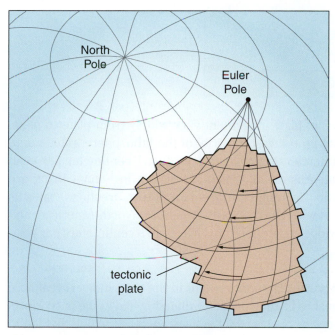

Figure 18.19 The geometry of plate movement follows Euler's theorem, a mathematical principle that describes the motion of a curved surface on a sphere. Each plate rotates about its own pole.

In any case, no matter the location of its pole, the arc of rotation of all plates is constant *but* the velocity of plate movement varies with distance from the pole. Therefore, as a plate departs from a mid-oceanic ridge, for example, the segment closest to the Euler pole will move least (slowest) and the segment farthest from the pole will move most (fastest). This affects the width and height of the mid-oceanic ridge. All things being equal, the ridge should be wider and lower at greater distance from its pole because the rate of movement is higher at greater distances.

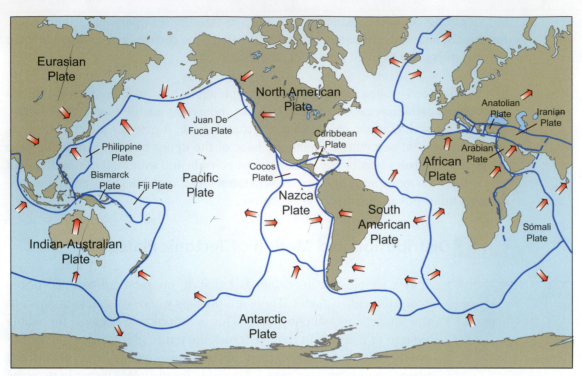

Figure 18.20 The distribution and names of Earth's tectonic plates. All major plates, except the Pacific, carry continents. The sets of arrows show the directions of movement at the plate boundaries. Where the movement is divergent, a mid-oceanic ridge is forming. Where it is convergent, subduction is taking place. Movement may also be parallel to a boundary, as along the California coast.

Rates of Movement: The rates of plate movement vary widely from plate to plate. Among the 15 plates that make up the vast majority of the lithosphere, rates range from as high as nearly 10 centimeters per year to less than 1 centimeter per year. The Pacific, Nazca, and Indian–Australian, which are shown in Figure 18.20, are fast moving with ridge-separation (combined movement in two directions) rates as high as 18 centimeters per year; whereas the South American, North American, and African are slow moving with ridge-separation rates less than 5 centimeters per year. Rates also vary over time.

Some plates, which showed a great deal of movement in the past, today reveal little detectable motion. Others appear to be moving faster than they did in the past. Moreover, there may be long-term cycles of spreading activity such as the one that led to the breakup of Pangaea and which, by the way, is still going on. The general conclusion we reach is that although individual plume systems may function for tens of millions of years, they are probably not smooth-flowing systems, but vary in velocity, force, and flow patterns over long periods. Combined with other factors such as resistance related to plate size and massive barriers related, for example, to continental collisions, the result is uneven movement rates from plate to plate and time to time.

Plate Types and Distribution: The Earth's lithosphere is made up of seven major plates, six or seven minor plates, and ten or more microplates or platelets (see the map in Figure 18.20). The latter are small fragments or splinters of larger plates that lie along plate contacts such as subduction zones. The largest plate in the world is the Pacific Plate and it is exceptional not only for its size, but because it is the only major plate that does not contain a continent. All the other major plates contain at least one continental platform. The Indian–Australian Plate contains two continental units; Australia on one end and India on the other. India is sutured onto Asia as a result of its collision with the Eurasian Plate after the breakup of Pangaea. The North American Plate includes all of one continent and a piece of another, northeastern Asia.

Plate Borders and Their Geographic Features: All the world's tectonic plates are moving in different directions, but only three types of borders are produced along the contacts:

- divergent , • convergent , • transform .

Divergent borders are found where the lithosphere is separating, as in seafloor spreading. They are also called *constructive borders* because new lithosphere is built as adjacent plates move away from each other. All major ocean basins – Pacific, Atlantic, and Indian – contain divergent plate borders; therefore, each of these basins is made up of two or more plates. A few smaller basins, most notably the Red Sea, also contains a divergent border as you can see in Figure 18.20. In fact, the Red Sea is growing as the African and Arabian plates separate along the center of the basin.

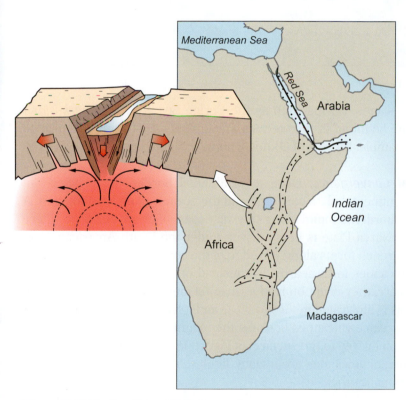

Figure 18.21 The East African Rift System can be traced from southeastern Africa to the Red Sea and beyond to the Sinai Peninsula of Egypt. The inset diagram shows a rift structure that has become the basin of a lake such as Lake Albert or Lake Nyasa.

Although most divergent borders are located in ocean basins, some also cut across continental platforms. When Pangaea split apart divergent borders formed between the major landmasses that became today's continents. In the modern world, the two most prominent divergent borders within a continent are the East Africa Rift System and the Lake Baikal (Siberian) Rift System. A rift is a fault formed when a large block of rock slips down between two larger sections of crust because the crust has pulled apart (see the inset diagram in Figure 18.21). In both rift systems, large lake basins mark the separation zone. The one in East Africa, which is over 5000 kilometers long and mapped in Figure 18.21, can be traced far northward where it joins the Red Sea Basin. Lake Baikal, the largest (based on volume) freshwater lake in the world, is a mile deep.

Convergent borders form where the lithosphere is being compressed and one plate is forced under the other. They are also called *destructive borders* because this is where lithosphere is lost to the mantle. There are several different types of convergent borders and all involve subduction. Which type forms depends on the motions and physical attributes of the plates involved in the contact. The three most common convergent borders are: oceanic–oceanic; oceanic–continental; and continental–continental.

Oceanic–Oceanic Convergence: In oceanic–oceanic convergent zones, one oceanic plate subducts under another oceanic plate as is shown in Figure 18.17. As in all types of subduction, which plate is driven down depends on which is heavier and rides at the lower elevation on the Earth's surface. A good example of an oceanic convergence is found at the northern end of the Pacific Ocean where the Pacific Plate is subducting under the Bering Sea, which is part of the North American Plate. Examine the map in Figure 18.22 and notice the Aleutian Islands running parallel to the Aleutian Trench. All of these islands are volcanic, scores are active, and together they form a prominent *island arc*, over 2000 kilometers long. On the eastern end of the arc, the volcanic islands have coalesced to form the Alaska Peninsula. Little by little the Bering Sea is being partitioned off from the Pacific as the Aleutian chain

Figure 18.22 The northern border of the Pacific Plate is marked by a convergence zone with deep trenches paralleling island chains. Behind the island chains lie back-island basins like the Bering Sea and the Sea of Japan.

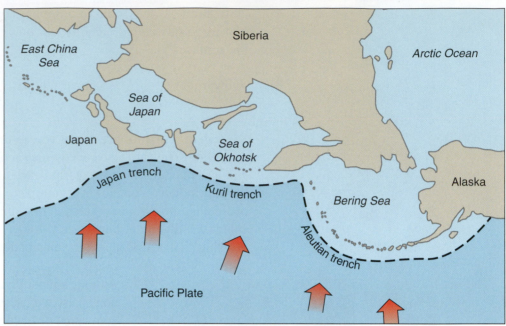

One of hundreds of active volcanoes in the island arcs of the northern Pacific. This one borders the Sea of Okhotsk.

fills in with volcanic islands. Basins such as the Bering Sea are called *back-island basins* and several can be found in the western Pacific that are strikingly similar to the Bering Sea, for example, the Sea of Okhotsk and the Sea of Japan.

Oceanic–Continental Convergence: Oceanic–continental borders form where an oceanic plate and a continental plate converge and the oceanic plate is subducted. A good example of this border is found along the west coast of South America where the eastward moving Nazca Plate is being driven under the South American Plate (see Figure 18.20). A trench forms along the edge of the continent and, as in oceanic subduction zones, volcanoes form nearby. But instead of the volcanoes erupting at sea to form an island arc, they erupt on land as magma melts through the overlying continental crust. This is illustrated in Figure 18.23 and it leads to a different magma composition than in oceanic volcanoes because the magma mixes with the granitic rock of the continent on its way to the surface as Figure 18.23 shows. Therefore, the lavas of these continental volcanoes are usually lighter, trending more to andesite than deep mafic (ultramafic) rock.

A second zone of activity is centered on the trench where sediment is accumulating giving rise to an accretionary wedge. The wedge is composed of mixed ocean and continental sediment and, when this mass is compressed with plate movement, it may be pushed up into a string of islands in the manner depicted in Figure 18.16. Further compression may force it even higher and drive it landward to become part of the coastal belt of mountains and the continental crust. Thus, the orogenic belt that ultimately forms is compositionally complex with volcanic, sedimentary, metamorphic, and granitic (continental) rocks mixed together in a mass of folded and faulted structures. Indeed, such is the character of the Andes.

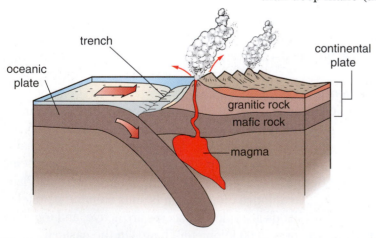

Figure 18.23 The main features of an oceanic–continental plate convergence. The trench forms offshore but the volcanoes form on land and incorporate continental rock into their compositions.

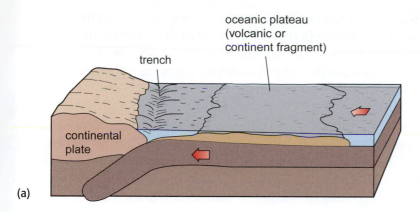

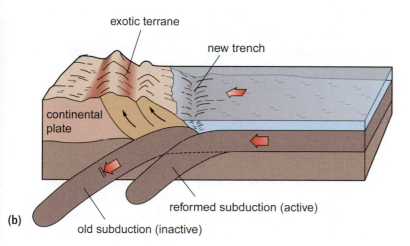

Figure 18.24 (a) Exotic terrane, such as an oceanic plateau, being added to a continental plate at a convergence border. (b) After it is added, a new subduction zone and trench form on the seaward edge of the terrane.

Throughout the oceans are relatively small areas of ocean floor that stand above the abyssal plains. These *oceanic plateaus* are "relict" terranes, the remnants of extinct volcanic clusters, old mid-oceanic ridges, or pieces of continental crust. Oceanic plateaus move with the ocean floor under them, which eventually transports them to a subduction zone. Here the smaller pieces are broken up and consumed, but the larger plateaus, such as the one in Figure 18.24, are too thick to be subducted, and instead are jammed into the continental crust. These pieces of crust are known as **exotic terranes**, and where they are driven landward, they often form mountain ridges along the coast. Once this new crust is in place, it adds a new piece of geography to the continent. The old subduction zone is abandoned and a new one forms on the seaward side of the new crust, as shown in Figure 18.24b.

Continental–Continental Convergence: This leads us to convergent borders involving two continental plates. Continental–continental convergences begin with a sea standing between two continents. In the simplest version of this convergence, each continent rests on a different plate with a common border at the subduction zone. As compressive forces drive the two toward each other, the sea shrinks. But closing the sea takes a long time, time for the sea to collect a huge amount of sediment, including a large continental shelf along each coast. As the continents draw near, this great mass of sediment is hardened and compressed, and thrown up into a large mountain mass. The rock formations are mangled, transformed into every imaginable folded and faulted configuration, and, at depth, heated and metamorphosed. Beneath this mass, magma forms and intrudes it, often melting its way to the surface and spilling out or erupting there, adding more mass and complexity.

The most superlative example of a continental convergence of this sort involved the collision of India with the southern edge of the Eurasian landmass. Take a look at Figure 18.2 showing the breakup of Pangaea and trace the movement of India. Notice the sea lying between India and Eurasia, called the Tethys Sea. It served as a huge reservoir of sediment, which became the building material for the highest mountain mass on Earth, the Himalayas and the Tibetan Plateau. This is illustrated in the sequence of diagrams in Figure 14.25. As India approached Asia, it took over 20 million years for the mountain mass to be heaved up and the subduction process to run its course. The subducted slab is still detectable on seismic charts deep beneath Tibet. The Himalayan orogeny was an event of monumental proportions with profound geographic effects including widespread changes in, among other things, the climate, drainage, and biota of Asia (Figure 18.25).

Figure 18.25 The Himalaya Front rising up from the plains of northern India. From here the great mountain mass stretches northward for nearly 1000 miles.

The third and final border type is neither constructive nor destructive but *conservative.* Called a **transform border,** it is characterized by plates slipping laterally along the contact and no lithosphere is lost or gained in the process. There are two classes of transform borders. One involves the sides of a plate and often takes the form of a long, straight segment or long series of linear segments. The most celebrated and scientifically scrutinized of such transform contacts is along the American West

Coast (see Figure 18.20) where the Pacific Plate slides along the edge of the North American Plate. The principal motion here is the northward movement of the Pacific Plate. Where the border runs through California, forming the famous *San Andreas Fault*, the actual line of contact and the pattern of displacement are vividly displayed in the desert landscape as the photograph in Figure 18.26 reveals.

Farther north, in Oregon, Washington, and British Columbia, the plate border runs offshore and a plate fragment or platelet, called the Juan de Fuca Plate, lies between the two larger plates. The Juan de Fuca Plate is being squeezed between the two giant slabs, causing it to subduct under the North American Plate as is shown in Figure 18.27. This has produced three distinct belts of terrain in the Pacific Northwest:

- The Cascade Mountains, a chain of volcanoes (a terrestrial volcanic arc) running north–south parallel to the subduction zone (see Figure 18.18);
- A long basin (called a *forearc basin* because it lies in front of the volcanic arc) occupied by the Georgia Strait, Puget Sound, and the Willamette Valley which extends from central Oregon into British Columbia;
- The mountains of Olympic Peninsula and Vancouver Island, an elevated accretionary wedge like the one shown in Figure 18.16.

The second class of transform border is a little more complicated. This border is found within the mid-oceanic ridges, such as in the Atlantic as represented in Figure 18.28, where the two plates pull apart forming a zigzag fracture pattern. The pattern has an interfingered configuration with the transform-border segments on the sides and divergent-border segments on the ends of the fingers. Notice the curved shape of the transform segments. This shape conforms to the arc of rotation of the plates and can be traced far beyond the ridge onto the abyssal plain.

Figure 18.26 California's premiere fault, the San Andreas, marks a transform border where the movement between the Pacific and North American plates is lateral.

Summary on Plate Tectonic Processes and Features: Plate tectonics is played out mainly along the plate borders. Most plates emerge in the ocean basins with seafloor spreading and disappear on the opposite border with subduction. In the course of these processes, most of Earth's volcanoes and earthquakes are generated, and most of the planet's major landforms are built in both the ocean basins and the continents. Of Earth's seven major plates, only the largest one does not carry a continent. As the Pacific Plate moves northward, several types of convergent borders are set up, each with different geographical implications in terms of landforms, coastal environments, ocean circulation, and other factors. This brings to mind other questions of a geographical nature. For example, how do ocean basins like the Atlantic originate, and if continents can grow can they also be reduced by plate tectonics, and does all this moving around of landmasses have anything to do with the geographic distribution of plants and animals?

18.4 Formation of Ocean Basins and Island Systems

For centuries the oceans have been a great mystery to humans. But until well into the twentieth century we really had no idea about their depths, bottom topography, and how they formed. Many scientists of the nineteenth and twentieth centuries speculated that the ocean basins were the oldest parts of the Earth's crust, formed long before the continents had taken shape. But as modern science unraveled the secrets of plate tectonics, it became apparent not only how the oceans formed, but why they are so much younger than the continents.

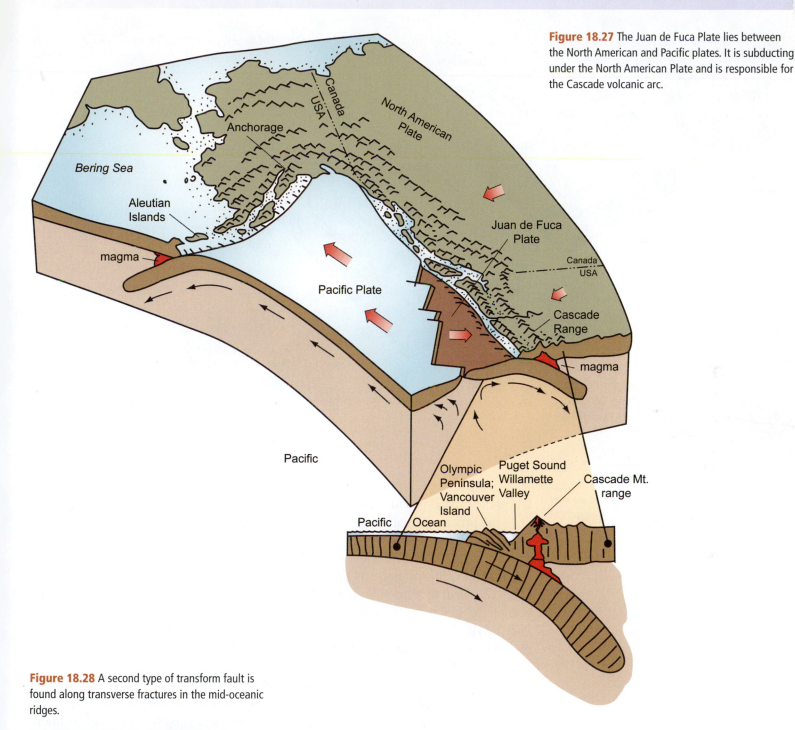

Figure 18.27 The Juan de Fuca Plate lies between the North American and Pacific plates. It is subducting under the North American Plate and is responsible for the Cascade volcanic arc.

Figure 18.28 A second type of transform fault is found along transverse fractures in the mid-oceanic ridges.

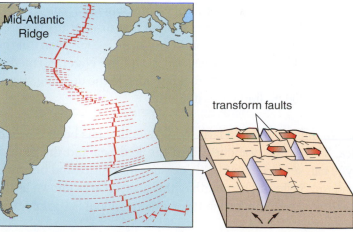

The Origin and Development of Ocean Basins: Let us review what we know about the ocean basins. First, considering the crust as a whole, they are very young indeed; nowhere is the ocean crust known to be more than 170 million years old (less than one-tenth the age of the interiors of the continents). Second, the ocean basins are made up essentially of one rock type, basalt, and this rock is produced from volcanic outpourings along the mid-oceanic ridges. Third, as the map in Figure 18.2 suggests, the basins grow from the inside out by seafloor spreading and are destroyed along the perimeter by subduction. Their lifetime is fixed by the time it takes to open and close a section of lithosphere between two landmasses.

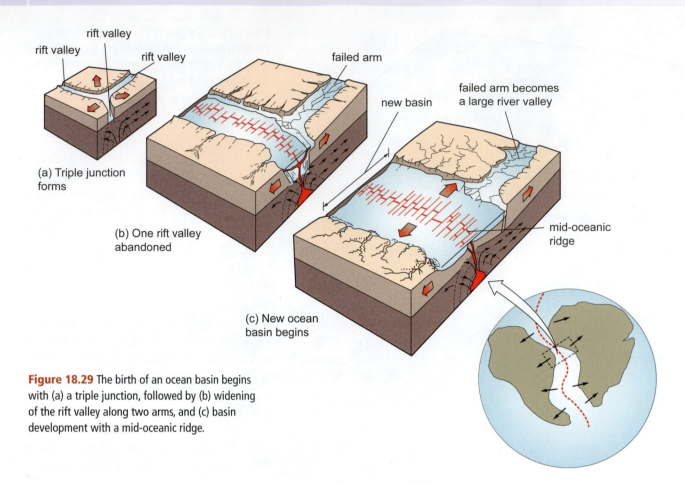

Figure 18.29 The birth of an ocean basin begins with (a) a triple junction, followed by (b) widening of the rift valley along two arms, and (c) basin development with a mid-oceanic ridge.

If oceanic crust is produced by seafloor spreading along divergent plate borders, how does the process actually begin? Evidence indicates that ocean basins begin with the splitting apart of a continent along the lines of what we observe with the African rift valleys and the Red Sea. The driving force is a mantle plume that rises up under a continental platform and rips the overlying lithosphere apart. Initially the rupture takes the form of a three-branched fracture system, called a **triple junction**, each branch marked by a rift valley as shown in Figure 18.29a. As the spreading advances and the rift valleys widen and deepen, one of the branches becomes inactive while the other two continue to grow.

The inactive branch forms a long valley called a **failed arm** (Figure 18.29b). Though it is removed from the picture in terms of ocean-basin formation, the failed arm will go on to play an important role in shaping the physical geography of the continent for these large lowland structures become major collection zones for runoff from the continent's interior. The product is a large river system that discharges through the failed arm and into the newly forming ocean basin (Figure 18.29c). Many of the world's largest rivers, including the Amazon, Mississippi, and the Niger, formed at various triple junctions as the Atlantic Basin was cracking open in the opening stages of the Pangaea breakup around 200 million years ago.

More recently (in the past 25 million years) a large triple junction has been developing in northeastern Africa. The active branches form the Red Sea and the Gulf of Aden. The failed arm, as you can see in Figure 18.30, extends into Africa and becomes a branch of the East African Rift System. The mouth of this arm, an area known as the Afar Triangle, has become the valley of the Awash River which is about 500 kilometers

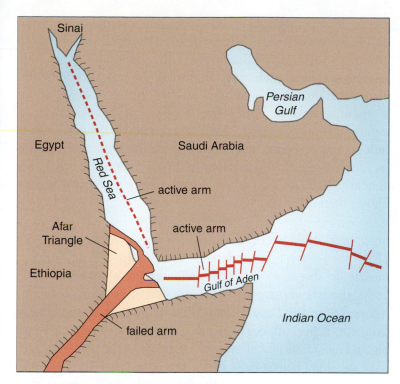

Figure 18.30 A modern triple junction in the East African Rift System. The two active arms form the Red Sea and the Gulf of Aden basins.

long. Will this valley develop into a great river system like the Niger has on the other side of Africa? It depends, of course, on the future of plate activity in this part of the world, and that depends on activity in the Earth's mantle, of which we know very little.

Back to the young ocean basin. The basin is now a narrow sea, like the Red Sea and the Gulf of Aden, and the failed arm is a connecting river valley. The spreading process is concentrated in the center of the basin, where a mid-oceanic ridge is forming, while faults on the sides of the basin have become largely inactive and are being buried under sediment with formation of continental shelves. Continued growth (widening) of the new basin means that somewhere else in the crust an ocean basin or two is shrinking.

Figure 18.31 shows shrinking ocean basins on the opposite sides of the two bordering plates. Taken to the extreme, if the expanding basin in the middle of the diagram doubles in size, the other two basins will be completely consumed in subduction. But this assumes that the expanding basin's growth is equal in both directions. Rarely do we find such symmetry in nature. Instead, the rate of seafloor spreading will usually favor one plate, sometimes dramatically so. In such cases the mid-oceanic ridge grows faster on one side than the other, meaning that just one basin shrinks – and may eventually disappear – while the other remains more or less stable.

Island Types and Formation: The ocean basins contain thousands of islands. We have already accounted for several types of islands, namely *ridge islands* in the ocean centers, *arc islands* along subduction zones, and *wedge islands* along selected segments of ocean–continental subduction zones. The latter are composed of compressed seafloor sediment, but ridge and arc islands are exclusively volcanic with compositions ranging from high-density basalts to porous lavas and volcanic ash.

Seamounts are another class of prominent volcanic islands. Some of our most celebrated islands are seamounts; for example, Hawaii, Bermuda, and the Galapagos. Seamounts form over hotspots in the lithosphere where plumes concentrate heat rising up from the mantle. The heat produces melting and the resultant magma

Figure 18.31 Expanding and shrinking ocean basins as a result of seafloor spreading and subduction.

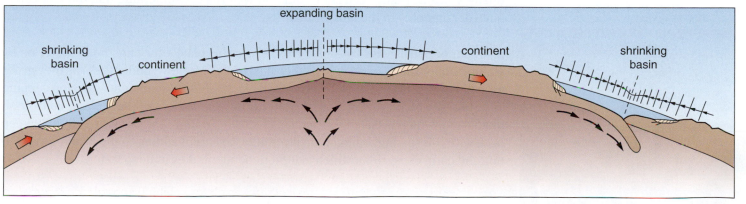

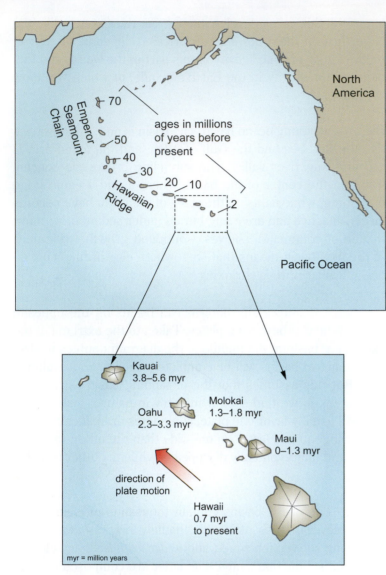

Figure 18.32 The Hawaiian chain of seamounts ranges in age from less than 1 to 70 million years and marks the directions of movement in the Pacific Plate. The older seamounts lie below sea level because of plate subsidence.

penetrates the lithosphere and pours out onto the ocean floor, building a volcanic rise. Highly active hotspots such as Hawaii produce enormous amounts of magma, as much as 1 cubic kilometer a year, and cover thousands of square kilometers of ocean floor. Although mostly obscured by ocean water, seamounts exhibit impressive vertical development, commonly 5000 meters (17,000 feet) or more above the ocean floor. Indeed, the Hawaiian seamount may be the highest mountain on Earth. From the ocean floor to the summit of Mauna Loa it is nearly 10,000m (33,000 feet) high, making it nearly a mile higher than Mt. Everest.

Seamounts often occur in chains. Chains form when an ocean plate passes over a stationary hotspot in the underlying mantle. In the case of the Hawaiian chain, shown in Figure 18.32, the islands stretch to the northwest in the direction of movement of the Pacific Plate. The seamounts farthest away from the active (modern) islands are about 70 million years old and most now lie below sea level because the plate has over time subsided isostatically under the added weight of the volcanoes and its own mass. At the other end (see the lower map in 18.32), the island of Hawaii is less than a million years old and continuing to spew out lava. Notice also that the Hawaiian chain also has a "dog-leg" alignment, revealing that the Pacific Plate changed direction by about 45 degrees. This happened about 40 million years ago.

There was a time in Earth's early years when islands were the only land on the planet. There must have been as many or more seamounts and ridge islands than there are today. But none of the islands we see today are much older than 100 million years. Therefore, thousands must have disappeared over the past 4 billion years. Where did they go? The life of an island can end in at least five different ways. In the case of ridge islands, most drown as the plate on which they are riding sinks as it moves across the ocean floor. Arc islands, on the other hand, are often lost when they are welded onto neighboring islands by lava flows and ash deposits or when they become attached to continental platforms as back-island basins fill in or as continental plates collide and shove the arc into a landmass. Islands can also be consumed in subduction or rammed into continents as parts of exotic terranes. And finally, volcanic islands can fall victim to wave erosion, something that the Icelanders have documented in the North Atlantic.

Seamounts usually die by sinking either as the plate carrying them sinks, or by sinking under their own weight as they push the crust down under them as they grow, as is illustrated in Figure 18.33. In tropical seas, sunken seamounts develop capping coral reefs, which grow in shallow water on the shoulders of the sinking volcano. If this island, which is called a *guyot*, sinks slowly enough, the coral growth may keep pace with the sinking, building up layer after layer and adding to the island's mass. If not, the coral ecosystem, which is light-dependent, dies in deep water.

A seamount in the South Pacific with its summit shrouded in clouds. The bulk of the volcano is below sea level.

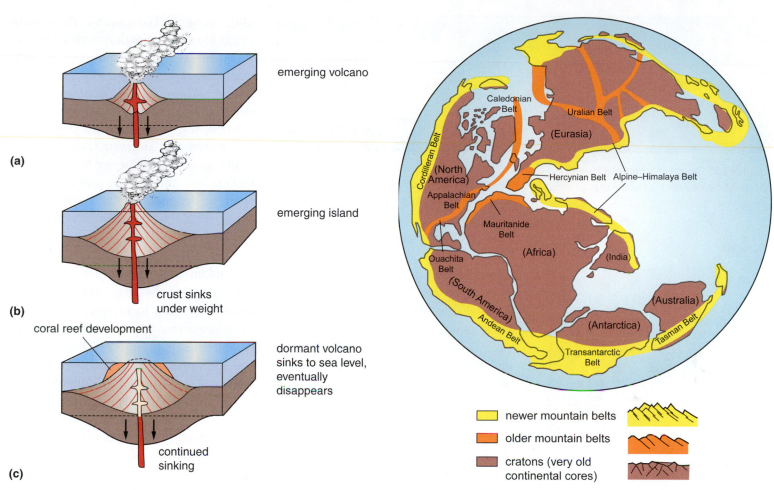

emerging volcano

(a)

emerging island

crust sinks
under weight

(b)

coral reef development

dormant volcano
sinks to sea level,
eventually
disappears

continued
sinking

(c)

Figure 18.33 As seamounts grow, they often sink under their own weight. In tropical seas, coral reefs may develop on the shoulders of the sinking volcano.

Figure 18.34 Continents grow with the formation of mountain belts (orogenies) on the margins. With the breakup of Pangaea, a worldwide system of mountains formed that includes the Alps, Andes and the Rockies.

newer mountain belts

older mountain belts

cratons (very old
continental cores)

18.5 Growth and Breakup of Continents

The continental crust throughout the world can be divided into two broad classes of rock formations, deformed and not deformed. Rock in the not-deformed class is mostly sedimentary rock found in the continental shelves and in thin sheets covering parts of the shields. It is relatively flat-lying, gently sloping, and/or slightly bent. Deformed rock is sedimentary, igneous, and metamorphic rocks in shields and mountain ranges that have been folded, fractured, faulted, and/or melted. Mine shafts and boreholes reveal the vast majority of the rocks making up the continents are decidedly deformed. The reason, of course, is that these rocks have been subject to mountain building, or orogenic processes. We conclude, therefore, that orogenies must be central to the growth of continents.

Orogeny is the term used to describe an episode of mountain-building. Orogenies produce orogenic belts, called *orogens*, which always develop on the margins of continents. Generally speaking, two groups of organic belts can be identified among the continents, newer ones and older ones. Newer orogenic belts, such as the Andean of South America and the Cordillera of North America, date from the time of Pangaea and formed on the perimeter of the great landmass as it was breaking up as the map in Figure 18.34 shows. Most date from around 250 million years ago and are still in the building process.

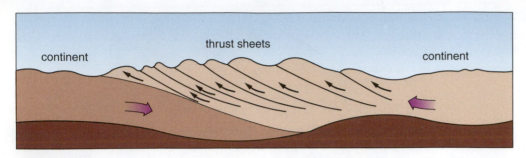

Figure 18.35 Growth of continental mass in a compression zone with the buildup of huge thrust sheets.

Older orogenic belts, such as the Appalachian of eastern North America, predate Pangaea, roughly before 500 million years ago. They formed along old continental margins when global geography was quite different and, when Pangaea formed, they ended up situated in the super-continent's interior. Older belts are distinguished from newer belts by their lower elevation and rounded land forms, the result of millions of years of erosion. Much older belts yet have been defined deep in the continental interiors among the ancient rocks of the cratons such as in the Canadian Shield, but today these are mere stumps of once larger mountains.

Growth by Orogeny: Orogenies thicken and extend the continental crust in three fundamental ways. The *first* involves the stacking of massive sheets of rock on the continental rim. When continental plates are rammed together in a convergence zone, such as when India collided with Asia, the compressional forces are so great that even thick, rigid crust is severely deformed. The crust is broken into huge sheets, called **thrust sheets**, which slide laterally and stack upon one another at low angles as is suggested in Figure 18.35. Individual sheets may be up to 20 kilometers thick and are made up of deformed rock of various origins, for example, wedges of sedimentary rock from the continental shelf and fragments of old crystalline rock detached from deeper parts of the crust.

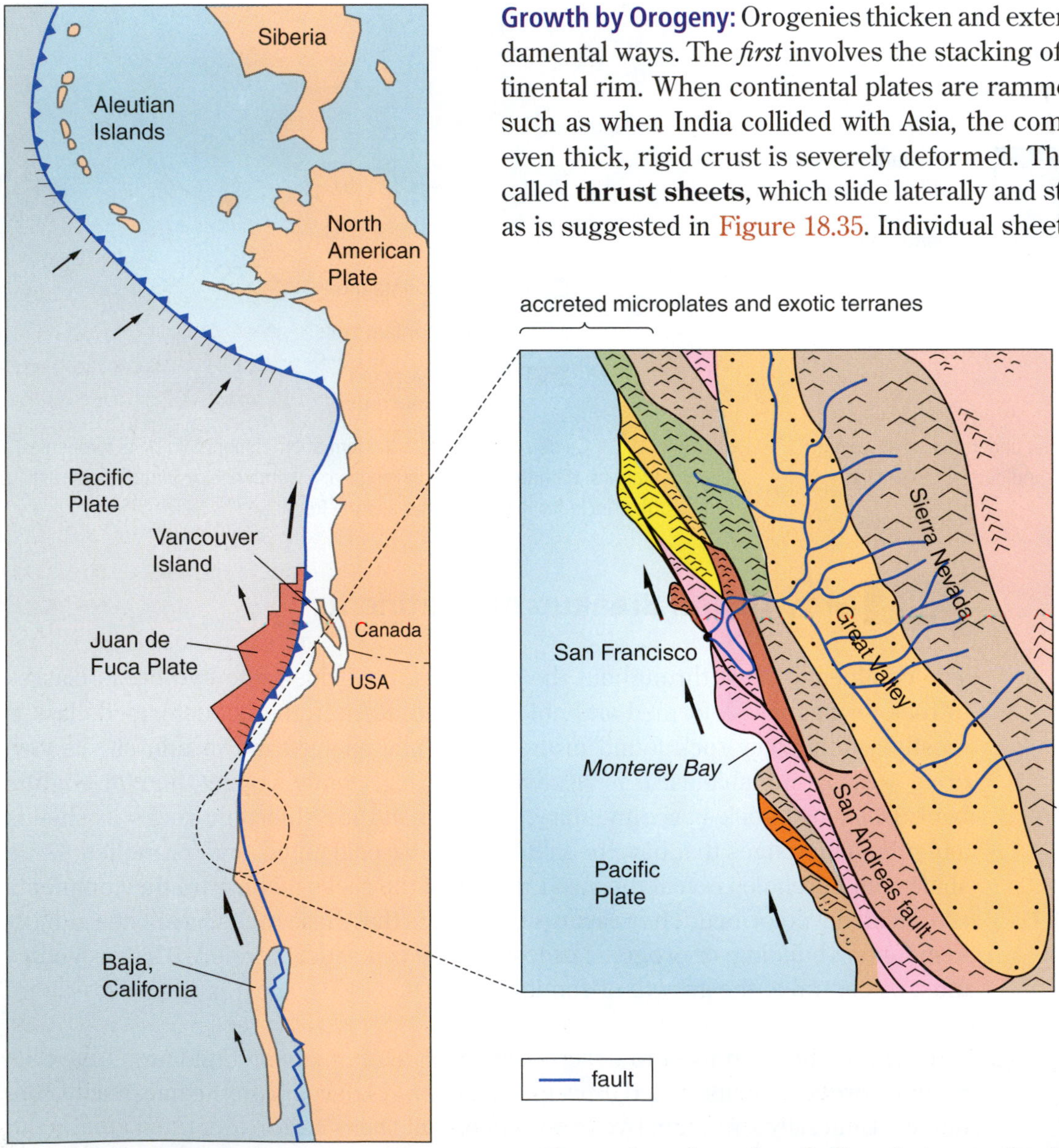

The *second* process is growth by accretion of microplates. **Microplates** are fragments of crust, including pieces of continents and exotic terranes that are drawn into convergence zones and spliced into the orogenic system. They may have traveled thousands of kilometers on a tectonic plate before being accreted onto a continent at a convergence border as is illustrated in Figure 18.24. Microplate accretion also takes place along transform borders, such as along the western edge of North America, where fragment after crustal fragment has been added to the continent (see the colored slices between the faults in Figure 18.36 inset). Each is marked by a mountain range or set of ridges and those that lie on the Pacific Plate, like those around Monterey Bay, are being displaced toward Alaska with the northward movement of the plate. Pieces of crust such as Vancouver Island, in southwest

Figure 18.36 Continental growth by microplates. Over millions of years microplates (the colored slices) have been added to North America along the Pacific Plate border.

British Columbia, and other islands and coastal mountain ranges north of it, appear to have moved northward several thousand kilometers and lodged on the edge of the North American Plate.

The *third* way in which orogenic belts grow and thereby build the continental crust involves the addition of magma. The forces that heave up the mountains also generate huge amounts of heat in the underlying mantle that melts vast amounts of rock. Much of the resultant magma builds up in deep chambers, called *batholiths*, but some intrudes the crust and discharges onto the surface as volcanoes and lava flows.

Given these three processes and the wide range of materials involved, it is not surprising that orogenic belts are exceedingly complex in rock composition. These processes also help explain how continents can over long periods of time grow from a succession of orogenies each of which adds a great swath of land to the continental platform. This model of growth is confirmed by the general pattern of rock ages on all the continents. The interiors are very old and the margins are much younger.

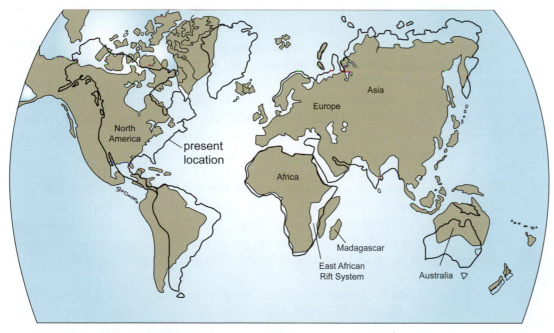

Figure 18.37 Possible global geography 100 million years in the future with Africa split apart along the East African Rift System.

Reduction by Breaking Up: Continents can also break up under the forces of plate tectonics. Strictly speaking, North America, South America, Australia, and Antarctica are remnant pieces of one massive continent's, Pangaea's, disintegration. We have already described some relatively recent examples of continents in the process of splitting apart. The most advanced is along the East African Rift System. Should this divergence continue, as the map in Figure 18.37 suggests, the eastern quarter or so of Africa will break off to form an island or two as Madagascar did about 75 million years ago (see Figure 18.1c). Asia also shows signs of splitting along the Lake Baikal Rift System, but the process is in its embryonic stages.

And North America is not to be outshone in this department. There is ample evidence of a separation zone extending from the Gulf of California northeastward across the American Southwest, and into the Rocky Mountains. It has produced a substantial rise in the crust, elevating the Colorado Plateau several thousand meters (shown in the chapter opening photograph), which in turn led to downcutting by the Colorado River and the formation of the Grand Canyon. In addition, it produced volcanic activity and increased numbers of earthquakes, as well as crustal fracturing and downfaulting over the large region in western United States called the Basin and Range. Yellowstone, in the Rocky Mountains, is a hotspot in this zone, the apparent product of a mantle-plume system that may be driving the splitting processes (Figure 18.38).

Figure 18.38 Yellowstone National Park marks a hotspot in the crust that may lie over a mantle plume.

We conclude that Earth has seen many different geographic arrangements of its landmasses in the past. Although the total area of continental platforms has gradually increased over the past 4 billion years, the shape, size, and location of individual landmasses has changed radically. The breakup of Pangaea is just one chapter in the plate tectonics story and it is still unfolding. As long as the Earth's internal energy system remains active new geographies will continue to emerge in the eons ahead. This means much more than simply changing the world map. Among other things it means new landforms and altered patterns of drainage on the continents, as well as changed patterns of ocean and atmospheric circulation leading to different climatic and biogeographical conditions for both Earth's lands and seas.

18.6 Geographic Significance of Plate Tectonics

Physical geography owes a great deal to plate tectonics. It strikes at the very heart of our theme of a dynamic Earth governed by great systems interacting on the planet's surface. Plate tectonics addresses some monumental scientific questions; in particular, how oceans open and close and how continents are constructed, break apart, change locations, and grow mountain ranges. But as fascinating as these questions are in terms of the forces, mechanics, and geologic processes involved, the geographic implications of plate tectonics may be even more scientifically exciting. Simply put, the changes induced by plate tectonics in surface systems such as ocean circulation, wind systems, storm patterns, precipitation, runoff, and biogeography have been literally world-shaping, especially when we consider their synergistic effects, that is, how they work in combinations to produce even greater effects than they would as a sum of the individual systems.

It would take an enormous effort to explore the full range of geographic changes brought about by plate tectonics but, for the present, let us take a look briefly at two. First, a puzzle in global biogeography, namely, the distribution of mammals among the continents, in particular, how Australia got its kangaroos; and second, the climate and drainage of south–central Asia, or how India got its summer monsoon.

How Australia Got Its Kangaroos: During the time of Pangaea, when Eurasia, North America, South America, Australia, and Antarctica, were clustered together around Africa, the geographic ranges of many plants and animals extended more or less uninterrupted over this vast landmass. But with the breakup of Pangaea, three major changes took place that shaped Earth's biogeography. First, formerly continuous populations of thousands of species became geographically separated by broad expanses of ocean allowing each population to take a different direction of evolution on each continent. Second, some landmasses after being isolated for millions of years joined or rejoined bringing different and sometimes competing species together. Third, landmasses changed locations, migrating great distances into different climatic zones resulting in new opportunities and constraints for their biota.

The evolution of mammals illustrates these effects. Marsupials (pouched mammals such as kangaroos and opossums) evolved about 100 million years ago when Africa, South America, Antarctica, and Australia were still connected or close to each other, and spread to all four landmasses. About 30 million years later, placental mammals evolved. The placentals competed with the marsupials for food, and rapidly drove the marsupials to extinction everywhere but in Australia and South America. Why? Because, as Figure 18.1c shows, Australia and South America had separated from Pangaea, or what was left of it, by the time placental mammals appeared.

Three of the nearly 200 marsupial species that inhabit Australia.

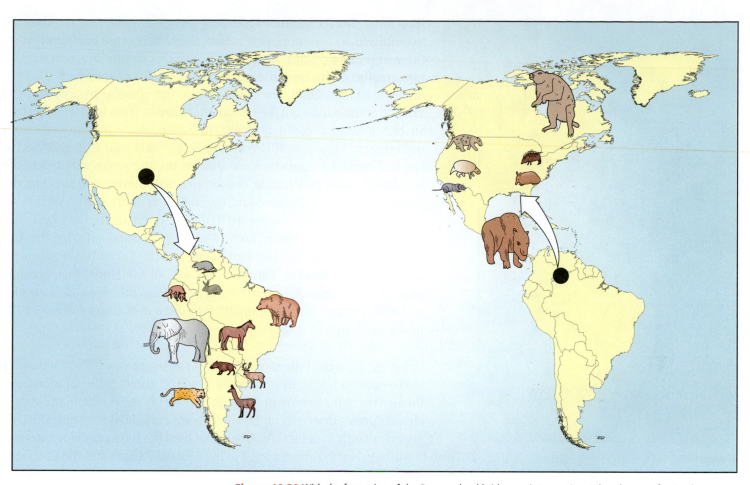

Figure 18.39 With the formation of the Panama land bridge, an intercontinental exchange of animals was initiated, which led to the extinction of most marsupials in South America.

The spread of placental mammals, therefore, was limited to North America, Europe, and Asia because these landmasses were still in contact with each other. Marsupials continued to thrive in both South America and Australia. But around 3 to 4 million years ago when North and South America became joined by the Panama land bridge. This led to an intercontinental exchange of animals, called the **Great American Interchange** (Figure 18.39), and when the placental mammals migrated into South America, the effects were the same as millions of years before, that is, marsupial extinction. Only the opossum survived and spread into North America.

This left only Australia with marsupial species and, without competition, they evolved into more than 180 species. However, geographic isolation from placental mammals was interrupted in the 1700s with European settlement. Among the placental mammals introduced by Europeans was the rabbit, which multiplied at astounding rates and had devastating effects on marsupial habitats (see Figure 11.21). As a result there have been significant population declines in many marsupial species.

Finally, what about Antarctica? It also split away from Pangaea at the same time as South America and Australia but, unfortunately for virtually all its animal and plant species, it moved into a deeply frigid polar zone, which proved far too stressful for these organisms. Today Antarctica supports only selected insect and bird species as terrestrial organisms.

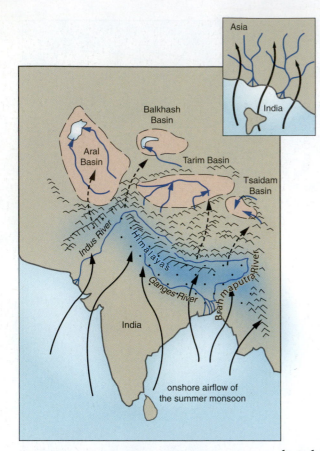

Figure 18.40 The Himalayan orogeny dramatically changed the drainage and climate of Asia giving rise to heavy monsoon precipitation over India and desert conditions in south–central Asia.

How India Got its Summer Monsoon: Orogenic belts take millions of years to build and, as they are being raised, they change the environment around them in profound ways. Besides the obvious changes in land elevation and topography, drainage and climate are radically altered as well. In southern Asia of 50 million years ago, before the Himalayan orogeny, streams drained southward from the continent's interior from land watered by moist air blown into the continent from the south, along the lines shown in the inset map in Figure 18.40). When India collided with Asia and the Himalayas and related mountains were thrown up, the mountains blocked moist air masses from penetrating the continental interior and feeding streams there. Streamflows declined and the direction of drainage was reversed so that streams now flow northward to the continental interior where they end in landlocked desert basins. Look at Figure 18.40 and note that the Aral Sea, the Balkhash Basin, the Tsaidam Basin, and the Tarim Basin of central Asia function in this fashion. On the other side of the Himalayas, precipitation increased dramatically, producing the famous monsoon climate of India, and with it came equally dramatic increases in the flows of rivers like the Indus and Ganges.

Orogenic belts can influence climate in two major ways. First, if they happen to rise across a system of prevailing winds, the transfer of moisture from the oceans to the continental interior is sharply reduced and the continental climate grows drier, streamflows decline, and inland seas shrink. This happened not only in central Asia, north of the Himalayas (as illustrated above), but also in western North America, east of the Coastal Ranges and the Cascades (see Figure 8.24), and in southern South America, east of the Andes. At the same time the climate on the ocean (windward) side of the orogenic belt grows much wetter and stream flows rise.

Second, climate changes as the land is elevated. For example, atmospheric temperature drops at an average rate of 6.4 °C per 1000 meters in altitude. Thus, with the growth of an orogenic belt, an elevation increase of 3000 meters can lower air temperature nearly 20 °C, producing a significant change in climate. For very high mountains, where the cold temperatures are coupled with increased precipitation on the windward side of the mountain belt, the result is heavy snowfall and the formation of glaciers. The glaciers and the mountain streams they help feed are powerful erosional agents which sculpt away large parts of the mountains quite literally as the mountains themselves are being upheaved by tectonic forces. The higher the mountains rise, the greater the force of erosional processes. On balance, it takes a lot of tectonic work to grow a truly big mountain, because there is a lot working against it.

Chapter Summary and Overview of The Formation and Geographic Organization of the Continents and Ocean Basins

Plate tectonics has emerged as one of the great advances in the long history of natural science. The theory holds that the Earth's lithosphere is subdivided into plate-like sheets that are moved around by slow motion in the mantle driven by Earth's internal energy system. Major plates function as input–output systems with seafloor spreading on the input side and subduction on the output side, but only the ocean-basin sections of plates are subject to subduction. Plate tectonics is critical to physical geography for many reasons. It determines the distribution of most volcanoes and earthquakes and, over the long term, it closes and opens ocean basins and influences the geographic arrangement of ocean currents, wind systems, climates, and the planet's plant and animal families.

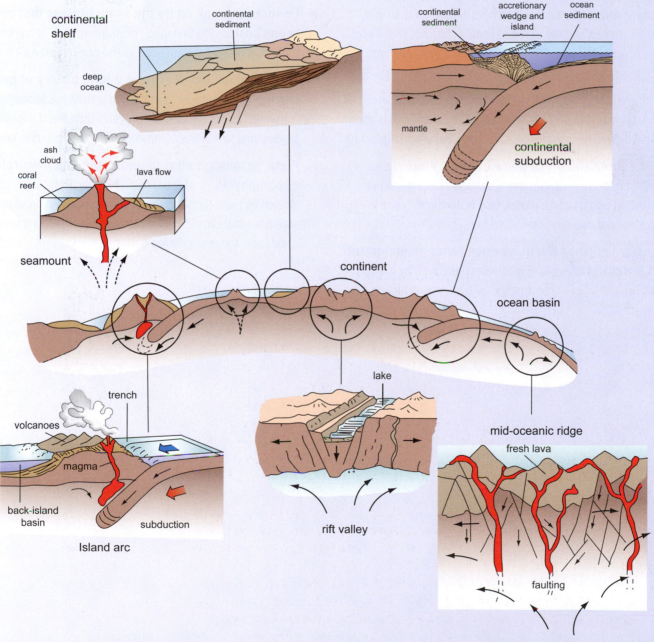

Summary diagram showing the essential features and processes associated with the global plate tectonic system and the building of continents and ocean basins.

▶ **The theory of plate tectonics ranks among the great advances in natural science.** It emerged from a concept called continental drift and the proposition that the Earth's continents were once grouped together in a single landmass.

▶ **Both the continents and the ocean basins are made up of three major subdivisions.** Each is marked by distinctive landforms and surface elevations and each is directly or indirectly related to plate tectonics.

▶ **Two major processes are critical to plate tectonics.** One is constructive, the other destructive and both processes generate earthquakes and volcanoes.

▶ **Each plate moves over the Earth's surface as a single, rigid unit.** The arc (angle) of movement is constant with distance from the Euler pole, but the velocity of motion increases with distance from the pole.

▶ **There are three geographic classes of tectonic plates.** All major plates except one contain at least one continental platform. Microplates often form in the contact zones between larger plates.

▶ **There are three classes of plate borders.** Which class a border belongs in depends on the directions of motion along the contact of two adjacent plates. One is destructive, one is constructive, and one is neither.

▶ **The western edge of North America forms an important plate border.** In California, it is marked principally by the San Andreas Fault. In Alaska, it is marked by a major subduction zone.

▶ **Ocean basins begin with the formation of a triple-junction rift system.** Two branches form a narrow ocean basin while the third branch becomes a failed arm and a potential influence on continental drainage.

▶ **Most ocean islands are volcanic in origin.** Three types of volcanic islands are found in the ocean basins and each originates from processes associated with tectonic plates.

▶ **Continents form and grow with repeated orogenies on their margins.** Orogenies thicken and extend the continental crust by ramming thrust sheets into the landmasses and by adding microplates and magma to the mountain mass.

▶ **Continents break up by the same process that creates ocean basins.** The first stage of disintegration is marked by the formation of a rift system such as the one in East Africa.

▶ **The Earth's landmasses have a long history of geographic change.** The total area of continental crust has increased since continents began about 4 billion years ago. But the locations, shapes, and size of the continents have changed dramatically.

▶ **Plate tectonics is vital to understanding physical geography.** Much of the world's biogeographical foundation was shaped by plate tectonics including the distribution of marsupial mammals and climates were changed by alterations in ocean circulation and precipitation patterns.

Review Questions

1 What is the theory of continental drift, and how did Alfred Wegener support it? What two key questions were not addressed by the theory, and how were these questions ultimately answered?

2 Referring to Figure 18.6, describe the relative geographical positions and elevations of the three major features of the continents: shields, orogenic belts, and continental shelves.

3 What are the relative locations and elevations of the major oceanic features: abyssal plains, oceanic ridges, and trenches?

4 Using Figures 18.11 and 18.12 and the accompanying text, list and briefly describe the sequence of processes involved in seafloor spreading.

5 How is it that subduction and seafloor spreading can be considered opposite, but equal, parts in a single system, what force(s) drives this system, and how does it apparently operate within the deep Earth?

6 Identify the relative changes in the temperature, density, and thickness of an oceanic plate as it moves from the mid-oceanic ridge to a subduction zone. What happens to the sediment that accumulates on the surface of the plate when it reaches the subduction zone?

7 What distinguishes the Pacific Plate from the other major plates and how fast and in what direction is this plate moving? What are microplates and what role do they play, if any, in the growth of continental crust?

8 Name the three types of convergent plate borders and identify which produced the Andes, the Himalayas, and the Aleutian Islands. Which of the three would you expect to be composed mainly of basaltic rock and why?

9 What is a triple junction and how does it lead to ocean basin formation? What is a failed arm and what role does in play in the geographic development of a continent?

10 Name the main types of islands produced in association with plate tectonics. Which is not associated with a plate border and explain why this island type forms where it does and why over geologic time it usually disappears.

11 Define the term orogeny and describe three ways that orogenies contribute to the growth of continental landmasses.

12 Are older orogenic belts physically different than younger ones as we see them today? How old are the newer ones considered to be, can you name some examples, and in general terms where are they located on the continents?

13 Continental crust is often described as relatively thick and stable. Is there any evidence that continents break up? In the past? When and where? In the present; and if so, can you cite an example location and some of the clues suggesting breaking up may be taking place?

14 What was the Great American Interchange, what did it have to do with plate tectonics, and what affect did it have on South American marsupials. How does this contrast with the story of marsupials in Australia?

Mountain Systems, Earthquakes, and Volcanoes

Chapter Overview

The Battle of Gettysburg, a pivotal event in American history, was strongly influenced by the lay of the land. But instances of landforms influencing battles and even shaping the development of countries and entire cultures are not unique in the annals of world history. Mountain ranges in particular have influenced where political boundaries are drawn, how religions and languages are distributed, and who trades with whom. So it behooves us to learn about mountains, both their geological and geographical aspects. In this chapter we want to learn first about their anatomies, that is, what do mountains look like on the inside? This will lead to a brief survey of mountain types and how rock is deformed by tectonic forces into folded, faulted, and volcanic structures. The remainder of the chapter is devoted to an examination of the two most studied and feared phenomena of mountain lands, earthquakes and volcanoes, with a glimpse at some of the most notorious of these natural villains including the infamous Mt. Pelee explosion of 1902 that killed 28,000 people, the more infamous Haitian earthquake of 2010 that killed more than 250,000 people, and the East Japan earthquake and tsunami of 2011 that killed over 25,000 people. The chapter ends on the question of the influence of volcanic eruptions on global weather and climate.

Introduction

We stopped on a low ridge near the town of Gettysburg, Pennsylvania, site of the famous Civil War battle, and silently surveyed the landscape around us. Bruce studied the horizon and said, "I tell you, Will, the whole battle was played out on a stage set by plate tectonics over 200 million years ago."

He pointed to a long mountain ridge just to the west. "That's the Blue Ridge, the main axis of the Appalachian Mountains, North America's eastern orogenic belt. It formed before Pangaea was assembled when a piece of continental crust rammed into North America from the east, thrusting up a great mass of crystalline rock. When Robert E. Lee brought the Army of Northern Virginia up here in the summer of 1863, he screened his movement from the Union scouts by staying west of the Blue Ridge." Bruce narrowed his focus and pointed to a notch in the

Above, the valley of Gettysburg, setting of the great battle, with the Blue Ridge in the background. Below, an artist's rendition of Pickett's charge against the Union lines on Cemetery Ridge.

ridge. "At that point there, the Chambersburg Notch, Lee's army turned east and entered the Gettysburg Valley. But they were quickly spotted by a unit of union cavalry under John Buford. Although the odds were overwhelming, Buford realized that unless he delayed Lee's advance the rebels would take the high ground on the valley floor before the main body of the Union Army could reach Gettysburg. That decision was crucial to the outcome of the great battle."

Turning south he extended his arms as if to embrace the valley and said, "The whole battlefield lies in an ancient rift valley. It formed about 200 million years ago, around the opening stage of Pangaea's breakup. One of the first breaks between Africa and North America began right here." He pointed to the front of the Blue Ridge. "A major fracture developed there. This side slipped down forming the broad structure of the Gettysburg Valley. What's important to the landscape we now see on the valley floor is that the rock here was intruded from beneath by magma and these intrusions formed many sills and dikes. It's the dikes that now stand slightly above the valley floor, forming the low ridges that together give the terrain its rolling topography."

We walked to a slightly higher point and looked down a long line of prominent war monuments along one ridge. "There," he said, "along that ridge, Cemetery Ridge, is where the Union Army made its stand. They positioned nearly 90,000 men behind a line extending all the way to those rock knobs to the south, Little Round Top and Big Round Top. Lee sent Pickett's division against the center of the Union line. The last 100 yards or so were up the front slope of the ridge directly into a massive barrage of Union musket and canon fire. Pickett's division suffered nearly 6000 casualties in 20 minutes. Topography meant everything here, and tectonic processes, driven by the gargantuan forces of moving plates, laid the foundation for these landforms and they, as much as anything else, determined the outcome of the battle."

19.1 Geographic Scale and Deformation of the Crust

The formation of an orogenic belt is a major planetary event. The process takes millions of years and its geographic coverage is broadly regional, typically covering an area of hundreds of thousands of square kilometers. Orogenic belts are structurally and compositionally complex and, as we saw in the last chapter, every one is built more or less in big pieces. These pieces represent the geographic subdivisions of a mountain belt, called **provinces**, and each province is characterized by certain types of rock structures and landforms.

In the Appalachian orogenic belt, shown in Figure 19.1, there are four provinces: Blue Ridge, Ridge and Valley, Piedmont, and Appalachian Plateau. The Civil War drama at Gettysburg was played out on the border of the Piedmont, a plateau-like surface composed of deformed metamorphic rocks, and the Blue Ridge, a mountainous spine of complexly folded igneous and metamorphic rocks. On the opposite side of North America, in the Cordilleran orogenic belt, which features the higher and more rugged Rocky Mountains, there are seven provinces. The map in Figure 19.2 gives you an idea of the marked differences in landforms among these western provinces such as between the Rockies and the Basin and Range Province.

Each mountain province, in turn, is made up of mountain ranges and, if we look still closer, each range is made up of individual mountains or groups of mountains. Finally, within each mountain there is a particular rock structure or complex of structures such as folds, faults, and/or some kind of solidified magma body. These structures are the architectural infrastructure of mountains; indeed, the term **tectonics**, which

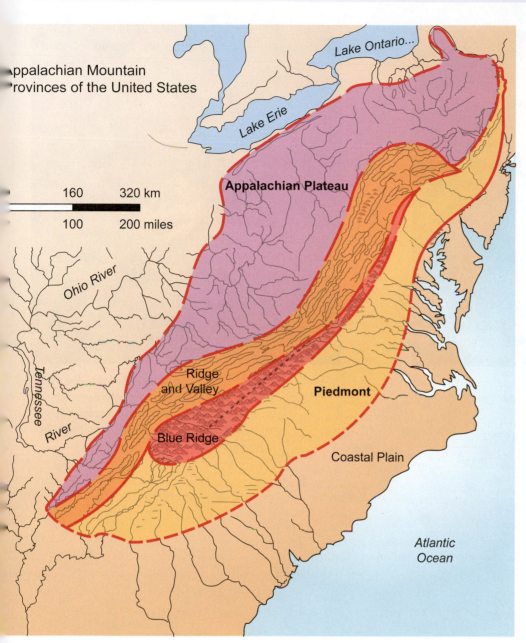

Appalachian Mountain Provinces of the United States

Lake Ontario...

Lake Erie

160   320 km

100   200 miles

Appalachian Plateau

Ohio River

Tennessee River

Ridge and Valley

Piedmont

Blue Ridge

Coastal Plain

Atlantic Ocean

Figure 19.1 The four provinces of the Appalachian orogenic belt, centering on the Blue Ridge.

we use to describe rock structures, is derived from the Greek word for architecture, *tecktonica*.

When we examine mountains as landforms, they may or may not reflect their inner rock structure, that is, the mountain's basic skeletal form is not always apparent from the outside. The reason for this is that most mountains take a long time to form and, as they are being raised up, erosional processes work on them, often changing their shape drastically. Glaciers, in particular, are capable of scouring the sides off a mountain, transforming a broad, rounded form into a much narrower, tower-like form which may show little resemblance to its original structure. The Matterhorn in the Alps on the Swiss–Italian border, shown in Figure 19.3, is the classical example of this mountain form, but there are thousands of other spectacular examples across the world.

The mountains which generally show the least erosional damage are young volcanoes, not because they are more resistant, but because they often form very quickly and erosional forces have not had much time to work on them. But the survival of a volcano can also depend on the environment into which it is born. Those that pop up in the ocean, as we noted in the last chapter, may be quickly obliterated by storm waves especially if exposed to the open sea. On the other hand, volcanoes in less rigorous settings, such as arid terrestrial environments, where runoff is weak and glaciation is absent, often retain their original structure for tens of thousands of years or more.

19.2 Folds, Faults, and Mountain Types

In the end, most mountain landforms represent some combination of structural and erosional influences. Like ancient buildings whose foundation and walls persist for thousands of years among architectural ruins in a weathered landscape, mountains also retain the influence of their original rock structure after millions of years of weathering and erosion as the photographs in Figure 19.4 suggest. The processes of natural decay work unevenly leaving the strongest and most resistant rock formations so that even after mountains have been worked down to mere hills, their internal structures are usually still apparent. Moreover, many mountains – the Himalayas are a good example – continue growing as they are being worn down.

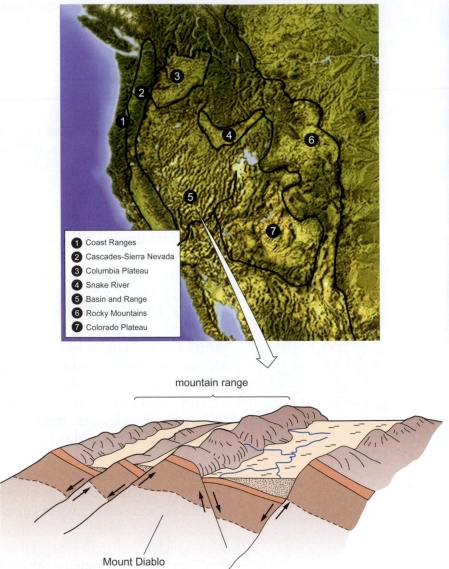

Figure 19.3 The Matterhorn in the Swiss–Italian Alps. The flanks of this great mountain have been cut away by the action of glaciers, changing the mountain to a steeper and narrower form that belies its once-broad structure.

Figure 19.2 The landform provinces of western North America with a section of one mountain range in the Basin and Range Province.

Figure 19.4 Ancient architectural ruins and tectonic ruins illustrating how differential weathering and erosion reduce rock formations to their basic and most resistant skeletal forms.

Earth's mountain ranges fall into three main classes: **folded**, **volcanic**, and **block-faulted**. Mountains of the ocean basins, including arcs, ridges, and seamounts, are almost exclusively volcanic whereas those of the continents are mainly folded. Both folded and volcanic mountains may also include fault structures. Faults come in all sizes and in some places involve great blocks of crust so large that the resultant structures themselves form individual mountain ranges. The mountains of the Basin and Range Province, shown in Figure 19.2, are an excellent example of block-faulted mountains. And it not uncommon for all three, folding, volcanism, and faulting, acting together in one big event or in separate events millions of years apart, to be responsible for a mountain range.

Folding is the process by which rock formations are bent into curved forms by forces within the crust. **Faulting** involves the displacement of rock along a fracture in the crust as a result of forces associated with plate movement or forces acting within a mountain range or section of crust, and **volcanism** involves the movement of magma through the crust and onto the Earth's surface where it is discharged as lava, ash, and gas.

Rock Deformation: In order to understand the processes of rock deformation, we should begin with a look at the behavioral characteristics of rocks when placed under stress. **Stress** is defined as a force that is acting on a body. If the force is not equal in all directions, it is referred to as *differential stress*. The diagrams in Figure 19.5 show the three main types of differential stress. Under tensional forces rock is pulled apart, as in seafloor spreading along the mid-oceanic ridges. Under compressional forces rock is squeezed together as in subduction zones where tectonic plates are moving against each other. And under shearing forces the stresses act parallel to each other but in opposite directions like the motion of scissors. Shearing forces produce transform faults, as we find along the California coast, for example, on the contact between the Pacific and North American plates (see the photograph in Figure 18.26).

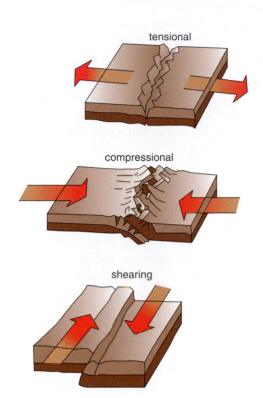

Figure 19.5 The three types of differential stress that produce rock deformation: tensional, compressional, and shear.

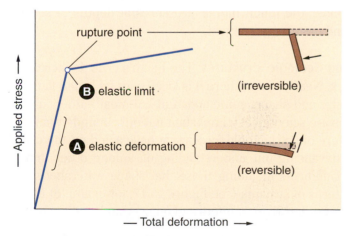

Figure 19.6 Elastic (reversible) deformation (A) in rock resulting from the application of stress. At B the rock reaches its elastic limit and ruptures.

When rock is subjected to differential stress, such as compression and shearing forces along the edge of tectonic plates, it may undergo three stages of deformation: elastic, plastic, and rupture. The first stage, marked by A in Figure 19.6, is **elastic deformation**. A simple test can be used to define elastic deformation; a slab of any solid substance, including rock, is bent but when the stress is withdrawn, it returns to its original shape and size. For each type of rock, however, there is an elastic limit (B in Figure 19.6), and if it is pushed beyond this limit it will either rupture or remain permanently bent. At this point it cannot return to its original shape when the stress is relaxed.

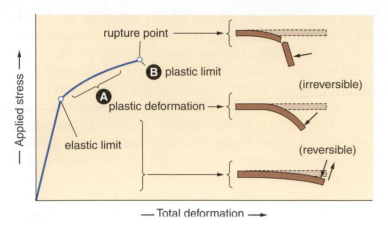

Figure 19.7 Plastic (irreversible) deformation (A) from the application of stress. At B the rock reaches its plastic limit and ruptures. Folding is a form of plastic deformation.

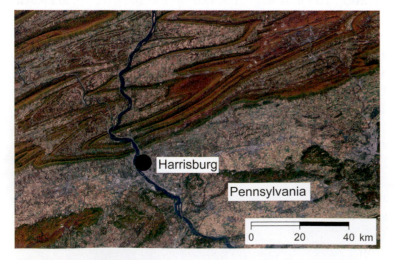

Figure 19.8 Plastic deformation in ice approximates the mechanics of folding in rock.

Figure 19.9 Mapping landforms such as these curious bedrock ridges in the Appalachian Mountains of southeastern Pennsylvania, is an important starting point in deciphering the nature of the larger rock formations underground.

Now move to Figure 19.7. The set-up is the same, but when the rock hits its elastic limit, it bends irreversibly. In other words, it is permanently bent. This constitutes **plastic deformation** or folding (A in Figure 19.7). If additional stress is applied to rock that is being folded, it may be pushed to its plastic limit, beyond which a fracture develops and it fails (B in Figure 19.7). The failure qualifies as a fault only if displacement takes place. How a rock behaves under differential stress, that is, whether it is apt to fracture and fail or bend into folds, depends on the rock type and many other factors.

These factors include certain internal properties, such as mineral composition and crystal structure, as well as the rate at which stress is applied, the rock's temperature, and the confining pressure exerted by the surrounding rock. It is very important to understand that stress in the crust builds up slowly over very long periods of time, millions of years. Because application is very long and gradual, rocks, such as granite and limestone, which can be very brittle in short-term laboratory tests, often behave as plastics in nature.

We can demonstrate this phenomenon with ice. Under sudden application of differential stress, ice will rupture, but when stress is applied gradually, it will bend appreciably as the photograph in Figure 19.8 illustrates. This accounts for the fact that glaciers are able to bend and flow. Thus, ice is both brittle and ductile (bendable), depending on the rate at which stress is applied. For most rock, confining pressure is also significant, because deformation usually takes place deep within the crust. Experiments show that rocks such as granite and limestone, which are brittle at the surface, are ductile if confined under the much higher pressures at depths of several thousand meters in the crust.

Mapping Geologic Structures: The Earth's crust is incredibly complex. Nearly everywhere it is laced with folds and faults of every size and shape, some apparent but most hidden. These features are scientifically important not only in understanding the nature of the tectonic forces that have acted on the crust, but in advancing our understanding of, among other things, major landforms including those that shape watersheds and the patterns of streams (see Figure 15.17, for example) and in locating groundwater aquifers and oil, gas, and coal deposits.

To figure out the broad patterns of crustal deformation several methods of investigation can be employed. One commonly used involves tracing the patterns of rock formations as they appear on topographic maps and aerial imagery such as that shown in Figure 19.9. This method may reveal certain shapes

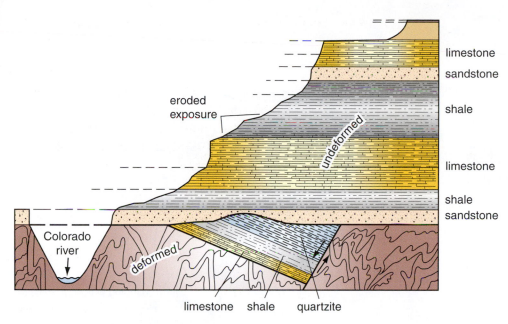

limestone
sandstone
shale

limestone

shale
sandstone

undeformed

Colorado river

deformed

limestone   shale   quartzite

Figure 19.10 A section of the Grand Canyon showing largely undeformed rock formations lying over highly deformed ones.

and geographic patterns in landforms that provide clues about geologic structures deep underground. But to verify such findings and provide detailed data, it is often necessary to go into the field and measure individual structures. The task can be difficult, for the bulk of most structures is hidden at depth and those portions that are exposed are often heavily altered by erosion. In non-mountainous regions, moreover, practically the entire upper surface of the bedrock is buried under thick mantles of soil and surface deposits. How, then, do we piece together the evidence that leads to the identification and classification of the geological structures of the crust?

The traditional method involves measuring the geometric attitude of the rock in outcrops. Outcrops are exposures of bedrock, and geometric attitude is their directional orientation. Outcrops range from small projections or ledges to huge exposures like those in the Grand Canyon that are more than a thousand meters high and many kilometers long. Inasmuch as most of the bedrock at or near the surface of the continental crust is sedimentary rock, most outcrops will exhibit distinct layers of strata, and knowing this is important in reconstructing geologic structures.

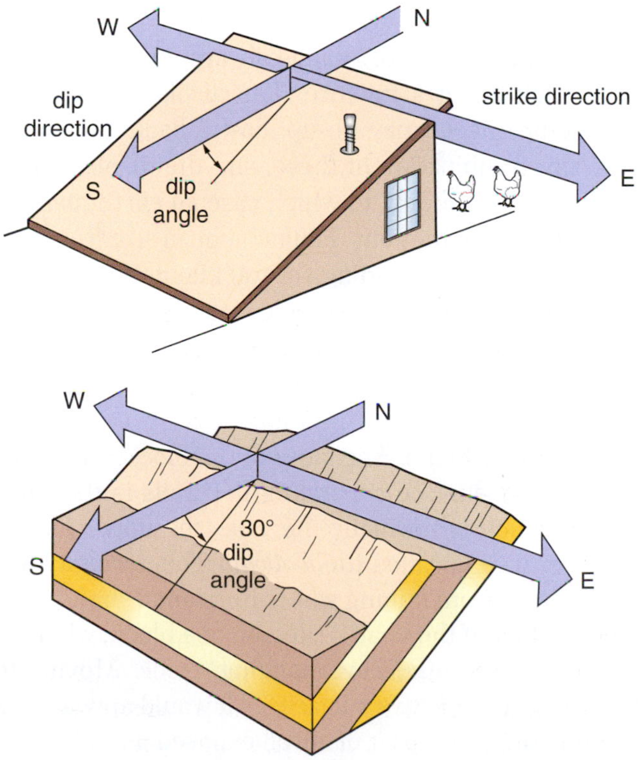

A first consideration in investigating an outcrop is whether the strata are horizontal or tilted. This is based on the **principle of original horizontality**, which holds that the layers of sediment that become rock strata were originally deposited in horizontal beds; thus, the bedding (strata) in undisturbed sedimentary rocks should be flat-lying. In addition, a sedimentary layer is deposited as a continuous sheet that usually ends by thinning away gradually or by changing gradually in composition with distance. Thus, where we find sedimentary rock has been tilted, bent, or broken as in the lower part of Figure 19.10, it means they have been deformed.

W   N

dip direction   strike direction

S   dip angle   E

W   N

S   30° dip angle   E

Figure 19.11 Strike and dip are illustrated here using a chicken coop (above) as a model for tilted sedimentary rock formations (below).

Strike and dip are the two directional properties measured in rock outcrops. Both are depicted in Figure 19.11. **Dip** is the angle of inclination of a bed from the horizontal. **Strike** is the direction perpendicular to dip where the broken or eroded exposures of the dipping beds intersect the surface. By marking the strike and dip directions of many outcrops on a map, it is possible to reconstruct individual folds and faults (Figure 19.12). They in turn can be combined with other fold and fault patterns to reconstruct larger structural patterns such as those of whole mountain ranges. The structural trends of mountain ranges in turn enable scientists to work out the relationship between crustal deformation and the tectonic forces associated with plate movement and mantle plumes.

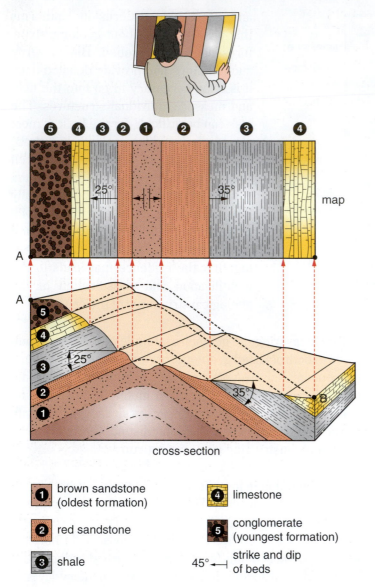

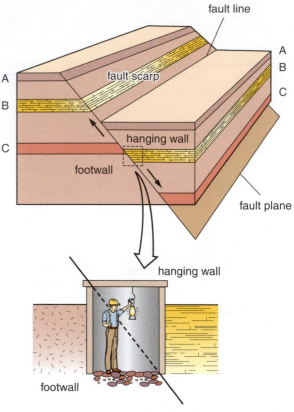

Figure 19.13 The basic properties and terminology of a fault structure including the hanging wall and footwall which are illustrated by the miner's shaft.

1 brown sandstone (oldest formation)

2 red sandstone

3 shale

4 limestone

5 conglomerate (youngest formation)

45° ⊣ strike and dip of beds

Figure 19.12 The application of strike and dip in mapping a fold which is not apparent based on the configuration of landforms visible on the surface.

Properties of Faults and Folds: Faults always involve the displacement of rock along a fracture. The displacement, relative to the fracture itself, may be up, down, back and forth, in or out, or any combination of these, and the displacement process itself may be sudden, in which case an earthquake is produced, or gradual, more like small slipping motions. Faults range in size from tiny ruptures only a few centimeters deep to those extending several kilometers into the crust. The longest faults are found along the edges of tectonic plates, where swarms of faults form branching systems that can be traced thousands of kilometers along plate borders (see Figure 18.36).

A standard terminology is used to describe faults. The faces of the blocks on either side of the fault are called the **walls**; the surface separating the walls is the **fault plane**. If the fault plane is inclined, which is normally the case, the upper face is called the *hanging wall*, and the lower face is called the *footwall*. This terminology, which is illustrated in Figure 19.13, comes from mining where horizontal shafts were often located along faults. The floor or foot of the shaft rested on one block; whereas the ceiling of the shaft was carved from the other, overhanging, block. Moving to ground level, the trend of the fault along the Earth's surface, as it would appear on a map, for example, is termed the **fault line**. The part of a wall exposed as a result of displacement is the **fault scarp**. In mountainous areas, fault scarps often appear as prominent slopes of cliffs or even as mountain fronts.

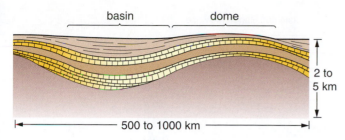

Figure 19.14 The broadest of folded structures, basins and domes, often measure hundreds of kilometers across and are common in the continental interiors.

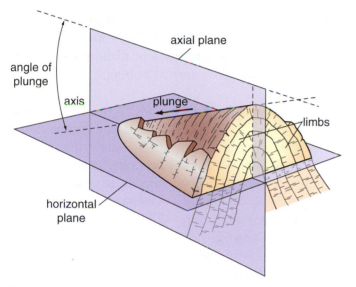

Figure 19.15 The basic properties of folds include limbs, axial plane, and angle of plunge.

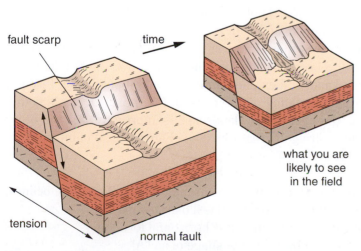

Figure 19.16 The basic structure of a normal fault where tensional stress has resulted in downward displacement of the hanging wall.

Folds are structures that resemble warps or wrinkles in rock. They are the most common geologic structure in orogenic belts and are often elegantly displayed in exposures of sedimentary and metamorphic rocks along mountain canyons, cliffs, and even in hillside cuts made in the construction of railroads and highways. Folds are also common to non-mountainous areas, though their configurations are far less dramatic. They usually take the form of broad upward and downward flexures, like those shown in Figure 19.14, called **domes** and **basins**. These structures are typically hundreds of kilometers in diameter and they often have a marked influence on the landscape, particularly on the sizes and shapes of watersheds and the locations of major streams. The Seine River watershed in northern France, for example, is shaped by the Paris Basin. Tributary streams such as the Oise, Marne, and Yonne drain toward the center of basin and meet at Paris to form the Seine, which flows westward to the English Channel.

The basic terms used to describe folds are illustrated in Figure 19.15. The two sides of a fold are the limbs; an imaginary plane drawn between the limbs, which divides the fold in half, is called the axial plane; and the crest of the fold is the axis. If the axis is inclined from the horizontal, the fold is said to plunge. Imagine a plunging fold to be somewhat like a submarine emerging on the ocean surface. When a fold is exposed at the Earth's surface it is usually torn apart by erosion – which has happened to most of those in the Appalachians – leaving only remnants of limbs visible as ridges in the landscape, as illustrated in Figure 19.9. Therefore, to determine the true size, dimensions, and geographic trends of the fold it is usually necessary to reconstruct the original structure on paper from strike and dip measurements.

Types of Faults and Folds: Faults are products of differential stress in the crust created by tensional, compressional, and shearing forces. The motion of displacement in a fault is dictated by which of these forces is dominant in a fracture zone. Faults are classed according to the direction of displacement of one wall relative to the other. If the displacement is up or down along the fault plane it is called a dip-slip fault. If the displacement is parallel to the fault line, as in the transform faults on the ocean floor, it is termed a strike-slip fault. Displacement that combine strike- and dip-slip are oblique-slip faults.

Common Fault Structures: Three basic types of dip-slip faults are recognized: normal faults, reverse faults, and thrust faults. In a **normal fault** the hanging wall is displaced downward relative to the footwall, exposing the upper part of the footwall in the form of a fault scarp as shown in Figure 19.16. Many mountain ranges in the American West, including the Wasatch of Utah and the Sierra Nevada of California, represent normal

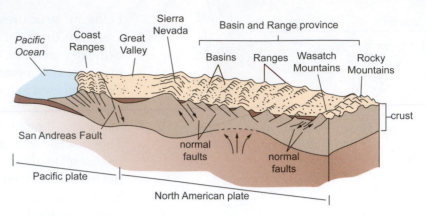

Figure 19.17 A generalized cross section from the Rockies to the Sierra Nevada showing the mountain ranges formed by normal faulting.

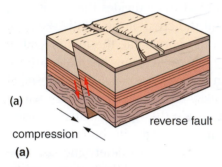

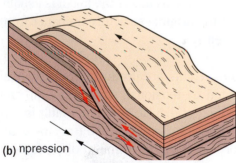

Figure 19.18 The basic structures of (a) a reverse fault where compressional stress has resulted in upward displacement of the hanging wall, (b) a thrust fault where extreme compressional stress produces lateral displacement in which the headwall overrides the hanging wall.

faults. Take a look at the diagram in Figure 19.17. In the Sierras the fault scarp faces east; in the Wasatch the fault scarp faces west. Between the Sierra Nevada and the Wasatch lies an extensive called the Great Basin – which we mentioned in the last chapter as part of the North America's zone of separation or divergence – where range after range of mountains has been formed by normal faulting.

The opposite displacement produces a **reverse fault** in which the hanging wall moves up relative to the footwall (see Figure 19.18a). If a reverse fault is subjected to great compressional force, and the fault plane is inclined at a low angle, the resultant movement would be mainly horizontal. This is a **thrust fault** and one is shown in Figure 19.18b. They are known to drive slabs of rock laterally for tens of kilometers over the surface. In some parts of the Rocky Mountains, rock formations were thrust great distances to form prominent front ranges that rise like a great wall along the western edge of the Great Plains. The Canadian Rockies in Alberta are a case in point; great slabs of rock were thrust eastward and piled against one another in an overlapping fashion to form an imposing front.

Finally, some dip-slip faults involve two fault planes. A **rift**, or **graben**, which we have already explored at some length, forms where a block is displaced downward between two normal faults. The opposite type of structure, called a *horst*, results in the elevation of a block between parallel normal faults. Rifts and horsts are common features in zones of tensional motion, and both are shown in Figure 19.19. The East African Rift System, which is shown in Figure 18.21, is the most extensive graben system on Earth, outside the ocean basins, of course. It is over 5000 kilometers long beginning in the Holy Land, and extending down the Red Sea Basin, through the Highlands of East Africa and ending on the coast of the Indian Ocean in southern Africa. Its influence on the East African landscape is impressive. Among other things, it forms the basin for a string of great lakes (see Figure 16.25) comparable to the North American Great Lakes.

The simplest folds can be described in two-dimensional terminology: monoclines, anticlines, synclines, and overturned folds. Each is illustrated in Figure 19.20.
- A **monocline** is a single bend in an otherwise horizontal formation;
- an **anticline** is a double bend upward in the shape of an "A";
- a **syncline** is the downward counterpart of an anticline.

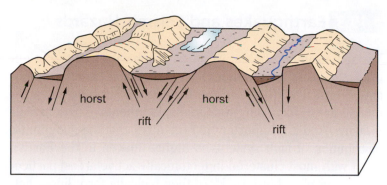

Figure 19.19 A system of rifts and grabens, common to areas of tensional motion in the crust such as East Africa.

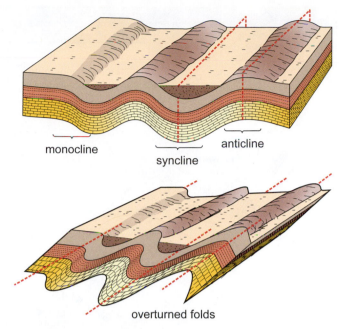

Figure 19.20 The four basic types of folds: monocline, syncline, anticline, and overturned.

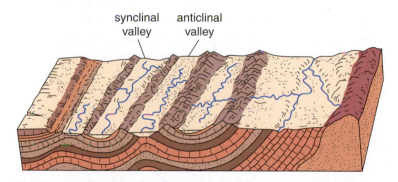

Figure 19.21 In the Ridge and Valley Province of the Appalachian Mountains valleys form in both synclines and anticlines.

They may be either symmetrical or asymmetrical depending on whether the axial plane is vertical or tilted. In extreme cases the asymmetry may be so severe that both limbs dip in the same direction (see (b) in Figure 19.20). This is called an *overturned* fold and if it is tipped far enough, the lower part of the fold may actually be turned upside down.

The Ridge and Valley section of the Appalachian Mountains contains spectacular examples of folding in sedimentary rocks. The ridges and valley trend north and south; when viewed in cross-section, they resemble a washboard terrain. On close examination, however, it is apparent that the landforms produced by the folding deviate from the simple structural pattern of anticlines and synclines in two ways. First, the folds plunge, like the section of the Ridge and Valley Province shown in Figure 19.9, so that the ridges formed by the exposed limbs form a zigzag geographic pattern. Second, valleys have formed not only in the synclines, where you would expect to find them, but in the anticlines as well, also shown in Figure 19.21. To figure out the origin of anticlinal valleys, you must infer that the tops of some anticlines have been eroded away, exposing the underlying, often weaker, formations to erosion. The ridges are formed from the limbs of adjacent anticlines and synclines.

Summary on Mountain Structures: Evidence of bending, breaking, and melting of the crust is found in abundance across the Earth, but it is in mountains that it is most conspicuous. Here stresses driven directly or indirectly by the motion of tectonic plates deform rock into folded, volcanic, and faulted structures. These structures can be mapped and analyzed to help us understand, first, the origin and nature of mountain ranges and orogenic belts, and second, the geographic significance of these features, because they set the stage upon which Earth's landscapes are built, including, among other things, major influences on drainage patterns and the distribution of groundwater and soil.

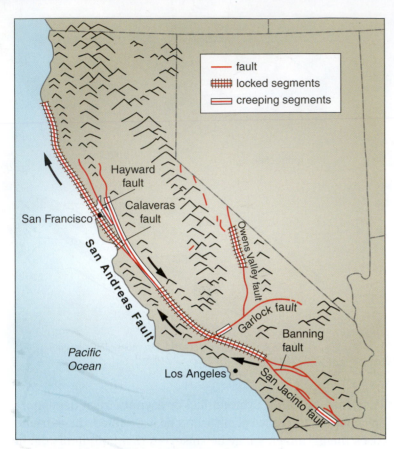

Figure 19.22 The San Andreas Fault System of western California. The main fault line marks the contact between the Pacific and North American plates and is made up of both locked and creeping segments.

19.3 Earthquakes and Seismic Hazards

We now examine the nature of fault behavior and the origin of earthquakes, and then go on to earthquake measurement and destructive effects on the landscape. Virtually all earthquakes are produced by sudden displacements of rock in fault systems. And it is important to remember that the geographic distribution of fault systems is highly uneven in the Earth's crust with most clustered in tectonically active areas, particularly along plate borders. Here they typically form linear networks, or systems, with branches leading from a main fault. The San Andreas Fault System, shown in Figure 19.22, is a prime example.

The Behavior of Fault Systems: Based on decades of continuous monitoring of selected fault systems, most notably those in California, we have learned that individual faults, especially large ones, do not behave as single, integrated units. Rather, it appears that they are divided into segments each with a different mode of motion or behavior somewhat like the fast and calm segments in a mountain stream. Some segments have fairly uniform, uninterrupted flows, whereas others, such as where a dam collects water and then bursts, have highly sporadic flow. In a fault system, segments are defined by the magnitude and frequency of seismic activity, that is, by the size and frequency of the earthquakes they produce.

Two types of segments have been defined in active fault systems: creep segments and locked segments. In *creep segments*, movement is slow and steady resulting in frequent, small earthquakes. In *locked segments*, motion is characterized by long periods of little or no action punctuated by infrequent, sudden bursts that result in big earthquakes. The locked segments behave like they are stuck in place and, as a result, stress, or *strain energy*, builds up in them until it is suddenly released in a massive jolt. Strain energy can build up for decades or centuries before it is unleashed in an earthquake which may shake the crust for hundreds of miles around. Notice in Figure 19.22 that San Francisco lies on a locked segment of the San Andreas Fault.

An **earthquake** is defined as shaking of the ground produced by a sudden release of stress when rock shifts in a fault. The earthquake's energy, called seismic energy, is released in waves that travel through the crust dissipating in strength with distance from the fault. Understanding earthquakes begins with the two main terms of reference, focus and epicenter, which are shown in the diagram in Figure 19.23. The earthquake's **focus** (also called **hypocenter**) is the precise underground center of the shift or rupture, whereas the **epicenter** is the geographic location on the surface directly above the focus. The locations of these two points are determined by measuring the seismic waves received at several distant seismic-recording stations and calculating back to the waves' points of origin. Not only can locked and creeping fault segments be discriminated using this technique, but hidden faults can be detected as

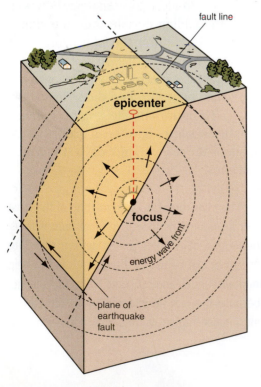

Figure 19.23 The key features of an earthquake: epicenter, focus, fault plane, and fault line. The epicenter is located directly above the focus.

well. The 1994 Northridge earthquake in the Los Angeles metropolitan region was produced by a hidden fault and resulted in $15 billion damage to bridges, highway overpasses, buildings, and other facilities.

Most earthquakes involve more than an isolated discharge of seismic energy. Some are preceded by a series of small quakes or "seismic noise" that is only detectable on seismic instruments. The main event, however, is nearly instantaneous, often lasting only seconds, and the energy waves it produces travel outward at more than 10,000 kilometers per hour (6000 miles per hour) with decreasing strength and destructive effect over distance. After a major earthquake, the fault usually continues to produce numerous small earthquakes, called *aftershocks*, as the rocks around the focus adjust to their new positions. Aftershocks can continue for a year of more, and some may be strong enough to cause damages to structures weakened by the main event.

Figure 19.24 The Haiti earthquake of 2010 was one of the most destructive in history not because its magnitude (7.0) was so great, but more because of its location and the nature of the landscape affected.

Measuring Earthquake Size and Destructiveness: It is estimated that every year the Earth produces over 30,000 earthquakes strong enough to be felt at the surface. Only about 75 of these could be called significant, that is, large enough to cause damage in the landscape, and of these most occur in the ocean basins and remote land areas where there are few people and settlements. Occasionally, however, one like the Haitian earthquake of 2010 hits a population center where it causes massive damage and alarming loss of life. An important question before us is just what is in an earthquake that enables it to cause damage?

The **seismic waves** produced by earthquakes fall into two general classes, surface waves and body waves. **Surface waves** travel within the outermost layer of rock and soil and are characterized by complex motion that includes wave-like ground rolls as well as a side-to-side ground motion. This lateral motion is like that of a snake crawling on the ground and it is responsible for most of the heavy damage in buildings and other structures such as that shown in Figure 19.24.

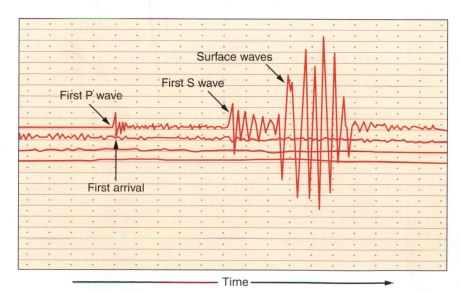

Time

Body waves are represented principally by primary (P) waves and secondary (S) waves both of which we discussed in Chapter 17 in connection with the exploration of the inner Earth. Recall that P waves produce push/pull, or compressional/decompressional motion in rock particles, whereas S waves produce a crossways or shearing motion in rock particles. When seismic waves are recorded on a seismograph, the recording pen makes a tracing on a moving sheet of paper scaled in minutes and seconds such as the one shown in Figure 19.25.

Figure 19.25 Earthquake energy waves recorded on a seismograph. The tracing shows the arrival times of P waves, S waves, and surface waves.

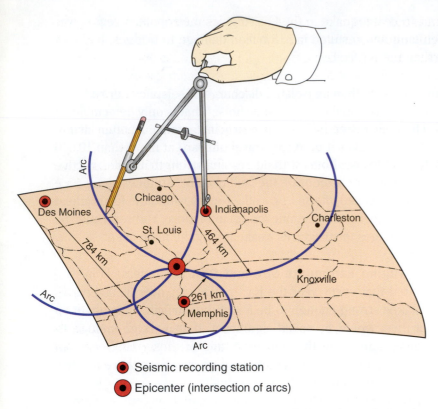

Figure 19.26 Finding the earthquake epicenter is a fairly simple exercise based on the difference in travel times between P waves and S waves received at recording stations in different geographic locations.

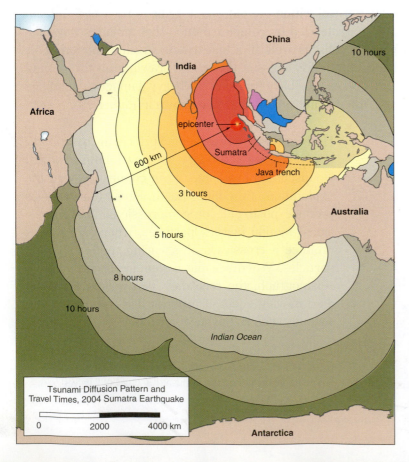

Tsunami Diffusion Pattern and Travel Times, 2004 Sumatra Earthquake

0   2000   4000 km

A typical seismograph tracing reveals that when an earthquake takes place, P waves are the first to be recorded followed by S waves and surface waves. Thus P waves are the fastest, nearly two times faster than S waves, and surface waves are the slowest, traveling at about nine tenths the S-wave velocity. The farther the waves travel from the earthquake, the greater the interval between their travel times. Therefore, if you work backwards from a distant recording (seismic) station until the time interval between the P and S waves is zero, you have defined the distance from the station to the epicenter. This distance can be plotted in an arc, as demonstrated in Figure 19.26, and the point where it intersects with arcs from other recording stations, marks the location of the earthquake epicenter.

One major benefit of this technique is that earthquakes on the ocean floor can be quickly located and the sea surface in the area monitored for huge waves, called **tsunamis**, which can cause serious damage to coastal areas. The Indonesian tsunami of 2004, one of the largest in recorded history, was produced by a huge earthquake in the ocean floor off the west coast of Sumatra. The giant wave killed more than 200,000 people on the perimeter of the Indian Ocean and many of these people lived in coastal areas more than 1000 miles away from the epicenter (Figure 19.27).

The second thing that can be learned from a seismograph is the **earthquake magnitude**, defined as the total energy released by an earthquake. This method of measuring magnitude is based on the amplitude of the largest peak recorded on a seismograph tracing during an earthquake. The *amplitude* is the actual height of the tracing (that is, the graph line, like the one in Figure 19.25), which is equal to one-half the distance between the peak and an adjacent trough. This method of measuring magnitude formed the basis for the standard scale for gauging the size of earthquakes, the Richter magnitude scale.

The Richter Magnitude Scale: This scale was originally proposed by the American seismologist Charles Richter (1900–1985) and today is generally known as the **Richter scale**. To interpret the number assigned to an earthquake, it is important to understand how this scale is calibrated. Because the difference in amplitude between small and large earthquakes may be huge, a thousand times or more, Richter set the scale so that a tenfold amplitude increase corresponds to one unit of magnitude. Thus, the amplitude of a surface wave produced by a magnitude 5 earthquake is 10 times greater than on for a magnitude 4 earthquake. In mathematical terms, the Richter scale is logarithmic because each unit represents a 10-fold change.

Figure 19.27 Tsunamis are one of the most destructive forces in coastal areas and they can be caused by both earthquakes and volcanoes. The Sumatra tsunami of 2004, which was generated by a huge earthquake and moved across the Indian Ocean in a matter of hours, killed over 200,000 people along the coast.

Even more important is the fact that in terms of earthquake energy each unit on the Richter scale is roughly equivalent to a 30-fold increase. Thus, the energy released in a magnitude 6 earthquake is 30 times greater than magnitude 5 and 900 (30 × 30) times greater than a magnitude 4. A major earthquake with a magnitude of, say, 8.5 produces millions of times more energy than a small one barely felt by humans.

Like all natural processes, earthquakes follow the magnitude and frequency principle in which frequency of occurrence declines with higher magnitudes. Table 19.1 relates Richter magnitudes to annual frequency of occurrence worldwide and to the general impacts on landscape and people associated with each magnitude. Notice that magnitudes 8 or larger are designated **great earthquakes**. Great earthquakes in the 8.0 to 8.9 magnitude range occur on the average of once a year for the whole planet. Larger ones occur far less often, but it would take many centuries of seismic monitoring to even approximate their average frequency. Since frequency is based on the average return period, it is important to remember that this number has nothing to do with the actual time of an earthquake's occurrence. In other words, it is possible that five to ten years may pass without a great earthquake on Earth or, conversely, two or three great earthquakes may strike in a single year. Seismologists estimate that the Los Angeles area is hit by a great earthquake every 160 years, but one has not struck the area in two centuries or more.

Table 19.1 Earthquake magnitude, frequency, and impacts

Magnitude	Annual frequency worldwide	Designation	Geographical effects
2–2.9	1,300,000	Very minor	Rarely felt by people but detected by instruments
3–3.9	130,000	Minor	Felt by people but no landscape damage
4–4.9	13,000	Light	Some landscape damage but no loss of life
5–5.9	1319	Moderate	Moderate landscape damage, some injuries, but no loss of life
6–6.9	134	Strong	Significant landscape damage with loss of life in populous areas
7–7.9	17	Major	Major landscape damage with large losses of life in populous areas
8–8.9	1	Great	Massive to total landscape destruction and heavy loss of life
9+	Unknown	Great	Total landscape destruction and massive loss of life.

Source: US Geological Survey, and others.

Moment Magnitude Scale: In 1979, scientists developed a second magnitude scale, called the **Moment magnitude scale**, to provide a more direct measure of an earthquake's total energy output. It, too, is ten-based (that is, logarithmic) and gives readings generally comparable to the Richter scale for small and medium-sized earthquakes. But it is different in how magnitude is measured. Instead of measuring it based on a seismograph recording, the Moment scale measures earthquake energy underground at the site or point of the displacement based on three factors:

- the amount of the rock displacement (slip), that is, the length or height of the offset;
- the extent or area of the displacement; and
- the rigidity of the rock involved.

Although the Richter scale is still the conventional international measure of earthquake magnitude, the Moment scale is considered more accurate for large quakes (greater than 7.0), and is the standard for the US Geological Survey, the agency most responsible for measuring and assessing earthquakes worldwide.

The Mercalli Scale: While the magnitude scales give us a measure of the raw energy produced by an earthquake, they may or may not reveal much about its destructive effect on the landscape, especially the built environment. Consider the occurrence of a great earthquake deep within the upper mantle far out in the ocean. Chances of it destroying cities, dams, and highways are remote because its energy level is greatly diffused (weakened) over the long travel distance to a populated land area. On the other hand, consider a more modest event, say, one of magnitude 6.5 or 7.0, with a shallow focus directly below a city. The destructive potential is enormous, as was clearly illustrated with the 2010 earthquake in Haiti with its focus near the heavily populated urban area of Port-au-Prince. Clearly, another scale is needed to measure the destructive power of an earthquake, or what is termed earthquake **intensity**.

The most widely used intensity scale, called the **Mercalli scale** (or the Mercalli modified scale), was developed in the late 1800s and early 1900s and is generally used worldwide today. The Mercalli scale uses 12 intensity levels ranging from those quakes that are hardly felt (level I) to those that are cataclysmic (level XII) in which the entire landscape is destroyed, that is, buildings are collapsed, forests are leveled, and landforms like cliffs and hillsides fail. The following discussion describes some high-intensity earthquakes.

Earthquake Magnitudes and Destructive Effects: Since the Richter scale has been in use (beginning in the 1930s), only one truly great earthquake has struck North America. It was the Alaskan quake of 1964. It measured 8.4 on the Richter scale (9.2 on the Moment scale), giving it global status among great earthquakes. The quake itself lasted nearly four minutes, damaging buildings over an area of 100,000 square kilometers, killed 131 people, and was so strong that it caused a 2 centimeter rise and fall in the ground as far away as Orlando, Florida. The photograph in Figure 19.28 shows some of the physical damage in Anchorage, Alaska, about 75 miles west of the epicenter, where the intensity reached levels X and XI on the Mercalli scale. Had the day of the event not been a holiday (Good Friday), the death toll would have been in the thousands. The cause of this earthquake, now known as the **Good Friday Earthquake**, was a large displacement associated with the subduction of the Pacific Plate under the North American Plate, the source of thousands of earthquakes every year (see Figure 18.36).

The Great East Japan (Sendai) Earthquake of 2011: If we move westward from Anchorage along the Pacific rim, we reach the Asian subduction zone of the Pacific Plate, where on March 11, 2011, Japan was hit by a magnitude 9.0 earthquake, one

Figure 19.28 Damage caused by the Alaska earthquake of 1964, considered the largest event in North America in the twentieth century.

(a)

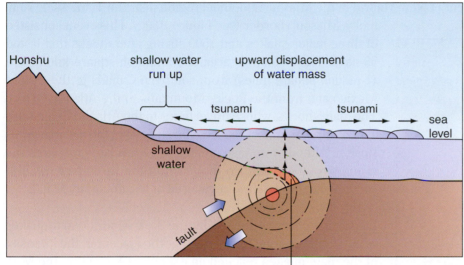

(b)

Figure 19.29 (a) The Great East Japan (Sendai) Earthquake, centered just off the east coast of Honshu Island, produced a large and destructive tsunami that killed more than 25,000 people. (b) The tsunami was produced by an upward bulge in the ocean floor and the water mass above it.

The leading edge of the Sendai Tsunami of March 11, 2011, rolling over coastal lowlands.

of the five largest earthquakes ever recorded in the world. It was centered along the contact between the Pacific and Eurasian plates 72 kilometers (45 miles) off the northeast coast of the main island of Honshu (see Figure 19.29a) at a depth of 32 kilometers (20 miles) below sea level (about 16 miles beneath the sea floor). Tremors radiated into both plates, taking only a few minutes to each Tokyo, about 300 kilometers to the south. The city experienced five minutes of heavy shaking, but damage was minor in large part because Tokyo has long practiced earthquake-resistant building construction. Not so in the coastal city of Sendai, where the shaking was followed by a large tsunami that washed ashore, causing massive damage.

The diagram in Figure 19.29b illustrates how the tsunami was generated. The crust on the ocean floor near the earthquake epicenter was displaced several meters, causing an upward displacement in the overlying mass of water and a huge bulge in the ocean surface. This mound of water formed a wave, the tsunami, which radiated in all directions across the ocean. To the west it took only 10 minutes (traveling at 500 kilometers per hour) to reach the Japanese coast. Typical of waves in general, as it crossed the shallow water offshore and was slowed by friction against the bottom, the tsunami rose higher, reaching over 10 meters (33 feet) by the time it hit land. The wave's destructive force was related not only to its height, but also to its great length. Because of this dimension, once the wave hit, it just kept on coming and coming (see the photograph below) until it pushed its way over 470 square kilometers (182 square miles) of coastal lowlands, killing over 25,000 people and destroying or damaging more than 100,000 buildings including three nuclear power plants.

The 2010 Haitian Earthquake: On the other side of the Earth, far away from the Pacific Plate, the heavily populated Caribbean country of Haiti, in January 2010, was struck by one of the most destructive earthquakes in world history. It registered a magnitude of 7.0 – well below the great earthquake class and 1000 times smaller than the Sendai event – but because it was centered in a crowded urban area of fragile architecture, the entire landscape near the epicenter more or less collapsed to the ground in a matter of minutes. Three million people were affected, 1 million were left homeless, 300,000 were injured, and 230,000 died. The quake registered a peak intensity of X on the Mercalli scale.

Haiti lies near the northern border of the Caribbean Plate where it abuts against the southern border of the North American Plate. It is an area with a long history of destructive earthquakes. The 2010 event produced a major displacement in a fault that scientists reason had been fully locked for the past 250 years during which stress had built up to massive levels. Following the initial shock there were 52 aftershocks registering magnitude 4.5 or greater that lasted for 15 days, including three at a magnitude of nearly 6.0.

Rubble from the 2010 Haitian Earthquake. Fragile buildings of unreinforced concrete and plaster crumbled under the force of tremors.

Why the great destruction? First, the location of a quake in an urban area of several million people. Second, the vulnerability of the built environment to seismic disturbance. Haiti is unable to build wooden houses (which are relatively pliable under earthquake motion), first, because most trees have been stripped from the land (for fuel and farmland), and second, because poverty is so extreme that the Haitians cannot afford to import lumber. As a result, a great many, perhaps most, houses and other buildings were made of weak concrete, which crumbled like plaster, burying people in the rubble.

The New Madrid, Missouri Earthquake: Not all the big earthquakes are on plate borders. The largest known earthquake event to strike the lower United States, known as the New Madrid quakes of 1811–12, was centered under the little frontier settlement of New Madrid in the Missouri Bootheel region near the Tennessee, Kentucky, Missouri border (see Figure 19.30). This event consisted of three major quakes and 1500 strong aftershocks that lasted 53 days. They shook an area of 2.5 million square kilometers (1 million square miles) from southern Canada to the Gulf of Mexico and from the Rocky Mountains to the Atlantic Ocean. The main shocks, whose travel times to the Atlantic Coast were a mere three minutes, caused church bells to toll as far away as Boston. Although no instruments were available to gauge the event, scientists estimate that each of the three New Madrid earthquakes would have registered at least magnitude 8.0.

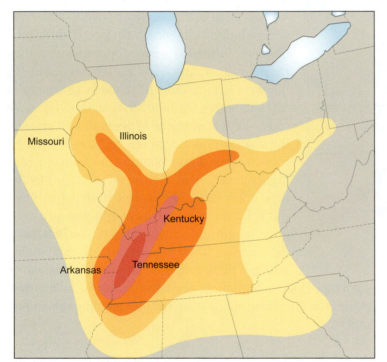

Figure 19.30 The estimated extent of the landscape damage caused by the New Madrid earthquakes of 1811–1812. Near the epicenter damage probably reached X to XI on the Mercalli Scale, nearly total destruction.

Landscape changes caused by the New Madrid earthquakes were dramatic to say the least. Near the epicenter, the intensity of the main quakes reached an estimated X to XI on the Mercalli scale and landscape damage extended over an astounding 600,000 square kilometers. Vast areas of forests were flattened as trees were snapped off by the ground roll driven by surface waves (Figure 19.31). Large tracts of land subsided forming broad lakes and swamps. Even the Mississippi River channel was altered as islands popped up in some places and disappeared in others while the channel itself shifted into a new alignment. Because the area was lightly populated with few settlements at the time, loss of life was insignificant. Today, the area between St. Louis and Memphis is home to more than 10 million people.

Figure 19.31 Archival drawing depicting the landscape chaos from the New Madrid earthquake. No photographs were available in 1812.

Why such a big earthquake in the middle of the continent? There are two hypotheses, both related to plate tectonics. One points to the breakup of Pangaea and the formation of a triple junction in the split between Africa and North America (see Figure 18.29). The lower Mississippi Valley, called the Mississippi Embayment, appears to have formed in the failed arm of the junction (the Atlantic Basin formed in the other two arms). The Mississippi arm was a huge rift structure and though the faults framing it are no longer detectable at the surface, the rift is sporadically active at depth. The other hypothesis proposes that in its westward movement, the North American Plate has passed over a major mantle plume and it heated up the upper Mississippi Embayment, triggering the New Madrid event. Was this earthquake a unique, one-time occurrence? Likely not. If the driver was a mantle plume, it is a virtual certainty that it is still there.

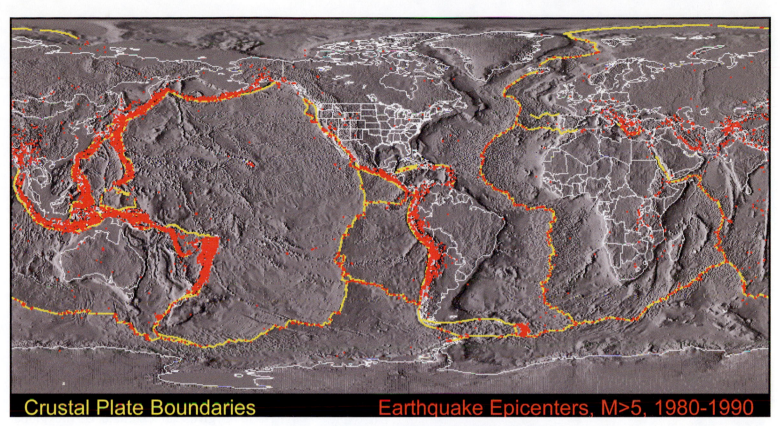

Crustal Plate Boundaries   Earthquake Epicenters, M>5, 1980-1990

Figure 19.32 The vast majority of earthquakes are found along the plate borders with the greatest number on the Pacific Rim (from NOAA).

The Distribution of Earthquake Hazard Zones: Despite the occurrence of major intraplate earthquakes such as the New Madrid, the vast majority of most large, destructive earthquakes fall along the borders of the tectonic plates, as the map in Figure 19.32 clearly shows. But among the three types of plate borders, there are significant differences in earthquake hazards. Divergent oceanic borders are decidedly the least hazardous because the earthquakes along mid-oceanic ridges, though shallow and frequent, tend to be small. The same is generally true for continental rift zones like the one in East Africa. In both zones, the crustal rock is hot and more pliable than cold crust and thus tends to deform more by bending rather than by rupture.

In subduction zones, on the other hand, the lithosphere is cold and brittle and earthquakes are large as illustrated by the east Japan earthquake of 2011. They begin at shallow depths as the plate starts to bend and continue with the descent of the plate to depths of hundreds of kilometers. Transform borders like the San Andreas Fault System also produce high magnitude earthquakes, although not as large and frequent as those in subduction zones. The dominance of these two borders in global earthquake production is underscored by the fact that every year 80 percent of the world's earthquake energy is released on the Pacific Rim. Most of the remaining energy is released in the long collision zone extending from southeast China and southeast Asia through the Himalayas westward to Turkey and Greece.

On the Pacific Rim, there are more than 40,000 kilometers (25,000 miles) of borderlands at high risk for large earthquakes. Within this belt, four areas are particularly active: on Japan, western North America, the southwestern Pacific, and Central and South America. The western Pacific is far and away the most hazardous segment. Each

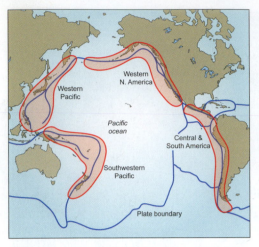

The four major areas of high-risk earthquake activity on the Pacific Rim.

year the Japanese endure more than 1000 earthquakes of magnitude 3.5 or greater. Historical records reveal that catastrophic quakes have struck Tokyo once every 69 years on the average and, in the twentieth century, Japan has suffered 25 earthquakes as powerful as the one that devastated San Francisco in 1906 (estimated at 7.8 or 7.9 on the Moment magnitude scale). Prior to the 2011 quake, Tokyo itself had not been struck by a major earthquake since 1923. Seismologists speculated that strain energy in the subduction zone along the east coast of Honshu had built up to disastrous levels, and as the Great East Japan Earthquake illustrated, they were correct.

Another vulnerable area to damaging earthquakes is western North America and within that area California stands out, way out, for two reasons. First, it is laced by many active fault systems, and second, because it has a huge population (over 40 million) and massive urban centers with heavy infrastructure development. Also vulnerable here are Oregon, Washington, and British Columbia, especially their big urban centers, Portland, Seattle, and Vancouver. On the favorable side, the North American West Coast is the most carefully studied and, along with Japan, the most seismically monitored area in the world, so if there is any place where earthquake warning and readiness are possible, it is here. Also, most jurisdictions enforce reasonably strict earthquake building codes.

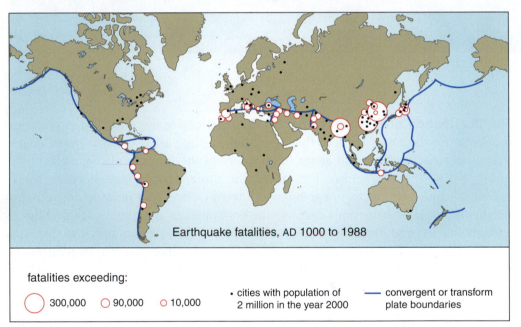

Earthquake fatalities, AD 1000 to 1988

fatalities exceeding:

○ 300,000 ○ 90,000 ○ 10,000

• cities with population of 2 million in the year 2000

— convergent or transform plate boundaries

Figure 19.33 The global distribution of earthquake deaths over the past 1000 years. Owing mainly to their traditionally heavy populations, the regions of eastern China and eastern India–Bangladesh have suffered most (adapted from Roger Bilham).

What has history recorded about earthquake disasters? The map in Figure 19.33 gives an estimate of the global distribution of earthquake deaths over the past 1000 years. Curiously, China far outweighs Japan in terms of loss of life. Eastern China has lost upwards of three quarters of a million people over the millennium. The explanation has more to do with China's historically heavy population and large cities than with earthquake tectonics, for China is less tectonically active than Japan. The same holds true for a second high-death area in Asia, the eastern India–Bangladesh region.

Summary on Earthquakes: Earth is continuously producing earthquakes, but only a few reach great-earthquake magnitudes, classed as magnitude 8.0 or larger. Magnitude is a measure of the energy released by a quake according to the Richter magnitude scale or the Moment magnitude scale. The destructiveness of an earthquake depends on its geographic location and depth in the crust as well as its magnitude. Where people and land-use facilities are concentrated along plate borders, prospects for disasters are high, as revealed by the 2010 Haitian event, which registered X on the Mercalli scale.

19.4 Volcanism, Volcanic Events, and History

A recent lava flow in Hawaii, similar to a basalt flood but much smaller.

Volcanism is the one other tectonic process capable of rendering sudden and massive change in the Earth's surface. Volcanism represents a wide range of processes and features related to the movement and solidification of magma both within the crust and on the surface. Here we are interested in surface processes, particularly the formation of volcanoes. **Volcanoes** are emissions of magma, airborne rock fragments called **pyroclastics**, and gases, released through openings in the Earth's crust. Most volcanoes are explosive in nature and produce more pyroclastics than lava.

The global distribution of volcanic activity is roughly the same as that of earthquake activity. The bulk of it occurs in the ocean basins along the plate borders, principally in mid-oceanic ridges and subduction zones. But volcanism is also prominently represented in the interiors of oceanic plates, especially by seamounts, which rise up from the abyssal plains. Volcanism is less common on the continents, but it is not uncommon. Two geographic settings are noteworthy for volcanic activity: the orogenic belts on the continental margins, such as the North American West Coast; and divergence zones, such as the East African Highlands, in the continental interior.

Types of Volcanoes: Although we tend to think of volcanoes as conical-shaped mountains, they actually come in a great variety of forms. Some of the largest volcanoes are not mountains at all, but are more like plateaus. Such volcanic landforms are the result of massive outflows of lava, called **basalt floods**, which issue from long, narrow openings, or fissures, in and around hotspots in the crust. The lava spills over the landscape in a relatively thin sheet, typically covering hundreds of square kilometers with each eruption. The fluid behavior of the lava is attributed to its high temperature, up to 1200°C. On sloping surfaces it can flow at velocities exceeding 50 kilometers per hour; however, in most basalt floods, lava moves at much slower rates, but always devouring the landscape in its path.

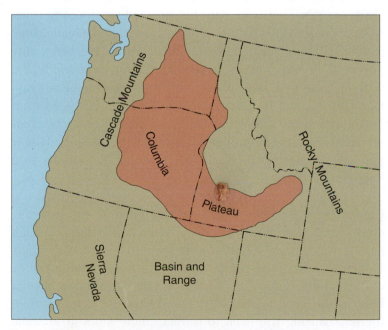

Figure 19.34 The Columbia Plateau (including the Snake River Province) contains North America's largest complex of lava fields resulting from basalt floods, covering more than 200,000 square kilometers.

The largest continuous area of flood basalts in North America is the Columbia Plateau, located just north of the Basin and Range Province in Oregon, Washington, and Idaho (Figure 19.34). It covers more than 200,000 square kilometers and appears to be related to divergence in the continental crust that fueled multiple lava emissions around 10 million years ago. The largest known areas of flood basalt in the world, more than 30 times larger than the Columbia Plateau, formed on Pangaea more than 200 million years ago. These outflows covered 7 million square kilometers and were generated by the same forces that produced the massive rift system that eventually widened to become the Atlantic Ocean. As Pangaea split apart, this great lava plateau was ripped apart and today large sections of it can be found in North America, South America, and Africa.

If a fluid magma is released from a cluster of tunnels, called vents, which are chronically active, the lava tends to pile up, forming a broad mound called a **shield volcano**. Although shield volcanoes may grow to heights of several thousand meters, most of their growth is lateral with lava issuing from the sides of the volcano in emissions called flank eruptions.

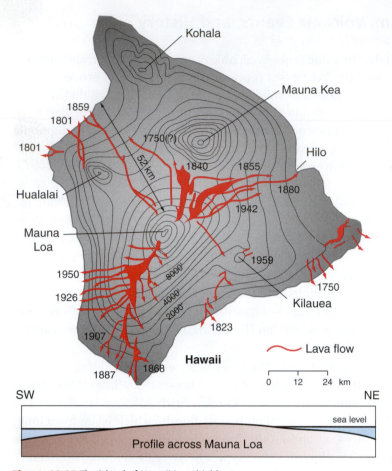

Figure 19.35 The island of Hawaii is a shield volcano that has built up about 10,000 meters from the ocean floor. This map shows the part above sea level. The red lines represent lava flows recorded over the past two centuries.

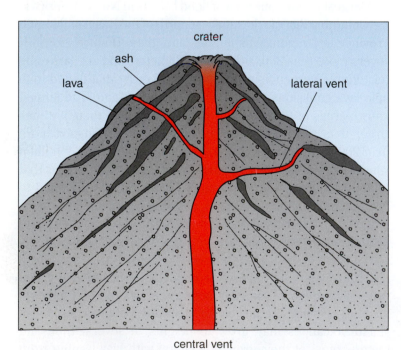

Figure 19.36 A cross-section of a composite (stratocone) volcano showing the interlayering of ash and lava, the central vent, lateral vents, and crater.

Side slopes are gentle, generally less than 5 degrees, and are laced with lava flows of various sizes. The Hawaiian Islands are excellent examples of shield volcanoes. The main island, Hawaii, was built from five shield volcanoes, shown in Figure 19.35, which coalesced into a single mass as they emerged from the sea. Like Hawaii, most shield volcanoes tend to form near the interiors of tectonic plates.

Composite volcanoes (also known as **stratocones**) are generally conical in shape and represent the popular stereotype of the "classical" volcano. Smaller and steeper than shield volcanoes, they also differ in composition; shield volcanoes are almost entirely basaltic lava, whereas composite volcanoes are made up of both lava and pyroclastic material. Pyroclastic material, which is mainly ash but can include all sizes of rock particles, is commonly ejected into the air in the opening stage of an eruption. The coarse particles rain back to the surface of the volcano forming a layer of some thickness whereas the fine particles are carried off in the atmosphere and eventually fall out elsewhere. Later in the eruption or in a subsequent eruption, lava flows may bury the pyroclastic layer. Thus, the lava and the pyroclastics become embedded in a layer-cake fashion as shown in Figure 19.36. Many of the famous volcanoes belong to the composite class, including Vesuvius and Etna in Italy, Fuji-san in Japan, Ararat in Turkey, and the well-publicized Mt. St. Helens and Mt. Rainier in Washington. Virtually all of the volcanoes in subduction zones are composites, which accounts for their abundance around the Pacific Plate.

Finally, there are volcanoes made up entirely of ash, called **cinder cones**. Similar in shape but smaller than the composite, cinder cones are clearly the weakest or least-resistant volcanic landform. Those formed at sea, for example, are commonly planed off by waves as fast as they form. Not surprisingly, the birth of some have been observed and documented firsthand. One cinder cone, Paracutin in Mexico, which emerged from a vent in a farm field in the 1940s, revealed a behavior characterized by intermittent periods of activity broken by longer periods of quiescence. By 1970, Paracutin had grown to a height of 1200 feet.

Composite and shield volcanoes, though different in scale, often develop similar anatomies, characterized by a main passageway (called a *central vent*), multiple secondary vents (which lead from the central vent to the sides), and a **crater** at the summit (Figure 19.36). The crater forms when magma held in the central vent is released through a lower vent, thereby causing the neck to retract. Such shifts in material can also break the hull of the volcano and create fissures that later serve as passageways for lava. Some volcanoes develop a much larger crater depression, called a **caldera**. Calderas form when a volcano collapses because a large mass of magma

Figure 19.37 The Santorini caldera. The eruption blew most of the island away leaving a huge water-filled caldera and a fringing rim of land.

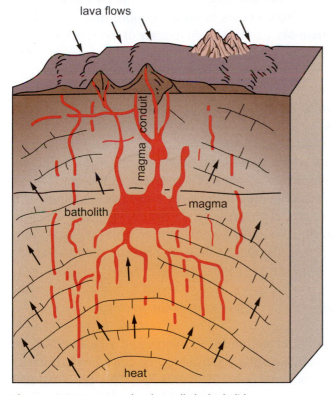

Figure 19.38 A magma chamber, called a batholith, at the base of the crust with conduits leading to the surface and volcanoes.

is drained away through a lower vent or because a massive amount of material is blown away in an explosive eruption. Explosive eruptions have been known to destroy the entire superstructure of a volcano and, as the photograph in Figure 19.37 reveals, some, such as Santorini in the Aegean Sea, obliterated most or all of an island and everyone on them. Santorini exploded around 3600 years ago.

Basic Mechanics of Volcanism: The underground mechanics of volcanism are in general poorly understood. For most volcanoes, the heat that creates the magma is supplied by convectional systems in the upper mantle. Where heat is concentrated, a pocket of magma can develop at the base of the lithosphere, which can melt its way through the overlying rock like a parcel of hot air slowly rising through the atmosphere, displacing its way upward by melting through zones of colder rock. Because of the huge quantities of heat needed to melt through rock, a great supply of magma must be available if the lithosphere is to be fully penetrated.

At selected spots in the lithosphere, huge chambers, called **batholiths**, form where magma is stored (Figure 19.38). From these chambers, large tunnels or conduits, ply their way into the crust and to the Earth's surface. Once a hot conduit to the surface has been created, the ascent of additional missiles of magma appears to take place with comparative ease. With their plumbing system in place, individual volcanoes tend to remain active for a long time, millions of years, after they are born.

Although most volcanic eruptions are explosive, most are not violent. Hundreds of volcanoes produce sputtering eruptions that rise and fall over short periods of time. Most basalt-flood eruptions involve little more than quiet outpourings of lava, but the birth of most composite volcanoes is usually heralded by discharges of gas and ash, which may continue for years before any lava appears. The cycle of ash and lava may be repeated many times before the volcano produces a truly explosive, sometimes violent, eruption. This pattern of behavior may catch people unawares as was the case at Pompeii, Italy, in AD 79 when Mt. Vesuvius exploded violently after decades of gurgling and burping, killing everybody in the city, an event we will look at in more detail later on.

Famous composite volcanoes. Left, Fuji-san; right, Mt. Rainier.

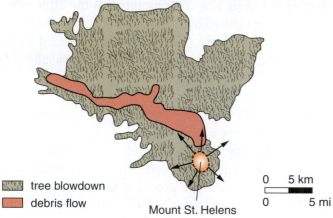

tree blowdown

debris flow

Mount St. Helens

0   5 km

0   5 mi

Figure 19.39 The Mt. St. Helens eruption of 1980, the most carefully documented volcanic eruption in history, with a map showing the area of destruction.

More typically, the most violent eruptions occur in established volcanoes that reactivate after a period of quiescence. During inactive periods, the superstructure (hull) and the internal passageways (neck and vents) become locked and rigid with cooling and magma solidification, and when new magma begins to push upward, it meets resistance from these capping materials. Pressure builds and, as in the case of Mt. St. Helens in May 1980, the volcano suddenly gives way at a weak point creating a huge explosion (Figure 19.39).

The Mt. St. Helens eruption was a major natural event for the United States. It was the first eruption to be monitored scientifically in the months leading to the explosion, and all the drama, including the explosion itself, was witnessed via television around the world. The destructive power of the eruption was shocking to people, but we should be reminded that it was a very modest event as volcanic explosions go. Judging from the size of ash deposits made by other volcanoes on the Pacific Rim, many eruptions in the past 20,000 years or so were more than a hundred times larger than Mt. St. Helens. And many of these occurred in areas that are still volcanically active with millions of people living there today. Japan is a good example.

Supervolcanoes: As we learn more about the Earth's crust and its dynamics, new findings come to light. One of these is the existence of volcanoes so big that they have earned the name **supervolcano** or **megacaldera**, and are defined as eruptions that produce more than 1000 cubic kilometers of ejecta (Mt. St. Helens produced about 1.5 cubic kilometers). They form over mantle plumes – 30 such plumes are currently operating on Earth – where the crust forms a confining cap preventing the release of magma. As a result, the magma builds up in a massive, pressurized chamber (batholith) that pushes the crust up until the whole thing bursts in a massive explosion.

One example serves to make the point. It is Yellowstone in the western United States, home of the famous national park (Figure 19.40). The Yellowstone supervolcano is the product of a magma chamber estimated to be 16 kilometers wide (the Yellowstone geysers are fueled by the chamber's heat). It has erupted 100 times in the past 16 million years with the last one occurring 640,000 years ago. This eruption, and most of the 99 others, so altered the atmosphere that it caused widespread extinctions. A new eruption, which volcanologists claim is way overdue, would discharge fire clouds over much of North America, killing millions of people, and send ash clouds far into the atmosphere leading to global cooling, maybe a freeze-down.

Volcanic Events and the Landscape: Volcanoes of all sizes can render sudden and massive changes in the landscape, both natural and human. But beyond the drama a much bigger and compelling story is unfolding, which the Yellowstone volcano more than hints at. This is the role of volcanoes in changing the global environment, for it appears that some of Earth's extreme changes of climate are volcano-driven. But first, let us look at some of what we know about violent volcanic eruptions on the landscape.

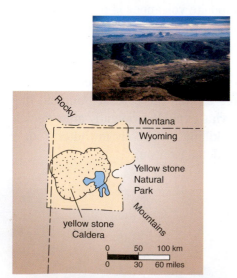

Rocky

Montana

Wyoming

Yellow stone
Natural
Park

yellow stone
Caldera

Mountains

0   50   100 km

0   30   60 miles

Figure 19.40 The Yellowstone area, site of a supervolcano. Note the size of the caldera. The photograph shows a section of the caldera rim.

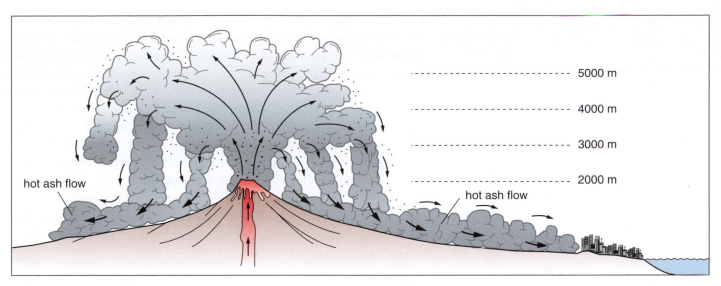

Figure 19.41 The processes associated with a pyroclastic cloud like the one discharged from Vesuvius in AD 79.

Many processes associated with volcanic eruptions influence land use and human lives including lava flows, forest fires, and big floods. But the most deadly appears to be eruptions that produce great masses of hot airborne ash, called **pyroclastic clouds**, that sear and bury the landscape in a matter of minutes. Ash can be discharged into the air laterally with a flank eruption like Mt. St. Helens or vertically from the top of the volcano. Until recently, the destructive processes associated with pyroclastic clouds were not fully understood.

It appears that the force of the blast and the rising, superheated air carry the ash to great heights in the atmosphere where it spreads out forming a great mushroom cloud. At some height this high density, airborne mass cools, and the cloud suddenly drops from the sky onto the landscape. In fact, observers report that the pyroclastics appear to pour from the sky as the diagram in Figure 19.41 portrays. The cloud hits the ground with great force and spreads over the surface at speeds in excess of 120 kilometers per hour (80 miles per hour). The depth of burial commonly exceeds 10 meters over tens of square kilometers or more. Those pyroclastic clouds, such as the one from Santorini, which occur at sea, are of course swallowed up mostly by the ocean. On land, the result can be horrendous. Perhaps the most famous of such events occurred in the Roman city of Pompeii.

Vesuvius, Italy, AD 79: Vesuvius, a composite volcano located on the Bay of Naples in Italy, was, and still is, notorious for its misbehavior. One morning in the summer of AD 79, it erupted with a modest burst of ash, a common sight to the residents of nearby Pompeii. Late the following day it became clear that this eruption was larger than usual, and the city's inhabitants began to flee. Before everyone could escape, however, Vesuvius suddenly released a huge pyroclastic cloud that ascended to great heights, then dropped to Earth, and rolled over Pompeii and the neighboring town of Herculanium literally burying people and animals in their tracks. Within several days Pompeii was buried to such a depth that excavation and reuse by the Romans were considered all but impossible. It remained buried and nearly forgotten until archeological excavations began in the 1700s. Mt. Vesuvius, on the other hand, has remained active over the centuries. Since 1800 it has erupted eleven times.

Excavated ruins of Pompeii with Mt. Vesuvius in the background. Ash completely buried the Roman city and most of its inhabitants to depths of 6m or more.

Mt. Pelee, Martinique, 1902: Another "hot-cloud" event that killed thousands of people and destroyed a city comes from the Caribbean island of Martinique. In 1902, Mt. Pelee, located inland of the port city of St. Pierre, erupted after a half-century of dormancy beginning with a series of mild explosions followed by a tremendous blast that sent a mass of incandescent gas and ash out the side of the volcano toward the city. The mass moved at a velocity approaching 200 kilometers per hour (120 miles per hour) engulfing the city and killing its 28,000 inhabitants instantly. Only one resident survived, a prisoner in a basement cell of the jail (a convicted murderer who was pardoned and later became a missionary), but a number of sailors aboard ships in the harbor watched the event. One offered this description:

> The mountain was blown in pieces. There was not warning. The side of the volcano was ripped out and there was hurled straight toward us a solid wall of flame. It sounded like a thousand cannon …

> The wave of fire was on us and over us like a flash of lightening. It was like a hurricane of fire. I saw it strike the cable steamship Grappler broadside on, and capsize her. From end to end she burst into flames and then sank. The fire rolled in mass straight down upon St. Pierre and the shipping. The town vanished before our eyes.
>
> (from K. H. Wilcoxon, *Chains of Fire – The Story of Volcanoes*, Chilton, 1966.)

19.5 Mountain Systems and the Global Environment

At many points throughout this book we have examined the influence of mountains on Earth systems and geographic conditions. In the opening chapters, for example, we learned that mountain slopes influence local climate by creating sunny slopes and shady slopes and that the differences in light and heat on these slopes induce differences in moisture, soil, and vegetation (see Figure 4.22). We also learned that mountain ranges can profoundly affect precipitation systems by intercepting moisture-bearing winds causing them to deposit massive loads of moisture on windward slopes and little on leeward slopes, giving rise to regional differentiation of the landscape in terms of ecology, water supplies, and land uses (see Figure 8.24).

Mountains influence the size and geographic organization of watersheds, and it is no fluke of nature that great rivers like the Mississippi, the Amazon, the Ganges, the Huang He and many others gain a principal share of their discharge from mountains (see Figure 15.37). And in the chapter ahead we will see that one of the major controls on atmospheric carbon dioxide is the building up and wearing down of mountain masses related to the weathering of limestone with the changes in moisture and erosion. As great mountain masses like the Himalayas are raised up, atmospheric carbon dioxide declines and global cooling takes place. This brings us to the broad and challenging subject of volcanic mountains and the global environment.

Volcanism, Atmospheric Systems, and Climate Change: There is little doubt in the scientific community that volcanic eruptions are capable of altering the global environment, mainly through their influence on weather and climate. The evidence is indisputable especially in light of measurements associated with modern eruptions and in light of historical records citing ancient eruptions and their effects. The connection between a volcano, representing a single point on the map, and the global environment, representing the entire 510,000,000 square-kilometer surface area of the planet, is another example where a system's perspective is called for, and the system that transforms a point into a global distribution is, of course, the atmosphere and its systems of prevailing winds.

Once the ejecta from a volcanic eruption are dispersed into the atmosphere, another system comes into play, the radiation system. The chief mechanism involves the reflection and scattering of incoming solar radiation off airborne volcanic particles. These particles, called aerosols, come from ash (pyroclastics) and minute liquid droplets produced when sulfur dioxide, the major gas from most eruptions, combines with water, oxygen, and tiny ash particles. The net effect of all these particles is to reduce the Sun's energy input to the lower atmosphere and the Earth's surface.

An example of a modern eruption that measurements show changed weather and climate is the Mt. Pinatubo event in 1991. Mt. Pinatubo is located in the Philippines well within the tropics (15 degrees north latitude) and the belt of the northeast tradewinds. When it exploded it drove millions of tons of ash and gas 34 kilometers (21 miles) into the atmosphere (Figure 19.42). Within a matter of weeks prevailing wind systems spread the ash throughout the world, measurably reducing the atmosphere's transparency to incoming solar radiation. The result was a drop in global temperature by 0.5 °C over the period 1991–1993.

In 1815, another tropical volcano, Mt. Tambora, erupted in what is generally regarded as the largest eruption in recorded history. It blew 160 cubic kilometers of ash into the atmosphere, 100 times more than Mt. St. Helens' output, which completely darkened the sky over a vast area. History records that 1816 was known in New England, eastern Canada and northwestern Europe as the year without a summer. Snowfalls extended well into July. The cold continued into 1817 and 1818, crops failed, and people suffered from malnutrition and starvation. Global temperature probably dropped by nearly 1 °C over this period.

History is dotted with similar events having similar effects. To name a few: Santorini (Greece, 1600 BC), Krakatoa (Indonesia, AD 535), Laki (Iceland, AD 1783), and Krakatoa (Indonesia, AD 1883). And when we reach back into the geologic record, there is abundant evidence of volcanic activity on Earth far in excess of anything humans have ever witnessed. The eruption of a supervolcano like Yellowstone is a case in point. Based on the size of its caldera (80 kilometers wide), this event may have been 10 or 100 times larger than Tambora with commensurate effects on the atmosphere, landscape, and life.

Figure 19.42 The Mt. Pinatubo eruption and the global pattern of ash in the atmosphere after only one or two weeks.

Chapter Summary and Overview of Mountain Systems, Earthquakes, and Volcanoes

The building of mountains and the attendant processes of earthquakes and volcanoes are enormously significant to physical geography. Among other things, mountains are the building blocks of the landmasses, the very stage upon which Earth's geography is played out, where, as the summary diagram highlights, they influence a host of landscape systems. We see these influences displayed in the patterns and dynamics of weather and climate, the size and shape of watersheds, the locations of streams, the composition of soil, and even in the distribution and makeup of Earth's biota. Increasingly the dynamics of mountain building as played out in earthquakes and volcanoes have caught our attention because these processes are very serious threats to humanity and increasingly so as global population grows and concentrates itself in vulnerable locations on the planet.

▶ **Orogenic belts are made up of provinces.** Each province contains mountain ranges made up of individual mountains.

▶ **Mountains are shaped by both internal structure and external erosional processes.** Most mountains are eroded and reconfigured as landforms while they are being built by tectonic processes.

▶ **Folding, faulting, and volcanism are the principal mountain-building processes.** One involves plastic deformation and one involves rupturing of rock formations.

▶ **Rock is deformed by differential stress.** The three basic types of differential stress are played out in the three main types of plate borders.

▶ **Strike and dip measurements are used to map rock structures.** Strike and dip patterns reveal folds, faults, and broad structural trends in orogenic belts.

▶ **A fault is any displacement of bedrock along a fracture.** There are three main classes of faults. Dip-slip faults include normal, reverse, and thrust faults.

▶ **Folds are described according to their geometric properties.** Common folds include monoclines, synclines, anticlines, and overturned folds.

▶ **Earthquakes are caused by a sudden shift of rock along a fault.** Locked segments of faults produce larger earthquakes than creep segments.

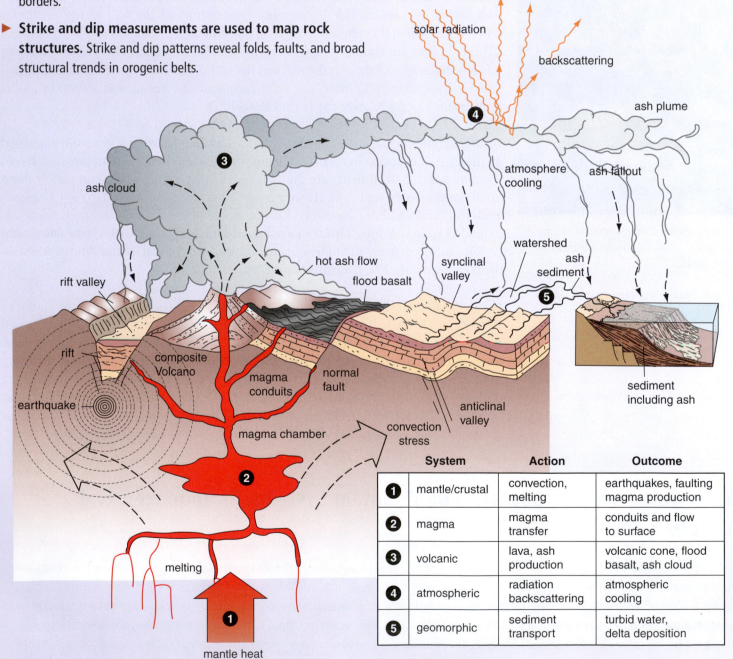

	System	Action	Outcome
❶	mantle/crustal	convection, melting	earthquakes, faulting magma production
❷	magma	magma transfer	conduits and flow to surface
❸	volcanic	lava, ash production	volcanic cone, flood basalt, ash cloud
❹	atmospheric	radiation backscattering	atmospheric cooling
❺	geomorphic	sediment transport	turbid water, delta deposition

The major systems and processes involved in mountain building. The scene is subdivided into five systems beginning with heat and convection in the mantle and ending with atmospheric cooling and ash returning to Earth as sediment in watersheds and the sea.

▶ **The epicenter of an earthquake lies directly above its focus.** Very large quakes may last several minutes and are usually followed by many aftershocks.

▶ **Earthquakes produce surface waves and body waves.** P waves and S waves are the two main types of body waves.

▶ **Magnitude is a measure of an earthquake's energy output.** Two scales are used to measure earthquake magnitude and both use logarithmic numbers.

▶ **Few earthquakes qualify as great earthquakes.** From the time magnitudes were first measured, only one North American earthquake registered as a great earthquake.

▶ **Most large earthquakes occur on the edges of tectonic plates.** The vast majority of the Earth's annual output of seismic energy is concentrated along the Pacific Rim.

▶ **The Mercalli scale is a measure of earthquake intensity.** The 2010 Haitian earthquake registered intensity X, but its magnitude was less than a great earthquake.

▶ **There are four basic types of volcanoes.** Only the composite has the tall, symmetrical form popularly associated with volcanic mountains.

▶ **Relatively few volcanic eruptions are violent.** Pyroclastic clouds are the most hazardous aspect of violent eruptions.

▶ **Supervolcanoes are driven by the heat of mantle plumes.** Their eruptions are capable of rendering massive and far-reaching environmental change.

▶ **Mountain systems are major players in the global environment.** This includes weather and climate changes related to alterations in atmospheric transparency and surface heating.

Review Questions

1 In terms of geographic organization, name the subdivisions of an orogenic belt, largest to smallest.

2 The form of a mountain often differs appreciably from the geologic structure of its origin. Explain.

3 What are the 3 basic types of stress operating in the crust and which of these is most common to which type of plate tectonic border?

4 Why is it that a rock that is brittle and breaks when we apply force to it with a laboratory machine can in nature behave as a plastic and bend into folds? Based on the graph in Figure 19.7, which part of the curve best describes isostasy? Faulting? Folding?

5 What is meant by the principle of original horizontality and what does it mean when rock formations do not conform to this principle? In the big picture of a continent, where would you generally expect rock formations to most conform and least conform to the horizontality principle?

6 What is the difference between the crustal movements that give rise to normal and reverse faults? And what differentiates a reverse fault from a thrust fault?

7 Describe the type of fault structure that led to the formation of (1) the valleys of the great lakes of Africa, (2) the mountain ranges of the Basin and Range Province of western North America. In terms of the forces responsible for the long term trends in plate tectonics, what do these two regions have in common?

8 Describe the origin of anticlinal and synclinal valleys; how the landforms are formed by the limbs; where in North America such valleys are found; and one major influence you would expect them to have on the geographic arrangement of the landscape.

9 Earthquakes emit both surface and body seismic waves. How do these waves differ and which is the most destructive in the landscape?

10 What is the difference between an earthquake's focus and epicenter? How is the location of the epicenter determined?

11 Someone is looking for an easy way to describe the global distribution of earthquakes. What feature(s) of the crust would you tell them to focus on and why?

12 What is meant by earthquake magnitude, how is it measured, and what term is given to earthquakes of 8.0 and higher magnitude? How common are such earthquakes?

13 What is the difference between the Richter and Mercalli scales? Is it possible for an earthquake to register a very high Mercalli number and a not so high Richter number and vice versa? How so? And how does the 2010 Haitian Earthquake fit into this discussion?

14 How do the four basic types of volcanoes (flood-basalt, shield, composite, and cinder cone) vary in terms of structure and geographic extent? What processes acting within each type contribute to their geographic form?

15 What is a supervolcano, what is the nature of the force that drives them, and should we consider them to be a phenomenon limited to the Earth's distant past? Explain.

16 In the summary diagram on page 496, trace the path of rock material from the mantle to the continental shelf and list the steps along the way. Does this constitute a system? And in the grand scheme of things is there any connection in the long run between the material buried in the continental shelf and the rock produced by the mantle?

Geomorphic Systems:
Rock Weathering, Hillslope Processes, and Slope Formation

Chapter Overview

Our planet's surface is a work of natural sculpture created by many sculptors, rivers, glaciers, waves, currents, wind, working in teams under a broad range of conditions and influences. But unlike human sculptors who aim to chisel and mold pieces of art for us to enjoy, unchanged, through time, Earth's sculptures, landforms, are works in progress, constantly changing in response to changing systems and the forces that drive them. Every sculptor begins with raw stone. So it is fitting that we begin this chapter with a brief look at the origin of the rocks that are brought to the Earth's surface as part of the rock cycle, the cycle driven by tectonic forces from the Earth's interior. These rocks are heavy and hard and, in order for them to be sculpted by water and air, they must first be weakened, broken up like an old sidewalk. This is the first phase of all geomorphic systems, one that changes solid rock into loose particles. The second phase involves the movement of those particles downhill by rolling, sliding, and falling under the force of gravity. The chapter concludes with an examination of how mass movements and related processes shape mountainsides and hillslopes and feed rock particles to stream, glacier, wave, current, and wind systems.

Introduction

Landscape is the term commonly applied to the complex of forms and materials that make up the terrestrial environment around us. It is the habitat of humanity, where we and a vast number of other organisms play out our lives. Landscape is a central topic of study in geography. Indeed, this book is largely devoted to an examination of the systems that operate in Earth's landscapes, the networks of water, biota, rock formations, soil, and climate, and how they interact and perform work. The one component of landscape we have yet to examine, although we have touched upon it at nearly every turn in previous chapters, is landform. **Landform** is the term used to describe the configuration of the land's surface. Landforms are terrestrial features such as plateaus, stream valleys, sea cliffs, and river deltas that provide the stage into which landscape is set and where its dynamics are played out.

More than perhaps any other aspect of the physical environment, landforms determine the geographic character of the Earth's surface. We reference places according to their landforms – the Great Plains, the Great Basin, the Central Valley – and they influence our sense of natural beauty, geographic intrigue, scientific curiosity, and even who we are as people. When we describe ourselves we often tell about where we come from in terms of the landforms of the area. It is no fluke of storytelling that in folklore distinctions are commonly made between mountain people and valley people and how they differ in their beliefs, attitudes, and behavior.

A sense of landforms is deeply embedded in the North American psyche. Much of the historical drive to explore and settle the continental interior was motivated by a sense of new lands and the hope to fulfill a utopian ideal of finding the perfect place. In the nineteenth century both Canada and the United States legislated systems of national parks, the first national parks in the world, aimed at preserving natural wonders. Virtually every major national park was landform-based: Yellowstone, Grand Canyon, and Banff are early examples. But it was more than just the grand landforms, for the fascination with these places also embraced processes, that is, landscape in action where the streams, glaciers, and geysers that created the landforms could be observed doing their work.

These processes are the tools of **geomorphic systems**, systems that operate on the Earth's surface shaping the land by wearing it down in some places, moving rock debris around, and leaving it as deposits in other places. Water in the form of runoff is the Earth's chief geomorphic system, because runoff in its various forms dislodges and moves more rock material than any other of the major system. And runoff, like all geomorphic systems, glaciers, wind, and waves and currents, functions as an open system with inputs and outputs of energy and matter and the capacity to do work.

Geomorphic systems also have the capacity for self adjustment, meaning that they can change the rate and character of their work in response to changes in the systems that feed and drive them. Among these, climate, water supply, and tectonics (most notably the elevation and structure of the crust) are the principal drivers of geomorphic systems. And very importantly, the different geomorphic systems occupy different geographic spaces on the Earth's surface, some wet, some dry, some cold, some windy, etc., and to these spaces the systems impart their own distinct imprints. These imprints are the landforms we see in the world around us and they in turn play a major role in shaping the larger landscape by influencing the distribution of water, soil, vegetation, animals, and land use.

Different systems in different places working at different rates? This means that the kinds of geomorphic processes and the forces they exert on the land are not uniformly distributed across the Earth's surface. Glaciers, for example, are limited to cold, snowy climates and big glaciers exert a lot more force than small ones. The same distinction holds for streams. The world's great rivers need a wet climate to feed their huge discharges and, within their valleys, they are capable of producing massive erosional power thousands of times greater than small streams. Not only that, but the resistance of Earth materials to the forces of streams, glaciers, wind, and waves and currents is not uniformly distributed over the Earth's surface. Think of the differences in the resistance of shorelines to storm waves. A seacoast composed of sand hardly holds a candle to one composed of hard granite. Thus, the geographic distribution of the work, represented by the erosion of land and the movement of sediment, is highly uneven geographically and knowing this is vitally important to understanding how Earth's landscapes are formed. The two landscapes shown in Figure 20.1 certainly bear this out.

Figure 20.1 (a) The Colorado Plateau is a high-energy, erosional landscape where streams have cut deeply into the crust. (b) The Gulf Coastal Plain is a low-energy landscape where streams have deposited sediment produced by erosion of the crust.

20.1 Geomorphic Systems and the Rock Cycle

Geomorphic systems are organized into three interconnected sets of processes: weathering, mass movement, and erosion. **Weathering** constitutes the processes that break down and disintegrate rock, **mass movement**, the processes such as landslides that drive weathered material downslope, and **erosion**, the processes such as streamflow that dislodge this material and transport it toward the sea.

Framing Concepts for Erosional Systems: Earth's erosional work is done by four systems: streams, glaciers, wind, and waves and currents. These systems, which are also called **erosional agents**, share similar attributes as systems. In their simplest form, each is made up of three basic parts or sequential components: first, *removal* of particles produced in the weathering of bedrock; second, *transportation* of this material as sediment; and third, *deposition* of the sediment in low-lying areas on the fringes of the continents or in the sea as shown in Figure 20.2.

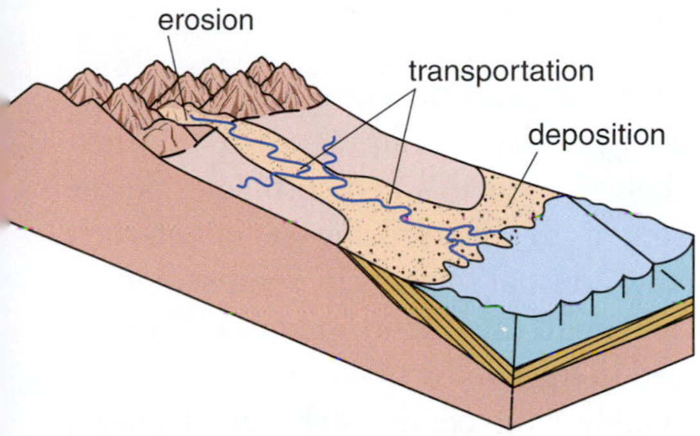

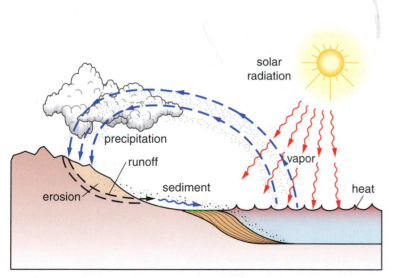

Figure 20.2 The three basic components of all erosional systems: erosion, transportation, and deposition.

As with most Earth systems, geomorphic systems involve flows of both energy and matter. Simply put, energy drives the forces that move matter through geomorphic systems. Matter comes in two forms: (1) water or air, and (2) sediment, and it is sediment, the residue of decayed rock, which attracts our attention in geomorphic studies. Most of the energy driving geomorphic systems owes its origin to solar radiation. In stream systems, for example, solar energy supplies the heat that drives ocean evaporation, which produces the vapor for precipitation, which supplies discharge to streams, which in turn erode rock and move sediment as summarized in Figure 20.3. A second source of energy is gravity because it sets the water from precipitation into motion and drives it downhill to the ocean.

Sediment in geomorphic systems is provided by surface or near-surface terrestrial rock. About 75 percent of the source rock is sedimentary rock, principally shale, limestone, and sandstone. In order for rock to pass through geomorphic systems, however, it must first be freed from its consolidated state and broken down into particles small enough to be mobilized. Weathering processes perform this task. The resultant residues, in the form of dissolved and non-dissolved (solid) particles of sediment, can then be moved by an erosional agent.

As geomorphic systems change the surface of the Earth through weathering, erosion, and the movement of sediment, landforms are created. Broadly speaking, landforms fall into two classes: erosional and depositional. **Erosional landforms** begin as rock masses elevated by tectonic processes, volcanism, isostatic uplift and/or by a drop in sea level. As these rock masses are worn down by geomorphic processes, rocky landforms take shape and sediment is produced. Most of the sediment is carried away by streams and, in the long run, deposited on the continental fringe as the photographs in Figure 20.1 illustrate. Trademark **depositional landforms** such as river deltas and coastal plains are created.

Figure 20.3 Solar energy drives geomorphic systems by powering the hydrologic cycle, wind systems, and wave and current systems. Here the hydrologic cycle and runoff via streams are featured.

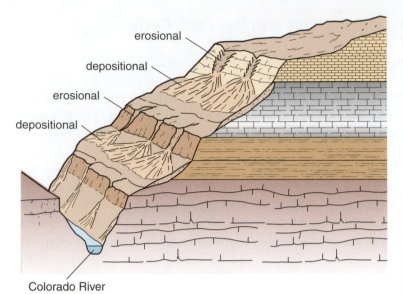

erosional

depositional

erosional

depositional

Colorado River

Figure 20.4 Canyon walls composed of alternating formations of sedimentary rock, some dominated by erosional and some by depositional landforms.

On the other hand, many of the Earth's landscapes are not purely depositonal or erosional. Many, especially in zones geographically intermediate between erosional and depositonal areas, are actually made up of a combination of erosional and depositional landforms. Even the Grand Canyon, widely recognized as one of Earth's premier erosional landforms, contains extensive depositional features where layers of sediment drape over the eroded ends of rock formations in the canyon walls in the manner depicted in Figure 20.4.

The Rock Cycle – Plate Tectonics Connection: Let us begin this section with a look at the diagram in Figure 20.5, which shows the connection between Earth's geomorphic cycle (or great system), represented by the erosion–sediment–rock sequence, and the Earth's internal rock cycle, represented by things like subduction and uplift (mountain building). The connection works like this: The sediment delivered to the continental fringe by geomorphic systems (principally streams) goes into building the continental shelves where, over geologic time, it is compressed and hardened into sedimentary rock. By virtue of its location on the edge of the continent, this rock, in time, is subject to deformation associated with the plate tectonics system and is the stuff from which new orogenic belts are built.

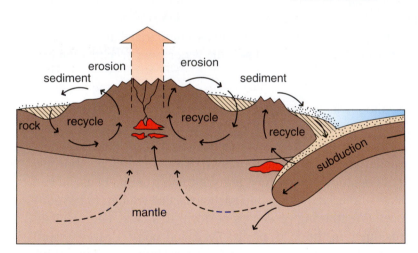

Figure 20.5 Cycles of sediment and rock associated with geomorphic systems (above) and plate tectonics/ mantle convention systems (below). Notice where the two systems connect.

Evidence of such recycling is abundant on Earth. Sedimentary rocks rich in delta sediment and marine fossils along with their metamorphic counterparts are the dominant rock types in the Rockies, Andes, Himalayas, and Alps, Earth's principal orogenic belts. From here the rock is subject to weathering and erosion, and a return trip to the sea, thus linking the geomorphic and rock-cycle systems into a grand recycling system for crustal rock. Some also goes back into the mantle with subduction of the crust as a part of the mantle convection system.

But major episodes of mountain building, *orogenies*, are widely spaced in geologic time, occurring every 500 million years or more on most continents. In the long time spans between orogenies, the landmasses are slowly worn down and the continental shelves are built. During these periods Earth's continents together give up an estimated 10 billion tons of sediment per year of which 3 billion tons are recycled as continental rock (see points A and B in Figure 20.6).

The remaining 7 billion tons (point C) go into the continental shelves. One billion tons (point D) of this shelf sediment are lost to subduction. This occurs along geologically active continental margins where trenches such as the Peru–Chile Trench in South America are located. Another 2 billion tons (point E) are incorporated into orogenic belts that are still forming such as the one along the North American West Coast. Eventually the remaining 4 billion tons (point F), after eons of buildup on passive (geologically inactive) continental margins, are incorporated into new orogenic belts as plates shift, plate borders change, and passive continental margins are transformed into active plate borders.

Figure 20.6 The quantities of sediment and rock moved by the two big systems illustrated in Figure 20.5. The values are in billions of tons per year. Notice that 70 percent of the sediment output goes into building continental shelves, part of which goes to subduction and the remainder to continental mountain building.

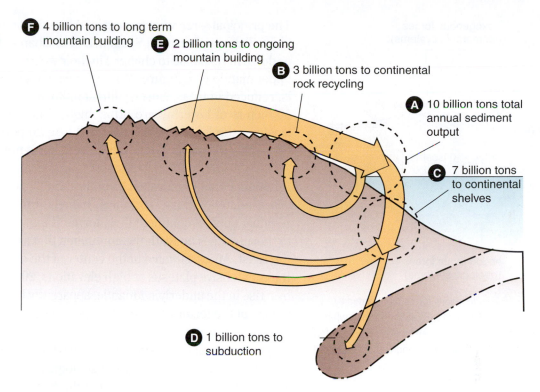

F 4 billion tons to long term mountain building

E 2 billion tons to ongoing mountain building

B 3 billion tons to continental rock recycling

A 10 billion tons total annual sediment output

C 7 billion tons to continental shelves

D 1 billion tons to subduction

20.2 The Denudation System and the Erosion Cycle

The general process of wearing down landmasses is referred to as **denudation**. Denudation is performed by a complex assortment of systems made up of all weathering and erosional processes that operate on the Earth's lands. We measure denudation rates by the total amount of sediment exported to the sea by streams throughout the world, most of which is carried by the world's 70 great rivers. Each of these rivers exports tens to hundreds of millions of tons of sediment annually. The largest, such as Huang He of China and the Ganges of India, each export a billion tons or more per year.

The denudation rate today is higher than any time since the origin of agriculture 12,000 years ago. Before farming, the world's rivers together exported a total of about 10 billion tons of sediment a year. Since that time, the rate has been rising because of increasing soil erosion brought on mainly by land clearing, deforestation, crop farming, and overgrazing and it now stands at more than 15 billion tons, and perhaps as high as 20 billion tons or more annually. This figure does not include the huge amounts of sediment trapped behind the hundreds of thousands of dams across the world.

Denudation versus Uplift: The actual rate of landmass lowering is small when measured over the short term, but enormous over geologic time (which, parenthetically, points up the significance of appreciating time scales when studying geomorphic phenomena). For the section of North America occupied by the United States, for example, the average denudation rate is 6.1 centimeters (2.4 inches) per 1000 years. At this rate the overall landscape would be lowered 60 meters (200 feet) in a million years. Assuming, for the sake of argument, that this rate is representative of the Earth as a whole, it would take less than 20 million years to lower all the continents to sea level. But the continents are several billion years old and none has yet been worn away, so there must be an opposing force or forces offsetting the effects of denudation.

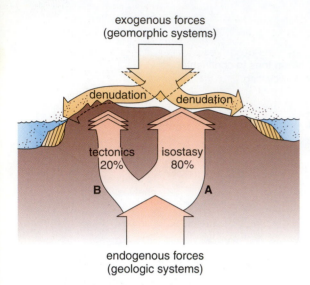

Figure 20.7 Endogenous forces represented by isostasy (A) and tectonics (B) uplift the crust while exogenous forces represented by geomorphic systems strip it away. The elevation of landmasses is a record of the balance between these systems.

The principal force working against denudation is isostasy. The principal of isostasy, you will remember from Section 17.3 in Chapter 17, holds that landmasses move up or down in response to changes in their weight as they float on the heavier rock of the upper mantle (see Figure 17.12). For each increment that a landmass is lowered, that is, reduced in mass, isostasy lifts it up by 80 percent of that amount. In other words, 10 meters of lowering are offset by 8 meters of uplift. But, given the full effects of isostacy, the landmasses would still lose 20 percent elevation with denudation and eventually disappear altogether. But this has not happened.

Additional uplift, equivalent to more than 20 percent of the loss, is made up by processes related to plate tectonics, namely, the formation of orogenic belts and the regional uplift of continental platforms (Figure 20.7). The latter appears to be related to thermal rises beneath the continents that are capable of elevating large sections of the crust. Africa, situated at the center of the old Pangaea landmass, is a case in point. It has extensive plateaus at high elevations, 2000 meters or more, which were uplifted by a rise in the underlying mantle, apparently induced by mantle convectional plumes, as part of the tectonic system that produced the East African Rift System.

Thus, a great contest is staged between denudational, or geomorphic, systems, driven mainly by exogenous forces and geological, or uplift, systems, driven by endogenous forces. The contest is played out at the Earth's surface and, in general, the elevation of the land is a record of the score or the relative balance. Obviously, the outcome is very uneven geographically.

In many mountainous regions, rates of uplift clearly exceed rates of denudation. In the great orogenic belts such as the Himalayas, uplift rates are on the order of 500 to 1000 meters per million years, roughly 12 times greater than the denudation rate for central North America. But there is another counterbalancing factor that must be also considered. Denudation rates rise substantially as land is uplifted.

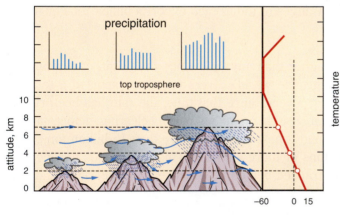

Figure 20.8 Precipitation and temperature changes with mountain elevation. In general, conditions for weathering and erosion increase, often dramatically, with elevation.

Precipitation increases and temperature decreases with higher elevations, a trend which is shown in Figure 20.8. The first results in higher runoff rates and the second in the formation of glaciers, both of which drive up erosion rates. In addition, the potential energy of erosional systems increases because mass (water) and elevation (distance above sea level) are increased. When water moves downhill this potential energy is converted into kinetic energy, enabling runoff and glaciers to denude massive amounts of mountain terrain rapidly. For large mountain belts like the Himalayas and the Andes, denudation rates are high, in the range of 200–300 meters over a million years. Yet this rate is less than half the rate of orogenic uplift in those areas.

How High Can a Mountain Grow?: Since uplift appreciably exceeds denudation in active orogenic belts, then why do not mountains go on rising up forever, reaching heights a hundred or more kilometers above sea level? Obviously there must be limits on the maximum elevation of mountains, for the tallest mountains on Earth are only 8 to 9 kilometers high. Chief among the limiting factors is the amount of internal (endogenous) energy available to lift great masses of rock against the force of gravity. It is the basic problem of the weight-lifter in Figure 20.9. To raise a mass he/she must overcome the resistance imposed by gravity. On a terrestrial planet smaller than Earth, where gravity is weaker (because planetary mass is smaller), a weight-lifter could raise a much larger mass overhead.

Figure 20.9 The height of mountains on a planet is controlled by the balance between the force of gravity and the force of uplift (endogenous) processes.

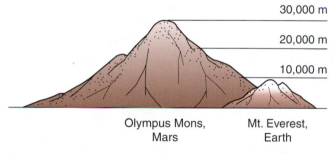

Figure 20.10 Olympus Mons is higher than Mt. Everest largely because Martian gravity is less than Earth's.

Mars is such a planet and it has much higher mountains than Earth. Mars once had an internal energy system with power probably comparable to Earth's. It was able to push a mountain up much higher than Earth simply because gravitational resistance was lower. Mars' biggest mountain, Olympus Mons, is about three times higher than Earth's Mt. Everest, which is nearly 30,000 feet (8800 meters) high (Figure 20.10). The other factor controlling the height of mountains on Earth is the power of erosional forces at high elevations. With its deep, moist atmosphere, erosional forces are much stronger on Earth than on Mars, thus land here is torn down faster as it gets higher – a form of positive feedback, that is, erosion leads to higher mountains, which lead to more erosion.

20.3 How Geomorphic Systems Operate

Eighty percent or more of the work in eroding the Earth's surface is performed by runoff systems. Most is done by the Earth's great rivers, their tributaries, and the runoff systems that feed them. These vast drainage systems, whose watersheds may cover more than a million square kilometers of land, begin with overland flow and groundwater seepage, all generated by precipitation, which feed streams linked together in great networks or drainage nets.

Stream Systems: As we noted earlier, drainage systems are powered mainly by solar energy via the hydrologic cycle. The hydrologic cycle feeds stream systems with the water that becomes discharge, which in turn produces erosion and carries sediment. But the energy of stream systems to erode the crust and move sediment is not determined only by the amount of precipitation and runoff they receive. It is also determined by the elevation of the land because elevation sets the steepness of the slope down which water flows: the steeper the slope, the greater the erosive force of streamflow. A watershed's power, as it were, is therefore a function of two sources of energy: water provided by the atmosphere, and elevated land provided by isostasy and tectonics.

These two energy sources behave quite differently and this affects the way streams operate as geomorphic systems. The atmosphere's input of water tends to be sporadic, that is, it comes in short spurts with storms of different magnitudes and frequencies and/or it varies seasonally. Uplift, on the other hand, tends to be *both* gradual and episodic but over long periods of time. In non-mountainous areas, uplift is more or less gradual with the continuous unloading of the land; whereas in mountainous areas, uplift tends to follow the erratic pattern of faulting and earthquake activity. Therefore, the work regime of stream systems over the broad interiors of the continents follows two modes at once, one dictated by frequent, short-term and seasonal runoff controlled by weather and climate, and the other dictated by long-term elevation changes controlled by Earth's interior forces. In mountainous areas, the work regime of streams is more complex, following two uplift modes – gradual regional uplift and episodic tectonic uplift – as well as a climatic mode in all its complexities.

When orogenic belts like the Rockies and Andes are being raised up, erosion and sediment production by stream systems increases as the land gets higher. We know this because the layers of sediment deposited in nearby river deltas get thicker as

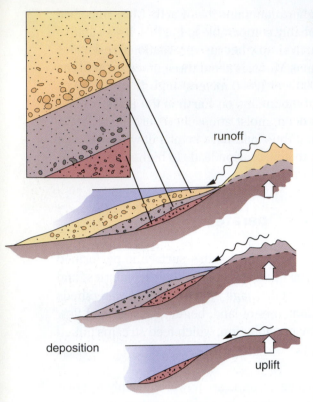

Figure 20.11 As mountains get higher erosion and sediment production increases, the thickness and composition of the depositional strata also change.

the mountains get higher, as shown in Figure 20.11. Not only that, but we often see thicker layers associated with sudden uplift related to episodes of earthquake and faulting activity. In addition, climate systems also change with uplift and mountain formation.

Almost invariably, precipitation increases with higher elevation, especially on windward slopes, as a result of orographic effects (see Figure 20.8). As total annual precipitation increases, streamflow also increases but variability decreases meaning that streamflows become more regular or steadier with greater precipitation. In contrast, leeward mountain slopes become drier as mountains get higher – the so-called rainshadow effect – and streamflow in turn becomes more variable. On balance, both sources of energy, namely, climate and elevation, are affected by mountain building not only in terms of the amount of energy available, but also in terms of how it is delivered. In the long run, a grand system, which is always trending toward equilibrium, a longterm condition that fits under the concept of *dynamic equilibrium*, is formed such as that illustrated in Figure 20.12. Mountain building leads to harsher climatic conditions, increased runoff and erosion, and loss of mountain mass. Reduced mountain mass, in turn, induces uplift.

The three remaining geomorphic systems – glaciers, wind, and waves and currents – do far less work in total than runoff systems. They are, however, very important in selected geographic environments, such as seashores and cold mountains, where they are the dominant systems. These systems are driven by atmospheric energy, and two of them, wind and waves, exhibit dramatic rises and falls with the vicissitudes of weather and climate. Table 20.1 attempts to put this concept into perspective by comparing how long it takes a system to respond to an atmospheric change like a change in air pressure. The second column gives a rough estimate of the system exchange time, another measure of response time.

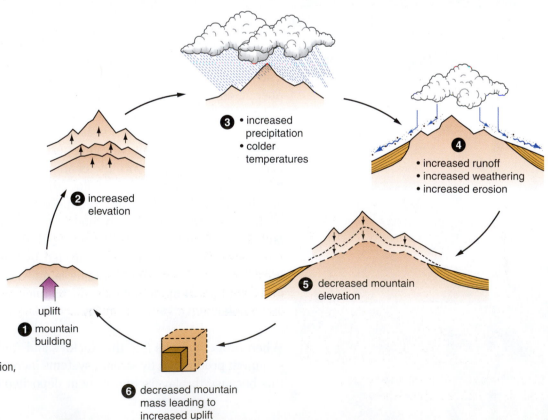

Figure 20.12 A system of mountain building and erosion in which uplift and increased elevation lead to climate change, increased erosion, reduced mountain mass, and further uplift and elevation increase as a result of feedback.

Table 20.1 System response times and exchange times

System	Response time to an atmospheric change*	Exchange time**
Wind	minutes	2–10 hours
Wave	1–2 hours	10–20 hours
Current	2–5 hours	20–40 hours
Glacier (mountain)	50–100 years	1000–10,000 years

* Approximate time to respond to a change in driving force such as wind's response to air pressure change

** Time taken to exchange the air or water mass in a system. Times will vary with the size of the system such as the area of a watershed.

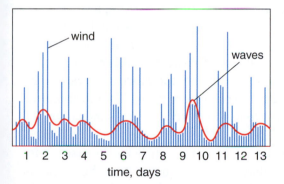

Figure 20.13 A schematic illustration of the relative variability of wind and wave magnitude over time. Waves are driven by wind but for each wind event the magnitude of wave change is smaller than that for wind.

Glaciers are slow-moving systems which vary gradually in response to climate change.

Wind Systems: Wind is the most variable of the geomorphic systems, changing speed and direction typically from minute to minute and blowing in directions not dictated by the slope of the land. But there are also longer-term patterns in wind such as those dictated by prevailing wind systems like the westerlies, or seasonal systems like the Asian monsoon, which are driven by global and regional pressure systems. The result is that in dry areas, small amounts of sand and dust are being moved around almost constantly. Massive movements, however, await major wind storms that cause dunes to shift and dust to blow great distances over continents and even beyond to the oceans.

Wave and Current Systems: Waves and currents are driven mainly by wind but they are less variable than the wind itself, as the graph in Figure 20.13 shows. They tend to vary with the magnitude and frequency of storms, rising and falling and changing direction over periods of several days, or with prevailing wind systems, driving rather steadily in one direction for months at a time. Although waves and currents can be treated as a single system, currents are not driven directly by wind but mostly by waves. Therefore, currents tend to have slower response times to wind because the energy must first pass through the wave system (Table 20.1). Waves of the tsunami class are an exception to the system described above. Tsunamis are generated not by wind but by earthquake and volcanic activity on the ocean floor, as we saw in Chapter 19. Their locations, magnitudes, and frequencies are governed by tectonic forces which, of course, have no relationship to the atmosphere.

Glacier Systems: Glaciers are the least variable geomorphic systems in terms of their response to short-term variations in the atmosphere. Like streams, they depend on precipitation for energy and, in the case of mountain glaciers, on land elevation as well. But there is a significant difference between streams and glaciers as energy systems. Glaciers are fed by snow, which is transformed into ice, which eventually becomes part of the glaciers. This process takes many years and it tends to buffer short-term variations in water supply, that is, weather-related variations in snow input (Table 20.1). Therefore, glaciers respond more to longer-term energy changes related to snowfall and temperature variations over decades and centuries. As they grow, glaciers expand their coverage, thicken, and exert greater erosive power on the land. When they shrink, erosional power declines rapidly and the water they lose with melting is added to the sea. The current world trend toward declining glaciers is related to a global warming trend that reaches back 50 years or more.

Summary on Geomorphic Systems: Although they are driven primarily by climate and water systems fueled by solar energy, geomorphic systems are intimately tied to Earth's rock cycle and mountain building driven by plate tectonics. Denudation increases with elevation as mountains grow and precipitation and stream gradients increase. Worldwide, most of the work in lowering the land is done by stream systems but, in selected geographic settings glaciers, waves and currents, and wind are far and away the principal workhorses. Although all four systems depend on the atmosphere for their energy, each responds differently to changes in the atmosphere. So important is the atmosphere in geomorphology that the first stage in the denudational system carries an atmospheric name, weathering.

20.4 Weathering Systems and the Breakdown of Rock

Weathering prepares rock for erosion by weakening and breaking it down. And the first question we should ask is why rock breaks down at all? Why should not granite, for example, which appears so durable, be practically invincible to nature's onslaught? To begin with, most of the rock we see on the Earth's surface did not form there. It formed within the crust, often at great depths, under conditions quite different from those at the surface and, over geologic time, was uplifted and exposed. This is helpful in understanding why solid rock is not invincible when exposed to surface or near surface conditions.

Rocks as Aliens: How different is the surface environment from the environment of rock formation? First, consider water, the most critical factor in weathering systems. It is abundant near the surface but scarce at depth. Groundwater, for example, typically constitutes 20 to 30 percent of near-surface rock and soil whereas at depths of 3000 meters or more, it is usually less than 1 percent. Second, near the surface the chemical environment is appreciably richer than at depth because, among other things, chemical compounds such as humic acids from vegetation are available there, often in great abundance.

Sandstone bedrock, once as solid as new concrete, breaking down with exposure and weathering.

Third, surface temperatures are considerably lower and more variable than temperatures at depth. Igneous rocks, for example, form at temperatures of 800 °C or more, whereas Earth's average surface temperature is only 15 °C. Moreover, temperature at depth varies little over millions of years but, as the graphs in Figure 20.14 reveal, at the surface, temperature may vary substantially from day to night and even more from winter to summer. Fourth, pressure is massive inside the crust, thousands of times greater than at the surface. Thus, when rock is uncovered near the surface it is subject to decompression, which causes it to expand. In combination, these contrasts in moisture, heat, chemistry, and pressure are severe and they place rock under much stress with surface exposure.

Water the Solvent: Water is Earth's most effective natural solvent. The power of water to affect weathering is controlled by three factors: (1) availability, (2) temperature, and (3) chemical additives. Climate is obviously a critical control on both the availability and temperature of water. Warm, wet environments are most conducive to high rates of chemical reactions. Consequently, in the wet tropics, weathering is much faster and, as Figure 20.15 shows, advances to much greater depths into the crust than it does in dry and cold environments.

Figure 20.14 Temperature (T) variations overtime in the atmosphere, at the Earth's surface, in the soil, and in the crust. Thermal variations are relatively extreme at ground level.

Temperature affects chemical weathering because heat influences the rate of chemical reactions. Most reactions double with each 10 °C rise in temperature. Therefore, we can expect that in a tropical environment with an average temperature around 27 °C, chemical weathering can decompose a rock as much as four times faster than in the midlatitudes at an average temperature of 7 °C or so. In addition, essentially all water in nature contains chemicals besides H_2O. Water takes on chemicals from the air, vegetation, soil, rock, and pollution sources and these set up reactions with the different minerals in rocks. For different combinations of chemicals and minerals, the rates, processes, and outcomes of weathering vary with bioclimatic conditions.

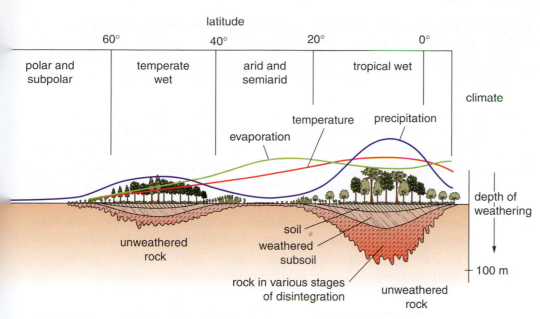

Figure 20.15 The depth of rock weathering and soil formation related to climate and vegetation from the tropical to the subpolar. Abundant water and warm temperatures advance weathering and soil formation.

Weathering Systems and Bioclimate: In general, wet, organically-rich environments such as tropical forest landscapes are the most active chemical environments and this is reflected in the great depth of weathering, among other things. By contrast, the depth of weathering in arid environments is slight (Figure 20.15).

A Lesson From the Mayans and Egyptians: In the tropical lowlands of Central America, the Mayans created a civilization that thrived for more than 1000 years and at times supported more than 5 million people. It was carved out of heavy forests and organized more or less into city states and at the center of each was an administrative—religious complex consisting of elegant temples, plazas, and royal palaces. All this the Mayans built from the most durable material available to them, limestone.

In AD 900, the entire Mayan civilization collapsed and the elegant cities were abandoned. Hundreds of limestone structures were left completely unattended and tropical forest began to retake the Mayan farmlands and cities. Plants crept over the structures and the limestone broke down giving rise to a deep mantle of soil and roots. By the time the Spanish crossed the region in the early 1500s, virtually no visible sign of the Mayan structures remained. The stepped pyramids were little more than cone-shaped hills in the forest. Not until the mid 1800s did a curious American dig into these hills and discover the remnants of the Mayan cities. Today archaeologists, aided by satellite images of vegetation, are clearing away the mantle of soil, forest, and rotten limestone revealing one of the world's truly spectacular ancient treasures (Figure 20.16a).

In the Nile Valley of Egypt, ancient builders erected structures similar to those of the Mayans. They began the process even earlier, built bigger structures, and in many places used the same material, limestone. Yet today, 5000 years later, most of these structures remain little changed by the elements (Figure 20.16b). Even

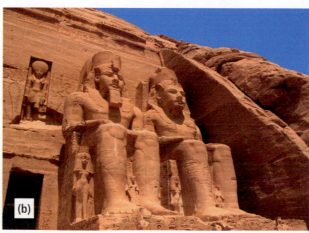

Figure 20.16 (a) A partially excavated Mayan temple, dating from AD 900. Weathering of the limestone was rapid under the wet, warm tropical climate. More than a meter of weathered debris covers most buildings. (b) An Egyptian sculpture little changed under an arid climate over the 5000 years since its construction.

the inscriptions carved in facing stones are still readable. Impressed by all this, the British, in the 1800s, decided to take a celebrated Egyptian monument, Cleopatra's Needle, back to London for outdoor display. In only a matter of years, in the damp polluted air of London, the inscriptions broke down and disappeared.

What all this points up is the tremendous differences in weathering rates across the world. The differences are tied directly to climate, in the Mayan and Egyptian examples mainly to moisture. But temperature is also important for if we go northward from Central America into the heart of North America, frost becomes an important player in the breakdown of rock. Added to heat and moisture is vegetation, a source of organic acids that accelerate weathering rates on most rock and, in the tropics, forests produce massive amounts of organic debris. In the desert there is of course little vegetation and thus few acids to add to sparse water. And in the case of Cleopatra's Needle in London, the hieroglyphics would have lasted longer had the city's atmosphere not been so polluted with carbon dioxide, sulfur dioxide, and other chemicals.

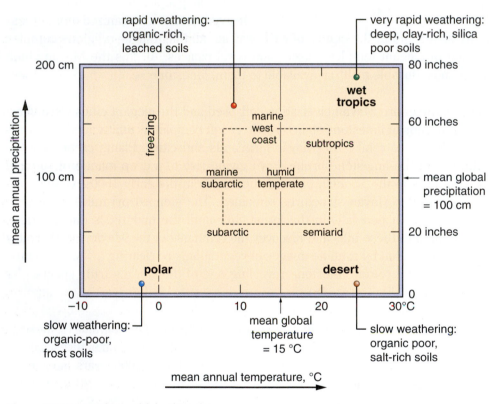

Figure 20.17 A graphic model depicting the relationship of weathering systems to annual temperature and precipitation. The major systems are wet tropics, desert, and polar. Several subsystems are also identified.

Global Variations: Earth is partitioned geographically into weathering systems. Broadly speaking, there are three extremes in these systems: tropical wet, desert, and polar. Each of these systems is labeled in the diagram in Figure 20.17 relative to mean annual precipitation and temperature. In addition, several other climatic zones, or subsystems, are shown, which are made up of various combinations of these. For example, the subarctic zone of North America and Eurasia, which covers an area approaching 10 million square kilometers, is notable for its long polar winters, with deep frost penetration and frozen soil water, and short summers warm enough to foster a vibrant organic system and abundant liquid water. Another example is the subtropical zone, part of which (the Mediterranean climate), is decidedly arid in summer but winters are wet enough to soak soils and produce leaching. And another is the marine West Coast zone with its long, wet winters, moderate temperatures, and highly active soil leaching and groundwater systems.

Add to this complexity climate change. As we discussed in Chapter 8, climate change has been the rule rather than the exception on our planet. Between 10,000 and 20,000 years ago, for example, climate zones shifted equatorward with the growth of the continental glaciers in North America and Eurasia. The American Southwest became fairly humid in places and water was abundant enough to advance cave formation in areas of limestone. The subarctic zone cited above extended southward well beyond the Great Lakes while polar conditions reached into the Great Lakes' basins. The point here is that most regions of the Earth bear the effects of two or more weathering systems, so that when we examine landscapes we must be aware of past conditions as well as present.

Mineral stains from groundwater seepage, oxides of iron, copper, and other minerals weathered from bedrock.

20.5 Basic Types of Weathering Processes

There are three basic types or classes of weathering processes: chemical, mechanical, and biological. **Chemical weathering** involves a number of processes, all associated with moisture, that disintegrate the minerals in rock. **Mechanical weathering** involves the breakdown or fragmentation of rock by processes that physically crack and wedge rocks apart, and **biological weathering** involves the effects of biota in disintegrating and breaking down rock. Biological effects include not only the role of acids produced, for example, by decaying vegetation in setting up chemical reactions but the mechanical effects of roots growing in cracks and forcing rocks to split apart.

Although we discuss weathering processes as three separate classes, we must recognize that in nature processes of all three classes actually work together, forming complex, intricate systems that weaken and destroy solid rock. In the end, four types of products are produced by weathering systems:

(1) physical particles ranging from huge boulders to tiny silt and clay sediments;
(2) chemicals dissolved in water in the form of ions of calcium, silica, and other elements;
(3) oxides, the residues of oxidation of metallic minerals such as iron and copper; and
(4) clay minerals produced by the chemical transformation of minerals weathered from rock such as feldspar into secondary, clay-type mineral forms.

Critical to our thinking is the fact that these products end up in different geographical environments – in the soil, on the continent shelves, in the deep ocean – but over geologic time all are transformed back into rock and reintegrated into the Earth's crust. In addition, weathering often produces distinctive landforms, including those characterized by deep holes and caverns where carbonic acid has dissolved limestone, a process known as carbonation.

Chemical Weathering: We begin with one of the most effective chemical weathering processes, carbonation. **Carbonation** begins in the atmosphere when water vapor condenses in a cloud and assimilates carbon dioxide (CO_2) forming a weak carbonic acid in rainwater. When rainwater infiltrates the ground, it takes on more carbonic acid. Topsoil is rich in organic matter, which is the food for bacteria, and these organisms in turn produce carbon dioxide. In moist, aerated soils, bacterial activity can raise the carbonic acid level of soil water and groundwater substantially.

When carbonic acid is introduced to limestone, it reacts with calcium carbonate, the rock's main constituent. From this reaction, a bicarbonate is formed that is soluble in water. In other words, as acidic water percolates through the limestone, the rock is leached away. The dissolved calcium enters the groundwater and is eventually released into a stream that ultimately discharges into the sea. Here it may be assimilated by shell creatures or precipitated directly from seawater and reconstituted as limestone.

Limestone is very abundant in the continental crust, and in many areas it weathers to form a distinctive type of terrain called **karst topography**. Karst landforms are produced by differential chemical weathering in which carbonation is focused where water underground concentrates in crack systems in the limestone. In time the cracks widen forming caverns that may link together in underground networks carrying

This majestic terrain of limestone towers in south China is a result of carbonation weathering in a warm, wet climate.

Figure 20.18 A small sinkhole, the result of cavern collapse. Groundwater fills the lower hole, forming a lake.

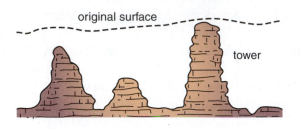

Figure 20.19 Profiles of two common karst landscapes. The upper one, dominated by towers, represents a more advanced state of weathering than the lower one which is dotted with sinkholes.

Figure 20.20 A bauxite mine in Brazil. Bauxite is associated with chemical weathering under wet tropical conditions and is the world's principal source of aluminum.

groundwater (see Figure 16.16). If the rock overlying a cavern is weak, it often sags or collapses, forming a **sinkhole** or **doline** like the one shown in Figure 20.18. As sinkholes enlarge and merge together, larger basins called *poljes*, are formed. If they fill with groundwater, a lake or wetland is formed, which has happened throughout the large karst area in central Florida, for example (see Figure 16.17).

Whole landscapes can be lowered by carbonation and karst processes. But the karstification process is typically uneven because of variations in rock resistance and/or groundwater distribution and flow. As a result, "islands" of rock are often left protruding above the landscape floor as isolated hills and towers. Such landscapes have the aspect of being inverted from those where sinkholes are the dominant karst landforms (Figure 20.19). Also see Section 16.4 in Chapter 16 for more on karst processes and cavern formation in connection with groundwater and subterranean streams.

Igneous rock also weathers by chemical processes involving water-carrying acids. It often starts when water penetrates the rock along the contacts between crystals and is absorbed by certain minerals. This process, called *hydration*, produces no chemical change itself, but sets up a sequence of chemical reactions that alter the minerals irreversibly. One set of these reactions is **hydrolysis**, a process involving the reaction of water and an acid on a mineral. Hydrolysis is considered to be most effective in weathering granite and related igneous rocks.

Feldspar is a key mineral in many igneous and metamorphic rocks. It is a major constituent of granite and when subjected to hydrolysis, it breaks down forming a clay mineral, kaolinite, as well as ions of silica in groundwater. Under the hot, wet conditions of the tropics, granite boulders, for example, can be observed in various states of disintegration as feldspar crystals rot away. The weathered feldspar yields **bauxite**, an oxide of aluminum. In some parts of the tropics, bauxite has accumulated in such quantities that it is commercially mined for aluminum ore (Figure 20.20). Bauxite is also the main constituent of **laterite**, the hard layer that forms within tropical soils.

Other Types of Chemical Weathering: Oxidation is common in the manufactured world. We see it as rusting metal in automobiles, for example. In nature, oxidation usually accompanies hydrolysis and is most apparent in rocks containing iron. Iron is a component in the minerals olivine and pyroxene and, in the presence of oxygen, it is oxidized to form ferric iron, which in turn is transformed into limonite, a mineral resembling rust. Oxidized minerals are abundant in surface and subsurface waters. Where groundwater seeps from rock faces, iron and other oxides often form colorful stains, and of course iron build up in water pipes is common in millions of homes.

Another chemical weathering process called **chelation** (pronounced key-la-tion) originates with plants. This process is characterized by bonding between mineral ions in rock and large organic molecules that are excreted from vegetation. Our knowledge of this process has developed in connection with research on agricultural fertilizers, and, although chelation is not well understood under natural conditions, some researchers regard it as an important weathering process. Lichens, for example, excrete chelating agents, and studies have shown that lichen-covered basalt, such as the rock shown in Figure 20.21, is often more deeply weathered than is lichen-free basalt.

Figure 20.21 A lichen-covered bedrock. Organic acids secreted by the lichens are a source of chelating molecules.

Table 20.2

Composition of Seawater	
Mineral	**Parts per 1000 Parts of Water**
Chloride	19.3
Sodium	10.6
Sulfate	2.7
Magnesium	1.3
Calcium	0.4
Potassium	0.4
Bicarbonate	0.1
Others	0.1
Total	**34.9**

Although rock seems to vanish when it is dissolved by chemical processes, we know that matter, like energy, cannot be destroyed. Most of the soluble ions produced in weathering are carried off in groundwater and streamflow, eventually reaching the sea. The composition of seawater (Table 20.2) attests to this unending process. On the other hand, lots of more resistant particles such as grains of quartz sand, are left behind in soils and surface deposits. But even they, in time, are corroded and leached away. This usually takes millions of years and there are ancient landscapes in the tropics where even quartz, a particularly resistant mineral, is almost non-existent in soil.

Mechanical Weathering: Mechanical weathering, the second major category, physically fragments rocks. Virtually everywhere mechanical weathering operates hand-in-hand with chemical processes, and in most places it is difficult to ascertain how much work should be ascribed to each. Mechanical weathering is apparently more effective in cold, dry environments, whereas chemical weathering is clearly more effective in warm, wet environments. There are many mechanical weathering processes operating in a wide variety of environments and here we describe a few of the most important ones.

Mechanical weathering begins with the formation of cracks in bedrock. Crack formation is initiated basically in three ways: (1) differential expansion of rock masses, (2) chemical decomposition along bedding planes or contacts between different rock types, and (3) expansion within bedrock by freezing water or plant roots. When cracks widen and deepen as chemical and mechanical weathering advances, they are called *joint lines*. Joint lines may extend tens or even hundreds of meters into bedrock and develop lateral offshoots. When horizontal and vertical joint lines intersect like those shown in Figure 20.22, blocks of rock are freed from the solid Earth.

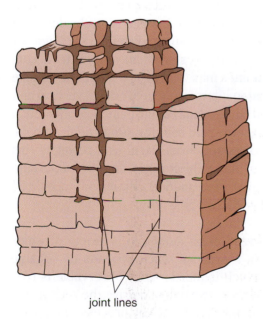

joint lines

Figure 20.22 Intersecting joint lines leading to the formation of blocks.

Where massive rock bodies, particularly granitic rocks, are exposed after millions of years deep within the crust, a type of mechanical weathering called **exfoliation** often develops. Exfoliation is caused by the differential expansion of a rock body as it decompresses with the stripping away of heavy rock overburden. The expansion produces a system of joint lines that yield large, scale-like sheets or slabs of rock. A somewhat similar expansion process can be observed in rock quarries where rock often breaks, or even explodes, with a "pop" as it decompresses after blasting operations.

In nature, the process is less dramatic, but as the sheets detach, they slide downslope or disintegrate into blocks and smaller particles that are subsequently eroded away. In any case, once the process starts, it forms a self-perpetuating system. Each

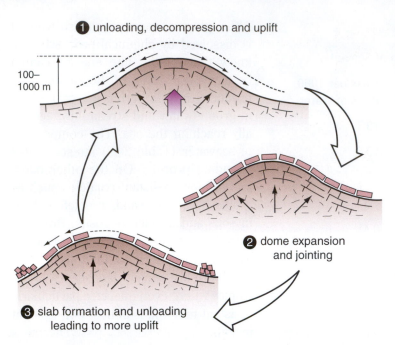

❶ unloading, decompression and uplift

100–1000 m

❷ dome expansion and jointing

❸ slab formation and unloading leading to more uplift

Figure 20.23 An exfoliation system described in three phases. The resultant landform is a dome-shaped rock knob.

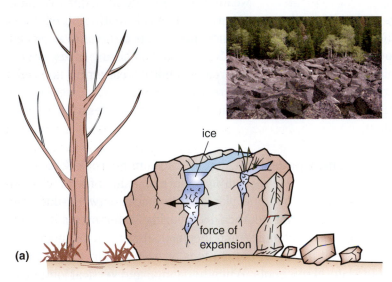

ice

force of expansion

(a)

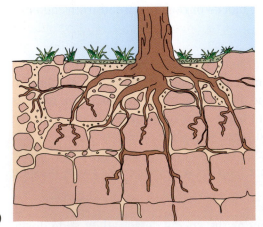

(b)

Figure 20.24 (a) Frost wedging leading to rock fragmentation. Inset shows felsenmeer, a product of frost wedging in high mountains. (b) Root wedging is another form of this process.

shedding of mass results in decompression, leading to expansion, more joints, more shedding, and so on – another example of positive feedback (Figure 20.23). The resultant landform is usually a rounded rock knob called a *dome*, a common landform in many areas, where great masses of granite rock are exposed.

At a much smaller scale, individual boulders may also be subject to scale-like disintegration called **spheroidal weathering**. Although this weathering pattern looks like a miniature version of exfoliation, it is attributed mainly to chemical processes and does not involve decompression and rock expansion.

Frost wedging (or **frost shattering**) is an effective weathering process in cold environments, especially polar and high mountain landscapes. The physical process of frost wedging is easy to demonstrate in laboratory experiments. When water freezes it expands by 9 percent of its liquid volume. The resultant force of crystallization can be tremendous, up to 30,000 pounds of pressure per square inch. At a temperature of −20 °C, a common surface temperature during winter in cold environments, the force exerted by the expansion of ice exceeds the resisting strength of rock by nearly 10 times. It is easily great enough to burst iron water pipes.

We know from the principles of soil heat flow that ground freezes from the top down. Therefore, the first ice to crystallize in water-filled joints and cracks forms a cap or plug over the liquid water. The ice plug resists some of the upward expansion produced by subsequent freezing, thereby directing the force of the crystallization downward and outward against the rock (Figure 20.24a). As a result, when water is truly confined, cracks and joint lines may enlarge and lengthen with ice wedging. This can lead to intersecting networks of vertical and horizontal cracks and joints, and the formation of individual blocks or boulders. In some cold settings, broad fields of boulders, called *felsenmeer* (stone seas), form as a result of this process. The boulders often exhibit sharp, angular edges suggesting that they formed from sudden splitting (Figure 20.24 inset).

Biological Weathering: Biota function as weathering agents in a variety of ways. We have already mentioned one, chelation, a process which depends on organic molecules secreted from plants. Another which is more readily observable is root wedging, where tree roots invade cracks in rock, expand, and drive the rock apart. Although we know relatively little about the mechanics of this process, the power of tree roots is certainly evident where sidewalks, streets, walls, and even building foundations have been lifted, separated, and/or broken by root growth. In addition, when plant roots invade rock, they promote chemical weathering because they contribute organic agents to soil water and groundwater (Figure 20.24b). Working together, mechanical and chemical processes appear to speed up the rate of breakdown as weathering advances for as rock is broken down into smaller pieces, the

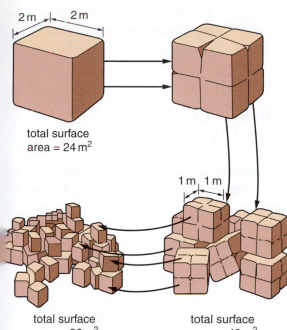

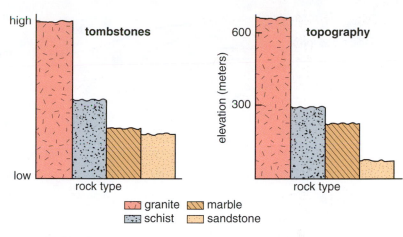

Figure 20.26 Relative resistance to weathering of four rock types in old tombstone (left) and elevations of the land (right) in New England (from P. Rahn).

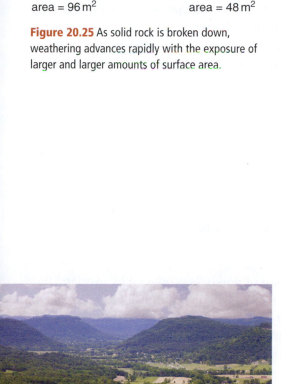

Figure 20.25 As solid rock is broken down, weathering advances rapidly with the exposure of larger and larger amounts of surface area.

Central Kentucky where differences in rock types and their resistance to weathering are often evident in landforms and their elevation. Sandstone often forms the higher ground.

total surface area exposed to weathering increases geometrically as the diagram in Figure 20.25 illustrates.

Weathering and Landforms: Weathering processes can produce distinctive landforms. In humid areas of limestone bedrock, karst processes produce landscapes like no others on Earth. Likewise, in areas of massive bedrock exposures, exfoliation produces rock domes whose symmetry seems to belie their natural origin. And where individual landforms are not so clearly related to weathering, the relief (relative elevation change) of the land, often is.

A study in New England, which is summarized in Figure 20.26, bears this out. This study related rock type (and its susceptibility to weathering) to land elevation by comparing weathering rates on tombstones of different rock compositions (granite, schist, marble, and sandstone) with elevations of nearby land of the same rock composition. Granite, which was the most resistant rock in tombstones, formed the highest-elevation land, whereas sandstone, which was the weakest rock in tombstones, formed the lowest elevation in landscape. On the other hand, a comparison of terrain developed in sandstone and limestone in central Kentucky shows sandstone to be more resistant. The difference is apparently related to differences in the susceptibility of Kentucky limestone and sandstone to dissolution under a weathering system driven by a humid climate with long, warm summers and a heavy vegetation cover.

Summary on Weathering Systems: Weathering operates on all terrestrial surfaces, but the processes, rates, and products (including landforms) may vary greatly from one geographic setting to another. In the broadest of terms, three extremes in weathering systems operate on Earth: tropical wet, arid, and polar, and these are defined in the chart in Figure 20.17. Where heat, moisture, and vegetation are abundant, high rates of chemical weathering can be expected for most rock types and the depth to which weathering advances is usually great. Where water is scarce, which includes polar as well as desert regions, weathering is generally much slower and often gives us different products including a relatively small output of

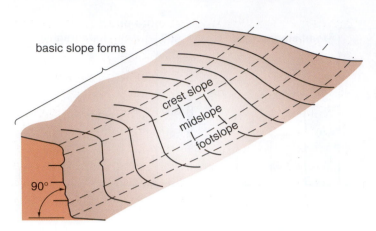

Figure 20.27 The three basic parts of a slope illustrated for a range of slope forms and inclinations.

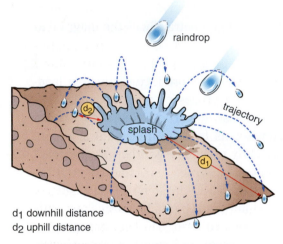

d₁ downhill distance
d₂ uphill distance

Figure 20.28 Upslope and downslope trajectories of particles released by the explosion from a raindrop impact on a slope.

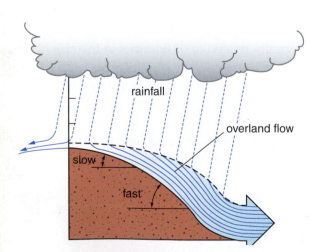

Figure 20.29 A conceptual model illustrating the increase in the depth of overland flow with distance downslope, assuming no water loss to infiltration. Velocity is greatest near midslope, the steep part of the slope, and least at the top and bottom.

dissolved minerals. Nevertheless, in all geographic settings weathering leaves the land with a residue of rock debris and, no matter where it occurs, it must ultimately be hauled downslope and eventually delivered to the sea.

20.6 The Hillslope Segment of the Denudational System

The second step in the denudational system is the downhill movement of weathered material or what we will call debris. The processes responsible for this movement are collectively referred to as **hillslope processes**. There are two classes of hillslope processes: (1) erosion generated by various forms of runoff, and (2) mass movement, which includes various gravitationally induced motions such as rock falls and landslides.

The term *hillslope* is taken to mean every sort of slope in the land from mountain cliffs to gentle inclines in farm fields. Irrespective of steepness, each slope is composed of three basic parts: a *crest slope*, which is the brow; a *midslope*, which is the central section; and a *footslope*, which is the base or the toe and each is shown in Figure 20.27. In general, the midslope is the steepest part of any slope; the surface of the midslope is also called the slope face.

Rainsplash and Rainwash: Erosion begins with the impact of raindrops on the soil. When a raindrop hits a wet, unprotected surface the impact sends out a circular splash of water and soil particles. If this occurs on a sloping surface, the downhill side of the splash travels further than the uphill side as the drawing in Figure 20.28 illustrates. The difference in the lengths of these trajectories multiplied times the millions of raindrops hitting a slope in a single rainstorm can account for appreciable downhill movement of soil on barren or sparsely vegetated slopes. The process is called **rainsplash**.

A related erosional process, called **rainwash** or **wash**, is produced by overland flow spilling over the slope face. Overland flow occurs when the intensity of precipitation exceeds the infiltration capacity of the surface. It is characterized by water moving slowly over the ground in thin sheets, and tiny threads. Near the top of the slope, overland flow is usually slight and incapable of effecting much erosion because the surface area available to gather water is small. But as it moves downslope, both volume and velocity increase, and at some point it generates enough force to displace small soil particles, especially organic matter, silt, and loose clay particles (Figure 20.29).

Together, rainwash and rainsplash can erode (lower) a slope by as much as 3 to 4 centimeters per year. Such rates are possible only on barren slopes, however. When vegetation is added to a slope, erosion rates decline significantly because the plants absorb much of the energy of raindrops, take up part of the rainwater through interception, increase soil intake of water through infiltration, slow velocity of overland flow, and increase the resistance of the soil to erosion (Figure 20.30). In fact, a dense forest may eliminate overland flow altogether and slopes suffer virtually no erosion by this form of runoff.

Gullying: As overland flow moves from the crest slope onto the midslope, the water merges into rivulets that can etch tiny channels, called **rills**, into the slope. Now the runoff velocity quickens because channel flow, even in the miniature channels, is a far more efficient means of flow than overland flow. A little further downslope, the rills merge together producing flows of sufficient force to erode ruts or gullies into the slope face, a process called **gullying** (Figure 20.31). As the gullies cut into the slope, they may intercept subsurface water, which seeps into the gully channels. As seepage water trickles out of the soil, it loosens and erodes small particles as shown in the inset in Figure 20.31. This process, which is termed **sapping**, causes the gully walls to be weakened and undermined along the seepage line. The gully walls may cave in, or slump, unless they are reinforced by the roots of plants, which usually slows the process down.

Figure 20.30 The important role of plants. Soil unprotected by vegetation is subject to pronounced erosion, especially on sloping ground.

Of all the erosional process that operate on hillslopes, gullying is decidedly the most effective in removing soil, so much so, in fact, that it is dreaded by farmers worldwide. Gullying can consume entire hillslopes in a matter of the years. Farmlands most vulnerable to gullying are found in humid, forested landscapes where land has been cleared and left unprotected from heavy rains and runoff. The American South was plagued by severe gullying throughout much of the nineteenth and twentieth centuries in the wake of land clearing for cotton, tobacco, corn, and other crops. Soil conservation practices and reforestation in the last 75 years have greatly reduced soil erosion in the South.

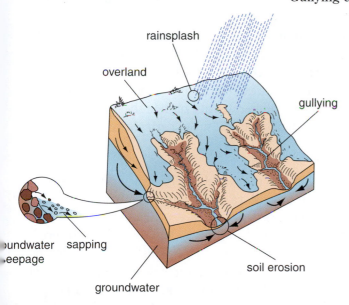

Figure 20.31 Runoff leading to gullying, the most severe form of slope erosion. Sapping accelerates gullying by weakening gully walls.

But gullying is spreading in other parts of the world. Today it is a serious problem in the wet tropics of Africa, Asia, and South America where forest is being removed and soil is left exposed to heavy rains and runoff. Gullying can also be prominent where land use is not involved. Landscapes called *badlands* are dominated by intensive gullying that leaves them with intricate networks of channels and finely textured, dissected topography. Most badlands develop on soft, erodible shale or deposits of volcanic ash in arid and semiarid areas where the plant cover is ineffective in stabilizing the surface. In the extreme, badlands are sometimes considered natural wonders and may be set aside as public reserves like the one shown in Figure 20.32.

Deposition and Alluvial-fan Formation: When runoff reaches the footslope and the ground begins to flatten out, flow velocity declines. With this drop in energy, the runoff deposits part of its sediment load. The largest particles are deposited first, followed sequentially by smaller particles as the water loses speed. The finest particles, mainly clays, are deposited last as the water soaks into the ground or draws to a halt near the footslope. In addition to slowing down, flows discharging from slopes also lose water to infiltration and evaporation.

Figure 20.32 The Badlands of Dinosaur Provincial Park, Alberta. A unique landscape in the Great Plains, carved from weak sedimentary rocks.

In arid regions, water loss is so pronounced that even sizable flows may vanish as they flow into valleys from mountain slopes. When this happens they give up all their sediment resulting in vast apron-shaped deposits called **alluvial fans**, which the streams wind their way across as they break down into distributary channels similar to those of a river delta. Most alluvial fans are broadly triangular or semiconical-shaped features up to several kilometers wide. Western North America sports thousands of magnificent fans along the flanks of mountain ranges and Figure 20.33 shows an example. Some of them fill in closed valleys but others feed sediment to stream systems that move it to the ocean via great rivers like the Missouri, Fraser, and Colorado.

Figure 20.33 An alluvial fan in Death Valley. These magnificent landforms are records of the work of erosional systems operating in the mountains above them. (For scale, see the highway skirting around the fan's perimeter.)

Stripping away tropical forest for farming results in accelerated soil erosion under the rainy climate of Amazonia.

20.7 Denudation Rates and Trends

Since runoff processes are the major agents of denudation, it is appropriate now to ask how much work they actually do and what controls their ability to produce sediment. Geographers are able to estimate rates of denudation by measuring, among other things, the amount of sediment delivered to streams by hillslopes. Generally, we estimate that continents are being lowered at rates of 4 to 6 centimeters per 1000 years or 40 to 60 meters per million years. As we noted earlier, however, these rates vary enormously depending on where you are related to three main controls: climate, topography, and land use.

Climate is critical because it is the main control on runoff and vegetation. In general, where precipitation is high, natural vegetation is abundant. There is great potential for high runoff and erosion rates, but both are held in check by plant cover. Thus in fully vegetated landscapes, denudation rates from soil erosion are typically very low. Things change when the vegetation is stripped away as Figure 20.30 clearly shows.

When vegetation is removed for agriculture, runoff and erosion rates rise dramatically. Soil loss to erosion typically increases more than 10 times, from less than one ton per hectare (2.5 acres) per year to more than 10 tons. Cropland occupies about 12 percent of Earth's land area and most is located in areas of moderate to relatively high precipitation. Not surprisingly, denudation rates are high in these areas: China, India, North America, Europe, and much of the tropics. But agriculture is not limited to cropland alone.

There are even larger areas of rangeland in the world where grazing has damaged the grass cover and led to heightened erosion rates not only from runoff but also wind. As erosion advances the remaining grass and soil are weakened, soil moisture declines and the landscape's general resilience to grazing and drought is lowered, eventually reaching the point where recovery is impossible. This cycle of landscape degradation transforms prairie (and marginal woodlands) into desert, a process that we know as **desertification**. As much as 200,000 square kilometers of grassland and tropical woodlands are being lost to desertification annually worldwide (Figure 20.34).

In deserts, denudation rates are naturally low despite a weak vegetative cover for the simple reason that runoff rates are very low. In semiarid lands, on the other hand, erosion rates appear to be higher than both deserts and forested lands. Two factors explain this: first, the vegetative cover is not dense enough to provide an effective defense against erosion, especially in dry grasslands; and second, runoff can be strong because it often comes in intensive spurts from thunderstorms. Where these lands are used as rangeland – which is almost everywhere they are found in the world – soil loss rates are even higher than under natural conditions.

The highest natural denudation rates are found in rugged, mountainous terrain. Mountains induce higher precipitation and runoff rates than lowlands, and runoff flows with much greater erosive force on mountains owing to the steepness of the slopes. We will discuss this in more detail in the next chapter. Suffice it to point out

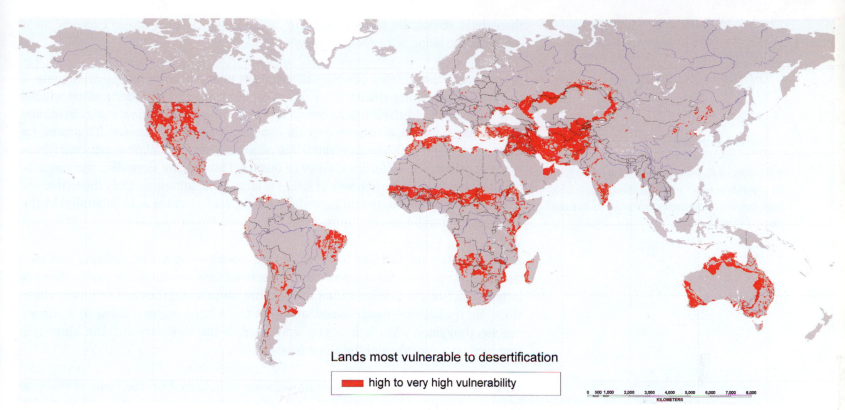

Lands most vulnerable to desertification

█ high to very high vulnerability

0 500 1,000   2,000   3,000   4,000   5,000   6,000   7,000   8,000
KILOMETERS

Figure 20.34 Areas of the world most subject to desertification, according to the US Department of Agriculture.

here that denudation rates are as much as 10 to 20 times higher in the Himalayas of Asia than in the far more modest terrain of north central North America, for example.

20.8 Mass Movement Processes and Features

Mass movement represents a second set of hillslope processes in the denudational system. When materials move by mass movement, they are drawn downslope under the influence of gravity rather than being carried by a transporting agent such as running water. Mass movement may be rapid, as in a landslide or avalanche, or so slow that it is barely detectable with sensitive instruments. Moreover, mass movement may involve gigantic amounts of debris, easily enough to bury towns and villages, or just individual grains of sand rolling down a streambank.

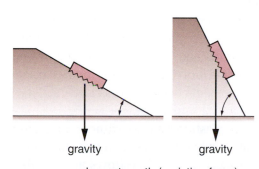

gravity   gravity

∿∿∿ = shear strength (resisting force)

Figure 20.35 The role of gravity in generating shear stress is governed mainly by slope steepness. Shear strength represents the holding forces that resist shear stress.

In any body of material situated on an incline, two sets of opposing forces determine whether it will move or remain in place. The forces drawing material downslope generate **shear stress**, the magnitude of which is primarily a function of slope steepness and the tendency for gravity to pull objects downhill as Figure 20.35 shows. The opposing forces impart **shear strength** to material and hold it together. Shear strength is governed by factors such as the frictional resistance between adjacent sand particles, the cohesive bonding among clay particles, and the capacity of plant roots to bind soil and rock particles together. The balance between shear stress and shear strength determines whether a slope is stable or unstable. When the two are equal, the slope is said to be at *critical threshold*; when shear strength is greater

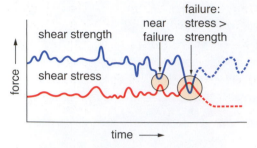

Figure 20.36 An illustration of the relationship between shear stress and shear strength over time. Both vary, but failure occurs only when shear stress exceeds shear strength.

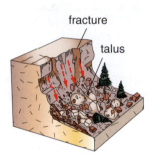

(a) fall
• cliff exposure with fractured bedrock
• angular blocks and slabs piled up at footslope as talus
• scarred and broken trees

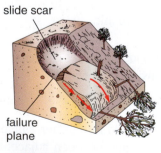

(b) slide
• concave scar on slope
• displacement one meter to a kilometer or more downslope
• swath of destruction

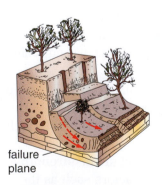

(c) slump
• bank or cliff exposure in soil or bedrock
• back rotational motion
• trees uprooted and tipped

Figure 20.37 Three types of mass movement involving a break in the slope: (a) fall, (b) slide, and (c) slump.

than shear stress, the slope is *stable* or *safe*; and when shear stress exceeds shear strength, the slope is *unstable* or *unsafe*.

In evaluating the stability or safety of a slope, it is important to realize that many of the factors governing shear stress and shear strength and whether a slope will fail (fall down) or hold itself up. These factors, which include groundwater, frost, and earthquakes, vary appreciably over time, as from season to season. To produce a mass movement such as a landslide, it is necessary for the failure-producing forces to exceed the resisting forces for only an instant. Thus, in any slope, the average state of affairs between these two sets of forces is largely meaningless. Only the extremes, that is, low shear strength and high shear stress, really matter as is illustrated by the overlapping graph lines in Figure 20.36.

Added to this is the fact that many slopes are *conditionally stable*, meaning that they would fail were it not for the presence of a special factor such as the root systems of large trees. Such slopes are called **metastable slopes** and they are common where slope angles become dangerously steep but trees have secured them by buttressing steep inclines with their trunks and roots. If the trees are cut, the kingpin is removed, so to speak, and the slope fails.

Types of Mass Movements: Mass movements are classed on the basis of the type of motion involved. Five basic types of motion seem to cover most movements: fall, slump, slide, flow, and creep, and the diagrams in Figures 20.37 and 20.38 illustrate the basic character of each. What type of motion prevails when a slope fails depends to a considerable degree on the physical properties of the materials in the slope. Bedrock, rock rubble, and sand behave as brittle elastic materials and tend to rupture (break) when they fail. The resulting movement takes the form of a fall, slide, or slump, shown as a, b, and c in Figure 20.37.

In a **fall**, pieces of rock break free and sail, topple, or tumble over the slope face. The resultant rock rubble, called *talus*, piles up at the footslope forming a steep ramp-like incline called a *talus slope*. In a **slide**, a sheet of rock or soil material slips downslope over a slippage surface called the failure plane. The area vacated is often marked by a distinct scar, often a denuded swath with exposed subsurface material. **Slumps** are characterized by a back rotational movement along the failure plane somewhat like that of a person slouching in a chair. They are often distinguishable by a scallop-shaped indentation left in the slope. Although slides and slumps may occur in one sudden, great movement, it is more common for them to occur in a series of small displacements over months or years.

Materials that behave as plastics deform without rupturing and follow a flowing motion when they fail and two such processes are illustrated in Figure 20.38. True **flows** are possible only with saturated materials that have a liquid or near-liquid consistency. Such consistencies are found in soft, wet, clayey materials or in sand or silt within which there is a flow of pressurized groundwater.

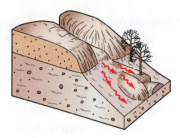

(a) flow
- lobate forms
- trees bent or upright
  flow-type movements

(b) creep
- absence of deformation in trees
  or soil surface.
- tree roots may reveal
  downslope bending

Figure 20.38 (a) Mudflow resulting from saturated soil material that flows like wet cement. (b) Soil creep, a very slow but widespread movement limited to the upper meter of soil.

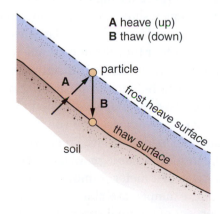

Figure 20.39 Soil creep by frost heaving. (a) Particles are lifted off the surface with freezing, and (b) settle back a little downslope with thawing.

The most common flow process is a **mudflow**, shown in Figure 20.38a, which is usually comprised of liquefied soil that has been soaked with rainwater and/or shallow groundwater. Mudflows are most likely where slopes have been stripped of stabilizing vegetation. They often begin as slides or slumps, emerging from the failure mass and moving surprisingly long distances downslope. They are a common hazard in California in wet years (particularly El Niño winters) on slopes weakened by forest fires, land development, and earthquakes.

A related, but much slower flow-type movement, called *solifluction*, is found mainly in periglacial (i.e. cold, permafrost) environments. It occurs in the wet surface layer of soil, called the *active layer*, which lies over permafrost in summer. Solifluction produces small, lobe-shaped features that flow gradually (as much as several centimeters per year) downslope under the combined influence of groundwater pressure, soil liquefaction, and frost action, which are described below. With climatic warming in the Arctic and subarctic, solifluction is expected to increase as the permafrost layer retreats beneath it.

Finally, there is the mass movement called **creep** or **soil creep** (see Figure 20.38b). It is characterized by a very slow (less than a few centimeters per year) downslope movement in the upper 0.5 to 1.0 meters of soil. Creep occurs almost everywhere hillslopes are found and, despite its slowness, it may be the most effective means of mass movement in terms of total tons of material moved worldwide. It can be driven by several factors, including ground frost, and is neither a flow nor a rupture-type movement.

Factors Influencing Slope Stability: Many factors influence slope stability and mass-movement processes. For example, ice wedging is an important cause of rock fall. Cliffs emitting seeps of groundwater are especially prone to frost wedging when water fills cracks, freezes, and expands. Frost is also a driver of soil creep because when moist soil freezes it expands upward and outward from the slope face, then, upon thawing, settles a little downslope, as shown in Figure 20.39. Repeated many times, the upper soil moves slowly downhill. A similar process may occur with changes in moisture in certain clayey soils as they expand with wetting and shrink with drying.

With the exception of gravity, water is probably the most important influence on slope stability. The addition of water to clayey soil reduces its shear strength because water changes soil consistency to one that is more liquid-like. As children we learned this principle while making mudpies. We found that beyond some critical amount of moisture the mixture simply got so runny that it would not stay in place and flowed away. In science, this critical moisture level is called the **liquid limit**. Once a soil material has reached or exceeded the liquid limit, it can flow downslope like viscous water.

Figure 20.40 The aftermath of a massive mudflow, which was brought on by clearing of slopes for residential development and the heavy rains of a hurricane.

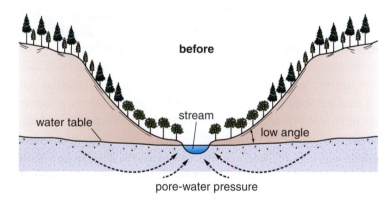

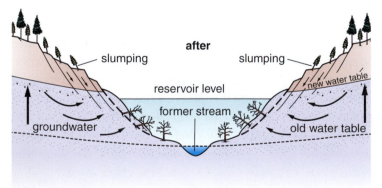

Figure 20.41 The change in water table with the raising of a reservoir resulting in failures of side slopes from pore-water pressure and soil liquefaction.

Such flows are classed as either **mudflows** or **debris flows**, depending on their particle composition. Mudflows consist primarily of clay and silt, whereas debris flows consist of clay and silt as well as a wide range of larger particles, even large boulders. Such flows are common in areas of heavy rainfall (we noted wet winters in California above) where steep slopes have been stripped of vegetation and they can be extremely dangerous. In recent decades countless people in poor tropical countries like Venezuela, Honduras, and the Philippines have been buried in mudflows induced by hurricane rains (Figure 20.40).

Another water influence on slope stability is groundwater pressure. All groundwater exists under pressure, which explains why it is able to flow through those minute pore spaces within soil and rock materials. If the pressure within pore spaces, that is, in the space between particles, is great enough to drive particles apart, they lose contact with each other and tend to float as if buoyant in a soup. This reduces the shear strength of the soil, causing it to fail, even in low angle slopes.

When groundwater rises under a slope, the base of the slope can weaken and give way. This can be caused by natural increases in groundwater, as from massive rainfalls driven, for example, by a hurricane or by man-made increases such as the raising of a reservoir. With a reservoir, the water table in a river valley can be raised 50 meters or more bringing groundwater high into side slopes where it never existed before. The combination of pore-water pressure and soil liquefaction coupled with wave erosion along the valley walls can quickly destabilize these slopes, causing them to fail in a manner similar to that shown in Figure 20.41.

Shear strength can also be reduced by removing vegetation from hillslopes. This is not, however, a simple cause-and-effect relationship because many slopes show no signs of instability even with complete deforestation. The reason has largely to do with tree root systems and whether they die or continue to live after trees have been cut. Those that live, such as redwood and many species of oak, continue to hold the slope in place while those that die rot away and leave the slope weakened. Added to this are the destabilizing effects of increased soil water and elevated water tables that often follow cutting. Trees are highly efficient water pumps and in some areas loggers have reported near-surface water-table rises of a meter or more after a clear cut. West Coast forests, for example, are able to extract from the soil as much as 40 percent of the total annual precipitation and release it through transpiration. This can amount to 100 centimeters or more water annually on some mountain slopes.

Figure 20.42 Celebrated landforms of the North American West. Each is an assemblage of a set of distinct slope forms.

Summary on Hillslope Processes: Hillslopes are held in place by a critical balance between forces that pull them down and those that hold them in place. Gravity and runoff belong to the former, and soil cohesion and vegetation belong to the latter. Slopes fail by erosion and mass movement when that balance is upset, and one of the chief causes of failure is removal of vegetation, something that is advancing in the world as agriculture, deforestation, and human population spread.

20.9 The Systems of Hillslope Form and Formation

The processes and rates of denudation of landmasses are not the only questions here. We are also interested in the form the hillslopes take and how they evolve over time. This is critical to understanding the essential character of landscapes because landforms, which are the foundations of landscapes, are merely an assemblage of slopes of different sizes, shapes, and angles. In the extreme, we are led to wonder about the origin of spectacular landforms such as the mesas, buttes, towers, and canyons in the American West (Figure 20.42). Fundamental to this discussion is an explanation of how slopes retreat.

Slope retreat is the process by which hillslopes are worked back over time. Generally, there are two basic modes of slope retreat: parallel and non-parallel (Figure 20.43). In parallel retreat, the slope is worn back and the angle remains comparatively constant over time. In non-parallel retreat, the slope is worn down and the angle grows gentler (lower) over time. Either mode is possible depending on the mass balance of the slope, but parallel retreat is generally regarded as the most prevalent mode in most landscapes.

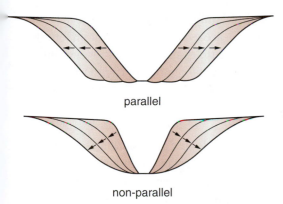

parallel

non-parallel

Figure 20.43 The two basic modes of slope retreat: parallel (wearing back), and non-parallel (wearing down).

Slopes as Mass-balance Systems: The *mass balance* of a slope refers to how the slope functions as a system, defined by the relative rate of production and removal of slope debris and its relationship to slope form. There are three components to a slope mass-balance system and they are illustrated in Figure 20.44:

(1) the production of debris (fragmented rock or soil particles) by weathering;
(2) the transfer of this debris downslope by hillslope processes; and
(3) the removal of debris from the foot of the slope by streams, glaciers, or waves.

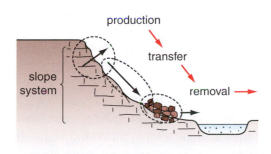

Figure 20.44 The three components of a slope system: production, transportation, and removal.

The relationship between the system and slope form is simple. If production and downslope transfer are faster than that rate of removal, then debris builds up on the lower slope. As it piles up there at the foot, features such as alluvial fans or talus slopes form, and the overall angle of the slope grows gentler. This is an example of non-parallel slope retreat because the overall angle, crest to toe, declines over time. On the other hand, if the debris produced and brought to the footslope is removed by, say, a stream or ocean waves as fast as it arrives, then the mode of retreat is parallel. If we take it one step further and give the removal agent, such as storm waves, lots of energy, it not only removes debris but cuts into the footslope as well. This may give the slope an oversteepened angle, causing it to fail. Examples of the latter are plainly evident along shorelines where waves and currents undercut cliffs or bluffs and then carry away the debris from failed slopes.

In order to maintain parallel retreat, one condition is essential: the slope must be part of an *open geomorphic system*. This system is defined by the through flow, or export, of rock debris from the slope environment to the margins of the continents. The vast majority of the export work is performed by stream systems, such as the Mississippi, Colorado, or Mackenzie whose thousands of tributaries are in contact with thousands of slope systems. In the case of the Mississippi, this great geomorphic system reaches from the upper slopes of the Rocky Mountains to the Gulf of Mexico. But what about those systems that stop short of the sea, that function as closed systems?

Where drainage is closed, rock debris cannot be transported out of its source area. Therefore, it builds up on valley floors and along footslopes, producing gentler slope angles. Over the long run, this results in non-parallel retreat especially on the lower parts of the slopes as the sequence of profiles in Figure 20.45 illustrates. About 20 percent of Earth's land area is served by closed or internal drainage systems. Most are located in arid, mountainous regions of Eurasia or Africa, but they are found on every continent, usually where mountains have blocked drainage to sea or the climate produces too little runoff to support through-flowing streams. Death Valley and the Dead Sea Basin are famous examples of small closed systems where basin infilling is readily apparent. But there are also very large closed systems, for example, the Tarim Basin of northern China, the Plateau of Iran, and the Great Salt Lake Basin. We discussed these in Chapter 15, Section 15.6.

System Trends and Slope Evolution: It is probably rare that a particular trend in slope development would continue undisturbed for millions of years. The Earth is too dynamic in most places for this to happen. In areas of active mountain building, as in parts of western North America, for example, faulting can re-establish slopes that were originally worn down by erosion, imparting new energy to the slope system thereby triggering a new episode of erosion and retreat. The drawings in Figure 20.46 show what this would look like.

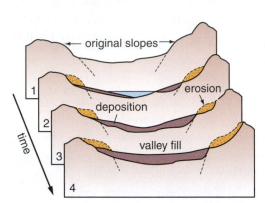

Figure 20.45 Slope retreat in a closed system such as landlocked valley. Lower slopes build out while upper slopes are worn back producing non-parallel retreat.

Another source of alteration in slope development is climatic change. Increased aridity in the American Southwest in the past 10,000 years, for example, has caused a reduction in the size of many lakes and streams and thus the loss of an erosional agent with sufficient power to remove debris from the footslopes of mountain ranges. As a result, former wave-cut and river-cut slopes in the Basin and Range country of Nevada and Utah, for example, have tended to grow gentler in the past several thousand years. The opposite is also possible, of course. Climate can grow wetter, as it did in the American Southwest during the last glaciation, giving rise to expanded lakes and bigger streams with more erosive power, accelerated debris removal, and steeper valley walls.

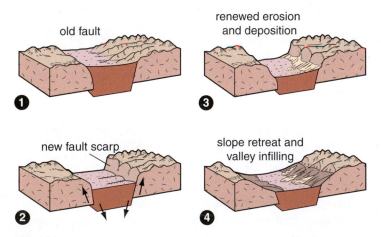

Figure 20.46 Slope system re-energized by renewed faulting leading to a new episode of slope erosion and retreat.

Common Slope Forms and Their Origins: If we examine the landforms in most localities, we are sure to notice some distinct similarities in the shapes of different hillslopes. The slopes in eastern North America, for example, tend to have long, smooth forms, whereas those in the mountainous west tend to be more angular and jagged. There are explanations for these differences. In some regions, slope form is controlled by bioclimatic factors such as soil cover and vegetation; in others by rock structures such as domes and faults; and in others by erosional forces such as waves beating against a rocky shore.

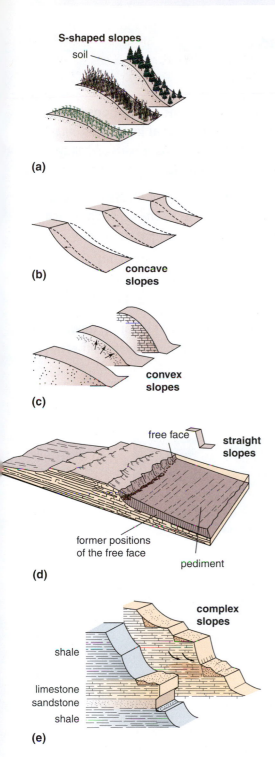

Figure 20.47 Basic slope forms: (a) S-shaped, (b) concave, (c) convex, (d) straight, and (e) complex.

There are five basic hillslope forms: S-shaped, concave, convex, straight, and complex and models of each are shown in Figure 20.47. Gentle **S-shaped slopes** are very common in non-mountainous regions where there is a thick soil overburden and heavy vegetative cover (Figure 20.47a). Such areas include much of eastern North America, northwestern Europe, and the wet tropics of South America, Africa, and southeastern Asia. The origin of the S-form appears to be tied to a particular combination of hillslope processes, notably runoff and soil creep, working on erodible soil held in place by a heavy plant cover. The role of vegetation is particularly evident when these slopes are cleared for land development and mass movement and erosion set in. The slope is quickly transformed from the S-shape into a concave shape as gullies are cut into it and large chunks of material are removed by slumps and slides.

Concave slopes also develop under natural conditions, though typically in areas of rough terrain. This slope form is usually the work of a powerful erosional agent or a mass movement such as a landslide or slump (Figure 20.47b). Glaciers, for example, are able to scour the sides of mountains into deep concavities called cirques, and where a mountain glacier flows down a stream valley it carves both side slopes into broadly concave forms. The inclination of such slopes increases upward and may approach 90 degrees beginning in the upper midslope.

Convex slopes, by contrast, are usually related to resistant rock formations or to uplift of the slope itself, and can be found in any climatic region (Figure 20.47c). In a slope composed of sedimentary rock, for example, resistant formations in the midslope may protrude to form a convexity. Convex slopes may also result from the uplift of mountains by tectonic forces, or from the expansion related to unloading and decompression in exfoliation domes.

Straight slopes form under various conditions and one of the most studied examples are pediments in arid landscapes. *Pediments* are broad, gentle slopes, usually with only several degrees inclination, which lie at the foot of cliffs, called *free faces*. They are composed of bedrock with a light veneer of gravelly rock debris strewn over them and form in the wake of a retreating free face. Because they are widespread in arid lands, pediments appear to be unique to dry climates and few places exhibit them better than the Canyon Country of the American Southwest (Figure 20.47d).

Finally, there are **complex slopes**. They are usually multifaceted with different angles and forms mixed together in a large incline. Most complex slopes are the result of varied bedrock structure in which rock formations of different thicknesses and resistances exert a variable control on slope processes (Figure 20.47e). In general, resistant formations form steep erosional segments whereas weak formations form gentle, depositional segments. Complex slopes may also result from different effects of processes. For example, groundwater seepage at different levels and locations on a slope can produce irregular patterns of failures resulting in the appearance of a chaotic slope form.

Chapter Summary and Overview on Geomorphic Systems:
Rock Weathering, Hillslope Processes, and Slope Formation

The work of wearing down of the Earth's landmasses and the creating of landforms is performed by a suite of geomorphic systems. These systems work hand-in-hand with the great system of tectonics and mountain building, changing work output with mountain elevation and related climatic conditions. Denudation begins with the weakening, breakdown, and disintegration of bedrock by weathering processes. These processes work at different rates and with different results depending on geographic conditions, particularly climate, vegetation, and rock types, and they produce an assortment of particles, which are moved downslope by runoff and mass-movement processes. These processes in turn are pivotal in shaping the landforms we see in the landscape around us.

▶ **Landforms are basic to the way we think about the geographic character of Earth's landscapes.** Among other things, they illustrate our sense of natural beauty and provide a basic infrastructure for the landscape.

▶ **Geomorphic processes and systems are responsible for wearing down and shaping terrestrial landforms.** All erosional systems are made up of three basic parts and all systems end with output in the form of deposition.

▶ **Landforms tend to fall into two major classes.** Erosional landforms begin as elevated rock masses and depositional landforms form in low areas mainly on the continental fringe.

▶ **Geomorphic systems are central to the Earth's rock cycle.** Most of the sediment produced from the landmasses is in time recycled back into the continental platforms as rock.

▶ **Denudation is the general process of wearing down landmasses.** Most of the rock mass lost in denudation is offset by isostatic uplift.

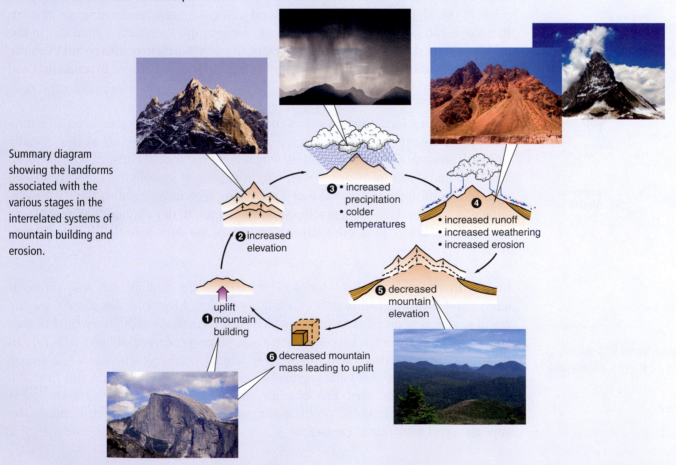

Summary diagram showing the landforms associated with the various stages in the interrelated systems of mountain building and erosion.

❸ • increased precipitation • colder temperatures

❹ • increased runoff • increased weathering • increased erosion

❷ increased elevation

❶ uplift mountain building

❺ decreased mountain elevation

❻ decreased mountain mass leading to uplift

▶ **Gravity and denudation limit the elevations of mountains.** As mountains get higher, denudation forces get stronger in response to greater potential energy from increased water mass and elevation above sea level.

▶ **Geomorphic systems are powered by both solar and gravitational energy.** Solar energy operates through the atmosphere whereas gravitational energy operates through land elevation.

▶ **Weathering is the first stage in the denudational system.** There are three types of weathering processes and they yield different products including particles of sediment and chemical ions.

▶ **Karst topography is a product of the decomposition of limestone.** It is noted for its distinctive landforms such as sinkholes and caverns.

▶ **Hillslope processes represent the second stage in the denudational system.** Vegetation is a critical control on erosion by runoff and, where it is removed for farming, gullying can be profoundly damaging, especially in rainy climates.

▶ **Mass-movement processes are downslope displacements of slope material under the force of gravity.** They are governed by the balance between the forces of shear stress and shear strength, and are usually identifiable by the slope forms they create.

▶ **Slope retreat is part of the dynamics of landform evolution.** Slopes function as mass-balance systems and exhibit two modes of retreat related to closed and open systems.

▶ **Landforms are made up of up of five basic slope forms.** Some forms are common to particular bioclimatic regions, whereas others are related to rock structures or to certain erosional processes.

Review Questions

1 What are the primary sources of energy that drive geomorphic systems and enable them to perform work? And what are the products of that work?

2 What is the nature of the relationship of geomorphic systems and the rock cycle? Of the 10 billion tons of sediment produced by geomorphic systems each year, where does most of it go? What about the rest of it?

3 Why has the amount of material exported to the sea by the Earth's major river systems increased over the past 12,000 years?

4 Describe the cycle of denudation and mountain uplift, and explain how feedback works to keep the system operating.

5 Geomorphic systems are driven by energy from different sources. What is the difference in the response of a stream system and a glacier system to inputs of precipitation?

6 Most rocks are alien to the Earth's surface. Does this statement explain why rocks weather so readily in surface environments? And what factors cause weathering rates to differ with geographic location on Earth?

7 Identify the principal chemical weathering processes and indicate how (and if) water plays a role in their operation. Which one is the major player in karst topography.

8 Why in most locations are the landforms produced by the weathering of sandstone so different than those produced by the weathering of limestone? Under what conditions would you expect these two to differ least?

9 Describe the steps or phases involved in the process of exfoliation and identify how feedback works in an exfoliation system.

10 Of the many processes responsible for the erosion of hillslopes, can you identify the most insidious one and some of the conditions that contribute to it?

11 If we look at slopes as systems, what do we identify as their three main parts? In what part(s) of the system would you expect to find alluvial fans and talus. What geographic conditions make for a closed sediment system in connection with slopes?

12 Identify the geographic settings where human activity has greatly accelerated desertification and hillslope erosion.

13 Define three ways in which water contributes to slope instability and name the types of mass movements that result.

14 Describe the relationship between parallel and non-parallel slope retreat with open and closed geomorphic systems.

15 Name a process, landscape feature, or geographic condition commonly associated with each of the following slope forms: S-shaped, concave, convex, straight, and complex.

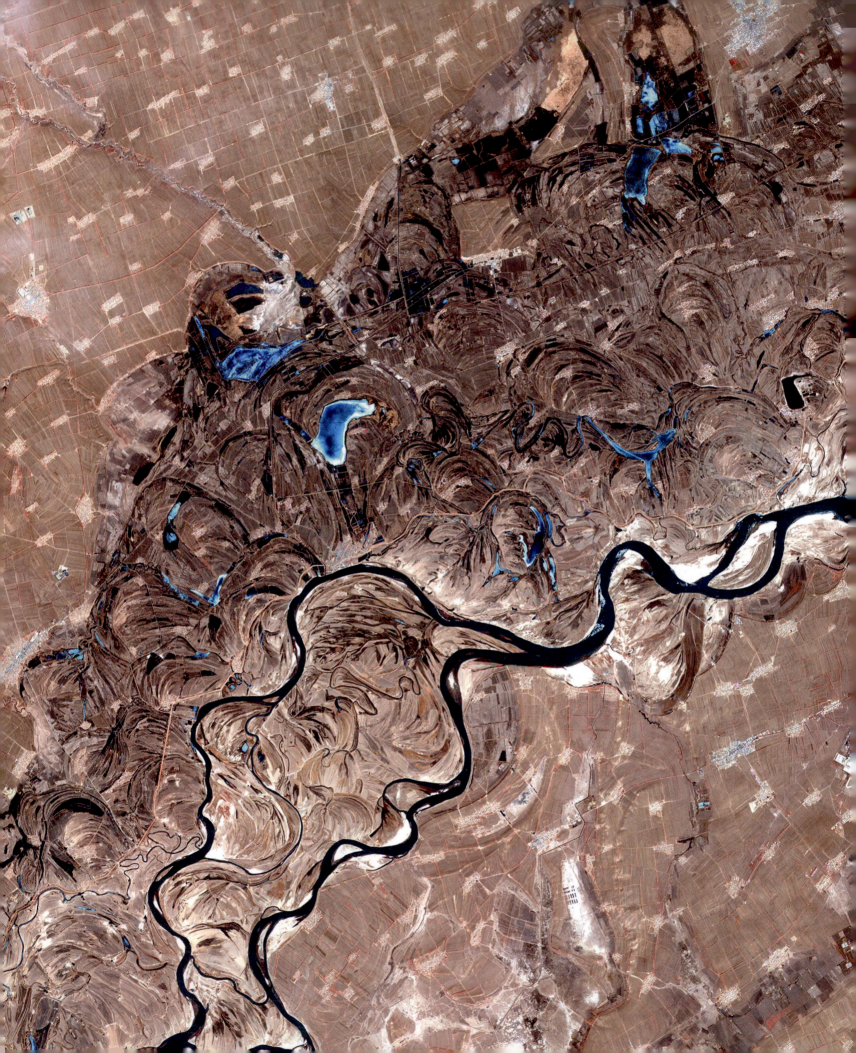

Stream Systems, Valley Formation, and Fluvial Landscapes

Chapter Overview

Mark Twain warned that we risk losing touch with the romance and beauty of a river by knowing about and attaching meaning to its parts. It appears that is a risk we will have to face in this chapter as we go in search of answers to some big questions about streams, beginning with why are they so effective in lowering the landmasses and shaping landscapes? Part of the answer has to do with their geographic coverage. They are one of Earth's most ubiquitous terrestrial features, operating in humid and dry and mountains and flat terrains alike. Part also has to do with the fact that they are organized into great branching systems, large enough to link the interiors of the landmasses to coastal lands and the sea. Accordingly, we begin this chapter with a brief discussion on streams as systems, the nature of streamflow, and the erosional processes of water running in open channels. This is followed by a look at the forms and dynamics of stream channels, how streams cut and shape their valleys, and develop landforms over entire watersheds. Next is a description of some of the grand ideas that have been advanced by scientists over the years to explain how stream systems shape the surfaces of continents. The chapter closes with some remarks about human impacts on stream systems.

Introduction

Before us lay a valley unlike any other on the Alaska Peninsula. It was completely barren, a desert landscape in the midst of immense mountain greenery. Bruce started out across it and said, "There it is, Will, the ash deposit from the 1912 Katmai eruption. The volcano was completely blown away. The blast came in this direction, filling the stream valley with a thick layer of volcanic ash." It was a startling scene, pastel tones of tan, gray, and red intermittently highlighted by solar spotlights shining through holes in the cloud cover.

We conferred with Dad on our plan. "Let's make our way along the east side, then cross over the upper deposit and take a look at what's left of the volcano. It'll take us about two days to make the round trip. There's no water on the ash deposit so maybe tonight we can make camp along the stream." "Good plan, boys," he replied, and stepped off with a stride that had long been familiar to us as boys in Northern Michigan. As we walked, I thought, what luck to be here in remote Alaska with two of the best outdoors men I know, brother Bruce, just returned from field studies in the Aleutian Islands, and Dad, fresh from the Lake Superior woods.

The stream ran along the east side of the valley, close to our route, and, late in the afternoon, we began looking for campsites. To our surprise, the stream was nowhere accessible, for it was confined in a small canyon about 50 feet deep with sharp, vertical walls. Although the stream was small, no more than 20 feet wide, it had cut completely through the ash deposit. This was no mean task, because below a thin layer of loose surface material, the ash was consolidated into low-density rock, called tuff. The stream – shown in Figure 21.1 – had eroded its way through this rock in less than 75 years. This was powerful testimony to the work of running water. We estimated that several hundred thousand cubic meters of ash and rock had been removed by the stream since the explosion.

That night we camped in a small side valley where the flanks of the ash deposit lapped upon the walls of the original valley. Near the head of a side valley, runoff from snowmelt and springs seeped into the ash and disappeared underground. Undoubtedly it and hundreds of small tributary valleys like it fed the stream in the canyon with groundwater and thus maintained its flow. The next day on the upper part of the ash field, we finally gained access to the stream channel and made a crossing. Although signs of stream erosion were plainly evident, the stream's flow here was smaller and apparently lacked the power to cut through the tuff. Later that day we climbed a ridge to see what remained of the volcano. Only a stump was left surrounded by snowfields that over time would supply the runoff that would eventually carry the mountain's shattered remains completely away.

Figure 21.1 Stream cut in the Katmai ashflow that took less than 75 years to erode.

21.1 Streams as Geomorphic Systems

Streams are among our most celebrated natural systems. And it is not hard to understand why. We have lived along them for thousands of years, used them for transportation and as sources of power, water supply, food, and pleasure. Most of us have a good idea about how streams work and, realize it or not, we intuitively think of streams as three-part systems. We look for a point of origin (headwaters), trace the downstream flow of the water in a channel, and mark a terminus (mouth) at the sea. Thus, the simplest model of a stream system is that of a thread connecting two points. This is how we often depict streams on maps.

But this model is sorely lacking as a scientific construct for, among other things, streams are actually complex, three-dimensional systems involving not just a thread of water but a network of threads connected to an overhead sprinkler system and underground plumbing. All of this is tied to a landscape within a watershed and involves more than water. Although water is indeed the defining matter of stream systems, here we are interested more in streams as erosional and sediment-transport systems. And the sediment they move includes much organic matter, both living and dead. So, streams are at once hydrologic systems, geomorphic systems, and ecosystems.

Phase One – Atmospheric Input: In terms of their dimensions, stream systems reach far into the atmosphere and deep into the ground. It can be argued that as systems streams begin with the evaporation of sea water. Most of the water that feeds the great Amazon system, for example, begins with evaporation over the Atlantic Ocean within a belt of about 40 degrees latitude along the Equator. The resultant vapor is driven eastward by the tradewinds into the heart of South America where it is dumped onto the Amazon watershed. This, then, is the first phase in the system and it is depicted in Figure 21.2. It takes only a few days to complete and, in addition to water, the atmospheric phase also brings sediment to the Amazon. The tradewinds

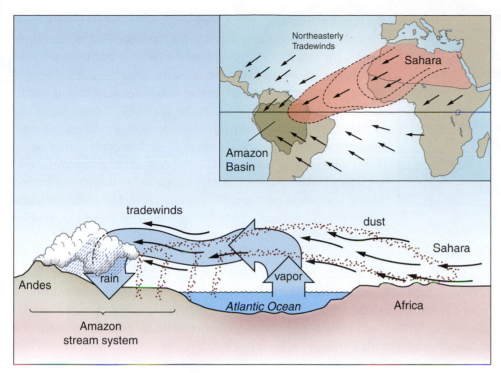

Figure 21.2 The first phase of the Amazon stream system. Atlantic moisture and Sahara dust are carried onto South America and the Amazon watershed.

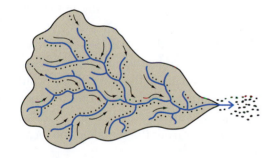

Figure 21.3 The surface phase of a stream system represented chiefly by a network of channels carrying water and sediment.

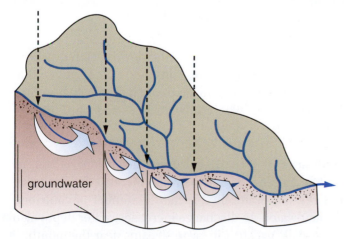

Figure 21.4 The groundwater phase of stream systems feeds streams with base flow of water and ions of sediment leached from soil and rock.

regularly stir up huge amounts of dust (clay and silt particles) in the dry, windy Sahara and carry much of it across the Atlantic to the Amazon Basin.

Phase Two – Surface Runoff: The second phase of the system is surface runoff. Rainfall first encounters tropical forest which breaks up the stream of precipitation, intercepting some, and giving up the remainder to the ground. On the ground most of the rainwater infiltrates the soil, or when the ground is wet, runs off as overland flow. In this segment the system receives its second input of sediment, tiny particles from the soil and forest floor, as it slowly makes its way to a stream channel. Once in the network of channels, the tempo of the system accelerates dramatically and with it the mass of water and sediment rises dramatically.

This is the heart of the stream system, the geomorphic workhorse most responsible for eroding the land, carrying sediment away, and shaping landscapes. We define it as a network of channels, along the lines of the one shown in Figure 21.3, organized by nature according to the principle of stream orders, and fed by a watershed. The time it takes to complete this, the surface-runoff phase of the stream system, is typically 10 or 15 days (longer in huge watersheds like the Amazon) and in the end most of the load of water and sediment is delivered to the ocean. That which is not, particularly the sediment, is stored in the system and moved months, years, or even millennia later.

Phase Three – Groundwater: The one remaining phase in the system is groundwater. For that rainwater which infiltrated the soil under the tropical forests, much of it percolates into aquifers, which in turn feed streams with baseflow, as the diagram in Figure 21.4 suggests. Though poor in sand, silt, and clay particles, groundwater is often rich in dissolved matter, especially in wet regions like the Amazon. Ions of silica, calcium, potassium, and many other chemicals leached from soil and rock are discharged with groundwater into streams to become part of the system's greater sediment load.

Driving Forces: Each of the phases in a stream system, atmospheric, surface runoff, and groundwater, is defined by flows of water and sediment. The driving forces for the system (see Figure 21.5) are solar energy and Earth's gravity. Without solar heating, which drives ocean evaporation and wind, there is no supply of water for watersheds. And without elevated landmasses and gravity there is no force to drive this water across the land toward the sea. And without flowing water there is no force to dislodge particles and carry them to the sea.

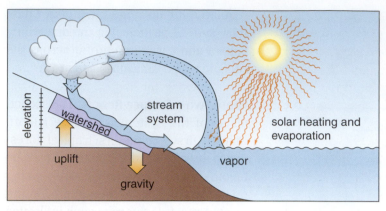

Figure 21.5 The driving forces of stream systems: solar heating, which powers the supply of the water system and gravity, which then pulls the water toward the sea.

System Connections and Change: As with all geographic phenomena, it is necessary to consider the nature of interconnections with other systems and how changes in these systems can bring about changes in a central system. Consider a change in climate such as global warming and how it might affect streams and their ability to do work. Global warming is expected to produce higher ocean temperatures and one outcome of a warmer sea surface is higher evaporation rates. This can lead to greater rainfall and runoff rates in watersheds such as the Amazon, which are downwind of large evaporation zones, leading in turn, possibly, to increased erosion rates and greater sediment output by the system. But there can also be other outcomes from such an increase in rainfall.

What if the increase in moisture improves the vigor and density of the watershed's forest cover? This would heighten the capacity of the landscape to take up rainwater while increasing its resistance to erosion, thereby offsetting the first change. Now let us go one step further and consider the changes brought on by another system, one not responding to a rainfall change, at least not initially.

Land use is advancing rapidly into the Amazon Basin and with it comes deforestation. (Almost 2 million hectares of forest (5 million acres) are being cleared every year.) The scene shown in Figure 21.6 is typical. Forest clearing reduces the landscape's capacity to hold water and resist erosion. Thus, while the intact forest may be more vigorous and resilient, the cleared areas are far less so and the net result is that, as a whole, the stream system may produce more work, depending on the extent and location of land-clearing activity. However, one thing will certainly have changed; namely, the configuration of the system measured by the geographic distribution of work and by the related development of landforms in the watershed. From here we could go on to explore many other connections, but let us get on to other matters and agree that streams are indeed complex systems with a host of ties to many geographic systems that operate at scales ranging from the global to the local.

Figure 21.6 Deforestation can lead to greater runoff and more soil erosion as well as changes in the distribution of work and landform development within a watershed.

21.2 Streamflow and the Energy of Running Water

In order to understand how streams do their work, we need first to examine a few principles about streamflow. Some of these principles are intuitively logical, others less so and demand a little more than intuitive thought.

Gradient and Discharge: First, streamflow is directed by the downstream **gradient** or slope of the channel. This is represented by curve A in Figure 21.7. Notice that channel gradient is steepest near the headwaters. Downstream it gradually becomes gentler, diminishing altogether at the mouth where flow declines and vanishes into the sea.

Second, for most streams, **discharge** (curve B) increases downslope. Thus, as the gradient declines the flow gets larger. The downstream increase in discharge is related to the downstream increase in watershed area, the number of contributing tributaries, and access to groundwater. As discharge increases, water depth also increases; thus channel depth is usually greatest in large streams near the mouth.

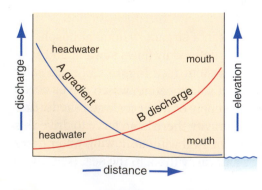

Figure 21.7 Changes in stream gradient (A) and discharge (B) from headwaters to mouth.

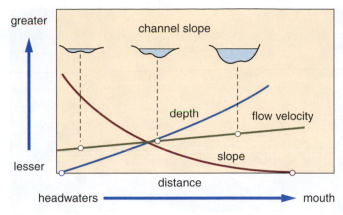

Figure 21.8 Flow velocity increases as channel gradient (slope) decreases, which is explained by the increase in channel depth downstream.

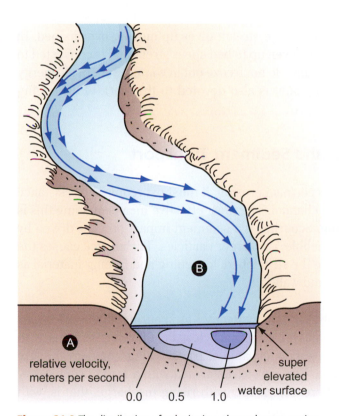

Figure 21.9 The distribution of velocity in a channel cross-section (A); and the pattern of fast-moving water in a meandering channel (B).

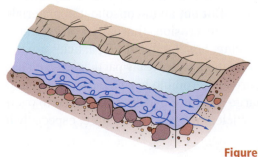

Velocity: Third, **flow velocity** in stream channels also changes from headwaters to mouth, but not as you might expect. Average velocity, represented by the green line in Figure 21.8, is often a little higher near the mouth where gradient is gentlest than near the headwaters where the gradient is steepest. The reason for this is related to three factors: the slope of the channel, the depth of the water, and the roughness of the channel. Increases in *channel slope* and/or *water depth* cause increase in velocity, whereas an increase in the channel *roughness* causes a decrease in velocity. Roughness decreases or remains constant downstream and depth, of course, increases. The effect of an increase in depth, represented by the blue line, is slightly greater than the effect of a decrease in slope, represented by the brown line. Hence, the fastest velocities in stream systems are found in the downstream reaches of larger streams where the depth and discharge are greatest. For example, the average flow velocity near the mouth of the Amazon, where slope is only a few centimeters per kilometer and the depth is 50 to 70 meters, is about 2.5 meters per second, whereas the average velocity in the Grand Canyon River of the Yellowstone National Park, where slope is 40 meters per kilometer and the depth is 1 to 2 meters, is only about one meter per second.

Flow velocity also changes with channel shape. Notice in Figure 21.9 that velocity is highest in the deepest part the channel and slowest where the water is shallow and contacts the channel. At channel bends, the faster-moving water responds more to centrifugal force than slower water does and slides toward the outside of the bend. In addition, this may also produce super elevation of the stream, in which the water surface near the outside of the bend is higher than the water surface near the inside. Where channels swing back and forth in meanders, the belt of fast-moving water is thrown from one side of the channel to the other in successive bends much as a bobsled team does in negotiating its run.

Motion: Fourth, streamflow is always characterized by **turbulent motion** even in streams with smooth surfaces and placid-looking flows. Turbulence is dominated by eddies, the rolling and swirling action of churning water (Figure 21.10). This mixing motion is important in the movement of sediment as eddies help dislodge particles and keep them aloft in the stream's flow. It is also a source of flow resistance as the slower-moving water on the streambed clashes with the faster-moving water above the bed, which reduces overall velocity.

Energy: Fifth, the stream functions as an **energy system**. Simply put, as water moves down the channel, *potential energy* (represented by the elevated mass of water in Figure 21.11) is continuously being converted to *kinetic energy*, the energy of water motion. Because the downhill force of the moving water mass does not result in an equivalent acceleration in stream velocity, there must be an equal and opposite force holding the water back. This resisting force is called **bed shear**

Figure 21.10 All streamflow is naturally turbulent, but turbulence is much greater with fast flows in rough channels.

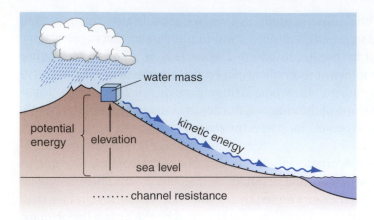

Figure 21.11 The stream as an energy system. Potential energy, represented by an elevated mass of water, is converted to kinetic energy when the mass is set into motion. Once set into motion, frictional resistance keeps the stream from becoming a run-away train of water.

Figure 21.12 An accumulation of boulders in a stream channel. During large flows, these huge particles are set into motion and become the instruments of channel scouring.

Table 21.1 Relative resistance of channel materials to stream erosion

Material	Relative resistance*
Fine sand	1.0
Sandy loam	1.13
Grass-covered sandy loam	1.7–4.0
Fine gravel	1.7
Stiff clay	2.5
Grass-covered fine gravel	2.3–5.3
Coarse gravel (pebbles)	2.7
Cobbles	3.3
Shale	4.0

* *Relative to fine sand*

stress, which is the amount of energy the stream gives up moving over its bed. In the end, most of a stream's energy is given up to bed shear stress and is converted to heat energy. This explains why streams do not freeze out in winter. But not all energy is converted to heat, for a small amount is also devoted to the work of dislodging particles and moving sediment.

21.3 Channel Erosion and Sediment Transport

Scouring: In order for a stream to perform erosion, it must overcome the resistance of the materials that make up the channel. The most effective means of doing this is by a process known as **scouring**. Scouring is a sediment impact process driven by turbulent flow armed with sand, pebble, and larger particles. Turbulence causes the particles to bounce and skid along the streambed, loosening and freeing material as they bang into the channel floor and walls.

Big streams are, understandably, far more effective at scouring than small ones because they can carry large loads of large particles. Fast-flowing streams with water depths of 2 meters or so are fully capable of transporting boulders the size of the ones shown in Figure 21.12. For any stream armed with a load of such heavy particles, erosion by scouring increases rapidly with rising discharge, that is, with flow velocity and depth.

Erodibility: Scouring is particularly effective in eroding unconsolidated (non-cemented) materials such as the sand and gravel that are found in most streambeds and banks. It is not uncommon, for example, for a large flow to scour away several feet of such material in a matter of only hours. But not all unconsolidated materials the stream encounters are equally susceptible to erosion, that is, are not equally erodible. Table 21.1 provides a resistance ranking. Small, sand-sized particles are the weakest, that is, least resistant to erosion by running water. Clay particles, which are a tiny fraction of the size of sand particles, are less erodible because they are bound together by a cohesive force. A relatively large amount of stress is required to overcome this force, and the smaller the clay particle, the greater the force, especially if the particles are compacted.

Potholes eroded in bedrock are indisputable evidence of the channel scouring.

Pebbles and larger particles are less erodible than sand, not because they are cohesive, but because they are much heavier. Thus, larger particles require faster and/or deeper water to be eroded than sand does. Such knowledge has practical application in erosion control because it tells us what size particles need to be placed in ditches, gullies, canals, and stream banks to stabilize them against erosion. According to many environmental experts, soil erosion and the resultant siltation of water systems may be the number one environmental problem facing humanity today.

Scouring is also capable of eroding bedrock, but the rate is miniscule compared to that in unconsolidated material. A centimeter or so per year would be a typical bedrock erosion rate. Nonetheless, the process leaves distinct marks of its work in the form of channel features called *potholes*. These are pits, often a meter or more deep, etched into solid rock on the streambed by the bumping action of stones swished around and around by currents.

Sources of Sediment: Although streams do derive sediment from the erosion of bedrock, especially in their upper reaches, most of their sediment comes from three other sources: (1) The erosion or re-erosion of alluvial deposits already on the valley floor; (2) inputs from the discharges of tributary streams; and (3) hillslope deposits along the sides of the valley from runoff, mass movements, and glaciers. This material is integrated into a great train of sediment, which the stream sorts out into two main classes of deposits: *channel deposits* such as sand bars and gravel beds that the stream lays down as it shifts back and forth across the valley floor; and *flood deposits* left when floodwaters spill across the valley floor. In terms of total volume of sediment, channel deposits make up the bulk of the alluvial material in most stream valleys (Figure 21.13).

The In-stream Sediment System: When the material from these sources is picked up by the stream and entrained with its flow, it is called **sediment load**. It usually begins as a more or less heterogeneous mix of particles, which the stream quickly sorts out according to grain size into three types of loads: bed load, suspended load, and dissolved load. **Bed load** consists of larger particles (sand, pebbles, cobbles, and boulders) that roll and bounce along in almost continuous contact with the streambed. (Bed load, of course, is responsible for channel scouring.) **Suspended load** consists of small-sized particles, principally clay and silt, that are held aloft in the stream by turbulent flow, and **dissolved load** consists of chemicals in the ionic form contributed mainly by weathering processes in the watershed. Because they are carried in the water column, suspended and dissolved loads move through the system much faster than bed load.

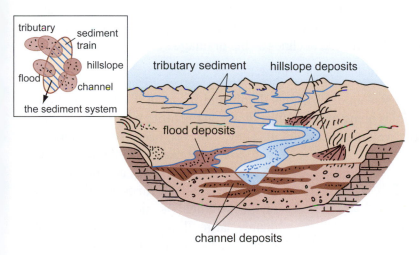

Figure 21.13 Sources of stream sediment: channel deposits, flood deposits, hillslope deposits, and tributary sediment. The inset illustrates these sources as subsets of the stream system. Airborne sediment is a fifth source that contributes throughout the system.

The mechanism of bed-load movement involves two forces: *drag*, a force directed along the bed stream, and *lift*, an upward force. The lift force is of primary importance in initiating particle motion. Laboratory flume experiments show that a lift force of about 70 percent of the submerged weight of a particle is sufficient to pivot it upward on its axis, whereupon the drag force can push it downstream. For any size class of particles there is a threshold of force, that is, a critical level of bed shear stress, below which no movement takes place, and above which the rate of sediment movement increases as some power (multiplier) of the force of bed shear stress.

Figure 21.14 The muddy appearance of the water on the left is produced by a heavy suspended sediment load and contrasts with the clear water on the right.

The lift component of turbulent flow is also important to suspended-load transport. Clay particles are relatively buoyant and, once suspended in the water column, will remain aloft even under the slowest of stream flows. Accordingly, clay sediments are extremely mobile in stream systems and, once entrained, tend to be moved great distances as part of the water mass. As the photograph in Figure 21.14 reveals, suspended load is easy to identify in most streams by the muddy appearance of the water.

Dissolved load is that material carried in solution. Ions of minerals produced in weathering are released into streams mainly through groundwater inflow. Total ionic concentrations in stream water are generally on the order of 200–300 milligrams per liter. However, in humid regions where soil leaching is strong, as in southeast Asia or eastern North America, ionic concentrations may reach several thousand milligrams per liter and represent 25–50 percent or more of a stream's total load. In dry areas, such as central Australia or southwestern United States, dissolved load is usually much lower, typically 10 percent or so of the total stream load. This is explained by the low rates of chemical weathering and soil leaching in arid-climate watersheds, which, of course, is a function of a moisture regime capable of supplying only a little infiltrating water.

Discharge and Sediment Transport: Sediment transport by streams varies radically with fluctuations in the magnitude of discharge as the examples in Figure 21.15 show. On September 15, discharge is low and the suspended sediment load is light, but when discharge rises on October 15, the sediment load increases dramatically. At the same time scouring cuts deeply into the streambed. Later, when discharge falls (October 25), the sediment load declines as the drainage system relaxes and the channel fills in with sediment.

We conclude that since stream discharge is dominated by small flows, most of the time very little sediment is being moved. Huge flows, on the other hand, such as 100-year floods, move great volumes of sediment. They, however, occur so infrequently that *in the long run* their total work adds up to a relatively modest amount of sediment movement. It turns out that most of the sediment is moved by intermediately large flows, those that happen every year or so or once every several years. These are mainly bankfull or near bankfull discharges and it appears they are responsible for around 90 percent of the total sediment transport in most streams. These flows are also responsible for the basic form of the channel, as well as the size of the meanders.

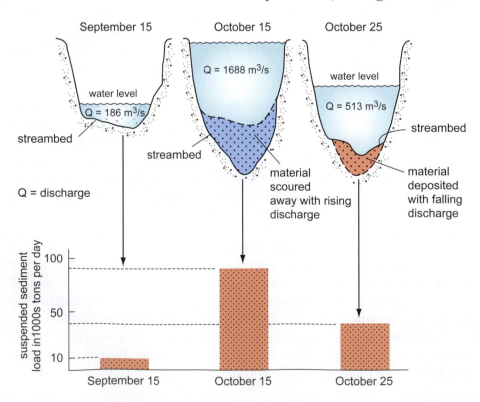

Figure 21.15 Channel and sediment load changes with fluctuations in discharge (Q). When water level rises in the channel (October 15), the stream deepens its bed and takes on much more sediment. The opposite occurs with falling discharge (October 25). (Adapted from Leopold *et al.*, 1964.)

Sediment, Channel Form, and Land Use: Channel geometry changes with fluctuations in both discharge and sediment supply, especially major fluctuations. When discharge rises, the channel deepens and enlarges as sediment is scoured from the streambed. Conversely, as discharge falls, the channel fills in with sediment and decreases in depth. These channel changes, both of which are illustrated in Figure 21.15, are referred to as **degradation** and **aggradation**, respectively. In the modern landscape, both urban and rural, land use is one of the common causes of such channel changes because it influences both stream discharge and sediment supply.

When land is cleared and cultivated, runoff and soil erosion rise, stream sediment loads increase, and channel aggradation usually sets in, especially in small streams. For Earth as a whole the major cause of this phenomenon is agriculture, principally crop farming (see Figure 21.16a and b). Locally, however, modern construction activity can have even greater impact than agriculture. When land is torn apart by heavy equipment, soil erosion and channel aggradation can reach extreme levels, resulting in sediment-glutted channels, elevated streambeds, and increased flooding (Figure 21.16c). As land-use advances and urban development overtakes a stream's watershed, soil is secured under pavement and buildings, soil erosion and sediment input to stream channels decline, while stormwater discharge rises abruptly. With larger and more frequent stormflows, streams scour and degrade their channels (Figure 21.16d). Channel sediment, such as gravel beds and sandbars, are removed, channels are deepened and widened, and most aquatic habitat is destroyed or damaged.

Channel form is also changed by other land-use practices. When a stream is dammed, its sediment load is captured in the reservoir behind the dam. Deprived of its sediment load, the stream, as it leaves the reservoir, generates a new sediment load by scouring its channel below the spillway. Under large spillway discharges, the channel degrades rapidly, stripping away sediment, downcutting, exposing boulders and bedrock, and damaging aquatic habitat.

At the opposite extreme are channels building up from excessive sediment loading. The Huang He of China suffers from such massive aggradation that much of the lower, main channel rides above the floodplain and the river is held in only by constructed levees as the diagram in Figure 21.17 illustrates. Heavy erosion of silty soils in tributary basins upstream is the chief cause of aggradation. The Huang He is huge and

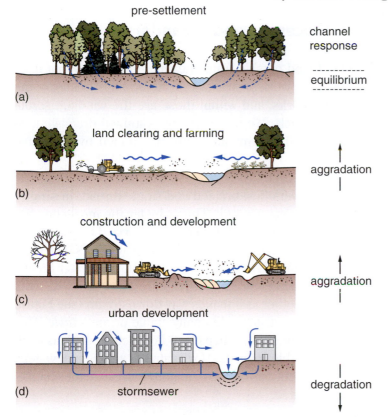

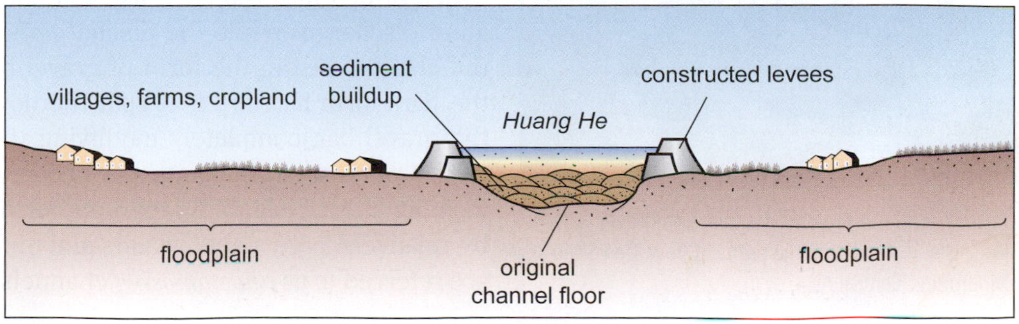

Figure 21.16 Land-use change leading to channel changes (aggradation and degradation) from pre-settlement (a) through farming (b), construction (c), and urban development (d) (after Wolman, 1967).

Figure 21.17 The relationship between the elevated (aggraded) channel of the Huang He and surrounding floodplain. For much of the year the river's surface lies above the floodplain, a very precarious arrangement.

millions of people live on its floodplain with hundreds of thousands below the elevation of the channel. This is a grievous situation, for eventually floodwater will break the levees, burying farmland, settlements, and people under a great mass of water and mud. In 1973 a flood on the Huang He resulting from burst dams killed over 250,000 people.

Summary on Stream Systems and their Work: Streams are complex systems linked to systems above and below the landscape that contribute water and sediment to a network of channels capable of eroding the land and moving sediment to the sea. The power of streams to do this work increases with discharge, but the total work accomplished is also related to the erodibility of rock and soil material and to the availability of sediment from other sources, especially tributaries. Transport and channel storage of sediment are related to flow events and to contributions from watershed sources including inputs from soil erosion brought on by land use.

21.4 Channel Forms, Processes, and Meanders

There is hardly a square mile of land outside Antarctica and Greenland that does not bear the distinctive imprint of the work of streams. At the broadest scale are the elaborate patterns of drainage networks, and within these are the winding, scrolling forms of their meandering channels. We have already examined drainage nets in Chapter 15. Now let us examine the forms and patterns created by individual channels.

Stream channels come in two basic varieties: braided and single-thread. In **single-thread** channels, water is confined to one conduit. This is by far the most common channel type in the world. They are relatively stable and characterized by one or two steep banks held in place by bedrock, and/or soil materials and plant roots. The main axis of the channel is marked by a zone of relatively deep, fast-moving water known as the **thalweg**. Most single-thread channels are sinuous or curving, within which the thalweg swings from one bank to the other as it makes way downstream (see Figure 21.9). Many single-thread channels are also conditionally stable or *metastable*. This means that they owe their stability to one (or perhaps two) key factors, such as bank vegetation, and the loss of decline or this factor (as, for example, when bank vegetation is broken down by heavy equipment during construction) can cause the channel form to break down and become braided.

Braided channels are made up of multiple conduits that weave in and out of contact with each other. They are often found in geomorphically active environments, such as near the front of a glacier or at construction sites, where massive amounts of sediment are being moved and stream banks are unstable or have broken down. Like the one shown in Figure 21.18, braided channels usually form in coarse, noncohesive bed material such as sand and gravel under conditions of sharply fluctuating discharge. While the discharge is decreasing, coarse debris is deposited on the channel bed in the form of gravel bars. When the discharge declines further, the bars form barriers that split the flow. Subsequent high discharges may erode the gravel bar, completely modifying the appearance of the channel. If a bar survives several seasons, plants may become established on it, adding to its stability and forming a small island. An extreme form of braided channels, which are dominated by relatively large stable islands and multiple channels winding among the islands, are referred to as *anastomosing* channels.

Figure 21.18 A braided channel with gravel bars, islands, and multiple paths of flow. Tien Shan Mountains, central Asia.

Meanders: Examine streams on maps or aerial photographs and it is hard not to be impressed by the curving, sometimes snake-like, path they take in crossing the land. The curves in stream channels are called **meanders**. Except for streams whose courses are confined to a relatively straight line by geologic structures such as a fault, virtually all streams contain meanders, though their geometry may vary from slightly wavy to excessively sinuous.

The process of meander formation is thought to begin with a disturbance in straight-line flow that causes the thalweg to shift from one bank to the other. Where this belt of fast-moving water strikes the bank, erosion takes place, the channel shifts later-ally, and a curve begins to form. In time, curves become bends and bends become meanders. As the meander system develops, the stream channel grows longer (that is, covers more ground), which in turn reduces its overall gradient. Meandering also appears to affect the stream as an energy system. Channel energy (bed shear stress) is more evenly distributed along a reach of meandering channel than along a similar reach of straight channel.

The size of meanders varies with the size of the stream. Huge rivers like the Missis-sippi build huge meanders, miles across. One measure of meander size is the width of the meander belt as defined in Figure 21.19. It appears to be related to the river's mean annual discharge; the larger this discharge, the greater the width. But while streams of the same size tend to develop meanders of the similar sizes, they often exhibit great differences in the curviness or sinuosity of the meander pattern.

Sinuosity is the ratio of channel length along the curve of the meanders to the length of the meander-belt axis. A line drawn down the center of the meander belt defines the meander-belt axis (Figure 21.19). Sinuosities of between 1.0 and 1.1 are essen-tially straight; those from 1.1 to 1.5 are wavy; and those greater than 1.5 are classed as meandering. Streams with extreme sinuosities, greater than 4.0, are classed as tortuous. **Tortuous channels** are characterized by a maze-like pattern of looping meanders that carry water in a variety of directions including up-valley in places.

Meander Movement: Meanders in most streams are continuously shifting or migrat-ing. For a given reach (segment) of a stream, the basic dimensions of the meander system in terms of scale and geometry, however, tend to remain relatively constant unless, of course, the stream's discharge changes, which can cause the meanders to grow or shrink. The mechanism of meander movement involves the lateral shifting of the channel as it is eroded on one side and filled in by deposition on the other. The pattern of erosion and deposition are governed by the shape of the channel meanders and the pattern of flow within it.

Water movement in a meander is characterized by two zones of contrasting flow. One is the belt of fast-moving water described above and the other is a belt of slow-moving water. These two belts occupy opposite sides of the channel in a meander bend. Centrifugal force pushes the thalweg to the outside of the meander where it causes bank erosion. Slow-moving water is left to the inside of the meander where it deposits sediment. These two processes take place simultaneously and the result is a lateral shift in the channel (Figure 21.20a).

Typical stream meanders. The belt of fast water shifts to the outside in each bend causing erosion there.

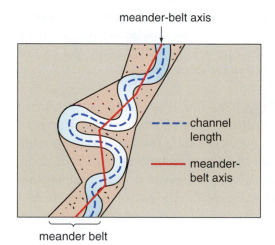

Figure 21.19 Meander-belt width is measured from the outside of one meander loop to the next on each side of the stream. Sinuosity is the ratio between the length of the meander-belt axis and the length of the channel.

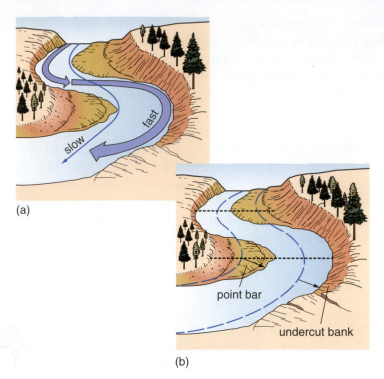

(a)

point bar

undercut bank

(b)

Figure 21.20 (a) Fast- and slow-moving water in meanders. (b) Corresponding lateral channel migration with resultant point bar and undercut bank. The broken line is the meander axis.

An extremely sinuous (tortuous) stream which has left oxbows of various sizes and shapes in its valley.

Two distinct landforms are created by this process. On the outside of the meander is an **undercut bank**, whereas on the inside is a **point bar**. Each year or so a new increment is added to the point bar while the stream erodes away a comparable amount on the opposite undercut bank. In this way the river shifts laterally, gradually changing its location in the valley (Figure 21.20b). Closer inspection reveals that the direction of change is actually both lateral and downstream, for erosion and deposition tend to be concentrated on the lower side of the meander axis. In some places, the rate of shift, or migration, can be measured by the age of trees on the valley floor. In Figure 21.21 the ages of cottonwood trees reveal that the Little Missouri River in North Dakota has shifted between 3 and 6 meters (10 to 20 feet) per year over the past 250 years or so.

Channel Dynamics, Riffles, and Pools: Streams can also make sudden changes in channel alignment. This usually happens where a stream breaches (cuts through) one of its own meanders. Breaching usually takes place in large looping meanders where the stream is eroding toward itself from opposite sides of the loop. A narrow neck is formed and, when the stream breaks through this ground, the larger meander is almost immediately abandoned because the new route is shorter and steeper and thus more efficient, that is, has greater energy in conducting the stream's flow. The abandoned channel segment forms a long thin lake, called an **oxbow**, which in time fills with sediment and organic debris, becoming a wetland. On aerial photographs of floodplains (see the one below), oxbow lakes and oxbow wetlands are easily identified by their distinctive crescent shapes.

For anyone who has waded or canoed down a stream, it is hard to overlook two prominent channel features: riffles and pools. **Riffles** are short runs of rapids where the channel gradient is steep, the flow relatively fast and turbulent, and the bottom material stony. **Pools**, on the other hand, are characterized by gentle gradient, deep, slower-moving water, and finer sediment. Riffles and pools occur in alternating sequences and the spacing between them increases with the size of the stream. As a rule of thumb, the distance between successive pools is five to seven times channel width.

Because of the stream's ability to adjust its channel by cutting (scouring) steep segments, and filling (depositing in) gentle segments, it seems that riffles and pools should eventually be obliterated as the stream smooths out its gradient. But observations on thousands of streams reveal that this is not so; streams have a way of maintaining these features. The explanation has to do with the variable nature of streamflow. When a stream rises to become a large flow, such as a bankfull discharge, both mass (water depth) and velocity increase across both riffles and pools. But the pools, which have the advantage of deeper water and slightly greater velocities, now have more kinetic energy and are flushed of their sediment and thus restored. There is an important lesson here: stream channels, and floodplains and valleys as well, are products of a wide range of flows and the flows we are apt to see on a casual visit may have little to do with the features we see in and around the channel.

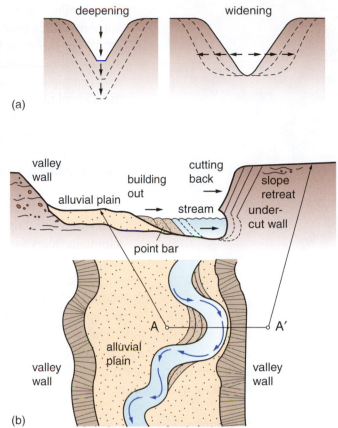

(a)

(b)

Figure 21.22 (a) Deepening and widening, the two processes of valley formation. (b) Valley widening by cutting back of the valley wall and building out of the alluvial plain.

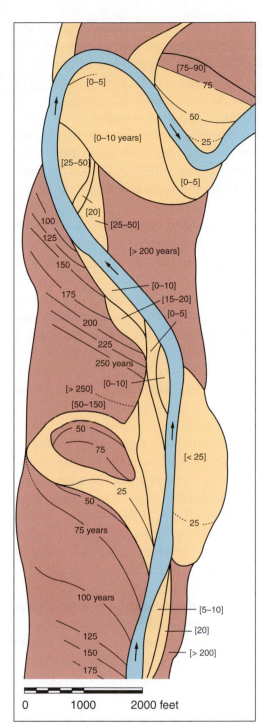

Figure 21.21 Changes in the location of the Little Missouri River as revealed by the ages (in years) of trees on the valley floor. Sudden changes in tree ages indicate sudden shifts in channel position as when a meander is breached.

The Formation of Valleys and Floodplains: As streams do their work eroding the land and transporting sediment, they also carve valleys. Valley formation involves two basic processes: downcutting, as streams work their way into the land; and widening, as streams erode laterally (Figure 21.22a). Downcutting is the dominant trend when a stream's gradient is steep and the stream is working toward a longitudinal profile adjusted to **base level**, the water elevation at its terminus. Steep gradients are usually irregular (rough) and the work of the stream is devoted to cutting down high spots and filling in low spots. As the stream approaches a smooth or graded profile, like the one shown by graph line A in Figure 21.7, more energy is devoted to lateral erosion and valley widening.

Lateral erosion and deposition produce a belt of flat ground on the valley floor. The process involves the stream tearing down high ground and building low ground in its place. It works as follows: when the stream flows against the edge of its valley, called the **valley wall**, it undercuts this slope, which fails and thereby retreats a short distance. At the same time, new ground is developing on the opposite bank with the formation of point-bar deposits. The new ground forms at a low elevation, near stream level. In time, the valley walls are cut back so far that a continuous ribbon of low ground, called **alluvial plain**, is formed along the valley floor (Figure 21.22b).

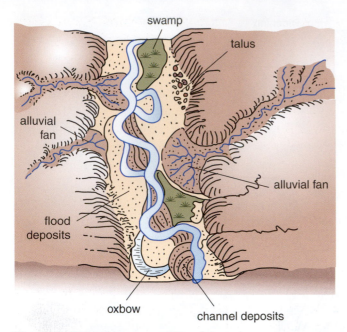

Figure 21.23 The diverse composition of stream valleys including alluvial fans, talus, flood deposits, and channel deposits.

The alluvial plain is composed principally of sands, gravels and other materials built from point bars and other channel deposits. When the river floods, the alluvial plain usually receives the overflow; therefore this ground is almost universally referred to as **floodplain**. Although floods alter the surface of the alluvial plain by eroding it and leaving deposits on it, they are clearly not the main cause of its formation. In this regard the term floodplain may be a little misleading for the alluvial plain, nonetheless because it is so widely used today, we will use floodplain for this part of the stream valley.

In addition to lateral cutting by the channel in widening the valley and building the floodplain, other processes also contribute to valley formation and composition. These are various slope processes such as slumping, overland flow, and gullying, which work on the valley walls, wearing them back and adding debris deposits to the floodplain. Especially significant are aprons of talus (rock debris) and alluvial fans that develop along the valley wall. The resultant composition of the floodplain is often very diverse indeed. Various types of channel deposits are laced together with deposits from floods, organic materials from forests and wetlands, and hillslope deposits from the valley walls (Figure 21.23).

Floodplain Features: Floodplains abound with distinctive topographic features. Oxbows and meander scars are the most salient features in many floodplains. As the term implies, **meander scars** are the imprints left by former channel locations. They often occur in regular series or sequences called *scrolls*, marking the progressive lateral shift of a channel such as the patterns shown in Figure 21.21. Oxbows are most abundant in the valleys of streams with tortuous meander patterns. Although meander breaching is a natural process in meandering streams, breaching is also performed by engineers to make streams straighter and more efficient as navigation channels.

Natural levees, scour channels, back-swamps, and terraces are also common floodplain features (Figure 21.24). **Levees** are mounds of sediment deposited along the riverbank by floodwaters. They occur on the bank because this is where flow velocity declines sharply as floodwater leaves the channel, which causes it to drop part of its sediment load. In the low areas behind levees, floodwater may become trapped and

Figure 21.24 Some of the prominent features found in floodplains.

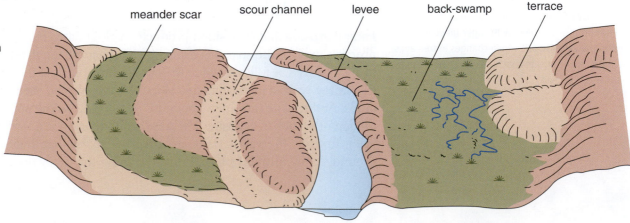

Table 21.2 Actions leading to changes in channel elevation

Action	Channel response	Elevation change
Land uplift	Degradation	−
Climate change (drier)	Aggradation	+
Farming	Aggradation	+
Dam construction: upstream	Aggradation	+
Dam construction: downstream	Degradation	−
Urban development	Degradation	−

linger for long periods, forming **back-swamps**. **Scour channels** are shallow channels etched into the floodplain by floodwaters. They often form across the neck of a meander loop and carry flow only during large floods. **Terraces** are elevated parts of a floodplain. They form when a river downcuts and establishes a lower valley floor leaving parts of the old valley floor (floodplain) slightly elevated. In mountainous areas, terraces often form when uplift associated with tectonic activity raises valley floors and induces streams to downcut.

Changes in the channel elevation are common in streams. Not only can streams lower themselves by cutting into their valley floors, but they can also raise their elevations by filling their channels in with sediment (see Figure 21.17). Which happens depends on the direction of change (increase or decrease) in three factors: discharge, sediment loads, and land elevation. Increases in land elevation and discharge and/or a decrease in sediment load can lead to downcutting whereas the opposite changes can lead to buildup and raising of the channel and the valley floor. These changes can take place naturally – with climate change, for example – or by human intervention as we demonstrated earlier with land clearing, soil erosion, urban development, and dam construction (Table 21.2).

Summary on Channels, Floodplains, and Valleys: Channels take on a variety of patterns, but almost all exhibit a meandering form of some degree, a form closely related to the winding nature of open-channel flow. By eroding on one side and depositing on the other, stream meanders shift laterally while sediment is moved from pool to pool, valley walls are carved back, and the larger valley takes shape. Valley widening leads to the formation of alluvial plains, or floodplains, and within this belt different landform features are formed with channel migration and flooding.

21.5 The Watershed as a Sediment System

Every watershed is a sediment system made up of many streams, which are fed particles by a host of processes. These processes operate throughout the watershed, moving sediment from upland surfaces, down hillslopes, and into stream valleys where it can be picked up by streams. Although this system moves huge amounts of sediment, it does not operate like a neatly engineered manufacturing system in a factory. Rather it operates in a sporadic manner over time and space within the watershed. Most sediment is produced during wet periods such as El Niño years or during certain events such as high-intensity rainstorms. In addition, the production of sediment is highly variable geographically depending on topography, surface materials (soil, deposits, and bedrock), vegetation, climate, and land use. As we noted in the previous chapter, mountainous terrain, semiarid lands, and cropland yield much larger amounts of sediment than flat lands, arid lands, and forested lands. For large watersheds such as the Mississippi, where topographic, bioclimatic, and land-use conditions vary radically over their vast areas, the output of sediment is typically very uneven from one part of the watershed to another.

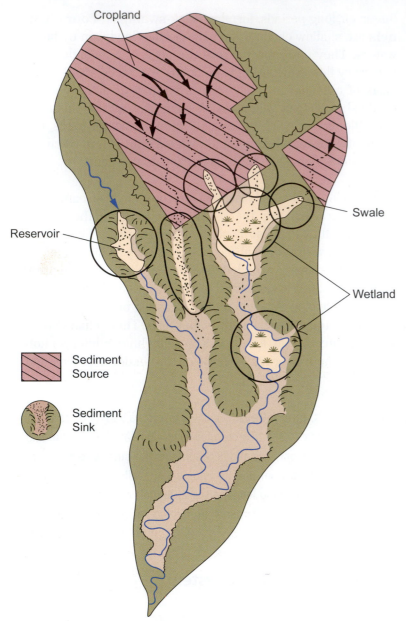

Cropland

Reservoir

Swale

Wetland

Sediment Source

Sediment Sink

Figure 21.25 A small watershed where cropland is the chief source of sediment and swales, wetlands, and reservoirs are the principal sinks, capturing 80 percent or more of the excess sediment.

Headwaters of the Brahmaputra; rugged mountain land, which yields huge amounts of sediment.

In its simplest form, the sediment delivery system moves sediment from hillslopes to stream channels and ultimately to the sea. But for most sediment the trip is not a non-stop ride. To begin with, the linkage between hillslopes and stream channels is often tenuous and much sediment gets trapped in various spots, called sinks, along the way. These sinks are wetlands, swales, wooded ravines, farm ponds, reservoirs, and the channels and valleys of small streams (Figure 21.25). For small basins at the heads of large watersheds where land had been cleared and farmed early in the twentieth century or before, as little as 10 to 20 percent of the sediment produced by runoff actually gets exported directly to larger streams. Sediment yield rates from such farm lands are typically high but studies in North America reveal that most of this sediment goes into storage before reaching large streams, where it probably remains for decades, centuries, or longer.

The sediment which does become entrained by streams moves through the system at irregular rates. That transported as bed load may move only short distances at a time, perhaps one or two meander lengths, before it is deposited in a point bar and buried in the floodplain as the stream channel shifts. Here it may remain for centuries or more until the stream migrates back over it and moves it again. Suspended and dissolved loads, on the other hand, usually move through the system much faster. Once entrained by the stream, they are carried entirely through the system to the sea, unless they are deposited in lakes, swamps, or reservoirs along the way. With the construction of tens of thousands of dams in the United States, Canada, and many other countries, sediment entrapment by reservoirs has greatly reduced export to the sea.

Sediment Transport by Great Rivers: This brings us to the question of how much sediment is actually moved by the various rivers of the world. Among the thousands of streams draining to the oceans, the 70 or so classed as great rivers do the vast majority of sediment export from the land. Among these, the great rivers of Asia take the prize for sediment output. Six Asian rivers (Brahmaputra, Yellow, Ganges, Yangtze, Indus, and Mekong) carry 50 percent or more of the world's total sediment to the ocean. Measured in tons of physical and chemical (dissolved) sediment produced *per square kilometer* of watershed per year, the Brahmaputra yields about 1850 tons per square kilometer per year, the Yellow about 1400 tons and the Ganges about 780 tons. By comparison, the Amazon produces about 175 tons, the Mississippi about 150 tons, the Mackenzie in Northern Canada about 75 tons, and the St. Lawrence less than 50 tons per square kilometer per year (Figure 21.26).

Why these great contrasts? The answer has mainly to do with topography and climate. Mountainous watersheds produce decidedly more sediment than low-relief watersheds (those dominated by plains) because runoff and erosion rates are higher in uplands. The great Asian stream systems emanate from the mountain ranges of the Himalayas and the Tibetan Plateau. The Brahmaputra, which drains the largest of these mountain ranges, yields 10 times more sediment per square kilometer of

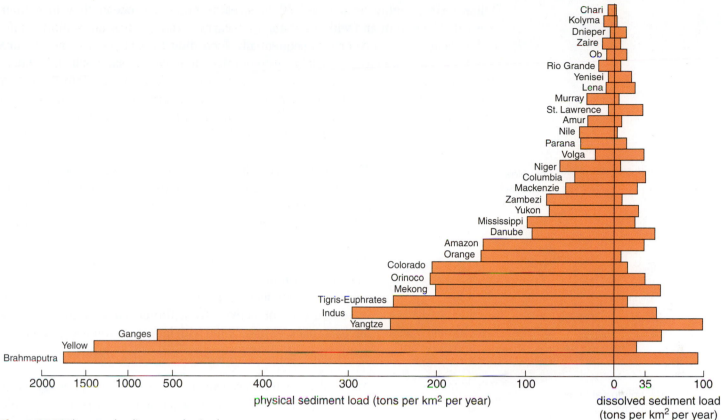

Figure 21.26 The annual sediment production by 32 of the world's largest rivers. Physical sediment is mainly sand, silt, and clay. Dissolved sediment is ions in solution.

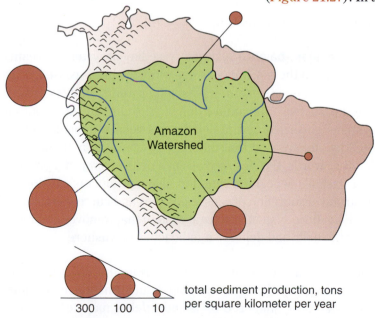

Figure 21.27 Contrast in sediment production between the mountainous parts of the Amazon watershed and the lower, flatter parts.

watershed than the Amazon. Both rivers are fed by a wet climate; in fact, the Amazon is the wetter of the two and has a much larger discharge. The big difference is that most of the Amazon watershed is low-relief terrain. Only the upper part of the watershed is mountainous and there we find a higher rate of sediment production (Figure 21.27). In addition, the Amazon watershed is mostly forested whereas much of the lower Brahmaputra and the other Asian watersheds are cropland, and cropland has high soil-erosion rates.

Another factor to consider is the seasonality of precipitation. The Asia watersheds are dominated by a monsoon climatic regime. They experience a season of drought, which weakens the plant cover, leaving the soil vulnerable to the torrential rains of the wet season. The Amazon's precipitation, on the other hand, is more evenly distributed throughout the year and the plant cover remains stronger and more resilient to erosional forces.

Deltas – The End of the Line: Eventually all streams come to an end. Most large streams end in the ocean where they drop their sediment load. This results in a great accumulation of sediment called a **delta**, a name we owe to the ancient Greeks who described the geographic feature at the end of the Nile as having the shape of the Greek capital letter Δ, delta. Although this term was appropriate for the Nile, most deltas actually resemble other things, including a bird's foot.

Deltas form roughly as follows. When streams enter the ocean they lose their forward momentum and with it the ability to carry sediment. Stream sediment falls to the ocean floor more or less sequentially according to particle size (see Figure 17.34). Pebbles and larger particles drop out first, followed by sand, silt, and clay at increasing distance from shore. Because it is so light, clay spreads into the ocean well beyond the river mouth, as much as tens of kilometers out to sea. As it settles to the ocean floor, it forms a base of sediment, called **bottomset beds**, upon which the coarser sediments are deposited as the delta builds seaward.

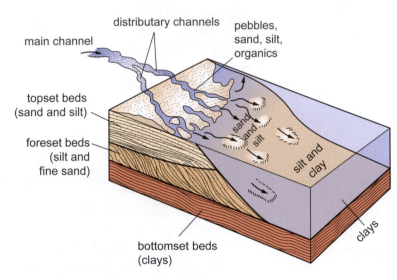

Figure 21.28 The basic structure of a delta beginning with bottomset beds and followed by foreset beds and topset beds.

The structure of a delta is made up of three sets of beds (Figure 21.28). First are the bottomset beds, which are flat-lying, thin layers of mud. Over these are thicker beds of fine sand and silt, called **foreset beds**, which are deposited as cross-beds (that is, at an angle), which slope gently seaward. At the top of the delta and nearer shore are horizontal beds composed mostly of sand, called **topset beds**. The delta surface is a very dynamic environment subject to the shifting action of stream channels, wave and current activity from the ocean, and assorted biological processes associated with lagoons, swamps, and related features.

As the delta extends seaward, the stream loses energy as its gradient is reduced to nearly a flat line. As a result the main channel becomes clogged with sediments and breaks down into smaller channels, called **distributaries**, which fan out over the delta. For large deltas like the Mississippi, the river and its distributary system periodically shift to a new location. This happens when an active part of the delta extends seaward so far that the river's route to the sea becomes so inefficient for the movement of sediment that the river shifts to a shorter and more efficient route. In the last 6000 years the Mississippi has made four major shifts (Figure 21.29).

Deltas as Mass-balance Systems: As sediment systems, streams do not end at the sea, but continue on as the ball is passed, so to speak, to another system, waves and currents. In this context, it is instructive to examine deltas as sediment input–output systems governed by the balance between sediment contributed by streams and that taken away by waves and currents. In other words, the size or mass of a delta is not just a product of a stream's deposition rate, but also a product of erosion by destructive shoreline processes. Waves and currents make up the longshore system, a coastal system that removes sediment from the delta and delivers it to beaches downshore.

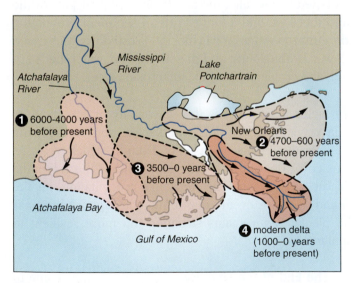

Figure 21.29 Shifting delta. Different locations of the Mississippi's active delta over the past 6000 years.

On high-energy coastlines, waves and currents can be so strong that river sediment in swept away as fast as it is brought to the coast, making delta formation impossible. On low-energy coasts, such as those sheltered by islands, deltas may develop with little interference from the sea. The Mississippi Delta is the product of massive input of river sediment and relatively modest removal rates by the sea. Over the past 150 million years the Mississippi has advanced its delta 1600 kilometers from where the Mississippi and Ohio rivers now join, meaning, of course, that the delta system has long been positively balanced.

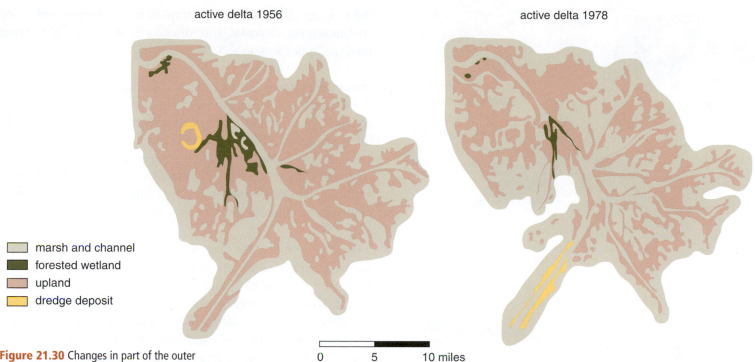

active delta 1956

active delta 1978

- marsh and channel
- forested wetland
- upland
- dredge deposit

0   5   10 miles

Figure 21.30 Changes in part of the outer Mississippi Delta from 1956 to 1978. Notice the large loss on the southwest side and the decrease in upland surface and wetland forest. Since this survey loss rates have accelerated. (From U.S. Geological Survey.)

This trend, however, has been reversed in the modern delta, first because sediment input from the Mississippi has declined (as a result of sediment entrapment in reservoirs) over the past century, and second, because of the destruction of delta wetlands with land-use development (Figure 21.30). Wetlands in and around the delta serve as effective buffers to storm waves and their loss exposes the delta to wave erosion. The Mississippi Delta is currently losing 20 square miles of wetland a year. The delta is increasingly vulnerable to storm surges, a point that was vividly illustrated by the surge of Hurricane Katrina which rolled over the weakened delta and flooded New Orleans in 2005. We will examine deltas further in connection with coastal geomorphology in the next chapter.

21.6 Watershed Systems, Landforms, and Landscape Development

All lands, wet and dry and cold and warm, are part of watersheds. Excluding glaciers and ice sheets, virtually all Earth's terrestrial surface is shaped by runoff processes. And since all runoff systems are organized into watersheds, the patterns of landforms and related landscape features like soils and vegetation should relate to the shapes of watersheds and their network of streams.

As geomorphic systems, watersheds are organized into three broad zones: upper, middle, and lower (Figure 21.31a). In the upper zone on the watershed perimeter, hillslope processes work in combination with glaciers, to shape erosional landforms and build hilly or mountainous landscapes. In the middle zone, these erosional processes are less powerful, depositional landforms begin to appear on valley floors, and massive amounts of sediment are transferred down the system. In the lower part of watershed systems, particularly those of large rivers, stream processes alone are dominant in shaping landforms. The landforms are principally depositional, usually complex associations of recent and ancient river-valley features including broad floodplains, terraces, alluvial soils, swamps, and deltas. The landscape in this part of

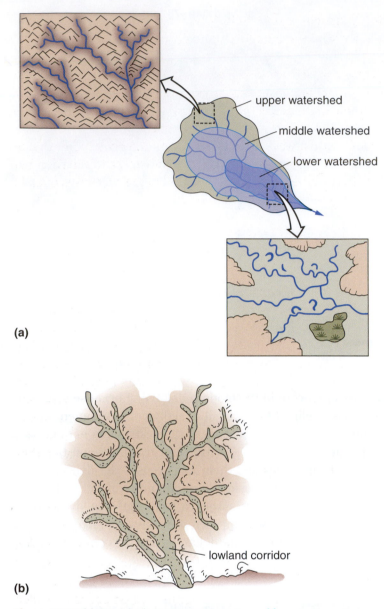

the system exhibits less topographic relief, deeper and more continuous soil deposits, and softer, i.e. less angular, landforms than the upper watershed.

Landform Patterns: Although streams shift about in their valleys, the skeletal pattern formed by the network of valleys remains constant for long periods in most watersheds. This network in turns shapes the basic pattern of landforms within watershed systems. Following the principle of stream orders, valleys grow larger as stream order gets higher. Thus, big valleys are more common in the lower watershed and small ones more common in the upper watersheds.

Taken as a whole, the fluvial lowlands form a *system of corridors* patterned after the drainage net (Figure 21.31b). The uplands, on the other hand, occupy the intervalley areas, or **interfluves**. This arrangement, in turn, strongly influences the distribution of moisture. Since uplands shed water, they tend to be drier than lowlands which collect water. In addition, lowlands lie closer to groundwater sources and usually have better nutrient supplies from stream and floodwater sediments. Not surprisingly, soils and ecosystems in watersheds follow this two-part division. The ecologically richest areas often follow the corridors formed by the networks of valleys. Uplands, on the other hand, are ecologically poorer with lower biodiversities and organic productivity rates.

Gorges, Canyons, and Water Gaps: Occasionally there are gorges and deep valleys, such as the Grand Canyon, in the lower parts of a major stream system where a large river cuts through a mountain range, plateau, or large ridge. River cuts in a single ridge or mountain range are called **water gaps** whereas longer river cuts are called **gorges** or **canyons**. While in most cases there is no question that rivers erode such features, the question does arise as to why a stream would cut through a mountain or plateau barrier rather than go around it. The answer is found in the geologic history of a region and is explained in one of two ways.

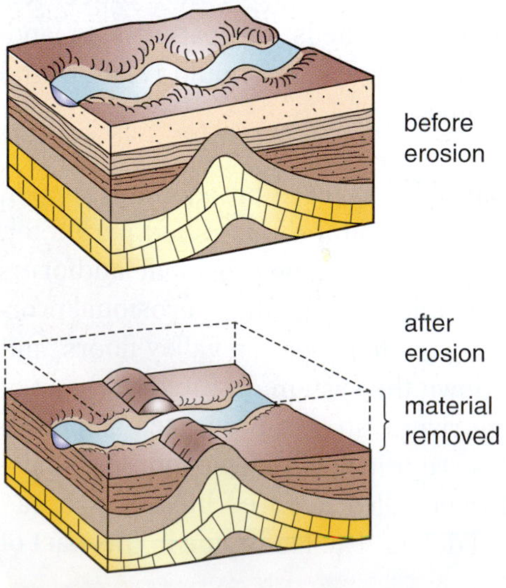

Figure 21.31 (a) Watersheds as geomorphic systems with processes and landforms organized roughly into three zones. (b) Drainage nets as a system of lowland corridors with distinctly different landscape attributes than adjacent interfluves.

The first explanation is that the stream existed in the region before the barrier formed, and the barrier subsequently rose up beneath it. As the land came up, the stream cut through the barrier to form the gap. Such a stream is called an **antecedent stream** because it existed before the ridge or mountain range formed.

In the second explanation, the barrier existed deep within the crust long before contact with the stream, while the stream flowed over it unimpaired. Then, as the stream eroded its way into the crust, it lowered itself onto the barrier, cutting into it (Figure 21.32). Such a stream is called a **superposed stream**, because it was

Figure 21.32 The concept of stream superimposition. The stream lowers itself onto an existing structure, such as a folded rock formation, creating a water gap, gorge, or canyon.

Beginning of the Grand Canyon as early scientists saw it in their epic trips down the great chasm.

Figure 21.33 The influence of plate tectonics on the formation of large river systems as illustrated by the opening of the Atlantic Rift. Notice the role of the domes and the failed arms in controlling the course of major rivers. On the opposite side of the continent, meanwhile, a mountain belt is thrown up which divides drainage between large interior watersheds and small coastal watersheds.

imposed onto a pre-existing structure as it cut its valley into the land. The Colorado River is a superposed stream where it cuts through the Colorado Plateau.

Stream Development, Watersheds, and Plate Tectonics: The cutting of the Grand Canyon over the past several million years was brought on principally by the uplift of the Colorado Plateau. Tectonic forces within the Earth – probably a huge plume rising from the mantle – pushed this part of North America up several thousand meters, which caused the Colorado River to cut deeply into the crust. We can cite many other examples of tectonic forces influencing not only stream behavior but the location and origin of stream systems. Most are related to the processes of plate tectonics and the formation of the continents themselves.

When large landmasses such as Pangaea break up into individual continents such as Africa, North America, and South America, the process begins with uplift of the crust, similar to that documented for the Colorado Plateau. Uplift is usually concentrated along a belt stretching thousands of kilometers across the landmass. Within this belt, large bulges or domes form in the crust that direct streams away from or around the uplift zone. Two such domes, labeled A and B in Figure 21.33, can be found along the uplift zone that formed between South America and Africa before they separated. The effects on drainage can still be seen in the patterns of streams on both continents. For example, notice how the Kasai River in Africa and the Uruguay River in South America follow the edge of these domes. This accounts for the indirect route these rivers take to the sea. Many other rivers including the Parana, Upper Niger, Sao Francisco, and the Cubango flow away from the sea in their upper reaches because they have been diverted by domes.

The second stage of the breakup process is marked by crustal separation along the uplift zone coupled with the formation of a narrow rift valley. In this rift system, several three-armed intersections, called **triple junctions**, form. As the rift widens, only two of the arms remain active, that is, continue to deepen and fill with ocean water (see Figure 18.29). The third arm fails or is aborted, that is, it ceases to deepen and widen. The **failed arm** always extends into one of the two bordering continents where it forms a large structural depression linking interior land to the sea. The depression serves as an outlet for runoff from the continental interior, eventually becoming the main valley of a major river. The Mississippi, Amazon, and Niger rivers have all formed in failed arms.

Senegal

Niger

Benue rift

Amazon rift

failed arm

Nile

failed arm

Amazon

Sao francisco

Parana

Zaire

Kasai

A

Cubango

Zambezi

Uruguay

B

plate movement

Orange

Atlantic rift

dome

A Angola–South Brazil dome

B Southwest Africa–Uruguay dome

Meanwhile, on the opposite side of the continent, the leading edge of the tectonic plate is being rammed against a neighbouring plate. The result is severe deformation as masses of sediment and fragments of crust are heaved up into mountain belts along the plate's edge. Streams flowing toward this coast are blocked by the mountains and drainage is diverted inland while new, smaller watersheds form on the ocean side of the mountain belts. The discharge diverted inland often joins the large interior watershed, which has already formed around the failed-arm depression described above. This process contributes to the formation of some of the world's largest watersheds including the Amazon and Mississippi. It also accounts for heavy sediment loading and large continental shelves on the trailing (rift) side of continents. And in the case of the Amazon, the formation of the Andes on the western rim of the basin helps account for its massive discharge because these huge mountains intercept the moisture-laden tradewinds which unload heavy precipitation on the great river's headwaters (see Figure 21.2).

21.7 The Geographic Cycle and Other Grand Ideas

For centuries, scholars have debated over the origin of the Earth's surface, how the landscapes we see around us came into being. What process or processes sculpted them? Streams? Waves and currents? Glaciers? Others? Were they shaped gradually, in small incremental changes, or suddenly, by catastrophic events? Is the Earth's surface old or relatively new? In the early 1800s, the questions about the rates of change and the age of the Earth's surface were largely put to rest with the introduction of the concept of **uniformitarianism**. This concept held that the types and rates of change taking place on Earth now are fundamentally the same as those that prevailed in the geologic past. Uniformitarianism proposed that Earth changed gradually over long periods of time (referred to as *gradualism*), and therefore its rocks and landforms must be millions of years old.

The answer to the first question on what process(es) sculpted the land surface did not become clear until the latter part of the nineteenth century with the American Western Surveys. Citing evidence no less compelling than the erosion of the Grand Canyon by the Colorado River, the Americans documented that runoff processes are Earth's primary erosional agent and that stream systems are capable of cutting into the crust and stripping massive amounts of rock off the continents.

The Geographic Cycle: From the North American work grew a grand idea or theory about continental erosion and landscape development. It was a called the **geographic cycle**. Its founder, American geographer William M. Davis, proposed that as stream systems erode landmasses, the landscape passes through a series of developmental stages. The stages take place sequentially like stages in a person's life, beginning with youth and then advancing into maturity and old age. Using this terminology, Davis proposed three basic stages of landscape development, shown in Figure 21.34, each marked by a distinctive valley form as streams deepen and widen their valleys.

The geographic cycle begins with an uplifted landmass and stream downcutting. The landforms are high and rugged – the product of tectonics – and the streams that form on them follow irregular patterns dictated largely by the configuration of the mountainous terrain. In this first stage of development, the **youthful stage**, streams downcut into the land forming V-shaped valleys while expanding their watersheds and integrating the landscape into large drainage systems. Streams continue to cut

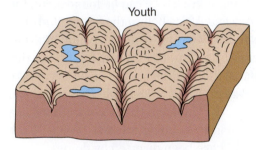

Youth

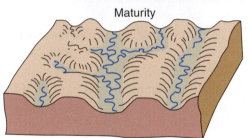

Maturity

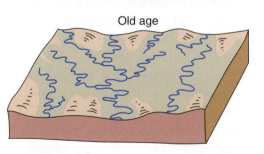

Old age

Figure 21.34 The three stages of the geographic cycle. As the land is lowered, it is transformed into a low plain, called a peneplain.

their valleys into the land until they establish **graded profiles**, that is, until their profiles are adjusted to base level, and slope gradually upward from mouth to headwaters as illustrated in Figure 21.7 (A).

In the second, or **mature stage**, valleys are widened into broad U-shapes. Valley walls are worn down to lower angles and stream-sculpted landforms have replaced tectonic or structural forms over much of the landscape. In the third stage, **old age**, valley sides are lowered to the point where the land between the valleys, the interfluves, is reduced to long gentle slopes. Extended over whole regions or continents, the result is a low, rolling plain, which Davis called a **peneplain**, meaning near-plain. At this point, tens of millions of years after the cycle began, the land is subject to renewed uplift which induces a new cycle of erosion. Davis called this **rejuvenation**.

The geographic cycle was immensely popular among educators and scientists in the first half of the twentieth century. Students found it easy to understand and for scientists it provided a model that helped them to explain the development of landforms in evolutionary terms. As with virtually all scientific formulations, however, the geographic cycle fell short in a couple of respects. First was the retreat of slopes. Davis described interfluves growing lower as the cycle advances, whereas scientific evidence reveals that most interfluves appear to grow narrower with parallel retreat of valley slopes. Second, there were problems with the notion that the uplift of the land does not take place until the cycle is completed. Abundant evidence indicates that isostasy takes place more or less continuously as denudation advances. Tectonics, on the other hand, is more episodic, but there is no evidence that it responds to the final stage of the geographic cycle. Not surprisingly, several alternatives to the geographic cycle were subsequently advanced.

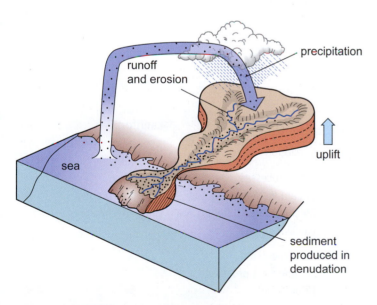

Figure 21.35 With dynamic equilibrium, if climate and uplift remain constant, then energy levels for runoff, erosion, and landform development should hold steady. Of course, the climate and runoff system cannot be expected to hold steady for long periods; therefore, the stream system will vary accordingly.

The Dynamic-equilibrium Concept: In the mid-twentieth century, geomorphology shifted from evolutionary-based models of landscape formation to models based on energy-system concepts. The idea was advanced that instead of going through developmental or evolutionary stages, river-eroded landscapes are constantly in a state of dynamic equilibrium. This means that the landscape functions as an ever-changing energy system, which is always tending toward a steady-state condition but because of the changeable nature of the stream's energy sources (represented by climate, runoff, and land uplift), rarely, if ever, achieves it. This is in sharp contrast to the assumptions underlying Davis's stage-wise development, in which the energy system winds down as the land gets lower. In the language of systems theory, the geographic cycle proceeds until the system approaches a state of entropy. *Entropy* is the state of minimum available energy in a system when a system falls into disorder.

According to the dynamic-equilibrium concept, if the climate were to remain steady while the land uplifted in response to denudation (unloading), the energy levels of stream systems would remain constant while they erode the land as Figure 21.35 attempts to show. This would be possible because the water available for runoff (mass) and elevation of land (which sets the stream's energy gradient) would hold steady with the passage of time (also see Figure 21.5). Furthermore, if the rock types and geologic structures remained constant as the stream system cut its way into the land, the valley forms would also remain constant. Thus, stage-wise development of stream valleys is untenable according to the dynamic-equilibrium concept, because as the land is worn down by erosion, the landforms remain essentially the same rather than changing from higher and steeper to lower and gentler.

Morphogenetic Region Concept: Neither the geographic cycle nor the dynamic equilibrium concept addressed the influence of Earth's varied climatic conditions on landscape development. Accordingly, around mid century, attempts were made to build broadly based concepts incorporating more or less the full range of climate types and the geomorphic processes and landscape types associated with each. From this emerged a concept called **morphogenetic regions**, which examined landscape as the product of different climatic regimes, including those dominated by stream systems.

According to this concept, landscapes differ with climate-driven processes emphasizing the differences, for example, between the wet tropics with strong chemical weathering and periglacial environments dominated by permafrost conditions and related processes. Morphogenetic regions or zones conform to the broad zonal patterns of world climate (Table 21.3). But climates change, of course, so some morphogenetic zones exhibit features of two or more climates and the borders between most zones are broadly transitional. Thus, the application of the morphogenetic-regions concept must recognize past climatic conditions, such as the periglacial conditions that existed over much of the midlatitudes only 10,000 to 12,000 years ago, as well as the inexact nature of climatic borders.

Table 21.3 Morphogenetic regions and their characteristics

Region	Present climate	Past climates	Active processes*	Landforms
Zone of glaciers	Glacial (cold)	Glacial	Glaciation	Glacial
Zone of pronounced valley formation	Polar, tundra (cool & wet)	Glacial, polar, tundra	Frost processes Stream erosion Mechanical weathering (Glaciation)	Box valleys Patterned ground Glacial forms
Extratropical zone of valley formation	Continental (cool, temperate, wet, dry)	Polar, tundra, continental	Stream erosion Ground frost processes (Glaciation)	Valleys
Subtropical zone of pediment and valley formation	Subtropical (warm, wet, dry)	Continental, subtropical	Pediment formation (stream action)	Planation surfaces and valleys
Tropical zone of planation surface formation	Tropical (hot, wet, wet-dry)	Subtropical, tropical	Planation** Chemical weathering	Planation surfaces and laterite

*and former processes (adapted from Budel, J., 1963)

**planation: processes leading to a low plain surface like a pleneplain

21.8 Human Impact on Stream Systems

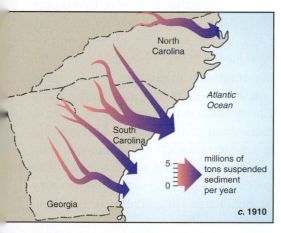

Landscape secured by vegetation against the erosive force of runoff, a condition common to less than 500 million years of Earth history.

Streams have been around as long as there has been land on Earth and during those billions of years they have seen nature render every conceivable sort of change in them. During the great global freeze-downs, stream systems were probably obliterated from the planet, and at other times, when climate grew warm and wet, stream systems expanded. The advent of terrestrial vegetation nearly a half a billion years ago had to have brought on one of the greatest changes in streams as erosional systems, for now the land was protected from erosion by a mantle of foliage and roots. No doubt rates of erosion fell precipitously as plants secured the ground. And channels must have changed as well as sediment loads declined and stream banks became stabilized with plants. These two changes probably caused a dramatic increase in single-thread channels as braided streams declined.

Today humans are the dominant change agent in stream systems. As we noted in the previous chapter, soil erosion and sediment export to the sea has more than doubled since the origins of agriculture about 12,000 years ago. However, in the last century or so, export of sediment by streams has increasingly been interrupted by dams and reservoirs. When a stream enters a reservoir, its velocity falls and it drops most of its load. When Hoover Dam was constructed on the Colorado River in 1934, for example, the river's downstream load of sediment dropped from around 150 million tons per year to less than 10 million tons. Similarly, in several large streams of eastern United States, discharge of suspended load to the sea dropped by as much as 75 percent with the construction of dams (Figure 21.36). In southern California, dams built for water supply in the1920s and 1930s caught so much sediment that beaches on the nearby Pacific shore, which depended on streams for their sand supply, declined dramatically.

When reservoirs deprive streams of their sediment loads, the energy that would normally be expended in moving that sediment is now available for additional work. The stream, in turn, uses this energy to erode its channel and pick up a new sediment load below the dam. As a result, in the reaches downstream from dams channels are often severely degraded, sand bars and gravel beds shrink, and the aquatic habitat, particularly benthic (bottom) community of plants and insects, is largely destroyed. In addition, the dams block the migration of fish and other animals. On the North American West Coast, dams have effectively destroyed entire salmon populations on many streams, including North America's largest Pacific stream, the Columbia.

In the United States and other developed countries enormous numbers of dams have been built for water supply, navigation, flood control, and other purposes. It is estimated that 80 percent of the major rivers of North America and Eurasia have been tampered with through the construction of barriers such as dams and lock systems. In the United States, 75,000 dams 2 meters or higher were constructed in the nineteenth and twentieth centuries. Dam construction in the United States and Canada has declined sharply in the past 25 years, but in less-developed countries like Brazil and China, it is still very active. The Chinese are now completing a project on the Yangtze, called the Three Gorges Project, which is being touted as the world's largest dam. Its reservoir will flood a huge area and capture a massive quantity of sediment.

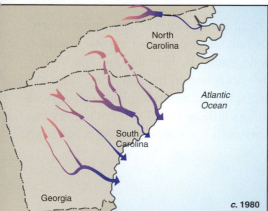

Figure 21.36 Changes in suspended-sediment output in several large streams in North Carolina, South Carolina, and Georgia as a result of dam construction between c.1910 and 1980. (From US Geological Survey.)

Managing Stream Systems: Human intervention in stream systems has come in three main forms. First, increased runoff in response to land-use development, both agriculture and urban. Second, increased sediment loads from soil erosion on expanding farmlands, and third, structural manipulation for flood control, navigation, water supply, and recreation. As a result, many stream systems no longer function as natural geomorphic systems in which water and sediment are moved to the sea by flows of various magnitudes and frequencies. They have in many respects become the wards of governments, as it were, and are the subjects of management programs designed to reduce their variability and natural function and/or make them serve some set of human needs (Figure 21.37).

Despite massive efforts to control streams with dams, levees, and other engineered facilities, flooding, property damage, and financial expenditure for flood control have continued to rise, leading many observers and policy makers to question the wisdom of the structural (engineering) approach. Added to this is the widespread concern that the ecological, historical, and esthetic character of streams and their valleys have been seriously damaged or lost. These concerns have led to alternative management approaches that are largely non-structural, including land-use planning and regulation, and restoration programs aimed at returning streams to more natural states. The twenty-first century may see more dams being programmatically removed than built in Canada and the United States.

In North America, habitat and history are among the most celebrated causes for stream restoration. In the eastern part of the continent, urban rivers – which for more than a century have been seriously abused by industrial, commercial, and transportation land uses – are the principal targets of restoration efforts. Most of these projects attempt to recover the stream's historical character by restoring (1) its natural hydrological and geomorphic function, and (2) its scenic value, by eliminating certain structures and land uses from the channel and floodplain. The result is a stream corridor or greenbelt that is reminiscent of the nineteenth century townscapes. Aquatic habitat is often improved, but channel degradation is usually so severe that full ecological restoration is virtually impossible.

In western North America, stream habitat is the principal target of restoration efforts. The most successful projects champion fish habitat, especially migratory salmon species whose spawning streams have been degraded by land use and blocked by dam and lock systems. Projects are typically controversial because they often call for the removal of dams and reduction of stormwater runoff from urban and agricultural lands. Studies show that watersheds with as little as 10–15 percent impervious cover may produce too much poor-quality stormwater to maintain salmon and trout populations.

Figure 21.37 Hoover Dam on the Colorado River. Such facilities severely alter streams as geomorphic systems, dramatically changing flow regimes and sediment transport.

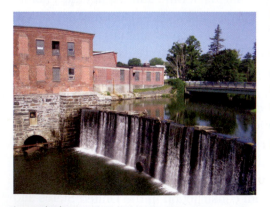

Increasingly stream restoration involves removal of defunct dams such as this one in Massachusetts. Thousands of such dams are found across North America.

Chapter Summary and Overview of Stream Systems, Valley Formation, and Fluvial Landscapes

More than any other geomorphic system, the great networks of freshwater streams create the stage upon which much of the landscape is arranged. Their work is driven by water delivered by the atmosphere and set into motion by gravity. En route to the sea, streams erode the land and move billions of tons of sediment annually to the continental shelves. By this process the continents are unloaded and landforms take shape, but at the same time more or less, the continent are raised up by isostasy and tectonics, thereby enervating the system and driving it to carve the land and produce more sediment. Thus, stream systems stand, as it were, with one arm in the atmosphere and one arm in the lithosphere, and function in response to both.

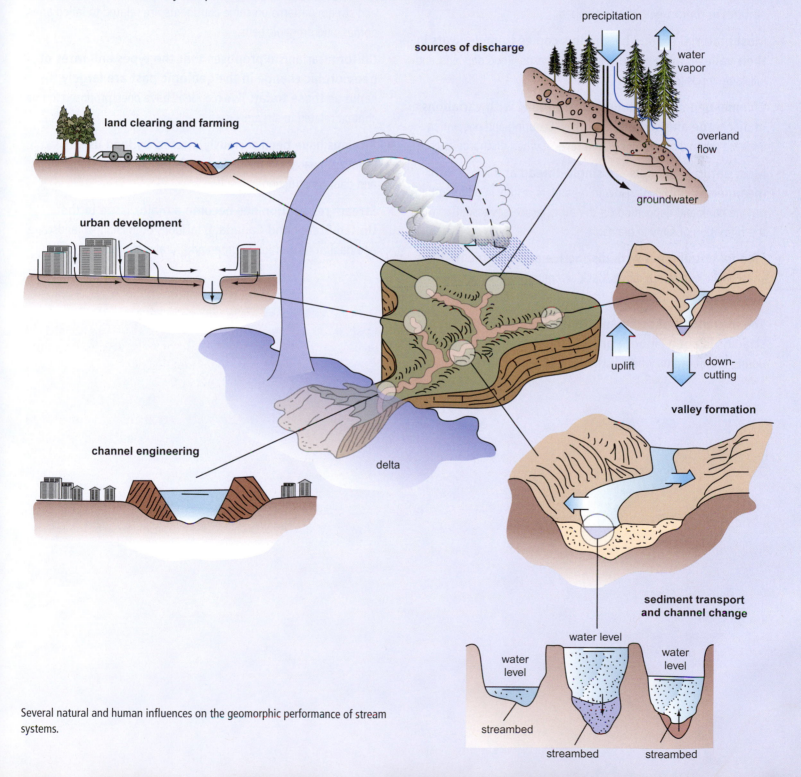

Several natural and human influences on the geomorphic performance of stream systems.

▶ **Streams are complex geomorphic systems comprised of three main phases.** Solar radiation and gravity are the primary sources of energy that drive these systems, but the landscape influences the distribution and amount of work they accomplish.

▶ **Stream discharge, velocity, and water depth change with distance downstream.** When water moves, potential energy is converted to kinetic energy, which produces bed shear stress.

▶ **Scouring accounts for the most stream erosion.** This process is closely related to turbulent flow and the motion of particles in contact with the streambed.

▶ **Most material eroded by streams comes from deposits in their valleys.** These include channel deposits, flood deposits, and hillslope deposits.

▶ **Channel geometry changes substantially with variations in discharge and sediment supply.** Scouring and deepening increase when discharge rises and reverse when discharge falls.

▶ **Most natural channels are single-thread and most form meandering patterns.** Flow in meanders results in undercutting on one bank and deposition on the other, processes essential to the formation of alluvial plains.

▶ **Floodplains abound with distinctive geomorphic features.** These include levees, oxbows, back-swamps, and terraces, all related to channel flow.

▶ **The movement of sediment from hillslopes to the sea via streams is not a smooth process.** Most sediment is moved during runoff events when discharge rises.

▶ **Most of the sediment exported from the continents is carried by the world's great rivers.** Of these rivers, those in Asia do the most work.

▶ **Watersheds are geomorphic systems whose landforms provide the basic infrastructure for the landscape.** The character and scale of landforms and landscape features generally changes from the upper to lower reaches of watersheds.

▶ **Plate tectonics and the breakup of large landmasses such as Pangaea can influence stream systems.** Watershed size and stream patterns on some continents are related to failed arms, domes, and orogenic belts.

▶ **Uniformitarianism proposes that the types and rates of geomorphic change in the geologic past are largely the same as those today.** Two big ideas have been proposed in this context related to stream systems.

▶ **Streams have been seriously altered by land use.** This includes impacts worldwide from soil erosion, sediment loading, and dam construction.

▶ **Stream restoration has become a major issue in the United States and Canada.** It includes both channel landscape and habitat restoration in urban and rural settings.

Review Questions

1 Using Figures 21.2, 21.3, and 21.4 briefly describe the three phases of stream systems in terms of inputs of water and sediment materials, and how this matter is transported through a stream system.

2 What are some of the potential impacts from global warming on stream systems? Why are we not sure about the exact outcomes on stream systems as a result of human activities?

3 Describe several of the downstream trends related to flow that are common to most streams. How do water depth and channel roughness influence streamflow velocity?

4 Using Figure 21.11, identify the different types of energy involved in a stream system, and offer an explanation for the fact that stream velocity does not accelerate wildly with distance downstream?

5 What is scouring and how does it generate sediment in a stream? What are the main sources of sediment in a stream system, how are bed load, suspended load, and dissolved load moved in streams, and where are these loads deposited?

6 What is the difference between aggradation and degradation, what natural and human activities contribute to each, and how is channel form affected by each?

7 Can you describe the pattern of streamflow in a meandering channel and how it relates to channel erosion and deposition and to the formation of an alluvial plain on the valley floor?

8 Referring to Figure 21.24, what are the common landforms found in floodplains? Which of these are produced by erosion? By deposition? How do you explain the origin of the terrace?

9 How do watersheds function as sediment systems, what are the sources of input, where is sediment stored and what are these sites (places) called? Is it accurate to think of the watershed sediment system operating in the fashion of a factory conveyor belt? Explain.

10 What is the general explanation for the huge contrasts in sediment production among the great rivers of the world? In particular, what accounts for the difference between the Amazon and Brahmaputra rivers? Where does the sediment from the great rivers end up and what is the nature of the resultant landforms?

11 Can you describe the character of the landforms and related landscape features in each of the three zones of a watershed system? Which of these is best described in terms of sediment production, sediment transport, and sediment deposition?

12 How has plate tectonics influenced the locations and patterns of large stream systems on the continents? In South America, how do you account for the fact that all the large rivers drain to the Atlantic Ocean. How does this relate to the difference in the geographic character in the continent's east and west coasts?

13 What are the main stages in the geographic cycle and how would you describe the general character of the landscape at each stage? In the dynamic-equilibrium concept, the landscape does not go through stages of development. What is the reasoning behind this?

14 How do you explain the assertion that in today's world humans are the dominant agent of change in stream systems? Describe some of the efforts now underway that are designed to lessen human impacts on stream systems.

Coastal Systems:
Waves, Currents, and Landforms

Chapter Overview

Our planet is laced with shorelines, more than a million miles of them in all. They are infinitely varied in form, composition, and geographic character, but all have one trait in common: a capacity for unending change. This change comes from a wide variety of sources including earthquakes, tides, volcanoes, glaciers, land use, and hurricanes, but one system stands above all these as the premiere coastal change agent: the geomorphic system of waves and currents. This system operates almost everywhere all the time and is responsible for doing the lion's share of the work in eroding coastal land and transporting and depositing sediment. Our mission here is to understand how that system works, what drives it, and how it is capable of shaping the landforms of this celebrated environment. We also want to know how all this relates to humanity, because humans are particularly fond of the sea coast. Each year more and more people crowd into coastal lands throughout the world. We begin with a brief examination of the various systems that move sediment along the coast and then go on to the master system of wind waves, currents, wave erosion, and coastal landforms.

Introduction

It was a glorious summer morning. We loaded our little boat for a trip along the Lake Superior shore. "What are we looking for, anyway?" Jim asked. "Shoreline features," I said without thinking. "Well, I took this geography course last semester and the professor talked about the way waves cut into shorelines leaving them with features like caves, cliffs, coves, and sea stacks (rock pillars). Do you remember that rugged piece of shore west of Five Mile Point where we used to hold family picnics? It's all sandstone with some cliffs, and I figure it might be fun to explore it by boat and see what we can find. What do you say?"

He glanced over at me and I knew what he was thinking, "Here we go on another of his wild excursions." But he never let his doubts show, which I considered a real credit to an 18-year-old who was also my brother.

Figure 22.1 (a) A section of the Lake Superior south shore showing the low sandstone cliffs, and (b) nearby sandy beaches where sediment collects.

We headed west around Powell's Point, and toward the open lake. By the time we could see Five Mile Point, a brisk wind had come up from the south and we began taking on water as white caps splashed over the stern. We headed for shore to bailout and make adjustments. Using some driftwood, we fashioned a splashboard over the back of the skiff. It worked, and within an hour, we reached the sandstone cliffs shown in Figure 22.1a.

The shore features were all I expected and more. Lake Superior's waves had cut deep into the sandstone strata, forming beautiful cliffs, irregular points and quaint coves. Beneath us the water revealed huge cracks running from the lake floor onto the shore. In places they had been eroded into wide chasms that disappeared into the cobalt blue of deep water offshore. But the treasures of the day were the caves. Hidden from view on the landward side of the shore by the overhang of the cliffs were two sets of caves. One was just above water level, the apparent work of storm waves, and the other was under water, clearly visible as we floated over it. The underwater caves were actually a network of small passages connecting barrel-sized pits, called potholes, that had been worn into the rock by stones whirled around in the turbulence of crashing waves. Later that summer we dived into the potholes and examined the cavities and the almost perfectly round stones that had done this work.

The adventure was exciting and finding textbook examples of shore features was immensely pleasing, but we failed to ask an important question. Where did all the eroded sandstone go? Obviously the huge sandstone blocks that littered the shore did not account for the great mass of rock that had been worn away. The answer was that most of the sandstone had been broken down into sand particles, which were carried away by the action of waves and currents and deposited in nearby coves and distant bays (Figure 22.1b). In other words, we were observing one part of a larger, three-part, geomorphic system involving erosion of bedrock followed by transportation and deposition of the resultant sediment.

22.1 The Geomorphic Systems of Earth's Coastlines

The coast is more than a line as the terms coastline and shoreline imply. It is band or zone of variable width where two sets of systems intersect, marine on one side and terrestrial on the other. On the marine side are four systems, wind waves, currents, tides, and tsunamis, each with the capacity to rub and pound against rock, soil materials, and river deltas, to transport and deposit sediment, and to create new landforms. But these systems vary radically in how much work they do and how and when they do their work.

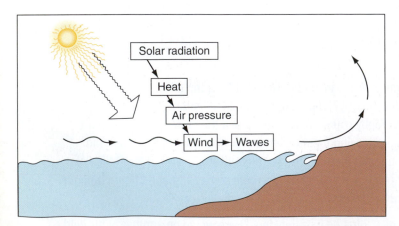

Figure 22.2 The system responsible for generating wind waves is powered by solar radiation.

Wind waves are far and away the most important system in terms of total work. Systems of wind waves operate on all shorelines, from small ponds and lakes to the largest ocean basins. As the name implies, wind waves are driven by wind, and since wind is a product of weather and climate systems and weather and climate are driven by solar energy, wind waves are also solar-powered (Figure 22.2). Where winds are strong and directionally steady, as within the great belts of the prevailing westerlies and the easterly tradewinds, wind waves are similarly strong and directionally steady. In zones frequented by the passage of turbulent weather systems such as midlatitude cyclones and tropical hurricanes, wave size, direction, and work output vary, and often radically so. Wind waves are the geomorphic workhorses of shorelines and a little later in the chapter we will say much more about them, including how they form, move, and do their work.

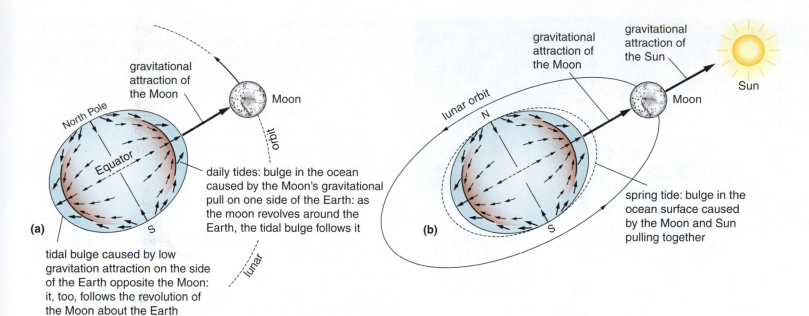

gravitational attraction of the Moon

North Pole

Equator

Moon

orbit

lunar orbit

(a)

daily tides: bulge in the ocean caused by the Moon's gravitational pull on one side of the Earth: as the moon revolves around the Earth, the tidal bulge follows it

tidal bulge caused by low gravitation attraction on the side of the Earth opposite the Moon: it, too, follows the revolution of the Moon about the Earth

lunar

gravitational attraction of the Moon

gravitational attraction of the Sun

Sun

N

Moon

(b)

spring tide: bulge in the ocean surface caused by the Moon and Sun pulling together

Figure 22.3 (a) A diagram illustrating the effect of the Moon's gravitational force in creating daily tides; (b) A similar diagram showing the combined effects of the Moon and Sun in creating spring tides.

Currents are another geomorphic system common to virtually all water bodies. Currents are river-like flows of water, capable of transporting sediment within the larger mass of a water body. They range in size from those as large as the Gulf Stream, which carries much more water than the Mississippi's biggest flows, to modest ribbons more or less the size of a two-lane road. Most currents are driven by wind waves. Therefore, we can expect large, strong current systems like the Gulf Stream within the belts of brisk, steady winds, a relationship which is particularly apparent in the belts of the prevailing wind systems. Where currents slide over the continental shelves and along landmasses, they work in concert with waves to pick up and transport huge amounts of sediment. We will also say much more about currents later in the chapter.

Tides are a global-scale system found only in the oceans. Tides take the form of very long, subtle waves characterized by gradual rises and falls in sea level. They are powered by the gravitational pull of the Moon and Sun and, unlike wind waves, tides occur with a regular and predictable periodicity related to the Moon's revolution about the Earth and the Earth's revolution about the Sun. *Lunar tides* peak simultaneously on two sides of the Earth as shown in Figure 22.3a, the side facing the Moon, where gravitational pull on the oceans is greatest, and the side facing away from the Moon, where the gravitational pull is weakest. It takes the Moon 24 hours and 50 minutes to orbit the Earth; therefore, lunar tides have an average period of 12 hours and 25 minutes.

The largest tides, called *spring tides*, occur semi-monthly during a full moon or a new moon when the Moon and Sun are aligned with the Earth and the two gravitational forces pull on the oceans together as shown in Figure 22.3b. Their counterparts, called *neap tides*, are the smallest tides, and they occur when the Moon and Sun are positioned at right angles to the Earth. In theory, the magnitude of tides should be greatest near the Equator and least near the poles, but in reality tide size varies considerably depending on water depth, coastal configuration, and land barriers. The primary work of tides in the coastal environment is that of pushing water and sediment in and out of bays and inlets.

Figure 22.4 The 2004 Sumatra earthquake produced a great tsunami wave shown here crashing onto the shore of India.

Tsunamis are waves generated by earthquakes when a big chunk of the ocean floor suddenly moves, and by volcanic eruptions when an explosion pushes up a great wall of water. Since the driving forces for earthquakes and volcanoes have nothing to do with the atmosphere or extraterrestrial gravity, tsunamis have no apparent connection to wind waves, currents, and tides. Unlike tides, there is no periodicity to their occurrence. And unlike wind waves, tsunamis have a relatively low frequency of occurrence, and we do not know when to expect them.

We do, however, have a general notion about where to expect them, and this is provided by the world map of tectonic activity (see Figure 19.32), which reveals that the prize for tsunami production goes to the Pacific Basin where earthquakes and volcanic activity are a continuous occurrence. When huge tsunamis, like the 2004 Sumatra tsunami, which produced waves 10 meters high, smash into coastal areas, they do an enormous amount of work in one sudden event (see Figure 22.4). But they, like massive floods, are infrequent and, in the long run, do less work in total than more frequent, moderate-sized wave events and certainly much less than wind waves.

System Interconnections and Work Habits: If we try to combine these four systems into a great working whole, the picture gets a bit complicated, but not impossibly so. First, wind-wave and current systems work together. They rise and fall and change direction with changes in wind velocity and direction. On most coasts, the wind-wave and current systems vary seasonally in both direction and magnitude. On the North American east and west coasts, for example, the winter systems are much stronger and do more work than the summer systems, and most large systems change direction from south-flowing in winter to north-flowing in summer.

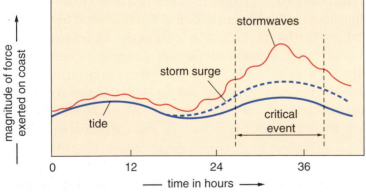

Figure 22.5 The concept of ocean systems combining to produce a high-magnitude event capable of exerting much greater force than any single system acting alone.

When large storms such as Atlantic hurricanes or Pacific cyclones cross the coast, the magnitude of waves and currents rises dramatically and the level of work output, measured by shores eroded and sediment moved, soars. Indeed, on many coasts the work of two or three large storms commonly exceeds several months' cumulative work by day-to-day wave and current activity. Add the force of a massive storm such as a hurricane coupled with elevated water levels from a high tide, and the work output is even greater as the graph in Figure 22.5 suggests.

This brings us to a critical concept in geomorphology, namely, the superimposition of multiple high-magnitude events at a single point in time and geographic space. Consider the probability of a huge hurricane striking during a spring tide. Or a hurricane and tsunami striking together. Granted, the probability is small, but the potential for work is enormous, especially if the point of intersection with the coast happens to be a soft, low-lying area such as the Mississippi Delta near New Orleans where the land's resistance to storm waves has been greatly reduced by the eradication of wetlands (see Figure 21.30) with the construction of pipeline corridors, navigation channels, and other facilities. And now take the concept one step further by imagining that the Mississippi River is at flood stage when the hurricane strikes, driving water levels in the delta even higher.

Time Trends in Coastal Systems: We know that coasts throughout the world have seen tremendous change over a wide range of time frames. Ocean basins have closed and opened over millions of years with plate tectonics, sea level has gone up and down by hundreds of feet over thousands of years with the rise and fall of glaciers, changing the entire configurations of coasts worldwide, and deltas and beaches have come and gone in as little as a few centuries with the actions of waves and currents. So it should come as no surprise that the coasts we know and the systems that shape them should change during our era on the planet, maybe even dramatically. And there are now some interesting prospects to consider. At the top of the list is the influence of global warming and here there are two major trends to mention: ocean warming and sea-level rise.

Hong Kong, an extreme in modern coastal development and crowding, but a trend being followed by scores of urban centers across the world.

We have already established that hurricanes (and other tropical storms) are fueled by water vapor from the sea, and that a rise in sea-surface temperature will drive more vapor into the air and fuel more and larger hurricanes (see Figure 8.36). Therefore, it is highly likely that coastlines traditionally visited by hurricanes and tropical storms, such as the American South and southeast Asia, will see more and larger storms as global warming advances. And while tropical storms are getting larger, sea level will also rise as the oceans expand with warming and the increased inflow of glacial meltwater, thereby allowing large waves to penetrate further onto shore than they are presently able to. The combination of more large waves with increased access to coastal lands is a sure formula for greater coastal flooding and erosion, increased sediment movement, and rapid change in the configuration of the shoreline. At the same time, world population is growing and crowding into coastal lands, creating the need for secure and stable coastlines. Coastal land uses and ecosystems will be dramatically affected as this scenario is played out and there is compelling evidence that it is already underway.

22.2 Wave Types, Origin, and Motion

To dislodge and move sediment, water must be set into motion and wind waves are the principal source of that motion in coastal waters. Accordingly, we devote most of the chapter to wind waves, the currents they generate, and the work they do together as a system.

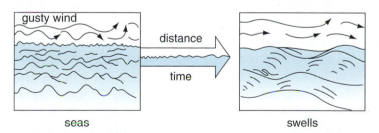

Figure 22.6 The basic difference in the topography of wave surfaces associated with seas and swells.

Generation of Wind Waves: The mechanism involved in wave generation by wind involves the transfer of momentum from moving air to the water surface. Though not well understood, the process of wave initiation probably involves surface-level air-pressure variations associated with wind gusts of various magnitudes and frequencies. The gusts differentially depress the sea surface producing a choppy sort of wave motion, called a *sea*, like that portrayed in Figure 22.6. These waves tend to be dragged along by the train of airflow gradually producing two simultaneous changes: a downwind motion and smoothing out of the wave form. The result is a more or less symmetrical wave called a *swell*, also shown in Figure 22.6. Swells commonly travel distances of hundreds of kilometers across the sea with little loss of size or change in form.

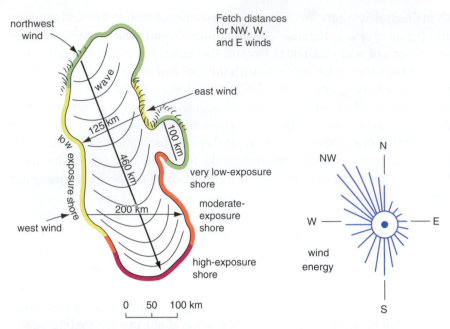

Figure 22.7 The concept of fetch related to wind direction. Here northwest winds not only have the greatest fetch but the greatest strength (as revealed by the wind rose), making the south end of the lake a high exposure shore.

Wave Size: The size that a wave can attain is controlled by four factors: wind velocity, wind duration, water depth, and fetch. **Fetch** is the distance of open water that a wind blows across a water body. Large values of all four factors are necessary to generate the largest waves: in other words, fast wind blowing from one direction for a long time over a great expanse of deep water. For any combination of these factors, there is a maximum wave size that can be generated. For small bodies of water like a lake or reservoir, fetch is the limiting factor on wave size as is demonstrated in Figure 22.7. Inland lakes with fetches less than 5 kilometers cannot generate waves much over a half-meter high even under the strongest winds. Large lakes such as Lake Superior with fetches of 300 kilometers can develop waves up to 8 meters high, whereas the oceans with fetches more than 10 times longer, can generate waves 30 meters high. For most water bodies, basin shape combines with regional wind patterns, especially storm winds, to create widely differing shore exposures to wave energy. High-exposure shores receive the full force of the largest waves, like the southern end of the water body in Figure 22.7, whereas low-exposure beaches are often protected by short fetches for strong winds.

Water depth influences wave size because a large part of a wave is actually under water. Every wave has a root, so to speak, defined by the motion of the water below the wave's surface form. When a wave rolls by, a large body of water under it rolls over, and the bigger the wave the greater the depth of water motion. The constraint imposed by water depth can be seen when a large wave crosses shallow water; its base drags on the bottom, and it slows down and loses energy. Therefore, a shallow water body is incapable of generating large waves and the shallower the water the smaller the maximum wave size.

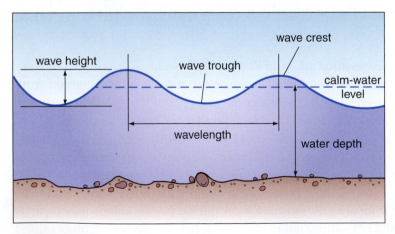

Figure 22.8 The basic terminology used in describing wave geometry.

Wave Forms and Motion: The description of a wave must include both its surface form and the fluid motion beneath it. First, the surface form (see Figure 22.8). The part of the wave that extends above the calm-water level is the *crest*, and that below the calm-water level is the *trough*. The *wavelength* is the distance from crest to crest, or trough to trough, and the *wave height* is the vertical distance between crest and trough. The typical slope of a wave, expressed as the ratio of wave height to the trough to crest distance, ranges from 1:12 to 1:25. Such long waveforms are stable – that is, do not break – and they are common in deep water. Nearshore the ratio changes as the wave drags on the bottom and its sides steepen. When a wave exceeds a slope of 1:7, it becomes unstable and fails, that is, falls over itself, or *breaks*.

Unlike the passage of water in a river, the passage of a wave in deep water does not result in the downwind transfer of mass, that is, in the flow of water. Rather, it is only the *wave form* that travels. Moreover, although water particles *do move* when a wave passes, the motion is only circular under the wave form, as illustrated

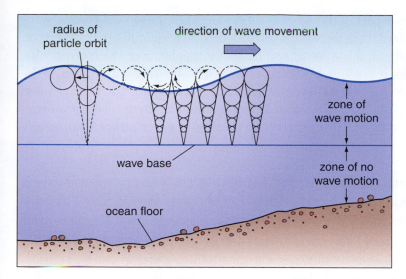

Figure 22.9 An illustration of the orbital motion of a wave as it passes from crest to trough. Below wave-base depth, water particles remain motionless with the passage of waves.

in Figure 22.9. This can be demonstrated by tracing the motion of a floating object such as a half filled bottle: with the passage of each wave it returns to the same point – if it is not pushed around by wind. Waves with such circular motion are termed **oscillatory waves**. The time it takes a wave to travel a distance of one wavelength is called the *wave period*. The *velocity* of a wave is equal to the distance traveled by a wave in one wave period.

Oscillatory Waves: The fluid motion of an oscillatory wave can be described as a series of circular orbits of water particles. The size (radii) of the orbits, shown in Figure 22.9, is greatest at the surface and decreases with depth. For larger waves, radii are larger and the wave motion extends to greater depths than for smaller waves. For all waves in deep water, there is a maximum depth, called **wave base**, below which they affect little or no water motion and are thus incapable of moving sediment. At depths below wave base, a submarine or diver, for example, should feel no rotating motion from waves passing overhead.

Various ratios relating wave geometry and wave motion have been worked out by scientists. The ratio most commonly used is related to wavelength. According to this measure, wave base is defined as one-half (0.5) wavelength, meaning that for waves spaced 10 meters apart, wave-base depth should be about 5 meters. Since wavelength increases with wave size, the huge storm waves on the deep ocean have the deepest wave bases, as much as 100 meters or more, but for most storm waves it is considerably less.

Translatory Waves: As a wave approaches the coast and first touches bottom, the water particles at its base, wave base, begin to rub on the sea floor. The shape of the rotational orbit of these particles now changes from a circle to an ellipse. Closer to shore in shallower water, the elliptical motion is compressed into linear motion as the water "slides" back and forth over the bottom with the passage of each wave as the inset diagram in Figure 22.10 illustrates. This produces friction, which slows the wave down and thereby compresses (shortens) the wavelength. But as the waves close their spacing, wave height increases.

Figure 22.10 The conversion of waves from oscillatory to translatory form as they pass from deep to shallow water. Nearshore the extreme tilt of the waves causes them to break and crash.

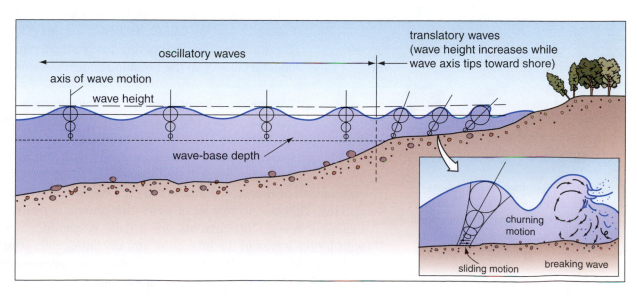

All the while the upper part of the wave is racing ahead of the slower-moving wave base causing a landward tilt in the wave's vertical axes. At a water depth between one to two times wave height, the wave tips so far that it topples over itself, or **breaks**. When a wave breaks it crashes headlong toward land heaving water against the bottom and the shore (once again see the inset in Figure 22.10). Since water mass is transferred toward shore, the wave no longer qualifies as an oscillatory wave. It is now called a **translatory** wave, meaning it transports water rather than merely rotating it.

Summary on Wave and Current Systems: The oceans are awash with waves and currents of a wide range of magnitudes and frequencies. All can be defined as systems, and they owe their origin and operation to several sources including volcanoes and earthquakes, but most are generated by wind. The size of wind waves increases with wind velocity, duration, and fetch and decreases with decreasing water depth. Waves and currents shape shorelines and their capacity to do this work changes over time with changes in climate, sea level, and other factors. How they do this work is the subject of the following sections in this chapter.

22.3 Wave-energy Distribution and Nearshore Circulation Systems

The energy of a wave consists of both potential and kinetic forms. Potential energy is represented by the mass of water displaced above the wave trough, and kinetic energy by the combined velocities of the water particles associated with wave motion. As waves grow and move faster, their energy increases and vice versa. When waves enter shallow water, they lose energy in erosion, in friction against the bottom, and in the turbulent motion of wave action. The energy of one 3-meter wave is roughly equivalent to a row of full-sized automobiles approaching shore side by side at full throttle. But how much force this wave exerts against the shoreline depends on the shape of the shore, especially offshore water depths.

Ramps and Cliffs: If the shore and offshore slope away gradually as in the form of a broad ramp, most of a wave's energy is spent in crossing this zone. The force of the wave is spread over the ramp from the point where the wave first "feels" bottom to the point where it finally washes up on the beach. This is illustrated in the area labeled *effective wave-energy zone* in Figure 22.11a. We can see evidence for this effect on swimming beaches. The wavelengths become gradually shorter with energy loss to friction (and turbulence in the breaking zone) as water depth decreases.

When a wave breaks part of its water mass is thrown toward shore. As wave after wave adds its water to the shoreline, the water level there literally becomes elevated, especially if the waves are large. This produces a seaward slope on the sea surface, which in turn sets up a gradient that forces the water back to sea, against the incoming waves. An important mechanism of this seaward flow is a curious current called a **rip current** (Figure 22.11a). This is a narrow jet of water that shoots seaward through the base of the incoming waves. Rip currents are a daily occurrence along some coasts, rising with larger waves of the afternoon and declining with smaller waves and offshore winds in the evening. In southern California, they are used by surfers to help propel themselves seaward against incoming waves.

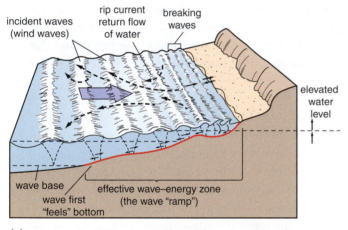

(a)

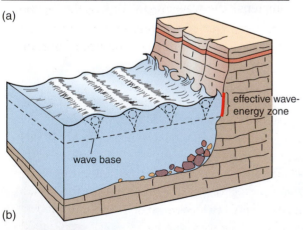

(b)

Figure 22.11 These diagrams illustrate the difference in the distribution of wave energy on (a) a shallow-water shore and (b) a deep-water shore. In (b) the full brunt of wave energy is more or less focused in a narrow zone near sea level.

The picture is quite different along coastal cliffs where deep water runs close to shore. Here waves undergo little energy attenuation before reaching shore, and instead smack the shore nearly full-force (see Figure 22.11b). As a result, a massive amount of energy can be focused on a relatively narrow zone at the shoreline. Because the surface area of this zone is so small, not all of a large wave's energy is expended and a lot is often left over. This excess energy is converted into new wave forms, such as reflected waves and edge waves that move along the shore or back toward sea.

Wave Refraction and Energy Redistribution: From a geographical standpoint, we need to know why certain parts of a coastline are the focus of erosion and others the sites of deposition. This leads us to the question of the spatial distribution of wave energy in coastal waters. To begin with, no coast has truly uniform offshore topography; therefore, as a wave approaches shore, some segments touch bottom before others do, and those that touch first lose energy first. Therefore, the velocity of the wave is reduced differentially with the parts over shallow water moving slower than the parts over deep water. This produces bending of the wave (measured along the crest line) resulting in a reorientation of wave energy relative to the shoreline.

The process of wave bending, called **wave refraction**, is analogous to refraction in light waves as they pass down through the atmosphere. As the atmosphere grows denser toward Earth, the more light refracts. And so it is with water waves, but the controlling variable is water depth. How far offshore a wave begins to refract depends on wave size and water depth. These two factors determine the location where the wave first touches bottom and begins to slow down. For big waves this location is farther off shore than it is for smaller waves because of the difference in the depth of wave base.

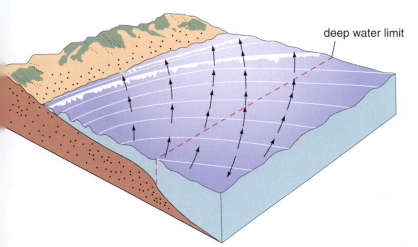

deep water limit

Figure 22.12 The concept of wave refraction. Refraction increases with decreasing water depth.

At depths greater than one-half of the wavelength, the wave functions as a deep-water wave, and no refraction takes place. At a water depth less than half of the wavelength, velocity begins to decline, and the wave refracts, first slightly, then increasingly so with decreasing water depth as the lines in Figure 22.12 show. This process defines the shallow-water zone where waves gradually change their direction of approach. As the wave gets closer to shore, it swings around until the direction of advance is nearly perpendicular to the contours of bottom topography, but not quite.

The redistribution of wave energy as a result of refraction can be defined by drawing *orthogonals* (lines perpendicular to the wave crest) for a group of waves traveling across the shallow zone. The relationship to the shape of the coastline is plainly evident as Figure 22.13 shows: wave energy is *convergent* on headlands and *divergent* in embayments, meaning that wave energy is greater than average where the land protrudes into the sea and less than average where the sea protrudes into the land. It follows that the distribution of coastal erosion and deposition follows this pattern with thin beaches (or none at all) on headlands and thick beaches in embayments. In other words, one is giving up sediment while the other is collecting it.

Along straight coasts, wave energy is more or less evenly distributed, but oriented in the direction of the approaching waves. Although refraction can reduce the angle between a wave crest and the shoreline to as little as 10 degrees, rarely does a wave

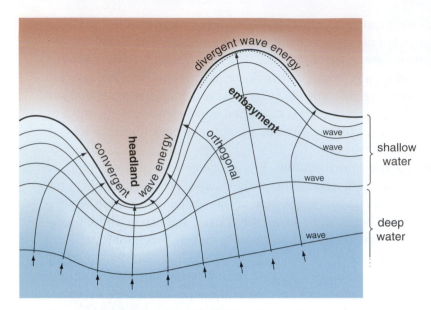

Figure 22.13 The pattern of wave refraction along an indented shoreline showing wave energy converging on the headland and diverging or weakening in the embayment.

Figure 22.14 Contrasting longshore-system configurations: (a) a long, continuous system on a long, smooth coast; (b) a more segmented system along a geographically irregular coast. The current is marked by the light-toned water.

approach the shore straight-on. The fact that the force of waves is exerted in a direction oblique to the shore is very important in terms of sediment transport because when particles are lifted off the bottom by wave turbulence, they tend to be carried along the shore with the flow of waves and associated coastal currents, called longshore currents. **Longshore currents** are shallow water currents that flow parallel to the shore in the direction of wave motion.

Current Systems Great and Small: With the exception of currents driven by tides, virtually all currents are the products of wind systems and their waves. The large current systems we see far offshore in the deep oceans, such as those that form the great gyres of the Atlantic and Pacific, are driven by the prevailing wind systems. Despite their size, they are less significant than longshore currents in terms of geomorphic work, but they are not insignificant. Deep-water currents are responsible for long-distance transport of fine sediment. They pick up suspended sediment, both organic and inorganic, near the mouths of certain large rivers and carry it far out to sea. Here the organic fraction feeds ocean ecosystems while the remainder settles out on the ocean floor (abyssal plains) to become part of the sea floor spreading system that we talked about in Chapter 18 in connection with plate tectonics. As sea floor sediment, this material is eventually moved into a subduction zone and at length transformed back into rock and added to the crust.

The Longshore System: Longshore currents are driven mainly by the force of wave action in the coastal waters. Since waves are powered by wind, longshore currents also respond to wind systems and vary with prevailing winds, seasonal winds, and winds associated with weather systems such as midlatitude cyclones. The magnitude of these longshore currents is governed by the velocity, direction, and duration of the coastal waves as well as by offshore topography and coastline configuration. Under a strong, steady flow of waves running oblique to a long, smooth coastline, like that in Figure 22.14a, longshore currents can reach velocities of 1 meter or more per second and extend more or less uninterrupted hundreds of kilometers along the coast.

Irregular coastlines, on the other hand, present a different picture. Where the coast is dominated by headlands, deep bays, and islands, the longshore system tends to be partitioned into many, smaller cells or subsystems is suggested by Figure 22.14b. These cells tend to expand and contract with the passage of storms and fair-weather systems. During periods of light winds, many small cells

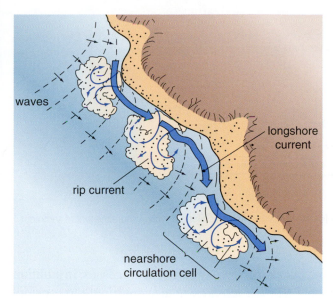

Figure 22.15 Nearshore circulation cells involve the interplay of three systems: waves, longshore current, and rip currents.

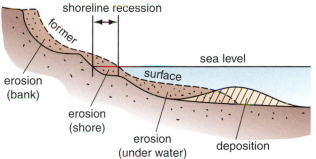

Figure 22.16 A cross-sectional diagram illustrating the three zones of coastal erosion: bank, shore, and underwater. Shoreline recession is the landward retreat of the water's edge.

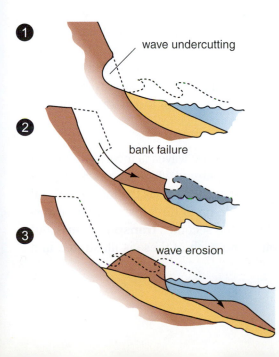

Figure 22.17 The process of bank erosion by wave undercutting and bank failure leading to shoreline recession.

dominate the coast, but during periods of heavy winds, these cells are overridden and integrated into a larger working whole.

Nearshore Circulation Cells: If we combine the various flow systems operating within the coastal zone, namely, waves, longshore currents, and rip currents, the result is a roughly circular pattern of water movement called a **nearshore circulation cell**. These cells are not a universal coastal phenomenon, but tend to occur on certain shores, usually long, sandy ones under the force of sizeable swells. At the heart of nearshore cells are rip currents. As we noted above, rip currents are driven by the pressure of elevated water along the shore and, when they shoot seaward, they entrain sediment from both incoming waves and longshore currents. When a rip current reaches a short distance offshore as shown in Figure 22.15, it slows down and fans out, giving a large part of its sediment load back to the wave and longshore systems. Thus, the motion of the sediment under such conditions is roughly circular, a combination of longshore, offshore, and onshore movements.

22.4 Wave Erosion, Sediment Transport, and Coastal Landforms

Wave erosion may take place underwater, at the waterline, or above it, and it may or may not result in *recession* of shoreline. **Recession** is defined as landward retreat of the shoreline and it can also be caused by a rise in water level, a subsidence of land, or both. Erosion is measured in terms of the volume of material removed, whereas recession is measured in terms of the distance of landward displacement of the shoreline (Figure 22.16).

Wave Erosion and Erosional Features: Wave erosion takes place when the hydraulic pressure of the moving water is sufficient to dislodge material. The hydraulic pressure applied by waves is greatest where the wave velocity and mass are greatest, which is in the zone where waves break. How much erosion a given wave can produce depends not only on wave force but also on the resisting strength on the material it strikes. Generally, bedrock is most resistant to waves, and loose (unconsolidated) sediment is least resistant. This difference is evident along shorelines composed of bedrock cliffs and sandy banks that are being attacked by storm waves. The cliff may be cut back little, if at all, whereas the bank can be cut back rapidly – as much as several feet per hour in big storms.

Bank erosion is a three-step process, which is illustrated in Figure 22.17. The first step is undercutting of the toe. The second step is slope failure as the bank collapses onto the beach. In the third step, waves over wash the heap of debris removing all but the heaviest particles, usually the large boulders. The fine particles are moved both downshore and offshore into deeper water.

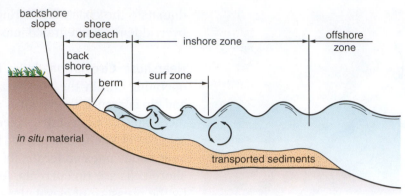

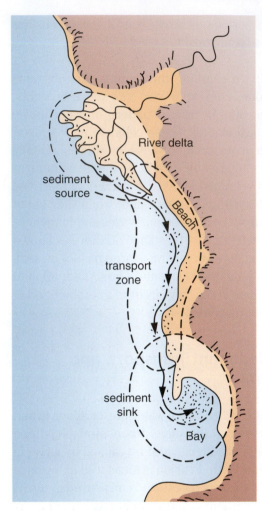

Figure 22.19 A longshore or drift sediment system is made up of three main parts: sediment source, transport zone, and sink.

Sea stacks. Remnant pillars of bedrock along an eroding shoreline.

Figure 22.18 The standard terminology for the zones and features along a typical shoreline where sediment like beach sand makes up the shore.

Erosion of bedrock is not only slow, but also involves additional processes. In addition to hydraulic pressure, *corrasion* and *solution weathering* also contribute to erosion. **Corrasion** is abrasion by stones – similar to scouring in stream channels – in which rocks of various sizes are rolled, bounced, and hurled by storm waves against solid rock, breaking fragments free. Solution weathering takes place when minerals are dissolved into seawater.

The shoreline features formed by wave erosion are familiar to most of us. Erosion of bedrock often results in the formation of a **sea cliff**, a vertical or near vertical rock face like the one in Figure 22.1. If a cliff is undercut near water level a *wave-cut notch* is formed. Where bedrock has variable resistance to wave erosion, an assortment of interesting features may form, including *sea caves* (such as those described in the Introduction) and *sea stacks*. Sea stacks are pillars of resistant rock left standing offshore.

Along soft (non-bedrock) shorelines, most of the features are formed in sediments that are in transit along the coast. Only the **backshore slope**, the steep bank or bluff behind the shore, is composed of *in situ* (in place or native) material (see Figure 22.18). The **shore**, or **beach**, stretches from the foot of the backshore slope to the shallow water just beyond shore. The width of the beach varies enormously from as much as 100 meters on depositional shores to as little as a meter or less on erosional shores. On wide shores the area between the *berm*, the highest part of the beach, and the foot of the backshore slope is designated the *backshore*. Seaward of the shore is the *inshore zone*, where waves break. Beyond the inshore is the *offshore zone*, where waves behave like deep-water waves. Longshore currents are strongest in the inshore zone, but they extend into the adjacent part of the offshore zone as well.

The Sediment Transport System: The material eroded by waves is incorporated into a train of sediment, called a **longshore** (or **drift**) **system,** which moves along the shore under the driving forces of wave and current action. This is the master coastal system and, like all geomorphic systems, it is made up of three parts: input, transport, and output. The first is a **sediment source**, the second a **transport zone**, and the third a **sediment sink** (depositional zone) and each of these is illustrated in a representative example in Figure 22.19.

The source of about 90 percent of the sediment in the longshore system comes from streams and rivers (the remainder comes from wave erosion of the land). Each year runoff systems deliver around 15 billion tons of sediment to the world's coastlines. Most comes from the great rivers, 10 of which produce more than 100 million tons each (see Figure 21.26). A large part of this material is picked up by waves and currents, transported along the coast, and is deposited in sediment sinks. Sinks are simply places conducive to deposition such as deep bays and submarine canyons. The movement of sediment along a coastline is called **littoral transport** or **longshore transport**, and the material moved, such as sand and silt, is known collectively as *littoral drift*. Many of the world's great beaches such as those of the US Gulf Coast are longshore transport zones and most are tied to one or more river sources.

Littoral transport involves several processes and one that virtually all of us have observed is **beach drift**. This component of longshore transport takes place on the beach itself and is driven by the swash and backwash of wave action. *Swash* is a thin sheet of wave water that slides up the beach face; *backwash* is its counterpart in return flow (Figure 22.20). Because most waves strike the shore at an angle, swash flows obliquely onto the beach. Backwash, on the other hand, flows more perpendicularly to the shoreline. Together, swash and backwash produce a ratchet like motion of water and sediment, resulting in a net downshore movement. Much of the sediment, especially the pebbles and larger particles, moved in this fashion is rolled along the beach and is referred to as **bed load**, the same term used for similar sediment in streams.

Beyond the beach, where waves are breaking, is the main body of the longshore system. Here the turbulence created by the breakers lifts sediment high into the water column whereupon it is carried downshore by longshore currents. Sediment moved in this fashion is called **suspended load** (because it is carried in the water column) and it constitutes the bulk of the littoral drift along most coasts. Most of it is sand and it moves in a leap-frog fashion as the churning motion of waves lifts it up, currents move it laterally, and gravity brings it back to the bottom. The fine sediments (silt and clay) raised by wave action or introduced to the system by rivers or tides are generally carried far beyond the breaking zone and into deeper water, maybe even into a system of deep-water currents and onto the abyssal plains.

Rates of Longshore Transport: Rates of longshore transport vary with wave energy, the angle at which waves approach the shore, the size and availability of sediments, and certain other factors including coastal ice and vegetation. Along a sandy shoreline that is free of ice, bedrock, and other controls, longshore sediment transport is directly proportional to the flux of wave and current energy in the longshore system. If wave energy is high, it is not uncommon to find a million or more cubic meters of sediment moved by the longshore system per year. But rates vary tremendously, even within relatively short distances on one coastline. One of the most striking differences in North America, for example, is found in southern California: at Oxnard Plain Shore, north of Los Angeles, longshore transport is ten times greater than it is at Camp Pendleton, south of Los Angeles.

swash

backwash

swash and backwash

swash zone

Figure 22.20 The processes of swash and backwash are responsible for the beach drift is part of the sediment movement in the longshore system.

Change in the color of breaking waves reveals the change in sediment content near the shore.

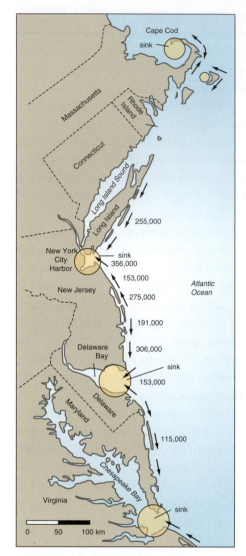

On the east coast of the United States, rates range from 100,000 to 380,000 cubic meters per year on shorelines exposed to the open sea (Figure 22.21). On sheltered shores, on the other hand, such as those behind the offshore islands that line much of the East Coast, transport often amounts to less than 25,000 cubic meters per year. In the upper Great Lakes, longshore transport averages between 50,000 and 100,000 cubic meters per year for most sandy shorelines. Along Arctic shorelines it is far less, on the order of 5000 to 10,000 cubic meters per year, owing to the presence of grounded ice in the shallow-water zone for most of the year, a condition, incidentally, which may change dramatically with warming of the seas and retreat of Arctic sea ice.

The longshore transport figures given in Figure 22.21 represent the balance of sediment transport in two directions along the coast: one the primary direction, the other the secondary direction. In Figure 22.22, these two directions are represented by arrows q_P and q_S. By subtracting the primary from the secondary transport value, we derive **net sediment transport**:

$$\text{Net sediment transport} = q_p - q_s.$$

If we add these two quantities together, we get **gross sediment transport**, which represents the sum total sediment moved by the longshore system in both directions over a year. Let us think for a moment about what these two values mean. Where net sediment transport is small but gross transport is large, sediment is merely being shifted back and forth. In other words, the same sediment mass, more or less, is being reworked year after year. This is often the case where wind systems shift seasonally. Where both net and gross are large, the longshore system tends to be more unidirectional. This may be the case in the northern Indian Ocean, shown in Figure 22.23, which is dominated by the strong northward flow of the summer monsoon.

Figure 22.21 Some longshore transport values and directions on the US East Coast. The values represent net annual rates in cubic yards of sediment. Chesapeake Bay, Delaware Bay, and New York City Harbor serve as major sinks.

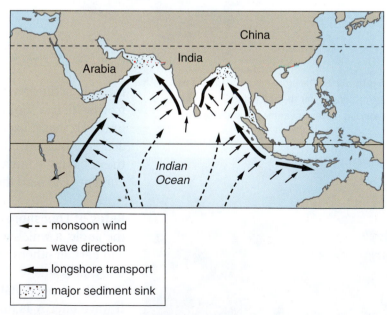

Figure 22.23 Longshore transport directions in the northern Indian Ocean are driven by monsoon circulation. In summer the flow is strongly northward with major sinks at the heads of the Bay of Bengal and the Arabian Sea.

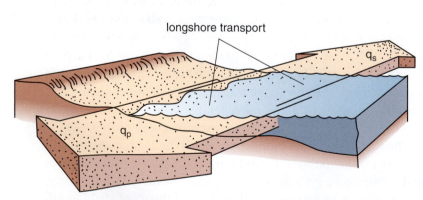

Figure 22.22 Measuring annual sediment movement along a shore. Gross transport ($q_p + q_s$) is total sediment.

Onshore-Offshore Transport: Although longshore transport is the dominant component of the coastal sediment system, there are several minor components or subsystems that operate on most coasts. One of these is onshore–offshore transport, that is, sediment transport perpendicular to the shore. The amount of sediment moved in this fashion is very small compared with longshore transport, but it is important to beach topography. The onshore movement is brought on by low-energy waves, usually the summer wave regime, and is characterized by a shoreward migration of sand bars. As sand is added to the beach, the beach expands seaward. With high-energy waves, which are typical of the winter wave regime in the midlatitudes, the trend reverses. Storm waves remove sediment, the beach face steepens, rip currents develop, and sediment shifts offshore.

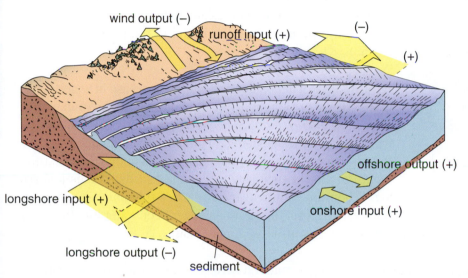

Figure 22.24 Detailed computation of the sediment mass balance for a segment of shoreline involves not only the longshore system but onshore–offshore sediment movement as well as inputs from and losses to the land.

Sediment Mass Balance: For any segment of coastline, such as a swimming beach or a park, we can determine whether it is gaining or losing mass by defining the total sediment inputs and outputs. Imagine the shore as a long, flat box into which and from which sediment is more or less constantly moving as shown in Figure 22.24. To calculate the mass balance, we need to measure all losses and gains, beginning with longshore transport in (+) and out (−) and then onshore (+) and offshore (−) migration of sand bars as well as losses to erosion by onshore winds (−) and input by runoff (+) and other processes, such as sediment pushed down the backshore slope by foot traffic. The single measure of the sediment balance for a segment of shore is the change in volume of sediment in the box over a year or longer time interval.

Examples of changes in sediment mass balance abound and most of the extreme ones are often related to some sort of human intervention in the coastal sediment system. For example, when water-supply reservoirs were built on certain streams in the Los Angeles region in the first-half of the twentieth century, sandy beaches on the Pacific began to shrink because the reservoirs were trapping the sand that naturally fed the beaches. This trend represented a negative sediment mass balance. The opposite has occurred in countless places with the construction of large piers or breakwaters near harbor entrances that block the outflow of longshore sediment causing beaches to grow on the upshore side as sediment builds up. On the downshore side, beaches dry up, of course, because they are deprived of sediment input. These examples and many others point to the need of seeing beaches as products of sediment systems in which the sand, like the water in a river, is on its way through and is not a permanent fixture.

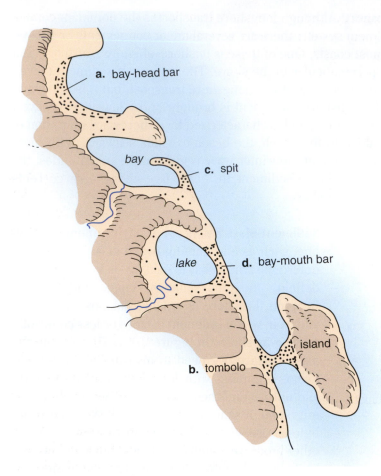

Figure 22.25 Common depositional landforms include spits, bars, and tombolos. All form where the longshore system loses energy and drops all or part of its load.

Here the stream of longshore sediment is interrupted by two large structures which keep waves and currents from mobilizing sediment.

Depositional Processes and Landforms Built of Sediment: In order for the longshore system to deposit its sediment load, it must lose energy. This can happen in several ways. One of the most common is *energy diffusion* related to wave refraction inside a bay. Sediment transported down the sides of the bay will accumulate at the landward end, or head, forming a broad bar of sediment called a **bay-head bar** (see Figure 22.25a).

Another cause of energy loss and sediment deposition is related to an island situated close to the shore. The island serves as a barrier to incoming waves shielding the shore from their full force. When the longshore train reaches this *energy shadow*, it slows down and drops part of its load. In time, enough sediment may collect behind the island to form a natural causeway, called a **tombolo**, linking the island to the mainland (Figure 22.25b).

A third cause of deposition is a sudden increase in water depth in the path of the longshore sediment train. As the train passes over this "*energy pitfall*," bed-load sediment falls below the wave base and is deposited. This commonly happens where the orientation of a shoreline changes abruptly such as at the mouth of a bay or estuary where the coastline takes a sharp jog. When the sediment train reaches this point, its momentum carries it past the shore and over deep water into the outer bay where it drops its sediment load.

Like a line of dump trucks building a roadbed across a valley, this process builds a slender finger of sand, called a **spit**, from the shore into the mouth of the bay (Figure 22.25c). If such deposits build all the way across the bay's mouth, a feature called a **bay-mouth bar** is formed (Figure 22.25d). Spit and bay-mouth-bar development are possible only where bays are fairly shallow and narrow and the rate of longshore transport is substantial. If a bay happens to be wide and deep and/or flushed by strong tidal currents, spit development is usually weak and bay-mouth-bar development very unlikely. Where bay-mouth-bar formation does take place, however, the bay is closed off from the ocean and transformed into a lake. Eventually, the lake's saltwater is displaced by freshwater from inflowing streams and groundwater.

The longshore train can also be diverted into deep water by an opposing current, or by a reef, pier, of breakwater, or the train may be intercepted by a submarine canyon that extends from deep water into the nearshore zone. Nearshore canyons are especially effective in capturing longshore sediment. Once sediment gets into a canyon head, it slides downslope to the outer edge of the continental shelf. The California coast has many such canyons and they are important conduits feeding longshore sediment to the deep ocean floor.

Summary on Circulation, Sediment, and Coastal Landforms: The motion of waves in shallow water sets up currents that flow along the coast. Together the waves and currents form longshore systems capable of moving huge amounts of sediment from source areas such as deltas to sinks such as bays. Whether a segment of shoreline in this system is retreating or advancing seaward depends on the relative balance of sediment inputs and outputs and these often change with the seasons. Longshore systems also reconfigure the coast, creating a wide variety of landforms that generally fall into two classes, erosional where energy is focused and deposition where energy is diffused. We now step back and examine a more sweeping question, the connection between coastlines and plate tectonics.

22.5 Coastal Development, Plate Tectonics, and Sediment Supply: The Big Picture

The interplay of the tectonic system and coastal geomorphic systems is staged at a grand scale with the formation of ocean basins and continental masses. It begins with the formation of new coastlines as a part of plate tectonics when the lithosphere and crust break into continents and ocean basins. As masses of continental lithosphere break apart, as happened with the breakup of Pangaea, new coastlines are created. For each continental landmass, two principal plate edges or coasts are formed: (l) a leading edge in the direction of plate movement; and (2) a trailing edge on the opposite side, and both are shown in Figure 22.26.

The leading edge forms an **active coast** so called because this is the side where the continental platform is being rammed against a neighboring oceanic plate. The western edges of North America and South America are excellent examples of active coasts. They are characterized by volcanic activity, subduction, thrust faulting, and related tectonic processes and it is these processes, rather than geomorphic processes, that dominate the coastline and determine much of its geographic character. Coastal erosion rates are typically high on active coast. But sediment buildup and continental-shelf development are modest because much of the sediment is lost in subduction and ongoing mountain building.

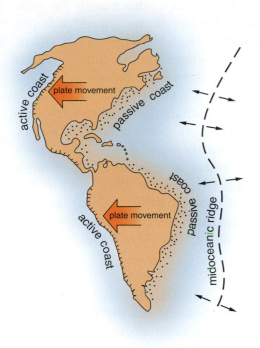

Figure 22.26 Coastlines at the continental scale are products of plate tectonics with active and passive coasts corresponding to the leading and trailing borders of the plates.

The trailing edge of the continent, by contrast, forms a **passive coast**, which is tectonically inactive (see Figure 22.27). Its geographic character is shaped largely by geomorphic systems and depositional landforms. It is helpful to review the origin of passive coasts. This coast begin as one side of a rift valley (see Figure 18.29) when the new ocean is just a narrow slit, as the Atlantic was between Africa and South America 200 million years ago. As the rift valley widens with seafloor spreading, the continental edge (for example, the east coast of South America) is moved further and further from the active separation zone along the mid-oceanic ridge. As sediment is added to the passive coast, the old rift structure is buried under thousands of meters of sediment.

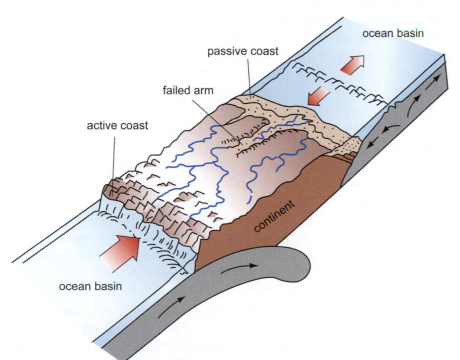

Figure 22.27 Active and passive coasts on the leading and trailing sides of a tectonic plate. Drainage and sediment systems favor the passive coast.

Big Rivers and Failed Arms: The sediment supply to passive coasts is often very large because some of the world's largest rivers terminate there. The combination of passive coasts and great rivers is not simply geographic coincidence. It is explained by the formation of failed arms in triple junctions in the opening stages of rifting and continental separation, which we examined in Chapter 18. The failed arm forms a structural trough, extending

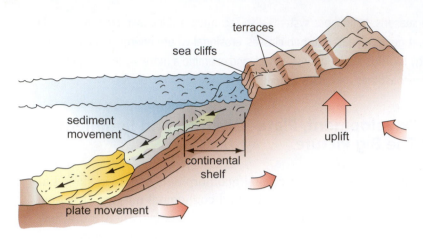

Figure 22.28 Common features of shorelines on the active edge of a continent including terraces marking upraised beaches.

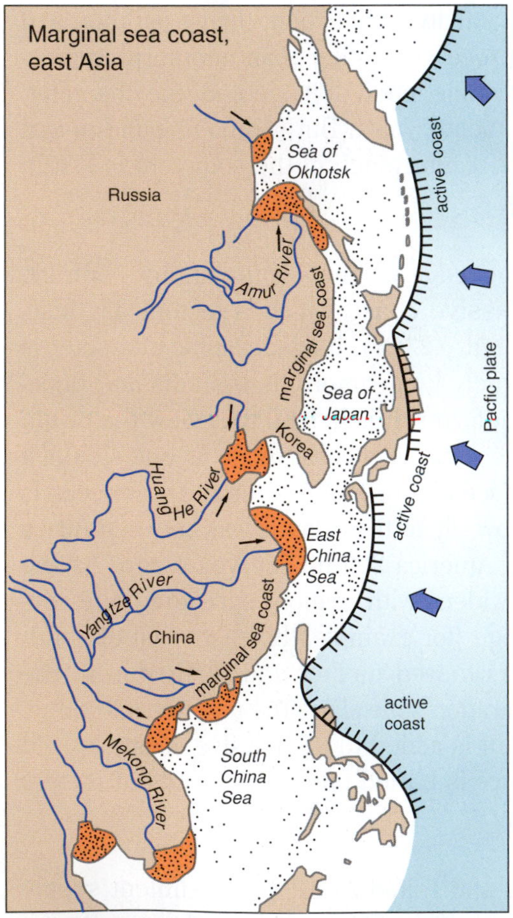

Figure 22.29 Marginal ocean basins such as the Sea of Japan and East China Sea form between the island arcs bordering subduction zones and the continental mainland. Over geologic time these basins fill with sediment building out the continental shelves.

into the continental platform, and in this trough the main valleys of some of the world's largest rivers form, including the Amazon, Niger, and Mississippi. As Figure 22.27 shows, these great rivers reach far into the continents collecting water and sediment and conducting them to the coast.

On the other hand, rivers discharging to active coasts are generally smaller because the orogenic belt along the leading side of the continent limits the size of their watersheds. South America illustrates this arrangement. Three great river systems, the Amazon, Orinoco, and Rio de la Plata, flow eastward draining over 75 percent of the continent, whereas no large rivers flow westward. These large rivers are fed by the easterly tradewinds whereas the much smaller west-slope rivers lie in the rain shadow of the Andes, an area of coastal desert.

Implications for Coasts at Continental Scales: Although the same geomorphic systems operate on active and passive coastlines, the landforms that develop along each are markedly different. On active coastlines, mountain ranges often rise more or less directly from the sea and, owing to deep water immediately offshore, they are subject to the nearly unmitigated force of large ocean waves (see Figure 22.11b). The result is a strongly erosional shoreline characterized by high topographic relief, as shown in Figure 22.28, and landforms such as sea cliffs. A large share of the sediment produced by coastal erosion, along with much of that contributed by streams, is lost to deep water rather than accumulating near shore.

Active coasts are also marked by marine terraces in many areas. These are bench-like landforms cut into the seaward slopes of coastal mountains by wave erosion when the land was positioned at a lower elevation. As the coast was uplifted by tectonic forces, the terraces were raised above sea level. Coasts that exhibit a series of terraces have been uplifted many times with each terrace corresponding to an episode of past tectonic activity.

Active coasts may also develop marginal basins that form between the continental coast and chains of arc islands offshore. The island arcs form from volcanic activity along subduction zones some distance seaward of the continental coast, giving rise to enclosed or partially enclosed basins between the chain of islands and the coast. The east coast of Asia is a clear example of this situation where a string of seas, notably, the Okhotsk, Japan, East China, and South China seas, have formed in marginal basins (Figure 22.29). On the outer (oceanic) side, these basins are active; on the continental side they are passive and the coast there is called a *marginal sea coast*. The passive sides are fed by some of Asia's largest rivers such as the Yellow and Yangtze, which eventually fill the basins with sediment.

Passive coasts are dominated by broad coastal plains leading offshore to wide continental shelves. The southeastern coast of the United States is a good example of such a coast. Both the coastal plains and the continental shelves are products of the large supply of sediment delivered to the coast over millions of years by rivers such as the Mississippi and which now hides the old rift structure that created the ocean basin. The diagram in Figure 22.30 shows the inactive (passive) faults that formed one wall of the rift. Far above it on the surface, water depths are shallow and depositional

Figure 22.30 Massive sediment accumulations on the passive side of continents produce large continental shelves, which bury the old rift structure that formed the original coastline.

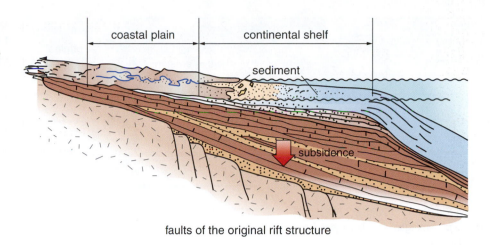

faults of the original rift structure

landforms abound on these sediment-rich coasts, including large river deltas, barrier islands, wide beaches, and sand dunes. Unlike active coasts, bedrock is mostly absent on passive coasts. As a result, these coasts tend to be softer and more susceptible to rapid change under the onslaught of storm waves such as those from hurricanes. Unlike active coasts, passive coasts often subside under the weight of sediment, which in turn drives ocean water up river mouths.

22.6 Classes of Ocean Coasts

Within the broad framework provided by active and passive continental edges, we can define different classes of coasts. These classes fall under two general headings, called emergent and submergent, based on long-term trends in the balance between changes in sea level and land elevation. **Emergent coasts** are characterized by land rising from the sea, resulting in elevated shorelines. **Submergent coasts** are characterized by falling land elevation and/or rising sea level, resulting in sunken or drowned shorelines. Below we describe nine of the more prominent coastal classes of the emergent and submergent types. Illustrations of each are provided in Figures 22.31 and 22.33. Each is referenced by letter in the discussion.

Emergent coasts: The most prevalent emergent coast is the *tectonic* class (Figure 22.31a). Tectonic coasts are formed mainly along active continental edges where mountain building by thrust faulting, volcanism, and related processes is driving the land up. Because uplift is usually episodic, tectonic coasts, as we noted above and in Figure 22.28, are often left with pronounced wave-cut terraces, marking the different stages of uplift. Tectonic coasts may also be *volcanic* along the mountainous edge of the continental and on islands such as the Aleutians and the Japanese archipelago (Figure 22.31b).

Most of the world's volcanoes form in the ocean basins as seamounts, island arcs, and ridge islands. Their growth takes place mainly under water where a huge base must develop

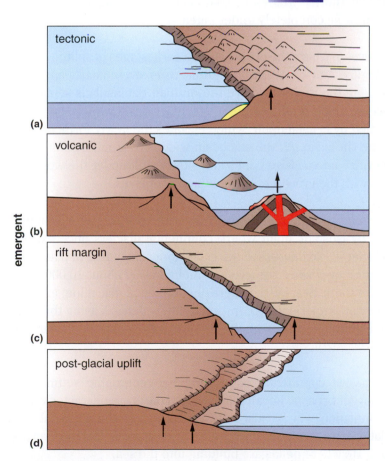

Figure 22.31 Principal types of emergent coastlines: tectonic, volcanic, rift margin, and post-glacial uplift.

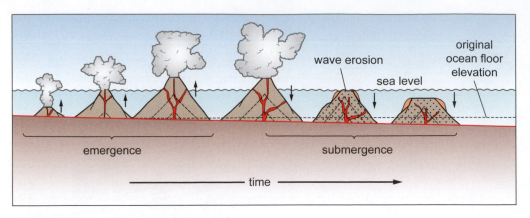

Figure 22.32 From emergent to submergent. The growth and decline of a volcanic island.

before they can reach the sea surface. Once an island forms and the volcano continues to grow, an emergent coastline forms, as shown on the left side of the diagram in Figure 22.32. But as the volcanic mass grows, it often becomes so heavy that the crust under it subsides isostatically and the island is transformed into a submergent coast, shown on the right side of Figure 22.32. This process often continues long after the volcano has become inactive, lowering it below sea level.

The Hawaiian volcanic ridge (chain) is a good example of island emergence followed by submergence. The Hawaiian chain has been forming for tens of millions of years from a volcanically active hotspot in the Pacific Plate. But as islands form, the plate, which is continually moving, drags them off to the northwest. We looked at this back in Chapter 18 in Figure 18.32. Thus there is a great string of islands and former islands stretching across the Pacific Basin. The newest islands, such as the big island of Hawaii, are volcanically active and emergent, in their early stages of development. As they gain mass they begin to subside, often while they are still emerging with the buildup of lava. When lava production stops, they continue sinking as the older islands have northwest of Hawaii. Many of these islands are more than 20 to 30 million years older and now lie completely underwater.

Emergent coasts also develop where the crust is in the early stages of splitting apart from plate tectonics (Figure 22.31c). We will call these *rift-margin* coasts and good examples are formed along the Red Sea where the Arabian Peninsula is separating from Africa. This process, which produces great rift structures in which the young sea forms, is accompanied by crustal uplift, which elevates the newly formed coastlines on each side of the rift basin. As the sea basin widens over millions of years, however, both coasts evolve into passive edges and slowly sink under the weight of accumulating sediment with the growth of the continental shelves.

The last class of emergent coast has nothing to do with tectonics or volcanism. Rather, it is dominated by isostatic uplift where continental glaciers once existed. The great masses of glacier ice so depressed the crust under them that in places the crust has been rebounding ever since the glaciers melted away about 10,000 years ago. Hudson Bay and the Baltic Sea, both centers of continental glaciation, are *postglacial-uplift* coasts. They are experiencing constant uplift, which is leaving the coast with a series of elevated shorelines that from the air look like rings around a partially drained bathtub (Figure 22.31d).

Submergent Coasts: Passive continental edges are the geographic settings for several classes of submergent coasts. Recall that passive coasts are characterized by massive sediment accumulation, which causes the coastal zone gradually to subside. The heaviest accumulations are formed at the mouths of great rivers like the Ganges, Orinoco, and the Huang He where deltas are building into the sea.

We covered the basic forms and features of deltas in the previous chapter (see Figure 21.28). Here we want to describe several types of deltas and their formation. Which type of delta develops is strongly influenced by the action of the sea at the river's mouth. Where the delta is exposed to strong waves and currents, it gets washed away as fast as the sediment is delivered. The Columbia River, which terminates on the Pacific coast, is an example of this arrangement. Where waves and currents are strong but not strong enough to remove all the sediment, the delta form is truncated and rounded on its seaward edge leaving it with a modified triangular shape, called an *arcuate delta*. The Nile Delta is the classic example of this type. Where ocean water extends up the lower valley of a river to form an estuary, the delta may take a long wedge shape as stream sediment fills in the drowned valley. Such a delta is called an *estuarine delta*, and the Rio de la Plata near Buenos Aires is an example of this type.

The Mississippi Delta is building directly into the sea and, for thousands of years, the river's deposition has exceeded the capacity of the ocean to carry most of the sediment away (see Figure 21.29). Because of its toe-like tentacles, the Mississippi's delta is called a *birdfoot delta*, resembling the one in Figure 22.33a. In recent decades, however, the growth trend has reversed in some parts of the delta because of a decline in sediment output by the great river and the loss of stabilizing wetlands. The decline in sediment is attributable mainly to sediment entrapment behind dams upstream in tributaries such as the Missouri, Ohio, and Tennessee rivers (see Figure 21.36).

Delta sediment transported downshore by waves and currents gives rise to wide beaches, spits, bay-mouth bars, barrier islands, and other depositional features. *Barrier-island* coasts are characterized by long, sandy islands that parallel the coastline (Figure 22.33b). The protected area between the island and the mainland is called a lagoon, an ecologically rich area with high rates of sedimentation. Barrier islands line most of the Atlantic coast between New York and Mexico. In the Gulf of Mexico, the Mississippi Delta is the primary source of sediment. In Texas, the system of barrier islands covers the entire coastline with occasional gaps between lagoons and the open sea through which tides drive inflows and outflows of water. As barrier-island coasts sink, the islands fragment and the lagoons fill with sediment.

Submergent coasts also form where the sea has risen and flooded low-lying areas, especially river mouths. *Ria* coasts are drowned coasts characterized by long fingers of the sea extending up river valleys (Figure 22.33c). Since the last glaciation, extensive areas of passive coasts have been flooded worldwide. Chesapeake Bay is an extreme example of a ria coast. Here the lower valley of the Susquehanna River and its tributaries form a network of estuaries following the pattern of the original drainage system.

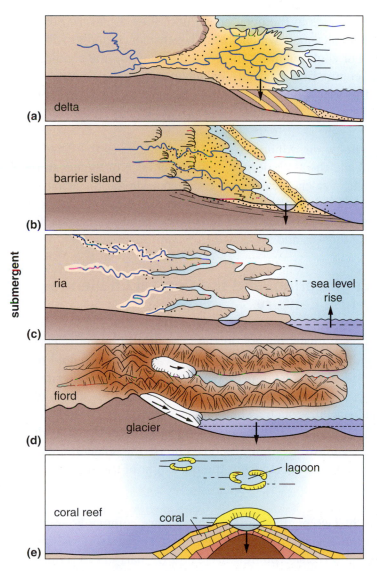

Figure 22.33 Principal types of submergent coastlines: delta, barrier island, ria, fiord, and coral reef.

A fjord coast is defined by long, deep glaciated valleys extending from the sea deep into the land.

Fiord coasts resemble ria coasts except that the valleys have been deeply scoured by glaciers (Figure 22.33d). As a result the valleys are deep, steep-sided and often very long, up to 100 kilometers from head to mouth in some areas. Norway's fiords are world famous, but equally spectacular ones are found in British Columbia, Alaska, southern Chile, and New Zealand. Water depths in fiords are so great that they are rarely altered by the buildup of shoreline deposits such as spits and bay-mouth bars.

Reef coasts are made up of coral reefs which form in tropical seas where the water is continually warm and less than 20 meters deep. Reefs grow from the buildup of massive amounts of coral, an organism that secretes calcium carbonate. The calcium cements the coral remains and other organisms together into a wave-resistant body of limestone.

Natural scientists of the eighteenth and nineteenth centuries were mystified by the presence of coral-reef islands in the deep ocean until they discovered that the islands had formed on old volcanoes, which through subsidence and erosion had been lowered to sea level or slightly lower. The coral builds up on the shoulders of the volcano to form a concentric ring of islands, called an atoll, and associated reefs, called fringing reefs. In the area of the crater, a small basin or lagoon forms (Figure 22.33e).

Some reefs have formed on *guyots*, which are flat-topped volcanoes that have been eroded down to sea level and then submerged with subsidence of the ocean floor. Much larger accumulations of calcium carbonate sediments, similar to ancient limestone bedrock formations found on the continents, also make up reef-like formations in tropical seas. These form in extensive shallow-water areas, such as the Great Bahama Bank southeast of Florida, and are called *carbonate platforms*.

22.7 Land Use, Engineering, and the Coastal Challenge of the Twenty-first Century

Among the many problems humans have with the sea, shore erosion is perhaps the most serious, and with global warming and rising sea-level, it is bound to get worse. When the sea erodes coastal land, people are faced with three alternatives: (1) abandon the area and move inland out of reach of the sea; (2) adjust land-use patterns so that vulnerable facilities such as buildings and roads are located in less-threatened areas; or (3) build defensive structures such as walls to try to divert waves and reduce erosion. Almost invariably, people choose the structural alternative and attempt to defend their property from the onslaught of the sea, especially where settlements are involved.

The Engineering Approach: Two engineering techniques are commonly employed to reduce shoreline erosion and both involve building structures: (1) walls, such as the one in Figure 22.34, to resist eroding waves; and (2) barriers to reduce the rate of longshore sediment transport. **Seawalls** (and similar engineered structures) are concrete, steel, wood, or rock rubble walls constructed along the beach face or the bank behind it. They are intended to function as defensive fortifications and are commonly used by individual property owners to hold their land in place against the onslaught of waves. Seawalls are very expensive and, in the short-term, often reduce erosion. In many instances, however, they actually increase erosion by deflecting wave energy downward and around them toward neighboring properties. In the long

Figure 22.34 A seawall built to protect property from shore erosion and storm damage. Where defended beaches suffer long-term sediment loss, seawall failure is inevitable.

Figure 22.35 A system of groins near Miami, Florida, designed to slow the movement of sand by the longshore system.

run they are bound to fail, especially if the coastline trends toward a negative mass balance, because they will eventually be undermined as sediment is carried away and the shore retreats.

Groins are generally the preferred structures for reducing sediment loss to longshore transport. A **groin** is a wall built into the sea perpendicular (or nearly so) to the shore for the purpose of slowing and capturing longshore sediment (Figure 22.35). Sediment accumulates on the upshore side of the groin and builds outward, eventually spilling around the end. A number of groins constructed in succession can slow the longshore transport sufficiently to build and maintain a beach in some locations. In other places, however, they have been known to increase erosion in beaches downshore because they interrupt the natural flow of sediment in the same manner as breakwaters do.

Breakwaters function like groins but they are usually much larger. They are constructed at harbor entrances to improve navigation safety by reducing wave energy and sediment accumulation. Because they reduce wave energy, breakwaters also interrupt the longshore sediment train. Sediments accumulate behind or on the upshore side of breakwaters, depriving the beach downshore of its sediment supply (Figure 22.36). This often results in beach erosion and recession downshore because, lacking sediments to move, wave energy is spent in the erosion of *in situ* material. At the same time, the area upshore of the breakwater becomes clogged with sediments, which eventually build out and spill around the breakwater into the harbor entrance. To solve the problem of erosion and sedimentation caused by breakwaters, a method called **sand bypassing** is used in which sand is mechanically moved from one side of the harbor entrance to the other.

less sediment, more erosion

sediment accumulates

Figure 22.36 Breakwaters at a harbor entrance showing the sharp contrast in beach conditions between the upshore (foreground) where sediment is accumulating and downshore (upper right) where it is not, and probably eroding.

In the end it is doubtful that the long-term benefits of most breakwaters outweigh the costs, especially if ecological and scenic resource losses are figured in. These factors, the costs of lawsuits from downshore property owners, and the increasing costs of energy related to sand-bypassing operations, are causing public decision-makers to pause over proposals to build new breakwaters and groin projects. But public pressure for property protection and safe harbors for recreational craft weighs heavily on politicians.

The Challenge of Rising Sea Level: Although we have frequently mentioned the effects of rising sea level related to global warming, it bears further consideration in the context of coastal geomorphology. In the past 15,000 years or so the level of Earth's oceans has risen by more than 100 meters. Most of that rise took place between 15,000 and 10,000 years ago, during a time when several million or so people lived along the coasts of the world. These people were hunter–gatherers who lived in shifting bands and as the sea came up they gradually, probably imperceptively, moved inland with it.

Humans in the twenty-first century face another rise in the sea to which they must adjust, but this time the circumstances are dramatically different.

First, more than 1000 million people now live on the world's coasts.

Second, coastal population is rising rapidly and by mid century may reach 5 billion or more.

Third, modern humans are sedentary, fixed in towns and cities dependent on massive, expensive, and embedded infrastructures in the form of roads, ports, buildings, piping systems, and so on.

Fourth, over much of the world, people and their governments have grown accustomed to resisting changes in the environment by constructing defenses against it. Seawalls and groins are simple examples. In river systems, it is dams and levees.

Fifth, as the atmosphere warms, it is capable of generating more and larger storms with far more power to generate strong winds and large waves (see Figure 8.36), and with water levels reaching 1 to 2 meters higher than present, the storm waves will strike land with greater force and frequency.

So, with humanity crowding against the sea and the sea pushing back with increased force, some major changes are forthcoming. Can we curtail the forces driving it all, namely, expanding humanity and global warming? Frankly, no. While the rate of population growth is slowing, we expect another 3 billion more people by mid century. With more people and higher rates of consumption overall, it will be very difficult to curb carbon dioxide loading of the atmosphere in this century, and even if the excess input could be stopped tomorrow, it would take the atmosphere about 200 years to adjust itself to pre-Industrial Revolution levels. It seems, therefore, that the struggle with the sea is inevitable. How will humanity adjust?

Kobe, Japan is not unlike large coastal cities across the world, with development pushing its way into the sea.

Undoubtedly the responses will vary widely across the Earth in much same way humanity has varied in its responses to other threats. Experience will be important. Countries like the Netherlands, which have wrestled with the sea for centuries, already have the knowledge, policies, and socio-economic systems to deal with ocean encroachment. Culture will play a part because some belief systems, religious, political, and others, will deny that change is taking place or that humanity has any power to control its destiny under such circumstances. Differences in geography will be extremely important because certain countries, Bangladesh, for example, are limited to coastal lowlands, while others have only modest exposure to the sea and/or their coasts and settlements lie at higher elevations.

And, of course, national wealth will be significant, especially when we consider that over 95 percent of the people added to the planet in the twenty-first century will live in less-developed countries where costly adaptations will be impossible to underwrite. Finally, how we plan and design our land-use facilities in the future will be critically important. If we resist the rigid, single-minded approaches of the past and instead build resilience in our coastal land-use systems, that is to say, incorporate flexibility and readiness into our land-use systems, humanity on the world's coasts may fare reasonably well.

Summary and Overview Coastal Systems: Waves, Currents, and Landforms

Some of the most interesting and meaningful geographic questions concern edge environments, and few if any rank higher than coastlines. As geomorphic systems, those of the coast are powerful and dynamic, for in the expanse of a lifetime or less it is possible to witness major geographic changes on the edges of landmasses large and small. And it appears that the magnitude and frequency of change are increasing with changes in the atmospheric energy, sea level, and land use. Once again we are reminded of the interconnected nature of geographic phenomena, not only in present times, but, as the summary diagram suggests, over the long term as well.

The passive coast is centered on a large drainage system that developed in a failed arm, which provides sediment to coastal systems and continental-shelf formation. On the opposite side of the continent is the active coast with deep water and mountain building.

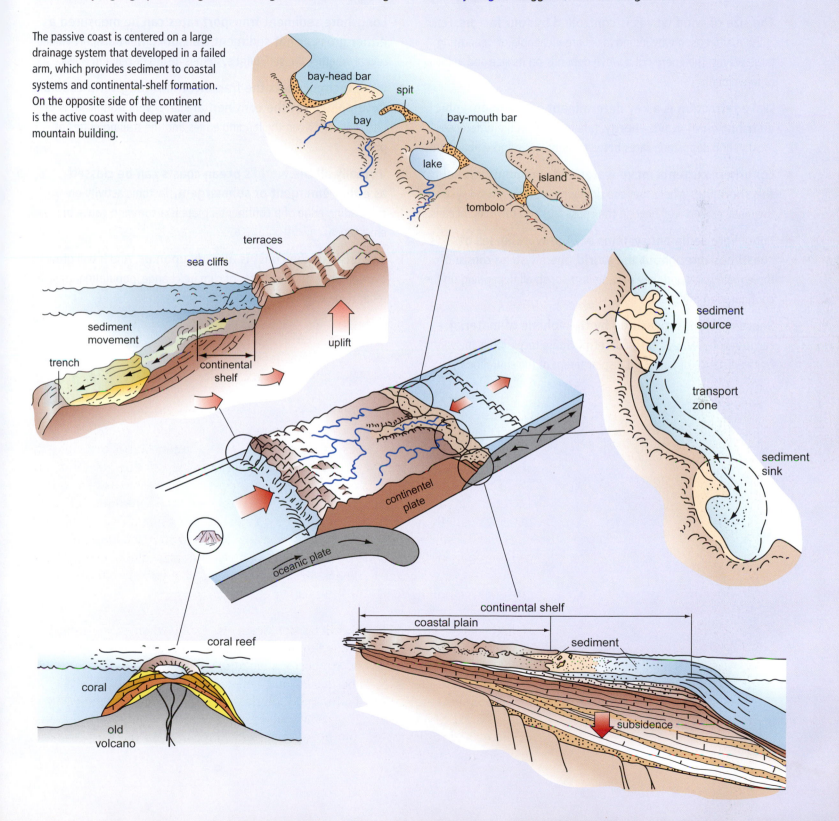

► **Four systems move sediment and shape the Earth's coasts.** Each operates at different magnitudes and frequencies and, when they work together, change can be catastrophic.

► **Coastlines are constantly changing with erosion and deposition and with variations in sea and land elevations.** During the last glaciations, sea level worldwide was much lower than today. In this century, sea level is rising with global warming.

► **The size of wind waves is controlled by four factors.** Fetch is the main reason why inland lakes are incapable of generating large waves. The energy of a wave depends on its size and velocity of movement.

► **Wave refraction is a key determinant of the geographic distribution of wave energy.** It helps explain the differences in erosion and deposition rates between headlands and embayments.

► **Longshore currents move water and sediment parallel to the shoreline.** Where they are driven by prevailing wind systems, they may extend well beyond the coast and into the deep oceans.

► **Longshore sediment systems are the workhorses of coastlines throughout the world.** These systems consist of three main parts and they may be large or small depending on the configuration of the shoreline.

► **Shore erosion is measured by the volume of material removed by waves and currents.** Shore recession is measured by the distance of landward displacement of the shoreline.

► **Shorelines are made up of three main zones each marked by different features.** Except for the backshore slope, these features are composed of sediment in transit.

► **The sediment transported parallel to the shore is called longshore drift.** The rate of longshore transport depends mainly on the availability of moveable sediment and wave and current energy.

► **Longshore sediment transport rates can be measured as either gross or net sediment transport.** Net transport values reveal whether the sediment system is losing or gaining mass.

► **Plate tectonics sets the framework for long-term coastal development at the continental scale.** Very significant is the difference between the leading edges and the trailing edges of plates.

► **Virtually all the world's ocean coasts can be classed as either emergent or submergent.** Tectonic activity on the leading edge of a continental plate is a common cause of emergence.

► **The crisis of the coast is already upon us.** And it will grow in this century with the effects of climate change, population growth, reliance on engineering, and other factors.

Review Questions

1 What are the four geomorphic systems operating on the marine side of coastlines? How do these systems differ in their origins and the type of work they perform?

2 Identify two events where at least two of the four marine geomorphic systems work together. What are the results of these interactions?

3 Figure 22.6 shows seas and swells. How do these products of wind differ in their form and movement?

4 What is the concept of fetch, and how does it change with direction and a geographic scale. And what does fetch have to do with shore exposure?

5 Referring to Figures 22.8 and 22.9, describe the motion within an oscillatory wave. Name 4 changes that take place in a wave's shape and motion when it changes from oscillatory to translatory.

6 What is meant by the term wave-base depth and how is it possible that this depth changes with different wave sizes?

7 What is wave refraction, and how does it influence the distribution of wave energy along a coast with bays and points?

8 What is a nearshore circulation cell? Using Figure 22.15 as a reference, name the three main flows of water drawing these cells and describe how they operate and move sediment.

9 Describe the three-step process of bank erosion. How are bedrock and softer shorelines eroded, and what shoreline features result in each setting?

10 What are sediment sources and sediment sinks and how are they linked together to form a system? What name is give to such systems and what does it mean if a system has a large net sediment transport value?

11 What explains the occurrence of deposition zones or sites, along a shoreline? Using Figure 22.25, identify the landforms associated with these sites: coastal island, mouth of a bay, head of a bay, and explain why they formed there.

12 At the global scale, how do plate tectonics influence coastline development? What are the differences between active and passive coasts in terms of their formative processes and landforms?

13 Using Figures 22.31 and 22.32 as a guide, identify the primary landforms associated with emergent and submergent coasts. What are the distinct roles of plate tectonics, glaciation, and deposition within each type of coast?

14 Name three structures commonly used in coastal engineering, what they are designed to do, and the nature of the problems they often create.

15 In the 21st century struggle between humanity and the sea, the outcomes will be different for different countries. Discuss several of the factors that will shape these outcomes.

Glacial Systems:
Growth, Motion, and Work of Glacial Ice

Chapter Overview

Glaciers have became media *causes celebres* because they are considered harbingers of a changing global climate, and rightly so. To form and survive, glaciers need cold temperatures and plenty of snow, and when one or both of these changes on an established glacier, it grows or shrinks, and lately glaciers the world over have been shrinking not because of too little snow but because of warmer temperatures. And when we look back over the past two million years or so in Earth history, we find concrete evidence of distinct patterns of glaciers growing and shrinking in response to changes in climate. This is meaningful, but equally meaningful is the work of glaciers as geomorphic systems. They have enormous erosional power, so great that they are capable of grinding the sides off mountains. So we also want to learn how they form, grow, move, erode the land, transport rock debris, and shape landforms. We begin with the types of glaciers found in the world and how they originate and function as systems. Next, we look at glaciers as geomorphic agents, first mountain glaciers and then their larger counterparts, continental glaciers. In this connection we are interested in the occurrence of ice ages and their relationship to climate change on Earth.

Introduction

We edged our way onto the ice shelf on Lake Superior's south shore. It was January and a frigid wind bore down on us from the northwest. Along the front of the shelf, about 200 meters offshore, storm waves rammed against the ice sending spray and chunks of loose ice high into the air. As it fell onto the shelf, the spray froze immediately in a slick glaze. With each storm the glaze and ice chunks built up along the shelf front eventually giving rise to a distinct ridge, like the one in Figure 23.1, reaching 6 or 7 meters high in places. Weighted down by this frozen superstructure, the shelf, like an overloaded ship, settled deeper into the lake until it rested firmly on the bottom.

We came to this forbidding place to study ice. On first glance, it seemed like an easy assignment, for ice was everywhere around us. But the glaze formed such a thick skin over the shelf that there was no easy way to examine its inner features. Len Bryan offered a proposal, "Why not dissect one the of the ice ridges so we can inspect its inner structure and

Figure 23.1 Lake Superior ice ridges form when storm waves throw spray and ice chunks onto grounded shelf ice.

Figure 23.2 The banded layering of ice in a glacier. Each band represents one year's snow accumulation after losses to melting and evaporation.

composition firsthand? It would be a big job, but we really have no other way to see what a ridge is like on the inside." John Koerner challenged him immediately. "What are you talking about? Look out there. You can watch the ridges forming. That'll tell you what's on the inside." "Yeah, I see that," Bryan replied, "but we really don't know what they're like at depth. Maybe there's a great slab of lake ice or a pile of sediment down there and the ridge is built on top of it?" Over the howling of the wind, Chuck Douthitt grumbled, "Let's do something, it's damned cold, and we won't last long standing around talking."

Cutting Bryan's trench was sheer grunt work, but we welcomed the exertion because it kept us warm. Like a gang of gold miners sensing an imminent strike, we competed for a turn in the trench. Hard work, and fast, cold air transfigured faces as moist breath froze on mustaches and beards. By midmorning, long ice sickles hung from Douthitt's whiskers and by midday he looked like a walrus in horn-rimmed glasses. By early afternoon, the trench was more or less complete.

The trench walls revealed the pattern of the ridge's growth. Larger pieces of ice, mostly cakes that had formed in deep water, made up the base. Over this was a sequence of thin layers interspersed with chunks of ice arranged in a pattern that matched the wave spray process we observed along the front of the shelf. Bands of snow could be seen between some layers of glaze ice. Lower layers appeared to be compressed and metamorphosed and we reflected on the similarities to layers of snow and ice in glaciers. Someone noted that many mountain glaciers also contain chunks of ice contributed by avalanches and that glaze ice often marks each year's summer melt.

But the differences between glacial ice and lake-shelf ice were far more striking than the similarities. First, each ice band in a glacier (see Figure 23.2) represents a whole winter of snow accumulation not just one storm. Second, the ice in glaciers begins as snow, not as liquid water, and is compressed and metamorphosed into ice over several decades. Third, glaciers represent centuries of snow and ice accumulation and they last thousands of years not just one winter season as lake ice does. Fourth, glaciers are able to move on their own, whereas shelf ice can move only under the influence of an outside force such as waves and currents.

23.1 Glacier Types, Environments, and Distribution

Glaciers are masses of ice formed from the accumulation and compression of large quantities of snow. They are capable of slow, flow-type movement in which the ice mass deforms as a plastic, somewhat like cold molasses or pudding moving down a gentle incline, and this is what distinguishes them from immobile heaps of ice. When glaciers move, they are capable of eroding the land under them and then transporting and depositing the eroded debris.

Glacial ice currently covers about 10 percent (15 million square kilometers) of the Earth's land surface. More than 95 percent of this area is occupied by two vast ice sheets or ice caps, Antarctica and Greenland. The remaining area is occupied by more than 150,000 alpine (mountain) glaciers, which are distributed across all belts of latitude including the equatorial. Not surprisingly, the altitude in a mountain range where alpine glaciers form, as Figure 23.3 shows, varies greatly from the high to the low latitudes. Near the Arctic Circle glaciers can form as low as 500 meters above sea level; at 45 degrees latitude they generally form only above 3000 meters; and in the tropics only mountains higher than 5000 meters elevation are cold enough to support glaciers.

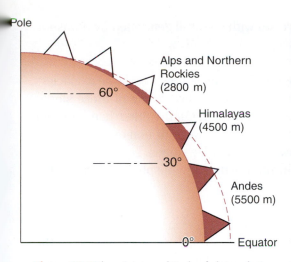

Figure 23.3 The minimum altitude of alpine glacier formation varies with latitude. Near the Equator it is 5500 meters (18,000 feet).

Figure 23.4 The Dry Valleys of Antarctica, a cold, arid environment in the midst of Earth's largest ice sheet that has not seen glaciers in millions of years.

Figure 23.5 Antarctica is almost completely covered by huge ice sheets, roughly the scale of continental glaciers.

But not all cold environments produce glaciers. Most mountains higher than the critical elevations cited above have no glaciers, and one of the world's coldest regions, northern Siberia, has no glaciers. And in Antarctica, the world's coldest place, there are barren valleys, called the Dry Valleys (shown in Figure 23.4), which have not produced glacial ice in more than 10 million years. Why? Because these places receive too little snowfall. Snow is the raw material for building glaciers and, unless a cold environment also produces large snowfalls, the formation and nourishment of glaciers is not possible. We conclude, therefore, that in order for Earth to produce an ice age, it must not only be colder, but also snowier.

Types of Glaciers: Glaciers can be classified according to their size, shape, and thermal characteristics. The largest are **continental glaciers**, huge sheets of ice covering large sections of continents. The Antarctic and Greenland ice caps are examples of continental glaciers, especially the Antarctic sheet (Figure 23.5). These two cover 90 percent or more of their respective landmasses, and lie almost entirely at polar latitudes, that is, above the Arctic and Antarctic circles. Only 18,000 years ago, during the last wave of continental glaciation, even larger ice sheets occupied Earth. These continental glaciers were truly massive in size, both in geographic coverage and thickness, warmer than today's polar ice sheet in Antarctica, and centered not in the polar zone, but in subpolar latitudes below the Arctic Circle. At their maximum the North American icefield covered the continent from the Atlantic to Pacific and from near the Arctic Coastal Plain to the valleys of the Ohio and Missouri rivers. We will say more about North America's continental glaciers later.

Alpine glaciers are the smallest glaciers. They range in size from ice patches only a few hundred meters across that form in pockets, called *cirques*, high on mountain slopes to long tongues of ice, known as *valley glaciers*, that extend tens of kilometers down mountain valleys. Most alpine glaciers are found in large orogenic belts, such as the Andes, on the continental margins. They are especially abundant where mountain ranges lie athwart prevailing wind systems and are fed by abundant orographic precipitation. The mountainous coasts of British Columbia, southern Chile, southern New Zealand, and Norway, for example, lie in the belt of the prevailing westerlies, which feed upper slopes with massive snowfalls that nourish thousands of glaciers in settings like the one in Figure 23.6. In Asia, most alpine glaciers are located in the

Figure 23.6 Alpine glaciers on the coast of Alaska are nourished by massive orographic snowfall fed by moist ocean air.

Heavy snowfall feeding large mountain glaciers high on windward slopes.

Himalayas and surrounding ranges and are fed with snowfall generated by monsoon airflow.

Wherever alpine glaciers are fed by orographic precipitation, the largest glaciers form on windward slopes because snowfall is heaviest there. Some of these slopes receive 10 to 20 meters of snowfall in an average winter. On the other hand, glacier development on the leeward mountain slopes is usually weak or non-existent except where storms and strong winds carry snow onto upper lee slopes. Lee-slope glaciers often favor shadow slopes (north-facing in the northern hemisphere) because the climate is colder there.

Another way of classifying glaciers is according to their thermal character. The class known as **temperate glaciers** can be thought of as relatively warm because the internal ice is at or very close to the melting threshold. These glaciers, which include most alpine glaciers, are characterized by profuse melting both within and beneath the ice, which apparently facilitates their slippage on the ground. **Polar glaciers**, by contrast, are frozen throughout and are distinctive for the near absence of meltwater (see Figure 23.7). The bottom of the ice is solidly frozen to the ground, which prohibits slippage at the base of the ice sheet. Most, but not all, of the vast areas of the Antarctic and Green-

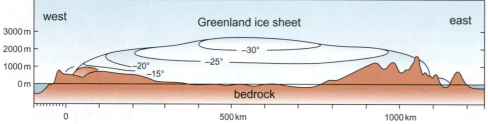

Figure 23.7 The Greenland ice sheet, like the Antarctic, is a polar glacier because it is deeply frozen and discharges little or no meltwater from most its ice mass.

land sheets are hard frozen in this fashion. Movement instead takes place more or less by plastic deformation within the ice mass, a motion involving internal shearing among planes of ice. Between these two classes is an intermediate thermal class, **subpolar**, which is polar throughout except for a temperate surface layer which produces meltwater much of the year. Much of the perimeter of the Greenland ice sheet where it descends to the sea is subpolar.

A third way of classifying glaciers is according to their shape and behavior related to topography under and around them. Alpine glaciers, particularly the valley variety, are designated **constrained glaciers** because their movement and shape are confined by the mountain terrain around them. They form in mountain valleys and, in order to move, must conform to the shape of the valley. These valleys are usually stream valleys and, as the glaciers flow down tributaries in the drainage net, they join like streams forging larger and larger ice streams with declining elevation.

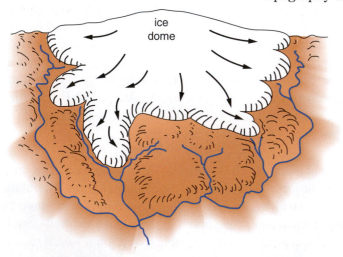

Figure 23.8 Unlike valley glaciers, continental glaciers are largely unconstrained by the landforms under and around them.

Ice caps, on the other hand, qualify as **unconstrained glaciers**, because they show little or no relationship to the topography under them. These glaciers radiate from great ice domes, like the one shown in Figure 23.8, moving over both upland and lowland surfaces. Only on their margins do they begin to conform to the lay of the land as large tongues of ice, called *lobes*, extend into broad lowlands such as the lobes of continental glaciers did in the basins of the Great Lakes 10,000 to 20,000 years ago.

23.2 Glaciers as Systems: Growth, Motion, and Decay

All glaciers operate as input–output systems. They are fed with matter in the form of frozen water, which moves through the system and is released mostly as liquid water and water vapor. In a perfectly balanced system, input and output are equal and the size (mass) of the system (glacier) remains constant. But no system in nature ever achieves perfect balance, and so it is with glaciers. Consequently, glaciers grow and shrink with changes in input and output. But because they are so large and slow-moving, glaciers are not very sensitive to short-term, that is, year-to-year, changes in weather and climate. Rather, they tend to respond to much longer-term variations, usually decades and centuries for small glaciers and much longer for larger ones. For this reason, glaciers can be used as indicators of past climates because they hold the records of atmospheric conditions over centuries or millennia, long before records were kept. The Antarctic ice sheet contains ice over 400,000 years old.

Input and Growth: Let us open this discussion with the sequence of diagrams in Figure 23.9. Glacial ice begins as snow and, through a process taking several decades or more, the snow is transformed into solid ice. The sequence of change goes roughly as follows. New-fallen snow (Figure 23.9a) is composed of loosely packed lacy ice crystals. It varies greatly in consistency from snowfall to snowfall, but generally has a density between 50 and 200 kilograms per cubic meter. (For comparison the density of pure ice is 917 kilograms per cubic meter.) Almost as soon as it has fallen, snow begins to metamorphose. The crystals are reduced by partial melting, evaporation, and compaction until they take on less angular forms. This reconstituted snow (Figure 23.9b) is called **névé** and it has a density of 400 to 500 kilograms per cubic meter. In most areas of the world where snow falls, névé melts completely away by the end of the summer, but in a few cold spots, part of it survives.

Névé that survives the summer season is called **firn** and it is the stuff that builds glaciers (Figure 23.9c). As each year's layer of firn is added to a glacier, it is compressed under the mass of succeeding years. Its density gradually increases and after 25 to 100 years, the firn is transformed into glacial ice with a maximum density around 850 kilograms per cubic meter (Figure 23.9d). Annual increments of firn accumulation range from less than a centimeter to more than 10 centimeters in thickness. After centuries of burial, individual increments can be identified as thin bands within the glacial mass (see Figure 23.2). These bands can be used as measures of past climatic conditions in as much as they reveal periods of heavy and light snowfall and, as we pointed out in Chapter 8, they also contain little pockets of trapped air that can be measured for indicators of past climatic conditions.

Glacier Movement: Movement takes place when an ice mass becomes too heavy to maintain its shape and literally squashes out near the base. In alpine glaciers, this begins when the ice reaches a thickness of about 20 meters (or 15 meters on steep slopes). As the glacier grows, it moves downslope gradually concentrating its flow where the topography offers the least resistance. This usually takes the glacier into a stream valley, which molds the ice into the streamlined form of a valley glacier.

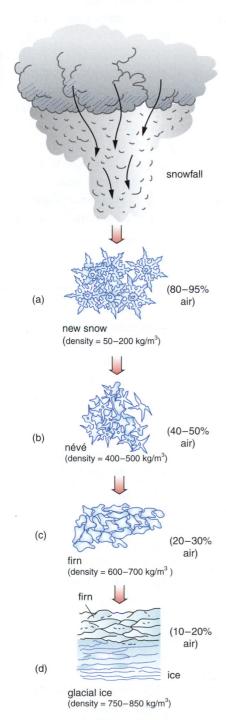

(a) new snow
(density = 50–200 kg/m³)
(80–95% air)

(b) névé
(density = 400–500 kg/m³)
(40–50% air)

(c) firn
(density = 600–700 kg/m³)
(20–30% air)

firn

(d) glacial ice
(density = 750–850 kg/m³)
(10–20% air)

ice

snowfall

Figure 23.9 The transformation firn (c) to glacial ice (d) with increasing density at each phase.

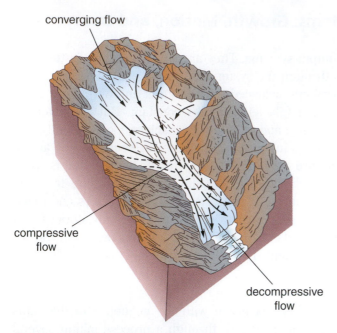

converging flow

compressive flow

decompressive flow

Figure 23.10 The pattern of flow in the surface of an alpine glacier, converging, compressive, and decompressive.

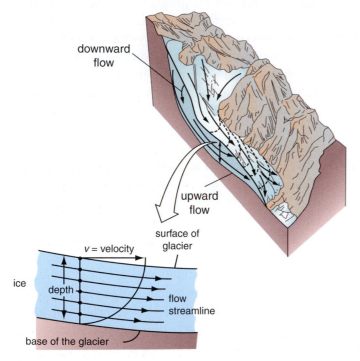

downward flow

upward flow

surface of glacier

v = velocity

ice

depth

flow streamline

base of the glacier

Figure 23.11 Inside the glacier, velocity is fastest near the surface and diminishes toward the base. In long view, flow tends to be downward in the upper reaches and upward in the lower reaches.

Within this stream of ice, the rate of flow varies greatly. Flow is fastest in the center and slowest on the margins, next to the valley walls. If we trace the flow lines in the surface of a glacier, represented by the arrows in Figure 23.10, flow converges at the head of a glacier where it is being fed with snow and ice from side slopes, compresses as it mounts up downslope, and then decompresses at the terminus where the ice is melting, spreading, and thinning out. Inside the glacier, velocity is greatest near the surface and diminishes to zero or nearly so at the base as shown in the inset in Figure 23.11.

Glaciers are very slow compared to other geomorphic systems. Typical velocities range from 5 to 300 meters per year. The Antarctic ice sheet at the South Pole is moving at a rate of 8 to 9 meters per year. Most alpine glaciers move faster, and some, called **surging glaciers**, move at rates as high as 50 meters per day. On the other hand, others have stopped moving altogether, mostly with the loss of mass as they enter the final phase of their existence, a trend accelerated in the past 50 years or more by global warming. In Glacier National Park in the Rocky Mountains, 110 alpine glaciers have melted away in the last 150 years (Figure 23.12).

What mechanisms enable a glacier to move and what factors control the rate of movement? There are two basic mechanisms of movement: plastic deformation and basal sliding. **Plastic deformation** takes place inside the ice mass and involves slippage within and between individual ice crystals. Each movement is microscopic, so it takes billions of slips to move the glacier a measurable distance. Also known as *creep*, this process is driven by pressure from the great weight of the ice mass; therefore it tends to be more pronounced deep within the glacier than near the surface. Plastic deformation is the commonest mode of movement in polar glaciers such as the Antarctic ice sheet. Within the body of the glacier, movement is often segregated into zones or thick planes that slide over each other.

Basal sliding involves slippage or skidding at the base of a glacier, such as the one shown in Figure 23.13. It accounts for 90 percent of the movement in temperate glaciers. The presence of liquid water at the base of the ice is central to the process of basal sliding. The base of the glacier is under tremendous pressure and this produces melting where the ice contacts the ground because the pressure drives up the ice's melting-point temperature. This process is similar to the mechanism involved in ice-skating. The immense pressure at the contact between runner (blade) and ice causes sudden melt-out and formation of a thin film of liquid water upon which the runner slides. In a glacier, melting detaches the ice from the ground by thawing frozen contacts, and the resultant water promotes slippage. Basal sliding increases with the thickness of the ice, the slope of the ground, and temperature at the base of the ice. Researchers recently learned that in some Antarctic glaciers meltwater becomes so concentrated at the base of the ice that it partially floats the glaciers, speeding up their movement.

glacier (1938)

glacier (1981)

glacier remnants (2009)

Figure 23.12 Grinnel Glacier, one of many declining glaciers in Glacier National Park of Montana. Given the current trend in atmospheric warming, the remaining 37 glaciers in the park will be gone by 2030.

Glacier Mass Budget: In order to understand glaciers as systems, we must examine their energy balances, or what in glaciology is termed **mass budget**. In brief, the mass budget of a glacier includes three main components, shown in Figure 23.14: (1) accumulation on the upper glacier; (2) forward movement of the glacier; and (3) ablation of the lower glacier. *Accumulation* is defined by the annual increment of firn, measured in tons of frozen water added to the glacier. *Ablation* is the term used to define processes of ice loss from a glacier, including melting, evaporation, and calving (breaking off into the sea). As an accounting problem, accumulation and ablation are deposits (input) and withdrawals (output) and forward movement is a transfer from the deposit to withdrawal column.

If over a year accumulation and ablation are equal, the glacier has neither lost nor gained any mass. If accumulation, forward movement, and ablation are all equal, the glacier has gained and pushed ahead an amount of ice equal to the amount lost. Given a number of years in which more ice accumulates than ablates, the glacier mass grows, and the terminus of the glacier, called the *snout* (shown in Figure 23.14), may advance downslope. The opposite of course means that the glacier is losing mass and probably retreating.

Most glaciers, no matter where they are found in the world, are currently retreating, and many are retreating rapidly (see Figure 23.12). This means that their mass balances are negative, which in turn means that climate in the last hundred years has

Figure 23.13 An exposed section of a mountain glacier revealing the contact between ice and rock, the surface of basal sliding.

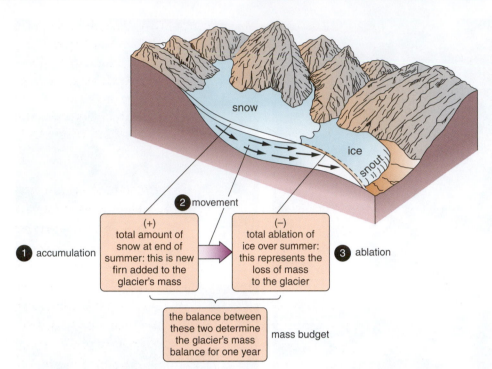

Figure 23.14 A glacier's mass budget represents the balance between (1) accumulation and (2) ablation in a system linked together by (3) forward movement of the ice.

Figure 23.15 An archival drawing from the Swiss Alps during the Little Ice Age, showing an advancing alpine glacier.

grown warmer, less snowy, or both. But this trend can reverse itself, as it did in the period 1400 to 1800, when, after a global-warming trend, glaciers over much of the world made such strong advances that the period has been branded as the Little Ice Age (Figure 23.15). But we must also recognize that the Little Ice Age occurred mainly before the Industrial Revolution (1750–1850), and since then carbon dioxide in the atmosphere has increased by 30 percent. Since 1975, the global atmospheric temperature has been rising dramatically and, since 1990, even the polar ice sheets have been thinning and retreating. See Chapter 8 for more details on climate change and ice ages.

Dynamics of the Glacier Terminus: To understand the true character of a glacier, we must realize that forward movement of the ice mass continues while accumulation is taking place at one end and ablation is taking place at the other. The rate of flow should be controlled by net accumulation, but the relationship is complex because of the time lag between the buildup and transformation of firn into ice and the corresponding movement of the glacier. The lag may amount to several decades, and for this reason it is not uncommon for a glacier's behavior to be out of phase with the climatic trends. This may help explain the occurrence of glacial surges, when there is no apparent reason for a glacier's frisky behavior. In other words, the glacier may be responding to a massive input of energy (net accumulation) some years before that slowly moved through the glacier and then bursts forth in a surge at the terminus.

Although forward movement is an important control on the behavior of the glacier, it is not the only control. The rate of ablation at the snout is also important. If snout ablation is equal to the rate of forward motion, the terminus may remain stationary while the glacier continues to move ahead. The glacier acts in this way much like the top-side of a conveyor belt. However, when the ablation rate exceeds forward movement, the snout retreats. Of course, the reverse can be expected when forward movement exceeds ablation. These motions – stability, advance, and retreat – are critical to understanding the glacier as a geomorphic agent; that is, how it produces erosion and deposition.

Eventually, all glacier ice is destroyed. In mountain glaciers most is lost to ablation by melting, but in large ice sheets huge amounts are also lost through **calving** (see Figure 8.40). This is the process that produces thick pieces of ocean ice called icebergs (Figure 23.16). Most icebergs today come from the Greenland and Antarctic glaciers. The Ross Ice Shelf, a great sheet of floating glacial ice on the edge of the Antarctic ice sheet, produces thousands of icebergs per year, which ocean currents carry around the continent as part of the west-wind drift. The world's largest recorded iceberg was produced in 1999 when a piece of ice the size of the state of Rhode Island broke away from the Ross Ice Shelf and floated slowly off.

Figure 23.16 Massive blocks of glacial ice calving into the ocean from the Hubbard Glacier, Alaska.

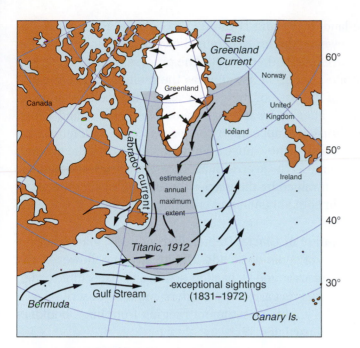

Figure 23.17 Icebergs from Greenland are carried southward by the Labrador Current and the East Greenland Current into the zone of the Gulf Stream and ocean shipping lanes.

In the North Atlantic, icebergs from Greenland are traditional menaces to transoceanic travel. Moved southward by the East Greenland Current and the Labrador Current, many icebergs reach well into the shipping lanes as far south as 40 degrees latitude (Figure 23.17). It was one of these pieces of ice that sunk the Titanic in 1912. Today icebergs can be detected with shipboard radar, making North Atlantic sailing much safer, though by no means failsafe.

Summary on Glacier Mass Budget: The concept of a glacier mass budget is summarized in Figure 23.18. Input is driven by the atmosphere and powered by solar radiation via the hydrologic cycle. Nourishment arrives as snow and that which survives the summer becomes part of the ice mass and part of the machinery of the glacier. At the output end, ablation consumes the ice, releasing its water as vapor, runoff, and icebergs. In general, if output (ablation) exceeds input (accumulation) for an extended period of time, the glacier will shrink and retreat, which is the trend of glaciers today across the world. Central to our interests at this point is the capacity of a glacier to move because it is movement that gives the ice its capacity for geomorphic work.

23.3 Glacial Erosion, Debris Transportation, and Erosional Landforms

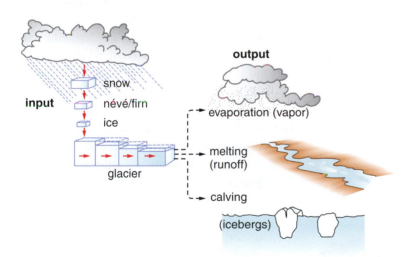

Figure 23.18 Summary figure representing a glacier system with input, movement, and output.

Glaciers are highly effective erosional agents. The amount of material eroded from a mountain valley by an alpine glacier over a year, for example, is substantially greater than a large stream could erode from the same valley. Measurements of sedimentation rates on the seafloor reveal that much more sediment is discharged to the oceans during ice ages that during non-glacial periods. Yet, in terms of total work output eroding land worldwide, glaciers are a poor second to runoff and streams, simply because glaciers cover only 10 percent of Earth's land area. Nevertheless, in those areas where glaciers are or have been, they have profoundly influenced the form and composition of landscape through erosion and deposition.

Erosional Processes of Glaciers: Glaciers erode by three main processes: abrasion, crushing, and quarrying. These are illustrated in Figure 23.19, but they need explanation. **Abrasion** (also called **scouring**) involves a rasping action in which rocks carried at the base of the moving ice grind away the underlying bedrock. The result is scratching, grooving, and polishing of bedrock which is accompanied by the production of a fine residue called *rock flour*, which clouds and discolors meltwater at the front of the glacier. **Crushing** is a related process involving the fragmentation of boulders under the weight of the ice as the glacier moves over obstacles at its base.

Quarrying involves plucking of large pieces of rock from underlying bedrock as the glacier moves over it. The glacier literally freezes onto a block of rock, pulls it out, and entrains it into the base of the ice. Because polar glaciers are frozen fast

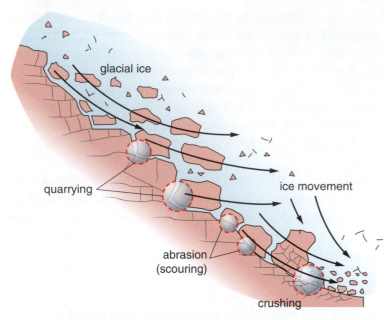

Figure 23.19 The three erosional processes at the base of a glacier, abrasion (scouring), crushing, and quarrying.

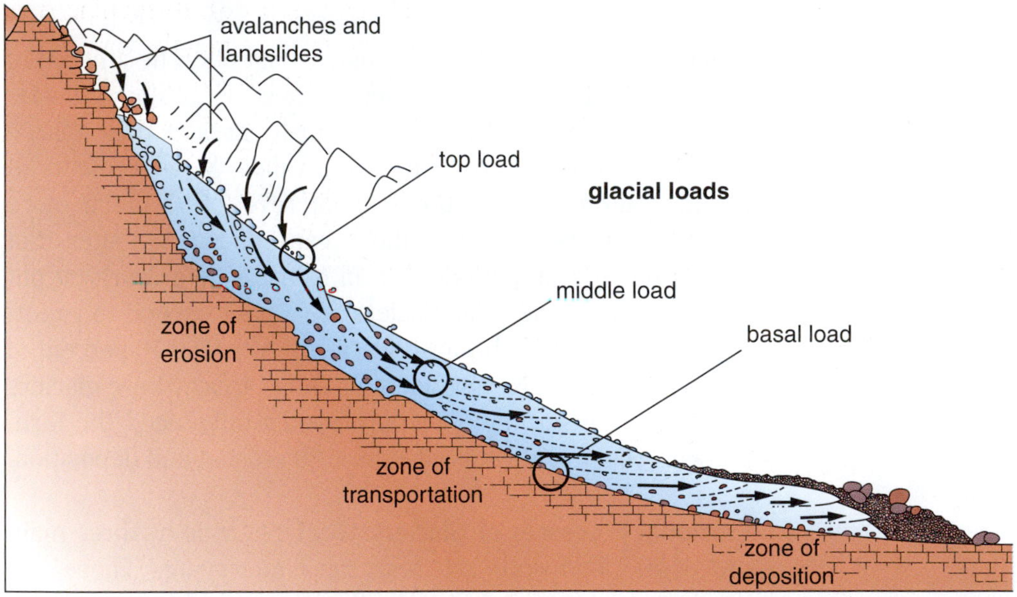

glacial ice

quarrying

ice movement

abrasion
(scouring)

crushing

avalanches and
landslides

top load

glacial loads

middle load

basal load

zone of
erosion

zone of
transportation

zone of
deposition

Figure 23.20 The transport system of an alpine glacier. Debris is carried as three loads, top, middle and basal and is deposited at or near the terminus.

to the land, quarrying is thought to operate under polar glaciers, however, the process must be very slow since polar glaciers move little at their bases. The process must also work in temperate glaciers in as much as the melting and refreezing process of the base of the glacier allows the ice to weld itself onto rock after meltwater has seeped into cracks and joints. As for abrasion, it appears to be more effective with temperate glaciers because they move mainly by sliding over the ground. And of course all three erosional processes are interrelated, for the rock entrained by the glacier through quarrying functions as a tool in both abrasion and crushing processes.

Transportation of Glacial Debris: Like streams, glaciers carry three types of sediment loads. Unlike streams, however, the loads are not arranged by particle size in any discernible order. Indeed the debris carried by glaciers could be described as a garbage-like jumble of sediment ranging from huge boulders to clay-sized particles.

Glaciers carry their loads of debris in three locations as illustrated in Figure 23.20, at the base of the ice, on the top, and more or less in the middle. The origin of the **basal**, or **subglacial**, debris we have described above. It comes from quarrying, abrasion, and crushing at the base of the glacier. The **top load** is derived from debris that is dumped onto the surface of the glacier by mass movements and erosional processes such as avalanches and streams discharging from mountainside slopes. Top loads are found only on alpine glaciers. Continental glaciers, on the other hand, have no top loads because they are not fed debris by surrounding mountain ranges. **Middle loads** are also limited to alpine glaciers because this debris is derived from top load debris that gets buried in firn in the upper part (accumulation zone) of the glacier. As the glacier moves downslope, this debris is dragged into the center of the glacier.

Erosional Landforms: When an alpine glacier expands from its source area high on a mountain slope, it inevitably finds its way into a stream valley. As the lobe of ice moves down the valley, it often merges with other valley glaciers emanating from neighboring mountain sides. With each merger the glacier, like a stream moving down its drainage network, grows larger, filling the stream valley from wall to wall.

Figure 23.21 A classic U-shaped valley in the Canadian Rockies, a product of erosion by a valley glacier.

Figure 23.22 The fiords on the west coast of Norway, British Columbia, and Alaska are, like fiords everywhere, the products of concentrated erosion by long and powerful valley glaciers.

Owing to the deep and steep-sided nature of mountain river valleys, the glacier must flow in a highly confined space. As a result, a great deal of stress is exerted against the walls of the valley. Talus and other loose deposits that line the walls of stream valleys are removed forthwith and carried down valley in the initial advance of the ice. The glacier then begins to erode the bedrock of the valley walls and floor. The valley is both widened and deepened, eventually emerging with a beautifully symmetrical U-shape like the one shown in Figure 23.21. **U-shaped valleys** contrast sharply with the V-shape of the original river valley and, for this reason, valley shape is a good indicator of how far down-valley glaciers once extended in mountain ranges.

In coastal mountains, such as Norway, British Columbia, and southern Chile, hundreds of glaciers protruded into the sea (and some still do), deepening and widening the valleys along the way. When they melted back, ocean water flooded the lower valley creating a **fiord** (Figure 23.22). Inland from the coast, beyond the long fingers of the fiords, narrow pre-glacial valleys (see Figure 23.23a), were widened with glaciation (b) and then with deglaciation are reoccupied by streams (c). But conditions now are much different than they were before glaciation. For one thing, the valleys are much larger and appear oversized relative to the scale of the streams. In other words, most streams are disproportionately

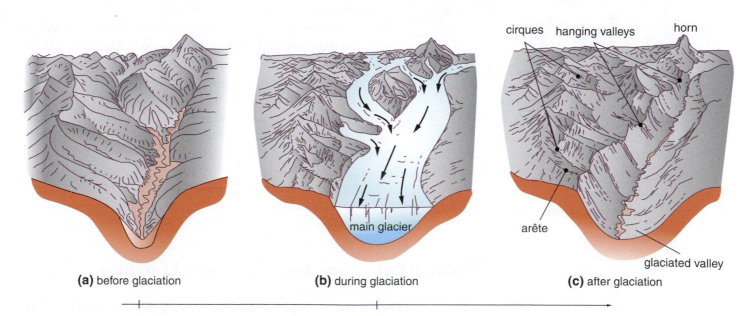

(a) before glaciation   **(b)** during glaciation   **(c)** after glaciation

Figure 23.23 Before (a), during (b), and after (c) diagrams illustrating the transformation of a mountain stream valley into a glaciated valley with its distinctive landforms.

A hanging valley and its waterfall on the rim of Yosemite Valley in eastern California.

small for their remodeled valleys. Geomorphologists give the term **misfit** to such streams. Second, valley gradients are severely altered. Glaciers gouged out the floors of the stream valleys, lowering the central sections far beyond the base of the original stream-eroded valley. This often leaves a small basin in the valley floor where a lake or wetland may form and sediments collect.

Third, after glaciation tributary valleys do not join trunk valleys at the same elevation, as they do in stream systems. Instead, tributary valleys lie at higher elevations, and appear to "hang" above the main valley and thus are given the name **hanging valleys** (see (c) in Figure 23.23. The explanation for this is related to the great difference in erosive power of large and small glaciers. Large glaciers are far more powerful and deepen their valleys much faster than smaller, tributary glaciers. Hence, the valley floors end up at unequal elevations and, as streams flow from tributary valleys into main valleys, they often tumble down the valley walls, sometimes forming spectacular waterfalls.

Other erosional features associated with alpine glaciers, also shown in Figure 23.23c, include cirques, horns, and arêtes. Where an ice field scours the side of a mountain for a long time, it often sculpts out a small basin called a **cirque**. Many cirques that formed during past glaciations no longer contain glaciers, and in their bottoms small lakes called *tarns*, have formed. Where several separate glaciers develop on different sides of a large mountain, they often erode the mountain's shoulders away, transforming if from a dome or box shape into a conical shape. Each glaciated side usually develops a beveled face, like a broad Eiffel Tower. Such mountains are called **horns** the most famous of which is the Matterhorn of Switzerland/Italy (see Figure 19.3). Each side or facet is called a *face* and faces are generally separated by sharp ridges termed **arêtes**.

Continental glaciers produce pronounced quarrying, grooving, scratching, and polishing but, with one exception, these great domes of ice produce none of the distinctive erosional landforms associated with alpine glaciers. The one exception is valley widening and deepening. This occurs mainly on the margin of the ice sheet where lobes of ice radiate from the main ice mass and enter river lowlands whose trend is in the direction of ice movement. The Finger Lakes of New York (see the inset in Figure 23.24), which have the form of U-shaped mountain valleys, are an excellent illustration of this process. At a larger scale are broad, shallow basins such as those of the Great Lakes. When the continental glaciers spread from the Canadian Shield about 20,000 years ago along the lines shown by the arrows in Figure 23.24, they gouged out basins in the less-resistant rocks on the margin of the shield. Geomorphologists speculate that before glaciation many of these basins were parts of river valleys.

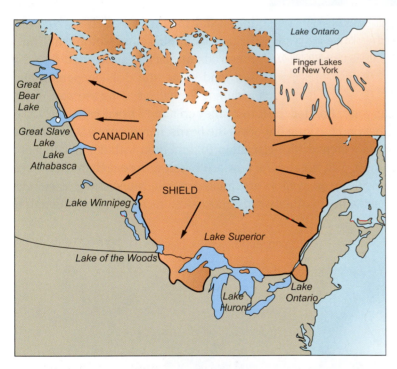

Figure 23.24 Shield margin lakes on the fringe of the Canadian Shield, the resistant core of the North American continent and a main source area for continental glaciers. The inset map shows the Finger Lakes of New York.

23.4 Glacial Deposition and Depositional Landforms

Since all the ice that passes through a glacier must eventually disintegrate, the debris carried in it must be dropped or deposited. For ice lost through calving, the debris is dropped to the bottom of the sea as icebergs melt away. This process accounts

Figure 23.25 Photographs showing the contrasting character of (a) glacial till and (b) glaciofluvial material. Till is an unstratified mix of particles, whereas glaciofluvial material is stratified and graded in composition.

for the occurrence of large stones on the ocean floor far from shore. But most glaciers do not end up in the sea, and instead deposit their loads of debris on land. Since ablation is concentrated in the lower part of the glacier, most deposition takes place near the snout. This holds for both alpine and continental glaciers. As the glacier moves ahead and melts, its load is deposited as though it was falling off the end of a conveyor belt.

There are two chief modes of glacial deposition. One is deposition directly from the ice as it melts, and the other is deposition by meltwater flowing from the glacier. Let us begin this section with the terms used to describe the materials and the deposits they form. The material deposited from the ice is called **till** and the landforms produced by such deposits are called **moraines**. All moraines are made up of till. Sediments laid down by meltwater are called **glaciofluvial deposits** and the same term is used for both the material and deposits. All material deposited by glaciers, both on land and at sea, is collectively known as **glacial drift**.

As the photographs in Figure 23.25 plainly show, till and glaciofluvial deposits are markedly different. Till is a heterogeneous mixture of unstratified materials ranging in size from massive boulders to clay-sized particles. Its composition, the manner of its deposition, and the resultant landforms vary widely, from hills and mounds arranged more or less in ridges to rolling plains. Glaciofluvial deposits, by contrast, are stratified and composed chiefly of sand and gravel. They vary in form from mounds to fan-like aprons and winding ridges. Glaciofluvial deposits are usually found in close proximity to moraines.

Till Deposits and Landforms: Moraines are named for the location where they form in and around the glacier, as illustrated in Figure 23.26. **End moraines** form at the ice front. **Terminal moraines** are end moraines marking the point of furthest advance by a glacier, whereas **recessional moraines** are end moraines laid down during halts as a glacier is melting back. Other types of moraines, include **lateral moraines**, which are deposited along the sides of a lobe or valley glacier; **medial moraines**, which are formed from the merger of two lateral moraines where two valley glaciers flow together; and **ground moraines**, which are deposited beneath the glacier as the ice

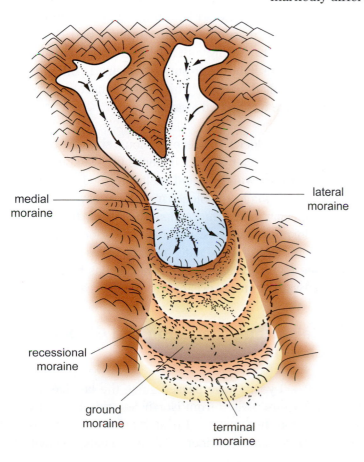

medial moraine

lateral moraine

recessional moraine

ground moraine

terminal moraine

Figure 23.26 The various types of moraines associated with an alpine glacier. Similar types and patterns of moraines form on and around the lobes of continental glaciers.

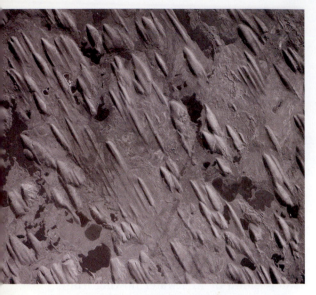

Figure 23.27 Drumlins, one of the truly curious landforms shaped by glaciers, mostly by continental glaciers.

melts along the bottom or by a "plastering-down" process as the glacier moves over the ground.

Drumlins are curious landforms composed of till in the shape of streamlined hills (Figure 23.27). They form parallel to the direction of the ice movement and are usually found in clusters. Drumlins are shaped somewhat like whales with a steeper front (snout) slope and a gentler back slope and measure in the range of 15 to 25 meters high and a kilometer or so long. Their origin is uncertain, but is generally thought to result from the reshaping of existing moraines by a re-advancing lobe of a continental glacier or by the sculpting of an accumulation of till under moving ice.

Erratics are another curious feature associated with till deposits. **Erratics** are large boulders, some the size of small houses, that are deposited with till. Because of their size, they cannot be moved by meltwater when the smaller particles are being washed away. Thus, they are often left protruding out of the ground like monoliths. Based on their bedrock composition, erratics can sometimes be traced to their place of geologic origin as a means of mapping the pattern of a glacier's movement.

Glaciofluvial Landforms: Glaciers release huge amounts of meltwater, which washes out sediment from within the ice and from the till deposits on its margin. Much of this sediment, particularly the pebbles and sand, is deposited in and around the ice front where it may build up in mounds, called **kames**, or spread out in broad fans, called **outwash plains**. Some is even deposited in the beds of the meltwater streams flowing under the ice. After the ice melts away, the beds are left as winding gravel ridges, called **eskers**.

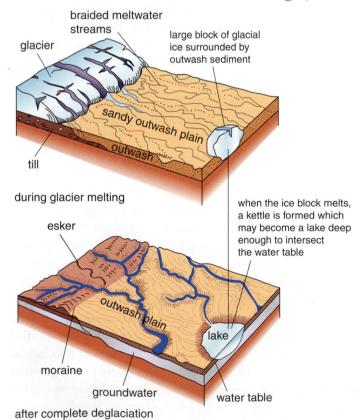

Figure 23.28 Meltwater processes and landforms near the glacial front. Lakes are formed in kettle deep enough to reach groundwater.

Outwash plains, like the one illustrated in Figure 23.28, are generally the largest meltwater landform. They are commonly found on the outer margins of end moraines and their flat topography contrasts sharply with the hilly topography of adjacent moraines. The flatness of outwash plains is often interrupted by pits, called **kettles**, which form where a partially buried block of ice has melted away.

It is common for blocks of ice of various sizes to become detached from the glacier. Stranded blocks (or "dead" ice) may last several decades while till and meltwater sediment fill in around them. Eventually, the blocks melt away leaving a kettle, which often appears as little more than a big hole in the ground. Kettles also form in moraines. Wherever they form, the deep ones usually fill with groundwater making lakes, ponds, and wetlands (see Figure 23.28). North America has hundreds of thousands of water features created by this process.

23.5 Pleistocene Glaciation and Global Change

Less than 10,000 years have passed since the last ice of the continental glaciers melted from North America and Eurasia. Prior to that time, humans lived near the fronts of the continental glaciers on both continents. They certainly witnessed the cold, stormy, glacial winters, and the watery summers with great meltwater streams pouring from ice canyons in the ice front. Did they understand the concept of a glacier as a great moving mass of ice? Probably not. That concept did not emerge until much later, in the eighteenth and nineteenth centuries, when naturalists began to make the connection between the landforms associated with valley glaciers in the Alps (see Figure 23.29) and similar features in the northern plains of Europe and the Great Lakes region of North America.

As the story of continental glaciation unfolded, we learned that only 18,000 years ago glacial ice covered 30 percent of the Earth's land area, three times as much as today. In Europe and North America, these ice sheets extended southward well into the midlatitudes. In addition, research has revealed more than one episode of glaciation, or **glacials**, with intervening warm periods, called **interglacials**. With each glacial, global temperatures fell by several degrees or more, and sea level dropped 100 meters or more in a pattern loosely illustrated by the graph in Figure 23.30. And with each interglacial, roughly the opposite occurred.

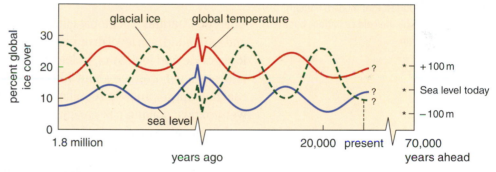

Figure 23.30 The general relationship among glacial ice, global temperature, and sea level during the Pleistocene Epoch.

Figure 23.29 Early clues. Glacial erratics found far beyond the areas of modern glaciers provided early scientists with evidence that glaciers were once much larger and more extensive.

Over the past 1.8 million years, Earth has been in an ice age (known as 'the Ice Age') during which there have been at least six major glacial stages (and probably several minor ones), each related to cooling of the global climate. (See Section 8.3 in Chapter 8.) This period of geologic time, which ended 14,000–12,000 years ago, is known as the **Pleistocene Epoch**. Currently, it appears that we are in an interglacial, called the **Holocene Epoch**, and if the climatic cycles of the Ice Age continue, we can expect another stage of glaciation in 70,000 years.

Biogeographical Change: The Pleistocene marked a time of profound geographic change on Earth. Besides the extensive geomorphic changes associated with the growth and decline of the continental glaciers, it was also a time of global biogeographical change, including the emergence and spread of modern humans, which we examined in Chapter 11. As climates cooled and the continental ice masses expanded, global bioclimatic zones shifted equatorward and generally compressed. Subtropical deserts such as the Sahara were cooler and drier, but some midlatitude deserts became wetter. In western North America, for example, the desert climate became wetter fostering heavier grass covers and more water features than today (see Figure 14.26). The core regions of tropical forests in Africa, South America, and Asia were smaller and, in the midlatitudes of North America and Europe, tundra stretched across areas that today support temperate forests (see Figure 8.8). In addition, massive volumes of silt were eroded from loose glacial deposits by wind and deposited in vast blankets of loess over parts of the midlatitudes.

Figure 23.31 Painted 17,000 years ago near the peak of the last glaciation, these cave paintings at Lasceaux in France suggest that modern humans were not vanquished by the cold conditions but lived and thrived under them.

Humans spread from Africa into Europe, Asia, North America and beyond during the Pleistocene. Instead of retreating from the advancing ice mass, human societies in the midlatitudes seemed to thrive there, they developed art and practiced painting on cave walls (see Figure 23.31), and progressed technologically despite the harsh conditions. As climate warmed at the end of the last glaciation, we invented agriculture, which, over the ensuing millennia, fostered the growth and spread of human population over most of the planet. In short, the Pleistocene appears to have given humans their start on Earth, and it can fairly be argued that we are children of the Pleistocene ice age. Other mammals also saw major changes at the close of the Pleistocene. As the graph in Figure 23.32 shows, there was a rash of extinctions in large mammals about 10,000 years ago. The woolly mammoth, the giant beaver, the giant wolf, and many others disappeared. Abrupt climate change was a significant contributor, but Ice Age hunters also probably played a role.

Oceanic Changes: As global climate cooled and the continental ice sheets thickened and expanded, sea level dropped correspondingly as water was taken out of the hydrologic cycle and locked up in ice. At the maximum of the last glaciation, 18,000 years ago, sea level dropped about 130 meters (400 feet) as the global volume of glacial ice reached approximately 70 million cubic kilometers. Vast areas of continental shelves were exposed along passive coasts and in marginal basins (or seas). Plants, animals, and humans expanded their ranges into these areas and, in some places, new land bridges were established between formerly disconnected land areas. One such connector called the Bering land bridge, shown in Figure 10.20, linked Asia and North America and was an important entryway to North America for humans and other mammals 15,000 to 20,000 years ago.

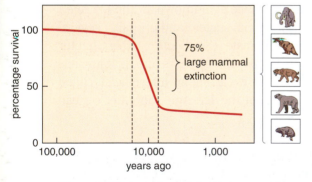

Figure 23.32 As the last glaciation drew to a close about 12,000 years ago, animal extinctions increased and many of the large mammals were especially hard hit.

Sea temperatures also changed during the Pleistocene, rising and falling by several degrees with interglacials and glacials. During glacials, ocean temperature zones shifted equatorward and sea ice expanded around Antarctica and in the North Atlantic and northern Pacific. In summer, cold water poured into the oceans from the melting ice masses. It descended to the ocean floor where it fed a massive cold current system that remains active today. (Turn to Chapter 5 for additional information on deep ocean circulation.)

North American Glaciation: Each of the major Pleistocene glaciations was marked by the buildup and expansion of massive domes of ice, 3 to 5 kilometers thick, like the one depicted in Figure 23.8. The centers, or *source areas*, of ice accumulation were located not in polar regions above the Arctic Circle, but in the subarctic zones, around 60 degrees latitude in North America. From two or three such areas in the Canadian Shield, the ice radiated outward in all directions, traveling as much as 2500 kilometers (1500 miles) before calving into the sea or melting away on land. In central North America, its southern limit is marked roughly by the Ohio and Missouri rivers.

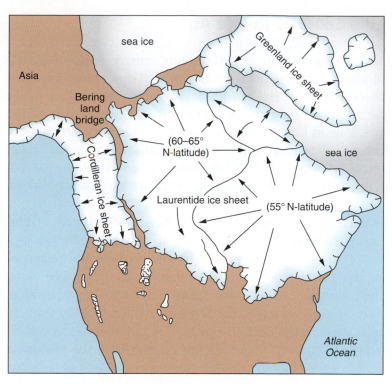

Figure 23.33 Ice coverage in North America about 18,000 years ago at the peak of the last glaciation. Notice the locations of the source areas of the Laurentide ice sheet south of the Arctic Circle.

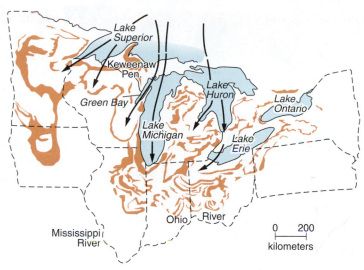

Figure 23.34 The paths of the lobes sent out by the Wisconsinan glaciation in the Great Lakes region and the patterns of the moraines they left behind.

At its maximum, each glaciation covered more than half the area now occupied by Canada and the United States (see Figure 23.33). In addition, the region of coverage by each stage of glaciation coincided roughly with that of its predecessor. As a result, in most locations, it is only the effects of the last glaciation, called the **Wisconsinan**, which can be found in the landscape. From the deposits left by the Wisconsinan glaciation in southern Canada and adjacent areas of the United States, we reason that the ice front advanced several times. Each of these re-advances is called a **substage** and, in most locations, substages took the form of glacial lobes moving from the main body southward along lowland corridors.

The Great Lakes Region provides an especially clear picture of the glacier's behavior. Here the pattern of advance and retreat of each substage followed the contour of the lake basins and connecting lowlands, as shown in Figure 23.34. One lobe followed the Lake Michigan Basin, another followed the Lake Huron Basin but split into sublobes when it reached the thumb of Michigan's Lower Peninsula. In Lake Superior, the Keweenaw Peninsula split the advancing ice into two lobes, one driving southwestward into central Minnesota and northern Wisconsin and the other driving southward along the Green Bay lowland deep into Wisconsin.

The pattern of movement of the late-Wisconsinan glacial substages is clearly marked by the distribution of the deposits they left behind. Especially pronounced are systems of moraines shown in Figure 23.34 that define the footprint of the individual lobes and their pattern of retreat. Row after row of end moraines mark the halting withdrawal of the ice as it slowly melted back toward its source areas in Canada.

These moraines and related landforms such as outwash plains, provide the basic geographic infrastructure for today's landscapes in the continent's central plains and lowlands. Many streams, for example, flow in the lowlands between moraines, and the outwash plains are associated with porous, infertile soils that in many areas correlate closely with the distribution of conifer forests and poor agricultural land. Till deposits, on the other hand, provide richer soils with far better agricultural potential.

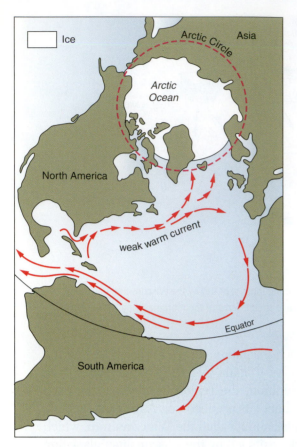

Figure 23.35 Before the closing of the Panama land bridge, most of the warm water from the North Equatorial Current in the Atlantic flowed westward between North and South America, leaving the northern Atlantic cold.

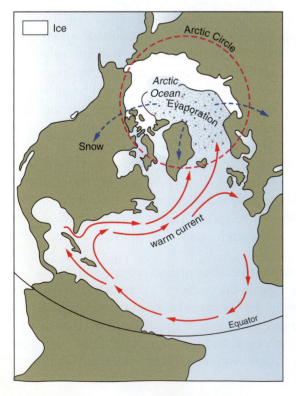

23.6 Ideas On the Cause of the Ice Age

Why after tens of millions of years without an ice age, did Earth freeze down about 2 million years ago? And why was the resultant ice age marked by a string of fluctuations that produced a series of glacial and interglacial periods? There is little doubt among scientists on the answer to the second question. The explanation is provided by two measurable variations in Earth's orbital geometry, called eccentricity and obliquity, which produce periodic changes in the receipt of solar radiation and which we discussed in Section 8.3 in Chapter 8. On the first question, as to why the Ice Age began in the first place, there is uncertainty. However, there have been plenty of ideas.

The ideas advanced on the cause of the Ice Age fall into three categories depending on which part of the Earth's radiation/heat system is involved: (1) the Sun's energy output; (2) the capacity of the Earth's atmosphere to conduct heat and radiation; and (3) the redistribution of heat by the oceans. The first two have received much serious attention but, for a variety of reasons, have been ruled out. Two of the most promising ideas from this set, namely, a reduction in solar radiation caused by a dirty atmosphere and reduced atmospheric heat absorption due to low levels of carbon dioxide have been discarded because ocean sediments dating from the beginning of the Ice Age show no evidence of an increase in atmospheric fallout of airborne particles or a decrease in carbon dioxide.

Changes in the pattern of heat redistribution in the ocean, however, hold promise. Ocean circulation has a major influence on air temperature and climate throughout the world and one of the key controls on ocean current patterns is the position of large landmasses in the ocean. When landmasses change position as a result of plate tectonics, ocean circulation often changes and climate is altered, sometimes dramatically. Antarctica is an example. About 35 million years ago, Antarctic split off from Australia and become positioned over the South Pole. When this took place, the continent lost touch with currents fed by warm, tropical waters and became dominated by cold, circumpolar ocean currents (see Figure 8.7). These currents, driven by the prevailing westerly wind system, flowed in a continuous loop around the continent. Without an influx of warm water, coupled with global cooling related to depressed CO_2 levels around this time, Antarctica went into a deep freeze where it remains to this day.

About 3 million years ago, a small but significant change took place in the geographic arrangement of the western hemisphere – the Panama land bridge formed between North and South America. This led to a major change in Atlantic Ocean circulation because prior to this time, as shown in Figure 23.35, the North Equatorial Current passed between the two continents. With the Panama closing, this tropical current was diverted northward to become the Gulf Stream, which then pumped massive amounts of warm water into the northern reaches of the Atlantic.

Two different ideas have been advanced to explain what happened next. The first idea proposes that warming extended into the Arctic Ocean, as Figure 23.36 illustrates, melting the ice cover there and opening this vast area of water to evaporation. Fed by water vapor from the Arctic Ocean, snowfalls increased to massive proportions

Figure 23.36 With the closing of the Panama isthmus, ocean currents warmed the northern Atlantic and Arctic oceans leading to melting of the Arctic sea ice, increased water-vapor production, heavy snowfalls, and glaciation.

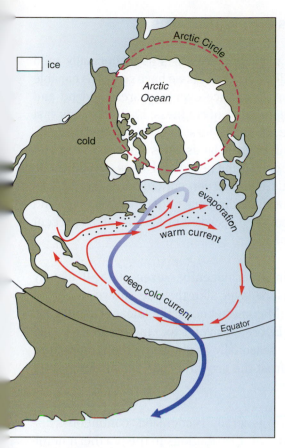

Figure 23.37 Heavy evaporation in the northern Atlantic leading to ocean cooling, atmospheric cooling, and initiation of the Ice Age.

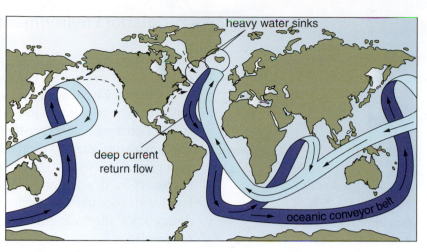

Figure 23.38 The grand scheme of ocean circulation with a deep water fed by heavy (high-density) water sinks in the northern Atlantic.

around the Arctic Basin in North America and Eurasia, giving rise to continental glaciers. As the glaciers grew, climate cooled, eventually affecting the entire globe by the time the ice reached its full 30 percent land coverage. When the climate got cold enough, the Arctic Ocean froze over again, cutting off the source of water vapor and the supply of snow.

The second idea, shown in Figure 23.37, proposes that the Arctic Ocean and surrounding lands got colder instead of warmer with the influx of the Gulf Stream into northern waters. The cooling process was driven by greatly increased evaporation in the northern Atlantic via a system that works like this. As evaporation rises, salt concentrates in surface water, driving its density up. The evaporation also cools surface water as heat is taken up with the vapor flux into the atmosphere. Together the salt and cool water result in surface water so heavy that it sinks into the ocean where it feeds the system of deep ocean currents (see Chapter 5 for a description of this system). Significant to this idea is the argument that dense surface water sinks in the northern Atlantic, well south of the Arctic Ocean. The Arctic Ocean is thus isolated from warm-water inflows and falls into a deep freeze, which drives temperatures in the region down initiating continental glaciation. The glaciers expand, feed more cold water to the deep current system, shown in Figure 23.38, which eventually spreads cold water throughout the oceans, bringing down temperatures worldwide.

Both ideas have merit. The first accounts for a source of water vapor at high latitude that is able to manufacture enough snow to grow an estimated million cubic kilometers of ice. On the other hand, it does not provide for cooling mechanism to start the Ice Age or a mechanism – beyond the cold ice sheets themselves – to extend cooling over the globe. The second idea provides for a cooling mechanism to start the Ice Age as well as a mechanism (deep ocean currents) to cool the globe. This system is very much active today, and some scientists speculate that it may expand with the influx of meltwater from the rapidly melting ice sheets in Antarctica and Greenland. Could this huge ocean cooling system offset twenty-first century global warming and trigger another continental glaciation?

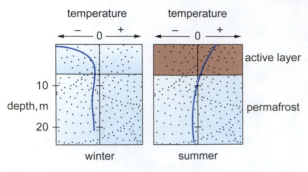

Figure 23.39 Permafrost capped by an active layer which freezes and thaws seasonally.

23.7 Periglacial Environments and Landforms

Earth is generally described by planetary scientists as a thermally mild planet, but in many ways it is actually quite cold. As we have seen, it is capable of producing glacial ages that can spread ice over vast areas of land and water. But ice coverage is just one measure of coldness. Earth also has vast areas of extreme cold where there are no glaciers. These are called **periglacial environments** and they are defined as landscapes of intense frost and frost-related processes. Together, glacial and periglacial areas cover 30 to 35 percent of Earth's land area. Of this, about 80 percent (26 percent of Earth's land area) is occupied by permafrost including that under the Greenland and Antarctic ice sheets. Permafrost covers 82 percent of Alaska and 50 percent of Canada.

Permafrost: Permafrost is cold ground in a permanently frozen state. It usually takes the form of a solidly frozen subsurface layer, like the one shown in Figure 23.39, ranging from 1 to 500 meters thick. (Maximum known thickness is about 1500 meters in Siberia). Above the permafrost layer is a layer 1 to 3 meters deep, called the **active layer**, which freezes and thaws seasonally as shown by the temperature profiles in Figure 23.39. Below the permafrost the ground is non-frozen and heated by geothermal energy from the Earth's interior.

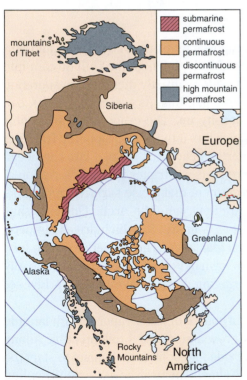

Figure 23.40 The distribution of permafrost in the northern hemisphere including the ground under the Greenland ice sheet.

Four geographic classes of permafrost are generally recognized, continuous, discontinuous, mountain, and submarine, and their distributions in the northern hemisphere are shown in Figure 23.40. **Continuous permafrost** extends uninterrupted over vast areas of polar lands in the coldest parts of North America, Siberia, and Antarctica. It generally exceeds 10 meters in thickness, and along its northern margin in Russia, Alaska, and a small area in Canada, it extends under the Arctic Ocean as **submarine permafrost**. This variety of permafrost is a remnant of the Wisconsinan glaciation when sea level was much lower and the land offshore was exposed to frost penetration.

Discontinuous permafrost is found mainly in subarctic lands on the southern margin of the continuous zone in North America and Asia. It is thinner than continuous permafrost, generally less than 5 meters thick, and distributed in a broad, patchy pattern. Patches favor locations with lower rates of solar heating and ground-heat penetration such as north-facing slopes and soil insulated by heavy layers of organic matter. **Mountain permafrost** has much the same pattern of distribution, favoring locations protected from the Sun such as cirques on the north faces in the Rockies, Alps, and Himalayas.

Periglacial Geomorphic Activity: Solidly frozen ground, bound together as it is by ice crystals, has the approximate strength of concrete. Thus, within the permafrost layer itself there is not much ground movement. The active layer, on the other hand *is* subject to movement as part of the seasonal freezing and thawing process. When water freezes, it expands by 9 percent of its liquid volume. Since the active layer is usually saturated with meltwater resting on the underlying permafrost table (surface), freezing causes it to expand. Two basic motions are produced: heaving and thrusting.

Figure 23.41 In periglacial landscapes, frost action produces curious surface features including (a) solifluction lobes on hillslopes here outlined by the pattern of snow accumulation; and (b) patterned ground in areas of stony soils.

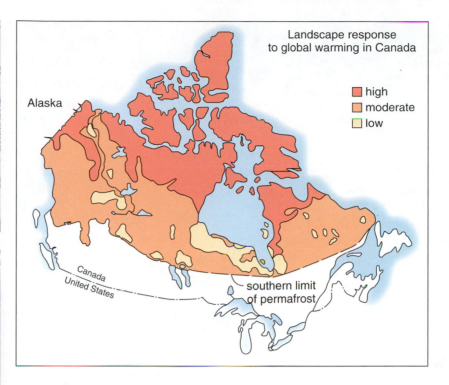

Figure 23.42 The anticipated distribution of the relative impacts on landscape in Canada with changes in permafrost related to global warming according to Environment Canada. Impacts include ground subsidence and slope failure.

Frost heaving (see Figure 20.39) lifts the surface up and **frost thrusting** pushes it sideways. On a hillslope, these motions drive several geomorphic processes that move soil material downhill. Among these, **solifluction** is probably the most pronounced. It produces a gradual flow-like movement in the upper active layer that results in the formation of elongated bulges or lobes on the slope face like those shown in Figure 23.41a. In tundra landscapes, solifluction lobes give the surface a distinctive bumpy texture.

Patterned ground is another familiar texture in tundra landscapes that is related to frost action in the active layer. Patterned ground is usually composed of stones arranged in various geometric shapes on the soil surface, for example, polygons and stripes (see Figure 23.41b). How stones become concentrated on the surface is explained mainly by frost heaving in the active layer. Heaving pushes stones out of the ground whereas thrusting moves them laterally causing them to accumulate and eventually form some sort of pattern on the surface.

Land Use in Periglacial Landscapes: Modern land use has affected periglacial environments at several scales beginning with global warming. The emerging picture of global warming is showing an uneven thermal pattern over the world with the most pronounced warming taking place in the periglacial regions of North America and Eurasia (see Figure 8.12). The effects on the periglacial landscape, as the map in Figure 23.42 suggests, are expected to range from slight to dramatic as permafrost retreats from shores and stream banks and melts to form deeper active layers. Among the outcomes are increased wave erosion in coastal areas, weakening of hillslopes leading to slides and other forms of failure, and changes in surface drainage leading to alterations in vegetation and ecosystems.

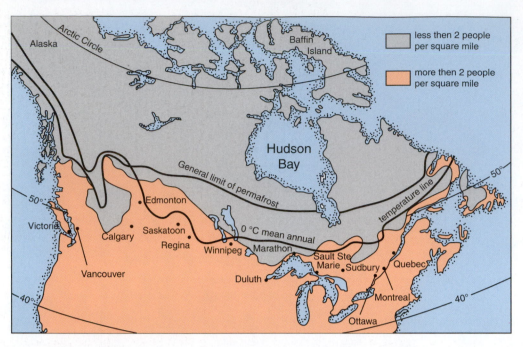

Figure 23.43 Periglacial environments support few people as this map of Canada reveals.

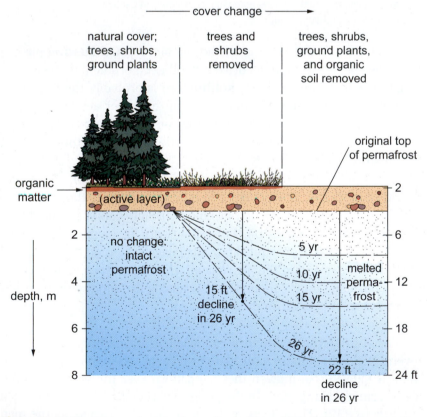

Figure 23.44 After 26 years, land clearing so changed the soil-heat balance that permafrost declined by 22 feet at one test location.

At the local scale, the interplay of permafrost and land use is well documented. As a whole, settlements are small and sparse in periglacial lands. In North America, the northern limits of urban development roughly follows the southern fringe of the periglacial region (see Figure 23.43). Beyond this border in Canada and Alaska, average population density is less than two persons per square mile (less than one person per square kilometer). The first point to note is that land-use facilities are difficult to construct and maintain in areas of permafrost. Among other things, land clearing upsets the soil-heat balance resulting, as Figure 23.44 shows, in permafrost decay and thickening of the active layer. Concrete, asphalt, pipes, and buildings can have similar effects and lead to sagging of the ground and breaking of foundations, pipes, and roads. If pipelines break, as they have in many places on the Russian tundra, environmental damage from oil spills can be severe. This is a serious concern because the tundra supports a rich complex of relatively undisturbed ecosystems and, as such, is one of the last great natural reserves on the planet.

Summary and Overview of Glacial Systems: Growth, Motion, and Work of Glacial Ice

Our species emerged and spread over the planet at a time in Earth history of changing climates and waxing and waning ice sheets. Increasingly, we have learned that glaciers and the geographic changes they brought on are not incidental to our existence, but in many ways central to it. They are clearly worthy of study, and we have learned what geographic conditions account for their formation and that they are powerful geomorphic agents, capable of obliterating and

rebuilding entire landscapes. And while we understand that glaciers and glaciations are intimately tied to the global atmospheric and oceanic systems, we do not fully understand how the connections function, especially the feedback mechanisms. The summary diagram highlights some of these uncertainties by showing that global warming could lead to markedly different outcomes for glaciers.

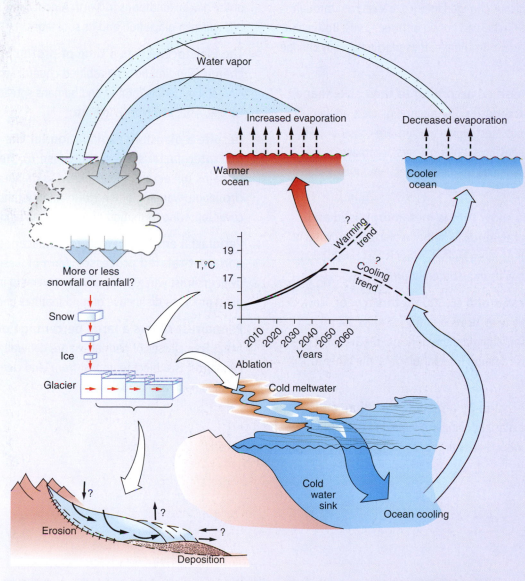

Feedback mechanisms and their possible effects on glacier mass budget (represented by the block diagram). One route follows a cooling trend related to the discharge of cold meltwater into the ocean, which leads to decreased ocean evaporation and the possibility of reduced snowfall and glacier nourishment. The other route follows a warming trend resulting in increased ocean evaporation and the possibility of greater snowfall and glacier nourishment.

► **Glaciers are masses of ice formed from the accumulation and compression of large quantities of snow.** They are capable of slow movement driven by the force of gravity pulling on the ice mass, and they are capable of eroding the land under them and transporting and depositing the eroded debris.

► **Glaciers currently cover about 10 percent of Earth's land area.** More than 95 percent of this area is taken up by two great ice sheets. The remainder is made up of thousands of alpine glaciers. During the Pleistocene Epoch, the global ice cover was much greater than today.

► **Glaciers operate as input–output systems.** Input comes in the form of snow, which is changed into ice, transferred through the system, and then discharged in two or three forms. Both input and output are dependent on climate; thus glaciers usually change with changes in climate.

► **Glaciers can be classified according to their size, shape, and thermal characteristics.** Continental glaciers are largely unconstrained, whereas most alpine glaciers are constrained by the shape of mountain valleys. Polar glaciers are frozen to the ground under them, whereas thermal glaciers are much warmer and highly fluid.

► **Glaciers are fed by snow that is metamorphosed and compressed into glacial ice.** Whether a glacier is growing or shrinking depends on its mass budget. Mass is lost by ablation which involves as many as three different processes.

► **Glaciers move at rates of 5 to 300 meters a year.** Flow rates vary significantly with location in the ice mass. Plastic deformation and basal sliding are the principal mechanisms of ice movement. One is common to polar glaciers, the other to temperate glaciers.

► **Glaciers produce erosion by abrasion, crushing, and quarrying.** In alpine glaciers, sediment loads are carried in three locations in the ice mass. Glacial drift is made up of stratified and non-stratified deposits.

► **Alpine glaciers create many distinctive erosional features.** Continental glaciers create less dramatic landforms but can carve out large basins like those of the Great Lakes. After an alpine glacier melts away, it leaves a valley with a distinctive shape.

► **Glacial drift is made up of till and meltwater deposits.** Deposits composed of till are called moraines and there are many types of moraines. Glaciofluvial deposits include outwash plains and eskers.

► **During the Pleistocene Epoch, there were at least six major glaciations.** Each was marked by global cooling and lowered sea levels and followed by a distinct interglacial. Most of the glacial landforms in North America are the products of Wisconsinan glaciation and its substages.

► **The Pleistocene was a time of profound geographic change.** Bioclimatic zones shifted equatorward and compressed. New soil materials formed and humans spread from Africa into Europe, Asia, and North America.

► **Despite a plausible explanation for the cause of glacial and interglacial stages, the cause of the Ice Age itself remains uncertain.** Ideas related to changes in ocean circulation, warming of the North Atlantic, and Arctic Ocean ice cover following formation of the Panama land bridge hold promise.

► **Periglacial environments are landscapes of intense frost and frost-related processes.** Most of these areas are occupied by permafrost where frost-driven movement in the active layer often produces distinctive ground features and landforms.

► **Permafrost covers a large percentage of Earth's land area.** Four classes of permafrost are generally recognized. Permafrost is subject to impact from land clearing, land-use development, and global warming.

Review Questions

1 Can you define four basic differences between glacial ice and lake ice?

2 What is the current percentage of Earth's land area covered by glacial ice and where is most of it found today? How about 18,000 years ago?

3 What are the key differences between temperate and polar glaciers? In which class would you place alpine glaciers? The continental glaciers of 18,000 years ago?

4 In our model of glaciers as systems, can you describe how the basic system works and how a glacier should behave when its mass budget is negative over an extended period of time. What happened to glacier mass budgets during the Little Ice Age?

5 Can you describe the processes involved in ablation and identify which one is dramatically displayed in Antarctica?

6 What is basal sliding, is it more common to polar or temperate glaciers, and what do you propose should be its connection to scouring or abrasion?

7 Where do glaciers carry their loads, where do most loads come from, and where in the glacier system are most deposited?

8 Based on the illustrations in Figure 23.23, can you describe the three stages of development in an alpine valley? What stage is associated with fiords and where in the world would we such features?

9 Alpine glaciers are famous for the distinctive landforms they impart to mountain landscapes. Can you define the features known as hanging valleys, cirques, horns, and arêtes and reflect on how they are formed?

10 What is the difference between material called till and that called glaciofluvial? And what landform features are associated with each, and what would they look like if you saw them in the field?

11 Humans have been described as the children of the Ice Age. Explain the logic behind this assertion.

12 What are glacials and interglacials, when did the last glacial reach its peak, and what were some of the major biogeographical changes associated with it?

13 Can you describe the general relationship between the increase and decrease in global temperature, glacial ice mass, and sea level?

14 Continental glaciations such as those that occurred during the Pleistocene begin in the sea. Is there any basis for this remark and why?

15 What is meant by the term periglacial environment, what is permafrost, and how are periglacial landscapes affected by global warming and land clearing?

Wind Systems:
Sand Dunes, Dust, and Deserts

Chapter Overview

Again and again we are reminded how critical the atmosphere is to understanding the geographic character of the Earth's surface. In earlier chapters we established that the atmosphere is, among other things, the medium for heat, radiation, and gas exchanges, the source of precipitation and freshwater, and the embodiment of the greenhouse that mitigates thermal extremes and nurtures all life. And now we take one last look at the atmosphere, this time as a geomorphic system, a fluid capable of eroding land, moving sediment and shaping landforms. Like most geographic phenomena, this system operates at several scales, ranging from the very large troposphere to the micro-world of ants on the ground. We explore the behavior of the airflow system with a brief look at wind velocity, direction, and patterns and then venture into wind erosion, sediment transport, and related landforms. The chapter goes on with a description of the processes involved in building and moving sand dunes, the names and the various classes of dunes found in desert and coastal zones, and ends with a review of wind deposits at the global scale and the big question of the relationship of the atmosphere as a geomorphic system to other global systems.

Introduction

Ray looked at the sky, thought for a moment, and said, "We'd best take a ride." He headed for the car behind the fish house. This was serious. It meant that the storm was getting bad and even these two old fishermen, veterans of howling gales, were hesitant about a trip to the center of Lake Superior. So the three of us rode down the coast to a high point where you could see over the open lake. It was ugly – a frothy mass of white against the gray backdrop of the early morning sky. No one spoke. Ray chewed his cigar. Vernie smoked a cigarette. I crossed my fingers and, with furtive glances at one and then the other, waited for a clue about whether we would make a run 30 miles into the lake to lift nets. None came. In 20 minutes we were back at the dock and without a word they both headed straight for the tug. In a few minutes the diesel was sputtering blue smoke and Ray signaled me to throw off the bowline.

The wind was hard from the northwest and with nearly 200 miles of open water to blow across, it could generate huge waves. And it did. As we cleared the north end of Grand Island, they hit us nearly head on. Great ridges of green water capped by smaller waves whose tops

Figure 24.1 The wind-beaten sandstone cliffs along the south shore of Lake Superior.

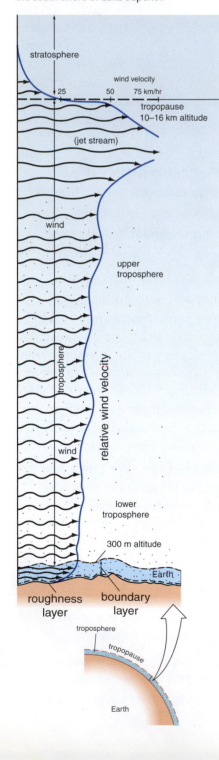

Figure 24.2 The three main airflow systems of the lower atmosphere. The troposphere at a thickness of 10 to 16 kilometers is the largest, but most geomorphic work is performed in the lower boundary layer and the roughness layer.

were being torn off by the wind. It was a magnificent ridge-and-valley topography of rolling liquid accentuated by the changing rhythm of the engine as it strained and relaxed with each crest and trough. Occasionally, a big one rose up under us enveloping the entire midsection of the boat, leaving the propeller partially spinning in the wind. Water sprayed in through cracks in the cabin and I remember thinking that small windows were a good thing in a boat in rough seas.

Years later, I thought about those magnificent waves and the force of the wind that drove them and how the great gusts tore over the cliffs and bluffs along the shore. No wonder the high sandstone cliffs, the very ones shown in Figure 24.1, were stripped clean and the trees along their crests tattered and flagged as though they'd survived a bomb blast. Winter and summer, the lake's powerful winds beat against the cliffs and the mantle of soil capping them. Particle by particle they wore them back, sending the tiny missiles of sand sailing inland to form sand dunes, which the winds relentlessly drove into the bordering forest. At length, the air lost its power to push the dunes against the forest. Had the forests not stood in their way, how far would the sand dunes have advanced? I imagined a different North American geography in which Lake Superior was bordered instead by a vast, treeless desert. What would subdue the wind? Would the sand dunes be unstoppable?

24.1 The System of Airflow over the Earth's Surface

Think of air as a light fluid and the atmosphere is an enormous ocean that rests on the Earth's surface. The currents of the atmosphere, its winds, are governed by the same principles of fluid motion that govern other fluids including water. But there are some important differences between air and water, chief among which is density. At a density of 1.29 kilograms per cubic meter, air is 775 times lighter than water (density equal to 1000 kilograms per cubic meter), and therefore exerts less force on the Earth's surface. Air is also much less viscous than water and therefore flows more readily when force is applied to it, which helps account for the atmosphere's nearly perpetual motion.

When we examine air as a geomorphic system, we need to focus on three scales of observation: the troposphere, the atmospheric boundary layer, and the roughness layer, each shown in Figure 24.2. The troposphere is the lower division of the atmosphere, the deep layer of active weather that lies between the base of the stratosphere and the Earth's surface (see Figure 4.5). The boundary layer is the lower layer of the troposphere where all the exchanges between the atmosphere and the Earth's surface take place, and the roughness layer is the very thin, lowermost layer of the boundary layer that rides directly on the Earth's surface. Wind velocity is fastest in the troposphere miles above the Earth's surface, and decreases toward the surface, reaching zero within a few centimeters of the ground.

The Troposphere: The **troposphere** is a great ocean of air 10 to 16 kilometers deep where all Earth's weather systems operate. It is the zone of pressure cells (which we explored at length in Chapter 5) and fast-moving, circumglobal airflow, called geostrophic winds, which travel up to several hundred kilometers per hour (see Figure 5.17). The entire troposphere is subject to mixing motion, which can lift particulate matter (or aerosols) from the Earth's surface high into the atmosphere where it enters the stream of geostrophic wind and the zone of long-distance transport, often around the world. This system is responsible for moving billions of tons of sediment picked up every year from volcanoes, forest fires, prairie fires, wars, cities, croplands, and deserts. Eventually it settles back to Earth, falling on the oceans and continents far from its place of origin.

The amount of sediment deposited on the land from tropospheric circulation is slight for most land areas, a mere dusting. But there are locations and times when it can be considerable. At the global scale there are two great corridors of long-distance transport of airborne sediment in each hemisphere, the easterly wind system in the tropics and the westerly wind system in the midlatitudes (Figure 24.3). These systems can move sediment thousands of kilometers in days, especially where upper atmospheric winds are strong. The strongest upper wind system is the polar-front jet stream, which courses west to east across the midlatitudes. In the tropics, jet-stream circulation is weaker, but beneath this system the west-flowing tradewinds are highly effective in moving sediment over long distances.

Long-distance transport of sediment by these wind systems can be substantial. Dust whipped up over the Sahara of North Africa is carried westward by easterly winds over the Atlantic to Central America, South America, and beyond. En route, millions of tons each year are dumped into the Atlantic Ocean and millions more are sprinkled over South America. Ash from tropical volcanoes such as the 1993 eruption of Mt. Pinatubo in the Philippines is often carried around the world, as shown in Figure 1.14, reaching high into the troposphere where it blocks out part of the incoming beam of solar radiation. In the midlatitudes, pollutants from urban–industrial centers and dust from cropland are prominent in long-distance transport. Plumes carrying aerosols from large cities such as New York have been observed thousands of kilometers downwind in the air stream of the westerly winds, and, in the Northern Pacific Basin, millions of tons of dust eroded from cropland in China have been measured raining down on the North American West Coast. We will say more about long-distance atmospheric transport at the end of the chapter.

The Atmospheric Boundary Layer: The **atmospheric boundary layer** forms the transition zone between the troposphere and the Earth's surface, roughly in the manner shown in Figure 24.4. Here the atmosphere drags over the Earth's surface and is slowed down by friction with the landscape and sea surface. The atmospheric boundary layer is about 300 meters (1000 feet) thick and more or less follows the

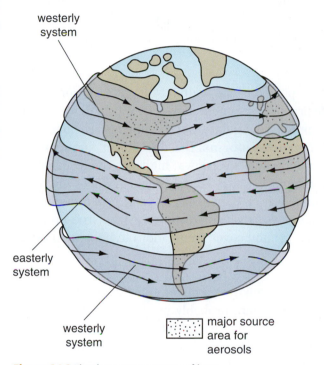

westerly system

easterly system

westerly system

[major source area for aerosols]

Figure 24.3 The three great systems of long-distance transport in the troposphere and some of the major source areas for aerosols.

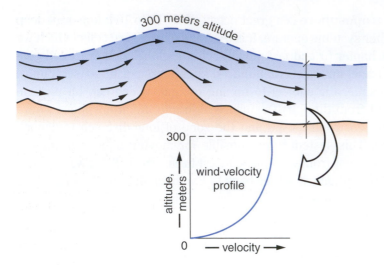

Figure 24.4 The atmospheric boundary layer is the lower 300 meters of the atmosphere where air is dragged over the Earth's surface.

contour of the Earth's surface over both land and water. Within the boundary layer wind speed is slowest at ground level and increases with elevation above the surface. At an elevation of 300 meters or so, however, the increase diminishes, beyond which airflow is variable but relatively fast for thousands of meters upward (Figure 24.2).

Airflow in the boundary layer is principally turbulent but the intensity of turbulence varies depending on surface roughness, thermal instability, wind speed, and height above the surface. In general, turbulence declines above ground level. Particles wafted upward by rising currents are carried downwind by streamlines of lateral flow. But to stay aloft for long periods, particles must be generally small, less than 0.01 millimeter in diameter.

Huge amounts of sediment are transported in the boundary layer. At ground level, in the roughness layer, sand is the principle sediment moved by wind. Above ground level, silt and clay particles are the primary boundary-layer sediments. The best illustration of boundary-layer transport are **dust storms**. In a dust storm, fine sediment completely fills the boundary layer and often "overflows" high into the troposphere above. Dust storms are most common in "less-dry" desert regions where the mean annual precipitation is 10–20 centimeters (4–8 inches) per year. It is thought that these environments are moist enough to advance weathering (and free up particles), but too dry to produce enough ground cover (vegetation) to hold the particle residues in place.

Figure 24.5 A desert dust storm which fills the boundary layer with dust of blinding density.

Dust storms are responsible for moving well over a billion tons of sediment worldwide annually. Satellites reveal that the Sahara probably produces the largest dust storms, but dust storms are common to all deserts and many grassland regions as well. In North America massive dust storms occurred in the Great Plains in the 1930s as wind eroded silt from dry, abandoned farm fields. The dust came from vast surface deposits, called **loess**, which were themselves laid down by wind thousands of years before, and which we will look at in more detail later in this chapter. As the photograph in Figure 24.5 reveals, the leading edges of dust storms are often marked by a well defined wall that more or less fills the atmospheric boundary layer.

The Roughness Layer: At the very bottom of the boundary layer, lying directly on the ground, is a shallow layer of nearly still air called the **roughness layer** (or **roughness length**). It is created by obstacles in the landscape such as trees and buildings, which force wind streamlines to rise above the ground, a process called *flow separation*. A forest, especially a dense one, can force wind several meters off the ground but, where the obstacles are shorter, as in a field of grass, the separation distance is only 2 or 3 centimeters as depicted in Figure 24.6. In both cases, however, wind is denied contact with soil particles, which explains why vegetation is such an important control on wind erosion. Over barren soil, by contrast, the roughness layer is microscopic, which allows eddies of fast air to strike the ground. Fine particles like silt are picked up by the eddies and hoisted into the wind stream. Sand particles, which are much heavier, can also be moved, but much closer to the ground in skidding and bouncing-like motions that we will describe a little later in connection with sand-dune formation and movement.

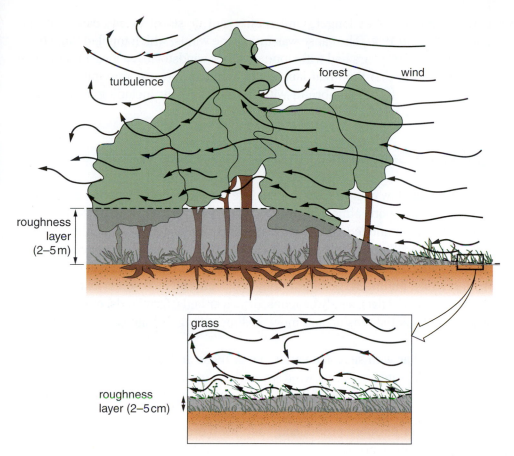

Figure 24.6 The roughness layer, found at the base of the boundary layer, is characterized as a zone of calm air which varies in thickness with vegetation and other landscape obstacles.

Wind Direction, Velocity, and Power: There are several characteristics of wind that make it quite different from other geomorphic agents. One is its capacity to move material uphill. Another is its pattern of behavior. Wind is extremely variable in terms of both direction and velocity. At ground level, wind in most places is very gusty, changing direction and velocity in a matter of minutes or seconds. Unlike watersheds where water and sediment are moved in essentially one direction, wind can move sediment in any direction. In addition, the erosive power of wind varies radically with changes in velocity.

In assessing the erosive power of wind, raw velocity readings can be somewhat misleading. This is because the actual force wind exerts on surface objects is not directly proportional to its velocity. Wind power increases with the cube of velocity. Thus a small increase in velocity can cause a large increase in power as the graph in Figure 24.7 illustrates. This means that fast winds are able to do tens of times more work than slow winds. Put in a geographic context, one powerful storm is capable of eroding more soil in a day or two than light winds can in several months or even several years. It follows that average or mean wind-velocity readings are not very meaningful in problems of wind erosion and sediment transport. It is the high magnitude ones that really count.

Wind Force and Resistance: Another way of looking at erosive power is in terms of the ability of wind to overcome the landscape's resistance to its force. That is, how strong must a wind be to damage the landscape by knocking down trees or eroding soil? Loose particles of tiny sediment such as clay particles need very little air movement to sweep them away; silt needs somewhat faster air; sand even faster; and pebbles need exceptionally fast wind to even nudge them forward. Large trees are highly resistant to wind, especially in forests, because a forest thickens the roughness layer, forcing wind to ride well above the ground. But forests are not totally exempt from wind damage.

In the massive tropical rainforests of the Amazon, patches of forest are commonly leveled by wind. The main cause of such blowdowns is *downbursts* of wind (extremely powerful downflows) from thunderstorms (Figure 24.8). In the midlatitudes, downbursts are less common, but hurricanes and tornadoes are major sources of tree blowdown. Such events are infrequent to be sure, for in most places such as New England, the Midwest, or Ontario, a wind of tree-toppling force may come only once

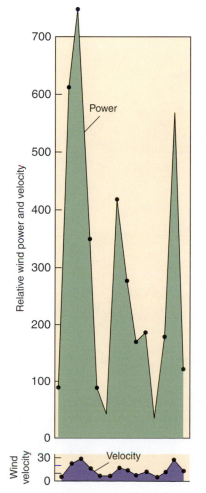

Figure 24.7 The relative change in wind power with changes in wind velocity, showing how power increases with the cube of velocity.

Figure 24.8 Trees shredded and uprooted by powerful winds, probably a downburst from a thunderstorm.

Figure 24.9 Famous wind systems and their trajectories. Most regional winds are seasonal and associated geographically with deserts and/or mountain ranges (compiled from various sources).

every 100 to 200 years or longer. One such event or set of events came to New England in 1937 and 1938. The cause was hurricane winds, which toppled huge trees in a forest that for 300 years had withstood all other windstorms, a forest that foresters and botanists of that time had labeled as a climax forest community, one supposed to be in a state of equilibrium with its environment including resilience to disturbances by big winds.

24.2 Some Geographic Patterns in Surface Winds

The patterns of airflow over the Earth's surface are exceptionally varied geographically. Many factors, both atmospheric and terrestrial, account for these patterns. We have already mentioned the broad variations associated with prevailing wind systems. The prevailing westerlies and the easterly tradewinds, in particular, dominate zonal swaths in the midlatitudes and tropics in both the northern and southern hemispheres (see Figure 24.3). Where they cross unprotected soil or exposed sediment, whether in tidal flats, sandy beaches, dry river beds, farmlands, or desert, they can produce substantial erosion and move sediment long distances downwind.

We are also familiar with seasonal variations in these and other wind systems. The world's most celebrated seasonal wind system is the monsoon of Asia (see Figure 5.23), which in the winter erodes soil as it blows across dry, unprotected farmlands of India, Bangladesh, and other Asian countries. But there are also other seasonal wind systems that figure far more prominently into wind erosion and sediment transport and a number of these are shown in Figure 24.9. In the Middle East, for example, there is a powerful winter wind, called the Shamal, which blows dust down the dry

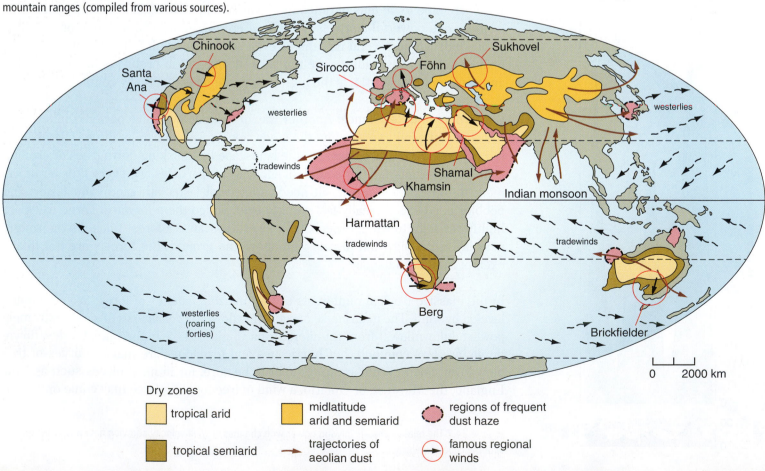

Dry zones

☐ tropical arid
☐ midlatitude arid and semiarid
⬭ regions of frequent dust haze

☐ tropical semiarid
→ trajectories of aeolian dust
◯ famous regional winds

Figure 24.10 Coastal sand dunes are found throughout the world in a wide range of geographic settings where there is an ample supply of sand and winds strong enough to blow it inland.

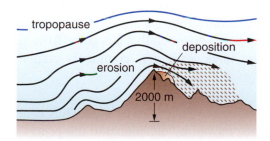

(a) mountain range

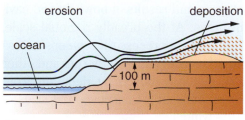

(b) coastal cliff

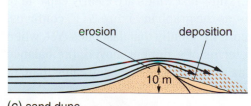

(c) sand dune

Tigris–Euphrates Valley and over the Persian Gulf, and in North Africa, the Khamsin (or Simoon) and the Sirocco drive great dust storms northward across the Sahara. These desert winds and others like them move millions of tons of sediment every year, and they can be so strong, persistent, and annoying in some places that they are also blamed for emotional disorders in humans such as depression and even suicide.

In North America, a monsoon-type wind develops over the southern Great Plains in late spring and early summer. As in Asia, this northward (southerly) flow is a response to the development of atmospheric low pressure from terrestrial heating in the continental interior. En route from the Gulf of Mexico it winnows dust from the dry southern plains and spreads it northward over the grasslands. Where loose sand is exposed, it too is moved by gusts of this wind and by other seasonal winds of the Great Plains, such as powerful winter northerlies (which some call Blue Northers), some of which ends up in fields of sand dunes. Dune fields, both active and inactive, dot the Great Plains in Canada and the United States. The largest of these is the Sandhills, a largely inactive (that is, grass-covered) mass of dunes in Nebraska, which formed several thousand years ago and today covers about one-third of the state.

In coastal areas, onshore winds are a daily occurrence in response to heating of land and the resultant differences in air pressure between land and water. The strongest of these winds are able to erode sand from beaches and move it short distances inland. Storms add significantly to this effect, and the combined result is sizable fields of coastal sand dunes along many of the world's windy shorelines (Figure 24.10).

Wind and Topography: As wind slides and gusts over the land, its patterns and rates of movement are altered by the lay of the land, by landforms. The most evident example is found where mountain chains like the Andes and the Rockies lie across the path of major wind systems like the prevailing westerlies. These mountains protrude well into the troposphere constricting its airflow and forcing wind to squeeze through a reduced space (between the mountain crest and the tropopause in Figure 24.11a). In order for the system to maintain continuity of flow, it must speed up as it crosses the mountain range. Velocity is greatest at the crest and it is often marked by erosion there as snow and rock and ice particles are driven downwind. The same generally holds true for smaller topographic barriers such as sea cliffs (Figure 24.11b), which also exhibit pronounced erosion where wind accelerates near their crests. In both instances, mountains and sea cliffs, wind decelerates beyond the crest and deposits most of the material eroded upwind.

This sequence of erosion and deposition on windward and leeward (downwind) slopes is the key to understanding the basic mechanism of how sand dunes work and how they migrate. As dune sand piles up, wind speed accelerates toward the crest of the pile, and at some point the wind gets strong enough to erode the sand (Figure 24.11c). Sand is swept off the upper dune, driven over the crest, and deposited on the lee slope as wind velocity declines. As sand is moved from the windward to leeward side of the dune, the dune moves, or migrates, downwind.

Figure 24.11 Influences of landform barriers to airflow at three scales: (a) mountain ridge, (b) coastal cliff, and (c) sand dune. Each induces windward slope erosion and lee-slope deposition.

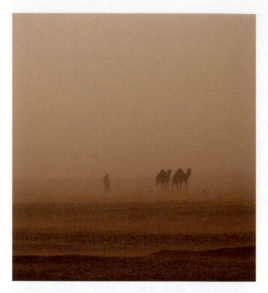

A sand and dust storm in the Sahara. Sand and finer particles, driven by powerful winds lasting for days, fill the boundary layer.

Summary on the Airflow System: This system of light fluid sweeps over the entire planet but varies greatly in its patterns, velocities, and power depending on location, topography, and our scale of observation. From ground level upward the system is organized into three layers, or elevation zones, each faster and deeper than the one under it and each associated with a particular kind of geomorphic activity. At landscape level, wind speed and power vary with weather conditions, the seasons, and topography. Obstacles such as mountain ranges and sea cliffs constrict airflow, forcing higher velocities and acceleration of erosional power.

24.3 Wind Erosion and Sediment Transport

Like all geomorphic systems, wind systems are made up of three basic parts: a source area where erosion takes place, a transport zone where sediment is carried over some distance, and a deposition area where sediment is laid down (Figure 24.12). Unlike streams or longshore systems, wind systems tend to be more flexible in terms of the direction of transport. Transport in all directions is possible, but in most wind systems one or two dominant directions produce most of the work. Distance of transport, on the other hand, tends to be similar to streams and longshore systems in that it varies with magnitude of flow and sediment particle size.

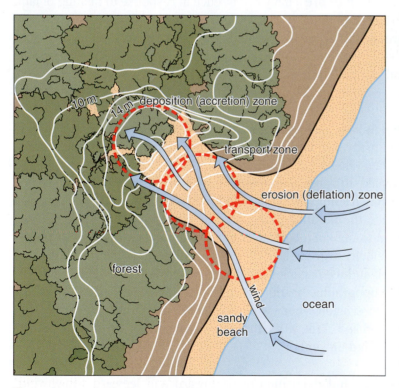

Figure 24.12 A simple three-part wind system where a sandy shore is the source area and onshore winds transport sand into a nearby forest which functions as a sink.

Wind erosion involves two processes: abrasion and deflation. **Abrasion** is the wearing away of a surface by bombardment of particles carried in the windstream. It is basically a sand-blasting process; however, the abrasive agent is not limited to sand, for silt and clay also appear to be effective agents. Wind abrasion is an effective erosional process. Automobiles caught in just one sandstorm can have their paint sandblasted down to metal and window glass left so pitted that it appears frosted. Where rock is exposed in deserts, it is commonly etched, grooved, and faceted by wind abrasion.

Deflation is the removal of loose particles by wind. Only sand and smaller particles are moved by deflation and the process is limited to unprotected (non-vegetated) surfaces. It occurs at a wide range of scales. At the local scale, in rangeland, dune fields, and beaches for instance, it is often related to peculiarities in wind micropatterns, and results in bowl-shaped cavities called **deflation hollows**. At the broadly regional scale, on the other hand, deflation is the process that sweeps across vast deserts removing residue from rock-weathering and other sources. The amount of material moved by deflation from any surface is dependent on both the supply of loose (free) particles, the sizes of particles, and, of course, the erosive force of the wind.

Because of the geometric relationship between wind velocity and erosive power, a fivefold velocity increase from, say, 2 meters per second to 10 meters per second produces a 125-fold increase in erosional power. Thus, fast winds are clearly capable of performing much more work than slow winds, as the graph in Figure 24.13 confirms. On the other hand, slow winds occur much more frequently than fast ones. Yet, when we compute the total work accomplished in terms of sediment moved, the balance strongly favors fast winds, defined roughly as those with velocities greater than 12

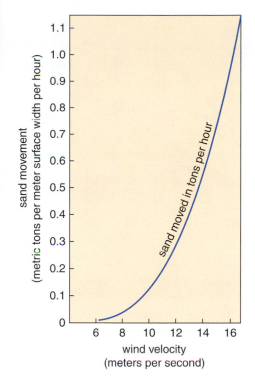

Figure 24.13 The relationship between wind velocity and sand movement showing a rapid rise in transport with increasing velocity.

meters per second (27 miles per hour). In fact, it appears that most work is done by wind storms that occur with an average frequency of several days or so a month.

Sand Transport: When wind rises over a sand surface, the first particle movements begin at a velocity, called **threshold velocity**, of about 4.5 meters per second (10 miles per hour). The initial movement of particles is characterized by a rolling motion called **traction** (or **surface creep**). In light winds, only small sand particles are moved, but in a strong windstorm, particles approaching small pebbles, even a bit larger, can be moved by this process.

A second mode of sand transport involves a jumping action, called **saltation**, which is driven by ground-level turbulence. Particles lift off the ground, become airborne, and sail in short trajectories, from several centimeters to more than 5 meters downwind. When a falling grain strikes the surface, it may either bounce and continue skidding along, or hit another grain whereupon part of its momentum is transferred to that grain and it goes sailing off in the manner illustrated in Figure 24.14. Saltation accounts for 75 to 80 percent of sand transport over sand dunes; the remaining 20 to 25 percent is moved by traction. Most sand grains travel within 20 centimeters (8 inches) of the surface, but very strong winds are able to lift grains as high as 2 meters above the ground.

Two factors help explain why saltation is so effective in sand transport. First, because of its low density, air offers little resistance to the movement of windblown particles. Second, because of its low viscosity, air is highly turbulent near the ground and eddies are able to produce a strong lifting force. In order for an airborne particle to settle back toward the surface, it must be heavy enough to overcome the lifting force of the eddies under it. If a particle's fall velocity is more than 20 percent of the upward velocity of the eddy, it cannot be held aloft. It is significant, therefore, that there are great differences in terminal settling velocities for sand, silt, and clay-sized particles. With slower fall velocities, clay and silt are held aloft much longer than sand are therefore subject to much longer travel distances.

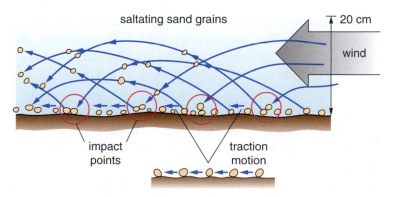

Figure 24.14 Sand transport by saltation, a leap-frog type motion responsible for as much as 80 percent of sand transport on sand dunes. Below, transport by traction, which is capable of moving particles much larger than sand.

Silt Transport: Silt and finer particles – let us call it all dust – do not move in contact with the ground but are lifted into the atmosphere by the wind's turbulent motion and transported in suspension. **Suspension** is a much faster means of transport than saltation and traction, because the particles travel in the airstream itself more or less as gas molecules do. Owing to this difference in the rate of particle movement, fine particles and coarse particles are quickly segregated, or sorted, into separate populations. The diagram in Figure 24.15, which reads from left to right, shows how this works with sand deposits at A and silt deposits at B. Although both are part of one wind system, dust is transported much farther than sand and, when it is deposited, dust forms a relatively homogeneous layer free of sand and larger particles.

This explains the occurrence of deposits called loess. **Loess** is wind-deposited silt that forms blankets 1 to 10 meters or more thick downwind from a wind-eroded source area such as a desert or fresh glacial deposits. Vast areas of loess cover the North American Great Plains, China, Argentina, and other areas of the world. In the Great Plains, the loess deposits were laid down by the same winds that created

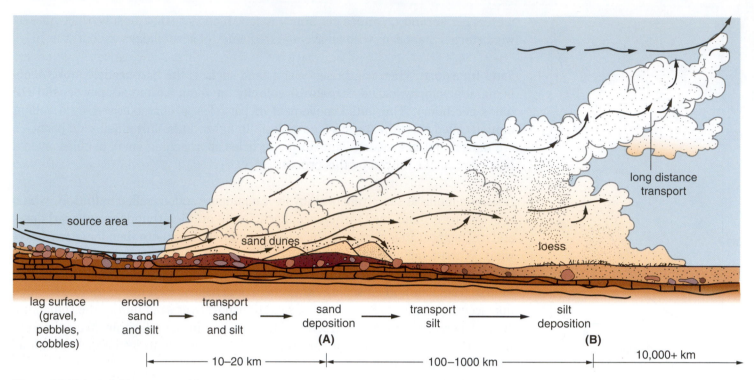

source area

lag surface
(gravel,
pebbles,
cobbles)

erosion
sand
and silt

transport
sand
and silt

sand
deposition
(A)

transport
silt

silt
deposition
(B)

sand dunes

loess

long distance
transport

|← 10–20 km →|← 100–1000 km →| 10,000+ km →|

Figure 24.15 A wind-driven geomorphic system illustrating sediment sorting and corresponding surface materials over distance downwind. Total travel distance for loess is hundreds of kilometers and much farther for dust carried higher in the troposphere.

the Nebraska Sandhills and similar high plain dune fields. Loess forms excellent agricultural soils, but it is highly prone to erosion by both wind and runoff. Wind erosion of loess was so heavy on abandoned farmland in the Great Plains in the 1930s that the region was branded as the Dust Bowl.

Because saltation cannot move sand particles as fast as suspension can move silt, the population of sand particles lags behind the dust, often concentrating in fields of sand dunes. And since traction is even slower that saltation, coarse sand and small pebbles lag even behind the dune sand. Lagging farthest behind are particles too large to be moved by wind at all. In desert areas, stones of various sizes are all that remain after the erodible particles have been blown away. Such material is called **lag** or **desert lag** (see the left side of Figure 24.15). Rocky surfaces, which include both lag and exposed bedrock, occupy 70 to 75 percent of the world's deserts. The remainder of Earth's deserts is covered by fields of sand dunes. Most loess deposits, on the other hand, are formed outside deserts, in the neighboring grasslands and beyond.

Landforms Sculpted by Wind: Landforms sculpted by wind erosion come in a wide variety of scales from small, wind-faceted stones, called ventifacts, to huge deflation basins more than 100 kilometers across. **Ventifacts** are stones shaped by abrasion from sandblasting and dustblasting. They come in all sorts of forms from those with flat bevelled surfaces to those with deeply hollowed-out cavities like the ones in Figure 24.16a from the Dry Valleys of Antarctica.

Yardangs are much larger wind-abraded landforms carved from rock or deposits of sediment. They often take the form of a streamlined ridge, with wind-eroded grooves, aligned in the direction of the prevailing wind. Yardangs are usually less than 10 meters high (see Figure 24.16b), and seem to be confined geographically to very dry inner-desert regions. **Deflation basins** are similar to deflation hollows but much larger, up to 100 kilometers across and 100 meters deep. Both deflation hollows and deflation basins are usually elongated in the direction of the prevailing wind.

(a)

(b)

Figure 24.16 Sculpture by wind: (a) ventifacts in the Dry Valleys of Antarctica; (b) yardangs in the Sahara.

The largest deflation basins, which average 250 meters deep, are found in Egypt west of the Nile Valley. The deepest is the Quttara Depression which reaches a depth to 134 meters below sea level and covers an area of about 20,000 square kilometers. Finally, there are **bedrock grooves** which are so large that they can be seen clearly on aerial imagery (Figure 24.17c). The largest, which are 30 to 40 kilometers long, are found in the Sahara and are cut into sandstone. The erosional process responsible is basically the same as in ventifact formation, namely, sandblasting.

Yardangs, deflation hollows, deflation basins, and bedrock grooves are important sources of sand and dust. These features are part of the vast, rocky arid landscapes, called **regs**, which make up most of desert surfaces worldwide and feature all manner of rock forms. The remainder of arid landscapes, sand deserts, called **ergs**, are dominated by great, sweeping masses of sand dunes, known as **sand seas**, which we examine in the next section.

Figure 24.16 (c) Bedrock grooves sculpted by wind in the American Southwest (aerial view).

24.4 Sand-dune Formation, Movement, and Forms

Sand dunes are among Earth's most intriguing landforms. Their shapes are as graceful as beautiful pieces of sculpture, yet they are very transient, changing shape and moving almost constantly (Figure 24.17). Although most sand dunes are found in the deserts, they are also common to seashores throughout the world in all sorts of bioclimatic environments (see Figure 24.10). Wherever they are found, two conditions are sure to exist: (1) an ample supply of free sand (i.e. loose particles without plant protection); and (2) strong, persistent winds capable of eroding and transporting this sand. But given a generous supply of moving sand, how does a dune actually get started?

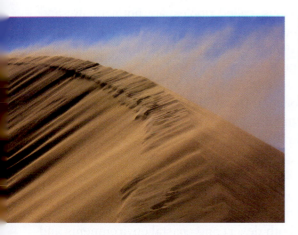

Figure 24.17 Earth art in progress. Wind sculpting the crest of a desert sand dune.

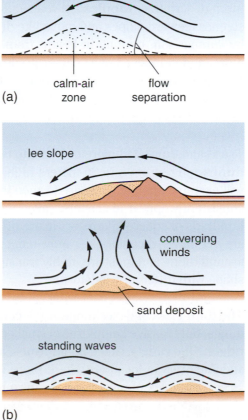

(a)   calm-air zone   flow separation

lee slope

converging winds

sand deposit

standing waves

(b)

Figure 24.18 (a) The process of flow separation where fast air lifts off the surface; and (b) three examples or sites of flow separation resulting in sand deposition.

There are probably several ways in which sand dunes originate, but one thing has to happen in all instances. The sand-transporting wind must slow down enough to drop its load of sand. This usually involves *flow separation* in which the fast air driving sand movement separates from the surface leaving slower air at ground level. Since sand moves mainly by saltation, and saltation requires contact with the ground, if fast air loses contact with the ground, then saltation ceases and sand drops out and is deposited (Figure 24.18a).

Flow separation can take place as air moves over an obstacle like a rock outcrop leaving a calm space in its lee; as air moves across a cliff (in either direction); where opposing winds meet and are driven upward; or where standing waves form (Figure 24.18b). A *standing wave* is a stationary rise or upward bend in wind streamlines. They form in all sorts of places even over flat surfaces.

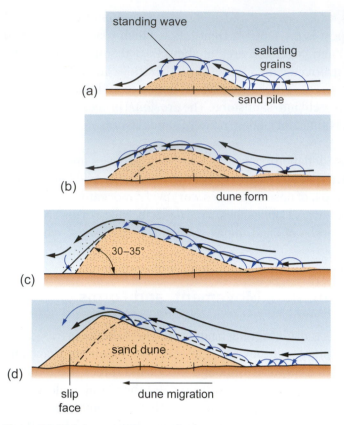

Figure 24.19 A four-step sequence of processes leading to dune formation and migration. As the dune moves ahead, it recycles its sand from the windward to the leeward slope.

Figure 24.20 Ripples are common microfeatures on wind-swept surfaces across which sand is being transported. They are very similar to the ripples formed by running water.

One explanation for them is that as wind sweeps over the land it tends to develop a rolling or wavy flow pattern. Beneath large waves a pocket of low-velocity air forms and inside this pocket saltating sand accumulates (see the lower diagram in Figure 24.18b).

At first the pile of sand is stationary (see (a) in Figure 24.19) but, as it grows higher, it reaches the level of faster and faster streamlines that move the sand across the top (b). Grains are moved from the windward to the leeward side of the pileup just over the crest. As the upper lee slope steepens to an angle approaching 35 degrees, it becomes unstable and small slides break loose, spilling sand down the slope (c). The sand pile has now become a dune, defined as a mobile heap of sand with a low-angle windward slope and a steep leeward slope, called the **slip face** (d). The **dune** migrates over the ground as sand is eroded from one side and deposited on the other as depicted in Figure 24.17. Rates of movement are highly variable depending on dune size and wind strength, but 10 to 20 meters per year is typical for desert dunes.

The size (volume) of the dune is limited by sand supply, whereas the height of the dune is limited by the velocity of the wind above ground level. As the dune grows higher and reaches faster streamlines, the streamlines become increasingly focused near the crest. At some point, usually at a height of less than 25 meters, vertical growth of most individual dunes must cease because wind stress becomes too great to allow further buildup.

Sand-dune Sizes and Forms: There is a wide variety of sand-dune features, sizes, and forms on Earth. The smallest dune forms, called ripples, are not actually sand dunes, but so similar to dunes and so common to dune environments that they merit description. **Ripples** are shaped like dunes, with gentle windward shapes and steeper leeward slopes and move like dunes (Figure 24.20). However, they are tiny by comparison, ranging in height from one centimeter or so to 50 centimeters or more, and composed of a coarser grade of sand than dune sand. They form on any surface where sand is being moved by wind, including the surfaces of dunes themselves. There is no evidence that ripples are baby sand dunes destined to eventually grow up to become full-blown sand dunes.

True sand dunes fall more or less into two size classes: small and very large. **Small dunes** range from 5 to 25 meters high and come in a variety of shapes, which we will examine shortly. They are formed in both desert and coastal environments and consistently appear in clusters, called *dune fields*, which vary from a few hectares to hundreds of square kilometers in area (Figure 24.21). Most dunes of the small class are *free dunes*, meaning their movement is unimpeded by vegetation growing on and around them, and most are highly transient, changing position, size, orientation, and shape with different windstorms.

The third class of desert dune is the complex case. Actually, they are not individual dunes, but massive complexes of smaller dunes heaped on top of one another in long ridges. One variety, called **megadunes**, are found in the great deserts of Africa, Asia, and Australia in locations where there is an abundant supply of sand. The world's largest complex of megadunes forms a huge sand sea in the Arabian Desert, called

Figure 24.21 A dune field in the Sahara region of North Africa region. Most dune fields migrate en masse and may intercept and merge with other migrating fields.

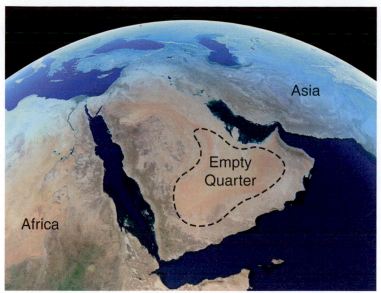

Figure 24.22 The Empty Quarter of the Arabian Peninsula, widely considered the world's largest area of large, complex sand dunes.

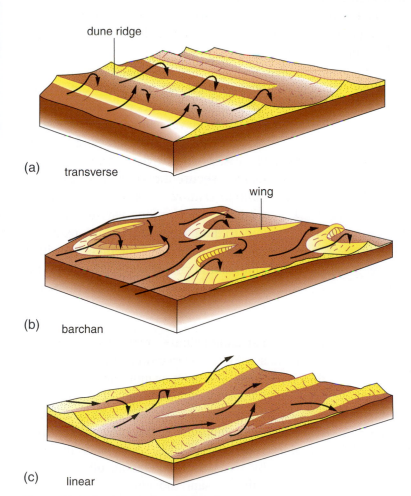

Figure 24.23 Common desert sand dune forms. Barchans belong in transverse class.

the Rub'al Khali or Empty Quarter. It covers more than 500,000 square kilometers, an area larger than the state of Montana (Figure 24.22).

We can group sand dunes into two general sets: desert dunes and coastal dunes. Most desert dunes are free dunes, whereas most coastal dunes are *impeded dunes*, that is, they are constrained in some way by vegetation. Plants not only inhibit mobility, but often alter patterns of erosion and deposition, which in turn influences the dune's shape. When sand dunes become fully overgrown by vegetation, they are immobilized and referred to as *stabilized dunes* or *relict dunes*. Vast areas of relict dunes are found on the margins of large deserts, and most are considered to be the result of climate change in the closing millennia of the last glaciation.

Desert Sand Dunes: Desert dunes come in three basic classes related to wind direction: transverse, linear, and complex, and examples of the first two are illustrated in Figure 24.23. **Transverse dunes** form perpendicular to the formative wind direction. They often take the form of long, wavy ridges, called *transverse ridges*, with a single slip face on the leeward side (Figure 24.23a). Another transverse dune, the **barchan** is distinctive for its graceful crescent shape. The points of the crescent, called *wings*, are curved downwind as if to partially envelop the slip face, as we can see in Figure 24.23b. Barchans are frequently found where sand supply is limited and they commonly travel in small fields or "schools" over hard, rocky surfaces. **Linear dunes** are sand ridges running more or less parallel to the transporting wind direction. They are apparently shaped not by one, but two or more winds blowing at slightly different angles to the ridge. The result, as Figure 24.23c shows, is linear dunes with multiple slip faces at different sites along both sides of the ridge.

Figure 24.24 A field of complex sand dunes, the result of some sort of combination or merger of different dunes and/or sand-transporting winds.

The third class of desert dune is the **complex class**. It covers dune masses of diverse forms and origins, including some resulting from mergers of different dune types, such as linear and barchan dunes. Others originate at intersections of several sand-transporting winds, and yet others appear to originate where newer dunes overtake masses of older (relict) dunes. Complex dunes may have dome shapes, star shapes, or irregular ridge shapes and they are common among megadunes in the great sand seas (Figure 24.24).

Coastal Sand Dunes: Outside deserts and fringing grasslands, the only significant areas of active sand dunes are found in coastal areas. Coastal dunes can form wherever there is an ample supply of beach sand and strong onshore winds to drive it inland. To foster dune development, beaches generally must be broad, sandy, and sufficiently agitated by wave action to keep them free of vegetation. Depositional shorelines such as those along barrier islands, bay-mouth bars, and bay-head bars are ideal sites for coastal-dune formation.

Coastal dunes start near shore and migrate inland. Most appear to begin with the formation of **blowouts** (small deflation hollows) in beach deposits behind the shore, similar to that shown in Figure 24.12. As onshore gusts drive sand winnowed from the blowout, it piles up just downwind forming a small dune. The dune is nourished by sand funneled through the blowout from the beach. As the dune moves inland, it grows in volume and elevation.

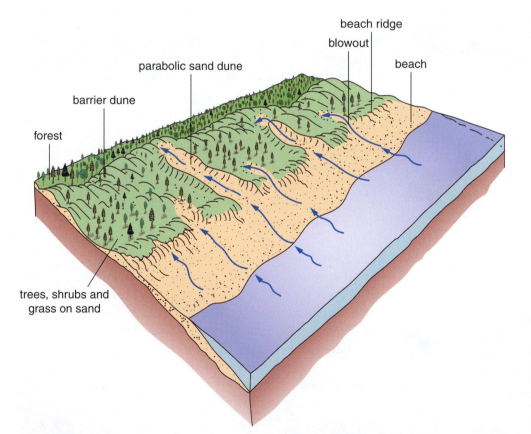

Figure 24.25 Coastal sand dunes of the parabolic variety. Sand is supplied by the beach and driven inland by onshore winds.

Where vegetation is part of the coastal environments, plants play an important role in the developing dune. Plants such as dune grass, sea oats, and various shrubs, which thrive in this sand-driven habitat, secure the sides or wings of the dune as it moves inland. As a result, the wings evolve into long narrow ridges or tails trailing seaward from the advancing mass of sand. The dune ends up with a long, slender form like a hairpin, and is called a **parabolic dune**, as shown in Figure 24.25.

Parabolic dunes can migrate inland a kilometer or more and reach elevations 50 meters or more above sea level. Landward advance usually stops when wind energy near the ground declines as streamlines of fast wind rise up over land, and when the resistance imposed by terrestrial vegetation, often forests, subdues sand movement. Near the landward limit of dune penetration parabolic dunes may bunch up to form an irregular ridge of sand hills paralleling the coast, called a **barrier dune** (Figure 24.25).

Coastal dunes are found throughout the world, but are especially abundant on passive continental coasts with huge supplies of beach sand and strong onshore winds. For example, they virtually line the Gulf of Mexico, especially on barrier

Figure 24.26 A field of perched sand dunes situated on top of a high coastal bluff. The dune field begins landward of the crest of the bluff.

islands, where southerly winds and tropical storms sweep sand off wide beaches. On active coasts – that is, those on the steeper, rockier sides of landmasses like the North American west coast – beach sand is often less abundant and high cliffs and bluffs often limit the landward advance of dunes that do form. However, high sea cliffs and bluffs sometimes yield enough sand to winds blowing over them to produce sand dunes along their crests. These elevated dunes are called **perched dunes**, and rank among the most spectacular coastal dunes (Figure 24.26).

Summary on Wind Erosion and Dune Forms: When wind erodes particles from soil and rock surfaces, it often leaves behind a curious array of forms and features with interesting names like lag, yardangs, and ventifacts. The particles themselves are entrained in the wind system, which then sorts them out according to size. Sand falls out first; silt later and farther away. From the sand, dunes may form that take on a variety of sizes (up to 400 meters high), shapes (including crescents and hairpins), and orientations (such as parallel and transverse), all depending more or less on sand supply, wind patterns, and geographic setting.

24.5 Global Distribution of Dune, Loess, and Related Wind Deposits

Sand dunes are not the only active (moving) sand deposits in the world. In fact, active dunes make up only about 75 percent of Earth's mobile sand deposits. The remaining area of shifting sand is covered mainly by thin blankets of sand called **sheets** and **streaks**. The volume of sand represented by sheets and streaks is relatively small, however, probably less than 10 percent of the global total. Like sand dunes, sheets and streaks are concentrated in the dry desert interiors and are easy to identify on satellite imagery such as that in Figure 24.27.

Relict dunes: In addition to active ergs (dune, sheet, and streak deposits) there are vast areas of *relict* or stabilized sand deposits in arid and semiarid areas. Their total coverage is nearly equal to that of Earth's active sand coverage. Most relict dunes are located in the humid parts of deserts and in grasslands near active ergs in Africa, Asia, and Australia (see the map in Figure 24.28). Their origin and distribution can be traced to climate changes associated with the rise and fall of Pleistocene glaciations. During periods of glaciation, global climate became cooler and drier. Forests and grasslands declined and deserts, with sand dunes, advanced over vast areas. With the decline of the Wisconsinan glaciers about 12,000 years ago and the return to warmer and wetter conditions, grasslands expanded leading to widespread sand-dune stabilization. As we noted earlier, one such area is the Nebraska Sandhills.

Dunes and Desertification: After several thousands of years of geographic decline in the size of Earth's deserts, the trend has reversed and most desert landscapes are now expanding. The driving force behind desert expansion is **desertification**, which, as we noted in earlier chapters, results from the combined effects of land-use pressure and drought. Simply put, when agriculture in grasslands (and dry woodlands) is pushed beyond its sustainable limits, the landscape is weakened making it highly vulnerable to drought. Drought, of course, is inevitable in grasslands, and under natural conditions plants and animals are usually able to weather it and rebuild after dry spells.

Atlantic Ocean

Africa

Sand sheets and streaks

0   250

kilometers

Figure 24.27 Sand sheets and streaks over the floor of the Sahara of North Africa.

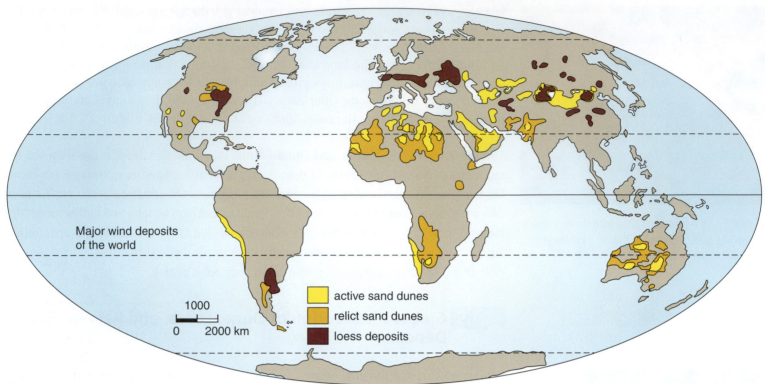

Figure 24.28 Major areas of wind deposits in the world, both sand and loess. Notice the large areas of relict (inactive) dunes in Africa, Asia, and Australia, and the midlatitude distribution of loess.

But when grassland is degraded by overgrazing, the plant cover is weakened and the soil damaged to the point that the landscape is unable to fully recover in the intervals between droughts. Soils are left unprotected from the drying and erosive effects of wind, relict dunes may be reactivated, and existing dune fields may begin to expand. It is a downward-spiraling system leading to increasingly negative effects, which some scientists argue also produces regional climate change because less moisture is fed to the atmosphere from the desiccated landscape. It is estimated that worldwide desertification is advancing at a rate in excess of 100,000 square kilometers a year, and global warming is expected to advance the process by forcing a greater magnitude and frequency of droughts.

Loess: Loess covers an area approaching that of active ergs, around 5 percent of Earth's land area (Figure 24.28). The distribution of loess, however, is quite different than the distribution of sand. Loess is found mainly in the central midlatitudes, whereas windblown sand is found mainly in the subtropics. The explanation for this difference is related to the principal source areas for silt and sand. Most loess was derived from glacial deposits near the front of the continental glaciers when they reached deep into the midlatitudes (see Figure 23.33). Silt was whipped up from moraines, outwash plains, huge floodways, and other deposits before they could be secured by plants. It was carried southward (in the northern hemisphere) several hundred kilometers or more, and deposited, eventually forming vast sheets over the landscape.

Figure 24.29 A scene from the North American Dust Bowl of the 1930s. This event played prominently into the drama of the Great Depression as farms declined, were abandoned, and loess-based soils were left to the ravages of the wind.

Loess is the parent material for some of the world's very best agricultural soils. These include the prairie soils of the central plains of the United States and the steppe of Russia. Loess-based soils are fertile with good moisture-holding and plowing qualities. But loess is also highly prone to wind erosion when it is exposed in farm fields as was vividly demonstrated during the dust storms of the North American Dust Bowl of the 1930s and documented in hundreds of photographs like the one in Figure 24.29.

24.6 Long-distance Transport and Relations to Other Earth Systems

Although we have touched on long-distance transport at various points in this and other chapters, it is necessary in the light of recent findings to add some summary remarks about this phenomenon for, among other things, it illustrates yet another aspect of the interconnected geography of our planet. First, long-distance transport appears to be much greater than we thought possible in the days before satellite imagery. It now appears that each year 1 to 3 billion tons of dust are wafted from the ground and moved around the globe by the atmosphere. This number does not include dust (aerosols) from pollution sources and volcanic eruptions. And second, two major wind systems, the westerlies and the easterlies, are responsible for the great majority of the transport (see Figure 24.3).

The source of the dust is Earth's great deserts and, among these, the Sahara, is the single largest source. Each year the western Sahara produces about 240 million tons of dust, most of which is carried westward over the Atlantic by the tropical easterly wind system. About 60 percent (140 million tons) of it is sprinkled on the ocean, with remainder going to the Caribbean and North America (60 million tons) and the Amazon Basin (40 million) (Figure 24.30). The consequences of this system may be nothing short of remarkable, for it is proposed that the dust falling on the Amazon is a principal, perhaps *the* principal, source of nutrients for the great rainforest ecosystem. Are similar systems at work in other parts of the planet fed by the Gobi Desert, the Great Australian Desert, the Atacama, and others? And what about effects of desertification on aerosol loading of the atmosphere?

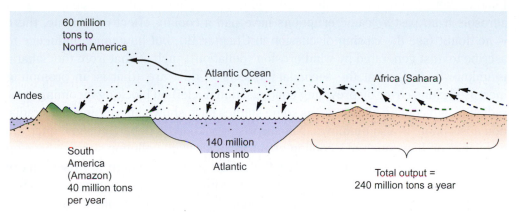

Figure 24.30 The long-distance transport system emanating from the Sahara of Africa produces 240 million tons of dust which are carried to the Atlantic Ocean, Amazon Basin, and southern North America.

Relations to Other Earth Systems: This brings us to the question of the influence of the atmosphere as a geomorphic system on other Earth systems. One of the most apparent is the influence on soil. The atmosphere has a tremendous effect on soil composition and fertility over vast areas of the world, a point illustrated here first, by the Sahara–Amazon connection cited above, second, by the role of dust storms in building the great loess beds of the midlatitudes, and third, by the vast sand-dune deposits of the subtropical and midlatitude deserts. The first two contribute to soil fertility and undoubtedly are significant to the character and productivity of the ecosystems grounded in them and, in the case of loess, to modern agriculture. Sand dunes, on the other hand, do not help soil fertility and represent some of the most biologically impoverished soil environments on the planet.

There is also a link to the hydrologic cycle. We turn again to dust, and begin with the concept of albedo, which we examined in Chapter 4. Albedo refers to the capacity of surface material to reflect solar radiation. Snow and ice have the very high albedos, which means they absorb little radiation and thus warm up very slowly. When dust is deposited on these materials, albedo can decline dramatically, inducing rapid melting and runoff, and in the extreme maybe even contribute to climate change. Overhead, aerosol particles from a host of surface sources (deserts, sea spray, volcanoes) are carried aloft where they contribute the vast majority of the condensation nuclei for

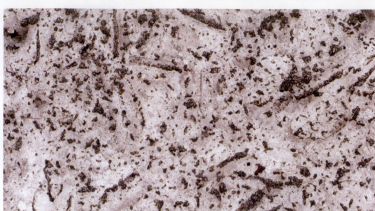

The albedo of dirty snow may be 30 percent lower than clean snow making a big difference in heating and melting rates.

precipitation particles. Without them, there would be far less precipitation on this planet. And while those infertile sand dunes contribute little to Earth's biological productivity, they have extremely high infiltration capacities, so high that they take in virtually every drop of water that lands on them, and thus can be very important in groundwater recharge.

Finally, there is the role of the geomorphic atmosphere in climate change. That aerosols from past volcanic eruptions have had a cooling effect on climate, there is no doubt (see the closing discussion in Chapter 19), but how much influence to ascribe to dust generated by wind erosion (deflation), such as that from the Sahara, is unknown. And, what the secondary effects of dust are (that is, in promoting increased cloud cover by providing condensation nuclei) on blocking incoming solar radiation and cooling climate or, as some studies have shown, increasing solar radiation absorption, is also unknown. Not surprisingly, there is debate in some quarters of science on how much of a cooling effect particles from pollution sources have on the modern atmosphere and whether that effect is great enough to offset part of the warming from carbon dioxide loading (see Figure 8.35). Confusing the matter further is evidence that an atmosphere darkened by aerosols may absorb more solar radiation than a cleaner one.

Chapter Summary and Overview of Wind Systems: Sand Dunes, Dust, and Deserts

As the diagrams below suggest, eventually the story of wind comes down to the Earth's surface, mostly in the deserts, whether as rocky surfaces where wind sweeps loose particles away or as sand surfaces, mainly as dunes, which shift about with changes in wind and climate. But to fully appreciate the work of wind as a geomorphic system, we have to stretch our thinking beyond these salient landforms and

consider also the broader effects of this great system. These include long-distance transport of particles from a host of sources and we should consider how particle transport at this scale takes place. We should also think about where this material is deposited and what influence it may have on other geographic systems, particularly soil, precipitation, and global climate.

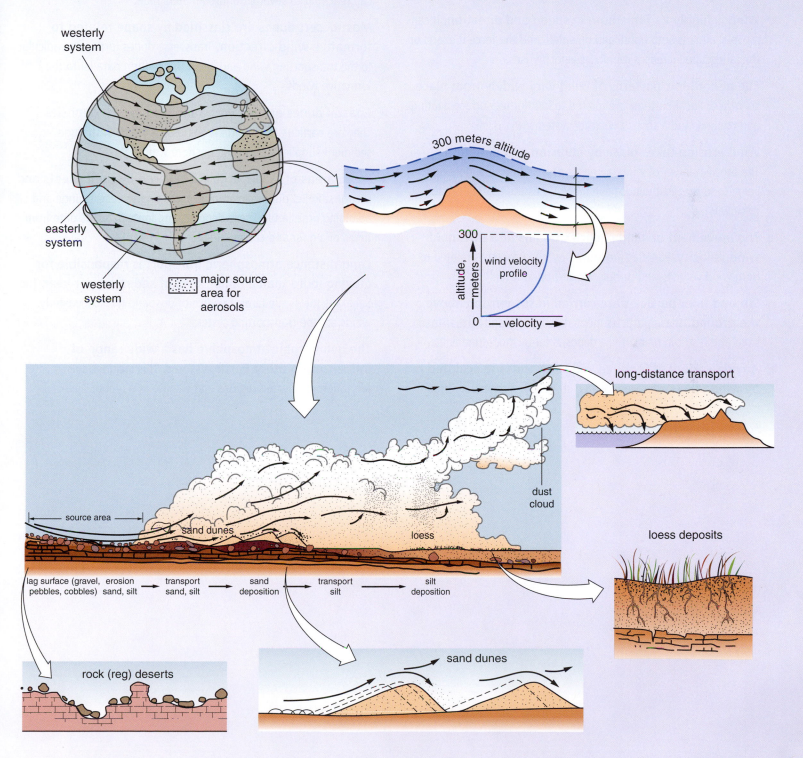

westerly system

easterly system

westerly system

major source area for aerosols

300 meters altitude

wind velocity profile

300

altitude, meters

0 — velocity →

long-distance transport

source area

sand dunes

dust cloud

loess

loess deposits

lag surface (gravel, pebbles, cobbles) → erosion sand, silt → transport sand, silt → sand deposition → transport silt → silt deposition

rock (reg) deserts

sand dunes

▶ **The work of the atmosphere as a geomorphic system occurs at three scales.** Long-distance transport takes place in the troposphere mostly within the corridors of prevailing winds and upper atmospheric winds. The atmospheric boundary layer is the zone of dust storms.

▶ **The roughness layer controls wind contact with the ground.** Where this layer is thick, fast wind cannot reach the ground, cause erosion, and move particles.

▶ **Wind is highly variable both in speed and direction.** It can move sediment both uphill and downhill, and the force it exerts on the landscape increases rapidly as velocity rises.

▶ **The geographic patterns of wind vary widely from place to place.** Local patterns are related to landforms, surface heating and cooling, and transient weather systems.

▶ **Wind erosion takes place by abrasion and deflation.** The erosive power of wind increases as a geometric function of velocity so that fast winds have many times the erosive capability of slow winds.

▶ **The movement of sand particles begins at a threshold wind velocity.** Particles are moved by two processes related to both wind velocity and turbulence at ground level.

▶ **Silt and finer particles are carried in suspension above the ground.** These particles move much faster and farther than sand and result in blanket-like deposits far beyond dune fields.

▶ **A wide variety of features and landforms are sculpted by wind.** The smallest are faceted stones. The largest are broad basins and long bedrock grooves.

▶ **Rocky surfaces dominate the world's deserts.** The largest areas of sand deserts are dune complexes called sand seas.

▶ **An ample supply of free sand and persistent, strong winds are needed to form sand dunes.** Flow separation in surface wind determines where dunes form in the landscape.

▶ **Sand dunes migrate downwind and most are limited to modest heights.** In sand seas, large dune complexes, however, can reach up to several hundred meters high.

▶ **Most desert dunes are classified by shape related to formative wind direction.** Transverse dunes form perpendicular to the transporting wind and linear dunes form parallel to the formative winds.

▶ **Coastal dunes are fed with sand from beach deposits.** They are particularly abundant on passive coasts with huge sediment supplies and strong onshore winds.

▶ **Large areas of active ergs are covered by sand sheets and streaks.** Relict dunes cover large areas on the desert fringe and adjoining grasslands. These deposits and loess deposits date from times of a drier and cooler climate.

▶ **Long-distance atmospheric transport is responsible for moving more than a billion tons of sediment per year.** The Sahara is the single largest source of dust and tropical easterly winds are the transporting system.

▶ **The geomorphic atmosphere has a wide range of influences on other Earth systems.** This includes soil formation in selected regions and climate at a global scale.

Review Questions

1 How does wind speed vary with altitude above the ground surface, and what effects do these variations in wind speed have on particle transport at the global, regional, and local geographic scales?

2 What landscape feature provides the best protection against wind erosion?

3 Considering the wide range of wind speeds that blow over the Earth's surface, what class of wind speed (for example, frequent light winds, average speed winds, or infrequent strong winds) is capable of rendering the most change in the landscape over the long term? Why?

4 Based on your knowledge of airflow and wind power, explain why auto and bike racers "draft" (stay close) behind the cars or bicycles in the front of the pack?

5 What are the Nebraska Sandhills and what processes are thought to be responsible for their formation?

6 Wind accelerates as it passes over an obstacle like a sea cliff or mountain range. What happens to the load of particles it carries once it has passed the obstacle?

7 Distinguish between saltation and traction in terms of the form of particle motion and the total work accomplished by each process.

8 The largest percentage of desert landscape is not covered by sand dunes. What type of desert is this and what are some of the key surface features you would expect to find there?

9 Describe the process in a wind system leading to the segregation of particles by size into distinctly different populations and deposits and why sand dune and loess deposits end up in different geographic locations.

10 Describe the sequence of processes that produce migration in a sand dune.

11 What are some key differences and similarities between coastal and desert sand dunes in terms of how they form, their size, and their shape?

12 What factors are contributing to desertification on Earth today and why are prospects high for an increase in the rate of desertification?

13 What is loess, what were the source environments of this material, and when and where were loess deposits laid down?

14 What wind systems are most responsible for long-distance atmospheric transport, what are the principal sources of atmospheric dust, and in this context what is the connection between the Sahara and the Amazon?

Appendix A
Units of Measurements and Conversions

Energy, Power, Force, and Pressure

Energy units and their equivalents

joule (abbreviation J); 1 joule = 1 unit of force (a newton) applied over a distance of 1 meter = 0.239 calorie

calorie (abbreviation cal); 1 calorie = heat needed to raise the temperature of 1 gram of water from 14.5°C to 15.5°C = 4.186 joules

British Thermal Unit (abbreviation BTU); 1 BTU = heat needed to raise the temperature of 1 pound of water 1°Fahrenheit from 39.4 to 40.4°F = 252 calories = 1055 joules

Power

watt (abbreviation W); 1 watt = 1 joule per second

horsepower (abbreviation hp); 1 hp = 746 watts

Force and Pressure

newton (abbreviation N); 1 newton = force needed to accelerate a 1-kilogram mass over a distance of 1 meter in 1 second squared

bar (abbreviated b); 1 bar = pressure equivalent to 100,000 newtons on an area of 1 square meter

millibar (abbreviation mb); 1 millibar = one-thousandth of a bar

pascal (abbreviation Pa); 1 pascal = force exerted by 1 newton on an area of 1 square meter

atmosphere (abbreviation Atmos.); 1 atmosphere = 14.7 pounds of pressure per square inch = 1013.2 millibars

Length, Area, and Volume

Length

1 micrometer (μm) = 0.000001 meter = 0.0001 centimeter

1 millimeter (mm) = 0.03937 inch = 0.1 centimeter

1 centimeter (cm) = 0.39 inch = 0.01 meter

1 inch (in.) = 2.54 centimeters = 0.083 foot

1 foot (ft.) = 0.3048 meter = 0.33 yard

1 yard (yd) = 0.9144 meter

1 meter (m) = 3.2808 feet = 1.0936 yards

1 kilometer (km) = 1000 meters = 0.6214 mile (statute) = 3281 feet

1 mile (statute) (mi.) = 5280 feet = 1.6093 kilometers

1 mile (nautical) (mi.) = 6076 feet = 1.8531 kilometers

Area

1 square centimeter (cm^2) = 0.0001 square meter = 0.15550 square inch

1 square inch (in^2) = 0.0069 square foot = 6.452 square centimeters

1 square foot (ft^2) = 144 square inches = 0.0929 square meter

1 square yard (yd^2) = 9 square feet = 0.8361 square meter

1 square meter (m^2) = 1.1960 square yards = 10.764 square feet

1 acre (ac) = 43,560 square feet = 4046.95 square meters

1 hectare (ha) = 10,000 square meters = 2.471 acres

1 square kilometer (km^2) = 1,000,000 square meters = 0.38 square mile

1 square mile ($mi.^2$) = 640 acres = 2.590 square kilometers

Volume

1 cubic centimeter (cm^3) = 1000 cubic millimeters = 0.0610 cubic inch

1 cubic inch (in^3) = 0.0069 cubic foot = 16.387 cubic centimeters

1 liter (1) = 1000 cubic centimeters = 1.0567 quarts

1 gallon (gal) = 4 quarts = 3.785 liters

1 cubic ft ($ft.^3$) = 28.31 liters = 7.48 gallons = 0.02832 cubic meter

1 cubic yard (yd^3) = 27 cubic feet = 0.7646 cubic meter

1 cubic meter (m^3) = 35.314 cubic feet = 1.3079 cubic yards

1 acre-foot (ac-ft) = 43,560 cubic feet = 1234 cubic meters

Mass and Velocity

Mass (Weight)

1 gram (g) = 0.03527 ounce* = 15.43 grains

1 ounce (oz) = 28.3495 grams = 437.5 grains

1 pound (1b) = 16 ounces = 0.4536 kilogram

1 kilogram (kg) = 1000 grams = 2.205 pounds

1 ton* (ton) = 2000 pounds = 907 kilograms

1 tonne = 1000 kilograms = 2205 pounds

Velocity

1 meter per second (m/sec) = 2.237 miles per hour

1 km per hour (km/hr) = 27.78 centimeters per second

1 mile per hour (mph) = 0.4470 meter per second

1 knot (kt) = 1.151 miles per hour = 0.5144 meter/second

Avoirdupois, i.e. the customary system of weights and measures in most English-speaking countries.

Appendix B
Global Climate Types and Descriptions

This appendix provides a description of each of the climate types generally recognized by geographers at the global scale. The climate names and letter codes, which are provided in Table B.1, follow a modified version of the Köppen–Geiger System. The descriptions that follow focus mainly on the seasonal patterns of temperature and precipitation as they relate to landscape composition, primarily vegetation and land use. The global distribution of the 12 climates examined is provided by the map at the end of the appendix. We begin with the tropical climates.

Table B.1 Climate zones, types, and letter codes

	Main zones	Climate types (Symbol)
A	Tropical rainy climates	Tropical rainforest (Af) Savanna (Aw) Monsoon (Am)
B	Dry climates	Steppe (BS) Desert (BW)
C	Temperate rainy climates	Mediterranean (Cs) Marine west coast (Cf) Humid subtropical (Cf)
D	Cold snow-forest climates	Humid continental (Df) Continental subarctic (Df, Dw) Marine subarctic (Df)
E	Polar climates	Tundra (ET) Perpetual snow and ice (EF) High mountain (ETH)

The Tropical Climates

Tropical Rainforest (Equatorial Wet) (Af): Along the equator and as much as 10 to 15° north and south of it, the climate is rainy, with high humidity and mean monthly temperatures around 30 °C throughout the year. The ITCZ is overhead or nearby in all seasons. As a result, every month has substantial precipitation, and clouds are prominent for a part of most days in the year.

Rainfall is generally due to convergent and convectional mechanisms, and over land it is associated with two distinct rhythmical patterns. The first is a daily rhythm characterized by rain showers in the afternoon of most days. The second is seasonal, characterized by heavy precipitation when the ITCZ is overhead and somewhat lower amounts when it is not. Over water, precipitation is predominantly convergent because surface heating is not intensive enough to initiate convection.

Near the Equator the seasonal pattern results in two rainfall peaks during the year as the ITCZ shifts from one hemisphere to the other with the migration of the sun. The graph in Figure B.1 for Georgetown, Guyana (latitude 6° 30′ N) shows this pattern. Notice, however, that the peaks of rainfall do not coincide with the equinoxes when the Sun is highest at the Equator but rather lag behind by a few months. This delay is caused by the time it takes for surface heat to build up and for the ITCZ to slide into place.

The principal land areas dominated by the tropical wet climate are found in South America, Africa, and insular Southeast Asia. There are, in addition, three narrow bands of this climate that extend poleward to about 25° latitude. Located on the coasts of Madagascar, southeastern Brazil, and Middle America, these bands are produced by the easterly tradewinds, which bring a steady supply of moisture to these coastlines. In months around the equinoxes, tropical storms may migrate onto these coasts, and some of the hurricanes are highly destructive.

The landscape of the equatorial-wet climate is predominantly heavy forest, called tropical forest, which is characterized by abundant species and rapid growth rates. In the wettest areas, such as the western Amazon Basin, the forest is especially large, productive, and dense, and is referred to as tropical rainforest. It is estimated that tropical forests support 50 percent or more of the Earth's species of plants, animals, and microorganisms. Because of land use pressure, mainly from agriculture, these forests are rapidly disappearing, and it some parts of the world will likely be eradicated by the middle of this century.

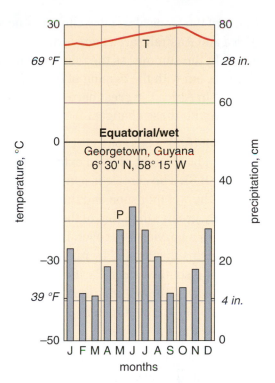

Figure B.1 Average monthly temperature and precipitation for a representative location in the tropical rainforest (equatorial wet) climate.

Tropical Wet-Dry (Tropical Savanna) (Aw): North and south of the equatorial-wet climate regions are areas that are in the ITCZ during the high-sun season, but under the subtropical high-pressure cell during the low-sun season. Thus, the areas have a wet season in the "summer" and a dry season during "winter." Temperatures are warm throughout the year, however (Figure B.2).

Understandably, the soil moisture available to plants is depleted in the first 1 to 2 months of the dry season, and vegetation declines until the landscape is brown. Streams also decline, and by midseason water holes (mainly deep stream pools and low spots where groundwater seeps out) are the only remaining surface water. With the return of summer rains, the soil-moisture reservoir is recharged, vegetation is revived, and streamflow is renewed.

The duration of the dry season is variable from year to year because of variations in the extent of the poleward migration of the ITCZ. Hence, areas on the poleward margins (the desert sides) of the tropical wet-dry climate are likely to experience variations in moisture that may have serious consequences for land and life. Both crop and pastoral farming in these zones are heavily dependent on summer rains, and when rains fall short of expected amounts, as they often do, crop failures and food shortages result. If drought conditions continue for many seasons, many of those inhabitants who do not migrate starve, as has happened more than once in the Sahel zone south of the Sahara in the past decade.

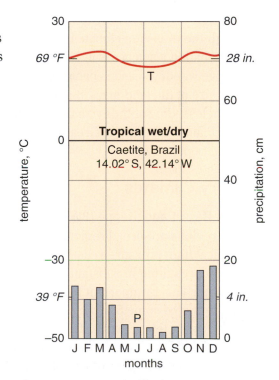

Figure B.2 Average monthly temperature and precipitation for a representative location in the tropical wet/dry (savanna) climate.

Much of the tropical wet-dry climate is occupied by savanna landscapes, which in Africa and Australia are characterized by scattered trees among large expanses of grass. In the summer as the rains sweep across the landscape, the area is vibrant with life, but with the coming of the dry season it grows brown and relatively lifeless. In Africa the great herds of grazing animals, such as the wildebeest, migrate with these changes.

Monsoon (Am): The monsoon climate, which is limited to South and Southeast Asia, can be thought of as an extreme version of the tropical wet-dry climate (Figure B.3). The climate is named for the seasonal winds that dominate this region, the wet monsoon and the dry monsoon. Summer rainfall here is heavy, generally greater than that of the tropical wet-dry landscape, and this is usually evident in a heavier tree cover. Precipitation is caused by convective and orographic mechanisms that are fed by massive amounts of moist air from the Indian Ocean and southwestern Pacific Ocean. During the peak months (June through September) of the wet monsoon, rainfall in the foothills of the Himalayas typically reaches 50 to 100 centimeters per month.

In winter, with the development of strong high pressure over Central Asia, airflow reverses. Now the air is dry and stable, producing a clear winter atmosphere. The duration of the dry monsoon season varies from year to year, and in years when it extends well into the spring, the resultant drought can be very damaging to agriculture. Marginal cropland in India is especially susceptible to this condition. Most of these croplands are located on the dry side of the monsoon zone in central and southwestern India. Conversely, excessive precipitation in the wet season can be equally damaging to agriculture because it causes flooding and soil erosion.

Although the monsoon climate is limited to South and Southeast Asia, monsoon-type circulation also occurs in North America; however, it does not produce the distinctive wet-dry seasons because the seasonal airflow is weaker owing to the smaller size of the land mass and lower magnitudes of the seasonal pressure cells. Nevertheless, the southerly airflow in summer from the Gulf of Mexico onto the Coastal Plains and into the Great Plains greatly increases precipitation between June and September.

The Dry Climates

Tropical Desert and Steppe (BW and BS): Poleward of the tropical wet-dry climate regions are broad regions dominated throughout the year by the stable subtropical high-pressure cells. The airflow near the centers of these cells is downward, resulting in substantial adiabatic heating by the time the air reaches the ground. This is important because this air starts out dry (by virtue of the fact that it originated some 5 to 8 km aloft) and gets even drier as its relative humidity falls with rising temperature. This condition, coupled with the fact that air masses with precipitable moisture rarely penetrate these regions, gives rise to the driest landscapes on earth (Figure B.4). In addition, cloud cover is scarce (in many areas, less than 30 days per year), with the result that intensive solar heating occurs at ground level. Therefore, any available surface moisture is rapidly driven into the atmosphere, giving rise to high potential evapotranspiration rates and extraordinary soil-moisture deficits.

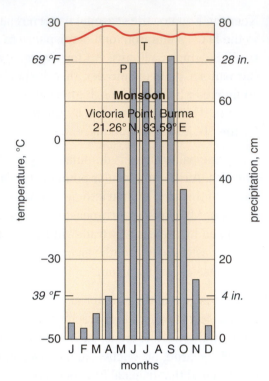

Figure B.3 Average monthly temperature and precipitation for a representative location in the monsoon climate.

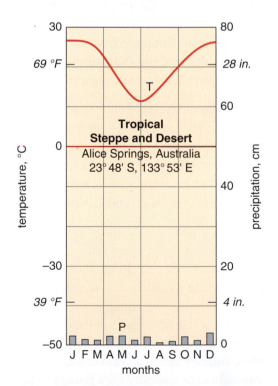

Figure B.4 Average monthly temperature and precipitation for a representative location in the tropical desert climate.

The world's major deserts, such as the Sahara (Africa), Rub'al Khali (Arabian Peninsula), Sonoran (North America), and Great Australian, are classical examples of deserts produced by the subtropical highs. Their life-threatening conditions for humans are legendary. Solar radiation on exposed skin can be lethal if prolonged; heat exhaustion from loss of body fluids and salts threatens those who risk physical exertion in the hottest hours of the day, which typically exceed 40 °C. The temperature falls drastically at night because the dry air has such a low capacity to retain longwave radiation.

The transition zones between the deserts and the tropical (A) and the midlatitude (C) climates are the semiarid steppes. Toward the & the steppes are characterized by a gradual change from perennially dry on one side to tropical wet-dry on the other. On the poleward border the steppe grades into the Mediterranean climate, which is also a wet-dry climate.

Along certain coastlines of the tropical steppe and deserts, the climate may be cool and foggy. This variant climate is associated with cold ocean currents, which not only set up conditions for ground-level condensation and the occurrence of frequent fog, but also add significantly to the stability of air near the ground. As a result, precipitation in this climate may be even less than it is in the large interior deserts. In the Atacama Desert of southern Peru, which is influenced by the cold Humboldt current, precipitation has never been recorded. Despite the paucity of rainfall, the combination of cool marine air and cool desert nights produces condensation on desert surfaces. This moisture is often the only source of water for plants and animals in these environments.

Patterns of traditional life in all the arid and semiarid climates have many similarities throughout the world. Overall population is very sparse. Settlements (villages, towns, and cities) are limited to those special locations, called *oases*, where a permanent supply of water is available. These exotic places are found where the desert is fed by a river such as the Nile or by springs at the foot of a mountain range. They are often places of heavy concentrations of population, especially the valleys of the great desert rivers such as the Nile, Tigris, Euphrates, and Indus, where geography is steeped in the annals of civilization.

Elsewhere over the broad reaches of the desert the traditional mode of land use is nomadic herding. It is characterized by small groups (tribes or large family units) and their grazing animals shifting about the desert in search of pasturage and often ranging great distances from winter to summer. Unfortunately, this way of life is rapidly being eliminated from the earth as national borders and central governments in Africa, the Middle East, and Asia become more and more restrictive in their policies toward non-sedentary populations, to say nothing of ethnic conflict and war.

Midlatitude Desert and Steppe (BW and BS): In North America, Eurasia, and the southern part of South America, precipitation declines toward the interior of the continents, eventually giving way to steppe and desert. In addition, these areas are shielded by mountain ranges from the maritime air driven landward by the westerlies, thereby creating a rainshadow effect. Midlatitude steppe and desert differs from its tropical counterparts in that it tends to be less severely arid, especially in winter, and more seasonal in temperature, with winter temperatures often falling below freezing (Figure B.5).

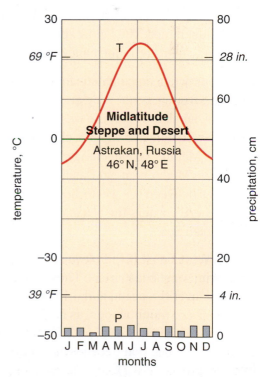

Figure B.5 Average monthly temperature and precipitation for a representative location in the midlatitude desert climate.

Some of the most damaging droughts in history have taken place in the midlatitude steppes. During average or better-than-average years these lands are extraordinarily alluring to grain farmers and grazers. But the good times are deceptive in dry regions because they are inevitably followed by drought, as the farmers of the American Great Plains found in the 1930s. In this instance, the effects of the drought were compounded by a national economic depression and a farm economy that had pushed itself well beyond the geographic limits of profitable grain farming. The costs of disasters such as the American Dust Bowl often include the permanent loss of native grass covers and severe soil erosion, and, in other regions, famine and death.

The Temperate Climates

Mediterranean (Cs): This climate is distinctive for its long sunny summers and short moist winters. This combination is caused by seasonal shifts in the subtropical high-pressure cells. During the summer, the cells expand poleward to cover the south-western coastal areas of the continents; in winter, the cells shift equatorward, and these areas fall under the influence of the westerlies and the midlatitude cyclones. Thus, you may think of the Mediterranean lands as belonging to the humid midlatitude climates in winter and the dry (B) climates in summer. Moreover, these areas tend to be thermally modified by the adjacent ocean, though summer temperatures are typically warm, averaging close to 30 °C in the warmest month (Figure B.6).

The soil-moisture balance is one of the most distinctive in the Mediterranean climate because of the seasonal extremes that occur there. In winter, evapotranspiration rates are low and the soil-moisture recharge is high. Summers are essentially the opposite; under the clear, sunny skies evapotranspiration rates are very high, and most of the available soil moisture is depleted by early summer or by midsummer. The landscape at this time turns brown, trees may drop leaves and go into a drought dormancy state, and streamflows decline or disappear.

The Mediterranean climate is found on all the continents except Antarctica, but its total coverage is not very great. In North and South America it is limited to relatively small areas on the west coast between the marine west coast climate on the poleward side and the dry climates on the equatorward side. Only in Eurasia does the Mediterranean climate extend eastward far beyond the continental west coast. The Mediterranean Basin (sea) accounts for this extension, of course, lying as it does between the westerlies and the subtropical high. The Mediterranean climate here borders the sea and extends beyond eastward to the southern end of the Caspian Sea.

Marine West Coast (Cf): Poleward of the Mediterranean climate regions are the marine west coast areas. These areas lie in the path of cyclonic storms that originate over the ocean and drift onto the continents with the westerlies. Where the coasts are mountainous, as in North and South America, the orographic mechanism induces heavy precipitation, often exceeding 300 cm per year. In the summer, precipitation declines as the cyclonic storms weaken and decrease in frequency (Figure B.7).

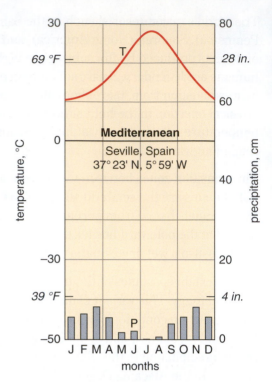

Figure B.6 Average monthly temperature and precipitation for a representative location in the Mediterranean climate.

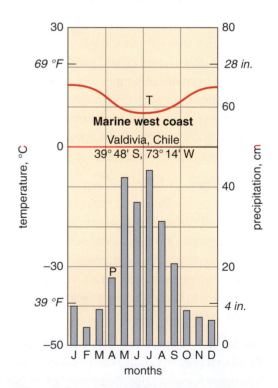

Figure B.7 Average monthly temperature and precipitation for a representative location in the marine west coast climate.

Seasonal temperatures in the marine west coast climates are very modest compared to locations of the same latitude in the continental interiors. The annual temperature range based on mean monthly temperatures is typically 10 to 15 °C, less than half that of continental counterparts. Summers are cooler, and the growing season is much longer than at inland locations. Coupled with the favorable moisture balance, these conditions are conducive to the formation of heavily forested landscapes. In fact, in the areas of heaviest rainfall, the forests are of such stature that they are termed rainforests.

The total land area classified as marine west coast is small, roughly comparable to that of its neighbor, the Mediterranean climate. This climate is not found in Asia, Africa, and, of course, Antarctica. The areas occupied by the marine west coast in North America, South America, Australia, and New Zealand are narrow coastal belts, 200 to 400 kilometers wide, which abut against high mountain ranges. Europe is the exception where the marine west coast climate extends from the British Isles and France 2000 kilometers or more inland into eastern Germany. The reason for this is related to the orientation of mountain ranges; in Europe the large mountain ranges (the Alps and the Pyrenees) trend east-west, thereby allowing the marine conditions to penetrate inland from the Atlantic. In Norway, on the other hand, the marine west coast situation is more like that of the Americas because the climate is limited by north-south trending mountains to a belt along the coast.

Humid Subtropical (Cf): In the subtropics on the southeastern sides of large continents, cyclonic and convectional storms are the principal sources of precipitation. Midlatitude cyclones are predominant in winter, but in spring and fall tropical cyclones occasionally cross the coastal areas. In North America and Asia, frequent incursions of polar air masses occur each winter.

As the polar front weakens in summer, the frontal precipitation declines. However, this is more than offset by convectional precipitation which is very pervasive throughout the warm months. In coastal areas, especially in China, summer monsoon type circulation also augments precipitation. In contrast to Mediterranean areas on the west coasts, these areas lack the thermal influence of the oceans; therefore, seasonal temperatures tend to be more extreme (Figure B.8). The largest areas of humid subtropical climate are found in China, Argentina and the American South.

A variation of the humid subtropical climate, characterized by a dry winter season, is found in five or six small areas of the world. These climates are found in or near regions generally dominated by tropical wet-dry or tropical monsoon climates, but are too cool to be given the tropical designation.

The Cold Forest Climates

Humid Continental (Df): Poleward of the humid subtropical climate, winters are more severe, with temperatures averaging below freezing in several months (Figure B.9). The polar front lies in this zone most of the year, and along it cyclonic storms are generated. Although winters are generally cold, they are subject to extremely frigid "cold snaps" with incursion of arctic air masses and warm spells ("thaws") associated with incursions of tropical air masses These incursions are associated with major shifts in the position of the midlatitude jet stream and the polar front.

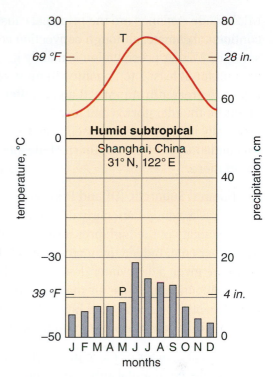

Figure B.8 Average monthly temperature and precipitation for a representative location in the humid subtropical climate.

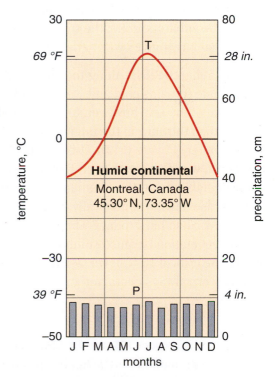

Figure B.9 Average monthly temperature and precipitation for a representative location in the humid continental climate.

Midlatitude cyclones are the principal cause of precipitation in the humid continental climate, although convection contributes appreciable rainfall in summer. Winter precipitation is mainly in the form of snow, and snow cover is present for at least 1 to 2 months in most year; however, monthly precipitation rates are greater in summer than in winter. Violent weather events in the form of intensive thunderstorms and tornadoes are not uncommon in this climatic zone, especially in the southern reaches of it. Because there are no sizable landmasses in the midlatitudes of the Southern Hemisphere, humid continental conditions are not found there.

Continental Subarctic (Df and Dw): In the northern interiors of North America and Eurasia, winters are fairly dry and very cold (Figure B.10). The dryness is due to the stability of the high-pressure cells that develop over the land in response to the winter cold. Indeed, in northeastern Siberia, the winters are so dry that large areas are given a dry winter designation (Dw). The subarctic zones are the source areas of continental polar air masses, which move southward and eastward into the humid continental zones and beyond. Summers may be mild, with long hours of sunlight, but there are only two to three months without frost.

At depths of a meter or more in the ground, permafrost can be found throughout much of the continental subarctic zone, but its distribution is discontinuous. Light and heat are sufficient, however, to support extensive boreal forests in both Eurasia and North America. These forests are floristically simple, usually made up of only four to six tree species (e.g., spruce, balsam fir, tamarack, and white birch), and growth rates are very slow not only because the growing season is short but because soils are typically saturated, especially over permafrost. In the northern belt of the subarctic zone, the trees decline in size and the forest cover thins out as the boreal forest gives way to the tundra landscape.

Marine Subarctic (Df): Along the shifting polar front, particularly on west coasts, the climate remains cold and wet throughout the year. Because of the thermal influence of the sea, however, temperatures in all but one or two months are higher than they are in the interiors and hover around freezing throughout much of the winter (Figure B.11). These coastal regions nevertheless have some of the cloudiest, windiest weather on Earth. Indeed, in the language of the Aleuts, the native inhabitants of the Aleutian Islands, there is no way to say "It's a nice day." The phrase used, when literally translated, corresponds more closely to "The wind isn't blowing so hard today." The marine subarctic climate is associated mainly with tundra vegetation because summer does not produce even a month of temperatures above the critical 10 °C level. Many tree species are usually found in this landscape, but they are always characterized by dwarf forms.

The Polar Climates

Tundra and Ice/snow (ET and EF): Poleward of the subarctic climates is the tundra climate (ET), named for the landscape comprised of diverse herbs, shrub-sized woody plants, and cold, wet soil conditions. On the average, only one month has an average temperature above 0 °C (but less than 10 °C). Most snow melts away in summer, and virtually the entire landscape is underlain by permafrost.

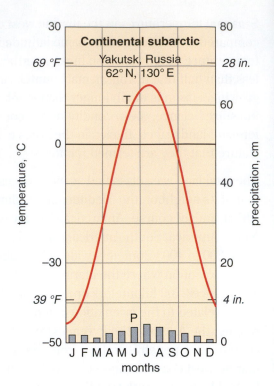

Figure B.10 Average monthly temperature and precipitation for a representative location in the marine subarctic climate.

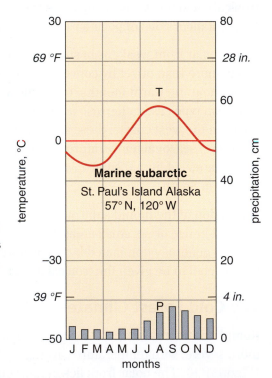

Figure B.11 Average monthly temperature and precipitation for a representative location in the marine subarctic climate.

Farther poleward is the earth's coldest climate overall (EF). It is characterized by permanent snowfields and ice caps. Average air temperatures do not exceed 0 °C in any month (Figure B.12). Despite the abundance of snow and ice, little snow is precipitated from the dry arctic air; the annual snowfall in liquid water is less than 25 centimeters. Greenland, Antarctica, and North America and Eurasia above 70°N latitude are the principal areas of arctic climate. A non-polar variant of the ET climate, denoted ETH, is found in high mountain areas outside polar regions.

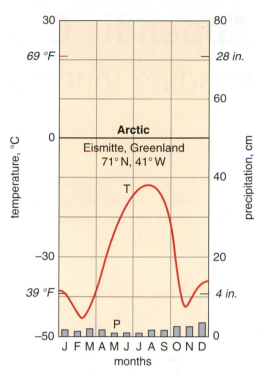

Figure B.12 Average monthly temperature and precipitation for a representative location in the ice/snow (Arctic) climate.

World Climates

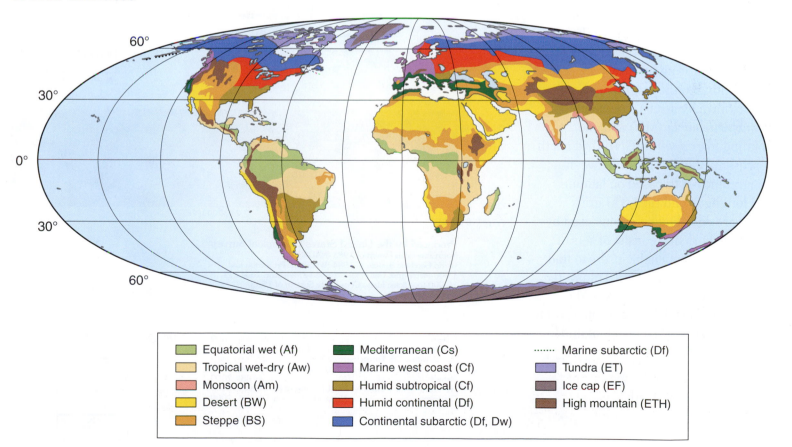

▉ Equatorial wet (Af)	▉ Mediterranean (Cs)	······ Marine subarctic (Df)
▉ Tropical wet-dry (Aw)	▉ Marine west coast (Cf)	▉ Tundra (ET)
▉ Monsoon (Am)	▉ Humid subtropical (Cf)	▉ Ice cap (EF)
▉ Desert (BW)	▉ Humid continental (Df)	▉ High mountain (ETH)
▉ Steppe (BS)	▉ Continental subarctic (Df, Dw)	

Appendix C
Reading Topographic Maps

Topographic maps for the United States are produced by the United States Geological Survey (USGS). These maps, also known as 7.5-minute quadrangles are the only uniform map series that covers the entire area of the United States in considerable detail.

Scale

The map scale is shown in two forms, a representative fraction (1:24,000) and bar scales (miles, feet, and kilometers), as shown in Figure C.1. Map scale indicates the ratio or proportion of the horizontal distance on the map to the corresponding horizontal distance on the ground.

Direction

It is common practice for maps to be oriented with true north at the top. Most USGS maps have a set of three arrows as shown in Figure C.2, one pointing to the *geographic North Pole* (shown by a star), one pointing to *magnetic north* (MN) and one pointing to *grid north* (GN). Grid north represents the difference between geographic north based on latitude and longitude and the UTM grid. Grids are coordinate systems used by cartographers that consist of sets of parallel lines that enable users to find locations. In the Universal Transverse Mercator (UTM) grid, the world is divided into 60 north-south zones, each covering a strip 6° wide in longitude.

In the Figure C.2, magnetic north is 14 degrees and 11 minutes east of geographic (true) north.

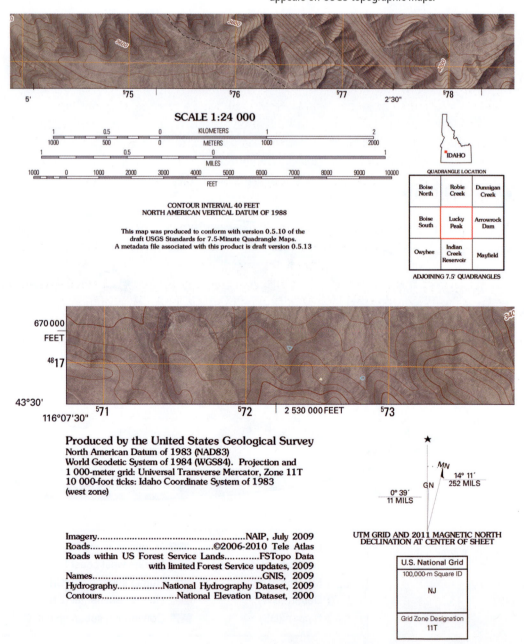

Figure C.1 Map scale and related information as it appears on USGS topographic maps.

Figure C.2 The three types of north as they appear on USGS topographic maps.

The difference between the geographic North Pole and magnetic north is the magnetic declination for that map.

Interpretation

A square mile (2.59 km²) portion of a topographic contour map is shown in Figure C.3. The curving, parallel lines are the *topographic contours*. Each contour connects points of equal elevation in the land; therefore, if you were walk along a contour your path would be perfectly level. As a result, contours never cross, and when many contours are viewed together, they reveal general shape of the terrain. To help the user determine elevations, selected contours, called *index contours* (usually every fourth or fifth contour) are labeled in feet or meters above sea level and printed in a heavier line weight. The narrower intermediate and supplementary contours found between the index contours help to show more details of the land surface shape.

Contours that are very close together represent steep slopes. Widely spaced contours, or an absence of contours, means that the ground slope is relatively level.

The elevation difference between adjacent contour lines, called the *contour interval*, determines the resolution of the map. Resolution is best, that is, most detailed, where the contour interval is small and this is possible only on large scale topographic maps. The largest scale topographic contour maps from the USGS are1:24,000 scale (about 240 meters on the ground to 1 cm on the map) and typically have a contour interval of 10 feet (3 m) or less. Smaller scale maps (representing larger areas) and those covering mountainous areas may have contour intervals of 100 feet or more.

In addition to elevation data and information, USGS topographic maps also address other geographic attributes of the land. These include the land cover and land uses such as forest, water features, agriculture, urban areas and roads (Figure C.4). Taken together these features reveal the extent of development, interrelationships between the local water and land resources, and the potential environmental hazards, as indicated in this map segment by the close proximity of the dam to a residential area in the northwest corner. Given these capabilities, topographic maps are very useful tools for urban/regional and natural resource planning applications.

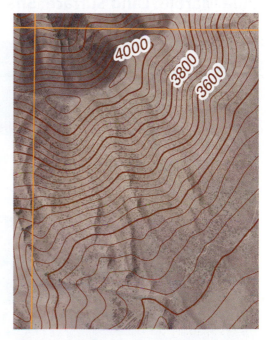

Figure C.3 Topographic contour map with a 40-foot contour interval showing intermediate and index contours.

Figure C.4 A topographic map excerpt of roughly one square mile with different land covers and land uses.

Appendix D

Blue Marble: Land Surface, Shallow Water, and Shaded Topography

Image Courtesy of NASA Goddard Space Flight Center. This spectacular "blue marble" image is the most detailed true-color image of the entire Earth to date. Using a collection of satellite-based observations, scientists and visualizers stitched together months of observations of the land surface, oceans, sea ice, and clouds into a seamless, true-color mosaic of every square kilometer (.386 square mile) of our planet.

Physical Map of the World, April 2005

Glossary

Ablation The wastage of ice or snow by melting and sublimation; in the case of glaciers, it also includes calving.

Abrasion The wearing away of a substance by rasping action; for example, the scouring of bedrock by the boulders carried in the base of a glacier.

Absolute humidity An expression for the water-vapor content of air usually expressed in grams of water vapor per cubic meter of air.

Absolute zero The zero point on the Kelvin temperature scale, which represents the state at which there is no molecular vibration in a substance and hence no heat. Corresponds to –273.15 on the Celsius scale.

Absorption The process by which incident radiation is taken up by a substance such as water vapor and converted to other forms of energy.

Abyssal plain The deep-ocean floor; the most extensive part of the ocean basin, which lies between the mid-oceanic ridges and the trenches, usually 5000 to 7000 meters below sea level.

Acidic soil A soil with a pH less than 7.0.

Acid rain Precipitation whose pH has been significantly lowered by air pollution, from a pH of 5.6 to one as low as 1.5. A serious problem in southeastern Canada, northeastern United States, and northwestern Europe.

Active layer The surface layer in a permafrost environment, which is characterized by freezing and thawing on an annual basis.

Actual evapotranspiration The true amount of moisture given up by the soil over some time period; it is equal to soil-moisture loss plus precipitation in the period of negative moisture balance.

Adaptation A change in an organism that brings it into better harmony with its environment; two types of adaptation are acquired and genetic.

Adaptation strategies A general means of plant adaptation. Three primary strategies are competition, stress toleration, and disturbance tolerance.

Adiabatic cooling A thermodynamic decline in a system, such as a parcel of air, in which there is no transfer of heat or mass from the system. In a rising parcel of air, decompression and expansion result in cooling.

Adsorption A chemical process by which ions are taken up by colloids.

Advection The transfer or exchange of energy in the atmosphere by the lateral movement of air as when cold air gains heat with movement over a warm surface.

Aggradation Filling in of a stream channel with sediment, usually associated with low discharges and/or heavy sediment loads.

Air pressure (See *Sea-level pressure*.)

Albedo The percentage of incident radiation reflected by a material. Usage in earth science is usually limited to shortwave radiation and landscape materials.

Alfisols Soils similar to mollisols but associated with forest covers. A whitish layer, called an argillic horizon, forms in the B horizon; upper soil is relatively rich in organic matter.

Alkaline soil A soil with a pH greater than 7.0.

Alluvial fan A fan-shaped deposit of sediment laid down by a stream at the foot of a slope; very common features in dry regions, where streams deposit their sediment load as they lose discharge downstream.

Alluvium Material deposited by a stream or river.

Alpine meadow A formation of grasses and forbs found in mountains above the treeline; similar to the tundra formation of the arctic.

Andesite An extrusive igneous rock comprised of intermediate amounts of dark minerals; the extrusive counterpart of diorite.

Andesite line The seaward extent of andesite lavas used in some places to delimit the geologic border of the continents.

Angiosperm A flowering, seed-bearing plant; the angiosperms are presently the principal vascular plants on Earth.

Angle of repose The maximum angle at which a material can be inclined without failing; in civil engineering the term is used in reference to clayey materials.

Angular momentum A measure of the momentum of an object with respect to its rotation about a point; in the atmosphere the farther poleward a particle of air goes, the closer it gets to the axis of Earth rotation and the faster its velocity becomes.

Anion A negatively charged ion.

Anticline A fold characterized by an upward bend, e.g. convex upward in cross-section.

Aphelion The position of the Earth in its orbit when it is farthest from the Sun – 152 million kilometers (94.25 million miles).

Aquiclude An impervious stratum or formation that impedes the movement of groundwater.

Aquifer Any subsurface material that holds a relatively large quantity of groundwater and is able to transmit that water readily.

Archipelago An arc-shaped group of islands, usually of volcanic origin; most archipelagos, e.g. the Japanese Islands, are associated with subduction zones.

Arête The term given to the sharp ridges that separate cirques or faces on a glaciated mountain.

Aridisols Soils of dry environments; poor in organic matter and often heavy in salts.

Artesian flow A pressurized flow of groundwater which causes the water to sometimes rise above ground level.

Artesian well (See *Artesian flow*.)

Asthenosphere The layer immediately under the lithosphere, where the rock appears to be in a plastic state and capable of slow-flowing motion.

Autumnal equinox (See *Equinox*.)

Azonal soil A soil order under the traditional USDA soil classification scheme; soils without horizons; those that are usually found in geomorphically active environments such as sand dunes and river valleys.

Backscattering That part of solar radiation directed back into space as a result of diffusion by particles in the atmosphere.

Backshore The zone behind the shore – between the beach berm and the backshore slope.

Backshore slope The bank or bluff landward of the shore that is comprised of *in situ* material.

Backswamps A low, wet area in the floodplain, often located behind a levee.

Backwash The counterpart to swash; the sheet of water that slides back down the beach face.

Bar A unit of force equal to 100,000 newtons per square meter; normal atmospheric pressure at sea level is slightly greater than 1 bar (1.0132 bars).

Barchan A type of sand dune that is crescent shaped, with its long axis transverse to the wind and its wings tipped downward.

Basalt floods A form of volcanism characterized by massive outflows of lava from long fissures in the crust.

Basaltic rock A general term applied to the rocks that form the ocean basins and the lower crust; relatively high-density, dark-colored igneous rock; basalt and related rocks of the sima layer.

Base In soil chemistry, a mineral that forms cations, e.g. magnesium and potassium.

Base level The lowest elevation to which a river can downcut its channel. For large rivers, this is controlled by sea level.

Baseflow The portion of streamflow contributed by groundwater; it is a steady flow that is slow to change even during rainless periods.

Basic soil (See *Alkaline soil*.)

Basin A rock structure formed by a large downward flexure, often hundreds of kilometers in diameter, e.g. Paris Basin of France.

Batholith A large accumulation of magma in a great chamber; upon cooling, it forms coarse-grained igneous rocks.

Bay-mouth bar A ribbon of sand deposited across the mouth of a bay.

Beam radiation Directed shortwave radiation that passes through the atmosphere without much scattering.

Bed load The stream-carried particles (sediment) that roll along the bottom and are in nearly continuous contact with the streambed.

Bed shear stress The force exerted against a streambed by moving water; it is a function of water density, gravitational acceleration, water depth, and the slope of the channel.

Berm A low mound that forms along sandy beaches.

Biochore A major region or division of vegetation defined on the basis of the structural and compositional characteristics of the plant cover; in a general sense, it is the vegetative version of a biome; see also *Biome*.

Biological diversity The number of species of plants, animals, and microorganisms per unit area of land or water.

Biomass The total weight of organic matter per unit area of landscape; also, the total weight of the organic matter in an ecosystem.

Biome A major division (region) of the Earth's surface, defined on the basis of the plants and animals inhabiting it; geographically, it may correspond to a climatic zone.

Black body A hypothetical body that is capable of absorbing all radiation incident on it and in turn is the most effective possible emitter of radiation.

Blowout (See **Deflation hollow**.)

Boreal forest Subarctic conifer forests of North America and Eurasia; floristically homogeneous forests dominated by fir, spruce, and tamarack; in Russia, it is called *taiga*.

Boundary layer The lower layer of the atmosphere; the lower 300 meters of the atmosphere where airflow is influenced by the Earth's surface.

Bowen ratio The ratio of sensible-heat flux to latent-heat flux between a surface and the atmosphere.

Braided channel A stream channel characterized by multiple threads or subchannels, which appear to weave in and out of one another.

Brittle rock Rock that ruptures with little or no plastic deformation.

Caatinqa (See *Thornbush*.)

Calcic layer A concentration of calcium at or near the surface of soils in arid regions.

Calcification A soil-forming regime of dry environments that results in the accumulation of calcium carbonate and, in grassy areas, a strong organic layer.

Caldera A large, circular depression in a volcano, resulting from an explosion or the loss of magma through a lower vent.

Caliche An accumulation of calcium carbonate at or near the soil surface in an arid environment.

Calorie A unit of energy; the amount of heat required to raise the temperature of water $1\,°C$, e.g. from $14.5\,°C$ to $15.5\,°C$.

Calving The process by which a glacier loses mass when ice breaks off into the sea.

Canopy The roof of foliage formed by the crowns of trees in a forest.

Capillarity The capacity of a soil to transfer water by capillary action; capillarity is greatest in medium textured soils.

Capillary fringe The transition in soil between the zone of aeration and the zone of saturation; where capillary water gives way to groundwater.

Cation A positively charged ion.

Cation adsorption The process by which cations become attached to colloids.

Cation-exchange capacity The total exchangeable cations that a soil can adsorb; the capacity per unit volume of soil increases with finer soil textures.

Cavitation A mechanism of stream and wave erosion brought about by the sudden increase in hydraulic pressure, which causes air bubbles to burst and exert force against rock.

Celsius A temperature scale, also known as centigrade, on which $0\,°C$ represents the normal freezing point of water, and $100\,°C$ represents the normal boiling point of water.

Centigrade (See *Celsius*.)

Central vent The main passageway by which magma ascends to the surface of a volcano.

Channel precipitation Rain or snow that falls directly on the stream channel and thereby contributes immediately to discharge.

Chapparal A vegetative formation of the Mediterranean climate of California, characterized by shrubs and small trees, often in shrubby thickets.

Chelation A weathering process involving the bonding of mineral ions to a large organic molecule, followed by the removal of the ions with the molecule.

Chemical stability A term referring to the tendency of a mineral to change to another form, such as another mineral; stable minerals are not readily transformed to other states and are usually those, such as quartz, that originated at relatively low temperatures.

Chemical weathering One of the two major types of weathering and generally considered to be the more effective one; it involves the chemical decomposition of rock by a variety of chemical processes, including dissolution, chelation, and hydrolysis.

Chernozem soils One of the great soil groups under the traditional USDA soil classification scheme; soil characterized by a heavy a horizon and calcium carbonate accumulation in the B horizon.

Chinook A dry, often warm, wind that descends the leeward side of a mountain range; in Germany, Austria, and Switzerland such a wind is called a *föhn*.

Circle of illumination The line dividing the illuminated half of the Earth from the dark (shadow) half; the alignment of this circle changes with the seasons, thereby changing the daily length of daylight hours at different latitudes.

Climate The representative or general conditions of the atmosphere at a place on Earth; it is more than the average conditions of the atmosphere, for climate may also include extreme and infrequent conditions.

Climatology The field of earth science devoted to the study of climate and climatic processes.

Climax community A group of organisms that represents an ecological equilibrium with its environment and therefore is capable of maintaining long-term stability.

Clone A genetically uniform group of plants regenerated by vegetative means (asexual) from a single parent.

Closed forest A forest structure with multiple levels of growth from the ground up; a forest in which undergrowth closes out the area between the canopy and the ground.

Coastal dune A sand dune that forms in coastal areas and is fed by sand from the beach.

Coefficient of runoff A number given to a type of ground surface representing the proportion of a rainfall converted to overland flow; it is a dimensionless number between 0 and 1.0 that varies inversely with the infiltration capacity; impervious surfaces have high coefficients of runoff.

Cold front A contact between a cold air mass and a warm air mass in which the cold air is advancing on the warm air, driving the warm air upward.

Colloid A small clay particle, less than 0.001 mm in diameter, that provides adsorption sites for ions.

Colluvium An unsorted mix of soil and mass-movement debris.

Community A group of organisms that live together in an interdependent fashion.

Community-succession concept A popular concept of vegetation change based on the idea that one plant community succeeds another in occupying a site ending with a climax community.

Composite volcano A cone-shaped volcano comprised mainly of pyroclastic material, e.g. Fuji-san of Japan and Vesuvius of Italy.

Compressional wave A type of seismic (elastic) wave that generates a back-and-forth motion along the line of energy propagation through a substance; also termed *primary wave*.

Concentration time The time taken for a drop of rain falling on the perimeter of a drainage basin to go through the basin to the outlet.

Condensation The physical process by which water changes from the vapor to the liquid phase.

Condensation nuclei Very small particles of dust or salt suspended in the atmosphere on which condensation takes place to initiate the formation of a precipitation droplet.

Conduction A mechanism of heat transfer involving no external motion or mass transport; instead, energy is transferred through the collision of vibrating molecules.

Cone of ascension The ascent of salt groundwater under a well in response to the development of a cone of depression in the overlying fresh groundwater.

Cone of depression A conical-shaped depression that forms in the water table or an aquifer surface immediately around a well from which groundwater is being rapidly pumped.

Conservation of angular momentum The principle that the angular momentum of a rotating mass, such as the Earth or a twirling skater, will not change unless torque (stress) is applied to it; if the radius of rotation decreases (arms in), rotational velocity increases, and vice versa.

Conservation-of-energy principle The principle that energy in an isolated system can be neither destroyed nor created; thus, total energy in the system remains constant.

Conservative zone A type of contact or border along a tectonic plate where lithosphere is neither destroyed nor created; the tectonic movement in conservative zones is often lateral, such as along the San Andreas fault of California.

Constructive zone A type of contact or border along a tectonic plate where new lithosphere is emerging; usually associated with mid-oceanic ridges.

Continental climate A climate characterized by a large annual temperature range and often relatively dry conditions.

Continental drift The term used to describe the wholesale movement of land masses over great distances on the surface of the Earth; the term is generally attributed to Alfred Wegener, who advanced the first coherent theory of continental drift.

Continental shelf The seaward-sloping margin of the continents under water to a depth of 200–300 meters; the shoulders of the continents where the rate sediment accumulation and sedimentary rock formation is high.

Continuity of flow A principle that describes the maintenance of flow in a system with changes in flow velocity and system capacity.

Core The innermost of the two major divisions of the solid Earth, the other being the mantle; the core includes the *outer core*, which is liquid, and the *inner core*, which is solid; the core is the densest part of the Earth.

Coriolis effect The effect of the Earth's rotation on the path of airborne objects; winds in the northern hemisphere are deflected to the right and those in the southern hemisphere are deflected to the left of their original paths; at the Equator it is negligible; the Coriolis is an effect apparent only to observers standing on the Earth and as such is not a force.

Corrasion Another term for abrasion.

Convection A mechanism of heat transfer which involves mass transport (mixing) of fluid, such as occurs with turbulence in the atmosphere; any mixing motion in a fluid.

Convectional precipitation A type of precipitation resulting from the ascent of unstable air; the instability may be caused by local heating in the landscape or by frontal activity; usually short-term, intensive rainfall.

Convergent precipitation A type of precipitation that takes place when air moves into a low-pressure trough or topographic depression and escapes by moving upward; precipitation in the ITC zone is at least partially convergent.

Cover density The percentage of areal coverage by vegetation of all types in an area.

Critical threshold The point at which shear stress (driving force) equals shear strength (resisting force) and beyond which change, such as slope failure, rock rupture, plant damage, or soil erosion, is imminent.

Crust The outermost zone of the lithosphere, which ranges from 8 to 65 kilometers (4–40 miles) in thickness and is bounded on the bottom by the Moho discontinuity.

Cumulus A class of clouds characterized by vertical development.

Cuspate foreland A large depositional feature along a coastline, often in the form of a triangular point.

Cyclone A large low-pressure cell characterized by convergent airflow and internal instability; the two main classes of cyclones are midlatitude cyclones and tropical cyclones; in some parts of the United States, tornadoes are also called cyclones.

Cyclonic/frontal precipitation A type of precipitation that results from a large cell of low pressure and the meeting of warm and cold air masses; most of the precipitation is concentrated along the fronts; see *Cold front* and *Warm front*.

Darcy's law The principle that describes the velocity of groundwater flow; velocity is equal to permeability times the hydraulic gradient.

Debris flow A type of mass movement characterized by the downslope flow of a saturated mass of heterogeneous soil material and rock debris.

Declination of the Sun The location (latitude) on Earth where the Sun on any day is directly overhead; declinations range from 23.27 degrees south latitude to 23.27 degrees north latitude.

Deflation hollow A topographic depression caused by wind erosion; also called a *blowout*.

Degradation Scouring and downcutting of a stream channel, usually associated with high discharges.

Denudation A term used to describe the erosion or wearing down of a land mass; also used to describe the process by which a site is stripped of its vegetative cover.

Denude (See *Denudation*.)

Deranged drainage Highly irregular drainage patterns in areas of complex geology, such as the Canadian Shield, which have been heavily glaciated.

Desert lag A veneer of coarse particles left on the ground after the fine particles have been eroded away, often by wind.

Desert soils Soils characterized by a weak 0 horizon and salt accumulation at or near the surface; in the traditional USDA classification scheme, these soils are classed as either Red or Gray desert soils.

Disseminule Any part of a plant from which another plant can be established; seeds, fruits, spores, or vegetative parts.

Destructive zone A type of contact or border along a tectonic plate where lithosphere is destroyed; see also *Subduction*.

Detrital rock Sedimentary rock composed of particles transported to their place of desposition by erosional processes.

Dew point The temperature at which a parcel of air becomes saturated.

Diagnostic horizon Soil horizons used to define the key conditions and processes of soil formation.

Diapir A conduit in the lithosphere through which magma moves to the surface.

Differential stress Force, directed on a body, that is not equal in all directions.

Diffuse radiation Solar radiation which has been scattered in the atmosphere; light rays moving in random patterns among molecules and particles in the air.

Dike A vertical fissure in the crust that serves as a passageway for magma.

Diorite An intrusive igneous rock that is both darker and of higher density than granite.

Dip (See *Strike and dip*.)

Dip-slip fault A fault in which the principal direction of displacement is up or down along the fault plane.

Discharge The rate of water flow in a stream channel; measured as the volume of water passing through a cross-section of a stream per unit of time, commonly expressed as cubic feet (or meters) per second.

Dispersal The process of distribution of a dissemmule from a parent plant to a new location.

Dissociation The process by which water molecules give up hydrogen ions.

Dissolved load Material carried in solution (ionic form) by a stream.

Disturbance Factors, other than the basic requirements, that affect a plant's well-being, e.g. floods, disease, and soil erosion.

Diurnal damping depth The maximum depth in the soil which experiences temperature change over a 24-hour (diurnal) period.

Doldrums The belt of calm and variable winds in the intertropical convergence zone; in the days of ocean sailing, a source of quiet.

Dolomite A sedimentary rock of chemical origin; it appears to be an altered form of limestone in which some of the calcium is replaced by magnesium.

Dome Large upward flexure of rock often hundreds of kilometers in diameter, e.g. the Ozark Plateaus of Missouri.

Dominant species A plant species that occupies the greatest amount of space in an area as measured by the extent of its foliage or root system.

Dominant stratum A number of plants whose combined foliage makes up the principal layer (story) in a vegetative formation.

Drainage basin The area that contributes runoff to a stream, river, or lake.

Drainage density The number of miles (or kilometers) of stream channels per square mile of land.

Drainage divide The border of a drainage basin or watershed where runoff separates between adjacent areas.

Drainage network A system of stream channels usually connected in a hierarchical fashion (see also *Principle of stream orders*).

Drift (See *Glacial drift* and *Littoral drift*.)

Ductile rock Rock that has a relatively large capacity for plastic deformation.

Dust Bowl The name given to the area of severe drought and wind erosion in the Great Plains during the 1930s.

Dwarfism The tendency for a plant to achieve less than full size at maturity because of environmental stress such as inadequate heat or light.

Dynamic equilibrium A term used to describe the behavior of a system, such as a river network, which is continually trending toward a state of equilibrium, but rarely reaches it. The trend may change with changes in the energy available to drive the system.

Earthquake intensity (See *Mercalli scale*.)

Earthquake magnitude (See *Richter scale*.)

Easterly wave A trough of low pressure in the trade wind belt. They form over water, move westward with the trade winds, and usually produce showers and thunderstorms.

Eccentricity A variation in the shape of the Earth's orbit from a more circular to a more elliptical shape over a period of 90,000 to 100,000 years.

Ecological amplitude The variation in tolerance and resource needs from member to member in a plant species.

Ecosystem A group of organisms linked together by a flow energy; also, a community of organisms and their environment.

Ecotone The transition zone between two groups, or zones, of vegetation.

Eddy The term given to a whirling or spiral motion in a fluid.

Edge wave A wave that moves parallel to the shore; usually a secondary wave of complex origins.

Elastic deformation Change in the shape of a body as a result of differential stress; on the release of the stress, the body returns to its original shape.

Elastic limit The maximum level of elastic deformation of a body, beyond which it ruptures.

Elastic wave An energy wave that causes motion in a material without permanently deforming it.

Electromagnetic spectrum The classification scheme used to describe the array of electromagnetic radiation; the various categories of radiation are distinguished on the basis of wavelength.

Eluviation The removal of colloids and ions from a soil or from one level to another in a soil.

Emissivity The ratio of total radiant energy emitted at a specified wavelength and temperature by a substance to that emitted by a *black body* under the same conditions.

Empirical approach An approach to scientific investigation based on direct observation and measurement.

Energy Generally, the capacity to do work; defined as any quantity that represents force times distance. Joules and calories are energy units commonly used in science.

Energy balance The concept or model that concerns the relationship among energy input, energy storage, work, and energy output of a system such as the atmosphere or oceans.

Energy flux The rate of energy flow into, from, or through a substance; also called radiant flux density and irradiance.

Energy pyramid The attenuation of organic energy in an ecosystem; the decline of energy in an ecosystem as organic matter is passed from one level of organisms to another.

Entisols Soils of recent origins with no horizons or weakly developed ones.

Ephemeral stream A stream without baseflow; one that flows only during or after rainstorms or snowmelt events.

Epiphyte A plant that grows in the superstructure of another plant without rooting in the soil; a non-parasitic aerial plant, e.g. Spanish moss and many orchids.

Equatorial zone The middle belt of latitude, extending 10° or so north and south of the Equator.

Equilibrium profile A gently sloping surface extending from the shore through the offshore zone across which wave energy is evenly distributed.

Equinox The dates when the declination of the Sun is at the Equator, March 20–21 and September 21–22, and the number of hours of dark and daylight in a 24-hour period is the same for all locations on Earth. These dates are known as the autumnal equinox and vernal equinox, but which is which depends on the hemisphere.

Erosion The removal of rock debris and soil by an agency such as moving water, wind, or glaciers; generally, the sculpting or wearing down of the land by erosional agents.

Euler's theorem The theorem that describes the movement of a plate on the surface of a sphere; such a plate moves about its own pole, called the Euler pole, along a small-circle path. This theorem helps explain the differences in the rates of sea floor spreading along the mid-oceanic ridges.

Eutrophication Accelerated biological productivity in a water body as a result of

the input of nutrients such as nitrogen and phosphorus.

Evaporite A rock or mineral formed from the evaporation of mineral-rich water, e.g. rock salt and gypsum.

Evapotranspiration The loss of water from the soil through evaporation and transpiration.

Event An episode of a process defined as some quantity of a variable such as a river discharge or wind velocity.

Evolution Biological change over time that results in new or changed relationships between organisms and the environment; irreversible biological change.

Exfoliation A mechanical weathering process involving the breaking off, or "shedding," of slabs of rock in response to the differential expansion of a rock mass.

Exotic river A river, such as the Nile or Colorado, that flows through an arid region after gaining its flow elsewhere; exotic streams usually lose much of their discharge to groundwater recharge and evaporation.

Extrusive rock Igneous rock that forms at the surface and cools quickly.

Fahrenheit A temperature scale on which 32 °F represents the normal freezing point of water and 212 °F represents the normal boiling point of water.

Fault A fracture in rock along which there has been displacement of one side relative to the other.

Fault line The linear trend of a fault along the Earth's surface as one would see it from the air.

Fault plane The plane representing the fracture surface in a fault.

Fault scarp The part of the fault plane exposed in a fault; in a normal fault it is the upper part of the footwall.

Feedback A return effect of a change; the consequences of a change have a feedback effect if they dampen or amplify the change or the causes of it.

Fell fields Areas of very light plant cover in polar and alpine regions; often rocky areas with scattered lichens, mosses, and small flowering plants.

Ferromagnesian minerals A subgroup of the silicate minerals; rock-forming minerals that are rich in iron and magnesium; dark-colored and relatively high density minerals, e.g. biotite, hornblende, and olivine.

Fetch The distance of open water in one direction across a water body; it is one of the main controls on wave size.

Firn Névé on a glacier that has survived the entire ablation season; the material that is transformed into glacial ice.

Firn line The lower edge of the accumulation zone on a glacier.

Flank eruption Volcanic eruption that breaks out on the side of a volcano.

Floristic system The principal botanical classification scheme in use today; under this scheme the plant kingdom is made up of divisions, each of which is subdivided into smaller and smaller groups arranged according to the apparent evolutionary relationships among plants.

Fluvioglacial deposits Materials deposited by glacial meltwaters, including outwash plains, kames, and eskers; usually stratified.

Fold A rock structure characterized by a bend in a rock formation.

Föhn (See *Chinook*.)

Foot ice An accumulation of grounded ice on and near shore in lakes, oceans, and rivers.

Footwall The lower surface of an inclined fault.

Forest fire (See *Ground fire*.)

Formation A structural unit of vegetation that may be considered a subdivision of a biochore; a formation may be made up of several communities. In the traditional terminology, it is called a physiognomic unit; in geology, a major unit of rock.

Free convection Mixing motion in a fluid caused by differences in density. In the atmosphere such differences are usually caused by differential heating of air near the surface.

Free-face A steep slope or cliff formed in bedrock.

Freeze-thaw activity Weathering and mass-movement processes associated with daily and seasonal cycles of freezing and melting.

Frequency The term used to express how often a specified event is equaled or exceeded.

Friable A term used to describe the tendency of a soil to crumble or break up when plowed.

Front (See *Cold front* and *Warm front*.)

Frost wedging A mechanical weathering process in which water freezes in a crack and exerts force on the rock, which may result in the breaking of the rock; a very effective weathering process in alpine and polar environments.

Fusion Another word for freezing.

Gabbro A coarse-grained, intrusive igneous rock that is dark and heavy owing to a relatively high percentage of ferromagnesian minerals.

Geodesy The science that measures the geoid and its major features.

Geographic cycle A concept developed by William Morris Davis on the formation of river-eroded landscapes. It describes three stages of landscape development (youth, maturity, and old age) and argues that rejuvenation takes place with uplift of the land, thereby renewing the cycle.

Geoid The term given to the true shape of the Earth, which deviates from a perfect sphere because of a slight bulge in the equatorial zone.

Geomechanical regime A soil environment where the formative processes are dominated by mechanical mixing associated with ground frost or wetting and drying of clay.

Geomorphic system A physical system comprised of an assemblage of landforms linked together by the flow of water, air, or ice.

Geomorphology The field of Earth science that studies the origin and distribution of landforms, with special emphasis on the nature of erosional processes; traditionally, a field shared by geography and geology.

Geophysics A field of Earth science devoted to the study of the Earth,

including the oceans and the atmosphere, through the application of models and techniques from physics.

Geostrophic winds Winds in the upper troposphere which generally flow parallel to isobars and often reach high velocities.

Geothermal energy Energy in the form of heat that is produced by the Earth's interior and flows through the crust mainly by conduction.

Glacial drift A general term applied to all glacial deposits, including moraine and fluvioglacial deposits.

Glacial flour (See *Glacial milk*.)

Glacial lake A natural impoundment of meltwater at the edge of a glacier.

Glacial milk Glacial meltwater of a light or cloudy appearance because of clay-sized sediment held in suspension.

Glacial polish Bedrock surfaces that have been made smooth and shiny by glacial abrasion.

Glacial surge A rapid advance of the snout of a glacier.

Glacial uplift Uplift of the crust following isostatic depression under the weight of the continental glaciers.

Gleization A soil-forming regime of poorly drained areas such as bogs and swamps; it results in a heavy organic layer over a layer of blue clay.

Global coordinate system The network of east–west and north–south lines (parallels and meridians) used to measure locations on Earth; the system uses degrees, minutes, and seconds as the units of measurement.

Gneiss A metamorphosed form of granite characterized by minerals arranged in bands.

Graben (See *Rift*.)

Graded profile The longitudinal profile of a stream representing an equilibrium condition toward which the stream adjusts in response to changes in discharge and sediment load.

Grafting The practice of attaching additional channels to a drainage network; in agricultural areas new channels appear as drainage ditches; in urban areas, as storm sewers.

Granite An intrusive igneous rock comprised mainly of quartz and feldspar; limited in its distribution to the continents.

Granitic rock A general term applied to the rocks that comprise the continental masses; low density, light-colored igneous rock; granite and related rocks of the sial layer.

Granodiorite An intrusive igneous rock that is intermediate in composition between granite and diorite.

Graupel A frozen precipitation particle comprised of a snow crystal and a raindrop frozen together.

Gravity water Subsurface water that responds to the gravitational force (in contrast to capillary water, which responds to molecular forces); the water that percolates through the soil to become groundwater.

Great circle Any circle that circumscribes the full circumference of the Earth and the plane of which passes through the center of the Earth; the Equator is a great circle; the shortest distance between any two points on the globe follows a great-circle route.

Greenbelt A tract of trees and associated vegetation in urbanized areas; it may be a park, nature preserve, or part of a transportation corridor.

Greenwich meridian The zero degree meridian from which east and west longitude are measured; it is named for the town of Greenwich, England, through which the line is drawn.

Groin A wall or barrier built from the beach into the surf zone for the purpose of slowing down longshore transport and holding sand.

Gross sediment transport The total quantity of sediment transported along a shoreline in some time period, usually a year.

Ground fire All-consuming fire that burns trees, ground plants, and topsoil.

Ground frost Frost that penetrates the ground in response to freezing surface temperatures.

Groundwater The mass of gravity water that occupies the subsoil and upper bedrock zone; the water occupying the zone of saturation below the soil water zone.

Gulf stream A large, warm current in the Atlantic Ocean that originates in and around the Caribbean and flows northwestward to the North Atlantic and northwest Europe.

Gullying Soil erosion characterized by the formation of narrow, steep-sided channels etched by rivulets or small streams of water. Gullying can be one of the most serious forms of soil erosion of cropland.

Guyot Volcanic island eroded to sea level and then submerged with subsidence of the ocean floor. Coral reefs may form on them.

Gymnosperm A plant that bears naked seeds; the most common group of gymnosperms are the conifers, needleleaf cone-bearing plants.

Gyre A large circular pattern of ocean currents associated with major systems of pressure and prevailing winds; the largest are the subtropical gyres.

Habitat The surrounding or local environment in which an organism gains its resources.

Habitat versatility (See *Ecological amplitude*.)

Hadley cells Large atmospheric circulation cells comprised of rising air near the equator, poleward flow aloft, descending air in the subtropics, and return flow on the surface in the form of trade winds; concept first proposed by George C. Hadley in 1735.

Hamada (See *Reg*.)

Hanging valley A tributary valley that enters the main valley at an elevation well above the valley floor; most common in areas of mountain glaciation, where hanging valleys are often the sites of spectacular waterfalls.

Hanging wall The upper surface of an inclined fault.

Hardpan A hardened soil layer characterized by the accumulation of colloids and ions.

Heat island The area or patch of relatively warm air which develops over urbanized areas.

Heat syndrome A health problem caused by some combination of extreme heat, exposure to solar radiation, physical exertion, and loss of body fluids and salts.

Heat transfer The flow of heat within a substance or the exchange of heat between substances by means of conduction, convection, or radiation.

High latitude The zones poleward of the Arctic and Antarctic circles.

Higher plants Generally, the larger and more advanced plants; the vascular plants; pteridophytes, gymnosperms, and angiosperms.

Hillslope processes The geomorphic processes that erode and shape slopes; mainly mass movements such as soil creep and landslides and runoff processes such as rainwash and gullying.

Histosols Soils dominated by a thick organic layer as a result of hydromorphic soil-forming conditions.

Horizon A layer in the soil that originates from the differentiation of particles and chemicals by moisture movement within the soil column; four major horizons are recognized in a standard soil profile: 0, A, B, and C.

Horst A fault characterized by a block displaced up, ward relative to adjacent rock formations.

Humboldt current A cold current that flows northward along the west coast of South America, contributing to the aridity there.

Humus Organic matter in the soil that has been broken down by physical, chemical, and biological processes into a granular form which is relatively stable.

Hurricane A large tropical cyclone characterized by convergent airflow, ascending air in the interior, and heavy precipitation.

Hydraulic gradient The slope of the water table or any body of groundwater; equal to the difference in the elevation of the water table at two points divided by the distance between them.

Hydraulic pressure (See *Cavitation*.)

Hydraulic radius The ratio of the cross-sectional area of a stream to its wetted perimeter.

Hydrograph A streamflow graph which shows the change in discharge over time, usually hours or days; see also *hydrograph method*.

Hydrograph method A means of forecasting streamflow by constructing a hydrograph that shows the representative response of a drainage basin to a rainstorm; the use of a "normalized" hydrograph for flow forecasting in which the size of the individual storm is filtered out; see also *Hydrograph*.

Hydrologic cycle The planet's water system, described by the movement of water from the oceans to the atmosphere to the continents and back to the sea.

Hydrologic equation The amount of surface runoff (overland flow) from any parcel of ground is proportional to precipitation minus evapotranspiration loss, plus or minus changes in storage water (groundwater and soil water).

Hydrolysis A complex chemical weathering process, or series of processes, involving the reaction of water and an acid on a mineral; it is considered to be the most effective process in the decomposition of granite.

Hydromorphic regime A soil environment where the formative processes are dominated by water, usually a swamp or a bog.

Hydrophyte Water-loving plants; aquatic plants such as water lily and water hyacinth.

Hydrostatic pressure The pressure exerted by elevated groundwater as in artesian flow.

Hygrophyte Water-tolerant plants, such as cattail, which are able to grow in saturated or lightly flooded sites.

Hygroscopic water Molecular water that resides directly on the surface of all materials; it is bound to surfaces under such great pressure that it is immobile and cannot be evaporated or used by plants.

Ice Age (See *Pleistocene Epoch.*)

Ice wedging (See *Frost wedging.*)

Illuviation The process of accumulation of ions and colloids in a soil.

Inceptisols Soils with horizons in the early phases of development; usually formed on young geomorphic surfaces such as midlatitude glacial deposits.

Individualistic concept A concept of vegetation change contrary to the community-succession concept, especially the idea of a climax community; it argues that the stability of the climax community is a matter of probability because those species in greatest abundance favor regeneration of their own kind.

Infiltration capacity The rate at which a ground material takes in water through the surface; measured in inches or centimeters per minute or hour.

Inflooding Flooding caused by overland flow concentrating in a low area.

Infrared radiation Mainly longwave radiation of wavelengths between 3.0–4.0 and 100 micrometers, but also includes near infrared radiation, which occurs at wavelengths between 0.7 and 3.0–4.0 micrometers.

In situ A term used to indicate that a substance is in place as contrasted with one, such as river sediment, that is in transit.

Instability A physical condition in which a fluid is gravitationally unstable; in the atmosphere it is one in which heavier (denser) air overlies lighter air.

Interception The process by which vegetation intercepts rainfall or snow before it reaches the ground.

Interflow Infiltration water that moves laterally in the soil and seeps into stream channels; in forested areas this water is a major source of stream discharge.

Interglacial A relatively warm and dry period in the Pleistocene Epoch during which most of the continental glaciers are thought to have melted away.

Intermittent A stream with baseflow in all but the dry season when the water table drops below the streambed.

Intertropical convergence zone (ITCZ) The belt of convergent airflow and low pressure in the equatorial zone (between the tropics) which is fed by the trade winds.

Intertropical zone The zone between the Tropic of Cancer (23.5 degrees north latitude) and the Tropic of Capricorn (23.5 degrees south latitude).

Intrazonal soils A soil order under the traditional USDA soil classification scheme: soils that form under conditions of impeded drainage.

Intrusive rock Igneous rock that forms within the Earth and cools slowly; see also *Plutonic rock.*

Inversion (See *Temperature inversion.*)

Ion A minute particle of a dissolved mineral; usually an atom or group of atoms that are electrically charged.

Isostatic depression Large-scale down-warping of the crust in response to an increase in mass (weight) on the surface; in areas of continental glaciation the crust was depressed by the weight of the ice.

Isostatic rebound The uplift or recovery of the Earth's crust following isostatic depression; elastic recovery of the crust from large-scale depression; see also *Isostatic depression.*

Jet stream Zone of concentrated geostrophic winds; in the midlatitudes it is called the polar front jet stream because it often coincides in location with the polar front.

Joint line An open crack or fracture in bedrock, usually a result of weathering.

Joule A unit of energy equal to one newton (a unit of force) applied over a one-meter length; in terms of heat, 4186 joules are needed to raise the temperature of 1 kilogram of water 1 °C, from 14.5 °C to 15.5 °C.

Kame A mound- or cone-shaped deposit of sand and gravel laid down by melting water in and around glacial ice.

Kaolinite A type of clay produced in the weathering of granite; it is especially widespread in tropical and subtropical regions.

Karst topography Irregular topography in areas of, carbonate rock, characterized by sinkholes, caverns, and underground drainage channels.

Katabatic wind Any wind blowing down a large incline such as a mountain slope; chinook winds are katabatic winds.

Kelvin A temperature scale based on absolute zero, the temperature at which a substance has no molecular vibration and thus generates no heat. Water freezes at 273.15 K and boils at 373.15 K.

Kilogram A metric unit of mass (weight) equal to 2.208 pounds.

Kinetic energy The energy represented by the motion of a substance; equal to mass times velocity squared, divided by two.

Kettlehole A pit in an outwash plain or moraine left from a buried block of glacial ice which melted away; see also *Pitted topography.*

Laminar flow Flow characterized by one layer of a fluid sliding over another without vertical (turbulent) mixing; the source offlow resistance is limited to intermolecular friction within the fluid.

Laminar sub layer In the atmosphere the layer of essentially calm air immediately adjacent to fixed surfaces such as vegetation and soil; in reality this air is not perfectly calm, but characterized by a faint laminar flow parallel to the surface.

Land cover The materials such as vegetation and concrete that cover the ground; see also *Land use.*

Landscape The composite of natural and human features that characterize the surface of the land at the base of the atmosphere; includes spatial, textural, compositional, and dynamic aspects of the land.

Landslide A type of mass movement characterized by the slippage of a body of material over a rupture plane; often a sudden and rapid movement.

Land use The human activities that characterize an area, e.g. agriculture, industry, and residential.

Latent heat The heat released or absorbed when a substance changes phase as from liquid to gas. For water at 0°C, heat is absorbed or released at a rate of 2.5 million joules per kilogram (597 calories per gram) in the liquid/vapor phase change.

Laterite A layer of iron and aluminum oxide accumulation in tropical soils, mainly the latosols.

Laterization A soil-forming regime of warm, moist environments that produces a strongly leached soil with light topsoil and heavy accumulations of iron and aluminum oxides; also called *ferratillization*.

Latosols One of the great soil groups under the traditional USDA soil classification scheme; soils characterized by a weak 0 horizon, heavy accumulation of laterite, and a deeply weathered profile.

Lava Molten rock that has reached the surface; see also *Magma*.

Law of plastic deformation A physical principle describing the deformational response of a substance to increasing shear stress.

Leaching The removal of minerals in solution from a soil; the washing out of ions from one level to another in the soil.

Levee A mount of sediment which builds up along a river bank as a result of flood deposition.

Life cycle The biological stages in the complete life of a plant.

Life form The form of individual plants or the form of the individual organs of a plant; in general, the overall structure of the vegetative cover may be thought of as life form as well.

Limb Term applied to the flanks of a fold when viewed in cross-section; in an anticline the limbs slope away from the axis.

Limestone A sedimentary rock of chemical and biological origins; calcium carbonate precipitated from seawater and deposited in the form of the shells and skeletons of sea creatures.

Limiting factors (See *Principle of limiting factors*.)

Lithosol An azonal soil comprised of large fragments of bedrock.

Lithosphere The upper layer of the mantle; the unit in which the tectonic plates are defined; it is about 100 kilometers thick and includes the crust.

Little Ice Age A period of climatic change in the Northern Hemisphere generally from the fourteenth through the eighteenth centuries; it was marked by a cooling trend and manifested by glacial advances in the seventeenth and eighteenth centuries.

Littoral drift The material that is moved by waves and currents in coastal areas.

Littoral transport The movement of sediment along a coastline; it is comprised of two components: longshore transport and onshore-offshore transport.

Loess Silt deposits laid down by wind over extensive areas of the midlatitudes during glacial and postglacial times.

Longshore current A current that moves parallel to the shoreline; velocities generally range between 0.25 and 1 meters per second.

Longshore transport The movement of sediment parallel to the coast.

Longwave radiation Radiation at wavelengths greater than 3.0–4.0 micrometers; includes infrared (thermal), radio waves, and microwaves.

Mafic lava Lava with a low quartz content in which silicon dioxide constitutes about 50 percent of the rock; this lava is prevalent in the ocean basins.

Magma Molten rock within the lithosphere which cools to become igneous rock; magma that reaches the surface is called lava.

Magnetic polarization The polarization of magnetized materials such as iron particles in volcanic or sedimentary rock.

Magnitude and frequency The concept concerning the behavior of processes and the resultant changes they produce individually and collectively in the landscape; it involves which events render the greatest change and what kinds of change different-sized events render.

Mantle One of the two major divisions of the solid Earth, the other being the core; the mantle includes the lithosphere, asthenosphere, mesosphere (the upper mantle), and the lower mantle; the mantle contains about two-thirds of the Earth's mass.

Maquis Shrubby vegetation of the Mediterranean lands of Europe; apparently a response to climate and long-term disturbance by various land uses; also called *macchia* and *garique*: see also *Chapparal*.

Maritime climate A climate characterized by a small annual temperature range and often high rainfall.

Mass balance The relative balance in a system, based on the input and output of material such as sediment or water; the state of equilibrium between the input and output of mass in a system.

Mass budget (See *Mass balance*.)

Mass movement A type of hillslope process characterized by the downslope movement of rock debris under the force of gravity; it includes soil creep, rock fall, landslides, and mudflows; also termed *mass wasting*.

Meander A bend or loop in a stream channel.

Meander belt The width of the train of active meanders in a river valley.

Mechanical weathering One of the two major types of weathering; it produces physical fragmentation of rock by ice wedging, rock expansion, and a variety of other mechanisms.

Megastorm A great cluster of thunderstorms covering an area of 10,000 square miles or more, and lasting 12 to 18 hours.

Mercalli scale A scale for rating the intensity of an earthquake; intensity is a measure of an earthquake's destructive effect in the landscape.

Meridians The north–south-running lines of the global coordinate system; meridians converge at the North and South poles; the Prime Meridian marks 0 degrees longitude.

Mesa A flat-topped mass of bedrock that rises sharply above the surrounding terrain; it is usually capped by a resistant formation of rock and has the general aspect of a broad table.

Mesophyll The inner tissue of a leaf where moisture is stored and from which it is released in transpiration.

Mesophyte Plants with intermediate water requirements, usually found in sites with well-drained soils but adequate soil moisture in most months.

Mesosphere A subdivision of the atmosphere that lies above the stratosphere, extending from 50 to 90 kilometers altitude.

Metastable state In chemical weathering the condition of a mineral when it is intermediate between stability and instability.

Meteorology The field of Earth science that studies the weather, with emphasis on forecasting short term changes and events.

Microflora Minute plant life in the soil, mainly bacteria, algae, and fungi, that consume vegetal matter and in turn help produce humus; the most effective consumers of the organic matter deposited on the soil by plants.

Middle latitude Generally, the zone between the pole and the equator in both hemispheres; usually given as 35 to 55 degrees latitude.

Mid-oceanic ridge The volcanic mountain chain located in the interior of an ocean basin along a zone of seafloor spreading.

Migration The successful growth and establishment of a plant in a new location.

Millibar A unit of force (or pressure) equal to one thousandth of a bar; normal atmospheric pressure at sea level is 1013.2 millibars (mb); see also *Bar*.

Mineral A naturally occurring inorganic substance with a characteristic crystal (molecular) structure that is fundamentally the same in all samples.

Mixing ratio An expression for the vapor content of air; the weight (mass) of water vapor relative to the weight of the dry air occupying the same space; usually measured in grams per kilogram.

Moho discontinuity The lower boundary of the crust, where seismic wave velocities show an appreciable increase; the exact nature of the Moho and its significance in the lithosphere is not known; also called the *M discontinuity*.

Moisture deficit A term in the soil-moisture balance; the difference between actual and potential evapotranspiration; the difference between the demand for and availability of soil moisture for evapotranspiration.

Moisture index The difference between precipitation and evapotranspiration; used in the Thornthwaite System of climate classification to distinguish different climatic zones.

Mollisols Soils of the grasslands, usually in semiarid zones. Rich in organic matter with calcium carbonate nodules at depth.

Monsoon A seasonal wind system in South Asia which blows from sea to land in summer, bringing moisture to the continent, and from land to sea in winter, bringing dry conditions to India and neighboring lands.

Montmorillonite A type of clay that is notable for its capacity to shrink and expand with wetting and drying.

Moraine The material deposited directly by a glacier; also, the material (load) carried in or on a glacier; as landforms, moraines usually have hilly or rolling topography.

Mudflow A type of mass movement characterized by the downslope flow of a saturated mass of clayey material.

Neap tide (See *Tide*.)

Net photosynthesis The energy balance of a plant; the balance between the energy produced in photosynthesis and that used in respiration.

Net sediment transport The balance between the quantites of sediment moved in two (opposite) directions along a shoreline.

Néve Partially melted and compacted snow; it generally has a density of at least 500 kg/m³.

Newton A measure of force; the force necessary to accelerate a 1-kilogram mass 1 meter in 1 second squared.

Nitrogen fixation The process by which gaseous nitrogen is converted by microorganisms living in association with certain plants to a form that can be stored in the soil and utilized by plants.

Non-ferromagnesian minerals A subgroup of the silicate minerals; light-colored, low-density, rockforming minerals, e.g., quartz and orthoclase feldspar; also called *Aluminosilicate* minerals.

Non-parallel slope retreat A mode of slope retreat in which the slope angle grows smaller as the slope is eroded back; see also *Parallel slope retreat*.

Non-sedentary A settlement type, such as nomadism, characterized by groups shifting about an area.

Normal fault A fault in which the hanging wall is displaced downward relative to the footwall.

Nuee ardente A dense, "glowing cloud" of hot volcanic gas and ash which moves downhill at high speeds, scorching the landscape.

Oblique-slip fault A fault that combines both strike-slip and dip-slip displacements.

Obliquity A variation in the angle of the Earth's axis from 21.8 to 24.4 degrees with a period of 40,000 years.

Occluded front A frontal condition in a midlatitude cyclone in which the cold and warm fronts have merged, forcing the warm-air sector upward.

Ocean trench A great trough in the ocean floor, between 7500 m and 11,000 m below sea level; trenches are associated with subduction and lie along island arcs or orogenic belts.

Open forest A forest structure with a strong upper one or two stories and limited undergrowth; a forest that is largely open at ground level.

Open system A system characterized by a throughflow of material and/or energy; a system to which energy or material is added and released over time.

Orogenic belt A major chain of mountains on the continents; one of the major geologic subdivisions of the continents.

Orographic precipitation A type of precipitation that results when moist air is forced to rise when passing over a mountain range; most areas of exceptionally heavy rainfall are areas of orographic precipitation.

Oscillatory wave A wave in which there is no mass transport of water; the motion of the wave is circular; thus, water particles return to their original position with the passage of each wave.

Out flooding Flooding caused by a stream or river overflowing its banks.

Outwash plain A fluvioglacial deposit comprised of sand and gravel with a flat or gentle sloping surface; usually found in close association with moraines.

Overland flow Runoff from surfaces on which the intensity of precipitation or snowmelt exceeds the infiltration capacity; also called Horton overland flow, for hydrologist Robert E. Horton.

Oxbow A crescent-shaped lake or pond formed in an abandoned segment of river channel.

Oxisols Soils of tropical, moist environments which have undergone intense weathering; weak in organic matter, but heavy in oxides of iron and aluminum.

Ozone One of the minor gases of the atmosphere; a pungent, irritating form of oxygen that performs the important function of absorbing ultraviolet radiation.

Paleosol A soil exhibiting relict features, the result of some past conditions and processes.

Pangaea An ancient supercontinent that was comprised of the world's major land masses packed together around Africa several hundred million years ago; the breakup of Pangaea led to the formation of today's continents.

Parallels The east–west-running lines of the global coordinate system; the equator, the Arctic Circle, and the Antarctic Circle are parallels; all parallels run parallel to one another.

Parallel slope retreat A mode of slope retreat in which the slope angle remains essentially constant as the slope is eroded back; see also *Non-parallel slope retreat*.

Parent material The particulate material in which a soil forms; the two types of parent material are *residual* and *transported*.

Parkland A savanna formation of the midlatitudes characterized by prairie or meadows, with patches and ribbons of broadleaf trees.

Patterned ground Ground in which vegetation, water features, or stones are arranged in a geometric pattern, e.g., circles or polygons; it is widespread in cold environments.

Peak annual flow The largest discharge produced by a stream or river in a given year.

Pedalfers A general class of soil characterized by accumulations of iron and aluminum; soils in areas that receive at least 60 cm of precipitation annually.

Pediment Long, gentle slope at the foot of a cliff or free-face; it is usually composed of bedrock with a light covering of rock debris; common in dry regions.

Pedocals A general class of soil characterized by calcium accumulations and found in areas that receive less than 60 cm of precipitation annually.

Pedon The smallest geographic unit of soil defined by soil scientists of the U.S. Department of Agriculture.

Percolation test A soil-permeability test performed in the field to determine the suitability of a material for wastewater disposal; the test most commonly used by sanitarians and planners to size soil-absorption systems.

Perennial stream A stream that receives inflow of groundwater all year; a stream that has a permanent baseflow.

Peridoite A coarse-grained igneous rock found at depth in the lithosphere; a very dark, high-density rock.

Periglacial environment An area where frost-related processes are -a major force in shaping the landscape.

Permafrost A ground-heat condition in which the soil or subsoil is permanently frozen; long-term frozen ground in periglacial environments.

Permeability The rate at which soil or rock transmits groundwater (or gravity water in the area above the water table); measured in cubic feet (or meters) of water transmitted through a specified cross-sectional area when under a hydraulic gradient of 1 foot per 1 foot (or 1 m per 1 m).

pH (See *Soil pH*.)

Phase change Reorganization of a substance at the atomic or molecular level resulting in a change in physical state as from liquid to vapor; also called *phase transition*.

Phase transition (See *Phase change*.)

Phenological adaptation A form of plant adaptation in which the stages in the life cycle (e.g., flowering, pollination, seed germination) are in phase (adjusted to) the seasons and the periodicity of certain events in the year.

Phloem (See *Xylem and phloem*.)

Photoperiod The duration of the daily light period when photochemical activity can take place in a plant.

Photosynthesis The process by which green plants synthesize water and carbon dioxide and, with the energy from absorbed light, convert it into plant materials in the form of sugar and carbohydrates.

Piedmont glacier A large glacier usually formed from the merger of many alpine glaciers.

Piezometric surface The theoretical elevation (datum) to which groundwater would adjust if released from the differential pressure under which it normally exists.

Pioneer One of the communities of the community succession concept; the first community of plants to occupy a new site.

Piping The formation of horizontal tunnels in a soil due to sapping, i.e. erosion by seepage water; piping often occurs in areas where gullying is or was active and is limited to soils resistant to cave-in.

Pitted topography Glacial terrain characterized by a pocked surface; the pits result from buried ice blocks that have melted away.

Plane of the ecliptic The plane defined by one complete revolution of the Earth around the Sun.

Plant production The rate of output of organic material by a plant; the total amount of organic matter added to the landscape over some period of time, usually measured in grams per square meter per day or year.

Plant stress Limitations placed on photosynthesis by too much or too little of the basic requirements, namely, light, heat, water, carbon dioxide, and certain minerals.

Plastic deformation Irreversible change in the shape of a body without rupturing.

Plate tectonics The geophysical theory or model in which the lithosphere is partitioned into great plates which move laterally on the surface of the Earth; plate tectonics emerged as a serious scientific proposition with the articulation of the theory of continental drift.

Playa A dry lake bed in the desert.

Pleistocene Epoch The present ice age which began 1.8 million years ago.

Plucking The process by which a glacier removes blocks of rock from the bedrock; an erosional process associated with melting and refreezing at the base of glacial ice.

Plunging fold A fold whose axis is inclined rather than horizontal to the Earth's surface.

Plutonic rock Deep intrusive rock; igneous rock that forms in large chambers well within the crust.

Podzolization A soil-forming regime of cool, moist environments that produces a strongly leached soil with a distinctive hardpan layer.

Podzols One of the great soil groups under the traditional USDA soil-classification scheme; soils characterized by a strong a horizon, a leached A horizon, and a B horizon containing oxides of iron and aluminum.

Point bar Deposit in a stream channel on the inside of a meander or bend.

Polar front The zone or line of contact in the midlatitudes between polar arctic air

and tropical air; it often coincides with the polar front jet stream.

Polar glacier A glacier characterized by frozen conditions throughout and no meltwater. Common in the Antarctic, these glaciers are frozen fast to the underlying ground.

Polar zone The upper high latitudes, 75 to 90 degrees latitude.

Polypedon A group of pedons having similar characteristics; also called a *soil body*.

Pools and riffles Features of stream channels; pools are quiescent places separated by riffles, or reaches of rapid flow.

Pore water pressure The pressure exerted by groundwater against the particles through which it is flowing.

Porosity The total volume of pore (void) space in a given volume of rock or soil; expressed as the percentage of void volume to the total volume of the soil or rock sample.

Potential energy The energy represented by the elevation of mass above a critical datum plane (elevation), e.g. the elevation of rainwater above sea level.

Potential evapotranspiration The projected or calculated loss of soil water in evaporation and transpiration over some time period given an inexhaustible supply of soil water.

Precession A variation in the Earth's orbit, which has a period of 21,000 to 23,000 years and results in the equinox date.

Precipitable water vapor The total amount of water in the atmosphere; the average depth of water added to the Earth's surface if all the moisture in the atmosphere were to condense and fall to Earth.

Precipitation The term used for all moisture-solid and liquid-that falls from the atmosphere.

Pressure cell A body of air, usually covering a large area, which is defined on the basis of air pressure, either high pressure or low pressure.

Pressure gradient The change in pressure over distance between two points; on weather maps the pressure gradient is

measured along a line drawn at right angles to the isobars.

Primary consumer An organism that eats plants as its sole source of substance; it may be either a plant (e.g. bacteria or algae) or an animal (e.g. deer or buffalo).

Prime meridian (See *Greenwich Meridian*.)

Principle of limiting factors The biological principle that the maximum obtainable rate of photosynthesis is limited by whichever basic resource of plant growth is in least supply.

Principle of stream orders The relationship between stream order and the number of streams per order; the relationship for most drainage nets is an inverse one, characterized by many low-order streams and fewer and fewer streams with increasingly higher orders; see also *Stream order*.

Productivity (See *Plant production*.)

Progradation A term used to describe a shoreline that builds seaward.

Pruning In hydrology the cutting back of a drainage net by diverting or burying streams; usually associated with urbanization or agricultural development.

Psychrometer An instrument for measuring atmospheric humidity, comprised of two thermometers, a wet bulb and a dry bulb; humidity is measured by the difference in readings between the two thermometers.

Pteridophyte A low, non-woody plant that reproduces via spores rather than seeds; the largest group of pteridophytes is the ferns.

Pyroclastic Materials Fragments of volcanic rock thrown out in a volcanic explosion.

Quickflow (See *Stormflow*.)

Quicksand Sand that is incapable of supporting overburden (added weight) because of high pore-water pressure.

Radiation The process by which radiant (electromagnetic) energy is transmitted through free space; the term used to describe electromagnetic energy, as in infrared radiation or shortwave radiation.

Radiation beam The column of solar radiation flowing into or through the atmosphere.

Rainfall intensity The rate of rainfall measured in inches or centimeters of water deposited on the surface per hour or minute.

Rainforest A forest formation dominated by a heavy cover of evergreen trees, with abundant secondary vegetation in the form of epiphytes and lianas; in addition to the equatorial and tropical rainforests, the conifer forests of the very humid portions of the marine west coast are often classed as rainforest.

Rainshadow The dry zone on the leeward side of a mountain range of orographic precipitation.

Rainsplash Soil erosion from the impact of raindrops.

Rainwash Soil erosion by overland flow; erosion by sheets of water running over a surface; usually occurs in association with rainsplash; also called *wash*.

Range The geographic area occupied by a species, genus, or family of organisms.

Rating curve A graph that shows the relationship between the discharge and stage of various flow events on a river; once this relationship is established it may be used to approximate discharge using stage data alone.

Rational method A method for computing the discharge from a small drainage basin in response to a given rainstorm; computation is based on the coefficient of runoff, rainfall intensity, and basin area.

Reach A stretch or segment of stream channel.

Recharge The replenishment of groundwater with water from the surface.

Recurrence The number of years on the average that separate events of a specific magnitude, e.g. the average number of years separating river discharges of a given magnitude or greater.

Reflected wave A wave that rebounds off the shore or an obstacle and is redirected seaward through shore-bound incident waves.

Reg A rocky desert landscape; also called a *hamada*.

Regolith The weathered material overlying the bedrock; usually coarse, unsorted.

Relative humidity An expression for the water vapor content of air at a given temperature; the vapor content of a body of air expressed as a percentage of the amount of vapor held by a parcel of air when it is saturated.

Relief The range of topographic elevation within a prescribed area.

Residual soil Soil formed in parent material derived from the underlying bedrock, i.e. from *in situ* material.

Respiration The internal cellular processes of a plant by which energy is used for biological maintenance.

Reverse fault A fault in which the hanging wall is displaced upward relative to the footwall.

Revolution The motion of a planet in its orbital path around the sun.

Ria coast A heavily indented coast marked by prominent headlands and deep reentrants.

Richter scale A scale for rating the magnitude of an earthquake, i.e. the amount of energy released at the focus of the earthquake. The Richter scale is a logarithmic scale; for each unit, magnitude increases about 32 times.

Riffles (See *Pools and riffles*.)

Rift (graben) A fault in a zone of tensional stress characterized by a block displaced downward relative to adjacent rock formations.

Rift valley A valley formed by a rift fault, e.g., the valleys of the large lakes of East Africa.

Rimed A term used to describe snow crystals on which condensation has taken place as they fall to Earth.

Rip current A relatively narrow jet of water that flows seaward through the breaking waves; it serves as a release for water that builds up near shore.

Riprap Rubble such as broken concrete and rock placed on a surface to stabilize it and reduce erosion.

Rockfall A type of mass movement involving the fall of rock fragments from a cliff or slope face.

Root wedging A mechanical weathering process in which a root grows inside a crack, placing stress on the rock and widening the crack.

Rotation The spinning motion of a sphere, such as that of the Earth about its axis.

Roughness length The height of the zone or envelope of calm air over a surface which marks the base of the zone of turbulent airflow.

Runoff In the broadest sense runoff refers to the flow of water from the land as both surface and subsurface discharge; the more restricted and common use, however, refers to runoff as surface discharge in the form of overland flow and channel flow.

Rupture Deformation of a substance by fracturing.

Salination Salt saturation of soil as a result of a rise in the water table due to irrigation.

Saltation The principal mode of transport by wind; it is characterized by particles "hopping" over the ground.

Sand bypassing A means of artificially feeding sand to the beach downshore from a barrier such as a breakwater across a bay-mouth.

Sand sea A huge field of sand dunes such as the Empty Quarter of Saudi Arabia.

Sapping An erosional process that usually accompanies gullying in which soil particles are eroded by water seeping from a bank.

Saturated adiabatic lapse rate The rate of decline in the temperature of a rising parcel of air after it has reached saturation; it is variable but averages -0.6 °C per 100 meters; this rate is less than the dry adiabatic lapse rate (0.98 °C per 100 meters), because of the heat released in condensation.

Saturation absolute humidity The maximum mass of water vapor that can be held in a cubic meter of air at a given temperature; see also *Absolute humidity*.

Saturation vapor pressure The maximum value that vapor pressure can attain in air at a given temperature; see also *Vapor pressure*.

Savanna A biochore characterized by trees and shrubs scattered among a cover of grasses and forbs; the tropical savanna is the most extensive savanna formation and is found in the areas of the tropical wet/dry climate.

Scattering The process by which minute particles suspended in the atmosphere diffuse incoming solar radiation.

Sclerophyll forest Forest of the Mediterranean climate, characterized by small, widely spaced evergreen hardwood trees; generally considered the least prominent of the world's forest formations.

Scouring A mechanism of erosion by streams and glaciers in which particles carried at the bed abrade underlying rock.

Sea A term used to describe the choppy sort of waves in an area of wave generation.

Sea breeze A local wind that blows from sea to land as a result of the differential heating of land and water in the coastal zone; usually a daily occurrence.

Sea-level pressure The pressure exerted by the atmosphere on the Earth's surface at sea level; measured by the height of a column of mercury in a mercurial barometer; normal (average) sea-level pressure is 29.92 inches, or 76 centimeters of mercury; 14.7 pounds of pressure per square inch; or 1013.2 millibars of force per square meter.

Seamount A volcanic mountain in an ocean basin whose origin is not connected with a mid-oceanic ridge or a subduction zone; volcanic rises in the abyssal plains such as Bermuda and Hawaii.

Secondary consumer An animal that preys on primary consumers; in the soil moles are secondary consumers.

Sediment sink A coastal environment, such as a baymouth, where massive amounts of sediment are deposited.

Seepage The process by which groundwater or interflow water seeps from the ground.

Seepage lake A lake that gains its water principally from the seepage of groundwater into its basin.

Seif A large sand dune that is elongated in the general direction of the formative winds; a dune formed by winds from more than one direction.

Seismology A branch of geophysics devoted to the study of earthquakes and the interpretation of seismic waves.

Semipermanent pressure cells Large pressure cells, such as the subtropical highs, which exist most of the time in all seasons within a zone of latitude.

Sensible heat Heat that raises the temperature of a substance and thus can be sensed with a thermometer. In contrast to latent heat, it is sometimes called the heat of dry air.

Septic system Specifically, a sewage system that relies on a septic tank to store and/or treat wastewater; generally, an on-site (small-scale) sewage-disposal system that depends on the soil to dispose of wastewater.

7th approximation The modern soil-classification system of the US Department of Agriculture; it uses six levels of classification, beginning with orders.

Shear stress Differential stress acting on a body in which the forces are directed at angles to one another.

Shear wave A type of seismic (elastic) wave that produces motion transverse (perpendicular) to the direction of seismic energy propagation; also termed *secondary wave*.

Shield A major geologic subdivision of the continents; the relatively low elevation interior of a geologically stable continent. The term is derived from the shape of a battle shield placed handle side down.

Shield volcano A volcano comprised mainly of lava, with the overall shape of a shield or dome. The Hawaiian Islands are shield volcanoes.

Shifting agriculture An agricultural practice in tropical and equatorial areas characterized by the movement of farmers from plot to plot as soil becomes exhausted under cultivation.

Sial layer The upper part of the crust; the part that forms the continents and is comprised of relatively light, granitic rocks. The term is a contraction of silicon and *aluminum*.

Silicate minerals The principal rock-forming group of minerals; minerals composed of a basic ion of silicon and oxygen, called the silicon oxygen tetrahedron, combined with one or more additional elements.

Silicic lava Lava with a high quartz content in which silicon dioxide constitutes 70 percent or more of the rock; this lava is prevalent on the continents.

Sima layer The lower part of the crust; the part that forms the ocean basin and is comprised of relatively heavy, basaltic rock. The term is a contraction of silicon and *magnesium*.

Single-thread channel A stream channel characterized by a single course; it may be straight or meandering.

Sinkhole A pit-like depression in areas of karst topography, caused by the removal of limestone or dolomite by underground drainage; also called a sink or doline.

Slip face The lee side of a sand dune where windblown sand accumulates and slides downslope.

Slope failure A slope that is unable to maintain itself and fails by mass movement such as a landslide, slump, or similar movement.

Slope form The configuration of a slope, e.g. convex, concave, or straight.

Sluiceway A large drainage channel or spillway for glacier meltwater.

Slump A type of mass movement characterized by a back rotational motion along a rupture plane.

Small circle Any circle drawn on the globe that represents less than the full circumference of the Earth; thus, the plane of a small circle does not pass through the center of the Earth. All parallels except the Equator are small circles.

Soil-absorption systems The term applied to sewage-disposal systems that rely on the soil to absorb wastewater; see also *Septic system*.

Soil creep A type of mass movement characterized by a very slow downslope displacement of soil, generally without fracturing of the soil mass; the mechanisms of soil creep include freeze-thaw activity and wetting and drying cycles.

Soil-forming factors The major factors responsible for the formation of a soil: climate, parent material, vegetation, topography, and drainage.

Soil-heat flux The rate of heat flow into, from, or through the soil.

Soil mantle A traditional term used to describe the composite mass of soil material above the bedrock.

Soil material Any rock or organic debris in which soil formation takes place.

Soil-moisture balance A model that describes the changes in the availability of soil moisture as a product of precipitation, evapotranspiration, and storage water in the soil.

Soil-moisture deficiency The amount of water needed to raise the moisture content of a soil to field capacity.

Soil-moisture recharge A term in the soil-moisture balance; the replenishment of soil moisture following a period of soil-moisture loss, i.e., following a period of negative moisture balance.

Soil-nutrient system The system defined by the flow of nutrients between the soil and the plant cover.

Soil order A major level of soil classification in the traditional USDA scheme; the three orders defined are zonal, azonal, and intrazonal.

Soil pH The degree of alkalinity or acidity of a soil; the ratio of hydrogen ions to hydroxyl ions. On the pH scale 7.0 is neutral.

Soil profile The sequence of horizons, or layers, of conditions, generally related to climate, that gives rise to certain soil processes and in turn a distinctive soil profile.

Soil regime A particular combination of soil-forming conditions, generally related to climate, that gives rise to certain soil processes and in turn a distinctive soil profile.

Soil structure The term given to the shape of the aggregates of particles that form in a soil; four main structures are recognized; blockly, platy, granular, and prismatic.

Soil texture The cumulative sizes of particles in a soil sample; defined as the percentage by weight of sand, silt, and clay-sized particles in a soil.

Soil-water balance (See *Soil-moisture balance*.)

Solar constant The rate at which solar radiation is received on a surface (perpendicular to the radiation) at the edge of the atmosphere. Average strength is 1353 joules per square meter per second, which can also be stated as 1.94 calories per square centimeter per minute.

Solifluction A type of mass movement in periglacial environments, characterized by the slow flowage of soil material and the formation of lobe-shaped features; prevalent in tundra and alpine landscapes.

Solstice The dates when the declination of the Sun is at 23.27 degrees north latitude (the Tropic of Cancer) and 23.27 degrees south (the Tropic of Capricorn) – June 21–22 and December 21–22, respectively.

Solum That part of soil material capable of supporting life; the true soil according to the agronomist; the upper part of the soil mass, including the topsoil and soil horizons.

Solution weathering A type of weathering in which a mineral dissolves on contact with water carrying a solvent such as carbonic acid.

Speciation The process by which new species originate.

Species A group, or taxon, of individuals able to freely interbreed among themselves, but unable to breed with other groups; the smallest taxon of the floristic system of plant classification.

Specific heat The relative increase in the temperature of a substance with the absorption of energy.

Specific humidity An expression for the vapor content of air based the weight (mass) of water vapor relative to the weight of the moist air (vapor plus dry air) to which it belongs; measured in grams per kilogram.

Spheroidal weathering A form of chemical weathering in which a boulder sheds thin plates of rock debris.

Spodosols Soils with pronounced zones of illuviation characterized by accumulations of iron and aluminum oxides. Formed in moist, cool climates.

Spore A reproductive cell in plants; generally any nonsexual reproductive cell; among the vascular plants, the pteridophytes are spore-bearing.

Spring tide (See *Tide*.)

Squall line The narrow zone of intensive turbulance and rainfall along a cold front.

Stand A floristically uniform growth of vegetation, often of similar size and age, that dominates an area.

Stefan-Boltzmann equation The intensity of energy radiation from a body increases with the fourth power of its temperature times a constant.

Stemflow Precipitation water that reaches the ground by running from the vegetation canopy down the trunk of a tree, shrub, or the stem of grass; see also *Interception*.

Steppe Short-grass prairie of the semiarid climatic zones; widespread in Eurasia and North America.

Stomata The openings in the foliage of a plant through which moisture is released during transpiration; the stomata open and close in response to air temperature, humidity, and other factors.

Stormflow The portion of streamflow that reaches the stream relatively quickly after a rainstorm, adding a surcharge of water to baseflow.

Story A layer or level of tree crowns in a forest.

Stratosphere The subdivision of the atmosphere that lies above the troposphere; it is characterized by stability and temperature that increases with altitude.

Stream order The relative position, or rank, of a stream in a drainage network. Streams without tributaries, usually the small ones, are first-order; streams with two or more first-order tributaries are second-order, and so on.

Stress A force acting on a body or substance; see also *Plant stress* or *Shear stress*.

Striation Scour line etched into bedrock by the rock debris on the base of a glacier.

Strike and dip Directional properties of a geologic structure such as a fault. Strike is the directional trend of a formation along the surface; dip is the angle of incline of the formation measured at a right angle to strike.

Strike-slip fault A fault in which the main direction of displacement is lateral, or along the fault line.

Subarctic zone The belt of latitude between 55 degrees and the Arctic and Antarctic circles.

Subduction The process by which a tectonic plate is consumed or destroyed as it slides into the Earth along the contact with an adjacent plate; subduction zones are places of frequent earthquakes and volcanism and are usually marked on the surface by ocean trenches.

Sublimation A physical process by which a solid is changed directly to a gas (or vice versa) without passing through the liquid phase.

Subtropical high-pressure cells Large cells of high pressure, centered at 25 to 30 degrees latitude in both hemispheres, which are fed by air descending from aloft; these cells are the main cause of aridity in tropics and subtropics.

Subtropical zone The zone of latitude near the tropics in both hemispheres; between 23.5 and 35 degrees.

Succulent habit A form of plant adaptation to arid conditions characterized by fleshy bodies and/or foliage with the capacity to store large amounts of water.

Summer solstice (See *Solstice*.)

Sun angle The angle formed between the beam of incoming solar radiation and a plane at the Earth's surface or a plane of the same attitude anywhere in the atmosphere.

Surge A large and often destructive wave caused by intensive atmospheric pressure and strong winds.

Suspended load The particles (sediment) carried aloft in a stream of wind by turbulent flow; usually clay- and silt-sized particles.

Swash The thin sheet of water that slides up the beach face after a wave breaks.

Swell A wave with a relatively smooth form, usually found at some distance from the area of wave generation.

Syncline A fold characterized by a downward bend, concave downward in cross-section.

Synoptic scale The scale of geographic coverage most commonly used on daily weather charts depicting air masses, winds, and storm cells.

System An interconnected set of objects or things; two or more components such as organisms, cities, or streams linked together in some fashion; e.g. energy systems, ecosystems, road systems.

Systeme International The preferred system of units according to an international consensus of scientists; the basic units of energy, time, and space are the joule, second, and square meter, respectively.

Taiga (See *Boreal forest*.)

Talus An accumulation of rock debris at the foot of a slope as a result of rockfall.

Taxon Any unit (category) of classification of organisms.

Tectonic plate A large sheet of lithosphere that moves as a discrete entity on the surface of the Earth; the lithosphere is subdivided into seven major plates and many smaller ones; each major plate, with the exception of the Pacific plate, contains a continent.

Temperate forest A forest of the midlatitude regions that could be described as climatically temperate, e.g. broadleaf deciduous forests of Europe and North America, comprised of beeches, maples, and oaks.

Temperate glacier A glacier characterized by abundant meltwater and an internal temperature slightly below the freezing threshold.

Temperature inversion An atmospheric condition in which the cold air underlies warm air; inversions are highly stable conditions and thus not conducive to atmospheric mixing.

Temperature profile The change in temperature along a line or transect through an environment, usually expressed in a graphical format.

Terrace A surface formed by wave erosion or river processes and elevated above the existing level of the ocean or floodplain.

Tertiary consumer A carnivore that preys on secondary as well as primary consumers; animals, such as birds of prey, that are near the ends of the food chains.

Theory of continental drift (See *Continental drift*.)

Thermal conductivity A thermal property of a substance describing its capacity to transmit heat given a thermal gradient of 10 K per meter (or 10 °C per meter).

Thermal diffusivity A thermal property of a substance that describes the rate at which a given temperature, represented, for example, by an isotherm, passes through a substance. It is defined as the ratio of thermal conductivity to volumetric heat capacity.

Thermal gradient The change in temperature over distance in a substance; usually expressed in degrees Celsius per centimeter or meter.

Thermal regime The annual or seasonal pattern of temperatures for a place or region; usually used in climatology.

Thornbush A vegetative formation of the tropical savanna regions, characterized by short, thorny trees and shrubs; called *caatinqa* in northeastern Brazil and *dornveld* in South Africa.

Threshold The level of magnitude of a process at which sudden or rapid change is initiated.

Thrust fault A fault in which the hanging wall is driven laterally over the footwall.

Thunderstorm An intensive convectional storm that produces heavy precipitation, strong local winds, as well as thunder and lightning.

Tide A large wave caused by bulges in the sea in response to the lunar and solar gravitational forces. The largest tides are *spring tides*, which occur when the Moon and Sun are aligned with the Earth; the smallest are *neap tides*, which occur when the Moon and Sun are positioned at a right angle relative to the Earth.

Tolerance The range of stress or disturbance a plant is able to withstand without damage or death.

Tombolo A depositional feature along some shorelines which forms a neck of land between an island and the mainland.

Topographic relief (See *Relief.*)

Topsoil The uppermost layer of the soil, characterized by a high organic content; the organic layer of the soil.

Township and range A system of land subdivision in the United States which uses a grid to classify land units; standard subdivisions include townships and sections.

Traction A mode of sediment transport by wind in which particles move in contact with the ground; also called *creep.*

Trade winds The system of prevailing easterly winds, which flow from the subtropical highs to the intertropical convergence zone (ITCZ) in both hemispheres; also called the *tropical easterlies.*

Transform fault A strike-slip fault; in particular, the term is applied to the faults that run transverse to the mid-oceanic ridges.

Transmission The lateral flow of groundwater through an aquifer; measured in terms of cubic feet (or meters) transmitted through a given cross-sectional area per hour or day.

Transpiration The flow of water through the tissue of a plant and into the atmosphere via stomatal openings in the foliage.

Transported soil Soil formed in parent material comprised of deposits laid down by water, wind, or glaciers.

Travel time (See *Concentration time.*)

Tree line The upper limit of tree growth on a mountain where forest often gives way to alpine meadow.

Tropical cyclone (See *Hurricane.*)

Tropics Correctly used, this term refers to the Tropic of Capricorn and the Tropic of Cancer; however, it is often used to refer to areas equatorward of the tropics.

Troposphere The lowermost subdivision of the atmosphere; the layer that contains the bulk of the atmosphere's mass and is characterized by convectional mixing and temperature that decreases with altitude.

Tsunami A large and often destructive wave caused by tectonic activity such as faulting on the ocean floor.

Tundra Landscape of cold regions, characterized by a light cover of herbaceous plants and underlain by permafrost.

Turbulent flow Flow characterized by mixing motion in which the primary source of flow resistance is the mixing action between slow-moving and faster-moving molecules in a fluid.

Turgor A term used to describe the status of water pressure in plant foliage; when a plant wilts, it loses its turgor, and the leaves become puckered and limp.

Typhoon (See *Hurricane.*)

Ultisols Soils in an advanced state of development in warm, moist climates. Pronounced eluviation; often poor in bases.

Ultraviolet radiation Electromagnetic radiation of wavelengths shorter than visible, but longer than X-rays.

Unified system A soil-classification scheme used in civil engineering, based on soil performance when it is placed under stress.

Uniformitarianism A concept attributed to eighteenth-century geologist James Hutton; the types of processes operating on the Earth today are the same ones that were active in the geologic past; often condensed to "the present is the key to the past."

US Geological Survey An agency of the US Department of Interior that is responsible for mapping and analyzing rock types, minerals, earthquakes, river flow, and related phenomena.

Urban climate The climate in and around urban areas; it is usually somewhat warmer, foggier, and less well lighted than the climate of the surrounding region.

Urbanization The term used to describe the process of urban development,

including suburban residential and commercial development.

Valley wall The side slope of a river valley where the floodplain gives way to upland surfaces.

Vapor pressure An expression for the water vapor content of air; it is the pressure exerted by the weight of the water vapor molecules independent of the weight of the other gases in the air; expressed in millibars or newtons per square meter.

Vascular plants Plants in which cells are arranged into a pipe-like system of conducting, or vascular, tissue; xylem and phloem are the two main types of vascular tissue.

Vegetative regeneration Asexual regeneration by plants in which some part of the plant, such as a root or a special organ such as a rhizome, is able to propagate new stems.

Vein Igneous rock or a deposit of minerals that form in a joint line or fracture.

Velocity The rate of movement in one direction, expressed as distance over time, e.g. meters per second, kilometers per hour, or miles per hour.

Vernal equinox (See *Equinox.*)

Vertisols Soils dominated by geomechanical mixing in areas of montmorillonite clay.

Viscosity A measure of the resistance to flow in a fluid due to intermolecular friction.

Visible light Electromagnetic radiation at wavelengths between 0.4 and 0.7 micrometer; the radiation that comprises the bulk of the energy emitted by the Sun.

Warm front A contact between a cold air mass and a warm air mass in which the warm air is moving against the cold air, sliding upward along the contact.

Water table The upper boundary of the zone of groundwater; in fine-textured materials it is usually a transition zone rather than a boundary line. The configuration of the water table often approximates that of the overlying terrain.

Watt A unit of power often used as an energy expression; equal to 1 joule per second.

Wave period The time it takes a wave to travel the distance of one wavelength.

Wave refraction The bending of a wave, which results in an approach angle more perpendicular to the shoreline.

Wave of translation A wave that produces a mass transport of water; in coastal areas it is often a breaking wave; see *Oscillatory wave*.

Weathering The breakdown and decay of Earth materials, especially rock; see also *Chemical weathering*.

Westerlies The prevailing eastward flow of air over land and water in the midlatitudes of both hemispheres; also called the *prevailing westerlies*.

West-wind drift Ocean current, or drift current, which flows eastward in the midlatitudes and subarctic, driven by the prevailing westerly winds.

Wetland A term generally applied to an area where the ground is permanently wet or wet most of the year and is occupied by water-loving (or tolerant) vegetation such as cattails, mangrove, or cypress.

Wetted perimeter The distance from one side of a stream to the other, measured along the bottom.

Wien's law A physical law stating that the wavelength of maximum-intensity radiation grows longer as the absolute temperature of the radiating body decreases; also called *Wien's displacement law*.

Wind chill Heat loss from the skin as a function of both air temperature and wind velocity.

Wind power The power generated by wind; proportional to the cube of speed or velocity.

Wind wave A wave generated by the transfer of momentum from wind to a water surface.

Winter solstice (See *Solstice*.)

Work A concept closely related to energy, work is the product of force and distance and is accomplished when the application of force yields movement of an object in the direction of the force.

Xerophyte Plants capable of surviving prolonged periods of soil drought, e.g. the cacti.

Xylem and phloem Conducting tissue in vascular plants through which the plant fluids are transmitted.

Zenith For any location on Earth, the point that is directly overhead to an observer. The zenith position of the Sun is the one directly overhead.

Zenith angle The angle formed between a line perpendicular to the Earth's surface (at any location) and the beam of incoming solar radiation (on any date).

Zonal soil A soil order under the traditional USDA soil-classification scheme; soils with well-developed horizons that reflect the climate conditions of the region in which they are found.

Zone of eluviation The level, or zone, in a soil losing materials in the form of colloids and ions; the zone of removal.

Zone of illuviation The level, or zone, in a soil where colloids and ions accumulate.

Zone of saturation (See *Groundwater*.)

Photographic credits

Chapter 1

Opener	Kapu/Shutterstock.
Figure 1.1	Tim Roberts Photography/Shutterstock.
Figure 1.2	Vladislav Gurfinkel/Shutterstock.
Figure 1.10	Attila Jandi/Shutterstock.
Figure 1.13	Cristi Matei/Shutterstock.
Figure 1.18	Denis Rozan/Shutterstock.
Page 4	iofoto/Shutterstock. Nelson Sirlin/Shutterstock. AISPIX/Shutterstock.
Page 10	NASA image courtesy of Jeff Schmaltz, MODIS Rapid Response Team at NASA GSFC.

Chapter 2

Opener	Frontpage/Shutterstock.
Figure 2.1	Image courtesy of Houghton Library, Harvard University, reference f*GC5. Ap34.533i2.
Figure 2.2	Map by Meriwether Lewis, William Clark, Nicholas Biddle and Paul Allen.
Figure 2.3	Elinag/Shutterstock. NASA/Johns Hopkins University Applied Physics Laboratory/Carnegie Institution of Washington, photograph number PIA13840.
Figure 2.4	Frontpage/Shutterstock.
Figure 2.5	Chris P/Shutterstock.
Figure 2.6	Image courtesy of NASA, Image Science and Analysis Laboratory, NASA, Johnson Space Center. *The Gateway to Astronaut Photography of Earth*, NASA image number ISS006-E-38632.
Figure 2.7	Chaikovskiy Igor/Shutterstock.
Figure 2.8a	Image courtesy of Image Science and Analysis Laboratory, NASA-Johnson Space Center.
Figure 2.8b	Image courtesy of NASA/GSFC/MITI/ERSDAC/JAROS, and U.S./Japan ASTER Science Team.
Figure 2.9	Image courtesy of Image Science and Analysis Laboratory, NASA-Johnson Space Center.
Figure 2.10	*Monthly snow cover for the Northern Hemisphere September 2005 and March 2006.* From D. K. Hall and J. L. Foster, 'Snow Cover: The Most Dynamic Feature on the Earth's Surface', pp. 110–115. In King, Parkinson, Partington and Williams (Eds.), *Our Changing Planet*, Cambridge Univ. Press, Cambridge, 2007. (Data from the MODIS instrument on the Terra satellite.)

Chapter 3

Opener	Brent Wong/Shutterstock.
Figure 3.1	Jeff Dozier.
Figure 3.2	Tom Bullock.
Figure 3.7	Witold Kaszkin/Shutterstock.
Figure 3.9	NASA Goddard Space Flight Center Scientific Visualization Studio and the Solar Dynamics Observatory.
Figure 3.17	Calee Allen, National Science Foundation.
Figure 3.24	Image courtesy MODIS Ocean Group, NASA GSFC, and the University of Miami.
Page 41	Copernicus in the tower at Frombork (oil on canvas), Matejko, Jan (1838–93)/Nicolaus Copernicus Museum, Frombork, Poland/ The Bridgeman Art Library. Image courtesy of North Wind Picture Archives/Alamy. Emilio Segre Visual Archives/American Institute of Physics/Science Photo Library.
Page 51	Natursports/Shutterstock.

Chapter 4

Opener	Lee Prince/Shutterstock.
Figure 4.1	David McDermott.
Figure 4.4	*Minimum total ozone content of the Southern Hemisphere for 1985 and 2006.* From Richard D McPeters, 'The Ozone Hole', pp. 78–80. In King, Parkinson, Partington and Williams (Eds.), *Our Changing Planet*, Cambridge Univ. Press, Cambridge, 2007. (Data from the TOMS instrument on the Nimbus 7, Meteor 3, and Earth Probe satellites, and the OMI instrument on Aura.)
Figure 4.8	MikLav/Shutterstock.
Figure 4.10	dampoint/Shutterstock.
Figure 4.19	Mares Lucian/Shutterstock. Yangchao/Shutterstock.

Figure 4.21   David Ionut/Shutterstock.
Figure 4.23   Image courtesy of NASA.
Figure 4.25   Brad Perks Lightscapes/Alamy.
Figure 4.30   Randy Schaetzl.
Figure 4.31   Dan Tautan/Shutterstock.
Page 74   leonid_tit/Shutterstock

Chapter 5

Opener   Image courtesy of NASA.
Figure 5.9   Duomo/Alamy.
Figure 5.27   Victor Shova/Shutterstock.
Figure 5.30   Image courtesy of Jacques Descloitres, MODIS Rapid Response Team at NASA GSFC.
Figure 5.39   Image courtesy of NASA. The sea surface temperature image was created at the University of Miami using the 11- and 12-micron bands, by Bob Evans, Peter Minnett, and co-workers.

Chapter 6

Opener   Zastol`skiy Victor Leonidovich/Shutterstock.
Figure 6.11   Ingo Arndt/Nature Picture Library.
Figure 6.12a   University Corporation for Atmospheric Research/Science Photo Library.
Figure 6.12b   Mark Romessa/Shutterstock.
Figure 6.17   Flagstaffotos/WMC.
Figure 6.27a   Media Union/Shutterstock. iofoto/Shutterstock. jaimaa/Shutterstock. Rafaifabrykiewicz/Shutterstock.
Figure 6.28   David P. Lewis/Shutterstock.
Figure 6.32   Image Science and Analysis Laboratory, NASA-Johnson Space Center. "The Gateway to Astronaut Photography of Earth".
Figure 6.39   Caitlin Mirra/Shutterstock.
Figure 6.40   Express Newspapers/Getty.
Figure 6.41   Image courtesy of Lawrence Ong, EO-1 Mission Science Office, NASA GSFC.
Figure 6.42   Image courtesy of National Oceanic and Atmospheric Administration NOAA.
Page 126   AJancso/Shutterstock.

Chapter 7

Opener   Gwoeii/Shutterstock.
Figure 7.4   Konstanttin/Shutterstock. kavram/Shutterstock.
Figure 7.6   *Figure to show Sea Surface Temperature of the Northerrn Hemisphere Spring 2006.* From Peter J. Minnett, 'Heat in the Ocean',

pp. 156–160. In King, Parkinson, Partington and Williams (Eds.), *Our Changing Planet*, Cambridge Univ. Press, Cambridge, 2007. (Data from the MODIS instrument on the Terra satellite.)
Figure 7.8   javarman/Shutterstock.
Figure 7.19   (inset) Alan Dykes/Alamy.
Figure 7.23   Philip Lange/Shutterstock.
Figure 7.24   Gusto Images/Science Photo Library.
Figure 7.25a   Mary Evans Picture Library/Alamy.
Figure 7.25b   Image courtesy of martinhartley.com.
Figure 7.26   Styve Reineck/Shutterstock.
Page 150   Patryk Kosmider/Shutterstock.
Page 161   Condor 36/Shutterstock.
Page 167   PavelSvoboda /Shutterstock. Paul Matthew Photography/Shutterstock.

Chapter 8

Opener   Galyna Andrushko/Shutterstock.
Figure 8.1   HABRDA/Shutterstock.
Figure 8.2   Felix Stampfi/icedrill.ch AG.
Figure 8.5   John Sibbick.
Figure 8.12   Figure redrawn from NASA Goddard Institute for Space Studies/Shutterstock.
Figure 8.19   Olsolya Harlsberg/Nature Picture Library.
Figure 8.33   Stuart Franklin/Magnum.
Figure 8.37   Image courtesy of the Met Office, Crown Copyright 2010.
Figure 8.38   Image courtesy of National Oceanic and Atmospheric Administration (NOAA).
Figure 8.39   Image courtesy NASA, aquired by the MODIS instrument on the Terra satellite.
Figure 8.41   Guy Edwards/Nature Picture Library.
Figure 8.42   Frans Lemmens/Alamy.
Page 187   Songquan Deng/Shutterstock.

Chapter 9

Opener   szefei/Shutterstock.
Figure 9.5a   Katerina Havelkova/Shutterstock.
Figure 9.5b   Ann Cantelow/Shutterstock.
Figure 9.7   Jeffrey T. Kreulen/Shutterstock.
Figure 9.9   Planetary Visions Limited /Science Photo Library.
Figure 9.10   Werner Otto/Alamy.
Figure 9.19   Data courtesy of United Nations World Food Programme.
Page 213   Nagel Photography/Shutterstock.
Page 218   Andrew Orlemann/Shutterstock.

Chapter 10

Opener	Pitugin Dmitri/Shutterstock.
Figure 10.2	Science Photo Library/Alamy.
Figure 10.9	Laurent Geslin/Nature Picture Library.
Figure 10.14a	Cindy Buxton/Nature Picture Library.
Figure 10.14b	Clément Phillipe/Alamy.
Figure 10.15	Radius Images/Alamy.
Figure 10.16	Stuart Franklin/Magnum.
Figure 10.19	Anson0618/Shutterstock.
Figure 10.22	Peter Oxford/Nature Picture Library.
Figure 10.27a	FrontPage/Shutterstock.
Figure 10.27b	Guentermanaus/Shutterstock.
Figure 10.29	Dinodia Photos/Alamy.
Figure 10.31	Ian Bracegirdle/Shutterstock.
Figure 10.32	Aurora Photos/Alamy.
Figure 10.33	Visuals Unlimited/Nature Picture Library. Brad Mitchell/Alamy. Peter Scoones/Nature Picture Library.
Figure 10.34	Graeme Shannon/Shutterstock. Oleg Znamenskiy/Shutterstock.
Figure 10.35a	Tom Bean/Alamy.
Figure 10.35b	Orientally/Shutterstock.
Figure 10.36	kavram/Shutterstock.
Figure 10.37a	Ecdoerner/Shutterstock.
Figure 10.37b	Julija Sapik/Shutterstock.
Figure 10.38a	Matthew Jacques/Shutterstock.
Figure 10.38b	Jakub Pavlinec/Shutterstock.
Figure 10.39	Jeff Schmaltz, MODIS Rapid Response Team, NASA/GSFC.
Figure 10.46	neelsky/Shutterstock.
Page 257	Roca/Shutterstock. Katie Dickinson/Shutterstock. Vladimir Melnik/Shutterstock.
Page 239	Karol Kozlowski/Shutterstock.
Page 254	Iakov Kalinin/Shutterstock.
Page 256	Iakov Filimonov/Shutterstock.

Chapter 11

Opener	Ocean/Corbis.
Figure 11.1a(i)	Cyril Hou/Alamy.
Figure 11.1a(ii)	Chawalit S./Shutterstock.
Figure 11.1b(i)	Lorenz Britt/Alamy.
Figure 11.1b(ii)	Clint Farlinger/Alamy.
Figure 11.2b	BlueOrange Studio/Shutterstock.
Figure 11.3	John Warburton-Lee Photography/Alamy.
Figure 11.4	imagebroker.net/Superstock.
Figure 11.11	Pichugin Dmitry/Shutterstock.
Figure 11.13a	Bon Appetit/Alamy.

Figure 11.13b	Nigel Cattlin/Alamy.
Figure 11.17	Image courtesy of David Beresford-Jones.
Figure 11.19a	Pete Oxford/Nature Picture Library.
Figure 11.19b	Bernard Castelein/Nature Picture Library.
Figure 11.20	Ricardo Beliel/Alamy.
Figure 11.25	Hung Chung Chih/Shutterstock.
Figure 11.29	The Art Archive/Alamy.
Figure 11.31	North Wind Picture Archives/Alamy.
Figure 11.32	Andrey N Bannov/Shutterstock.
Figure 11.33	trekandshoot/Shutterstock.
Figure 11.35	testing/Shutterstock.
Page 283	Chris Howey/Shutterstock.
Page 287	Delmas Lehman/Shutterstock.

Chapter 12

Opener	Vitaly Titov & Maria Sidelnikova/Shutterstock.
Figure 12.19a	Photolinc/Shutterstock.
Figure 12.19b	Nickolay Stanev/Shutterstock.
Figure 12.19c	2009fotofriends/Shutterstock.
Figure 12.26	Caitlin Mirra/Shutterstock.
Page 294	Image courtesy of Alex Maclean/Courtesy of the City of New York.

Chapter 13

Opener	Barnaby Chambers/Shutterstock.
Figure 13.1	Simon Fraser University/Roy Carlson.
Figure 13.2a	curved light USA/Alamy.
Figure 13.2b	Vasiliy Koval/Shutterstock.
Figure 13.9	Image courtesy of Delta Waterfowl.
Figure 13.10	Robert Harding Picture Library Ltd/Alamy.
Figure 13.11	Vlad Ageshin/Shutterstock.
Figure 13.12	Image courtesy of Randy Schaetzl.
Figure 13.14	FloridaStock/Shutterstock.
Figure 13.15b	Image redrawn courtesy of United States Department of Agriculture.
Figure 13.17	Image courtesy of Randy Schaetzl.
Figure 13.18	Image courtesy of Randy Schaetzl.
Figure 13.19	Image courtesy of Randy Schaetzl.
Figure 13.20	Patrick Cunningham/Alamy.
Figure 13.22	John Martin as on Victorian Resources Online www.dpi.vic.gov.au/vro. Copyright Victoria, Department of Primary Industries.
Figure 13.23	Image courtesy of Randy Schaetzl.
Figure 13.24	Image courtesy of National Soils Database, Agriculture and Agri-Food Canada.
Figure 13.25	Vladimir Melnik/Shutterstock.
Figure 13.26	Vivienne Sharp/Alamy.

Figure 13.27 Image by Pep Fuster/Shutterstock.
Figure 13.28 Image by Suzanne Long /Shutterstock.
Figure 13.30 Nigel Cattlin/Alamy.
Figure 13.31 Xico Putini /Shutterstock.
Figure 13.32 Image by Yurikr/Shutterstock.
Figure 13.33 Eye Ubiquitous/Alamy.
Page 328 Vladimir Sazonov/Shutterstock.

Chapter 14
Opener Ana de Sousa/Shutterstock.
Figure 14.6 Image courtesy of NASA, Image Number STS110-330-019.
Figure 14.9 Sarah Theophilus/Shutterstock.
Figure 14.19 Natalia Bratslavsky/Shutterstock.
Figure 14.20b Elzbieta Sekowska/Shutterstock.
Figure 14.21 AMA/Shutterstock.
Figure 14.24 Victor Englebert/Science Photo Library.
Figure 14.28b BDR/Alamy.
Figure 14.30 iofoto/Shutterstock.
Page 336 Patrick Poendl/Shutterstock.
Page 349 Alecia Scott/Shutterstock.
Page 355 Image courtesy of United States Geological Survey.

Chapter 15
Opener Image by Johnny Lye/Shutterstock.
Figure 15.1 Image by Christian Jegou Publiphoto Diffusion/Science Pot Library.
Figure 15.12a Image courtesy of Austin History Centre.
Figure 15.12b Image courtesy of AP Photo/ABC.
Figure 15.16 Image courtesy of NASA/JPL/University of Arizona.
Figure 15.24 Image courtesy of Jim Wark, Airphoto North America.
Figure 15.25a NASA Earth Observatory image created by Robert Simmon, using EO-1 ALI data provided courtesy of the NASA EO-1 team.
Figure 15.25b aAianet-Pakistan/Shutterstock.
Figure 15.27 Donald P. Schwert, North Dakota State University.
Figure 15.28 Gary Blakeley/Shutterstock.
Figure 15.30 Bob Srenco, ©Srenco1993.
Figure 15.31 Photo from China Images/Alamy.
Figure 15.33 Image courtesy of NASA.

Chapter 16
Opener Sander van der Werf/Shutterstock.
Figure 16.6 morphart/Shutterstock.
Figure 16.11 B. Brown/Shutterstock.

Figure 16.14 Ria Novosti/Science Photo Library.
Figure 16.17 Image courtesy of SRTM Team NASA/JPL/NIMA.
Figure 16.20 Image courtesy of NASA.
Figure 16.24 Milevshi/Shutterstock..
Figure 16.26 Nicholas Rjabow/Shutterstock.
Figure 16.28 Image courtesy of U.S. Fish and Wildlife Service.
Figure 16.34 Fabienne Fossez/Alamy.
Page 386 Robert Harding Picture Library Ltd/Alamy.
Page 386 Alta Oosthuizen/Shutterstock.
Page 400 hinm/Shutterstock.
Page 400 Steffen Foerster Photography/Shutterstock.

Chapter 17
Opener beboy/Shutterstock
Figure 17.4 Image courtesy of NASA.
Figure 17.5 Image courtesy of NASA.
Figure 17.6 Image courtesy of NASA.
Figure 17.7b Image courtesy of NASA.
Figure 17.8 Natalia Lukiyanova/Shutterstock.
Figure 17.13 Sue Flood/Nature Picture Library.
Figure 17.19b Alan Majchrowicz/Alamy.
Figure 17.24a S_E/Shutterstock.
Figure 17.24b Borislav Dopudja/Alamy.
Figure 17.25a Tyler Boyes/Shutterstock.
Figure 17.26a Tyler Boyes/Shutterstock.
Figure 17.26b Sergey Lavrentev/Shutterstock.
Figure 17.26c Tyler Boyes/Shutterstock.
Figure 17.29a Jorg Hackemann/Shutterstock.
Figure 17.29b Quing Ding/Shutterstock, Jose Gil/Shutterstock, Kevin Schafer/Alamy.
Figure 17.32a senk/Shutterstock.
Figure 17.32b Image courtesy of The Detroit News Archives.
Figure 17.35 mary416/Shutterstock.
Page 433 Gustavo Toledo/Shutterstock, LesPalenik/Shutterstock.

Chapter 18
Opener Joseph Hardy/Shutterstock.
Figure 18.3 Image courtesy of UNEP/GRID-Arendal Maps and Graphics Library. February 2008. Hugo Ahlenius/UNEP/GRID-Arendal.
Figure 18.18 David Cobb/Alamy.
Figure 18.25 Attila Jandi/Shutterstock.
Figure 18.26 Image courtesy of Robert E. Wallace, USGS.
Figure 18.38 Sebastien Burel/Shutterstock.

Page 454   Attila Jandi/Shutterstock.

Page 460   Xavier Marchant/Shutterstock.

Page 465   ale_rizzo/Shutterstock, Rafael Ramirez Lee/Shutterstock, Max Earey/Shuterstock.

Chapter 19

Opener   Nikita Rogul/Shutterstock.

Figure 19.2   Map courtesy of NASA/JPL.

Figure 19.3   Phillipe Clement/Nature Picture Library.

Figure 19.4   ahmad80/Shutterstock, photoBeard/Shutterstock

Figure 19.8   Image courtesy of Josephine Mewett.

Figure 19.9   NASA image by Robert Simmon, based on Landsat data from the University of Maryland Global Land Cover Facility.

Figure 19.24   Mark Pearson/Alamy.

Figure 19.28   Everett Collection Inc/Alamy.

Figure 19.31   Image from *American Progress: OrGreat Events of the Greatest Century*, R. H. Deren (1877).

Figure 19.32   Image courtesy of NOAA.

Figure 19.37   S. R. Lee Photo Traveller/Shutterstock.

Figure 19.39   Image courtesy of USGS.

Figure 19.40   Smith, R. B., and L. Siegel, *Windows into the Earth: The geologic story of Yellowstone and Grand Teton National Parks*, Oxford Univ. Press, New York, 2000.

Figure 19.42   Photo by Karin Jackson, U.S. Air Force, June 12, 1991.

Page 470   Attila Jandi/Shutterstock, Stock Montage/SuperStock.

Page 485   Press Association Images/AP Photo/Kyodo News.

Page 486   Arindambanerjee/Shutterstock.

Page 489   Nina B/Shutterstock.

Page 491   Lazar Mihai-Bogdan/Shutterstock, neelsky/Shutterstock.

Page 493   Sailor/Shutterstock.

Chapter 20

Opener   A Jaye/Shutterstock.

Figure 20.1a   iofoto/Shutterstock.

Figure 20.1b   Image courtesy of USGS/Kathryn Smith.

Figure 20.16a   kschrei/Shutterstock.

Figure 20.16b   Nestor Noci/Shutterstock.

Figure 20.18   Caitlin Mirra/Shutterstock.

Figure 20.20   Eye Ubiquitous/Alamy.

Figure 20.21   TTphoto/Shutterstock.

Figure 20.24   Jason Vandehey/Shutterstock.

Figure 20.30   Neil Bradfield/Shutterstock.

Figure 20.32   Elena Elisseeva/Shutterstock.

Figure 20.33   Visuals Unlimited/Nature Picture Library.

Figure 20.34   Redrawn from image courtesy of US Department of Agriculture.

Figure 20.40   U.S. Geological Survey photo/Tom Casadevall.

Figure 20.42   Nickolay Stanev/Shutterstock.

Page 507   Christopher Kolaczan/Shutterstock.

Page 508   John S. Sfondilias/Shutterstock.

Page 511   Dean Pennala/Shutterstock, Henry Tsui/Shuterstock/Shutterstock.

Page 515   Alexey Stiop/Shutterstock.

Page 518   Frontpage/Shutterstock.

Page 526   Sam DCruz/Shutterstock, Isabella Pfenninger/Shutterstock, David R. Frazier Photolibrary, Inc./Alamy, funkyfood London – Paul Williams/Alamy, drewthehobbit/Shutterstock, Jose Gil/Shutterstock.

Chapter 21

Opener   Image courtesy of NASA/GSFC/METI/ERSDAC/JAROS, and U.S./Japan ASTER Science Team.

Figure 21.1   Image courtesy of Bruce Marsh.

Figure 21.6   guentermanaus/Shutterstock.

Figure 21.12   Ishbukar Yalilfatar/Shutterstock.

Figure 21.14   Jim Wark, Airphoto North America.

Figure 21.18   Image courtesy of Marli Miller.

Figure 21.30   Image courtesy of US Fish and Wildlife Services.

Figure 21.37   Andy Z/Shutterstock.

Page 535   dp Photography/Shutterstock.

Page 539   Qba from Poland/Shutterstock.

Page 540   Lloyd Homer, GNS Science/Shutterstock.

Page 544   Pichugin Dmitry/Shutterstock.

Page 549   Angel's Gate Photography/Shutterstock.

Page 553   Vladimir Melnik/Shutterstock.

Page 554   drewthehobbit/Shutterstock.

Chapter 22

Opener   photoneye/Shutterstock.

Figure 22.1a   William Marsh.

Figure 22.1b   Philip O'Brien/Shutterstock.

Figure 22.4   Sadatsugu Tomizawa/AFP/Getty Image.

Figure 22.14a   Pichugin Dmitry/Shutterstock.

Figure 22.14b   Mircea Bezergheanu/Shutterstock.

Figure 22.34a   John Wollwerth/Shutterstock.
Figure 22.35   U.S. Geological Survey Circular 1075.
Figure 22.36   photosilta/Alamy.
Page 563   leungchopan/Shutterstock.
Page 570   Oliver Taylor/Shutterstock.
Page 571   William Marsh.
Page 580   Rene van Rijn/Shutterstock.
Page 582   Laitr Keiows/Shutterstock.

Chapter 23

Opener   blickwinkel/Alamy.
Figure 23.1   Image courtesy of Jeff Dozier.
Figure 23.2   Image courtesy of Lonnie G. Thompson, The Ohio State University.
Figure 23.4   Kevin Schafer/Alamy.
Figure 23.5   Graphic Science/Alamy.
Figure 23.6   Mary Lane/Shutterstock.
Figure 23.12   Photos by T. J. Hileman, courtesy of Glacier National Park Archives.
Figure 23.13   Chris Howarth/Chile/Alamy.
Figure 23.15   Image courtesy of Samuel Birman, *La mer de Glace, vue du Montanvert*, Inv. SG-Birmann-2/18, The Gugelmann Collection, Prints and Drawings Department, Swiss National Library.
Figure 23.16   Hubbard Glacier Calving/Shutterstock.
Figure 23.21   H. Mark Weidman Photography/Alamy.
Figure 23.22   Photo by oleandra /Shutterstock.
Figure 23.25   Neil F. Glasser.
Figure 23.27   Image courtesy of National Resources Canada.
Figure 23.29   Steve Estvanik/Shutterstock.
Figure 23.31   Ray Roberts/Alamy.

Figure 23.41   Neil F. Glasser.
Page 590   Rechitan Sorin/Shutterstock.
Page 598   k45025/Shutterstock.

Chapter 24

Opener   apdesign/Shutterstock.
Figure 24.1   SNEHIT/Shutterstock.
Figure 24.5   Image courtesy of US Government.
Figure 24.8   Gina Sanders/Shutterstock.
Figure 24.10   photoshot holdings ltd/Alamy, Luiz Claudio Marigo/Nature Picture Library.
Figure 24.16a   Image courtesy of Bruce Marsh.
Figure 24.16b   Xidong Luo/Shutterstock.
Figure 24.16c   iofoto/Shutterstock.
Figure 24.17   Maxim Petrichuk/Shutterstock.
Figure 24.20   Nuno Miguel Duarte Rodrigues Lopes/ Shutterstock.
Figure 24.21   Nuno Miguel Duarte Rodrigues Lopes/ Shutterstock.
Figure 24.22   Joao Virissimo/Shutterstock.
Figure 24.24   Martin Harvey/Alamy.
Figure 24.26   Dean Pennala/Shutterstock.
Figure 24.27   Image courtesy of Jacques Descloitres, MODIS Rapid Response Team at NASA GSFC.
Figure 24.29   Science and Society/Superstock.
Figure 24.30   Image courtesy of NASA/Goddard Space Flight Center, The SeaWiFS Project and GeoEye, Scientific Visualization Studio.
Page 620   tomashlavac/Shutterstock.
Page 630   maukun/Shutterstock, Artsem Martysiuk/ Shutterstock.

Index

Notes